Progress in Nonlinear Differential Equations and Their Applications
Volume 16

Editor

Haim Brezis
Université Pierre et Marie Curie
Paris
and
Rutgers University
New Brunswick, N.J.

Editorial Board

A. Bahri, Rutgers University, New Brunswick
John Ball, Heriot-Watt University, Edinburgh
Luis Cafarelli, Institute for Advanced Study, Princeton
Michael Crandall, University of California, Santa Barbara
Mariano Giaquinta, University of Florence
David Kinderlehrer, Carnegie-Mellon University, Pittsburgh
Robert Kohn, New York University
P. L. Lions, University of Paris IX
Louis Nirenberg, New York University
Lambertus Peletier, University of Leiden
Paul Rabinowitz, University of Wisconsin, Madison

Alessandra Lunardi

Analytic Semigroups and Optimal Regularity in Parabolic Problems

1995

Birkhäuser
Basel · Boston · Berlin

Alessandra Lunardi
Dipartimento di Matematica
Università di Parma
Via D'Azeglio 85/A
I-43100 Parma

A CIP catalogue record for this book is available from the Library of Congress, Washington D.C., USA

Deutsche Bibliothek Cataloging-in-Publication Data
Lunardi, Alessandra:
Analytic semigroups and optimal regularity in parabolic
problems / Alessandra Lunardi. – Basel ; Boston ; Berlin :
Birkhäuser, 1995
 (Progress in nonlinear differential equations and their applications ;
 Vol. 16)
 ISBN 3-7643-5172-1 (Basel ...)
 ISBN 0-8176-5172-1 (Boston)
NE: GT

This work is subject to copyright. All rights are reserved, whether the whole or part of the material is concerned, specifically the rights of translation, reprinting, re-use of illustrations, broadcasting, reproduction on microfilms or in other ways, and storage in data banks. For any kind of use permission of the copyright owner must be obtained.

© 1995 Birkhäuser Verlag, P.O. Box 133, CH-4010 Basel, Switzerland
Printed on acid-free paper produced from chlorine-free pulp
Printed in Germany
ISBN 3-7643-5172-1
ISBN 0-8176-5172-1

9 8 7 6 5 4 3 2 1

Contents

Introduction IX

0 Preliminary material: spaces of continuous and Hölder continuous functions
  0.1 Spaces of bounded and/or continuous functions 1
  0.2 Spaces of Hölder continuous functions 3
  0.3 Extension operators 8

1 Interpolation theory
  1.1 Interpolatory inclusions 12
  1.2 Interpolation spaces 15
    1.2.1 The K-method 15
    1.2.2 The trace method 20
    1.2.3 The Reiteration Theorem 25
    1.2.4 Some examples 28
  1.3 Bibliographical remarks 32

2 Analytic semigroups and intermediate spaces
  2.1 Basic properties of e^{tA} 35
    2.1.1 Identification of the generator 40
    2.1.2 A sufficient condition to be a sectorial operator 43
  2.2 Intermediate spaces 44
    2.2.1 The spaces $D_A(\alpha, p)$ and $D_A(\alpha)$ 45
    2.2.2 The domains of fractional powers of $-A$ 54
  2.3 Spectral properties and asymptotic behavior 56
    2.3.1 Estimates for large t 56
    2.3.2 Spectral properties of e^{tA} 60
  2.4 Perturbations of generators 64
  2.5 Bibliographical remarks 67

3 Generation of analytic semigroups by elliptic operators
- 3.1 Second order operators 70
  - 3.1.1 Generation in $L^p(\Omega)$, $1 < p < \infty$ 72
  - 3.1.2 Generation in $L^\infty(R^n)$ and in spaces of continuous functions in R^n 76
  - 3.1.3 Characterization of interpolation spaces and generation results in Hölder spaces in R^n 84
  - 3.1.4 Generation in $C^1(R^n)$ 90
  - 3.1.5 Generation in $L^\infty(\Omega)$ and in spaces of continuous functions in $\overline{\Omega}$ 94
- 3.2 Higher order operators and bibliographical remarks 112

4 Nonhomogeneous equations
- 4.1 Solutions of linear problems 123
- 4.2 Mild solutions 129
- 4.3 Strict and classical solutions, and optimal regularity 133
  - 4.3.1 Time regularity 133
  - 4.3.2 Space regularity 142
  - 4.3.3 A further regularity result 150
- 4.4 The nonhomogeneous problem in unbounded time intervals 153
  - 4.4.1 Bounded solutions in $[0, +\infty[$ 153
  - 4.4.2 Bounded solutions in $]-\infty, 0]$ 160
  - 4.4.3 Bounded solutions in R 164
  - 4.4.4 Exponentially decaying and exponentially growing solutions 168
- 4.5 Bibliographical remarks 170

5 Linear parabolic problems
- 5.1 Second order equations 175
  - 5.1.1 Initial value problems in $[0,T] \times R^n$ 178
  - 5.1.2 Initial boundary value problems in $[0,T] \times \overline{\Omega}$ 192
- 5.2 Bibliographical remarks 208

6 Linear nonautonomous equations
- 6.1 Construction and properties of the evolution operator 212
- 6.2 The variation of constants formula 228
- 6.3 Asymptotic behavior in the periodic case 234
  - 6.3.1 The period map 234
  - 6.3.2 Estimates on the evolution operator 236
  - 6.3.3 Asymptotic behavior in nonhomogeneous problems 240
- 6.4 Bibliographical remarks 250

7 Semilinear equations

- 7.1 Local existence and regularity 257
  - 7.1.1 Local existence results 257
  - 7.1.2 The maximally defined solution 265
  - 7.1.3 Further regularity, classical and strict solutions 268
- 7.2 *A priori* estimates and existence in the large 271
- 7.3 Some examples 275
  - 7.3.1 Reaction-diffusion systems 275
  - 7.3.2 A general semilinear equation 278
  - 7.3.3 Second order equations with nonlinearities in divergence form 283
  - 7.3.4 The Cahn-Hilliard equation 284
- 7.4 Bibliographical remarks for Chapter 7 286

8 Fully nonlinear equations

- 8.1 Local existence, uniqueness and regularity 290
- 8.2 The maximally defined solution 298
- 8.3 Further regularity properties and dependence on the data 301
  - 8.3.1 C^k regularity with respect to (x, λ) 302
  - 8.3.2 C^k regularity with respect to time 306
  - 8.3.3 Analyticity 308
- 8.4 The case where X is an interpolation space 309
- 8.5 Examples and applications 313
  - 8.5.1 An equation from detonation theory 313
  - 8.5.2 An example of existence in the large 315
  - 8.5.3 A general second order problem 319
  - 8.5.4 Motion of hypersurfaces by mean curvature 329
  - 8.5.5 Bellman equations 332
- 8.6 Bibliographical remarks 334

9 Asymptotic behavior in fully nonlinear equations

- 9.1 Behavior near stationary solutions 340
  - 9.1.1 Stability and instability by linearization 341
  - 9.1.2 The saddle point property 345
  - 9.1.3 The case where X is an interpolation space 347
  - 9.1.4 Bifurcation of stationary solutions 350
  - 9.1.5 Applications to nonlinear parabolic problems, I 355
  - 9.1.6 Stability of travelling waves in two-phase free boundary problems 358

- 9.2 Critical cases of stability 363
  - 9.2.1 The center-unstable manifold 363
  - 9.2.2 Applications to nonlinear parabolic problems, II 371
  - 9.2.3 The case where the linear part generates a bounded semigroup 373
  - 9.2.4 Applications to nonlinear parabolic problems, III 375
- 9.3 Periodic solutions 378
  - 9.3.1 Hopf bifurcation 379
  - 9.3.2 Stability of periodic solutions 387
  - 9.3.3 Applications to nonlinear parabolic problems, IV 397
- 9.4 Bibliographical remarks 398

Appendix: Spectrum and resolvent 399
- A.1 Spectral sets and projections 402
- A.2 Isolated points of the spectrum 404
- A.3 Perturbation results 408

Bibliography 411

Index 423

Introduction

This book gives a systematic treatment of the basic theory of analytic semigroups and abstract parabolic equations in general Banach spaces, and of how such a theory may be used in parabolic PDE's. It takes into account the developments of the theory during the last fifteen years, and it is focused on classical solutions, with continuous or Hölder continuous derivatives. On one hand, working in spaces of continuous functions rather than in Lebesgue spaces seems to be appropriate in view of the number of parabolic problems arising in applied mathematics, where continuity has physical meaning; on the other hand it allows one to consider any type of nonlinearities (even of nonlocal type), even involving the highest order derivatives of the solution, avoiding the limitations on the growth of the nonlinear terms required by the L^p approach. Moreover, the continuous space theory is, at present, sufficiently well established. For the Hilbert space approach we refer to J.L. LIONS – E. MAGENES [128], M.S. AGRANOVICH – M.I. VISHIK [14], and for the L^p approach to V.A. SOLONNIKOV [184], P. GRISVARD [94], G. DI BLASIO [72], G. DORE – A. VENNI [76] and the subsequent papers [90], [169], [170].

Many books about abstract evolution equations and semigroups contain some chapters on analytic semigroups. See, e.g., E. HILLE – R.S. PHILLIPS [100], S.G. KREIN [114], K. YOSIDA [213], A. PAZY [166], H. TANABE [193], PH. CLÉMENT ET AL. [54]. However, none of them is entirely devoted to analytic semigroups, which are seen as particular cases of the general theory of semigroups rather than an independent topic. On the contrary, we develop the subject without invoking theorems from the general theory of semigroups. This is done not only to construct a self-contained theory, but mainly because associating analytic semigroups with general C_0 semigroups is, in our opinion, misleading. For instance, the classical Hille-Yosida Theorem may be applied as well to hyperbolic equations, which are substantially different from the parabolic ones.

Moreover, we emphasize the optimal regularity results, which are not treated in the above mentioned books — except, partially, in [54]. They are, in fact, a typical feature of abstract parabolic equations, and yield old and new optimal regularity results for parabolic partial differential equations and systems.

No special prerequisites are requested to read this book, except the fundamental notions of functional analysis and some familiarity with PDE's. To help the reader, elements of spectral theory are collected in an appendix. In Chapter 1, dealing with interpolation, Bochner integrals are used. We refer to the appendix of [37] for a simple approach to the basic theory of Bochner integration.

The starting point of our analysis is the study of the linear problem

$$u'(t) = Au(t), \ t > 0, \ u(0) = u_0, \tag{1}$$

where $A : D(A) \subset X \mapsto X$ is a closed operator in a general Banach space X, and $x \in X$. Motivated by the applications to PDE's, we do not assume that $D(A)$ is dense in X. We only need to assume that A is sectorial, i.e. that the resolvent set $\rho(A)$ of A contains a sector

$$S = \{\lambda \in \mathbf{C} : \lambda \neq \omega, \ |\arg(\lambda - \omega)| < \theta\},$$

with $\omega \in \mathbb{R}$, $\theta > \pi/2$, and there exists $M > 0$ such that

$$\|\lambda(\lambda I - A)^{-1}\|_{L(X)} \leq M, \ \lambda \in S. \tag{2}$$

Problem (1) leads to the construction of the analytic semigroup e^{tA} generated by A, defined through the Dunford integral

$$e^{tA} = \frac{1}{2\pi i} \int_\gamma e^{t\lambda} (\lambda I - A)^{-1} d\lambda, \ t > 0, \tag{3}$$

where γ is a suitable curve with support in the resolvent set of A.

Condition (2) may be interpreted as an abstract ellipticity assumption on A. It is indeed satisfied by the realizations of elliptic partial differential operators in many functional spaces.

Under appropriate hypotheses, the variation of constants formula

$$u(t) = e^{tA} u_0 + \int_0^t e^{(t-s)A} f(s) ds, \ t \geq 0, \tag{4}$$

gives the unique solution of the nonhomogeneous problem

$$u'(t) = Au(t) + f(t), \ t > 0; \ u(0) = u_0. \tag{5}$$

A number of regularity results, according to the regularity of f and u_0, are proved. Among them, of particular importance are the above mentioned optimal regularity results, i.e., the cases when u' and Au enjoy the same regularity properties of f. Unfortunately, optimal regularity results are not always available. For instance, if $f \in C([0,T]; X)$ (the space of the continuous functions in the interval $[0,T]$ with values in X) and $u_0 = 0$, then, in general, u' and Au do not belong necessarily to $C([0,T]; X)$.

Two optimal regularity results are stated below. They are due to E. SINESTRARI [177]. To formulate them, one has to introduce the real interpolation spaces $D_A(\alpha, \infty)$. For $0 < \alpha < 1$, $D_A(\alpha, \infty)$ may be easily defined in terms of the semigroup e^{tA}, as the set of all $x \in X$ such that $t^{1-\alpha}\|Ae^{tA}x\|_X$ is bounded near $t = 0$

(recall that $t\|Ae^{tA}x\|_X$ is bounded for every $x \in X$, and $\|Ae^{tA}x\|_X$ is bounded for every $x \in D(A)$).

Theorem 1. *If $f \in C^\alpha([0,T];X)$ (the space of all α-Hölder continuous functions in the interval $[0,T]$ with values in X, $0 < \alpha < 1$), then u' and Au belong to $C^\alpha([0,T];X)$ if and only if the compatibility conditions*

$$u_0 \in D(A), \quad Au_0 + f(0) \in D_A(\alpha,\infty)$$

hold. In this case, u' is bounded in $[0,T]$ with values in $D_A(\alpha,\infty)$.

The last statement of Theorem 1 concerns a further regularity property which is very important in the applications; see below. Since $Au_0 + f(0) = u'(0)$, it may be restated by saying that the condition $u'(0) \in D_A(\alpha,\infty)$ implies that $u'(t) \in D_A(\alpha,\infty)$ for every t.

Theorem 2. *If $f \in C([0,T];X) \cap B([0,T];D_A(\alpha,\infty))$ (the space of all bounded functions in the interval $[0,T]$ with values in $D_A(\alpha,\infty))$, then u' and Au belong to $C([0,T];X) \cap B([0,T];D_A(\alpha,\infty))$ if and only if*

$$u_0 \in D(A), \quad Au_0 \in D_A(\alpha,\infty).$$

In this case, Au belongs to $C^\alpha([0,T];X)$.

One guesses from Theorems 1 and 2 that interpolation spaces are of crucial importance in the theory of regularity for abstract parabolic equations. For this reason, Chapter 1 is devoted to interpolation spaces. Interpolation theory, in connection with evolution equations, is indeed rather wide, and giving an exhaustive treatment is out of the realm of this book. We refer to the books of H. TRIEBEL [200], P.L. BUTZER – H. BERENS [42], J. BERGH – J. LÖFSTRÖM [32]. In Chapter 1 we introduce concisely and in a self-contained way what is necessary for the development of the rest of the book: equivalent definitions of real interpolation spaces, some elementary properties, interpolatory inclusions, and the important Reiteration Theorem. The reader with some knowledge of interpolation may skip it.

The optimal regularity results allow the solving of more complicated problems, such as nonautonomous, delay, or nonlinear problems, by "simple" perturbation arguments. An example is given in Chapter 6, through the construction of the parabolic evolution operator relevant to nonautonomous equations,

$$u'(t) = A(t)u(t), \quad t > 0; \quad u(0) = u_0, \qquad (6)$$

where all the operators $A(t)$ are sectorial and have the same domain $D \subset X$.

Another example is a class of parabolic fully nonlinear problems, such as

$$u'(t) = F(u(t)), \quad t > 0; \quad u(0) = u_0, \qquad (7)$$

where F is a regular (say, twice continuously differentiable) function defined in a Banach space $D \subset X$, with values in X. See below for motivating examples. The main assumption, which gives a parabolic character to the problem, is that for every $u_0 \in D$ the linear operator $A = F'(u_0) : D(A) \simeq D \mapsto X$ is sectorial.

Let us explain briefly the argument used in the proof of local existence. First of all, problem (7) is rewritten as

$$u'(t) = Au(t) + G(u(t)), \quad t > 0; \quad u(0) = u_0, \tag{8}$$

where $G : D \mapsto X$ satisfies $G'(u_0) = 0$. To avoid technicalities, we consider here the simplest situation, when $u_0 \in D$, $Au_0 + G(u_0) = F(u_0) \in D_A(\alpha, \infty)$ for some $\alpha \in {]}0,1{[}$. A local solution, defined in a small interval $[0, \delta]$, is sought as a fixed point of a nonlinear operator Γ in a ball

$$B = \{u \in C^\alpha([0,\delta]; D) : u(0) = u_0, \, \|u(\cdot) - u_0\|_{C^\alpha([0,\delta];D)} \leq r\}.$$

The operator Γ is defined by $\Gamma u = v$, where v is the solution of

$$v'(t) = Av(t) + G(u(t)), \quad 0 \leq t \leq \delta; \quad v(0) = u_0.$$

If $u \in B$ and δ is small enough, then $G(u(\cdot))$ belongs to $C^\alpha([0, \delta]; D)$. Since $u_0 \in D$, $Au_0 + G(u(0)) = Au_0 + G(u_0) \in D_A(\alpha, \infty)$, then Theorem 1 yields $\Gamma u \in C^\alpha([0, \delta]; D)$. After, one sees that if r and δ are suitably chosen, then Γ maps B into itself, and it is a contraction on B in the norm of $C^\alpha([0, \delta]; D)$. Consequently, Γ has a unique fixed point u in B, which is in fact the unique solution of (7) belonging to $C^\alpha([0, \delta]; D)$. Theorem 1 gives also $u'(\delta) = Au(\delta) + G(u(\delta)) = F(u(\delta)) \in D_A(\alpha, \infty) = D_{F'(u(\delta))}(\alpha, \infty)$. This makes it possible to repeat the same procedure in an interval $[\delta, \delta + \delta_1]$ and to extend the solution to the interval $[0, \delta + \delta_1]$, and then to a maximal time interval $[0, \tau[$.

The compatibility condition $F(u_0) \in D_A(\alpha, \infty)$ is somewhat restrictive. The natural condition on the initial value u_0 to get a local strict solution (i.e., $u \in C([0,\delta]; D) \cap C^1([0,\delta]; X))$ is $u_0 \in D$, $F(u_0) \in \overline{D}$. This is because $F(u_0) = u'(0) \in \overline{D}$ if $u(t) \in D$ for $t \geq 0$. The difficulty is overcome by working in a suitable space of weighted (near $t = 0$) Hölder continuous functions, and the above procedure is still effective, although the proof is rather long. See Chapter 8.

Other situations in which optimal regularity has been successfully used are integrodifferential and delay equations, with delay term defined in $D(A)$, such as

$$u'(t) = Au(t) + \int_0^t k(t,s)Au(s)ds, \quad t \geq 0; \quad u(0) = u_0,$$

where k is a real kernel, and

$$u'(t) = Au(t) + Au(t-r), \quad t \geq 0; \quad u_{|[-r,0]} = \varphi_0,$$

with $r > 0$ and $\varphi_0 : [-r, 0] \mapsto D(A)$ given.

Also in these equations, the delay terms are considered as perturbations of $u' - Au$. Under suitable assumptions on k, u_0, φ_0, the solutions are constructed step by step, by a procedure similar to the one described above. See [176], [144], [155], [27], [74], [75], [73].

Optimal regularity is needed also in certain highly degenerate quasilinear systems. See H. AMANN [23].

The connection between abstract theorems and concrete PDE's is made by generation theorems, and characterizations of interpolation spaces.

At this time, many results about generation of analytic semigroups by strongly elliptic operators are available. We present here the basic ones, concerning generation in L^p spaces, in spaces of continuous and Hölder continuous functions, and in C^1 spaces.

The problem consists in proving existence and estimates for the solutions of elliptic equations or systems depending on a complex parameter λ,

$$\lambda u - \mathcal{A}u = f \quad in \ \Omega, \tag{9}$$

supported with the usual homogeneous boundary conditions. Here Ω is a regular, possibly unbounded, open set in $\mathbb{R}^n$, $\mathcal{A}$ is a strongly elliptic operator of order $2m$, with uniformly continuous and bounded coefficients, and f belongs to a Banach space X of functions defined in Ω. Existence and uniqueness results are provided by the classical theory of elliptic PDE's. Then, *a priori* estimates of the type

$$\|u\|_X \leq const. \frac{\|f\|_X}{|\lambda|} \tag{10}$$

are proved by different techniques, depending on the choice of the space X. In the case where X is a Lebesgue space $L^p(\Omega)$, as shown by S. AGMON[12], the main tools are the Agmon – Douglis – Nirenberg estimates. The case $X = L^\infty(\Omega)$, or $X = C(\overline{\Omega})$, follows from the case $X = L^p(\Omega)$ through a suitable localization procedure due to H. B. STEWART [187, 188]. In the case where X is a Hölder space, the resolvent estimate (10) is found by interpolation, after the characterization of Hölder spaces as interpolation spaces between $C(\overline{\Omega})$ and the domain of the realization of $\mathcal{A}$ in $C(\overline{\Omega})$. Such a characterization is also of primary importance in the applications to parabolic equations. In particular, it lets one prove in a simple way the Schauder theorem for elliptic equations in $\mathbb{R}^n$. Moreover, it lets one apply the abstract results about problem (5) to find several Hölder regularity results in parabolic equations, among which are the well known Ladyzhenskaya – Solonnikov – Ural'ceva theorems.

Let us explain what we mean by the simplest significant example,

$$\begin{cases} u_t = \Delta u + \varphi & in \ [0,T] \times \overline{\Omega}, \\ u(0,x) = u_0(x) & in \ \overline{\Omega}, \\ \dfrac{\partial u}{\partial \nu} = 0 & in \ [0,T] \times \partial\Omega, \end{cases} \tag{11}$$

where Ω is a bounded open set in $\mathbb{R}^n$ with regular boundary $\partial\Omega$, and $\partial/\partial\nu$ denotes the normal derivative at the boundary.

The realization of the Laplace operator Δ in $X = C(\overline{\Omega})$ with homogeneous Neumann boundary condition,

$$D(A) = \{\xi \in C(\overline{\Omega}) : \Delta\xi \in C(\overline{\Omega}), \partial\xi/\partial\nu = 0\}, \quad A\xi = \Delta\xi,$$

is sectorial in X, and

$$D_A(\alpha, \infty) = \begin{cases} C^{2\alpha}(\overline{\Omega}), & \text{if } \alpha < 1/2, \\ \{\xi \in C^{2\alpha}(\overline{\Omega}) : \partial\xi/\partial\nu = 0\}, & \text{if } 1/2 < \alpha < 1 \end{cases}$$

(see Chapter 3). Problem (11) is seen as an evolution equation of the type (5) in X, by setting

$$u(t) = u(t, \cdot), \quad f(t) = \varphi(t, \cdot).$$

Let us apply Theorems 1 and 2, with $\alpha < 1/2$ for simplicity. In our case, $f \in C^\alpha([0,T];X)$ means that $\varphi \in C^{\alpha,0}([0,T]\times\overline{\Omega})$ (the space of all continuous functions in $[0,T]\times\overline{\Omega}$ which are α-Hölder continuous in t, uniformly with respect to x). The condition $u_0 \in D(A)$, $Au_0 + f(0) \in D_A(\alpha, \infty)$ means that

$$u_0, \Delta u_0 \in C(\overline{\Omega}), \quad \partial u_0/\partial\nu = 0, \quad \Delta u_0 + \varphi(0, \cdot) \in C^{2\alpha}(\overline{\Omega}). \tag{12}$$

If $\varphi \in C^{\alpha,0}([0,T]\times\overline{\Omega})$ and (12) holds, Theorem 1 guarantees that the solution u of (11) is such that u_t and Δu belong to $C^{\alpha,0}([0,T]\times\overline{\Omega})$.

The condition $f \in C([0,T];X) \cap B([0,T];D_A(\alpha, \infty))$ means that φ belongs to $C^{0,2\alpha}([0,T]\times\overline{\Omega})$ (the space of all continuous functions in $[0,T]\times\overline{\Omega}$ which are 2α-Hölder continuous in x, uniformly with respect to t). The condition $u_0 \in D(A)$, $Au_0 \in D_A(\alpha, \infty)$ means – via the Schauder regularity theory for elliptic PDE's – that

$$u_0 \in C^{2+2\alpha}(\overline{\Omega}), \quad \partial u_0/\partial\nu = 0. \tag{13}$$

If $\varphi \in C^{0,2\alpha}([0,T]\times\overline{\Omega})$ and (13) holds, Theorem 2 guarantees that the solution u of (11) is such that u_t and Δu belong to $C^{0,2\alpha}([0,T]\times\overline{\Omega})$. Consequently, all the second order space derivatives of u are 2α-Hölder continuous with respect to x.

Patching together the above results, we find that if $\varphi \in C^{\alpha,2\alpha}([0,T]\times\overline{\Omega})$, $u_0 \in C^{2+2\alpha}(\overline{\Omega})$, and $\partial u_0/\partial\nu = 0$, then u_t and Δu belong to $C^{\alpha,2\alpha}([0,T]\times\overline{\Omega})$. In its turn, this implies that $u \in C^{1+\alpha,2+2\alpha}([0,T]\times\overline{\Omega})$.

More details may be found in Chapter 5, where general linear parabolic problems are studied.

Together with regularity results, we consider also asymptotic behavior. As in ordinary differential equations, the asymptotic behavior of e^{tA} depends heavily on the spectral properties of A. The study of asymptotic behavior in linear problems is a preliminary step to study asymptotic behavior in nonlinear problems.

In the second part of the book we consider nonlinear problems.

Semilinear abstract and concrete parabolic problems have been thoroughly studied, even recently. See the monographs of R.H. MARTIN [158], H. TANABE

[193], A. FRIEDMAN [83, 84] A. PAZY [166], D. HENRY [99], E. ROTHE [172], W. VON WAHL [206], R. TEMAM [196], J. SMOLLER [180], D. DANERS – P. KOCH MEDINA [61], C. COSNER – J. HERNANDEZ – E. MITIDIERI [56]. We give a general treatment, which does not pretend to be exhaustive, but unifies and synthetizes several approaches, providing the basic results about existence, regularity, continuous dependence, and criteria for existence in the large. We consider equations of the type

$$u'(t) = Au(t) + F(t, u(t)), \quad t \geq 0; \quad u(0) = u_0, \tag{14}$$

with nonlinearity $F(t,x)$ defined for $t \geq 0$ and x in a general intermediate space between X and $D(A)$.

We have chosen to skip over asymptotic behavior and geometric properties of the solutions to semilinear equations, referring the interested reader to D. HENRY [99]. He considered nonlinearities defined for x in $D(-A)^\alpha$, the domain of a fractional power of $-A$, but most of his techniques may be adapted without important modifications to the more general setting considered here.

The last two chapters of the book are concerned with fully nonlinear problems. The interest of the mathematicians in fully nonlinear problems grew significantly in the last years, also because of an increasing number of fully nonlinear equations and systems arising from the applications. We mention some of them here.

(i) If a smooth initial hypersurface Γ_0 evolves by mean curvature into a family of smooth hypersurfaces $\{\Gamma_t\}_{t\geq 0}$, the signed distance $v(t,x)$ from Γ_t, before the possible appearance of singularities, satisfies

$$\begin{cases} v_t = \sum_{i=1}^n \dfrac{\lambda_i}{1 - \lambda_i v}, & t \geq 0, \ x \in \overline{\Omega}, \\ |Dv|^2 = 1, & t \geq 0, \ x \in \partial\Omega, \end{cases}$$

λ_i being the eigenvalues of the symmetric matrix $D^2 v$, and Ω being a thin regular neighborhood of Γ_0. The above problem is parabolic near sufficiently small initial data.

(ii) Bellman equations, arising in Stochastic Control,

$$\begin{cases} v_t(t,x) = \inf_{\alpha \in \Lambda} [L^\alpha v(t,\cdot)(x) + f(x,\alpha)] \\ v(T,x) = v_0(x), \ x \in \mathbb{R}^n, \end{cases}$$

are in some cases fully nonlinear parabolic problems. For instance, in the (physically meaningful) case in which $\Lambda = \mathbb{R}^n$,

$$L^\alpha u(x) = \frac{1}{2} \sum_{i,j=1}^n (a_{ij}^0(x) + \alpha_i \alpha_j) D_{ij} u(x) + \sum_{i=1}^n (\alpha_i + b_i^0(x)) D_i u(x),$$

with
$$\sum_{i,j=1}^{n} a_{ij}^0(x)\xi_i\xi_j \geq c|\xi|^2, \quad x,\xi \in \mathbb{R}^n,$$

and
$$f(x,\alpha) = \frac{1}{2}|\alpha|^2,$$

the problem reduces to
$$\begin{cases} v_t = \frac{1}{2}\left(\langle (I+D^2v)^{-1}Dv, Dv\rangle - Tr(A_0 D^2 v)\right), & t \leq T, \ x \in \mathbb{R}^n, \\ v(T,x) = v_0(x), & x \in \mathbb{R}^n \end{cases}$$

which, after reversing time, becomes a forward parabolic fully nonlinear problem.

(iii) Certain free boundary problems with jumps at the interface, such as
$$\begin{cases} u_t = u_{yy} + uu_y, \ t \geq 0, \ y \neq \xi(t), \\ u(t,\xi(t)) = u^*, \ [u_y(t,\xi(t))] = u_y(t,\xi(t)^+) - u_y(t,\xi(t)^-) = -1, \ t \geq 0, \end{cases}$$

may be reduced to fully nonlinear problems. In a frame attached to the free boundary $\xi(t)$, the systems reads as
$$\begin{cases} u_t = u_{xx} + uu_x + \dot{\xi}(t)u_x, \ t \geq 0, \ x \neq 0, \\ u(t,0) = u^*, \ [u_x(t,0)] = u_x(t,0^+) - u_x(t,0^-) = -1, \ t \geq 0. \end{cases}$$

The simplest way to decouple the system is to differentiate with respect to time the equality $u(t,\xi(t)) = u^*$, getting $u_t(t,\xi(t)^\pm) + u_y(t,\xi(t)^\pm)\dot{\xi}(t) = 0$, so that $\dot{\xi}(t) = [u_t(t,\xi(t))] = [u_{yy}(t,\xi(t))] - u^*$. Replacing the latter in the equation, we get
$$u_t(t,x) = u_{xx}(t,x) + (u(t,x) + [u_{xx}(t,0)] - u^*)u_x(t,x), \quad x \neq 0,$$

which may be seen as a fully nonlinear equation, with nonlinearity depending on the nonlocal term $[u_{xx}(t,0)]$.

These examples, together with others, are considered in Chapter 8, where also the fundamental results about existence, uniqueness, regularity, and dependence on the data are stated.

Another important class of nonlinear problems are the quasilinear ones,
$$u'(t) = A(t,u(t))u(t) + F(t,u(t)), \ t > 0; \quad u(0) = u_0. \tag{15}$$

The simplest case is when the linear operators $A(t,x)$ and the function $F(t,x)$ are defined for $t \geq 0$ and for x in an intermediate space X_α between X and D. D is a Banach space, continuously embedded in X; all the operators $A(t,x)$ are sectorial and have domain D, and the initial datum u_0 belongs to D. Such problems may be seen as particular fully nonlinear equations.

The case where the linear operators $A(t,x)$ have variable domains, depending on t and x, requires much longer discussion, which goes beyond the aims of this book. We refer to the monograph of H. AMANN [24], which is entirely devoted to quasilinear problems, and to the classical treatment of parabolic quasilinear equations, which may be found in the books of O.A. LADYZHENSKAYA – V.A. SOLONNIKOV – N.N. URAL'CEVA [124], A. FRIEDMAN [83, 84], N.V. KRYLOV, and [122].

Chapter 10 is devoted to the study of geometric properties of the solutions of fully nonlinear problems. First we consider existence and stability of stationary solutions. Existence of stationary solutions in equations depending on parameters is proved by bifurcation arguments. Then we show that the Principle of Linearized Stability holds, and we construct (when possible) invariant manifolds near stationary solutions: the stable manifold, the unstable manifold, and the center-unstable manifolds. The center manifold also allows one to solve some critical cases of stability, reducing the problem of the stability of a fully nonlinear, infinite dimensional problem to the problem of the stability of a system of ordinary differential equations. However, other typically infinite dimensional critical cases of stability, arising in problems in unbounded domains, are also studied. Then we consider periodic solutions, Hopf bifurcation, and the Principle of Linearized Stability for periodic solutions.

Geometric theory of fully nonlinear equations is not an easy generalization of geometric theory of semilinear equations, even if the final results sound similar. For instance, *a priori* estimates on the D-norm of the solutions are not sufficient, in general, to prove existence in the large. When $u' = Au + f$, estimates like

$$\|u(t)\|_D \leq c(\|u(0)\|_D + \int_0^t k(t-s)\|f(s)\|_X ds)$$

with $k \in L^1$, are not available, so that the Gronwall Lemma cannot in general be used. Usually, bootstrap arguments to get regularity properties or estimates, as well as compactness arguments, do not work.

The geometric theory of fully nonlinear equations may serve also for quasilinear equations with constant domains.

The main part of this book was written during a sabbatical year spent at Scuola Normale Superiore, Pisa, which I would like to thank for generous support and hospitality.

Chapter 0

Preliminary material: spaces of continuous and Hölder continuous functions

0.1 Spaces of bounded and/or continuous functions

Let X be a real or complex Banach space with norm $\|\cdot\|$, and let $I \subset \mathbb{R}$ be a (possibly unbounded) interval. We consider the functional spaces $B(I;X)$, $C(I;X)$, $C^m(I;X)$ ($m \in \mathbb{N}$), $C^\infty(I;X)$, consisting respectively of the bounded, continuous, m times continuously differentiable, infinitely many times differentiable, functions $f: I \mapsto X$. $B(I;X)$ is endowed with the sup norm

$$\|f\|_{B(I;X)} = \sup_{t \in I} \|f(t)\|.$$

We set also

$$C_b(I;X) = B(I;X) \cap C(I;X), \quad \|f\|_{C(I;X)} = \|f\|_{B(I;X)},$$

$$C_b^m(I;X) = \{f \in C^m(I;X) : f^{(k)} \in C_b(I;X),\ k=0,\ldots,m\},$$

$$\|f\|_{C_b^m(I;X)} = \sum_{k=0}^{m} \|f^{(k)}\|_{B(I;X)}.$$

We denote by $C_0^\infty(I;X)$ the subset of $C^\infty(I;X)$ consisting of the functions with support contained in the interior of I.

Where there is no danger of confusion, we shall write $\|f\|_{B(X)}$ or simply $\|f\|_\infty$ instead of $\|f\|_{B(I;X)}$ for any bounded function f; moreover, if $X = \mathbb{R}$ or $\mathbb{C}$, we shall write $B(I)$, $C(I)$, etc., instead of $B(I;X)$, $C(I;X)$, etc.

We now give some approximation results which will be used throughout.

Definition 0.1.1 *Let $a < b \in \mathbb{R}$, let $f \in C([a,b]; X)$, and set, for $n \in \mathbb{N}$, $a \leq t \leq b$,*

$$P_n(t) = (b-a)^{-n} \sum_{k=0}^{n} \binom{n}{k} (t-a)^k (b-t)^{n-k} f\left(a + \frac{k}{n}(b-a)\right). \tag{0.1.1}$$

The functions P_n are called Bernstein polynomials *of f.*

Later we will use the following consequences of (0.1.1).

$$P_n(a) = f(a), \quad P_n(b) = f(b), \quad \|P_n\|_\infty \leq \|f\|_\infty, \quad \forall n \in \mathbb{N}. \tag{0.1.2}$$

Proposition 0.1.2 *Let $f \in C([a,b]; X)$, and let P_n be defined by (0.1.1). Then $P_n \to f$ in $C([a,b]; X)$ as $n \to \infty$.*

The proof given in K. YOSIDA [213, p.8] for the case $X = \mathbb{R}$ may be extended to the case of general X. By Proposition 0.1.2 it follows that $C^\infty([a,b]; X)$ is dense in $C([a,b]; X)$. (0.1.2) will be used in the sequel.

Another important consequence is the following.

Corollary 0.1.3 *Let $Y \subset X$ be a subspace dense in X. Then $C^\infty([a,b]; Y)$ is dense in $C([a,b]; X)$.*

Proof — Fix $f \in C([a,b]; X)$ and $\varepsilon > 0$. Let n be so large that, denoting by P_n the n-th Bernstein polynomial of f,

$$\|f - P_n\|_\infty \leq \varepsilon.$$

For $k = 0, .., n$ let $y_k \in Y$ be such that

$$\|f(a + k(b-a)/n) - y_k\|_X \leq \varepsilon/(\dot n + 1),$$

and set

$$Q_n(t) = (b-a)^{-n} \sum_{k=0}^{n} \binom{n}{k} (t-a)^k (b-t)^{n-k} y_k.$$

Then $\|f - Q_n\|_\infty \leq 2\varepsilon$, and the statement follows. ∎

We shall consider also functions defined in $\mathbb{R}^n$ or in subsets of $\mathbb{R}^n$, particularly in $\mathbb{R}^n_+ = \{x = (x_1, \ldots, x_n) \in \mathbb{R}^n : x_n \geq 0\}$ and in domains with uniformly C^m boundary. If Ω is an open set in $\mathbb{R}^n$, and $m \in \mathbb{N}$ we say that the boundary $\partial\Omega$ is uniformly C^m if there exists a (at most countable) collection of open balls $U_j = \{x \in \mathbb{R}^n : |x - x_j| < r\}$, $j \in \mathbb{N}$, covering $\partial\Omega$ and such that there exists an integer k with the property that $\cap_{j \in J} U_j = \emptyset$ for all $J \subset \mathbb{N}$ with more than k elements. Moreover, we assume that there is $\varepsilon > 0$ such that the balls centered at x_j with radius $r/2$ still cover an ε-neighborhood of $\partial\Omega$, and that there exist coordinate transformations φ_j such that $\varphi_j : \overline{U_j} \mapsto B(0,1) \subset \mathbb{R}^n$ is a C^m diffeomorphism, mapping $\overline{U_j} \cap \Omega$ onto the upper half ball $B_+(\mathbb{R}^n) = \{y \in B(0,1) : y_n > 0\}$, and mapping $U_j \cap \partial\Omega$ onto the basis $\Sigma_n = \{y \in B(0,1) : y_n = 0\}$. All the coordinate transformations φ_j and their inverses are supposed to have uniformly bounded derivatives up to the order m,

$$\sup_{j \in \mathbb{N}} \sum_{1 \leq |\alpha| \leq m} (\|D^\alpha \varphi_j\|_\infty + \|D^\alpha \varphi_j^{-1}\|_\infty) \leq M. \tag{0.1.3}$$

For every $x \in \partial\Omega$ we denote by $\nu(x)$ the unit exterior normal vector to $\partial\Omega$ at x.

For further properties of domains with uniformly C^m boundary see [38].

If Ω is any open set in $\mathbb{R}^n$, $C(\overline{\Omega})$ (respectively, $UC(\overline{\Omega})$) denotes the Banach space of all the continuous (respectively, uniformly continuous) and bounded functions in $\overline{\Omega}$, endowed with the sup norm, and $C^m(\overline{\Omega})$ (respectively, $UC^m(\overline{\Omega})$) denotes the set of all m times continuously differentiable functions in Ω, with derivatives up to the order m bounded and continuously (respectively, uniformly continuously) extendable up to the boundary. It is endowed with the norm

$$\|f\|_{C^m(\overline{\Omega})} = \sum_{|\alpha| \leq m} \|D^\alpha f\|_\infty.$$

0.2 Spaces of Hölder continuous functions

The Banach spaces of Hölder continuous functions $C^\alpha(I; X)$, $C^{k+\alpha}(I; X)$ ($k \in \mathbb{N}$, $\alpha \in\,]0,1[$), are defined by

$$C^\alpha(I; X) = \{f \in C_b(I; X) : [f]_{C^\alpha(I;X)} = \sup_{t,s \in I,\ s<t} \frac{\|f(t) - f(s)\|}{(t-s)^\alpha} < +\infty\},$$

$$\|f\|_{C^\alpha(I;X)} = \|f\|_\infty + [f]_{C^\alpha(I;X)};$$

$$C^{k+\alpha}(I; X) = \{f \in C_b^k(I; X) : f^{(k)} \in C^\alpha(I; X)\},$$

$$\|f\|_{C^{k+\alpha}(I;X)} = \|f\|_{C_b^k(I;X)} + [f^{(k)}]_{C^\alpha(I;X)}.$$

We shall also consider spaces of Lipschitz continuous functions,

$$Lip(I; X) = \{f \in C_b(I; X) : [f]_{Lip(I;X)} = \sup_{t,s \in I,\ s<t} \frac{\|f(t) - f(s)\|}{t-s} < +\infty\},$$

$$\|f\|_{Lip(I;X)} = \|f\|_\infty + [f]_{Lip(I;X)}.$$

Again, where there is no danger of confusion, we shall write $[f]_{C^\alpha(X)}$ or $[f]_{C^\alpha}$, $\|f\|_{C^\alpha(X)}$ or $\|f\|_{C^\alpha}$, etc., instead of $[f]_{C^\alpha(I;X)}$, $\|f\|_{C^\alpha(I;X)}$ etc.

The spaces of the little-Hölder continuous functions are defined by

$$h^\alpha(I;X) = \{f \in C^\alpha(I;X) : \lim_{\delta \to 0} \sup_{t,s \in I, |t-s|<\delta} \frac{\|f(t)-f(s)\|}{|t-s|^\alpha} = 0\},$$

$$h^{k+\alpha}(I;X) = \{f \in C_b^k(I;X) : f^{(k)} \in h^\alpha(I;X)\}.$$

One checks immediately that $C^\theta(I;X) \subset h^\alpha(I;X)$ for $\theta > \alpha$. Moreover, the following proposition holds.

Proposition 0.2.1 *Let $0 < \alpha < 1$, and let $\theta > \alpha$. Then $h^\alpha(I;X)$ is the closure of $C^\theta(I;X)$ in $C^\alpha(I;X)$.*

Proof — Since $C^\theta(I;X) \subset h^\alpha(I;X)$, to show that the closure of $C^\theta(I;X)$ is contained in $h^\alpha(I;X)$ it is sufficient to prove that $h^\alpha(I;X)$ is closed in $C^\alpha(I;X)$. If a sequence $\{f_n\}_{n \in \mathbb{N}} \subset h^\alpha(I;X)$ converges to f in $C^\alpha(I;X)$, then

$$\|f(x) - f(y)\| \leq \|f(x) - f_n(x) + f_n(y) - f(y)\| + \|f_n(x) - f_n(y)\|$$
$$\leq [f - f_n]_{C^\alpha(I;X)}|x-y|^\alpha + \|f_n(x) - f_n(y)\|.$$

Fix $\varepsilon > 0$, and let n be such that $[f - f_n]_{C^\alpha} \leq \varepsilon/2$, let $\delta > 0$ be such that $\|f_n(x) - f_n(y)\| \leq |x-y|^\alpha \varepsilon/2$ for $|x-y| \leq \delta$. So, for $|x-y| \leq \delta$ it holds that $\|f(x) - f(y)\| \leq \varepsilon |x-y|^\alpha$. Therefore, f is α-little Hölder continuous.

Concerning the other inclusion, we first consider the case $I = \mathbb{R}$.

Let $f \in h^\alpha(\mathbb{R};X)$. For each $n \in \mathbb{N}$ let $\rho_n \in C^\infty(\mathbb{R})$ be such that

$$\operatorname{supp} \rho_n \subset \,]-1/n, 1/n[, \quad \|\rho_n\|_{L^1(\mathbb{R})} = 1,$$

and set

$$f_n(x) = \int_\mathbb{R} f(x-y)\rho_n(y)dy, \quad x \in \mathbb{R}, \; n \in \mathbb{N}.$$

It is not hard to see that $[f_n - f]_{C^\alpha}$ goes to 0 as $n \to \infty$. This implies that $h^\alpha(\mathbb{R};X)$ is contained in the closure of $C^\eta(\mathbb{R};X)$ for every $\eta > \alpha$.

Let now $I = [a,b]$, and let $f \in h^\alpha([a,b];X)$. To show that there is a sequence of smooth functions converging to f in $C^\alpha([a,b];X)$, first we extend f to a function F defined in the whole $\mathbb{R}$, and then we use the previous step. We extend f to the whole of $\mathbb{R}$ by setting

$$\tilde{f}(x) = \begin{cases} f(a) & \text{for } x \leq a, \\ f(x) & \text{for } a \leq x \leq b, \\ f(b) & \text{for } x \geq b. \end{cases}$$

Then $\tilde{f}$ is little-Hölder continuous in $\mathbb{R}$, and

$$[\tilde{f}]_{C^\alpha(\mathbb{R};X)} = [f]_{C^\alpha([a,b];X)}, \quad \|\tilde{f}\|_\infty = \|f\|_\infty.$$

2. Spaces of Hölder continuous functions

We have seen that $\widetilde{f}$ can be approximated in $C^\alpha(\mathbb{R}; X)$ by a sequence of functions $\{\widetilde{f_n}\}_{n\in\mathbb{N}}$ belonging to $C^\eta([a,b]; X)$ for every $\eta > \alpha$. The restrictions of the $\widetilde{f_n}$'s to $[a,b]$ are smooth functions approximating f in $C^\alpha([a,b]; X)$.

The case where I is a halfline is similar and is left to the reader. ∎

We now introduce another class of spaces. For $0 \leq \alpha < 2$ we set

$$\mathcal{C}^\alpha(I; X) = \{f \in C_b(I; X) : [f]_{\mathcal{C}^\alpha(I;X)} = \sup_{x,y\in I;\ x\neq y} \frac{\|f(x) - 2f\left(\frac{x+y}{2}\right) + f(y)\|}{|x-y|^\alpha} < +\infty\},$$

$$\|f\|_{\mathcal{C}^\alpha(I;X)} = \|f\|_\infty + [f]_{\mathcal{C}^\alpha(I;X)}.$$

The interest of such spaces relies on the fact that for $\alpha \neq 1$ they are equivalent to Hölder spaces, as the following proposition shows.

Proposition 0.2.2 *For $0 < \alpha < 2$, $\alpha \neq 1$, we have*

$$\mathcal{C}^\alpha(I; X) = C^\alpha(I; X),$$

with equivalence of the respective norms.

Proof — The embedding $C^\alpha(I; X) \subset \mathcal{C}^\alpha(I; X)$ is trivial; indeed if $f \in C^\alpha(I; X)$ with $0 < \alpha < 1$, then

$$\begin{aligned}\|f(x) - 2f(\tfrac{x+y}{2}) + f(y)\| &\leq \|f(x) - f(\tfrac{x+y}{2})\| \\ &+ \|f(y) - f(\tfrac{x+y}{2})\| \leq 2\left(\tfrac{|x-y|}{2}\right)^\alpha [f]_{C^\alpha},\end{aligned} \quad (0.2.1)$$

whereas if $f \in C^\alpha(I; X)$ with $1 < \alpha < 2$, then

$$\begin{aligned}\|f(x) - 2f(\tfrac{x+y}{2}) + f(y)\| &= \Big\|\tfrac{x-y}{2}(f'(x) - f'(y)) \\ &+ \int_{\frac{x+y}{2}}^x (f'(\sigma) - f'(x))d\sigma - \int_y^{\frac{x+y}{2}}(f'(\sigma) - f'(y))d\sigma\Big\| \\ &\leq \tfrac{|x-y|^\alpha}{2}[f']_{C^{\alpha-1}} + 2\left(\tfrac{|x-y|}{2}\right)^\alpha [f']_{C^{\alpha-1}}.\end{aligned} \quad (0.2.2)$$

Let us show that $\mathcal{C}^\alpha(I; X) \subset C^\alpha(I; X)$. Fix $x \in I$. For every τ such that $x + \tau \in I$, set

$$g(\tau) = f(x+\tau) - f(x).$$

Then

$$\|g(\tau) - 2g(\tau/2)\| = \|f(x+\tau) - 2f(x+\tau/2) + f(x)\| \leq [f]_{\mathcal{C}^\alpha}\tau^\alpha,$$

and replacing successively τ by $\tau/2$, $\tau/4$ etc., it is easy to see by recurrence that

$$\|2^{n-1}g(\tau/2^{n-1}) - 2^n g(\tau/2^n)\| \leq 2^{(n-1)(1-\alpha)}\tau^\alpha [f]_{\mathcal{C}^\alpha}, \quad \forall n \in \mathbb{N}.$$

From this we get the estimate

$$\|g(\tau) - 2^n g(\tau/2^n)\| \leq \tau^\alpha \sum_{k=0}^{n-1} 2^{k(1-\alpha)} [f]_{C^\alpha}, \quad \forall n \in \mathbb{N}. \tag{0.2.3}$$

Let us consider the case $0 < \alpha < 1$, and $I = [0,1]$. From (0.2.3) we get

$$\|g(\tau) - 2^n g(\tau/2^n)\| \leq \tau^\alpha \frac{2^{n(1-\alpha)}}{2^{1-\alpha} - 1} [f]_{C^\alpha}. \tag{0.2.4}$$

Now we are able to estimate $\|f(x_1) - f(x_2)\|$, for $0 \leq x_1 \leq x_2 \leq 1$, and $x_2 - x_1 \leq 1/4$. If $x_1 \leq 1/2$, set $x = x_1$ and $h = x_2 - x_1$; if $x_1 > 1/2$, set $x = x_2$ and $h = x_1 - x_2$. Let $\tau = 2^n h$, where $n \in \mathbb{N}$ is such that $1/4 \leq 2^n |h| \leq 1/2$. Therefore, both x and $x + \tau$ belong to $[0,1]$, and from (0.2.4) we get

$$\|f(x_1) - f(x_2)\| = \|g(h)\| \leq \frac{1}{2^n} \left(\|g(\tau)\| + \tau^\alpha \frac{2^{n(1-\alpha)}}{2^{1-\alpha} - 1} [f]_{C^\alpha} \right)$$

$$\leq 8 \|f\|_\infty |h| + (2^{1-\alpha} - 1)^{-1} [f]_{C^\alpha} |h|^\alpha.$$

For $x_2 - x_1 > 1/4$, we have obviously

$$\|f(x_1) - f(x_2)\| \leq 2 \|f\|_\infty \leq 2^{1+2\alpha} |x_1 - x_2|^\alpha \|f\|_\infty,$$

so that $f \in C^\alpha([0,1]; X)$ and

$$[f]_{C^\alpha([0,1];X)} \leq 8 \|f\|_\infty + (2^{1-\alpha} - 1)^{-1} [f]_{C^\alpha([0,1];X)}. \tag{0.2.5}$$

Let us now consider the case $1 < \alpha < 2$, in which estimate (0.2.3) gives

$$\|g(\tau) - 2^n g(\tau/2^n)\| \leq \tau^\alpha \frac{2^{\alpha-1}}{2^{\alpha-1} - 1} [f]_{C^\alpha}, \quad \forall n \in \mathbb{N}. \tag{0.2.6}$$

To show that f is differentiable in $[0, 1/2]$, for every $x \in [0, 1/2]$ we set

$$F_k(x) = 2^k \int_x^{x+1/2^k} f(\sigma) d\sigma, \quad k \in \mathbb{N}.$$

Since f is continuous, then F_k converges uniformly to f in $[0, 1/2]$ as $k \to \infty$. Moreover,

$$F_k'(x) = \frac{f(x + 1/2^k) - f(x)}{1/2^k} = \frac{g(1/2^k)}{1/2^k}.$$

From (0.2.6), taking $\tau = 1/2^k$, we get

$$\|F_k'(x) - F_{k+n}'(x)\| \leq \frac{2^{\alpha-1}}{2^{k(\alpha-1)}(2^{\alpha-1} - 1)} [f]_{C^\alpha}, \quad \forall x \in [0, 1/2], \, k, \, n \in \mathbb{N},$$

2. Spaces of Hölder continuous functions

so that F'_k is a Cauchy sequence in $C([0, 1/2]; X)$. It follows that f is differentiable in $[0, 1/2]$, and using again (0.2.6), with $\tau = 1/2$, $n = k$, we get

$$\|f'(x)\| = \lim_{k \to +\infty} \frac{\|g(1/2^k)\|}{1/2^k} \leq \frac{\|g(1/2)\|}{1/2} + \frac{1}{4^{\alpha-1}}[f]_{C^\alpha} \leq 2\|f\|_\infty + \frac{1}{4^{\alpha-1}}[f]_{C^\alpha} \quad (0.2.7)$$

To show that f is differentiable in $[1/2, 1]$ it is sufficient to replace $1/2^k$ by $-1/2^k$ in the definition of F_k. Estimate (0.2.7) still holds for $x \in [1/2, 1]$.

Let us show that f' is $(\alpha - 1)$-Hölder continuous. First we estimate the difference between the incremental ratio and the derivative, using (0.2.6).

$$\left\| \frac{f(x+\tau) - f(x)}{\tau} - f'(x) \right\| = \lim_{k \to +\infty} \left\| \frac{f(x+\tau) - f(x)}{\tau} - \frac{f(x+\tau/2^k) - f(x)}{\tau/2^k} \right\|$$

$$= \lim_{k \to +\infty} \left\| \frac{g(\tau)}{\tau} - \frac{g(\tau/2^k)}{\tau/2^k} \right\| \leq \tau^{\alpha-1} \frac{2^{\alpha-1}}{2^{\alpha-1} - 1}[f]_{C^\alpha}, \quad \forall k \in \mathbb{N}.$$

It follows that for x, $x + \tau \in [0, 1]$ we have

$$\|f'(x+\tau) - f'(x)\| \leq \left\| f'(x+\tau) - \frac{f(x+2\tau) - f(x+\tau)}{\tau} \right\|$$

$$+ \left\| \frac{f(x+2\tau) - 2f(x+\tau) + f(x)}{\tau} \right\| + \left\| \frac{f(x+\tau) - f(x)}{\tau} - f'(x) \right\|$$

$$\leq \frac{2^{\alpha-1}}{2^{\alpha-1} - 1} \tau^{\alpha-1}[f]_{C^\alpha} + \tau^{\alpha-1}[f]_{C^\alpha},$$

so that f' is $(\alpha - 1)$-Hölder continuous.

The same proof works in the case where I is a halfline or $I = \mathbb{R}$. In the case where $I = [a, b]$, it is sufficient to map $[a, b]$ into $[0, 1]$ by the transformation $x \mapsto (x - a)(b - a)^{-1}$. ∎

The proof of the inclusion $\mathcal{C}^\alpha([0, 1]; X) \subset C^\alpha([0, 1]; X)$ is adapted from A. ZYGMUND [216, p. 44]. By the same proof of the case $0 < \alpha < 1$ one obtains

$$\|f(x+h) - f(x)\| \leq \text{const.} \, |h \log |h|| \, \|f\|_{C^1(I;X)}$$

for any $f \in \mathcal{C}^1(I; X)$ and x, $x + h \in I$.

If Ω is an open set in $\mathbb{R}^n$, and $0 < \theta < 1$, the Banach space $C^\theta(\overline{\Omega})$ and its subspace $h^\theta(\overline{\Omega})$ are defined by

$$C^\theta(\overline{\Omega}) = \{ f \in C(\overline{\Omega}) : [f]_{C^\theta} = \sup_{x, y \in \overline{\Omega}, x \neq y} \frac{|f(x) - f(y)|}{|x - y|^\theta} < \infty \},$$

$$\|f\|_{C^\theta(\overline{\Omega})} = \|f\|_\infty + [f]_{C^\theta},$$

$$h^\theta(\overline{\Omega}) = \{ f \in C^\theta(\overline{\Omega}) : \lim_{\tau \to 0} \sup_{x, y \in \overline{\Omega}, 0 < |x-y| < \tau} \frac{|f(x) - f(y)|}{|x - y|^\theta} = 0 \},$$

whereas if $\theta > 1$, $\theta \notin \mathbb{N}$, we set

$$C^\theta(\overline{\Omega}) = \{f \in C^{[\theta]}(\overline{\Omega}) : D^\beta f \in C^{\theta-[\theta]}(\overline{\Omega}), \ |\beta| = [\theta]\},$$
$$\|f\|_{C^\theta(\overline{\Omega})} = \|f\|_{C^{[\theta]}(\overline{\Omega})} + \sum_{|\beta|=[\theta]} [D^\beta f]_{C^{\theta-[\theta]}(\overline{\Omega})},$$

$$h^\theta(\overline{\Omega}) = \{f \in C^\theta(\overline{\Omega}) : D^\beta f \in h^{\theta-[\theta]}(\overline{\Omega}), \ |\beta| = [\theta]\}.$$

Here $[\theta]$ is the greatest integer $\leq \theta$.

The definition of uniformly $C^{m+\alpha}$ boundary and uniformly $h^{m+\alpha}$ boundary ($m \in \mathbb{N}$, $0 \leq \alpha < 1$) for open sets $\Omega \subset \mathbb{R}^n$ is analogous to the definition of uniformly C^m boundary (see Section 0.1).

Arguments similar to the ones of Proposition 0.2.1 show that $h^\theta(\mathbb{R}^n)$ is the closure of $C^k(\mathbb{R}^n)$ in $C^\theta(\mathbb{R}^n)$, for every $k \in \,]\alpha, +\infty]$, and that if Ω is an open set with regular boundary, then $h^\theta(\overline{\Omega})$ is the closure of $C^k(\overline{\Omega})$ in $C^\theta(\overline{\Omega})$, for every $k \in \,]\alpha, +\infty]$.

0.3 Extension operators

In the next chapters we shall need to extend functions defined on a smooth hypersurface, boundary of an open set, to the whole open set. As a first step, one extends functions defined in $\mathbb{R}^{n-1} \times \{0\}$ to the whole $\mathbb{R}^n$.

Proposition 0.3.1 *Let $n \geq 2$. Fix a function $\varphi \in C^\infty(\mathbb{R}^{n-1})$, with compact support, and such that $\int_{\mathbb{R}^{n-1}} \varphi(\xi) d\xi = 1$. For each $k \in \mathbb{N}$ and $f \in C(\mathbb{R}^{n-1})$ set*

$$F_k(y_1, \ldots, y_n) = \frac{y_n^k}{k!} \int_{\mathbb{R}^{n-1}} \varphi(\xi) f(y_1 + \xi_1 y_n, \ldots, y_{n-1} + \xi_{n-1} y_n) d\xi. \quad (0.3.1)$$

Then $F_k \in C^k(\mathbb{R}^n)$, and for every multi-index β such that $0 \leq |\beta| \leq k$, for every $(y_1, \ldots, y_{n-1}) \in \mathbb{R}^{n-1}$ it holds

$$D^\beta F_k(y_1, \ldots, y_{n-1}, 0) \begin{cases} = 0, & \text{if } \beta \neq (0, \ldots, 0, k) \\ = f(y_1, \ldots, y_{n-1}), & \text{if } \beta = (0, \ldots, 0, k). \end{cases} \quad (0.3.2)$$

Moreover, if $f \in C^{m+\theta}(\mathbb{R}^{n-1})$, with $m \in \mathbb{N}$ and $\theta \in [0, 1[$, then $F_k \in C^{k+m+\theta}(\mathbb{R}^n)$, and there is $C > 0$, independent of f, such that

$$\|F_k\|_{C^{k+m+\theta}(\mathbb{R}^n)} \leq C\|f\|_{C^{m+\theta}(\mathbb{R}^{n-1})} \quad (0.3.3)$$

The proof of Proposition 0.3.1 is omitted.

3. EXTENSION OPERATORS

With the aid of Proposition 0.3.1, one constructs extension operators in arbitrary smooth domains. We do not state a more general result, for which we refer to A. LUNARDI, E. SINESTRARI, W. VON WAHL [156, Theorem 6.3], but only what will be used in the sequel.

Theorem 0.3.2 *Let Ω be an open set in $\mathbb{R}^n$ with uniformly $C^{2+\alpha}$ boundary $\partial\Omega$, $0 \leq \alpha < 1$. Then:*

(i) *there exists an extension operator $\mathcal{D} \in L(C^\theta(\partial\Omega), C^\theta(\overline{\Omega}))$ for each $\theta \in [0, 2+\alpha]$, such that*
$$\mathcal{D}f_{|\partial\Omega} = f, \quad \forall f \in C(\partial\Omega).$$

(ii) *Let $\beta_i, \gamma \in C^{1+\alpha}(\overline{\Omega})$ be such that*
$$\left|\sum_{i=1}^n \beta_i(x)\nu_i(x)\right| \geq \varepsilon > 0, \quad x \in \partial\Omega, \tag{0.3.4}$$

and for every $u \in C^1(\overline{\Omega})$ set
$$\mathcal{B}u(x) = \sum_{i=1}^n \beta_i(x)D_i u(x) + \gamma(x)u(x), \quad x \in \overline{\Omega}.$$

There exists an operator $\mathcal{N} \in L(C^\theta(\partial\Omega), C^{\theta+1}(\overline{\Omega}))$ for each $\theta \in [0, 1+\alpha]$, such that
$$\mathcal{B}(\mathcal{N}f)_{|\partial\Omega} = f, \quad \forall f \in C(\partial\Omega).$$

Proof — Let $\{\xi_j\}_{j\in\mathbb{N}}$ be a partition of 1 associated with the covering $\{U_j : j \in \mathbb{N}\}$ of $\partial\Omega$, with the following properties: every ξ_j is a C^∞ function with support contained in U_j, $\sum_{j=1}^\infty \xi_j(x) = 1$ for every x in a neighborhood of $\partial\Omega$, and $\sup_{j\in\mathbb{N}} \|\xi_j\|_{C^{2+\alpha}} < \infty$.

Let us prove statement (i). For $f \in C^\theta(\overline{\Omega})$, $0 \leq \theta \leq 2+\alpha$, define
$$\widetilde{f}_i(y') = f(\varphi_i)^{-1}(y', 0), \quad y' \in \mathbb{R}^{n-1}, \ |y'| \leq 1, \tag{0.3.5}$$

where φ_i are the functions which locally straighten the boundary. The function $\widetilde{f}_i$ has an obvious extension to $B_{n-1}(0,1) \times [0, +\infty[$, defined by
$$\widetilde{F}_i(y', y) = \widetilde{f}_i(y') = f(\varphi_i)^{-1}(y', 0), \quad y' \in B_{n-1}(0,1), \ y \geq 0.$$

Then $\widetilde{F}_i \in C^\theta(B_{n-1}(0,1) \times [0, +\infty[)$. Extend the function $x \mapsto \xi_i(x)\widetilde{F}_i(\varphi_i(x))$ to the whole of $\overline{\Omega}$, by setting
$$F_i(x) \begin{cases} = \xi_i(x)\widetilde{F}_i(\varphi_i(x)), & x \in \overline{U_i} \cap \overline{\Omega}, \\ = 0, & x \in \overline{\Omega}\setminus \overline{U_i}. \end{cases}$$

Since the support of ξ_i is contained in U_i, then F_i belongs to $C^\theta(\overline{\Omega})$. Due to the properties of the functions ξ_i and φ_i, it holds

$$\|F_i\|_{C^\theta(\overline{\Omega})} \leq C\|f\|_{C^\theta(\partial\Omega)},$$

with C independent of i.

Define an extension $\mathcal{D}f$ of f by

$$\mathcal{D}f(x) = \sum_{i=1}^\infty \xi_i(x) F_i(x), \quad x \in \overline{\Omega}.$$

Then the operator $\mathcal{D}$ enjoys the claimed properties.

Let us prove statement (ii). For $f \in C^\theta(\partial\Omega)$, $0 \leq \theta \leq 1 + \alpha$, and for every $i \in \mathbb{N}$, the function $\widetilde{f}_i$ defined by (0.3.5) belongs to $C^\theta(B_{n-1}(0,1))$. By statement (i), it has an extension to the whole $\mathbb{R}^{n-1}$, which we still denote by $\widetilde{f}_i$, satisfying

$$\|\widetilde{f}_i\|_{C^\theta(\mathbb{R}^{n-1})} \leq C\|f\|_{C^\theta(\partial\Omega)},$$

with C independent of i. By Proposition 0.3.1, applied with $k = 1$, for every $i \in \mathbb{N}$ there exists a function $F_i \in C^{\theta+1}(\mathbb{R}^n_+)$ such that

$$F_i(y',0) = D_k F_i(y',0) = 0, \quad k = 1,\ldots,n-1; \quad D_n F_i(y',0) = \frac{\widetilde{f}_i(y')}{\psi_i(y')},$$

where

$$\psi_i(y') = \sum_{k=1}^n \beta_k(\varphi_i^{-1}(y',0)) \frac{\partial \varphi_i^{(n)}}{\partial x_k}(\varphi_i^{-1}(y',0)).$$

Note that the gradient of φ_i^n (the n-th component of φ) is orthogonal to $\partial\Omega$, so that the uniform nontangentiality assumption (0.3.4) implies that $|\psi_i(y')| \geq \varepsilon$ for every y'. So, there is $C > 0$ such that

$$\|F_i\|_{C^{\theta+1}(\mathbb{R}^n_+)} \leq C\|f\|_{C^\theta(\partial\Omega)}.$$

We define now the operator $\mathcal{N}$ by

$$\mathcal{N}f(x) = \sum_{i=1}^\infty \xi_i(x) F_i(\varphi_i(x)), \quad x \in \overline{\Omega}.$$

It is easy to check that $\mathcal{N}$ enjoys the properties stated. ∎

Chapter 1

Interpolation theory

In this chapter we give a self-contained exposition of the part of interpolation theory which will be used in the following. We do not attempt to give an exhaustive overview on interpolation theory and its applications. We refer the interested reader to the bibliography in Section 1.3.

To describe the contents of the chapter, let us give some definitions. If X, Y, D are Banach spaces such that[1]

$$D \subset Y \subset X$$

we say that Y is an *intermediate space* between X and D. If, in addition, for every linear operator $T \in L(X)$ such that $T_{|D} \in L(D)$ it holds $T_{|Y} \in L(Y)$, then Y is called *interpolation space* between X and D.

Section 1.1 deals with interpolatory estimates. There we study the spaces belonging to the class J_α ($0 \leq \alpha \leq 1$) between X and D, that is the intermediate spaces Y for which there is, $c > 0$ such that

$$\|x\|_Y \leq c \|x\|_D^\alpha \|x\|_X^{1-\alpha}, \quad \forall x \in D.$$

We provide important examples of such spaces. For instance, we prove that for every regular open set $\Omega \subset \mathbb{R}^n$, the space $C^k(\overline{\Omega})$ belongs to the class $J_{k/m}$ between $C(\overline{\Omega})$ and $C^m(\overline{\Omega})$, for $0 < k < m$. We prove also inclusions between spaces of functions with values in Banach spaces, such as

$$C^\theta([a,b]; D) \cap C^{\theta+1}([a,b]; X) \subset C^{1-\alpha+\theta}(I; Y),$$

where $0 < \alpha, \theta < 1$, $\alpha \neq \theta$, and Y belongs to the class J_α between X and D.

In Section 1.2 we describe two methods to construct a family of interpolation spaces: the *K-method* and the *trace method*. Such spaces are called *real interpolation spaces*. Then we prove the Reiteration Theorem, a very useful tool which will be employed frequently in the following.

---
[1]The symbol $\subset$ denotes continuous embedding.

At the end of the chapter we give some examples. In particular, we prove that for every regular open set $\Omega \subset \mathbb{R}^n$, the Hölder and little-Hölder spaces $C^{k+\sigma}(\overline{\Omega})$, $h^{k+\sigma}(\overline{\Omega})$ ($k \in \mathbb{N} \cup \{0\}$, $0 < \sigma < 1$) are real interpolation spaces between $C(\overline{\Omega})$ and $C^m(\overline{\Omega})$, for $m > k + \sigma$.

Other elements of interpolation theory will be given in Section 2.2, where we shall consider real interpolation spaces between Banach spaces and domains of sectorial operators. Some explicit characterizations will be given in Chapter 3.

1.1 Interpolatory inclusions

Let X, D be Banach spaces, with norms $\|\cdot\|$, $\|\cdot\|_D$ respectively, and assume that D is continuously embedded in X.

Definition 1.1.1 *Let $0 \leq \alpha \leq 1$. A Banach space Y such that $D \subset Y \subset X$ is said to belong to the class J_α between X and D if there is a constant c such that*

$$\|x\|_Y \leq c \|x\|^{1-\alpha} \|x\|_D^\alpha, \ \forall x \in D.$$

In this case we write $Y \in J_\alpha(X, D)$.

Some important examples are given in the next propositions.

Proposition 1.1.2 *Let k, m be positive integers such that $k < m$. Then:*

(i) *$C_b^k(\mathbb{R}; X)$ belongs to the class $J_{k/m}$ between $C_b(\mathbb{R}; X)$ and $C_b^m(\mathbb{R}; X)$;*

(ii) *$C^k(\mathbb{R}^n)$ belongs to the class $J_{k/m}$ between $C(\mathbb{R}^n)$ and $C^m(\mathbb{R}^n)$; $C^k(\mathbb{R}^n_+)$ belongs to the class $J_{k/m}$ between $C(\mathbb{R}^n_+)$ and $C^m(\mathbb{R}^n_+)$;*

(iii) *if Ω is an open set in $\mathbb{R}^n$ with uniformly C^m boundary, then $C^k(\overline{\Omega})$ belongs to the class $J_{k/m}$ between $C(\overline{\Omega})$ and $C^m(\overline{\Omega})$.*

Proof — Let us prove statement (i). It is sufficient to show that for every $m \in \mathbb{N}$ there is $c_m > 0$ such that if $f \in C_b^m(\mathbb{R}; X)$ then

$$\|f^{(k)}\|_\infty \leq c_m (\|f\|_\infty)^{1-k/m} (\|f^{(m)}\|_\infty)^{k/m}, \ \forall k = 1, .., m-1. \quad (1.1.1)$$

First we prove that (1.1.1) holds for $m = 2$ and $k = 1$. From the inequality

$$\|f(x+h) - f(x) - f'(x)h\| \leq \frac{1}{2} \|f''\|_\infty h^2, \ \forall x \in \mathbb{R}, \ h > 0,$$

we get

$$\|f'(x)\| \leq \frac{\|f(x+h) - f(x)\|}{h} + \frac{1}{2} \|f''\|_\infty h, \ \forall x \in \mathbb{R}, \ h > 0,$$

so that

$$\|f'\|_\infty \leq \frac{2\|f\|_\infty}{h} + \frac{1}{2} \|f''\|_\infty h, \ \forall h > 0.$$

1. INTERPOLATORY INCLUSIONS

Taking the minimum on h over $(0, +\infty)$ we get

$$\|f'\|_\infty \leq 2(\|f\|_\infty)^{1/2}(\|f''\|_\infty)^{1/2}, \quad \forall f \in C_b^2(\mathbb{R}; X). \tag{1.1.2}$$

Assume now by recurrence that (1.1.1) holds for some $m > 2$. Then for every $f \in C_b^{m+1}(\mathbb{R}; X)$ we get

$$\|f'\|_\infty \leq c_m(\|f\|_\infty)^{1-1/m}(\|f^{(m)}\|_\infty)^{1/m}$$

$$\leq c_m(\|f\|_\infty)^{1-1/m}\left[c_m(\|f'\|_\infty)^{1-\frac{m-1}{m}}(\|f^{(m+1)}\|_\infty)^{\frac{m-1}{m}}\right]^{1/m}.$$

It follows that

$$\|f'\|_\infty \leq c_m^{\frac{m+1}{m}}(\|f\|_\infty)^{1-\frac{1}{m+1}}(\|f^{(m+1)}\|_\infty)^{\frac{1}{m+1}},$$

and, for $k = 2, ..., m$,

$$\|f^{(k)}\|_\infty = \|(f')^{(k-1)}\|_\infty \leq c_m(\|f'\|_\infty)^{1-(k-1)/m}(\|f^{(m+1)}\|_\infty)^{(k-1)/m}$$

$$\leq c_m^{1+\frac{m-k+1}{m-1}}(\|f\|_\infty)^{1-\frac{k}{m+1}}(\|f^{(m+1)}\|_\infty)^{\frac{k}{m+1}}.$$

Therefore, (1.1.1) holds with m replaced by $m+1$, and statement (i) is proved.

Statement (ii) can be shown similarly, replacing h by he_i, where e_i is the vector whose k-th component is 0 if $k \neq i$, 1 if $k = i$.

Statement (iii) follows from (ii) by localizing and straightening the boundary. ∎

Similar arguments lead to the following generalization.

Proposition 1.1.3 *Let $0 < \theta < \alpha$. Then:*

(i) $C^\theta(\mathbb{R}; X)$ *belongs to the class* $J_{\theta/\alpha}$ *between* $C_b(\mathbb{R}; X)$ *and* $C^\alpha(\mathbb{R}; X)$;

(ii) $C^\theta(\mathbb{R}^n)$ *belongs to the class* $J_{\theta/\alpha}$ *between* $C(\mathbb{R}^n)$ *and* $C^\alpha(\mathbb{R}^n)$; $C^\theta(\mathbb{R}^n_+)$ *belongs to the class* $J_{\theta/\alpha}$ *between* $C(\mathbb{R}^n_+)$ *and* $C^\alpha(\mathbb{R}^n_+)$;

(iii) *if Ω is an open set in $\mathbb{R}^n$ with uniformly C^α boundary, then $C^\theta(\overline{\Omega})$ belongs to the class $J_{\theta/\alpha}$ between $C(\overline{\Omega})$ and $C^\alpha(\overline{\Omega})$.*

Now we are able to state some interpolatory inclusions between spaces of functions defined in an interval I. Such inclusions will be widely used throughout the book.

Proposition 1.1.4 *Let $X_\alpha \in J_\alpha(X, D)$. Then*

(i) $B(I; D) \cap Lip(I; X) \subset C^{1-\alpha}(I; X_\alpha)$,

(ii) $B(I; D) \cap C^\theta(I; X) \subset C^{\theta(1-\alpha)}(I; X_\alpha), \quad \forall \theta \in]0, 1[$,

(iii) $B(I; D) \cap C_b(I; X) \subset C_b(I; X_\alpha)$.

Proof — Let us prove statement (i). For every $u \in B(I;D) \cap Lip(I;X)$ and for $s,t \in I$ we have

$$\|u(t) - u(s)\|_{X_\alpha} \le c\|u(t) - u(s)\|^{1-\alpha}\|u(t) - u(s)\|_D^\alpha$$
$$\le c[u]_{Lip(I;X)}^{1-\alpha}|t-s|^{1-\alpha}(2\|u\|_{B(I;D)})^\alpha,$$

so that u is $(1-\alpha)$-Hölder continuous with values in X_α, and

$$\begin{aligned}\|u\|_{C^{1-\alpha}(I;X_\alpha)} &\le c\|u\|_{B(I;X)}^{1-\alpha}\|u\|_{B(I;D)}^\alpha + 2^\alpha c[u]_{Lip(I;X)}^{1-\alpha}\|u\|_{B(I;D)}^\alpha \\ &\le (2^\alpha + 1)c\|u\|_{Lip(I;X)}^{1-\alpha}\|u\|_{B(I;D)}^\alpha.\end{aligned} \qquad (1.1.3)$$

The proof of statements (ii) and (iii) is similar. ∎

Proposition 1.1.5 *Let $X_\alpha \in J_\alpha(X,D)$. Then for every $\theta \in \,]0,1[$*

$$C^\theta(I;D) \cap C^{\theta+1}(I;X) \subset C^{\theta+1-\alpha}(I;X_\alpha)$$

In particular, if $\theta \ne \alpha$, then

$$\mathcal{C}^\theta(I;D) \cap \mathcal{C}^{\theta+1}(I;X) \subset \mathcal{C}^{\theta+1-\alpha}(I;X_\alpha).$$

Proof — Fix $u \in C^\theta(I;D) \cap C^{\theta+1}(I;X)$. Then for $t,s \in I$

$$\left\|u(t) + u(s) - 2u\left(\tfrac{t+s}{2}\right)\right\|_{X_\alpha}$$
$$\le c\left\|u(t) + u(s) - 2u\left(\tfrac{t+s}{2}\right)\right\|_D^\alpha \left\|u(t) + u(s) - 2u\left(\tfrac{t+s}{2}\right)\right\|^{1-\alpha},$$

so that from estimates (0.2.1) and (0.2.2) we get

$$\left\|u(t) + u(s) - 2u\left(\tfrac{t+s}{2}\right)\right\|_{X_\alpha}$$
$$\le c\left(2^{1-\theta}[u]_{C^\theta(D)}|t-s|^\theta\right)^\alpha \left((1/2 + 2^{1-\theta})[u']_{C^\theta(X)}|t-s|^{\theta+1}\right)^{1-\alpha}$$
$$= c'([u']_{C^\theta(X)})^{1-\alpha}([u]_{C^\theta(D)})^\alpha |t-s|^{\theta+1-\alpha},$$

so that $u \in \mathcal{C}^{\theta+1-\alpha}(I;X_\alpha)$, and

$$\|u\|_{\mathcal{C}^{\theta+1-\alpha}(I;X_\alpha)} \le c(\|u\|_{C_b(I;D)})^\alpha (\|u\|_{C_b(I;X)})^{1-\alpha}$$
$$+c'([u]_{C^\theta(I;D)})^\alpha([u']_{C^{\theta+1}(I;X)})^{1-\alpha} \le c''(\|u\|_{C^\theta(I;D)})^\alpha(\|u\|_{C^{\theta+1}(I;X)})^{1-\alpha}.$$
$$(1.1.4)$$
∎

2. Interpolation spaces

Remarks 1.1.6

(i) We shall see in Chapter 2 that the inclusions given by Propositions 1.1.4 and 1.1.5 are sharp. In particular, in the case $\alpha = \theta$ the functions belonging to $C^{\theta+1}(I;X) \cap C^{\theta}(I;D)$ are not necessarily Lipschitz continuous or differentiable with values in X_θ, but they belong only to $C^1(I;X_\theta)$. See the remarks after Example 2.2.11.

(ii) If $I = [a,b]$, the embedding constants of estimates (1.1.3) and (1.1.4) do not depend on a, b. But the embedding constant of the inclusion $C^{\theta+1-\alpha}([a,b];X) \subset \mathcal{C}^{\theta+1-\alpha}([a,b];X)$ (see Proposition 0.2.2), for $\alpha \neq \theta$, blows up as $b - a \to 0$. Indeed, consider the function $f(t) = (t-a)x$, where $x \neq 0$ is any element of X. Then, for every $\beta \in]0,1[$, $[f]_{\mathcal{C}^\beta([a,b];X)} = 0$, so that $\|f\|_{\mathcal{C}^\beta([a,b];X)} = (b-a)\|x\|$, whereas $\|f\|_{C^\beta([a,b];X)} = (b-a)\|x\| + (b-a)^{1-\beta}\|x\|$. Therefore,

$$\lim_{b-a \to 0} \frac{\|f\|_{C^\beta([a,b];X)}}{\|f\|_{\mathcal{C}^\beta([a,b];X)}} = +\infty.$$

1.2 Interpolation spaces

1.2.1 The K-method

Let X, Y be Banach spaces, with $Y \subset X$, and let $c > 0$ be such that

$$\|y\|_X \leq c\|y\|_Y, \quad \forall y \in Y.$$

We describe briefly the construction of a family of intermediate spaces between X and Y, called *real interpolation spaces*, and denoted by $(X,Y)_{\theta,p}$, $(X,Y)_\theta$, with $0 < \theta \leq 1$, $1 \leq p \leq \infty$. We follow the so-called K-method.

Throughout this and the next section we set $1/\infty = 0$.

Definition 1.2.1 *For every $x \in X$ and $t > 0$, set*

$$K(t,x,X,Y) = \inf_{x=a+b,\ a\in X,\ b\in Y} (\|a\|_X + t\|b\|_Y). \quad (1.2.1)$$

If there is no danger of confusion, we shall write $K(t,x)$ instead of $K(t,x,X,Y)$.

From Definition 1.2.1 it follows immediately that for every $t > 0$ and $x \in X$ we have

$$\begin{cases} (i) & \min\{1,t\}K(1,x) \leq K(t,x) \leq \max\{1,t\}K(1,x), \\ (ii) & K(t,x) \leq \|x\|_X. \end{cases} \quad (1.2.2)$$

Now we define a family of Banach spaces by means of the function K.

Definition 1.2.2 *Let $0 < \theta \leq 1$, $1 \leq p \leq \infty$, and set*

$$\begin{cases} (X,Y)_{\theta,p} = \{x \in X : t \mapsto t^{-\theta - 1/p} K(t,x,X,Y) \in L^p(0,+\infty)\}, \\ \|x\|_{(X,Y)_{\theta,p}} = \|t^{-\theta - 1/p} K(t,x,X,Y)\|_{L^p(0,+\infty)}; \end{cases} \quad (1.2.3)$$

$$(X,Y)_\theta = \{x \in X : \lim_{t \to 0} t^{-\theta} K(t,x,X,Y) = 0\}. \quad (1.2.4)$$

The mapping $x \mapsto \|x\|_{(X,Y)_{\theta,p}}$ is easily seen to be a norm in $(X,Y)_{\theta,p}$. Where there is no danger of confusion, we shall write $\|x\|_{\theta,p}$ instead of $\|x\|_{(X,Y)_{\theta,p}}$.

Since $t \mapsto K(t,x)$ is bounded, it is clear that only the behavior near $t = 0$ of $t^{-\theta} K(t,x)$ plays a role in the definition of $(X,Y)_{\theta,p}$ and of $(X,Y)_\theta$. Indeed, one could replace the half line $(0, +\infty)$ by any interval $(0, a)$ in Definition 1.2.2, obtaining equivalent norms.

For $\theta = 1$, from the first inequality in (1.2.2)(i) we get

$$(X,Y)_1 = (X,Y)_{1,p} = \{0\}, \quad p < \infty.$$

Therefore, from now on we shall consider the cases $(\theta, p) \in \,]0,1[\,\times [1,+\infty]$ and $(\theta, p) = (1, \infty)$.

If $X = Y$, then $K(t,x) = \min\{t,1\}\|x\|$. Therefore, as one can expect, $(X,X)_{\theta,p} = (X,X)_{1,\infty} = X$ for $0 < \theta < 1$, $1 \leq p \leq \infty$, and

$$\|x\|_{(X,X)_{\theta,p}} = \left(\frac{1}{p\theta(1-\theta)}\right)^{1/p} \|x\|_X, \quad 0 < \theta < 1, p < \infty,$$

$$\|x\|_{(X,X)_{\theta,\infty}} = \|x\|_X, \quad 0 < \theta \leq 1.$$

Some inclusion properties are stated below.

Proposition 1.2.3 *For $0 < \theta < 1$, $1 \leq p_1 \leq p_2 \leq \infty$ we have*

$$Y \subset (X,Y)_{\theta,p_1} \subset (X,Y)_{\theta,p_2} \subset (X,Y)_\theta \subset (X,Y)_{\theta,\infty} \subset \overline{Y}. \quad (1.2.5)$$

For $0 < \theta_1 < \theta_2 \leq 1$ we have

$$(X,Y)_{\theta_2,\infty} \subset (X,Y)_{\theta_1,1}. \quad (1.2.6)$$

Proof — From the inequality $K(t,x) \leq \min\{c,t\} \|x\|_Y$ for every $x \in Y$ it follows immediately that Y is continuously embedded in $(X,Y)_{1,\infty}$ and in $(X,Y)_{\theta,p}$ for $0 < \theta < 1$, $1 \leq p \leq \infty$.

Let us show that $(X,Y)_{\theta,\infty}$ is contained in $\overline{Y}$ and it is continuously embedded in X. For $x \in (X,Y)_{\theta,\infty}$ and for every $n \in \mathbb{N}$ there are $a_n \in X$, $b_n \in Y$ such that $x = a_n + b_n$, and

$$n^\theta \left(\|a_n\|_X + \frac{1}{n}\|b_n\|_Y\right) \leq 2\|x\|_{\theta,\infty}.$$

2. INTERPOLATION SPACES

In particular, $\|x - b_n\|_X = \|a_n\|_X \le 2\|x\|_{\theta,\infty} n^{-\theta}$, so that the sequence $\{b_n\}$ goes to x in X as $n \to \infty$. This implies that $(X,Y)_{\theta,\infty}$ is contained in $\overline{Y}$. Moreover, from the inequality

$$\|x\|_X \le \|a\|_X + \|b\|_X \le \|a\|_X + c\|b\|_Y, \quad \text{if } x = a+b,$$

we get

$$\|x\|_X \le K(c,x) \le c^\theta \|x\|_{\theta,\infty}, \quad \forall x \in (X,Y)_{\theta,\infty},$$

so that $(X,Y)_{\theta,\infty}$ is continuously embedded in X.

The inclusion $(X,Y)_\theta \subset (X,Y)_{\theta,\infty}$ is trivial, since $K(\cdot,x)$ is bounded.

Let us show that $(X,Y)_{\theta,p}$ is contained in $(X,Y)_\theta$ and it is continuously embedded in $(X,Y)_{\theta,\infty}$ for $p < \infty$. Note that $K(\cdot,x)$ satisfies

$$K(t,x) \le \frac{t}{s} K(s,x) \text{ for } x \in X,\ 0 < s < t.$$

Therefore, for each $x \in (X,Y)_{\theta,p}$ and $t > 0$

$$\begin{aligned}
t^{1-\theta} K(t,x) &= [(1-\theta)p]^{1/p} \left(\int_0^t s^{(1-\theta)p-1} ds \right)^{1/p} K(t,x) \\
&\le [(1-\theta)p]^{1/p} \left(\int_0^t s^{-\theta p-1} t^p K(s,x)^p ds \right)^{1/p},
\end{aligned}$$

so that

$$t^{-\theta} K(t,x) \le [(1-\theta)p]^{1/p} \left(\int_0^t s^{-\theta p-1} K(s,x)^p ds \right)^{1/p}.$$

Letting $t \to 0$ it follows that $x \in (X,Y)_\theta$. The same inequality yields

$$\|x\|_{\theta,\infty} \le [(1-\theta)p]^{1/p} \|x\|_{\theta,p}. \tag{1.2.7}$$

Let us prove that $(X,Y)_{\theta,p_1} \subset (X,Y)_{\theta,p_2}$ for $p_1 < p_2$. For $x \in (X,Y)_{\theta,p_1}$ we have

$$\begin{aligned}
\|x\|_{\theta,p_2} &= \left(\int_0^{+\infty} t^{-\theta p_2 - 1} K(t,x)^{p_2} dt \right)^{1/p_2} \\
&\le \left(\int_0^{+\infty} t^{-\theta p_1 - 1} K(t,x)^{p_1} dt \right)^{1/p_2} \left(\sup_{t>0} t^{-\theta} K(t,x) \right)^{(p_2-p_1)/p_2} \\
&= (\|x\|_{\theta,p_1})^{p_1/p_2} (\|x\|_{\theta,\infty})^{1-p_1/p_2},
\end{aligned}$$

and using (1.2.7) we find

$$\|x\|_{\theta,p_2} \le [(1-\theta)p_1]^{1/p_1 - 1/p_2} \|x\|_{\theta,p_1}. \tag{1.2.8}$$

Let us prove that (1.2.6) holds. If $0 < \theta_1 < \theta_2 \leq 1$ and $x \in (X,Y)_{\theta_2,\infty}$, we have

$$\|x\|_{\theta_1,1} = \int_0^1 t^{-\theta_1-1} K(t,x) dt + \int_1^{+\infty} t^{-\theta_1-1} K(t,x) dt$$
$$\leq \int_0^1 t^{-\theta_1-1} \|x\|_{\theta_2,\infty} t^{\theta_2} dt + \int_1^{+\infty} t^{-\theta_1-1} \|x\|_X dt \leq \frac{1}{\theta_2-\theta_1} \|x\|_{\theta_2,\infty} + \frac{1}{\theta_1} \|x\|_X. \quad (1.2.9)$$

The statement is so completely proved. ∎

Proposition 1.2.4 $(X,Y)_{\theta,p}$ *is a Banach space.*

Proof — Let $\{x_n\}_{n\in\mathbb{N}}$ be a Cauchy sequence in $(X,Y)_{\theta,p}$. Due to the continuous embedding of $(X,Y)_{\theta,p}$ in X, $\{x_n\}_{n\in\mathbb{N}}$ is a Cauchy sequence in X too, so that it converges to an element $x \in X$.

Let us estimate $\|x_n - x\|_{\theta,p}$. Fix $\varepsilon > 0$, and let $\|x_n - x_m\|_{\theta,p} \leq \varepsilon$ for $n, m \geq n_\varepsilon$. Since $y \mapsto K(t,y)$ is a norm, for every $n, m \in \mathbb{N}$ and $t > 0$ we have

$$t^{-\theta} K(t, x_n - x) \leq t^{-\theta} K(t, x_n - x_m) + t^{-\theta} \|x_m - x\|_X. \quad (1.2.10)$$

Let $p = \infty$. Then for every $t > 0$ and $n, m \geq n_\varepsilon$

$$t^{-\theta} K(t, x_n - x) \leq \varepsilon + t^{-\theta} \|x_m - x\|_X.$$

Letting $m \to +\infty$, we find $t^{-\theta} K(t, x_n - x) \leq \varepsilon$ for every $t > 0$. This implies that $x \in (X,Y)_{\theta,\infty}$ and that $x_n \to x$ in $(X,Y)_{\theta,\infty}$.

Let now $p < \infty$. Then

$$\|x_n - x\|_{\theta,p} = \lim_{\delta \to 0} \left(\int_\delta^{1/\delta} t^{-\theta p-1} K(t, x_n - x)^p dt \right)^{1/p}.$$

Due again to (1.2.10), for every $\delta \in \,]0,1[$ we get, for $n, m \geq n_\varepsilon$,

$$\left(\int_\delta^{1/\delta} t^{-\theta p-1} K(x_n - x)^p dt \right)^{1/p} \leq \|x_n - x_m\|_{\theta,p}$$
$$+ \|x_m - x\|_X \left(\int_\delta^{1/\delta} t^{-\theta p-1} dt \right)^{1/p} \leq \varepsilon + \|x_m - x\|_X \left(\frac{1}{\theta p \delta^{\theta p}} \right)^{1/p}.$$

Letting $m \to \infty$ and then $\delta \to 0$ we get $x \in (X,Y)_{\theta,p}$ and $x_n \to x$ in $(X,Y)_{\theta,p}$. ∎

Corollary 1.2.5 *For $0 < \theta \leq 1$, $(X,Y)_\theta$ is a Banach space, endowed with the norm of $(X,Y)_{\theta,\infty}$.*

2. INTERPOLATION SPACES

Proof — It is easy to see that $(X,Y)_\theta$ is a closed subspace of $(X,Y)_{\theta,\infty}$. Since $(X,Y)_{\theta,\infty}$ is complete, then also $(X,Y)_\theta$ is complete. ∎

The spaces $(X,Y)_{\theta,p}$ and $(X,Y)_\theta$ enjoy an important interpolation property, stated in the next proposition. It implies that they are in fact interpolation spaces.

Proposition 1.2.6 *Let X_1, X_2, Y_1, Y_2 be Banach spaces, such that Y_i is continuously embedded in X_i, for $i = 1, 2$. If $T \in L(X_1, X_2) \cap L(Y_1, Y_2)$, then $T \in L((X_1,Y_1)_{\theta,p}, (X_2,Y_2)_{\theta,p}) \cap L((X_1,Y_1)_\theta, (X_2,Y_2)_\theta)$ for every $\theta \in \,]0,1[$ and $p \in [1,\infty]$, and for $(\theta,p) = (1,\infty)$. Moreover,*

$$\|T\|_{L((X_1,Y_1)_{\theta,p},(X_2,Y_2)_{\theta,p})} \leq (\|T\|_{L(X_1,X_2)})^{1-\theta}(\|T\|_{L(Y_1,Y_2)})^\theta. \tag{1.2.11}$$

Proof — If $T = 0$ the statement is trivial, so that we can assume that $T \neq 0$. Let $x \in (X_1, Y_1)_{\theta,p}$: then for every $a \in X_1$, $b \in Y_1$ such that $x = a + b$ and for every $t > 0$ we have

$$\|Ta\|_{X_2} + t\|Tb\|_{Y_2} \leq \|T\|_{L(X_1,X_2)}\left(\|a\|_{X_1} + t\frac{\|T\|_{L(Y_1,Y_2)}}{\|T\|_{L(X_1,X_2)}}\|b\|_{Y_1}\right),$$

so that

$$K(t, Tx, X_2, Y_2) \leq \|T\|_{L(X_1,X_2)} K\left(t\frac{\|T\|_{L(Y_1,Y_2)}}{\|T\|_{L(X_1,X_2)}}, x, X_1, Y_1\right). \tag{1.2.12}$$

Setting $s = t\frac{\|T\|_{L(Y_1,Y_2)}}{\|T\|_{L(X_1,X_2)}}$ we get $Tx \in (X_2, Y_2)_{\theta,p}$, and

$$\|Tx\|_{(X_2,Y_2)_{\theta,p}} \leq \|T\|_{L(X_1,X_2)}\left(\frac{\|T\|_{L(Y_1,Y_2)}}{\|T\|_{L(X_1,X_2)}}\right)^\theta \|x\|_{(X_1,Y_1)_{\theta,p}},$$

and (1.2.11) follows. From (1.2.12) it follows also that

$$\lim_{t\to 0} t^{-\theta} K(t, x, X_1, Y_1) = 0 \Rightarrow \lim_{t\to 0} t^{-\theta} K(t, Tx, X_2, Y_2) = 0,$$

that is, T maps $(X_1, Y_1)_\theta$ into $(X_2, Y_2)_\theta$. ∎

Corollary 1.2.7 *For $0 < \theta < 1$, $1 \leq p \leq \infty$ and for $(\theta, p) = (1, \infty)$ there is $c(\theta, p)$ such that*

$$\|y\|_{(X,Y)_{\theta,p}} \leq c(\theta, p) \|y\|_X^{1-\theta} \|y\|_Y^\theta \quad \forall y \in Y. \tag{1.2.13}$$

Proof — Set $\mathbb{K} = \mathbb{R}$ or $\mathbb{K} = \mathbb{C}$, according to the fact that X is a real or a complex Banach space. Let $y \in Y$, and define $T : \mathbb{K} \mapsto X$, by $T(\lambda) = \lambda y$ for each $\lambda \in \mathbb{K}$. Then $\|T\|_{L(\mathbb{K},X)} = \|y\|_X$, $\|T\|_{L(\mathbb{K},Y)} = \|y\|_Y$, and $\|T\|_{L(\mathbb{K},(X,Y)_{\theta,p})} = \|y\|_{(X,Y)_{\theta,p}}$. The statement follows now from Proposition 1.2.6, through the equality $(\mathbb{K}, \mathbb{K})_{\theta,p} = \mathbb{K}$. ∎

The statement of Corollary 1.2.7 can be rephrased saying that every $(X,Y)_{\theta,p}$ belongs to $J_\theta(X,Y)$. In particular, $(X,Y)_{\theta,1}$ belongs to $J_\theta(X,Y)$. We will see later (Proposition 1.2.13) that in fact a space E belongs to the class $J_\theta(X,Y)$ if and only if $(X,Y)_{\theta,1}$ is continuously embedded in E.

1.2.2 The trace method

We describe now another construction of the real interpolation spaces, which is one of the most common in the literature and which will be useful for proving other properties.

Definition 1.2.8 *For $0 \leq \theta < 1$ and $1 \leq p \leq \infty$ set*

$$\begin{aligned} V(p,\theta,Y,X) &= \{u : \mathbb{R}_+ \mapsto X : t \mapsto u_\theta(t) = t^{\theta-1/p}u(t) \in L^p(0,+\infty;Y), \\ &\quad t \mapsto v_\theta(t) = t^{\theta-1/p}u'(t) \in L^p(0,+\infty;X)\}, \\ \|u\|_{V(p,\theta,Y,X)} &= \|u_\theta\|_{L^p(0,+\infty;Y)} + \|v_\theta\|_{L^p(0,+\infty;X)}. \end{aligned} \quad (1.2.14)$$

Moreover, for $p = +\infty$ we define a subspace of $V(\infty,\theta,Y,X)$, by

$$V_0(\infty,\theta,Y,X) = \{u \in V(\infty,\theta,Y,X) : \lim_{t \to 0} \|t^\theta u(t)\|_X = \lim_{t \to 0} \|t^\theta u'(t)\|_Y = 0\}. \quad (1.2.15)$$

It is not difficult to see that $V(p,\theta,Y,X)$ is a Banach space endowed with the norm $\|\cdot\|_{V(p,\theta,Y,X)}$, and that $V_0(\infty,\theta,Y,X)$ is a closed subspace of $V(\infty,\theta,Y,X)$. Moreover, if $\theta < 1$, any function belonging to $V(p,\theta,Y,X)$ has a X-valued continuous extension at $t = 0$. Indeed, for $0 < s < t$ from the equality $u(t) - u(s) = \int_s^t u'(\sigma)d\sigma$ it follows, for $1 < p < \infty$,

$$\begin{aligned} \|u(t) - u(s)\|_X &\leq \left(\int_s^t \|\sigma^\theta u'(\sigma)\|_X^p \frac{d\sigma}{\sigma}\right)^{1/p} \left(\int_s^t \sigma^{-(\theta-1/p)q} d\sigma\right)^{1/q} \\ &\leq \|u\|_{V(p,\theta,Y,X)}[q(1-\theta)]^{-1/q}(t^{q(1-\theta)} - s^{q(1-\theta)})^{1/q}, \end{aligned}$$

with $q = p/(p-1)$. Arguing similarly, one sees that if $p = 1$ or $p = \infty$, then u is Lipschitz continuous (respectively, $(1-\theta)$-Hölder continuous) near $t = 0$.

In this and in the next section we shall use the Hardy-Young inequalities, which hold for every positive measurable function $\varphi : (0,a) \mapsto \mathbb{R}$, $0 < a \leq \infty$, and every $\alpha > 0$, $p \geq 1$. See [97, p.245-246].

$$\begin{cases} (i) & \int_0^a t^{-\alpha p} \left(\int_0^t \varphi(s) \frac{ds}{s}\right)^p \frac{dt}{t} \leq \frac{1}{\alpha^p} \int_0^a s^{-\alpha p} \varphi(s)^p \frac{ds}{s}, \\ (ii) & \int_0^a t^{\alpha p} \left(\int_t^a \varphi(s) \frac{ds}{s}\right)^p \frac{dt}{t} \leq \frac{1}{\alpha^p} \int_0^a s^{\alpha p} \varphi(s)^p \frac{ds}{s} \end{cases} \quad (1.2.16)$$

We shall use the following consequence of inequality (1.2.16)(i).

Corollary 1.2.9 *Let u be a function such that $t \mapsto u_\theta(t) = t^{\theta-1/p}u(t)$ belongs to $L^p(0,a;X)$, with $0 < a \leq \infty$, $0 < \theta < 1$ and $1 \leq p \leq \infty$. Then also the mean value*

$$v(t) = \frac{1}{t}\int_0^t u(s)ds, \quad t > 0 \quad (1.2.17)$$

has the same property, and setting $v_\theta(t) = t^{\theta-1/p}v(t)$ we have

$$\|v_\theta\|_{L^p(0,a;X)} \leq \frac{1}{1-\theta}\|u_\theta\|_{L^p(0,a;X)} \tag{1.2.18}$$

With the aid of Corollary 1.2.9 we are able to characterize the real interpolation spaces as trace spaces.

Proposition 1.2.10 *For $(\theta, p) \in \,]0,1[\,\times [1,+\infty] \cup \{(1,\infty)\}$, $(X,Y)_{\theta,p}$ is the set of the traces at $t = 0$ of the functions in $V(p, 1 - \theta, Y, X)$, and the norm*

$$\|x\|_{\theta,p}^T = \inf\{\|u\|_{V(p,1-\theta,Y,X)} : x = u(0),\ u \in V(p, 1 - \theta, Y, X)\}$$

is an equivalent norm in $(X,Y)_{\theta,p}$. Moreover, for $0 < \theta < 1$, $(X,Y)_\theta$ is the set of the traces at $t = 0$ of the functions in $V_0(\infty, 1 - \theta, Y, X)$.

Proof — Let $x \in (X,Y)_{\theta,p}$. For every $n \in \mathbb{N}$ let $a_n \in X$, $b_n \in Y$ be such that $a_n + b_n = x$, and

$$\|a_n\|_X + \frac{1}{n}\|b_n\|_Y \leq 2K(1/n, x).$$

For $t > 0$ set

$$u(t) = \sum_{n=1}^\infty b_{n+1}\chi_{]\frac{1}{n+1}, \frac{1}{n}]}(t) = \sum_{n=1}^\infty (x - a_{n+1})\chi_{]\frac{1}{n+1}, \frac{1}{n}]}(t),$$

where χ_I is the characteristic function of the interval I, and

$$v(t) = \frac{1}{t}\int_0^t u(s)ds.$$

Since $(X,Y)_{\theta,p} \subset (X,Y)_{\theta,\infty}$, then $\lim_{t\to 0} K(t,x) = 0$. In particular, $x = \lim_{n\to\infty} b_n$, so that $x = \lim_{t\to 0} u(t) = \lim_{t\to 0} v(t)$. Moreover,

$$\|t^{1-\theta}u(t)\|_Y \leq t^{1-\theta}\sum_{n=1}^\infty \chi_{]\frac{1}{n+1}, \frac{1}{n}]}(t)2(n+1)K(1/(n+1), x) \leq 4t^{-\theta}K(t,x),$$
$$\tag{1.2.19}$$

so that $t \mapsto t^{1-\theta-1/p}u(t) \in L^p(0,+\infty;Y)$. By Corollary 1.2.9, $t \mapsto t^{1-\theta-1/p}v(t)$ belongs to $L^p(0,+\infty;Y)$, and

$$\|t^{1-\theta-1/p}v\|_{L^p(0,+\infty;Y)} \leq 4\theta^{-1}\|x\|_{\theta,p}.$$

On the other hand,

$$v(t) = x - \frac{1}{t}\int_0^t \sum_{n=1}^\infty \chi_{]\frac{1}{n+1}, \frac{1}{n}]}(s)a_{n+1}ds,$$

so that v is differentiable almost everywhere with values in X, and

$$v'(t) = \frac{1}{t^2}\int_0^t g(s)ds - \frac{1}{t}g(t),$$

where $g(t) = \sum_{n=1}^\infty \chi_{]\frac{1}{n+1},\frac{1}{n}]}(t)a_{n+1}$ is such that

$$\|g(t)\|_X \le t^{-\theta}\sum_{n=1}^\infty \chi_{]\frac{1}{n+1},1/n]}(t)2K(1/(n+1),x) \le 2K(t,x).$$

It follows that

$$\|t^{1-\theta}v'(t)\| \le t^{-\theta}\sup_{0<s<t}\|g(s)\| + \|t^{-\theta}g(t)\| \le 4t^{-\theta}K(t,x). \tag{1.2.20}$$

Then $t \mapsto t^{1-\theta-1/p}v'(t)$ belongs to $L^p(0,+\infty;X)$, and

$$\|t^{1-\theta-1/p}v'\|_{L^p(0,+\infty;X)} \le 4\|x\|_{\theta,p}.$$

Therefore, x is the trace at $t=0$ of a function $v \in V(p,1-\theta,Y,X)$, and

$$\|x\|_{\theta,p}^T \le 2(2+1/\theta)\|x\|_{\theta,p}.$$

If $x \in (X,Y)_\theta$, then, by (1.2.19), $\lim_{t\to 0} t^{1-\theta}\|u(t)\|_Y = 0$, so that $\lim_{t\to 0} t^{1-\theta}\|v(t)\|_Y = 0$. By (1.2.20), $\lim_{t\to 0} t^{-\theta}\|g(t)\|_X = 0$, so that $\lim_{t\to 0} t^{1-\theta}\|v'(t)\|_X = 0$. Then $v \in V_0(\infty, 1-\theta, Y, X)$.

Conversely, let x be the trace of a function $u \in V(p, 1-\theta, Y, X)$. Then

$$x = x - u(t) + u(t) = -\int_0^t u'(s)ds + u(t) \quad \forall t > 0,$$

so that

$$t^{-\theta}K(t,x) \le t^{1-\theta}\left\|\frac{1}{t}\int_0^t u'(s)ds\right\|_X + t^{1-\theta}\|u(t)\|_Y. \tag{1.2.21}$$

Corollary 1.2.9 implies now that $t \mapsto t^{-\theta-1/p}K(t,x)$ belongs to $L^p(0,+\infty)$, so that $x \in (X,Y)_{\theta,p}$, and

$$\|x\|_{\theta,p} \le \frac{1}{\theta}\|x\|_{\theta,p}^T.$$

If x is the trace of a function $u \in V_0(\infty, 1-\theta, Y, X)$, then, by (1.2.21), $\lim_{t\to 0} t^{-\theta}K(t,x) = 0$, so that $x \in (X,Y)_\theta$. ∎

Remark 1.2.11 By Proposition 1.2.10, if $x \in (X,Y)_{\theta,p}$ or $x \in (X,Y)_\theta$, then x is the trace at $t = 0$ of a function u belonging to $L^p(a,b;Y) \cap W^{1,p}(a,b;X)$ for $0 < a < b$. But it is possible to find a more regular function $v \in V(p, 1-\theta, Y, X)$

2. INTERPOLATION SPACES

(or $v \in V_0(\infty, 1-\theta, Y, X)$) such that $v(0) = x$. For any $u \in V(p, 1-\theta, Y, X)$ (or $u \in V_0(\infty, 1-\theta, Y, X)$) such that $u(0) = x$, set

$$v(t) = \frac{1}{t}\int_0^t u(s)ds, \quad t \geq 0.$$

Then $v \in W^{1,p}(a,b;Y) \cap W^{2,p}(a,b;X)$ for $0 < a < b$, and $v(0) = x$. By Corollary 1.2.9, $t \mapsto t^{\theta-1/p}v(t)$ belongs to $L^p(0, +\infty; Y)$; moreover

$$v'(t) = -\frac{1}{t^2}\int_0^t (u(s) - u(t))ds = -\frac{1}{t^2}\int_0^t ds \int_s^t u'(\sigma)d\sigma,$$

so that

$$\|v'(t)\|_X \leq \frac{1}{t}\sup_{0<s<t}\left\|\int_s^t u'(\sigma)d\sigma\right\|_X \leq \frac{1}{t}\int_0^t \|u'(\sigma)\|_X d\sigma,$$

and again by Corollary 1.2.9, $t \mapsto t^{1-\theta-1/p}v'(t)$ belongs to $L^p(0, +\infty; X)$, and

$$\|v\|_{V(p,\theta,Y,X)} \leq \frac{1}{\theta}\|u\|_{V(p,\theta,Y,X)}. \tag{1.2.22}$$

Moreover, $v'(t) = (u(t) - v(t))/t$, so that $t \mapsto t^{2-\theta-1/p}v'(t)$ belongs to $L^p(0, +\infty; Y)$, and

$$\|t^{2-\theta-1/p}v'\|_{L^p(0,+\infty;Y)} \leq (1 + 1/\theta)\|t^{1-\theta-1/p}u\|_{L^p(0,+\infty;Y)}. \tag{1.2.23}$$

If $u \in V_0(\infty, 1-\theta, Y, X)$, it is easy to see that $v \in V_0(\infty, 1-\theta, Y, X)$, and that $\lim_{t \to 0} t^{2-\theta}\|v'(t)\|_Y = 0$. ∎

By means of the trace method it is easy to prove some important density properties.

Proposition 1.2.12 *Let $0 < \theta < 1$. For $1 \leq p < \infty$, Y is dense in $(X, Y)_{\theta, p}$. For $p = \infty$, $(X, Y)_\theta$ is the closure of Y in $(X, Y)_{\theta, \infty}$.*

Proof — Let $p < \infty$, and let $x \in (X, Y)_{\theta,p}$. By Remark 1.2.11, $x = v(0)$, where $v \in V(p, \theta, Y, X)$, and moreover $t \mapsto t^{2-\theta-1/p}v' \in L^p(0, +\infty; Y)$. Set

$$x_\varepsilon = v(\varepsilon), \quad \forall \varepsilon > 0.$$

Then $x_\varepsilon \in Y$, and we shall show that $x_\varepsilon \to x$ in $(X, Y)_{\theta,p}$.

We have $x_\varepsilon - x = z_\varepsilon(0)$, where

$$z_\varepsilon(t) = (v(\varepsilon) - v(t))\chi_{[0,\varepsilon]}(t).$$

It is not hard to check that $z_\varepsilon \in W^{1,p}(a,b;X)$ for $0 < a < b < \infty$, and that $z'_\varepsilon(t) = -v'(t)\chi_{]0,\varepsilon[}(t)$. It follows that

$$\lim_{\varepsilon \to 0}\|t^{1-\theta-1/p}z'_\varepsilon(t)\|_{L^p(0,+\infty;X)} = 0.$$

Moreover, due to the equality

$$z_\varepsilon(t) = \int_t^{+\infty} \chi_{]0,\varepsilon[}(s) v'(s) ds,$$

we get, using the Hardy-Young inequality (1.2.16)(ii),

$$\|t^{1-\theta-1/p} z_\varepsilon(t)\|_{L^p(0,+\infty;Y)} \leq \left(\int_0^{+\infty} t^{(1-\theta)p} \left(\int_t^{+\infty} \chi_{]0,\varepsilon[}(s) s \|v'(s)\|_Y \frac{ds}{s} \right)^p \frac{dt}{t} \right)^{1/p}$$

$$\leq \frac{1}{1-\theta} \left(\int_0^{+\infty} \chi_{]0,\varepsilon[}(s) s^{(2-\theta)p} \|v'(s)\|_Y \frac{ds}{s} \right)^{1/p},$$

so that $t \mapsto t^{1-\theta-1/p} z_\varepsilon(t) \in L^p(0,+\infty; Y)$ for every ε, and

$$\lim_{\varepsilon \to 0} \|t^{1-\theta-1/p} z_\varepsilon(t)\|_{L^p(0,+\infty;Y)} = 0.$$

Therefore, $z_\varepsilon \to 0$ in $V(p, 1-\theta, Y, X)$ as $\varepsilon \to 0$, which means that $\|x_\varepsilon - x\|_{\theta,p}^T \to 0$ as $\varepsilon \to 0$. From Proposition 1.2.10 we get $\lim_{\varepsilon \to 0} \|x_\varepsilon - x\|_{\theta,p} = 0$.

Let now $x \in (X,Y)_\theta$. Due again to Remark 1.2.11, x is the trace at $t=0$ of a function $v \in V_0(\infty, 1-\theta, Y, X)$, such that $t \mapsto t^{2-\theta} v'(t) \in L^\infty(0,+\infty; Y)$ and $\lim_{t \to 0} t^{2-\theta} \|v'(t)\|_Y = 0$. Let x_ε, z_ε be defined as above. Then $\lim_{t \to 0} t^{1-\theta} \|v'(t)\|_X = 0$, so that

$$\sup_{t>0} t^{1-\theta} \|z'_\varepsilon(t)\|_X = \sup_{0<t<\varepsilon} t^{1-\theta} \|v'(t)\|_X \to 0, \quad \text{as } \varepsilon \to 0,$$

and

$$\sup_{t>0} t^{1-\theta} \|z_\varepsilon(t)\|_Y = \sup_{0<t<\varepsilon} t^{1-\theta} \|v(\varepsilon) - v(t)\|_Y \leq 2 \sup_{0<s<\varepsilon} s^{1-\theta} \|v(s)\|_Y \to 0,$$

as $\varepsilon \to 0$. Arguing as before, it follows that $\|x_\varepsilon - x\|_{\theta,\infty} \to 0$ as $\varepsilon \to 0$. ∎

In the previous subsection we have seen that every $(X,Y)_{\theta,p}$ belongs to $J_\theta(X,Y)$. In particular, $(X,Y)_{\theta,1}$ belongs to $J_\theta(X,Y)$. Now we can characterize all the spaces in the class $J_\theta(X,Y)$.

Proposition 1.2.13 *Let $0 < \theta < 1$, and let E be a Banach space such that $Y \subset E \subset X$. The following statements are equivalent:*

(i) E belongs to the class J_θ between X and Y,

(ii) $(X,Y)_{\theta,1} \subset E$.

Proof — The implication (ii) ⇒ (i) is a straightforward consequence of Corollary 1.2.7, with $p = 1$. Let us show that (i) ⇒ (ii). For every $x \in (X,Y)_{\theta,1}$, let $u \in V(1, 1-\theta, Y, X)$ be such that $u(t) = 0$ for $t \geq 1$, $u(0) = x$, and set

$$v(t) = \frac{1}{t} \int_0^t u(s) ds.$$

Then $v(0) = x$, $v(+\infty) = 0$, so that
$$x = -\int_0^{+\infty} v'(t)dt.$$

Let c be such that $\|y\|_E \leq c\|y\|_Y^\theta \|y\|_X^{1-\theta}$ for every $y \in Y$. Then
$$\|v'(t)\|_E \leq c\|v'(t)\|_Y^\theta \|v'(t)\|_X^{1-\theta} = ct^{-1}\|t^{2-\theta}v'(t)\|_Y^\theta \|t^{1-\theta}v'(t)\|_X^{1-\theta}.$$

By Remark 1.2.11, $t \mapsto t^{1-\theta}v'(t)$ belongs to $L^1(0, +\infty; Y)$, and $t \mapsto t^{-\theta}v'(t)$ belongs to $L^1(0, +\infty; X)$. By the Hölder inequality, v' belongs to $L^1(0, +\infty; E)$, and
$$\|x\|_E \leq c(\|t^{1-\theta}v'(t)\|_{L^1(0,\infty;Y)})^\theta (\|t^{-\theta}v'(t)\|_{L^1(0,\infty;X)})^{1-\theta} \leq \text{const.}\,\|x\|_{\theta,1}.\;\blacksquare$$

1.2.3 The Reiteration Theorem

We need some preliminaries about certain classes of intermediate spaces between X and Y. We have introduced the classes J_θ in Section 1.1, and we have shown that a Banach space E such that $Y \subset E \subset X$ belongs to $J_\theta(X, Y)$ if and only if $(X, Y)_{\theta,1}$ is continuously embedded in E. Now we define another class of intermediate spaces.

Definition 1.2.14 *Let E be a Banach space such that $Y \subset E \subset X$, and let $0 \leq \theta \leq 1$. E is said to belong to the class K_θ between X and Y if there is $k > 0$ such that*
$$K(t, x) \leq kt^\theta \|x\|_E, \quad \forall x \in E,\; t > 0.$$

In other words, E belongs to the class K_θ if and only if it is continuously embedded in $(X, Y)_{\theta, \infty}$. In this case, we write $E \in K_\theta(X, Y)$.

By Definition 1.2.14 and Proposition 1.2.12, a space E belongs to $K_\theta(X, Y) \cap J_\theta(X, Y)$ if and only if
$$(X, Y)_{\theta, 1} \subset E \subset (X, Y)_{\theta, \infty}.$$

Now we are able to state the Reiteration Theorem.

Theorem 1.2.15 *Let $0 \leq \theta_0 < \theta_1 \leq 1$. Fix $\theta \in \,]0, 1[$ and set $\omega = (1-\theta)\theta_0 + \theta\theta_1$. The following statements hold true.*

(i) *If E_i belong to the class K_{θ_i} $(i = 0, 1)$ between X and Y, then*
$$(E_0, E_1)_{\theta, p} \subset (X, Y)_{\omega, p}, \;\; \forall p \in [1, \infty], \quad (E_0, E_1)_\theta \subset (X, Y)_\omega.$$

(ii) *If E_i belong to the class J_{θ_i} $(i = 0, 1)$ between X and Y, then*
$$(X, Y)_{\omega, p} \subset (E_0, E_1)_{\theta, p}, \;\; \forall p \in [1, \infty], \quad (X, Y)_\omega \subset (E_0, E_1)_\theta.$$

Consequently, if E_i belong to $K_{\theta_i}(X,Y) \cap J_{\theta_i}(X,Y)$, then

$$(E_0, E_1)_{\theta,p} = (X,Y)_{\omega,p}, \quad \forall p \in [1, \infty], \quad (E_0, E_1)_\theta = (X,Y)_\omega,$$

with equivalence of the respective norms.

Proof — Let us prove statement (i). Let k_i, be the embedding constants of the inclusions $E_i \subset (X,Y)_{\theta_i, \infty}$, $i = 0, 1$. For each $x \in (E_0, E_1)_{\theta,p}$, let $a \in E_0$, $b \in E_1$ be such that $x = a + b$. Then

$$K(t, x, X, Y) \leq K(t, a, X, Y) + K(t, b, X, Y) \leq k_0 t^{\theta_0} \|a\|_{E_0} + k_1 t^{\theta_1} \|b\|_{E_1}.$$

Since a and b are arbitrary, it follows that

$$K(t, x, X, Y) \leq \max\{k_0, k_1\} t^{\theta_0} K(t^{\theta_1 - \theta_0}, x, E_0, E_1).$$

Consequently,

$$t^{-\omega} K(t, x, X, Y) \leq \max\{k_0, k_1\} t^{-\theta(\theta_1 - \theta_0)} K(t^{\theta_1 - \theta_0}, x, E_0, E_1). \tag{1.2.24}$$

By the change of variable $s = t^{\theta_1 - \theta_0}$ we see that $t \mapsto t^{-\omega - 1/p} K(t, x, X, Y)$ belongs to $L^p(0, +\infty)$, which means that x belongs to $(X,Y)_{\omega,p}$, and

$$\begin{cases} \|x\|_{(X,Y)_{\omega,p}} \leq \max\{k_0, k_1\}(\theta_1 - \theta_0)^{-1/p} \|x\|_{(E_0, E_1)_{\theta,p}}, & \text{if } p < \infty, \\ \|x\|_{(X,Y)_{\omega,\infty}} \leq \max\{k_0, k_1\} \|x\|_{(E_0, E_1)_{\theta,p}}, & \text{if } p = \infty. \end{cases}$$

If $x \in (E_0, E_1)_\theta$, by (1.2.24) we get

$$\lim_{t \to 0} t^{-\omega} K(t, x, X, Y) \leq \max\{k_0, k_1\} \lim_{s \to 0} s^{-\theta} K(s, x, E_0, E_1) = 0,$$

so that $x \in (X,Y)_\omega$.

Let us prove statement (ii). By Proposition 1.1.2 and Remark 1.2.11, every $x \in (X,Y)_{\omega,p}$ is the trace at $t = 0$ of a regular function $v : \mathbb{R}_+ \mapsto Y$ such that $v(+\infty) = 0$, $t \mapsto t^{1-\omega-1/p} v'(t)$ belongs to $L^p(0, +\infty, X)$, $t \mapsto t^{2-\omega-1/p} v'(t)$ belongs to $L^p(0, +\infty, Y)$, and

$$\|t^{1-\omega-1/p} v'(t)\|_{L^p(0,+\infty,X)} + \|t^{2-\omega-1/p} v'(t)\|_{L^p(0,+\infty,Y)} \leq k \|x\|^T_{(X,Y)_{\omega,p}},$$

with k independent of x and v. We shall show that v belongs to $V(p, 1-\theta, E_0, E_1)$: this will imply, through Proposition 1.1.2, that $x \in (E_0, E_1)_{\theta,p}$. Let c_i be such that

$$\|y\|_{E_i} \leq c_i \|y\|_X^{1-\theta_i} \|y\|_Y^{\theta_i} \quad \forall y \in Y, \quad i = 0, 1.$$

Then

$$\|v'(s)\|_{E_i} \leq \frac{c_i}{s^{\theta_i + 1 - \omega}} \|s^{1-\omega} v'(s)\|_X^{1-\theta_i} \|s^{2-\omega} v'(s)\|_Y^{\theta_i}, \quad i = 0, 1,$$

2. INTERPOLATION SPACES

so that from the equalities

$$\theta_0 + 1 - \omega = 1 - \theta(\theta_1 - \theta_0), \quad \theta_1 + 1 - \omega = 1 + (1 - \theta)(\theta_1 - \theta_0),$$

we get

$$\begin{cases} (i) & \|s^{1-\theta(\theta_1-\theta_0)-1/p}v'(s)\|_{L^p(0,+\infty;E_0)} \leq c_0 k \|x\|^T_{(X,Y)_{\omega,p}}, \\ (ii) & \|s^{1+(1-\theta)(\theta_1-\theta_0)-1/p}v'(s)\|_{L^p(0,+\infty;E_1)} \leq c_0 k \|x\|^T_{(X,Y)_{\omega,p}}. \end{cases} \quad (1.2.25)$$

From the equality $v(t) = -\int_t^\infty v'(s)ds$ and 1.2.25(ii), using the Hardy-Young inequality (1.2.16)(ii) if $p < \infty$, we get

$$\|t^{(1-\theta)(\theta_1-\theta_0)-1/p}v(t)\|_{L^p(0,+\infty;E_1)} \leq \frac{c_0 k}{(1-\theta)(\theta_1-\theta_0)} \|x\|^T_{(X,Y)_{\omega,p}}.$$

Set now

$$g(t) = v(t^{1/(\theta_1-\theta_0)}), \quad t > 0.$$

Then $t \mapsto t^{1-\theta-1/p}g(t) \in L^p(0,+\infty;E_1)$, and

$$\|t^{1-\theta-1/p}g(t)\|_{L^p(0,+\infty;E_1)} \leq (\theta_1 - \theta_0)^{-1/p} \|t^{(1-\theta)(\theta_1-\theta_0)-1/p}v(t)\|_{L^p(0,+\infty;E_1)},$$

moreover $g'(t) = (\theta_1 - \theta_0)^{-1} t^{-1+1/(\theta_1-\theta_0)} v'(t^{1/(\theta_1-\theta_0)})$, so that, by (1.2.25)(i), $t \mapsto t^{1-\theta-1/p}g'(t) \in L^p(0,+\infty;E_0)$, and

$$\|t^{1-\theta-1/p}g'(t)\|_{L^p(0,+\infty;E_0)} \leq (\theta_1 - \theta_0)^{-1-1/p} \|t^{1-\theta(\theta_1-\theta_0)-1/p}v'(t)\|_{L^p(0,+\infty;E_0)}.$$

Therefore, $g \in V(p, 1-\theta, E_0, E_1)$, so that $x = g(0)$ belongs to $(E_0, E_1)_{\theta,p}$, and

$$\|x\|^T_{(E_0,E_1)_{\theta,p}} \leq (\theta_1 - \theta_0)^{-1-1/p} k \|x\|^T_{(X,Y)_{\omega,p}}.$$

If $x \in (X, Y)_\omega$, then (1.2.25)(i) has to be replaced by

$$\lim_{s \to 0} s^{1-\theta(\theta_1-\theta_0)} \|v'(s)\|_{E_0} = 0,$$

so that

$$\lim_{t \to 0} t^{1-\theta} \|g'(t)\|_{E_0} = \lim_{t \to 0} \frac{t^{-\theta+1/(\theta_1-\theta_0)}}{\theta_1 - \theta_0} \|v'(t^{1/(\theta_1-\theta_0)})\|_{E_0} = 0.$$

Similarly, (1.2.25)(ii) has to be replaced by

$$\lim_{s \to 0} s^{1+(1-\theta)(\theta_1-\theta_0)} \|v'(s)\|_{E_1} = 0.$$

Using the equality

$$t^{1-\theta} g(t) = t^{1-\theta} \int_{t^{1/(\theta_1-\theta_0)}}^{\varepsilon^{1/(\theta_1-\theta_0)}} v'(s)ds + \frac{t^{1-\theta}}{\varepsilon^{1-\theta}} \left(\varepsilon^{1-\theta} \int_{\varepsilon^{1/(\theta_1-\theta_0)}}^{+\infty} v'(s)ds \right),$$

which holds for $0 < t < \varepsilon$, one deduces that $\lim_{t \to 0} t^{1-\theta} \|g(t)\|_{E_1} = 0$. ∎

Remark 1.2.16 By Proposition 1.2.3, $(X,Y)_{\theta,p}$ and $(X,Y)_\theta$ belong to $K_\theta(X,Y)$ $\cap J_\theta(X,Y)$ for $0 < \theta < 1$ and $1 \leq p \leq \infty$. The Reiteration Theorem yields

$$((X,Y)_{\theta_0,q_0},(X,Y)_{\theta_1,q_1})_{\theta,p} = (X,Y)_{(1-\theta)\theta_0+\theta\theta_1,p},$$
$$((X,Y)_{\theta_0},(X,Y)_{\theta_1,q})_{\theta,p} = (X,Y)_{(1-\theta)\theta_0+\theta\theta_1,p},$$
$$((X,Y)_{\theta_0,q},(X,Y)_{\theta_1})_{\theta,p} = (X,Y)_{(1-\theta)\theta_0+\theta\theta_1,p},$$

for $0 < \theta_0, \theta_1 < 1$, $1 \leq p, q \leq \infty$. Moreover, since X belongs to $K_0(X,Y) \cap J_0(X,Y)$, and Y belongs to $K_1(X,Y) \cap J_1(X,Y)$ between X and Y, then

$$((X,Y)_{\theta_0,q},Y)_{\theta,p} = (X,Y)_{(1-\theta)\theta_0+\theta,p}, \quad ((X,Y)_{\theta_0},Y)_\theta = (X,Y)_{(1-\theta)\theta_0+\theta},$$

and

$$(X,(X,Y)_{\theta_1,q})_{\theta,p} = (X,Y)_{\theta_1\theta,p}, \quad (X,(X,Y)_{\theta_1})_\theta = (X,Y)_{\theta_1\theta},$$

for $0 < \theta_0, \theta_1 < 1$, $1 \leq p, q \leq \infty$.

1.2.4 Some examples

We give here the characterization of some important interpolation spaces.

Theorem 1.2.17 *For $0 < \theta < 1$, $m \in \mathbb{N}$, it holds*

$$(C(\mathbb{R}^n), C^m(\mathbb{R}^n))_{\theta,\infty} = \mathcal{C}^{\theta m}(\mathbb{R}^n),$$

with equivalence of the respective norms. In particular, if θm is not an integer, then

$$(C(\mathbb{R}^n), C^m(\mathbb{R}^n))_{\theta,\infty} = C^{\theta m}(\mathbb{R}^n),$$

with equivalence of the respective norms, and

$$(C(\mathbb{R}^n), C^m(\mathbb{R}^n))_\theta = h^{\theta m}(\mathbb{R}^n).$$

Proof — The last statement is an obvious consequence of the others, since, due to Proposition 1.2.12, $(X,Y)_\theta$ is the closure of Y in $(X,Y)_{\theta,\infty}$, and $h^{\theta m}(\mathbb{R}^n)$ is the closure of $C^m(\mathbb{R}^n)$ in $C^{\theta m}(\mathbb{R}^n)$.

Let us prove the first part of the theorem. The proof is in three steps: first we show that the statement holds for $m = 2$, then we show that for $m > 2$ we have $(C^k(\mathbb{R}^n), C^{k+2}(\mathbb{R}^n))_{\theta,\infty} = \mathcal{C}^{k+2\theta}(\mathbb{R}^n)$ for $k = 0, \ldots, m-2$, and then we use the Reiteration Theorem to conclude.

Let $\rho \in C_0^\infty(\mathbb{R}^n)$ be even with respect to all variables, with support contained in $B(0,1)$, and such that $0 \leq \rho(x) \leq 1$, $\int_{\mathbb{R}^n} \rho(x)dx = 1$. For every $f \in C(\mathbb{R}^n)$ set

$$f_t(x) = \int_{\mathbb{R}^n} f(x-y)\rho_t(y)dy = \int_{\mathbb{R}^n} f(y)\rho_t(x-y)dy, \quad x \in \mathbb{R}^n, \quad (1.2.26)$$

where $\rho_t(x) = t^{-n}\rho(x/t)$. Set moreover

$$c_{k,\theta} = \int_{\mathbb{R}^n} |y|^\theta \sum_{|\alpha|=k} |D^\alpha \rho(y)| dy, \quad k \in \mathbb{N}, \ \theta \geq 0.$$

Step 1 : $m = 2$. First we show that $\mathcal{C}^{2\theta}(\mathbb{R}^n)$ is continuously embedded in $(C(\mathbb{R}^n), C^2(\mathbb{R}^n))_{\theta,\infty}$. For every $f \in \mathcal{C}^{2\theta}(\mathbb{R}^n)$ we have

$$\begin{aligned}
f_t(x) - f(x) &= \int_{\mathbb{R}^n} (f(x-y) - f(x))\rho_t(y) dy \\
&= \int_{\mathbb{R}^n} (f(x-y) - f(x)) \frac{\rho_t(y) + \rho_t(-y)}{2} dy \\
&= \frac{1}{2} \int_{\mathbb{R}^n} (f(x-y) - 2f(x) + f(x+y))\rho_t(y) dy,
\end{aligned}$$

so that

$$\|f_t - f\|_\infty \leq [f]_{\mathcal{C}^{2\theta}} \frac{1}{2} \int_{\mathbb{R}^n} |2y|^{2\theta} \rho_t(y) dy \leq 2^{2\theta-1} c_{0,2\theta} t^{2\theta} [f]_{\mathcal{C}^{2\theta}}.$$

Moreover, $\|f_t\|_\infty \leq \|f\|_\infty$, and

$$D_{ij} f_t(x) = \frac{1}{t^{n+2}} \int_{\mathbb{R}^n} f(y) D_{ij}\rho((x-y)/t) dy.$$

Since $\int_{\mathbb{R}^n} D_{ij}\rho((x-y)/t) dy = 0$, we get

$$D_{ij} f_t(x) = \frac{1}{t^{n+2}} \int_{\mathbb{R}^n} (f(x-y) - f(x)) D_{ij}\rho(y/t) dy.$$

Since $D_{ij}\rho$ is even, arguing as above we find

$$\|D_{ij} f_t\|_\infty \leq [f]_{\mathcal{C}^{2\theta}} \frac{1}{2} \int_{\mathbb{R}^n} |2y|^{2\theta} D_{ij}\rho_t(y) dy \leq 2^{2\theta-1} c_{2,2\theta} t^{2\theta-2} [f]_{\mathcal{C}^{2\theta}}.$$

From Proposition 1.1.2 it follows

$$\|D_i f_t\|_\infty \leq c 2^{\theta-1/2} c_{2,2\theta}^{1/2} ([f]_{\mathcal{C}^{2\theta}})^{1/2} (\|f\|_\infty)^{1/2} t^{\theta-1}, \quad i = 1, \ldots, n,$$

so that there exists $k_1 = k_1(\theta)$ such that for $0 < t \leq 1$

$$\|f_t\|_{C^2} \leq k_1 t^{2\theta-2} \|f\|_{\mathcal{C}^{2\theta}}.$$

We split now f as $f = (f - f_{t^{1/2}}) + f_{t^{1/2}}$. Then for $0 < t \leq 1$ we get

$$K(t,f) \leq \|f - f_{t^{1/2}}\|_\infty + t\|f_{t^{1/2}}\|_{C^2} \leq k_0 t^\theta [f]_{\mathcal{C}^{2\theta}} + k_1 t^\theta \|f\|_{\mathcal{C}^{2\theta}},$$

with $k_0 = k_0(\theta) = 2^{2\theta-1}c_{0,2\theta}$. Recalling that $K(t,f) \leq \|f\|_\infty$ for $t \geq 1$, we find that $f \in (C(\mathbb{R}^n), C^2(\mathbb{R}^n))_{\theta,\infty}$, and

$$\|f\|_{\theta,\infty} \leq \|f\|_\infty + (k_0 + k_1)\|f\|_{C^{2\theta}}.$$

We show now that $(C(\mathbb{R}^n), C^2(\mathbb{R}^n))_{\theta,\infty}$ is continuously embedded in $\mathcal{C}^{2\theta}(\mathbb{R}^n)$. Let $f \in (C(\mathbb{R}^n), C^2(\mathbb{R}^n))_{\theta,\infty}$, and $x, y \in \mathbb{R}^n$. There are $f_0 \in C(\mathbb{R}^n)$, $f_1 \in C^2(\mathbb{R}^n)$, such that $f = f_0 + f_1$ and

$$\|f_0\|_\infty + |x-y|^2 \|f_1\|_{C^2} \leq 2\|f\|_{\theta,\infty}|x-y|^{2\theta}.$$

Then

$$\|f(x) + f(y) - 2f\left(\tfrac{x+y}{2}\right)\|$$
$$\leq \|f_0(x) + f_0(y) - 2f_0\left(\tfrac{x+y}{2}\right)\| + \|f_1(x) + f_1(y) - 2f_1\left(\tfrac{x+y}{2}\right)\|$$
$$\leq 4\|f_0\|_\infty + \tfrac{1}{2}|x-y|^2\|f_1\|_{C^2} \leq (8|x-y|^{2\theta} + |x-y|^{2\theta})\|f\|_{\theta,\infty},$$

so that f belongs to $\mathcal{C}^{2\theta}(\mathbb{R}^n)$, and

$$[f]_{\mathcal{C}^{2\theta}(\mathbb{R}^n)} \leq 9\|f\|_{\theta,\infty}.$$

Step 2: $(C^k(\mathbb{R}^n), C^{k+2}(\mathbb{R}^n))_{\theta,\infty} = \mathcal{C}^{k+2\theta}(\mathbb{R}^n)$.

The embedding $\mathcal{C}^{k+2\theta}(\mathbb{R}^n) \subset (C^k(\mathbb{R}^n), C^{k+2}(\mathbb{R}^n))_{\theta,\infty}$ can be proved in the same way as in the case $k = 0$; it is sufficient to recall that if $f \in C^k(\mathbb{R}^n)$ then for $|\alpha| = k$ we have

$$D^\alpha f_t(x) - D^\alpha f(x) = \int_{\mathbb{R}^n} (D^\alpha f(x-y) - D^\alpha f(x))\rho_t(y)dy,$$

$$D^\alpha D_{ij} f_t(x) = \frac{1}{t^{n+2}}\int_{\mathbb{R}^n} (D^\alpha f(x-y) - D^\alpha f(x))D_{ij}\rho(y/t)dy,$$

and to argue as in Step 1.

Let now $f \in (C^k(\mathbb{R}^n), C^{k+2}(\mathbb{R}^n))_{\theta,\infty}$. Then $f = u(0)$, with $u \in V(\infty, 1-\theta, C^k(\mathbb{R}^n), C^{k+2}(\mathbb{R}^n))$. In particular, for $|\alpha| = k$, $D^\alpha f = D^\alpha u(0)$, and $D^\alpha u$ belongs to $\in V(\infty, 1-\theta, C(\mathbb{R}^n), C^2(\mathbb{R}^n))$. Due to Step 1, $D^\alpha f \in \mathcal{C}^{2\theta}(\mathbb{R}^n)$, that is $f \in \mathcal{C}^{k+2\theta}(\mathbb{R}^n)$.

Step 3: conclusion. For $m = 1$ the statement follows from Step 1 via the Reiteration Theorem. Indeed, by Proposition 1.1.2(i), $C^1(\mathbb{R}^n) \in J_{1/2}(C(\mathbb{R}^n), C^2(\mathbb{R}^n))$, and from Step 1, since $C^1(\mathbb{R}^n)$ is continuously embedded in $\mathcal{C}^1(\mathbb{R}^n)$, it follows that $C^1(\mathbb{R}^n) \in K_{1/2}(C(\mathbb{R}^n), C^2(\mathbb{R}^n))$. (In fact, the equality $(C(\mathbb{R}^n), C^1(\mathbb{R}^n))_{\theta,\infty} = \mathcal{C}^\theta(\mathbb{R}^n)$ could be easily proved directly, arguing as in Step 1, with obvious modifications).

For $m > 2$ the statement follows from the Reiteration Theorem, provided we show that, for $k = 1, \ldots, m-1$, $C^k(\mathbb{R}^n)$ belongs to the class $K_{k/m}$ between $C(\mathbb{R}^n)$ and $C^m(\mathbb{R}^n)$ (we know already from Proposition 1.1.2(i) that it belongs

2. INTERPOLATION SPACES   31

to the class $J_{k/m}$). Indeed, once we know that $C^k(\mathbb{R}^n) \in K_{k/m} \cap J_{k/m}$, then for $k < m\theta < k+2$ we get

$$(C(\mathbb{R}^n), C^m(\mathbb{R}^n))_{\theta,\infty} = (C^k(\mathbb{R}^n), C^{k+2}(\mathbb{R}^n))_{m\theta/2-k,\infty} = C^{m\theta}(\mathbb{R}^n),$$

thanks to Step 2. So, the rest of the proof is devoted to show that for $k = 1, \ldots, m-1$, $C^k(\mathbb{R}^n)$ belongs to $K_{k/m}(C(\mathbb{R}^n), C^m(\mathbb{R}^n))$.

First we show that $C^k(\mathbb{R}; X)$ belongs to $K_{1/(m-k+1)}(C^{k-1}(\mathbb{R}; X), C^m(\mathbb{R}; X))$. The proof is quite similar to that of Step 1. Let $f \in C^k(\mathbb{R}; X)$, and define f_t by (1.2.26). Then, for $0 < |\alpha| \le k-1$, we have, denoting by D the gradient,

$$\|D^\alpha f - D^\alpha f_t\|_\infty \le \|D(D^\alpha f)\|_\infty \int_{\mathbb{R}^n} |y|\rho_t(s) ds = c_{0,1} t \|f\|_{C^k},$$

so that $\|f - f_t\|_{C^{k-1}} \le k_0 t \|f\|_{C^k}$. For $|\alpha| = 0, \ldots, m-k$, $|\beta| = k$, we have

$$D^\alpha D^\beta f_t(x) = \frac{1}{t^{n+|\alpha|}} \int_{\mathbb{R}^n} D^\beta f(x-y) D^\alpha \rho(y/t) dy$$

so that

$$\|D^{\alpha+\beta} f_t\|_\infty \le \frac{1}{t^{|\alpha|}} c_{k,0} \|D^\beta f\|_\infty,$$

which implies that $\|f_t\|_{C^m} \le k_1 t^{k-m} \|f\|_{C^k}$ for $0 < t \le 1$. Therefore, splitting f as $f = (f - f_{t^{1/(m-k+1)}}) + f_{t^{1/(m-k+1)}}$, for $0 < t \le 1$ we get

$$K(t, f) \le \|f - f_{t^{1/(m-k+1)}}\|_{C^k} + t\|f_{t^{1/(m-k+1)}}\|_{C^m} \le (k_0 + k_1) t^{1/(m-k+1)} \|f\|_{C^k}.$$

So, $C^k(\mathbb{R}^n)$ belongs to the class $K_{1/(m-k+1)}$ between $C^{k-1}(\mathbb{R}^n)$ and $C^m(\mathbb{R}^n)$ for $k = 1, \ldots, m-1$. In particular, if $k \ge 2$, the space $C^{k-1}(\mathbb{R}^n)$ belongs to $K_{1/(m-k+2)}(C^{k-2}(\mathbb{R}^n), C^m(\mathbb{R}^n))$, so that, from part (ii) of the Reiteration Theorem it follows that $C^k(\mathbb{R}^n) \subset (C^{k-2}(\mathbb{R}^n), C^m(\mathbb{R}^n))_{2/(m-k+2),\infty}$. Again, if $k \ge 3$, since $C^{k-2}(\mathbb{R}^n) \in K_{1/(m-k+3)}(C^{k-3}(\mathbb{R}^n), C^m(\mathbb{R}^n))$, from the second part of the Reiteration Theorem it follows that $C^k(\mathbb{R}^n) \subset (C^{k-3}(\mathbb{R}^n), C^m(\mathbb{R}^n))_{3/(m-k+3),\infty}$. Arguing in this way, after $k-1$ steps we find that $C^k(\mathbb{R}^n)$ is continuously embedded in $(C(\mathbb{R}^n), C^m(\mathbb{R}^n))_{k/m,\infty}$, which means that $C^k(\mathbb{R}^n)$ belongs to the class $K_{k/m}$ between $C(\mathbb{R}^n)$ and $C^m(\mathbb{R}^n)$. This completes the proof. ∎

Corollary 1.2.18 *For $0 \le \theta_1 < \theta_2$, $0 < \sigma < 1$ it holds*

$$(C^{\theta_1}(\mathbb{R}^n), C^{\theta_2}(\mathbb{R}^n))_{\sigma,\infty} = C^{\theta_1+\sigma(\theta_2-\theta_1)}(\mathbb{R}^n).$$

Proof — If θ_1 and θ_2 are not integers, the statement is an immediate consequence of Theorem 1.2.17 and the Reiteration Theorem. Let one of them be integer, say $\theta_i \in \mathbb{N}$. Due to Proposition 1.1.3(i), $C^{\theta_i}(\mathbb{R}^n)$ belongs to the class J_{m/θ_i} between $C(\mathbb{R}^n)$ and $C^m(\mathbb{R}^n)$, for every integer $m > \theta_2$. Due to Theorem 1.2.17, it belongs also to the class K_{m/θ_i}, since it is continuously embedded in $\mathcal{C}^{\theta_i}(\mathbb{R}^n)$. The Reiteration Theorem may be applied also in this case, and it yields the statement. ∎

Theorem 1.2.17 and Corollary 1.2.18 yield the characterization of other interpolation spaces between spaces of functions defined in arbitrary smooth domains.

Corollary 1.2.19 *Let $0 \leq \theta_1 < \theta_2$, and $0 < \sigma < 1$. If Ω is an open set in $\mathbb{R}^n$ with uniformly C^{θ_2} boundary, then*

$$(C^{\theta_1}(\overline{\Omega}), C^{\theta_2}(\overline{\Omega}))_{\sigma,\infty} = \mathcal{C}^{\theta_1+\sigma(\theta_2-\theta_1)}(\overline{\Omega}),$$

with equivalence of the respective norms, and if $\theta_1 + \sigma(\theta_2 - \theta_1)$ is not integer, then

$$(C^{\theta_1}(\overline{\Omega}), C^{\theta_2}(\overline{\Omega}))_\sigma = h^{\theta_1+\sigma(\theta_2-\theta_1)}(\overline{\Omega}).$$

Proof — The proof of the statement in the case where $\Omega = \mathbb{R}^n_+$ is quite similar to the proof of Theorem 1.2.17, with obvious modifications. The general case follows from the cases $\Omega = \mathbb{R}^n_+$ and $\Omega = \mathbb{R}^n$, through locally straightening the boundary. ∎

1.3 Bibliographical remarks

Extensive treatments of interpolation theory may be found in the paper by J.L. LIONS – J. PEETRE [129], and in the books by H. TRIEBEL [200], P.L. BUTZER – H. BERENS [42], J. BERGH – J. LÖFSTRÖM [32], YU. BRUDNYI – N. KRUGLJAK [39], S.G. KREIN – YU. PETUNIN – E.M. SEMENOV [115], B. BEAUZAMY [30], J. PEEETRE [167].

The K-method has been introduced by J. PEETRE, and the trace method by J.-L. LIONS. See [42, Ch. 3] for historical and bibliographical references. The spaces $(X,Y)_\theta$ have been defined by G. DA PRATO – P. GRISVARD in [63] and called *continuous interpolation spaces* .

Chapter 2

Analytic semigroups and intermediate spaces

Let X be a complex Banach space, with norm $\|\cdot\|$. This chapter deals with the solution of an initial value problem in X,

$$u'(t) = Au(t), \ t > 0; \ u(0) = x,$$

where $A : D(A) \subset X \mapsto X$ is a linear operator, with not necessarily dense domain.

Definition 2.0.1 A *is said to be* sectorial *if there are constants* $\omega \in \mathbb{R}$, $\theta \in]\pi/2, \pi[$, $M > 0$ *such that*

$$\begin{cases} (i) & \rho(A) \supset S_{\theta,\omega} = \{\lambda \in \mathbb{C} : \lambda \neq \omega, |\arg(\lambda - \omega)| < \theta\}, \\ (ii) & \|R(\lambda, A)\|_{L(X)} \leq \dfrac{M}{|\lambda - \omega|} \ \forall \lambda \in S_{\theta,\omega}. \end{cases} \quad (2.0.1)$$

The fact that the resolvent set of A is not void implies that A is closed, so that $D(A)$, endowed with the graph norm

$$\|x\|_{D(A)} = \|x\| + \|Ax\|,$$

is a Banach space. For every $t > 0$, (2.0.1) allows us to define a linear bounded operator e^{tA} in X, by means of the Dunford integral

$$e^{tA} = \frac{1}{2\pi i} \int_{\omega + \gamma_{r,\eta}} e^{t\lambda} R(\lambda, A) d\lambda, \ t > 0, \quad (2.0.2)$$

where $r > 0$, $\eta \in]\pi/2, \theta[$, and $\gamma_{r,\eta}$ is the curve $\{\lambda \in \mathbb{C} : |\arg \lambda| = \eta, |\lambda| \geq r\} \cup \{\lambda \in \mathbb{C} : |\arg \lambda| \leq \eta, |\lambda| = r\}$, oriented counterclockwise. We also set

$$e^{0A} x = x, \ \forall x \in X. \quad (2.0.3)$$

Since the function $\lambda \mapsto e^{t\lambda} R(\lambda, A)$ is holomorphic in $S_{\theta,\omega}$, the definition of e^{tA} is independent of the choice of r and η. We shall show in Section 2.1 that the mapping $t \mapsto e^{tA}$ is analytic from $]0, +\infty[$ to $L(X)$, and moreover it enjoys the semigroup property
$$e^{tA} e^{sA} = e^{(t+s)A}, \quad \forall\, t,\, s \geq 0.$$
These properties motivate the following definition.

Definition 2.0.2 *Let $A : D(A) \subset X \mapsto X$ be a sectorial operator. The family $\{e^{tA} : t \geq 0\}$ defined by (2.0.2)-(2.0.3) is said to be the analytic semigroup generated by A in X.*

Section 2.1 is devoted to the main properties of e^{tA}. All of them are deduced from Definition 2.0.2, without invoking results from the general theory of semigroups.

We recall that a family of linear operators $\{T(t)\}_{t \geq 0} \subset L(X)$ is said to be a *semigroup* if
$$\begin{cases} T(t)T(s) = T(t+s), \quad t,\, s \geq 0, \\ T(0) = I. \end{cases}$$

A semigroup $T(t)$ is said to be *analytic* if the function $t \mapsto T(t)$ is analytic in $]0, +\infty[$ with values in $L(X)$. It is said to be *strongly continuous* if for each $x \in X$ the function $t \mapsto T(t)x$ is continuous on $[0, +\infty[$. We shall show that if A is sectorial, then $\{e^{tA}\}_{t \geq 0}$ is analytic, so that it is strongly continuous if and only if
$$\lim_{t \to 0} e^{tA} x = x \quad \forall x \in X.$$

We shall see in Proposition 2.1.4 that
$$\lim_{t \to 0} e^{tA} x = x \iff x \in \overline{D(A)}.$$

Therefore, $\{e^{tA}\}_{t \geq 0}$ is strongly continuous if and only if the domain $D(A)$ is dense in X.

Section 2.2 deals with intermediate spaces between X and $D(A)$, which are of fundamental importance in the following chapters. In Section 1.1 we introduced the class J_α, with $0 < \alpha < 1$. Examples of spaces belonging to the class J_α between X and $D(A)$ are the interpolation spaces $D_A(\alpha, p)$, with $1 \leq p \leq \infty$, the continuous interpolation spaces $D_A(\alpha)$, the complex interpolation spaces $[D(A), X]_\alpha$, and the domains of fractional powers $D(-A)^\alpha$. We do not pretend to give a systematic treatment of intermediate spaces: we only present the basic results and those results which will be used in the sequel.

We shall give several characterizations and equivalent norms in the spaces $D_A(\alpha, p)$ and $D_A(\alpha)$, which will be shown to be real interpolation spaces between X and $D(A)$. Their elements x will be characterized in terms of the behavior of $t \mapsto e^{tA} x$, $t \mapsto Ae^{tA} x$ near $t = 0$ and of $\lambda \mapsto AR(\lambda A)x$ as $\lambda \to \infty$. Moreover we

1. Basic properties of e^{tA}

shall study the properties of the function $t \mapsto e^{tA}x$ when x belongs to any of these spaces.

In Section 2.3 we shall study the spectrum and the asymptotic behavior of e^{tA}, in connection with the spectral properties of A.

Finally in Section 2.4 we shall prove some perturbation results for sectorial operators.

2.1 Basic properties of e^{tA}

Let $A : D(A) \mapsto X$ be a sectorial operator, and let e^{tA} be the analytic semigroup generated by A. In the following proposition we state the main properties of e^{tA}.

Proposition 2.1.1 *(i) $e^{tA}x \in D(A^k)$ for each $t > 0$, $x \in X$, $k \in \mathbb{N}$. If $x \in D(A^k)$, then*
$$A^k e^{tA} x = e^{tA} A^k x, \quad \forall t \geq 0.$$

(ii) $e^{tA} e^{sA} = e^{(t+s)A}$, $\forall\, t, s \geq 0$.
(iii) There are constants M_0, M_1, M_2, ..., such that
$$\begin{cases} (a) & \|e^{tA}\|_{L(X)} \leq M_0 e^{\omega t}, \quad t > 0, \\ (b) & \|t^k(A-\omega I)^k e^{tA}\|_{L(X)} \leq M_k e^{\omega t}, \quad t > 0, \end{cases} \quad (2.1.1)$$

where ω is the constant of assumption (2.0.1). In particular, from (2.1.1)(b) it follows that for every $\varepsilon > 0$ and $k \in \mathbb{N}$ there is $C_{k,\varepsilon} > 0$ such that
$$\|t^k A^k e^{tA}\|_{L(X)} \leq C_{k,\varepsilon} e^{(\omega+\varepsilon)t}, \quad t > 0. \quad (2.1.2)$$

(iv) The function $t \mapsto e^{tA}$ belongs to $C^\infty(]0, +\infty[; L(X))$, and
$$\frac{d^k}{dt^k} e^{tA} = A^k e^{tA}, \quad t > 0, \quad (2.1.3)$$

moreover it has an analytic extension in the sector
$$S = \{\lambda \in \mathbb{C} : \lambda \neq 0,\ |\arg \lambda| < \theta - \pi/2\}.$$

Proof — Let us prove that statement (i) holds. By using several times the identity $AR(\lambda, A) = \lambda R(\lambda, A) - I$, which holds for all $\lambda \in \rho(A)$, it follows that, for each $x \in X$, $e^{tA}x$ belongs to $D(A^k)$ for all $k \in \mathbb{N}$, and that
$$A^k e^{tA} = \frac{1}{2\pi i} \int_{\omega + \gamma_{r,\eta}} \lambda^k e^{t\lambda} R(\lambda, A) d\lambda.$$

If $x \in D(A)$, the equality $Ae^{tA}x = e^{tA}Ax$ follows from Definition 2.0.2 through the obvious equality $AR(\lambda, A)x = R(\lambda, A)Ax$. For $k > 1$, statement (i) follows by recurrence.

To prove statements (ii), (iii), it is convenient to introduce the operator

$$B : D(A) \mapsto X; \quad Bx = Ax - \omega x. \tag{2.1.4}$$

Then the resolvent set of B contains the sector $S_{\theta,0}$, where θ is given by (2.0.1), and $R(\lambda, B) = R(\lambda + \omega, A)$, so that

$$\|\lambda R(\lambda, B)\|_{L(X)} \leq M, \quad \lambda \in S_{\theta,0}. \tag{2.1.5}$$

From Definition 2.0.2 we get easily

$$e^{tB} = e^{-\omega t} e^{tA}, \quad t \geq 0. \tag{2.1.6}$$

Now we can prove that (ii) holds. For t, $s > 0$, $r > 0$, $\pi/2 < \eta' < \eta$ we have

$$\begin{aligned}
e^{tB} e^{sB} &= \left(\frac{1}{2\pi i}\right)^2 \int_{\gamma_{r,\eta}} e^{t\lambda} R(\lambda, B) d\lambda \int_{\gamma_{2r,\eta'}} e^{s\mu} R(\mu, B) d\mu \\
&= \left(\frac{1}{2\pi i}\right)^2 \int_{\gamma_{r,\eta} \times \gamma_{2r,\eta'}} e^{t\lambda + s\mu} \frac{R(\lambda, B) - R(\mu, B)}{\mu - \lambda} d\lambda d\mu \\
&= \left(\frac{1}{2\pi i}\right)^2 \int_{\gamma_{r,\eta}} e^{t\lambda} R(\lambda, B) d\lambda \int_{\gamma_{2r,\eta'}} e^{s\mu} (\mu - \lambda)^{-1} d\mu \\
&\quad - \left(\frac{1}{2\pi i}\right)^2 \int_{\gamma_{2r,\eta'}} e^{s\mu} R(\mu, B) d\mu \int_{\gamma_{r,\eta}} e^{t\lambda} (\mu - \lambda)^{-1} d\lambda,
\end{aligned}$$

so that from the equalities $\int_{\gamma_{2r,\eta}} e^{s\mu}(\mu - \lambda)^{-1} d\mu = 2\pi i\ e^{s\lambda}$ for $\lambda \in \gamma_{r,\eta}$ and $\int_{\gamma_{r,\eta}} e^{t\lambda}(\mu - \lambda)^{-1} d\lambda = 0$ for $\mu \in \gamma_{2r,\eta'}$ we get $e^{tB} e^{sB} = e^{(t+s)B}$. Statement (ii) follows now from (2.1.6).

Let us prove (iii) and (iv). For $t > 0$ it holds

$$\begin{aligned}
e^{tB} &= \frac{1}{2\pi i} \int_{\gamma_{r,\eta}} e^{t\lambda} R(\lambda, B) d\lambda = \frac{1}{2\pi i t} \int_{\gamma_{tr,\eta}} e^{\xi} R(\xi/t, B) d\xi \\
&\quad \frac{1}{2\pi i t} \int_{\gamma_{r,\eta}} e^{\xi} R(\xi/t, B) d\xi,
\end{aligned}$$

so that

$$\|e^{tB}\|_{L(X)} \leq \frac{M}{2\pi}\left(2\int_r^{+\infty} \rho^{-1} e^{\rho \cos \eta} d\rho + \int_{-\eta}^{\eta} e^{r \cos \theta} d\theta\right),$$

and (2.1.1)(a) follows.

Due to statement (i), $e^{tB}x$ belongs to $D(B) = D(A)$ for every $x \in X$, and

$$\begin{aligned} Be^{tB} &= \frac{1}{2\pi i}\int_{\gamma_{r,\eta}} \lambda e^{t\lambda} R(\lambda, B)d\lambda = \frac{1}{2\pi i}\int_{\gamma_{tr,\eta}} t^{-2}\xi e^{\xi} R(\xi/t, B)d\xi \\ &= \frac{1}{2\pi i}\int_{\gamma_{r,\eta}} t^{-2}\xi e^{\xi} R(\xi/t, B)d\xi, \end{aligned}$$

so that

$$\|Be^{tB}\|_{L(X)} \le \frac{M}{2\pi t}\left(2\int_{r}^{+\infty} e^{\rho\cos\eta}d\rho + r\int_{-\eta}^{\eta} e^{r\cos\theta}d\theta\right),$$

and (2.1.1)(b) follows for $k = 1$. Moreover we get easily, for $t > 0$,

$$\frac{d}{dt}e^{tB} = \frac{1}{2\pi i}\int_{\gamma_{r,\eta}} \lambda e^{t\lambda} R(\lambda, B)d\lambda = Be^{tB}, \tag{2.1.7}$$

so that

$$\frac{d^k}{dt^k}e^{tB} = B^k e^{tB},$$

which implies that (2.1.3) holds. From the equality $Be^{tB} = e^{tB}B$ on $D(B)$ it follows $B^k e^{tB} = (Be^{\frac{t}{k}B})^k$ for all $k \in \mathbb{N}$, so that

$$\|B^k e^{tB}\|_{L(X)} \le (M_1 k t^{-1})^k \le (M_1 e)^k k! t^{-k},$$

and (2.1.1)(b) holds for every k, with $M_k = (M_1 e)^k k!$. Now (2.1.2) follows.

Let us prove statement (iv). Let $0 < \varepsilon < \theta - \pi/2$, and choose $\eta = \theta - \varepsilon$. The function

$$z \mapsto e^{zA} = \frac{1}{2\pi i}\int_{\omega + \gamma_{r,\eta}} e^{z\lambda} R(\lambda, A)d\lambda$$

is well defined and holomorphic in the sector

$$S_\varepsilon = \{z \in \mathbb{C} : z \ne 0, \ |\arg z| < \theta - \pi/2 - \varepsilon\}.$$

The union of the sectors S_ε, for $0 < \varepsilon < \theta - \pi/2$, is S, and statement (iv) follows. ∎

The semigroup e^{tA} may be approximated by a sequence of semigroups generated by bounded operators. This may be useful in several circumstances, such as in next Corollary 2.1.3.

Proposition 2.1.2 *For every integer $n > \omega$ set*

$$A_n : X \mapsto X, \quad A_n = nAR(n, A). \tag{2.1.8}$$

Then $\rho(A) \subset \rho(A_n)$, and $R(\lambda, A_n) \to R(\lambda, A)$ in $L(X)$ as $n \to \infty$, for every $\lambda \in \rho(A)$. Moreover, $e^{tA_n} \to e^{tA}$ in $L(X)$ as $n \to \infty$, for every $t > 0$.

Proof — Without loss of generality we may assume that $\omega < 0$. Then it is not hard to see that $\rho(A) \subset \rho(A_n)$, and that

$$R(\lambda, A_n) = \frac{n^2}{(\lambda+n)^2} R\left(\frac{\lambda n}{\lambda+n}, A\right) + \frac{1}{\lambda+n} I,$$

for every $n \in \mathbb{N}$ and $\lambda \in \rho(A)$. It follows that $R(\lambda, A_n) \to R(\lambda, A)$ for every $\lambda \in \rho(A)$, and since $\|R(\lambda, A_n)\|_{L(X)} \leq c|\lambda|^{-1}$ for $|\arg \lambda| < \theta_A$, with constant c independent of n, it follows that $e^{tA_n} \to e^{tA}$ in $L(X)$ as $n \to \infty$, for every $t > 0$. ■

The family $\{A_n : n \in \mathbb{N}\}$ is said to be the *Yosida approximation* of A.

Corollary 2.1.3 *Let X be a real Banach space, and let $A : D(A) \subset X \mapsto X$ be a linear operator such that the complexification*

$$\widetilde{A} : D(\widetilde{A}) = D(A) + iD(A) \mapsto \widetilde{X} = X + iX, \quad \widetilde{A}(x+iy) = Ax + iAy,$$

is a sectorial operator in $\widetilde{X}$. Then $e^{t\widetilde{A}}(X) \subset X$.

Proof — For $n > \omega$, let $\widetilde{A}_n = n\widetilde{A}R(n, \widetilde{A})$. By Proposition 2.1.2, $e^{t\widetilde{A}_n}x \to e^{t\widetilde{A}}x$ for every $x \in X$. On the other hand, since $\widetilde{A}_n$ is a bounded operator, then $e^{t\widetilde{A}_n} = \sum_{k=0}^{\infty} t^k \widetilde{A}_n^k / k!$, so that $e^{t\widetilde{A}_n}$ maps X into itself. The statement follows. ■

Corollary 2.1.3 will be used in Chapters 5, 7, 8, 9, where we will consider differential operators with real coefficients, acting in spaces of real functions.

The following proposition deals with the behavior of $e^{tA}x$ near $t = 0$. The behavior of $e^{tA}x$ for large t will be the subject of Section 2.3.

Proposition 2.1.4 *The following statements hold true.*

(i) *If $x \in \overline{D(A)}$, then $\lim_{t\to 0^+} e^{tA}x = x$. Conversely, if there exists $y = \lim_{t\to 0^+} e^{tA}x$, then $x \in \overline{D(A)}$, and $y = x$.*

(ii) *For every $x \in X$ and $t \geq 0$, the integral $\int_0^t e^{sA}x\,ds$ belongs to $D(A)$, and*

$$A \int_0^t e^{sA}x\,ds = e^{tA}x - x.$$

If in addition the function $s \mapsto Ae^{sA}x$ belongs to $L^1(0, t; X)$, then

$$e^{tA}x - x = \int_0^t Ae^{sA}x\,ds.$$

(iii) *If $x \in D(A)$ and $Ax \in \overline{D(A)}$, then $\lim_{t\to 0^+}(e^{tA}x - x)/t = Ax$. Conversely, if there exists $z = \lim_{t\to 0^+}(e^{tA}x - x)/t$, then $x \in D(A)$ and $z = Ax \in \overline{D(A)}$.*

1. Basic properties of e^{tA}

(iv) If $x \in D(A)$ and $Ax \in \overline{D(A)}$, then $\lim_{t \to 0+} Ae^{tA}x = Ax$. Conversely, if there exists $v = \lim_{t \to 0+} Ae^{tA}x$, then $x \in D(A)$ and $v = Ax \in \overline{D(A)}$.

Proof — (i) Let $\xi > \omega$, $0 < r < \xi - \omega$. For every $x \in D(A)$, let $y = \xi x - Ax$. We have

$$e^{tA}x = e^{tA}R(\xi, A)y = \frac{1}{2\pi i}\int_{\omega+\gamma_{r,\eta}} e^{t\lambda}R(\lambda, A)R(\xi, A)y\, d\lambda$$

$$= \frac{1}{2\pi i}\int_{\omega+\gamma_{r,\eta}} e^{t\lambda}\frac{R(\lambda, A)}{\xi - \lambda}y\, d\lambda - \frac{1}{2\pi i}\int_{\omega+\gamma_{r,\eta}} e^{t\lambda}\frac{R(\xi, A)}{\xi - \lambda}y\, d\lambda$$

$$= \frac{1}{2\pi i}\int_{\omega+\gamma_{r,\eta}} e^{t\lambda}\frac{R(\lambda, A)}{\xi - \lambda}y\, d\lambda.$$

Then $\lim_{t \to 0+} e^{tA}x = \frac{1}{2\pi i}\int_{\omega+\gamma_{r,\eta}} e^{t\lambda}\frac{R(\lambda, A)}{\xi - \lambda}y\, d\lambda = x$. Since $D(A)$ is dense in $\overline{D(A)}$, then $\lim_{t \to 0+} e^{tA}x = x$ for every $x \in \overline{D(A)}$. Conversely, if $y = \lim_{t \to 0} e^{tA}x$, then $y \in \overline{D(A)}$ because $e^{tA}x \in D(A)$ for $t > 0$, and $R(\xi, A)y = \lim_{t \to 0+} R(\xi, A)\, e^{tA}x = \lim_{t \to 0+} e^{tA}R(\xi, A)x = R(\xi, A)x$ because $R(\xi, A)x \in D(A)$. Therefore $y = x$.

(ii) Let $\xi \in \rho(A)$ and $x \in X$. For every $\varepsilon \in \,]0, t[$ we have

$$\int_\varepsilon^t e^{sA}x\, ds = \int_\varepsilon^t (\xi - A)R(\xi, A)e^{sA}x\, ds$$

$$= \xi \int_\varepsilon^t R(\xi, A)e^{sA}x\, ds - \int_\varepsilon^t \frac{d}{ds}(R(\xi, A)e^{sA}x)\, ds$$

$$= \xi \int_\varepsilon^t R(\xi, A)e^{sA}x\, ds - e^{tA}R(\xi, A)x + e^{\varepsilon A}R(\xi, A)x.$$

Since $R(\xi, A)x$ belongs to $D(A)$, letting $\varepsilon \to 0$ we find

$$\int_0^t e^{sA}x\, ds = \xi R(\xi, A)\int_0^t e^{sA}x\, ds - R(\xi, A)(e^{tA}x - x). \quad (2.1.9)$$

Hence, $\int_0^t e^{sA}x\, ds \in D(A)$, and $(\xi - A)\int_0^t e^{sA}x\, ds = \xi \int_0^t e^{sA}x\, ds - (e^{tA}x - x)$, so that (ii) follows.

(iii) If $x \in D(A)$ and $Ax \in \overline{D(A)}$, then

$$\frac{e^{tA}x - x}{t} = \frac{1}{t}A\int_0^t e^{sA}x\, ds = \frac{1}{t}\int_0^t e^{sA}Ax\, ds.$$

Since $s \mapsto e^{sA}Ax$ is continuous in $[0, t]$ by (i), then $\lim_{t \to 0}(e^{tA}x - x)/t = Ax$. Conversely, if there exists $z = \lim_{t \to 0}(e^{tA}x - x)/t$, then $\lim_{t \to 0+} e^{tA}x = x$, so that $x, z \in \overline{D(A)}$. Moreover, for every $\xi \in \rho(A)$,

$$R(\xi, A)z = \lim_{t \to 0} R(\xi, A)\frac{e^{tA}x - x}{t},$$

so that, by statement (ii),

$$R(\xi, A)z = \lim_{t \to 0} \frac{1}{t} R(\xi, A) A \int_0^t e^{sA} x\, ds = \lim_{t \to 0} (\xi R(\xi, A) - I) \frac{1}{t} \int_0^t e^{sA} x\, ds.$$

Since $x \in \overline{D(A)}$, then $s \mapsto e^{sA} x$ is continuous near $s = 0$, so that

$$R(\xi, A)z = \xi R(\xi, A)x - x.$$

Therefore, $x \in D(A)$, and $z = \xi x - (\xi - A)x = Ax$.

(iv) The first part of the statement is a consequence of (i). Moreover, if there exists $v = \lim_{t \to 0} Ae^{tA} x$, then the function $s \mapsto Ae^{sA} x$ is continuously extendible at $s = 0$, and $v = \lim_{t \to 0} \frac{1}{t} \int_0^t Ae^{sA} x\, ds$. Using (ii) we get $v = \lim_{t \to 0} (e^{tA} x - x)/t$, so that, by (iii), $x \in D(A)$ and $v = Ax \in \overline{D(A)}$. ■

If X_0 is a subspace of X, the *part of A in X_0* is defined by

$$\begin{cases} D(A_0) = \{x \in D(A) : Ax \in X_0\}, \\ A_0 : D(A_0) \mapsto X_0, \ A_0 x = Ax. \end{cases}$$

Remark 2.1.5 Let $X_0 = \overline{D(A)}$, and let A_0 be the part of A in X_0. Then $D(A_0)$ is dense in X_0. Moreover, A_0 is sectorial, so that it generates the analytic semigroup e^{tA_0} in X_0, and we have $e^{tA_0} x = e^{tA} x$ for $x \in X_0$. Due to Proposition 2.1.4(i), e^{tA_0} is strongly continuous in X_0.

2.1.1 Identification of the generator

Now we consider the problem of identifying the generator of a given analytic semigroup.

We ask two (related) questions: is it possible that two different sectorial operators generate the same analytic semigroup? Or, given an analytic semigroup $\{T(t)\}_{t \geq 0}$, is it possible to find a sectorial operator A such that $T(t) = e^{tA}$? The answers are below. We begin with a simple lemma.

Lemma 2.1.6 *Let $A : D(A) \subset X \mapsto X$ satisfy (2.0.1). Then for every $\lambda \in \mathbb{C}$ such that $\mathrm{Re}\, \lambda > \omega$ we have*

$$R(\lambda, A) = \int_0^{+\infty} e^{-\lambda t} e^{tA} dt. \tag{2.1.10}$$

Proof — Let $0 < r < \mathrm{Re}\, \lambda - \omega$ and $\eta \in\,]\pi/2, \theta[$. Then

$$\int_0^{+\infty} e^{-\lambda t} e^{tA} dt = \frac{1}{2\pi i} \int_{\omega + \gamma_{r,\eta}} R(z, A) \int_0^{+\infty} e^{-\lambda t + zt} dt\, dz$$

$$\frac{1}{2\pi i} \int_{\omega + \gamma_{r,\eta}} R(z, A)(z - \lambda)^{-1} dz = R(\lambda, A). \quad ■$$

1. Basic properties of e^{tA} 41

Corollary 2.1.7 *For every $t \geq 0$, e^{tA} is one to one.*

Proof — $e^{0A} = I$ is obviously one to one. Assume that there are $t_0 > 0$, $x \neq 0$ such that $e^{t_0 A} x = 0$. Then, for $t \geq t_0$, $e^{tA} x = e^{(t-t_0)A} e^{t_0 A} x = 0$. Since $t \mapsto e^{tA} x$ is analytic, then $e^{tA} x \equiv 0$ in $]0, +\infty[$. By Lemma 2.1.6 we get $R(\lambda, A) x = 0$ for all $\lambda > \omega$, so that $x = 0$, a contradiction. ∎

Corollary 2.1.8 *If $A : D(A) \subset X \mapsto X$ and $B : D(B) \subset X \mapsto X$ are sectorial operators such that $e^{tA} = e^{tB}$ for every $t > 0$, then $D(A) = D(B)$ and $A = B$.*

Proof — Let A and B satisfy (2.0.1) with constant $\omega = \omega_A$, $\omega = \omega_B$, respectively. By Lemma 2.1.6, for $\operatorname{Re} \lambda > \max\{\omega_A, \omega_B\}$ we have $R(\lambda, A) = R(\lambda, B)$, and the statement follows. ∎

Proposition 2.1.9 *Let $\{T(t) : t > 0\}$ be a family of linear bounded operators such that $t \mapsto T(t)$ is differentiable with values in $L(X)$, and*

(i) $T(t)T(s) = T(t+s)$, for every $t, s > 0$;

(ii) there are $\omega \in \mathbb{R}$, M_0, $M_1 > 0$ such that $\|T(t)\|_{L(X)} \leq M_0 e^{\omega t}$, $\|t T'(t)\|_{L(X)} \leq M_1 e^{\omega t}$ for $t > 0$;

(iii) either (a) there is $t > 0$ such that $T(t)$ is one to one, or (b) for every $x \in X$, $\lim_{t \to 0} T(t) x = x$.

Then $t \mapsto T(t)$ is analytic in $]0, +\infty[$ with values in $L(X)$, and there exists a unique sectorial operator $A : D(A) \subset X \mapsto X$ such that $T(t) = e^{tA}$ for every $t \geq 0$.

Proof — The function
$$F(\lambda) = \int_0^{+\infty} e^{-\lambda t} T(t) \, dt$$
is well defined and holomorphic in the half-plane $\Pi = \{\lambda \in \mathbb{C} : \operatorname{Re} \lambda > \omega\}$. To prove the statement, it is clearly sufficient to show that

(a) $F(\lambda)$ has an analytic extension to a sector $S_{\beta, \omega}$, with $\beta > \pi/2$, and $\|(\lambda - \omega) F(\lambda)\|_{L(X)}$ is bounded in $S_{\beta, \omega}$;

(b) there exists a linear operator $A : D(A) \subset X \mapsto X$ such that $F(\lambda) = R(\lambda, A)$ for $\lambda \in S_{\beta, \omega}$.

To prove (a), we show by recurrence that $t \mapsto T(t)$ is infinitely many times differentiable, and
$$T^{(n)}(t) = (T'(t/n))^n, \quad t > 0, \ n \in \mathbb{N}. \tag{2.1.11}$$

Equality (2.1.11) holds for $n = 1$. If (2.1.11) holds for $n = n_0$, from the identity $T(t+s) = T(t)T(s)$ we get $T^{(n_0)}(t+s) = T^{(n_0)}(t)T(s) = T^{(n_0)}(s)T(t)$ for every t, $s > 0$, and moreover

$$\lim_{h \to 0} \tfrac{1}{h} \left(T^{(n_0)}(t+h) - T^{(n_0)}(t) \right)$$
$$= \lim_{h \to 0} \tfrac{1}{h} T^{(n_0)} \left(\tfrac{tn_0}{n_0+1} \right) \left(T\left(\tfrac{t}{n_0+1} + h \right) - T\left(\tfrac{t}{n_0+1} \right) \right)$$
$$= \left(T'\left(\tfrac{t}{n_0+1} \right) \right)^{n_0} T'\left(\tfrac{t}{n_0+1} \right) = \left(T'\left(\tfrac{t}{n_0+1} \right) \right)^{n_0+1},$$

so that $T^{(n_0+1)}$ exists, and (2.1.11) holds for $n = n_0 + 1$. Therefore, (2.1.11) holds for every n, and it implies that

$$\|T^{(n)}(t)\|_{L(X)} \le (nM_1/t)^n e^{\omega t} \le (M_1 e)^n t^{-n} n! e^{\omega t}, \quad t > 0, \; n \in \mathbb{N}.$$

So, the series

$$\sum_{n=0}^{\infty} \frac{(z-t)^n}{n!} \frac{d^n}{dt^n} T(t)$$

converges for all $z \in \mathbb{C}$ such that $|z - t| < t(M_1 e)^{-1}$. Therefore, $t \mapsto T(t)$ is analytically extendible to the sector $S_{\beta_0, 0}$, with $\beta_0 = \arctan(M_1 e)^{-1}$, and denoting the extension by $T(z)$, we have

$$\|T(z)\|_{L(X)} \le (1 - (eM_1)^{-1} \tan \theta)^{-1} e^{\omega \operatorname{Re} z}, \quad z \in S_{\beta_0, 0}, \; \theta = \arg z.$$

Shifting the halfline $[0, +\infty[$ to the halfline $\{\arg z = \beta\}$, with $|\beta| < \beta_0$, we see that (a) holds, with any $\beta \in]\pi/2, \beta_0[$.

Let us prove statement (b). It is easy to see that F satisfies the resolvent identity in the half-plane Π: indeed, for $\lambda \ne \mu$, $\lambda, \mu \in \Pi$, we have

$$F(\lambda)F(\mu) = \int_0^{+\infty} e^{-\lambda t} T(t) dt \int_0^{+\infty} e^{-\mu s} T(s) ds$$
$$= \int_0^{+\infty} e^{-\mu \sigma} T(\sigma) d\sigma \int_0^{\sigma} e^{-(\lambda+\mu)t} dt = \int_0^{+\infty} e^{-\mu \sigma} T(\sigma) \frac{e^{-(\lambda-\mu)\sigma} - 1}{\lambda - \mu} d\sigma$$
$$= \frac{1}{\lambda - \mu} (F(\lambda) - F(\mu)).$$

Let us show that $F(\lambda)$ is one to one for every $\lambda \in \Pi$. Assume that there are $x \ne 0$, $\lambda_0 \in \Pi$ such that $F(\lambda_0)x = 0$. From the resolvent identity it follows easily that $F(\lambda)x = 0$ for every $\lambda \in \Pi$. Therefore, for every x^* in the dual space X',

$$\langle F(\lambda)x, x^* \rangle = \int_0^{+\infty} e^{-\lambda t} \langle T(t)x, x^* \rangle dt = 0, \quad \forall \lambda \in \Pi.$$

Since $\langle F(\lambda)x, x^* \rangle$ is the Laplace transform of the scalar function $t \mapsto \langle T(t)x, x^* \rangle$, then $\langle T(t)x, x^* \rangle \equiv 0$ in $]0, +\infty[$. Since x^* is arbitrary, then $T(t)x \equiv 0$ in $]0, +\infty[$,

1. Basic properties of e^{tA}

which is impossible if (iii)(a) or (iii)(b) holds. Therefore, $F(\lambda)$ is one to one for every $\lambda \in \Pi$. By Proposition A.0.2, there exists a linear operator $A : D(A) \subset X \mapsto X$ such that $\rho(A) \supset \Pi$ and $R(\lambda, A) = F(\lambda)$ for every $\lambda \in \Pi$. Since F is holomorphic in $S_{\beta_0, \omega}$, by Proposition A.0.2 we get $\rho(A) \supset S_{\beta_0, \omega}$, and $R(\lambda, A) = F(\lambda)$ for $\lambda \in S_{\beta_0, \omega}$. Statement (b) is so proved. ∎

Remark 2.1.10 From the proof of Proposition 2.1.9 it follows that if A is a sectorial operator such that $\|e^{tA}\|_{L(X)} \leq M_0 e^{\alpha t}$ and $\|tAe^{tA}\|_{L(X)} \leq M_1 e^{\alpha t}$ for every $t > 0$, then A satisfies (2.0.1) with $\omega = \alpha$. The converse is true for $\omega = 0$, thanks to Proposition 2.1.1(iii).

2.1.2 A sufficient condition to be a sectorial operator

Proposition 2.1.11 *Let $A : D(A) \subset X \mapsto X$ be a linear operator such that $\rho(A)$ contains a half plane $\{\lambda \in \mathbb{C} : \operatorname{Re} \lambda \geq \omega\}$, and*

$$\|\lambda R(\lambda, A)\|_{L(X)} \leq M, \quad \operatorname{Re} \lambda \geq \omega, \qquad (2.1.12)$$

with $\omega \in \mathbb{R}$, $M > 0$. Then A is sectorial.

Proof — By Proposition A.0.3, for every $r > 0$ the resolvent set of A contains the open ball centered at $\omega + ir$ with radius $|\omega + ir|/M$. The union of such balls contains the sector $S = \{\lambda \neq \omega : |\arg(\lambda - \omega)| < \pi - \arctan M\}$. Moreover, for $\lambda \in V = \{\lambda : \operatorname{Re} \lambda < \omega, |\arg(\lambda - \omega)| \leq \pi - \arctan 2M\}$, $\lambda = \omega + ir - \theta r/M$ with $0 < \theta \leq 1/2$, formula (A.0.4) gives

$$\|R(\lambda, A)\| \leq \sum_{n=0}^{\infty} |\lambda - (\omega + ir)|^n \frac{M^{n+1}}{(\omega^2 + r^2)^{(n+1)/2}} \leq \frac{2M}{r}.$$

On the other hand, for $\lambda = \omega + ir - \theta r/M$ it holds

$$r \geq (1/(4M^2) + 1)^{-1/2} |\lambda - \omega|,$$

so that $\|R(\lambda, A)\| \leq 2M(1/(4M^2) + 1)^{-1/2} |\lambda - \omega|^{-1}$. The statement follows. ∎

Proposition 2.2.1 and the Reiteration Theorem imply that for all $0 < \theta < 1$ and $1 \leq p \leq \infty$ such that $k\theta/n$ is not integer we have

$$\begin{aligned}(X, D(A^k))_{\theta, p} &= (X, D(A^n))_{k\theta/n, p}, \\ (X, D(A^k))_{\theta} &= (X, D(A^n))_{k\theta/n}.\end{aligned} \qquad (2.1.13)$$

2.2 Intermediate spaces

Through the whole section we set

$$C_n = \sup_{0<t\leq 1} \|t^n A^n e^{tA}\|_{L(X)}, \quad \forall n \in \mathbb{N}. \tag{2.2.1}$$

Then $C_n < +\infty$ for every n, thanks to estimates (2.1.1).

Proposition 2.2.1 *For $0 < k < n$, $D(A^k)$ belongs to the class $J_{k/n} \cap K_{k/n}$ between X and $D(A^n)$.*

Proof — Since $D(A^k) = D((A - (\omega+1)I)^k)$ for every k, we may assume without loss of generality that $\omega < 0$. Then the graph norm on $D(A^k)$ is equivalent to the norm $x \mapsto \|A^k x\|$, which will be used here.

Let us prove that $D(A^k) \in J_{k/n}(X, D(A^n))$. First we consider the case $k = 1$, $n = 2$. We claim that there is $C > 0$ such that

$$\|Ax\| \leq C \|x\|^{1/2} \|A^2 x\|^{1/2}, \quad x \in D(A^2). \tag{2.2.2}$$

Let $x \in D(A^2)$. Since $\omega < 0$, then $0 \in \rho(A^2)$, so that if $A^2 x = 0$ then $x = 0$, and in this case (2.2.2) holds. Moreover, for every $t > 0$ we have

$$Ae^{tA}x - Ax = A^2 \int_0^t e^{sA}x\, ds = \int_0^t e^{sA} A^2 x\, ds,$$

so that

$$\|Ax\| \leq \|Ae^{tA}x\| + \int_0^t \|e^{sA} A^2 x\|\, ds \leq \frac{M_1}{t}\|x\| + M_0 t \|A^2 x\|, \quad t > 0.$$

If $A^2 x \neq 0$, taking the minimum of the right hand side for $t > 0$ we get (2.2.2) with constant $C = 2\sqrt{M_0/M_1}$. (2.2.2) implies that $D(A) \in J_{1/2}(X, D(A^2))$. Arguing by recurrence as in the proof of Proposition 1.1.2(i) one can see that $D(A^k) \in J_{k/n}(X, D(A^n))$ for $0 < k < n$.

To prove that $D(A^k) \in K_{k/n}(X, D(A^n))$ we show preliminarly that $D(A) \in K_{1/n}(X, D(A^n))$. If $x \in D(A^n)$ split $x = e^{t^{1/n}A}x + (x - e^{t^{1/n}A}x)$, where

$$\|e^{t^{1/n}A}x\|_{D(A^n)} \leq M_0\|x\| + M_{n-1} t^{-1+1/n} \|Ax\|,$$

and

$$\|x - e^{t^{1/n}A}x\| = \left\|\int_0^{t^{1/n}} Ae^{sA}x\, ds\right\| \leq M_0 t^{1/n} \|Ax\|.$$

It follows that

$$K(t, x; X, D(A^n)) = \inf_{x=a+b} \|a\| + t\|A^n b\| \leq M_0 t^{1/n} \|Ax\| + M_{n-1} t^{1/n} \|Ax\|, \quad t \geq 0.$$

2. Intermediate spaces

Therefore, $t^{-1/n}K(t,x;X,D(A^n))$ is bounded in $]0,+\infty[$, and $x \in (X,D(A^n))_{1/n,\infty}$, with $\|x\|_{(X,D(A^n))_{1/n,\infty}} \leq (M_0+M_{n-1})\|Ax\|$. This means that $D(A)$ is in the class $K_{1/n}$ between X and $D(A^n)$.

Let us argue by recurrence. Assume that for some $n \geq 3$ we have $D(A^h) \subset (X,D(A^n))_{h/(n-1),\infty}$ for $h = 1, \ldots, n-2$. We have just proved that this is true for $n = 3$. Set $Y = D(A)$. The part A_Y of A in Y is sectorial, so that $D(A_Y^h) \subset (Y,D(A_Y^{n-1}))_{h/(n-1),\infty}$ for $h = 1, \ldots, n-2$. On the other hand, $D(A_Y^h) = D(A^{h+1})$, $D(A_Y^{n-1}) = D(A^n)$, so that $D(A_Y^{h+1}) \subset (D(A),D(A^n))_{h/(n-1),\infty}$. Since $D(A)$ belongs to $K_{1/n}(X,D(A^n))$, by part (i) of the Reiteration Theorem 1.2.15 we get $(D(A),D(A^n))_{h/(n-1),\infty} \subset (X,D(A^n))_{(h+1)/n,\infty}$ for $1 \leq h \leq n-2$. Setting $h+1 = k$ we get

$$D(A_Y^k) \subset (X,D(A^n))_{k/n,\infty}, \quad 2 \leq k \leq n-1,$$

and the statement follows. ∎

2.2.1 The spaces $D_A(\alpha,p)$ and $D_A(\alpha)$

There are several equivalent definitions of the spaces $D_A(\alpha,p)$. The most useful in the study of regularity of the solutions to abstract equations comes out from the behavior of $\frac{d}{dt}e^{tA}x = Ae^{tA}x$ near $t = 0$. We have seen in Section 2.1 that, for each $x \in X$, $\|tAe^{tA}x\|$ is bounded in $]0,1[$ (and it goes to 0 as $t \to 0$ if $x \in \overline{D(A)}$), whereas for every $x \in D(A)$, $\|Ae^{tA}x\|$ is bounded in $]0,1[$. This leads naturally to the definition of a class of intermediate spaces between X and $D(A)$ ($0 < \alpha < 1$, $1 \leq p \leq \infty$, and $(\alpha,p) = (1,\infty)$), by

$$\begin{cases} D_A(\alpha,p) = \{x \in X \,:\, t \mapsto v(t) = \|t^{1-\alpha-1/p}Ae^{tA}x\| \in L^p(0,1)\}, \\ \|x\|_{D_A(\alpha,p)} = \|x\| + [x]_{D_A(\alpha,p)} = \|x\| + \|v\|_{L^p(0,1)}; \end{cases} \quad (2.2.3)$$

$$D_A(\alpha) = \{x \in D_A(\alpha,\infty) \,:\, \lim_{t \to 0} t^{1-\alpha}Ae^{tA}x = 0\}. \quad (2.2.4)$$

(We set as usual $1/\infty = 0$.) As easily seen, for every $x \in D_A(\alpha,p)$ and $T > 0$ the function $s \mapsto \|Ae^{sA}x\|$ belongs to $L^1(0,T)$, so that by Proposition 2.1.4(ii) we have

$$e^{tA}x - x = \int_0^t Ae^{sA}x\,ds \;\forall t \geq 0, \quad x = \lim_{t \to 0} e^{tA}x.$$

In particular, all the spaces $D_A(\alpha,p)$ and $D_A(\alpha)$ are contained in the closure of $D(A)$. Moreover we have

$$D_A(\alpha,p) = D_{A_0}(\alpha,p), \quad D_A(\alpha) = D_{A_0}(\alpha),$$

where A_0 is the part of A in $\overline{D(A)}$. See Remark 2.1.5.

Some characterizations

We state below several characterizations of the spaces $D_A(\alpha, p)$ and $D_A(\alpha)$. All of them will be used in the sequel.

First, the spaces $D_A(\alpha, p)$ and $D_A(\alpha)$ are real interpolation spaces between X and the domain of A.

Proposition 2.2.2 *For $0 < \alpha < 1$ and $1 \leq p \leq \infty$, and for $(\alpha, p) = (1, \infty)$ we have*

$$D_A(\alpha, p) = (X, D(A))_{\alpha, p},$$

with equivalence of the respective norms. Moreover, for $0 < \alpha < 1$,

$$D_A(\alpha) = (X, D(A))_\alpha.$$

Proof — Let $\varphi : [0, +\infty[\mapsto \mathbb{R}$ be a C^∞ function such that

$$\begin{cases} 0 \leq \varphi(t) \leq 1, \ |\varphi'(t)| \leq 2 \ \forall \, t > 0, \\ \varphi(t) = 1 \text{ for } 0 \leq t \leq 1/3, \ \varphi(t) = 0 \text{ for } t \geq 1. \end{cases} \quad (2.2.5)$$

Let $x \in D_A(\alpha, p)$. Then, choosing $u(t) = \varphi(t)e^{tA}x$, we get $x = u(0)$, $u(t) = 0$ for $t \geq 1$, and for $0 < t \leq 1$

$$\begin{cases} \|t^{1-\alpha}u(t)\|_{D(A)} \leq \|t^{1-\alpha}Ae^{tA}x\| + \|t^{1-\alpha}e^{tA}x\|, \\ \|t^{1-\alpha}u'(t)\|_X \leq \|t^{1-\alpha}Ae^{tA}x\| + \|\varphi'\|_\infty \|t^{1-\alpha}e^{tA}x\|, \end{cases}$$

so that $u \in V(p, 1-\alpha, D(A), X)$, and

$$\|u\|_{V(p,1-\alpha,D(A),X)} \leq 2[x]_{D_A(\alpha,p)} + 3\|t^{1-\alpha-1/p}e^{tA}x\|_{L^p(0,1;X)}.$$

Due to Proposition 1.2.10, $x \in (X, D(A))_{\alpha, p}$, and

$$\|x\|_{\alpha,p}^T \leq 2[x]_{D_A(\alpha,p)} + 3C_0 c_p \|x\|, \quad (2.2.6)$$

where c_p is a suitable constant. Therefore, $D_A(\alpha, p) \subset (X, D(A))_{\alpha, p}$. From the above considerations it is clear that if $x \in D_A(\alpha)$, then u belongs to $V_0(\infty, 1-\alpha, D(A), X)$, so that $x \in (X, D(A))_\alpha$.

Conversely, let $x \in (X, D(A))_{\alpha, p}$. Then $x = u(0)$, with $u \in V(p, 1-\alpha, D(A), X)$. It follows that

$$\begin{aligned} \|t^{1-\alpha}Ae^{tA}x\| &\leq \|t^{1-\alpha}Ae^{tA}u(t)\| + \left\|t^{1-\alpha}Ae^{tA}\int_0^t u'(s)ds\right\| \\ &\leq C_0\|t^{1-\alpha}Au(t)\| + C_1 \left\|t^{1-\alpha}\frac{1}{t}\int_0^t u'(s)ds\right\|. \end{aligned} \quad (2.2.7)$$

Due to Corollary 1.2.9, $t \mapsto \|t^{1-\alpha-1/p}Ae^{tA}x\|$ belongs to $L^p(0, 1)$, and

$$\begin{aligned} \|t^{1-\alpha-1/p}Ae^{tA}x\|_{L^p(0,1)} &\leq C_0\|t^{1-\alpha-1/p}Au(t)\|_{L^p(0,1)} \\ +\alpha^{-1}C_1\|t^{1-\alpha-1/p}u'(t)\|_{L^p(0,1)} &\leq \max(C_0, \ \alpha^{-1}C_1)\|x\|_{\alpha,p}^T. \end{aligned} \quad (2.2.8)$$

2. INTERMEDIATE SPACES   47

Estimate (2.2.8) also holds for $p = \infty$, if we set $1/\infty = 0$. Therefore, $D_A(\alpha, p)$ is continuously embedded in $(X, D(A))_{\alpha, p}$ for $1 \leq p \leq \infty$.

If $x \in (X, D(A))_\alpha$, then $\|t^{1-\alpha} u(t)\|_{D(A)}$ and $\|t^{1-\alpha} u'(t)\|$ go to 0 as $t \to 0$, and from estimate (2.2.8) it follows that $\lim_{t \to 0} t^{1-\alpha} A e^{tA} x = 0$, so that $x \in D_A(\alpha)$, and the statement is completely proved. ∎

The above characterization yields immediately several properties of the spaces $D_A(\alpha, p)$ and $D_A(\alpha)$.

Corollary 2.2.3 *The following statements hold true.*

(i) *The spaces $D_A(\alpha, p)$ and $D_A(\alpha)$ do not depend explicitly on the operator A, but only on $D(A)$ and on the graph norm of A. Precisely, if $B : D(B) = D(A) \mapsto X$ is a sectorial operator such that*

$$c^{-1} \|Ax\| \leq \|Bx\| \leq c \|Ax\| \quad \forall x \in D(A)$$

for some $c \geq 1$, then we have (with equivalence of the respective norms)

$$D_B(\alpha, p) = D_A(\alpha, p), \quad D_B(\alpha) = D_A(\alpha).$$

(ii) *For $0 < \alpha_1 < \alpha_2 < 1$, $1 \leq p \leq \infty$, and for $(\alpha_2, p) = (1, \infty)$, we have*

$$D_A(\alpha_2, p) \subset D_A(\alpha_1, p).$$

For $0 < \alpha < 1$, $1 \leq p_1 < p_2 < \infty$,

$$D_A(1, \infty) \subset D_A(\alpha, p_1) \subset D_A(\alpha, p_2) \subset D_A(\alpha) \subset D_A(\alpha, \infty) \subset \overline{D(A)}.$$

(iii) *The spaces $D_A(\alpha, p)$ and $D_A(\alpha)$ belong to the class J_α between X and $D(A)$. Moreover, if E is a Banach space such that $D(A) \subset E \subset X$, then E belongs to the class J_α between X and $D(A)$ if and only if $D_A(\alpha, 1) \subset E$.*

(iv) *$D(A)$ is dense in $D_A(\alpha, p)$ for $p < \infty$. $D_A(\alpha)$ is the closure of $D(A)$ in $D_A(\alpha, \infty)$.*

Proof — Statement (i) is an obvious consequence of Proposition 2.2.2. Statements (ii), (iii), and (iv) follow from Proposition 2.2.2 through Proposition 1.2.3, Corollary 1.2.7 and Proposition 1.2.12, respectively. ∎

The next proposition gives a characterization of the spaces $D_A(\alpha, p)$ and $D_A(\alpha)$ in terms of the behavior of the function $u(t) = e^{tA} x$ (instead of its derivative $A e^{tA} x$) near $t = 0$.

Proposition 2.2.4 *It holds*

$$D_A(\alpha, p) = \{x \in X : t \mapsto w(t) = t^{-\alpha - 1/p} \|e^{tA} x - x\| \in L^p(0, 1)\},$$

and, setting $[[x]]_{D_A(\alpha,p)} = \|w\|_{L^p(0,1)}$, the norm

$$x \mapsto \|x\| + [[x]]_{D_A(\alpha,p)}$$

is equivalent to the norm of $D_A(\alpha, p)$. Moreover,

$$D_A(\alpha) = \{x \in X : \lim_{t \to 0} t^{-\alpha}(e^{tA}x - x) = 0\}.$$

Proof — Let $x \in D_A(\alpha, p)$. Then

$$t^{-\alpha}(e^{tA}x - x) = t^{1-\alpha} \frac{1}{t} \int_0^t Ae^{sA}x \, ds, \tag{2.2.9}$$

so that, by Corollary 1.2.9,

$$[[x]]_{D_A(\alpha,p)} \leq \alpha^{-1}[x]_{D_A(\alpha,p)}, \tag{2.2.10}$$

Conversely, let $[[x]]_{D_A(\alpha,p)} < \infty$. From the equality

$$Ae^{tA}x = Ae^{tA} \frac{1}{t} \int_0^t (x - e^{sA}x) \, ds + e^{tA} \frac{1}{t} A \int_0^t e^{sA}x \, ds$$

we get

$$\|t^{1-\alpha} Ae^{tA}x\| \leq C_1 t^{1-\alpha} \frac{1}{t} \int_0^t \frac{\|x - e^{sA}x\|}{s} ds + C_0 t^{-\alpha} \|e^{tA}x - x\|. \tag{2.2.11}$$

Since the function $s \mapsto s^{1-\alpha-1/p} \frac{\|x - e^{sA}x\|}{s}$ belongs to $L^p(0,1)$, using Corollary 1.2.9 we deduce that $t \mapsto t^{1-\alpha-1/p} Ae^{tA}x$ belongs to $L^p(0,1)$, and

$$\|t^{1-\alpha-1/p} Ae^{tA}x\|_{L^p(0,1)} = [x]_{D_A(\alpha,p)} \leq (C_1 \alpha^{-1} + C_0)[[x]]_{D_A(\alpha,p)} \tag{2.2.12}$$

Hence the seminorms $[\,\cdot\,]_{D_A(\alpha,p)}$ and $[[\,\cdot\,]]_{D_A(\alpha,p)}$ are equivalent. Finally, the characterization of $D_A(\alpha)$ follows by letting $t \to 0$ in (2.2.9) and in (2.2.11).
∎

Remark 2.2.5 We remark that, in the case $p = +\infty$, Proposition 2.2.4 states that $x \in D_A(\alpha, \infty)$ if and only if the function $t \mapsto e^{tA}x$ belongs to $C^\alpha([0,1]; X)$, if $0 < \alpha < 1$, or to $Lip([0,1]; X)$, if $\alpha = 1$. This is clearly equivalent to $t \mapsto e^{tA}x \in C^\alpha([0,T]; X)$ (respectively, $Lip([0,T]; X)$) for all $T > 0$.

As we have seen in Section 2.1, the properties of $t \mapsto e^{tA}x$ are strongly connected with the behavior of $R(\lambda, A)x$ for large $|\lambda|$. The following proposition gives a characterization of $D_A(\alpha, p)$ and $D_A(\alpha)$ in terms of the behavior of $AR(\lambda, A)x$ as $\lambda \to +\infty$.

2. Intermediate spaces

Proposition 2.2.6 *Let $a > \max(1, \omega)$. Then*
$$D_A(\alpha, p) = \{x \in X : t \mapsto z(t) = \|t^{\alpha-1/p} AR(t, A)x\| \in L^p(a, +\infty)\},$$
and, setting $[x]^*_{D_A(\alpha,p)} = \|z\|_{L^p(a,+\infty)}$, the norm
$$x \mapsto \|x\| + [x]^*_{D_A(\alpha,p)}$$
is equivalent to the norm of $D_A(\alpha, p)$. Moreover,
$$D_A(\alpha) = \{x \in X : \lim_{t \to +\infty} t^\alpha AR(t, A)x = 0\}.$$

Proof — From the identity $x = e^{tA}x + (x - e^{tA}x)$ we get, for $0 < t < 1/a$,
$$\|t^{-\alpha} AR(t^{-1}, A)x\| \leq \|t^{-\alpha} R(t^{-1}, A) A e^{tA} x\| + \|t^{-\alpha} AR(t^{-1}, A)(x - e^{tA}x)\|$$
$$\leq \frac{M}{1/t - \omega} \|t^{1-\alpha} A e^{tA} x\| + \left(\frac{M}{1 - \omega t} + 1\right) \|t^{-\alpha}(x - e^{tA}x)\|$$
$$\leq M'(\|t^{1-\alpha} A e^{tA} x\| + \|t^{-\alpha}(x - e^{tA}x)\|).$$
(2.2.13)

Let $x \in D_A(\alpha, p)$. Then
$$[x]^*_{D_A(\alpha,p)} = \|z\|_{L^p(a,+\infty)} = \|t^{-\alpha-1/p} AR(t^{-1}, A)x\|_{L^p(0,1/a)}$$
$$\leq M'(\|t^{1-\alpha-1/p} A e^{tA} x\|_{L^p(0,1)} + \|t^{-\alpha-1/p}(x - e^{tA}x)\|_{L^p(0,1)}) \quad (2.2.14)$$
$$= M'([x]_{D_A(\alpha,p)}[[x]]_{D_A(\alpha,p)}).$$

Conversely, let x be such that $t \mapsto \|t^{\alpha-1/p} AR(t, A)x\| \in L^p(a, +\infty)$. Using estimates (2.2.1) and the equality $x = tR(t,A)x - AR(t,A)x$ we find, for every $t \geq a$,
$$\|t^{\alpha-1} A e^{t^{-1}A} x\| \leq \|t^{\alpha-1} e^{t^{-1}A} tAR(t,A)x\| + \|t^{\alpha-1} A e^{t^{-1}A} AR(t,A)x\|$$
$$\leq C_0 \|t^\alpha AR(t,A)x\| + C_1 \|t^\alpha AR(t,A)x\|,$$
(2.2.15)

whereas for $1 \leq t \leq a$
$$\|t^{\alpha-1} A e^{t^{-1}A} x\| \leq C_1 t^\alpha \|x\|.$$
Then $t \mapsto \|t^{1-\alpha-1/p} A e^{tA} x\|$ belongs to $L^p(0,1)$, since
$$\|t^{1-\alpha-1/p} A e^{tA} x\|_{L^p(0,1)} = \|t^{\alpha-1-1/p} A e^{t^{-1}A} x\|_{L^p(1,+\infty)}$$
$$\leq \|t^{\alpha-1-1/p} A e^{t^{-1}A} x\|_{L^p(1,a)} + \|t^{\alpha-1-1/p} A e^{t^{-1}A} x\|_{L^p(a,+\infty)}.$$

If $p < \infty$ we find
$$[x]_{D_A(\alpha,p)} \leq C_1 \left(\frac{a^{\alpha p} - 1}{\alpha p}\right)^{1/p} \|x\| + (C_0 + C_1)[x]^*_{D_A(\alpha,p)}, \quad (2.2.16)$$
and, if $p = \infty$,
$$[x]_{D_A(\alpha,\infty)} \leq C_1 a^\alpha \|x\| + (C_0 + C_1)[x]^*_{D_A(\alpha,p)}. \quad (2.2.17)$$

The characterization of $D_A(\alpha)$ is a consequence of estimates (2.2.13) and (2.2.15). ∎

Behavior of e^{tA} in real interpolation spaces

For $k \in \mathbb{N}$, $\alpha \in \,]0,1[$, $p \in [1,\infty]$, set

$$D_A(\alpha+k,p) = \{x \in D(A^k): A^k x \in D_A(\alpha,p)\}, \quad \|x\|_{\alpha+k,p} = \|x\| + \|A^k x\|_{\alpha,p},$$

$$D_A(\alpha+k) = \{x \in D_A(\alpha+k,p): A^k x \in D_A(\alpha)\}.$$

As a consequence of Proposition 2.2.4, the function $t \mapsto e^{tA}x$ belongs to $C^{k+\alpha}([0,1]; X)$ if and only if $x \in D_A(\alpha+k,\infty)$. Moreover, since $D(A)$ is dense in $D_A(\alpha,p)$ for $p < \infty$, then $D(A^k)$ is dense in $D_A(\alpha+k,p)$ for $p < \infty$; since $D_A(\alpha)$ is the closure of $D(A)$ in $D_A(\alpha,\infty)$, then $D_A(\alpha+k)$ is the closure of $D(A^{k+1})$ in $D_A(\alpha+k,\infty)$.

A very important fact is that the parts of A in $D_A(\alpha,p)$ and in $D_A(\alpha)$, defined by

$$A_{\alpha,p}: D_A(\alpha+1,p) \mapsto D_A(\alpha,p), \quad A_{\alpha,p}x = Ax,$$

$$A_\alpha: D_A(\alpha+1) \mapsto D_A(\alpha), \quad A_\alpha x = Ax,$$

are sectorial operators in $D_A(\alpha,p)$ and in $D_A(\alpha)$, respectively.

Proposition 2.2.7 *$\rho(A)$ is contained in $\rho(A_{\alpha,p})$, and $\|R(\lambda, A_{\alpha,p})\|_{L(D_A(\alpha,p))} \le \|R(\lambda,A)\|_{L(X)}$ for every $\lambda \in \rho(A)$. Consequently, $A_{\alpha,p}$ is a sectorial operator in $D_A(\alpha,p)$ and A_α is a sectorial operator in $D_A(\alpha)$.*

Proof — Let $\lambda \in \rho(A)$. Obviously, $R(\lambda,A)_{|D_A(\alpha,p)}: D_A(\alpha,p) \mapsto D_A(\alpha,p)$ is one to one. Moreover, for every $x \in D_A(\alpha,p)$ and $0 < t \le 1$ it holds

$$\|t^{1-\alpha} A e^{tA} R(\lambda,A)x\| = \|R(\lambda,A) t^{1-\alpha} A e^{tA} x\| \le \|R(\lambda,A)\|_{L(X)} \|t^{1-\alpha} A e^{tA} x\|.$$

Therefore,

$$[R(\lambda,A)x]_{D_A(\alpha,p)} \le \|R(\lambda,A)\|_{L(X)} [x]_{D_A(\alpha,p)},$$

and the statement follows. ∎

We study now the behavior near $t = 0$ of the function $t \mapsto e^{tA}x$, when $x \in D_A(\alpha,p)$ or $x \in D_A(\alpha)$.

Proposition 2.2.8 *For $0 < \alpha < 1$ and $1 \le p < \infty$,*

$$\lim_{t \to 0} \|e^{tA}x - x\|_{D_A(\alpha,p)} = 0, \quad \forall\, x \in D_A(\alpha,p). \tag{2.2.18}$$

For $0 < \alpha < 1$ and $x \in D_A(\alpha,\infty)$,

$$\lim_{t \to 0} \|e^{tA}x - x\|_{D_A(\alpha,\infty)} = 0 \iff x \in D_A(\alpha). \tag{2.2.19}$$

2. INTERMEDIATE SPACES

Proof — Since $D_A(\alpha,p) \subset \overline{D(A)}$, then by Proposition 2.1.4(i) for every $x \in D_A(\alpha,p)$ we have $\lim_{t\to 0} e^{tA}x - x = 0$ in X. If $p < \infty$, from the inequality

$$\|t^{1-\alpha}Ae^{tA}(e^{sA}x - x)\| = \|t^{1-\alpha}(e^{sA} - I)Ae^{tA}x\| \leq (C_0 + 1)\|t^{1-\alpha}Ae^{tA}x\|,$$

which holds for $t > 0$ and $0 < s < 1$, one deduces that $\lim_{s\to 0}[e^{sA}x - x]_{D_A(\alpha,p)} = 0$, thanks to Lebesgue Dominated Convergence Theorem.

For every $x \in D_A(\alpha)$ and $0 < s < 1$ we have

$$[e^{sA}x - x]_{\alpha,\infty} \leq \left(\sup_{0<t<s} + \sup_{s\leq t\leq 1}\right) \|t^{1-\alpha}Ae^{tA}(e^{sA}x - x)\|$$
$$\leq \|e^{sA} - 1\|_{L(X)} \sup_{0<t<s} \|t^{1-\alpha}Ae^{tA}x\|$$
$$+ \sup_{s\leq t\leq 1} \|tAe^{tA}\|_{L(X)} s^{-\alpha} \left\|\int_0^s \sigma^{1-\alpha}Ae^{\sigma A}x \frac{d\sigma}{\sigma^{1-\alpha}}\right\|$$
$$\leq (C_0 + 1)\sup_{0<t<s} \|t^{1-\alpha}Ae^{tA}x\| + C_1\alpha^{-1} \sup_{0<\sigma<s} \|\sigma^{1-\alpha}Ae^{\sigma A}x\|.$$

Then $\lim_{s\to 0} e^{sA}x = x$ in $D_A(\alpha, \infty)$.

Conversely, if $\lim_{t\to 0} \|e^{tA}x - x\|_{D_A(\alpha,\infty)} = 0$, then x belongs to the closure of $D(A)$ in $D_A(\alpha, \infty)$, and hence to $D_A(\alpha)$ thanks to Corollary 2.2.3(iv). ∎

Now we give some estimates for the function $t \mapsto A^n e^{tA}$ when $t \to 0$ and $t \to \infty$. For convenience, in the next proposition we set

$$D_A(0,p) = X, \quad \forall p \in [1, \infty].$$

Other estimates (for large t) will be given in Section 3.

Proposition 2.2.9 *Let $(\alpha,p), (\beta,p) \in\ ]0,1[\ \times [1,+\infty] \cup \{(1,\infty)\}$, and let $n \in \mathbb{N}$. Then there are $C = C(n,p;\alpha,\beta)$, $C' = C'(n,p;\alpha,\beta)$ such that*

$$\begin{cases} (i) & \|t^{n-\alpha+\beta}A^n e^{tA}\|_{L(D_A(\alpha,p),D_A(\beta,p))} \leq C, \quad 0 < t \leq 1; \\ (ii) & \|t^n(A - \omega I)^n e^{tA}\|_{L(D_A(\alpha,p),D_A(\beta,p))} \leq C' e^{\omega t}, \quad t \geq 1. \end{cases} \quad (2.2.20)$$

The statement holds also for $n = 0$, provided $\alpha \leq \beta$.

Proof — First we prove that the estimates (2.2.20)(i) hold.

Let $\alpha = 0$. We have to estimate $\|t^n A^n e^{tA}x\|_{D_A(\beta,p)}$ for $x \in X$ and $n \in \mathbb{N} \cup \{0\}$. By Corollary 2.2.3(iii), $D_A(\beta,p)$ belongs to the class J_β between X and $D(A)$. Let c be such that

$$\|y\|_{D_A(\beta,p)} \leq c\|y\|_{D(A)}^\beta \|y\|_X^{1-\beta} \quad \forall y \in D(A).$$

For $0 < t \leq 1$ we get, using (2.2.1),

$$\|t^n A^n e^{tA}x\|_{D_A(\beta,p)} \leq c\|t^n A^n e^{tA}x\|_{D(A)}^\beta \|t^n A^n e^{tA}x\|^{1-\beta} \leq c't^{-\beta}\|x\|,$$

and (2.2.20)(i) is proved for $\alpha = 0$, $n \in \mathbb{N} \cup \{0\}$.

Let $x \in D_A(\alpha, p)$, with $0 < \alpha < 1$, or $x \in D_A(1, \infty)$ and let $n \in \mathbb{N}$. Then

$$\|t^n A^n e^{tA} x\|_{D_A(\beta,p)} \leq 2^n \|(t/2)^{n-1+\alpha} A^{n-1} e^{\frac{t}{2}A}\|_{L(X, D_A(\beta,p))} \|(t/2)^{1-\alpha} A e^{\frac{t}{2}A} x\|$$
$$\leq 2^{n+\beta-\alpha} t^{\alpha-\beta} C(n-1, p, 0, \beta) \|x\|_{D_A(\alpha,\infty)},$$

and (2.2.20)(i) holds for $0 < \alpha < 1$, $n \in \mathbb{N}$.

Let now $n = 0$, and let $x \in D_A(\alpha, p)$, $\beta \leq \alpha$. Then $\xi^{1-\beta} \leq \xi^{1-\alpha}$ for $0 < \xi \leq 1$, so that

$$\|e^{tA}x\|_{D_A(\beta,p)} = \|\xi^{1-\beta-1/p} A e^{\xi A} e^{tA} x\|_{L^p(0,1;X)} + \|e^{tA}x\|$$
$$\leq C_0 \|\xi^{1-\alpha-1/p} A e^{\xi A} e^{tA} x\|_{L^p(0,1;X)} + C_0 \|x\| = C_0 \|x\|_{D_A(\alpha,p)}.$$

On the other hand, if $\beta > \alpha$ then

$$\|e^{tA}x\|_{D_A(\beta,p)} \leq \|e^A x\|_{D_A(\beta,p)} + \left\| \int_t^1 A e^{sA} x \, ds \right\|_{D_A(\beta,p)}$$
$$\leq C(0, p, 0, \beta) \|x\| + C(1, p, \alpha, \beta) \int_t^1 s^{\alpha-\beta-1} ds \, \|x\|_{D_A(\alpha,\infty)}$$
$$\leq C(0, p, 0, \beta) \|x\| + C(1, p, \alpha, \beta)(\beta-\alpha)^{-1} t^{-\beta+\alpha} \|x\|_{D_A(\alpha,\infty)},$$

so that (2.2.20)(i) holds for $0 < \alpha < \beta \leq 1$, $n = 0$.

Estimates (2.2.20)(ii) follow easily from (2.2.20)(i) and (2.1.1): it is sufficient to recall that for $t \geq 1$

$$\|t^n(A - \omega I)^n e^{tA}\|_{L(D_A(\alpha,p), D_A(\beta,p))} \leq \|t^n(A - \omega I)^n e^{tA}\|_{L(X, D_A(\beta,p))}$$
$$\leq \|e^{\frac{1}{2}A}\|_{L(X, D_A(\beta,p))} M_n \left(\frac{t}{t-1/2}\right)^n e^{\omega(t-1/2)} \leq \|e^{\frac{1}{2}A}\|_{L(X, D_A(\beta,p))} 2^n e^{-\omega/2} M_n e^{\omega t}.$$

∎

Remark 2.2.10 In the proof of Proposition 2.2.9 we have shown in fact that

$$\sup_{0 < t \leq 1} \|t^{n-\alpha+\beta} A^n e^{tA}\|_{L(D_A(\alpha,\infty), D_A(\beta,p))} < \infty, \quad \forall n \in \mathbb{N},$$

for $0 < \alpha, \beta < 1$, $1 \leq p \leq \infty$. In particular, for $\alpha = \beta$

$$\sup_{0 < t \leq 1} \|t^n A^n e^{tA}\|_{L(D_A(\alpha,\infty), D_A(\alpha,p))} < \infty, \quad \forall n \in \mathbb{N}, \ p \in [1, \infty].$$

This is not true in general for $n = 0$, as the next counterexample shows.

Example 2.2.11 Let $X = L^\infty(0, +\infty)$, and set

$$A: \ D(A) = \{f \in X : \sup_{x > 0} x^2 |f(x)| < \infty\} \mapsto X, \quad (Af)(x) = -x^2 f(x).$$

2. INTERMEDIATE SPACES

Then $(e^{tA}f)(x) = e^{-x^2 t}f(x)$ for every $f \in X$, so that $D_A(\alpha, \infty)$ is the set of all functions f such that $\sup_{0 < \xi < 1, x > 0} \xi^{1-\alpha} x^2 e^{-x^2 \xi} f(x) < \infty$. In particular, the function f defined by $f(x) = 0$ for $0 \leq x \leq 1$, $f(x) = x^{-2\alpha}$ for $x \geq 1$ belongs to $D_A(\alpha, \infty)$. However, $\lim_{t \to 0}[e^{tA}f]_{D_A(\alpha,p)} = +\infty$ if $p < \infty$, so that $t \mapsto \|e^{tA}\|_{L(D_A(\alpha,\infty), D_A(\alpha,p))}$ is not bounded.

From Proposition 2.2.9 a number of results on the behavior of $e^{tA}x$ near $t = 0$ when x belongs to some interpolation space may be deduced. We do not list them here, but we shall show them when needed.

We end this section by some sharp interpolatory inclusions. Since the spaces $D_A(\alpha,p)$ belong to $J_\alpha(X, D(A))$, it follows from Proposition 1.1.5 that $C^\alpha(I; D(A)) \cap C^{1+\alpha}(I; X)$ is continuously embedded in $C^1(I; D_A(\alpha, p))$, for every interval $I \subset \mathbb{R}$ and for every $p \geq 1$. Example 2.2.11 proves that $C^1(I; D_A(\alpha, p))$ cannot in general be replaced by $C^1(I; D_A(\alpha, p))$ or $Lip(I; D_A(\alpha, p))$: indeed, if $x \in D_A(\alpha + 1, \infty)$ then the function $t \mapsto e^{tA}x$ belongs to $C^\alpha(I; D(A)) \cap C^{1+\alpha}(I; X)$ but in general it is not Lipschitz continuous wih values in $D_A(\alpha,p)$ for $p < \infty$. However, for $p = \infty$ the result of Proposition 1.1.5 may be improved.

Proposition 2.2.12 *Let $0 < \alpha < 1$, and let I be a (possibly unbounded) interval. The following statements hold.*

(i) *If $u \in C^\alpha(I; D(A)) \cap C^{1+\alpha}(I; X)$, then $u'(t) \in D_A(\alpha, \infty)$ for every $t \in I$, and*

$$\|u'(t)\|_{D_A(\alpha,\infty)} \leq C(\|u\|_{C^\alpha(I;D(A))} + \|u'\|_{C^{1+\alpha}(I;X)}), \quad \forall t \in I. \quad (2.2.21)$$

In addition, $C^\alpha(I; D(A)) \cap C^{1+\alpha}(I; X) \subset Lip(I; D_A(\alpha, \infty))$.

(ii) $h^\alpha(I; D(A)) \cap h^{1+\alpha}(I; X) \subset C^1(I; D_A(\alpha))$.

Proof — (i) First we prove that (2.2.21) holds. Let $u \in C^\alpha(I; D(A)) \cap C^{1+\alpha}(I; X)$. For $t, t + h \in I$, split $u'(t)$ as

$$u'(t) = \int_0^1 (u'(t) - u'(t + \sigma h))d\sigma + \frac{u(t+h) - u(t)}{h}.$$

Then for every $\xi \in \,]0, 1]$ it holds

$$\|\xi^{1-\alpha} A e^{\xi A} u'(t)\| \leq \xi^{1-\alpha} \|A e^{\xi A}\|_{L(X)} \int_0^1 \|u'(t) - u'(t + \sigma h)\| d\sigma$$
$$+ \xi^{1-\alpha} \|e^{\xi A}\|_{L(X)} \left\| \frac{Au(t+h) - Au(t)}{h} \right\| \quad (2.2.22)$$
$$\leq \frac{C_1}{\alpha + 1} \xi^{-\alpha} [u']_{C^\alpha(X)} |h|^\alpha + C_0 \xi^{1-\alpha} [u]_{C^\alpha(D(A))} |h|^{\alpha-1}.$$

If I is unbounded, for every $\xi \in \,]0,1]$ and $t \in I$ there is $h \in \mathbb{R}$ such that $|h| = \xi$ and $t + h \in I$. Replacing in (2.2.22) we find

$$\|\xi^{1-\alpha} Ae^{\xi A} u'(t)\| \leq \frac{C_1}{\alpha + 1}[u']_{C^\alpha(X)} + C_0 \xi^{1-\alpha}[u]_{C^\alpha(D(A))}, \quad (2.2.23)$$

and estimate (2.2.21) holds. If $I = [a,b]$, and $\xi \leq (b-a)/2$, for every $t \in [a,b]$ there is h such that $|h| = \xi$ and $t + h \in [a,b]$. For such values of ξ, (2.2.23) holds. On the other hand, if $\xi > (b-a)/2$ then

$$\|\xi^{1-\alpha} Ae^{\xi A} u'(t)\| \leq \xi^{-\alpha}\frac{C_1}{\alpha + 1}\|u'(t)\| \leq \left(\frac{2}{b-a}\right)^\alpha \frac{C_1}{\alpha + 1}\|u'\|_{C(X)},$$

so that $[u']_{D_A(\alpha,\infty)}$ is bounded, and (2.2.21) is proved.

Let us show now that u is Lipschitz continuous with values in $D_A(\alpha, \infty)$. For $0 < \xi \leq 1$ and $s \leq t \in I$ it holds

$$\|\xi^{1-\alpha} Ae^{\xi A}(u(t) - u(s))\| \leq \int_s^t \|\xi^{1-\alpha} Ae^{\xi A} u'(\sigma)\| d\sigma$$
$$\leq (t-s) \sup_{\sigma \in I}[u'(\sigma)]_{D_A(\alpha,\infty)},$$

so that

$$[u(t) - u(s)]_{D_A(\alpha,\infty)} \leq (t-s) \sup_{\sigma \in I}[u'(\sigma)]_{D_A(\alpha,\infty)}.$$

Let us prove that statement (ii) holds. If $u \in h^\alpha(I; D(A)) \cap h^{1+\alpha}(I; X)$, the sequence of approximating functions u_n given by Proposition 0.2.1 belong to $C^\infty(I; D(A))$ and converge to u in $C^\alpha(I; D(A)) \cap C^{1+\alpha}(I; X)$. By statement (i), $\{u'_n\}$ converges to u' in $L^\infty(I; D_A(\alpha, \infty))$. Since $C(I; D_A(\alpha))$ is closed in $L^\infty(I; D_A(\alpha, \infty))$, then $u' \in C(I; D_A(\alpha))$. $\blacksquare$

2.2.2 The domains of fractional powers of $-A$

Throughout the subsection we assume that for $t > 0$

$$\|e^{tA}\|_{L(X)} \leq M_0 e^{-\omega t}, \quad \|tAe^{tA}\|_{L(X)} \leq M_1 e^{-\omega t},$$

with $M_0, M_1, \omega > 0$. For every $\alpha > 0$ we define

$$(-A)^{-\alpha} = \frac{1}{\Gamma(\alpha)} \int_0^{+\infty} t^{\alpha - 1} e^{tA} dt,$$

where the Euler Γ function is defined by

$$\Gamma(\theta) = \int_0^{+\infty} e^{-\theta t} t^{\theta - 1} dt. \quad (2.2.24)$$

2. Intermediate spaces

For $\alpha = 1$, $(-A)^{-1}$ defined above coincides in fact with the inverse of $-A$, thanks to (2.1.10). For $\alpha = n \in \mathbb{N}$, $n > 1$, integrating by parts several times and using Proposition 2.1.4(ii), it follows that $(-A)^{-n}$ defined above coincides with the inverse of $(-A)^n$.

Lemma 2.2.13 *Let α, $\beta > 0$. Then $(-A)^{-\alpha}(-A)^{-\beta} = (-A)^{-(\alpha+\beta)}$.*

Proof — We have

$$\Gamma(\alpha)\Gamma(\beta)(-A)^{-\alpha}(-A)^{-\beta} = \int_0^{+\infty} \int_0^{+\infty} t^{\alpha-1} s^{\beta-1} e^{(t+s)A} ds\, dt$$

$$= \int_0^{+\infty} t^{\alpha-1} \left(\int_t^{+\infty} (\sigma - t)^{\beta-1} e^{\sigma A} d\sigma \right) dt$$

$$= \int_0^1 \xi^{\alpha-1}(1-\xi)^{\beta-1} d\xi \int_0^{+\infty} \sigma^{\alpha+\beta-1} e^{\sigma A} d\sigma$$

$$= \Gamma(\alpha)\Gamma(\beta)(-A)^{-(\alpha+\beta)}. \quad \blacksquare$$

Lemma 2.2.13 has several consequences. The first one is the fact that $(-A)^{-\alpha}$ is one to one. Indeed, if $(-A)^{-\alpha} x = 0$, then for every integer $n > \alpha$ we have $(-A)^{-n} x = (-A)^{-n+\alpha}(-A)^{-\alpha} x = 0$, so that $x = 0$.

Definition 2.2.14 *For $\alpha > 0$, we set $D(-A)^\alpha = \mathrm{Range}\,(-A)^{-\alpha}$, and*

$$(-A)^\alpha = ((-A)^{-\alpha})^{-1}.$$

Moreover, we set $(-A)^0 = I$.

$D(-A)^\alpha$ is endowed with the norm $\|x\|_{D(-A)^\alpha} = \|(-A)^\alpha x\|$. Since $(-A)^\alpha$ is one to one, such norm is equivalent to the graph norm.

The operators $(-A)^\alpha$, with α not integer, are usually called *fractional powers* of $-A$.

The following properties are easy consequences of the definition and of Lemma 2.2.13.

(i) If $\alpha > \beta > 0$, then $D((-A)^\alpha) \subset D((-A)^\beta)$,

(ii) For every $\alpha, \beta \in \mathbb{R}$, we have $(-A)^\alpha (-A)^\beta = (-A)^{\alpha+\beta}$ on $D((-A)^\gamma)$, with $\gamma = \max\{\alpha, \beta, \alpha+\beta\}$.

Note that, since $(-A)^{-\alpha}$ commutes with e^{tA}, for $\alpha > 0$, then $(-A)^\alpha$ commutes with e^{tA} on $D(-A)^\alpha$. It follows that

$$\|e^{tA}\|_{L(D((-A)^\alpha))} \le M_0 e^{-\omega t}, \quad t > 0.$$

For $0 < \alpha < 1$, the domains of the operators $(-A)^\alpha$ are intermediate spaces between X and $D(A)$, as the following proposition states.

Proposition 2.2.15 *For $0 < \alpha < 1$, $D((-A)^\alpha)$ belongs to the class $J_\alpha(X, D(A)) \cap K_\alpha(X, D(A))$. In other words,*

$$D_A(\alpha, 1) \subset D((-A)^\alpha) \subset D_A(\alpha, \infty), \quad 0 < \alpha < 1.$$

Proof — First we show that $D((-A)^\alpha)$ belongs to the class J_α.

For every $x \in D(A)$ it holds $(-A)^\alpha x = (-A)^{-(1-\alpha)}(-Ax)$, so that for every $\lambda > 0$ we have

$$\|(-A)^\alpha x\| = \frac{1}{\Gamma(1-\alpha)} \left\| \left(\int_0^\lambda + \int_\lambda^{+\infty} \right) t^{-\alpha} A e^{tA} x \, dt \right\|$$

$$\leq \frac{1}{\Gamma(1-\alpha)} \left(\frac{M_0}{1-\alpha} \|Ax\| \lambda^{1-\alpha} + \frac{M_1}{\alpha} \|x\| \lambda^{-\alpha} \right).$$

Taking $\lambda = \|x\|/\|Ax\|$ we get

$$\|(-A)^\alpha x\| \leq c \|Ax\|^\alpha \|x\|^{1-\alpha},$$

so that $D((-A)^\alpha) \in J_\alpha(X, D(A))$ thanks to Corollary 2.2.3(iii).

Let us prove that $D((-A)^\alpha)$ belongs to the class K_α, that is, it is continuously embedded in $D_A(\alpha, \infty)$. For $x \in D((-A)^\alpha)$ set $y = (-A)^\alpha x$. Then, for $0 < \xi \leq 1$,

$$\|\xi^{1-\alpha} A e^{\xi A} x\| = \|\xi^{1-\alpha} A e^{\xi A} (-A)^\alpha y\| \leq \frac{\xi^{1-\alpha}}{\Gamma(\alpha)} \left\| \int_0^{+\infty} t^{\alpha-1} A e^{(\xi+t)A} y \, dt \right\|$$

$$\leq \frac{\xi^{1-\alpha} M_1}{\Gamma(\alpha)} \int_0^{+\infty} \frac{t^{\alpha-1}}{\xi+t} dt \|y\| \leq \frac{M_1}{\Gamma(\alpha)} \int_0^{+\infty} \frac{s^{\alpha-1}}{1+s} ds \, \|(-A)^\alpha x\|,$$

so that $x \in D_A(\alpha, \infty)$, and the statement follows. ∎

2.3 Spectral properties and asymptotic behavior

2.3.1 Estimates for large t

One of the most useful properties of analytic semigroups is the so called *spectral determining condition*: roughly speaking, the asymptotic behavior (as $t \to +\infty$) of $A^n e^{tA}$ is determined by the spectral properties of A. Set

$$\omega_A = \sup\{\operatorname{Re} \lambda : \lambda \in \sigma(A)\} \tag{2.3.1}$$

Then $\omega_A \leq \omega$, where ω is the number in Definition 2.0.1.

Proposition 2.3.1 *For every $\varepsilon > 0$ and $n \in \mathbb{N} \cup \{0\}$ there is $M_{n,\varepsilon} > 0$ such that*

$$\|t^n A^n e^{tA}\|_{L(X)} \leq M_{n,\varepsilon} e^{(\omega_A + \varepsilon)t}, \quad t > 0. \tag{2.3.2}$$

3. Spectral properties and asymptotic behavior

Moreover, for every $\varepsilon > 0$, and $\alpha, \beta \in\,]0,1[$, $p \in [1, \infty]$, we have

$$\sup_{t>0} t^{n+\beta} e^{-(\omega_A+\varepsilon)t} \|A^n e^{tA}\|_{L(X, D_A(\beta,p))} < \infty, \quad n \in \mathbb{N} \cup \{0\}, \tag{2.3.3}$$

$$\sup_{t>0} t^{n-\alpha+\beta} e^{-(\omega_A+\varepsilon)t} \|A^n e^{tA}\|_{L(D_A(\alpha,p), D_A(\beta,p))} < \infty, \quad n \in \mathbb{N}, \tag{2.3.4}$$

and (2.3.4) holds also for $n = 0$, provided $\alpha \leq \beta$.

Proof — Let us show that (2.3.2) holds. For $0 < t \leq 1$, estimate (2.3.2) is an obvious consequence of (2.1.1). If $t \geq 1$ and $\omega_A + \varepsilon \geq \omega$, it is again a trivial consequence of estimates (2.1.1). So, we consider the case where $t \geq 1$ and $\omega_A + \varepsilon < \omega$. Since $\rho(A) \supset S_{\theta,\omega} \cup \{\lambda \in \mathbb{C} : \operatorname{Re}\lambda > \omega_A\}$, then, setting $a = (\omega - \omega_A - \varepsilon)|\cos\theta|^{-1}$, $b = (\omega - \omega_A - \varepsilon)|\tan\theta|$, the path

$$\begin{aligned}\Gamma_\varepsilon &= \{\lambda \in \mathbb{C} : \lambda = \xi e^{-i\theta} + \omega,\ \xi \geq a\} \cup \{\lambda \in \mathbb{C} : \lambda = \xi e^{i\theta} + \omega,\ \xi \geq a\} \\ &\cup\ \{\lambda \in \mathbb{C} : \operatorname{Re}\lambda = \omega_A + \varepsilon,\ |\operatorname{Im}\lambda| \leq b\}\end{aligned}$$

is contained in $\rho(A)$, and $\|R(\lambda, A)\|_{L(X)} \leq M_\varepsilon |\lambda - \omega_A|^{-1}$ on Γ_ε, for some $M_\varepsilon > 0$. Since for every t the function $\lambda \mapsto e^{\lambda t} R(\lambda, A)$ is holomorphic in $\rho(A)$, then the path $\omega + \gamma_{r,\eta}$ can be replaced by Γ_ε in formula (2.0.2), getting, for every $t \geq 1$,

$$\begin{aligned}\|e^{tA}\| &= \left\|\frac{1}{2\pi i}\int_{\Gamma_\varepsilon} e^{t\lambda} R(\lambda, A) d\lambda\right\| \leq \frac{M_\varepsilon}{\pi} \int_a^{+\infty} \frac{e^{(\omega+\xi\cos\theta)t}}{|\xi e^{i\theta} + \omega - \omega_A|} d\xi \\ &+ \frac{M_\varepsilon}{2\pi}\int_{-b}^{b} \frac{e^{(\omega_A+\varepsilon)t}}{|iy+\varepsilon|} dy \leq \frac{M_\varepsilon}{\pi}\left(\frac{1}{b|\cos\theta|} + \frac{b}{\varepsilon}\right) e^{(\omega_A+\varepsilon)t}.\end{aligned}$$

Then (2.3.2) follows for $n = 0$. Arguing similarly, we get, for every $t \geq 1$,

$$\begin{aligned}\|Ae^{tA}\| &= \left\|\frac{1}{2\pi i}\int_{\Gamma_\varepsilon} e^{t\lambda}\lambda R(\lambda, A) d\lambda\right\| \\ &\leq \frac{M_\varepsilon}{2\pi} \sup_{\lambda \in \Gamma_\varepsilon} |\lambda(\lambda-\omega_A)|^{-1}\left(2\int_a^{+\infty} e^{(\omega+\xi\cos\theta)t} d\xi + \int_{-b}^{b} e^{(\omega_A+\varepsilon)t} dy\right) \\ &\leq \frac{M_\varepsilon}{\pi}(|\cos\theta|^{-1} + b) e^{(\omega_A+\varepsilon)t} \leq \widetilde{M_\varepsilon} e^{(\omega_A+2\varepsilon)t} t^{-1}.\end{aligned}$$

Since ε is arbitrary, (2.3.2) follows also for $n = 1$.

From the equality $A^n e^{tA} = (Ae^{\frac{t}{n}A})^n$ we get, for $n \geq 2$,

$$\|A^n e^{tA}\|_{L(X)} \leq (M_{1,\varepsilon} n t^{-1} e^{\frac{t}{n}(\omega_A+\varepsilon)})^n \leq (M_{1,\varepsilon} e)^n n!\, t^{-n} e^{(\omega_A+\varepsilon)t},$$

and (2.3.2) is proved. (2.3.3) and (2.3.4) are now easy consequences of (2.3.2) and (2.2.20). ∎

We remark that in the case $\omega_A = \omega = 0$, estimates (2.1.1) and (2.2.20) are better than (2.3.2), (2.3.3), (2.3.4) for t large.

A natural question to ask is whether (2.3.2) holds for $\varepsilon = 0$. The answer is yes in some important cases: the first is when $\omega_A = \omega = 0$ (see estimates (2.1.1)), and the second will be discussed below (see Corollary 2.3.5).

The *type* of e^{tA} is usually defined by

$$\inf\{\omega \in \mathbb{R} : \exists M > 0 \text{ such that } \|e^{tA}\|_{L(X)} \leq Me^{\omega t}, \ \forall t > 0\}.$$

Proposition 2.3.1 implies that the type of e^{tA} is less or equal to ω_A. On the other hand, if $\|e^{tA}\|_{L(X)} \leq Me^{\omega t}$ for every $t > 0$, then $R(\lambda, A) = \int_0^\infty e^{-\lambda t} e^{tA} dt$ (see Lemma 2.1.6) exists for $\text{Re}\,\lambda > \omega$, so that $\omega_A \leq \omega$. We have so proved the following Corollary.

Corollary 2.3.2 ω_A *is the type of* e^{tA}.

Let us assume now that the spectrum of A may be decomposed as

$$\sigma(A) = \sigma_1 \cup \sigma_2, \tag{2.3.5}$$

where σ_1, σ_2 are nonempty spectral sets, and σ_1 is bounded. Set

$$\alpha_i = \inf\{\text{Re}\,\lambda : \lambda \in \sigma_i\}, \ \omega_i = \sup\{\text{Re}\,\lambda : \lambda \in \sigma_i\}, \ i = 1, 2. \tag{2.3.6}$$

Let P be the projection associated to σ_1,

$$P = \frac{1}{2\pi i} \int_\gamma R(\lambda, A) d\lambda, \tag{2.3.7}$$

where γ is a suitable curve around σ_1. According to (A.1.2), we may decompose X as

$$X = X_1 \oplus X_2, \ X_1 = P(X), \ X_2 = (I - P)(X). \tag{2.3.8}$$

We recall that X_1 is contained in $D(A^n)$ for each $n \in \mathbb{N}$, and that AP is a bounded operator. The decomposition (2.3.8) induces a splitting of the operator A: indeed, according to (A.1.3), we define

$$\begin{cases} A_1 : X_1 \mapsto X_1, \ A_1 x = Ax \ \forall x \in X_1, \\ A_2 : D(A_2) = D(A) \cap X_2 \mapsto X_2, \ A_2 x = Ax \ \forall x \in D(A_2). \end{cases}$$

By Proposition A.1.2, $\sigma(A_1) = \sigma_1$, $\sigma(A_2) = \sigma_2$, and $R(\lambda, A_1) = R(\lambda, A)_{|X_1}$, $R(\lambda, A_2) = R(\lambda, A)_{|X_2}$ for $\lambda \in S_{\theta,\omega}$. Then A_1 and A_2 generate analytic semigroups in X_1 and in X_2 respectively, and

$$e^{tA_1} = e^{tA}{}_{|X_1} = e^{tA} P_{|X_1}, \ e^{tA_2} = e^{tA}{}_{|X_2} = e^{tA}(I - P)_{|X_2}. \tag{2.3.9}$$

Let us state estimates for $e^{tA}P$ and $e^{tA}(I - P)$.

3. Spectral properties and asymptotic behavior

Proposition 2.3.3 *Let (2.3.5), (2.3.6) hold. Then $e^{tA}P = Pe^{tA}$, so that $e^{tA}(X_2) \subset X_2$, $e^{tA}(X_1) \subset X_1$.*
For every $\varepsilon > 0$ and $n \in \mathbb{N} \cup \{0\}$ there is $M_{n,\varepsilon} > 0$ such that

$$\|t^n A^n e^{tA}(I-P)\|_{L(X)} \leq M_{n,\varepsilon} e^{(\omega_2+\varepsilon)t}, \quad t > 0. \tag{2.3.10}$$

Moreover, for every $\varepsilon > 0$ and $\alpha, \beta \in \,]0,1[$, $p \in [1,\infty]$,

$$\sup_{t>0} t^{n+\beta} e^{-(\omega_2+\varepsilon)t} \|A^n e^{tA}(I-P)\|_{L(X,D_A(\beta,p))} < \infty, \quad n \in \mathbb{N} \cup \{0\}, \tag{2.3.11}$$

$$\sup_{t>0} t^{n-\alpha+\beta} e^{-(\omega_2+\varepsilon)t} \|A^n e^{tA}(I-P)\|_{L(D_A(\alpha,p),D_A(\beta,p))} < \infty, \quad n \in \mathbb{N}, \tag{2.3.12}$$

and (2.3.12) holds also for $n = 0$, provided $\alpha \leq \beta$.
Moreover, $t \mapsto e^{tA}P$ has an analytic extension to $\mathbb{R}$, and for every $\varepsilon > 0$, $n \in \mathbb{N} \cup \{0\}$ there is $N_{n,\varepsilon} > 0$ such that

$$\begin{aligned}(i) \quad & \|A^n e^{tA}P\|_{L(X)} \leq N_{n,\varepsilon} e^{(\omega_1+\varepsilon)t}, \quad \forall t \geq 0, \\ (ii) \quad & \|A^n e^{tA}P\|_{L(X)} \leq N_{n,\varepsilon} e^{(\alpha_1-\varepsilon)t}, \quad \forall t \leq 0.\end{aligned} \tag{2.3.13}$$

Proof — The equality $e^{tA}P = Pe^{tA}$ follows easily from the equality $PR(\lambda, A) = R(\lambda, A)P$: see Proposition A.1.2.

Since $\sigma(A_2) = \sigma_2$, then estimates (2.3.2), (2.3.3), (2.3.4) hold for e^{tA_2}, with ω_A replaced by ω_2. Using (2.3.9)(ii), estimates (2.3.10), (2.3.11), (2.3.12) follow.

Since A_1 is a bounded operator, then $t \mapsto e^{tA_1}$ may be extended to $\mathbb{R}$, by setting

$$e^{tA_1} = \frac{1}{2\pi i} \int_\gamma e^{t\xi} R(\xi, A) d\xi. \tag{2.3.14}$$

To prove (2.3.13), choose γ such that $\inf\{\operatorname{Re}\lambda : \lambda \in \gamma\} \geq \alpha_1 - \varepsilon$, and $\sup\{\operatorname{Re}\lambda : \lambda \in \gamma\} \leq \omega_1 + \varepsilon$. Then we have, for each $t \in \mathbb{R}$,

$$A^n e^{tA_1} = \frac{1}{2\pi i} \int_\gamma e^{t\xi} \xi^n R(\xi, A) d\xi,$$

so that, denoting by ℓ the lenght of γ, we get

$$\|A^n e^{tA_1}\| \leq \frac{1}{2\pi} \sup\{\|\xi^n R(\xi, A)\|_{L(X)} : \xi \in \gamma\} \ell e^{(\omega_1+\varepsilon)t}, \quad t \geq 0,$$

$$\|A^n e^{tA_1}\| \leq \frac{1}{2\pi} \sup\{\|\xi^n R(\xi, A)\|_{L(X)} : \xi \in \gamma\} \ell e^{(\alpha_1-\varepsilon)t}, \quad t \leq 0. \quad \blacksquare$$

Proposition 2.3.4 *Let $\sigma_1 = \{\lambda_1, \ldots, \lambda_N\}$, where all the λ_j's are isolated elements of $\sigma(A)$. Let P_j, D_j be defined by (A.2.2) with λ_0 replaced by λ_j, $j = 1, \ldots, N$, and let P be defined by (2.3.7). Then for every $t \in \mathbb{R}$ we have*

$$e^{tA}P = \sum_{j=1}^{N} e^{\lambda_j t}\left(P_j + \sum_{n=1}^{\infty} \frac{t^n}{n!} D_j^n\right). \tag{2.3.15}$$

Proof — Let γ_j, $j = 1, ..., N$, be a circle centered at λ_j, with radius so small that no other element of $\sigma(A)$ than λ_j lies in the interior of γ_j or in γ_j, and let us plug equality (A.2.1) in (2.3.14). We get

$$e^{tA}P = \sum_{j=1}^{N} \frac{1}{2\pi i} \int_{\gamma_j} e^{\lambda t} \left(P_j(\lambda - \lambda_j)^{-1} + \sum_{n=1}^{\infty} D_j^n(\lambda - \lambda_j)^{-n-1} \right) d\lambda,$$

and (2.3.15) follows. ∎

Corollary 2.3.5 *Let $\sigma(A) \cap \{\lambda \in \mathbb{C} : \operatorname{Re} \lambda = \omega_A\}$ be a spectral set consisting of a finite number of semisimple eigenvalues $\lambda_1, \ldots, \lambda_n$. Then*

$$e^{tA} = e^{tA}(I - P) + \sum_{j=1}^{N} e^{\lambda_j t} P_j, \quad t \geq 0, \tag{2.3.16}$$

so that estimates (2.3.2), (2.3.3), (2.3.4), with $n = 0$, hold also for $\varepsilon = 0$.

2.3.2 Spectral properties of e^{tA}

As one can expect, for each $T > 0$ the spectral properties of e^{TA} are strongly connected with the spectral properties of A.

In the following proposition we study the invertibility of $I - e^{TA}$. To do this, we define a curve $\gamma_\# = \gamma_1 - \gamma_2$, where

$$\gamma_1 = \{r \mid \rho e^{i\eta} : \rho \geq 0\} \cup \{r + \rho e^{-i\eta} : \rho \geq 0\}$$

and $r \neq 0$, $\eta \in \,]\pi/2, \pi[$ are such that $\sigma(A)$ lies on the left hand side of γ_1, $r + \rho e^{i\eta} \neq 2k\pi i/T$, for every $\rho \geq 0$, $k \in \mathbb{N}$. Moreover,

$$\gamma_2 = \emptyset, \text{ if } r < 0, \quad \gamma_2 = \bigcup_{|k| \leq K} C\left(\frac{2k\pi i}{T}, \varepsilon\right), \text{ if } r > 0,$$

where $K = \max\{k \in \mathbb{N} : 2k\pi/T < r\tan(\pi - \eta)\}$, and $\varepsilon < \pi/T$ is a sufficiently small number. See the figure below.

Proposition 2.3.6 *Let $T > 0$. If $2k\pi i/T \in \rho(A)$ for every $k \in \mathbb{Z}$, then $1 \in \rho(e^{TA})$, and*

$$(1 - e^{TA})^{-1} = \frac{1}{2\pi i} \int_{\gamma_\#} \frac{e^{Tz}}{1 - e^{Tz}} R(z, A) dz + I = \Gamma + I \tag{2.3.17}$$

where $\gamma_\#$ is the curve defined above, with $\varepsilon > 0$ so small that for each $k \in \mathbb{Z}$ the balls centered at $2k\pi i/T$ with radius ε are contained in $\rho(A)$ and do not intersect γ_1.

Conversely, if $1 \in \rho(e^{TA})$, then $2k\pi i/T \in \rho(A)$ for every $k \in \mathbb{Z}$.

3. Spectral properties and asymptotic behavior

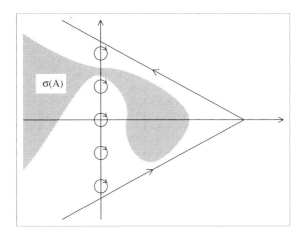

Proof — Define Γ by (2.3.17), and define $\gamma'_\#$ as $\gamma_\#$, with r replaced by $r' \in \,]r, 2(K+1)\pi/\tan(\pi-\eta)[$ and ε replaced by $\varepsilon/2$. Then

$$\begin{aligned}
\Gamma e^{TA} &= e^{TA}\Gamma = \left(\frac{1}{2\pi i}\right)^2 \int_{\gamma'_\# \times \gamma_\#} \frac{e^{T\lambda} e^{Tz}}{1-e^{Tz}} \frac{R(\lambda,A) - R(z,A)}{z-\lambda} dz\, d\lambda \\
&= \left(\frac{1}{2\pi i}\right)^2 \int_{\gamma'_\#} e^{T\lambda} R(\lambda, A) \left(\int_{\gamma_\#} \frac{e^{Tz}}{(1-e^{Tz})(z-\lambda)} dz\right) d\lambda \\
&\quad - \left(\frac{1}{2\pi i}\right)^2 \int_{\gamma_\#} \frac{e^{Tz}}{1-e^{Tz}} R(z,A) \left(\int_{\gamma'_\#} \frac{e^{T\lambda}}{z-\lambda} d\lambda\right) dz \\
&= \frac{1}{2\pi i} \int_{\gamma_\#} \frac{e^{2Tz}}{1-e^{Tz}} R(z,A) dz.
\end{aligned}$$

On the other hand, we have

$$e^{TA} = \frac{1}{2\pi i} \int_{\gamma_\#} e^{Tz} R(z,A) dz = \frac{1}{2\pi i} \int_{\gamma_\#} \left(\frac{e^{Tz}}{1-e^{Tz}} - \frac{e^{2Tz}}{1-e^{Tz}}\right) R(z,A) dz$$

Therefore, $e^{TA}(\Gamma+I) = (\Gamma+I)e^{TA} = \Gamma$, so that $(I - e^{TA})(\Gamma+I) = (\Gamma+I)(I - e^{TA}) = I$, and the first part of the statement follows.

From the identity

$$e^{T\lambda} - e^{TA} = (\lambda - A)\int_0^T e^{\lambda(T-s)} e^{sA} ds \left(= \int_0^T e^{\lambda(T-s)} e^{sA} ds (\lambda - A) \text{ on } D(A)\right), \tag{2.3.18}$$

which holds for every $\lambda \in \mathbb{C}$, it follows that if $e^{T\lambda} - e^{TA}$ is invertible then $\lambda I - A$ is invertible. Taking $\lambda = 2k\pi i/T$ with $k \in \mathbb{Z}$ we find that if $1 \in \rho(e^{TA})$ then $2k\pi i/T \in \rho(A)$. ∎

Proposition 2.3.6 yields easily the following corollary, which is known as *Spectral Mapping Theorem*.

Corollary 2.3.7 *For every $T > 0$*

$$\sigma(e^{TA}) \setminus \{0\} = e^{T\sigma(A)}.$$

Proof — For every $\lambda \neq 0$ consider the rescaled semigroup $e^{-\lambda t} e^{tA}$, generated by $A - \lambda I$. The number $e^{T\lambda}$ belongs to $\sigma(e^{TA})$ if and only if 1 belongs to $\sigma(e^{-\lambda T} e^{TA})$. By Proposition 2.3.6, 1 belongs to $\sigma(e^{-\lambda T} e^{TA})$ if and only if there exists $k \in \mathbb{Z}$ such that $2k\pi i/T$ belongs to $\sigma(A - \lambda I)$. Since $\sigma(A - \lambda I) = \sigma(A) - \lambda$, the statement follows. ∎

We know from Corollary 2.1.7 that 0 is not an eigenvalue of e^{TA}. However, in the most interesting examples 0 belongs to $\sigma(e^{TA})$. Indeed, if e^{TA} is invertible, then $X = \text{Range } e^{TA} \subset D(A)$, so that $D(A) = X$.

A spectral mapping theorem holds also for the point spectrum σ_p.

Proposition 2.3.8 *For every $T > 0$*

$$\sigma_p(e^{TA}) = e^{T\sigma_p(A)}.$$

Proof — Fix $\lambda \in \mathbb{C}$. From the identity (2.3.18) it follows that if λ is an eigenvalue of A, then $e^{T\lambda}$ is an eigenvalue of e^{TA}. So, $e^{T\sigma_p(A)} \subset \sigma_p(e^{TA})$.

To prove the converse, we note that 0 is not an eigenvalue of e^{TA}, which is one to one thanks to Corollary 2.1.7. Therefore, for every eigenvalue z of e^{TA} there exists $\lambda \in \mathbb{C}$ such that $z = e^{\lambda T}$. By rescaling the semigroup e^{tA} as in Corollary 2.3.7 (that is, replacing e^{TA} by $e^{-\lambda T} e^{TA}$), we may assume $z = 1$, $\lambda = 0$. So, we have to show that for some $k \in \mathbb{Z}$, $2k\pi i/T$ is an eigenvalue of A.

The eigenspace E of e^{TA} corresponding to the eigenvalue 1 is contained in $D(A^n)$ for every $n \in \mathbb{N}$, and for every eigenvector x the function $t \mapsto e^{tA}x$ is either constant, or periodic with minimum period $\leq T$. From equality (2.1.10) it follows, after easy computation, that for Re λ large it holds

$$R(\lambda, A)x \begin{cases} = x/\lambda, & \text{if } e^{tA}x \text{ is constant,} \\ = \dfrac{1}{1-e^{-\lambda T}} \displaystyle\int_0^T e^{-\lambda t} e^{tA} x \, dt, & \text{if } e^{tA}x \text{ is T-periodic.} \end{cases} \quad (2.3.19)$$

In particular, the spectrum of the part of A in E is contained in $\{2k\pi i/T : k \in \mathbb{Z}\}$, so that it is either empty or it consists of a finite number of elements, each of them being a semisimple eigenvalue, thanks to (2.3.19). Setting

$$P_k x = \frac{1}{2\pi i} \int_{C(2k\pi i/T, \pi/T)} R(\lambda, A) x \, d\lambda, \quad x \in E,$$

then all except at most a finite number of P_k's vanish, and

$$P_k x = \frac{1}{T} \int_0^T e^{-2k\pi i t/T} e^{tA} x \, dt, \quad x \in E.$$

Let $x \in E$, $x \neq 0$, and set $y = \sum_{k \in \mathbb{Z}} P_k x$. Then for every x' in the dual space E' the Fourier coefficients of the T-periodic functions $t \mapsto \langle e^{tA} x, x' \rangle$ and $t \mapsto \langle e^{tA} y, x' \rangle$ coincide, so that $e^{tA} x = e^{tA} y$ for every t, and since e^{tA} is one to one, then $x = y$. In particular, there exists $k \in \mathbb{Z}$ such that $P_k x \neq 0$. Therefore, $P_k x$ is an eigenvector of A with eigenvalue $2k\pi i/T$, and the statement follows. ∎

The following proposition deals with the case where 1 is an isolated eigenvalue of e^{TA}, for some $T > 0$. It will be used in the study of the T-periodic solutions of evolution equations.

Proposition 2.3.9 *Let k_j, $j = 1, \ldots, n$ be integers such that for every $j = 1, \ldots, N$, $2k_j \pi i/T$ is a semisimple eigenvalue of A, and $\rho(A) \supset \{2k\pi i/T : k \in \mathbb{Z}, k \neq k_1, \ldots, k_N\}$. Then 1 is a semisimple eigenvalue of e^{TA}.*

Fix $\varepsilon > 0$ so small that the sets $\{\lambda \in \mathbb{C} : 0 < |\lambda - 2k_j \pi i/T| \leq \varepsilon\}$ are contained in $\rho(A)$ for each $j = 1, \ldots, N$, and set

$$P = \sum_{j=1}^N P_j, \quad P_j = \frac{1}{2\pi i} \int_{C(2k_j \pi i/T, \varepsilon)} R(\lambda, A) d\lambda, \quad (2.3.20)$$

Then P is a projection on $\operatorname{Ker} e^{TA}$, $1 \in \rho(e^{tA}_{|(I-P)(X)})$, and

$$\left(I - e^{TA}_{|(I-P)X}\right)^{-1} = \frac{1}{2\pi i} \int_{\gamma_\# + \bigcup_{j=1}^N C(2k_j \pi i/T, \varepsilon)} \frac{e^{Tz}}{1 - e^{Tz}} R(z, A) dz + I - P, \quad (2.3.21)$$

where $\gamma_\#$ is the curve defined above.

Proof — We know from Propositions 2.3.6 and 2.3.8 that 1 is an isolated eigenvalue of e^{TA}. We have only to show that it is semisimple.

Since $2k_j \pi i/T$ are semisimple eigenvalues of A, then $X = P(X) \oplus (I - P)(X)$ (see Remark A.2.4). One can verify that $(I - e^{TA})$ is invertible on $(I - P)(X)$, by constructing explicitly the inverse of $I - e^{TA}_{|(I-P)(X)}$. Indeed, arguing as in the proof of Proposition 2.3.6, one can show that (2.3.21) holds. It follows that $P(X) = \operatorname{Ker}(I - e^{TA})$, and $(I - P)(X) = \operatorname{Range}(I - e^{TA})$. Then, $X = \operatorname{Ker}(I - e^{TA}) \oplus \operatorname{Range}(I - e^{TA})$, so that 1 is a semisimple eigenvalue of e^{TA}. ∎

2.4 Perturbations of generators

We give here some simple, although very useful, results about perturbations of sectorial operators. Throughout the section, $A : D(A) \subset X \mapsto X$ is a linear operator satisfying (2.0.1), with constants $\omega(A)$, $\theta(A)$, $M(A)$.

We begin with the case of a perturbation "dominated" by A, in the sense that either its domain contains an intermediate space between X and $D(A)$, or its range is contained in an intermediate space between X and $D(A)$.

Proposition 2.4.1 *(i) Let X_α be a space belonging to the class J_α between X and $D(A)$, with $0 \leq \alpha < 1$, and let $B \in L(X_\alpha, X)$. Then $A + B : D(A) \mapsto X$ is sectorial.*
(ii) Let X_α be a space belonging to the class K_α between X and $D(A)$, with $0 < \alpha \leq 1$, and let $B \in L(D(A), X_\alpha)$. Then $A + B : D(A) \mapsto X$ is sectorial.

Proof — For $\lambda \in S_A = \{\lambda \in \mathbb{C} : \lambda \neq \omega(A), |\arg(\lambda - \omega(A))| < \theta(A)\}$ and $f \in X$, consider the resolvent equation

$$\lambda u - Au - Bu = f. \qquad (2.4.1)$$

(i) Setting $v = \lambda u - Au$, (2.4.1) is equivalent to

$$v = BR(\lambda, A)v + f. \qquad (2.4.2)$$

Let c_α be such that $\|x\|_{X_\alpha} \leq c_\alpha \|x\|^{1-\alpha} \|x\|_{D(A)}^\alpha$ for each $x \in D(A)$. Then

$$\|BR(\lambda, A)\|_{L(X)} \leq \|B\|_{L(X_\alpha, X)} \frac{c_\alpha(M(A) + 1)}{|\lambda - \omega(A)|^{\alpha - 1}},$$

so that $BR(\lambda, A)$ is a 1/2-contraction, provided $|\lambda|$ is sufficiently large. In this case, (2.4.2) has a unique solution v, with $\|v\| \leq 2\|f\|$, and (2.4.1) has a unique solution u, with $\|u\| \leq 2M(A)|\lambda - \omega(A)|^{-1}\|f\|$. It follows that $A + B$ is sectorial.
(ii) Since X_α belongs to the class K_α between X and $D(A)$, then it is continuously embedded in $D_A(\alpha, \infty)$. We will see that there exist $c, r > 0$ such that

$$\|AR(\lambda, A)x\| \leq \frac{c}{|\lambda|^\alpha} \|x\|_{X_\alpha}, \quad \lambda \in S_A, \ |\lambda| > r, \ x \in X_\alpha.$$

Indeed, from the equality

$$AR(\lambda, A) = AR(|\lambda|, A) + A(\lambda - |\lambda|)R(\lambda, A)R(|\lambda|, A),$$

which holds for every $\lambda \in \rho(A)$ such that $|\lambda| \in \rho(A)$, we get, from Proposition 2.2.4,

$$\|AR(\lambda, A)x\| \leq C \left(\frac{1}{|\lambda|^\alpha} + \frac{2M_A}{|\lambda|^\alpha} \right) \|x\|_{D_A(\alpha, \infty)} \leq \frac{c}{|\lambda|^\alpha} \|x\|_{X_\alpha},$$

4. PERTURBATIONS OF GENERATORS

for every $x \in X_\alpha$, $\lambda \in S_A$, with $|\lambda|$ large enough. It follows

$$\|R(\lambda, A)B\|_{L(D(A))} \leq \|R(\lambda, A)\|_{L(X_\alpha, D(A))} \|B\|_{L(D(A), X_\alpha)}$$
$$\leq \left(\frac{c}{|\lambda|^\alpha} + \frac{M(A)}{|\lambda - \omega(A)|} \right) \|B\|_{L(D(A), X_\alpha)}.$$

Therefore, there is $r > 0$ such that if $\lambda \in S_A$, and $|\lambda| \geq r$, then $\|R(\lambda, A)B\|_{L(D(A))} \leq 1/2$. For such λ's, the resolvent equation (2.4.1) is equivalent to

$$u = R(\lambda, A)(Bu + f),$$

and it is uniquely solvable, with

$$\|u\|_{D(A)} \leq 2\|R(\lambda, A)\|_{L(X, D(A))} \|f\| \leq 2 \left(M(A) \frac{|\lambda| + 1}{|\lambda - \omega(A)|} + 1 \right) \|f\|.$$

From (2.4.1) we get

$$|\lambda| \|u\| \leq \|u\|_{D(A)} (1 + \|B\|_{L(D(A), X)}) + \|f\|,$$

and the statement follows. ∎

Now we consider the case where B is not dominated by A, but its norm is suitably small.

Proposition 2.4.2 *Let $B \in L(D(A), X)$, with $\|B\|_{L(D(A), X)} < (M(A)+1)^{-1}$. Then $A + B : D(A) \mapsto X$ is sectorial.*

Proof — For every $\lambda \in S_A$

$$\|R(\lambda, A)\|_{L(X, D(A))} \leq \|AR(\lambda, A)\|_{L(X)} + \|R(\lambda, A)\|_{L(X)} \leq M(A) \frac{|\lambda| + 1}{|\lambda - \omega(A)|} + 1. \tag{2.4.3}$$

If $|\lambda|$ is sufficiently large, then $\|B\|_{L(D(A), X)} \|R(\lambda, A)\|_{L(X, D(A))} < 1$, so that $BR(\lambda, A)$ is a contraction, and the statement follows as in the proof of Proposition 2.4.1(i). ∎

The next proposition deals with the case where B is not dominated by A, and its norm is not necessarily small, but it is a compact operator.

Proposition 2.4.3 *Let $D(A)$ be dense in X, and let $B \in L(D(A), X)$ be compact. Then $A + B : D(A) \mapsto X$ is sectorial.*

Proof — For every $x \in D(A)$, we have $\|R(\lambda, A)x\|_{D(A)} \leq M_A|\lambda - \omega_A|^{-1} \|x\|_{D(A)}$, so that $\|R(\lambda, A)x\|_{D(A)}$ goes to 0 as $|\lambda|$ goes to $+\infty$, $\lambda \in S_A$. Estimate (2.4.3) implies that $\|R(\lambda, A)\|_{L(X, D(A))}$ is bounded for large $\lambda \in S_A$. Since $D(A)$ is dense in X, we get
$$\forall x \in X, \quad \lim_{|\lambda| \to +\infty, \, \lambda \in S_A} \|R(\lambda, A)x\|_{D(A)} = 0.$$
Since $B : D(A) \mapsto X$ is compact, it follows easily that
$$\lim_{|\lambda| \to +\infty, \, \lambda \in S_A} \|R(\lambda, A)Bx\|_{D(A)} = 0,$$
uniformly for $x \in B(0,1) \subset D(A)$. Therefore, there is $r > 0$ such that if $\lambda \in S_A$, $|\lambda| \geq r$, then $\|R(\lambda, A)B\|_{L(D(A))} \leq 1/2$. The statement follows now as in the proof of Proposition 2.4.1(ii). ∎

The following proposition concerns the powers of generators.

Proposition 2.4.4 *Let $n \in \mathbb{N}$. Let $A : D(A) \subset X \mapsto X$ be a sectorial operator, satisfying (2.0.1) with angle $\theta > \pi(1 - 2^{-n})$. Then the operator*
$$(-1)^{n+1}A^n : D(A^n) \mapsto X$$
is sectorial, and it satisfies (2.0.1) with angle $\beta = n[\theta - \pi(1 - 1/n)]$. In particular, if A satisfies (2.0.1) with $\theta = \pi$, then $(-1)^{n+1}A^n$ is sectorial for every $n \in \mathbb{N}$.

Proof — By the usual translation argument, it is sufficient to consider the case where $\omega = 0$. Let $\lambda \in S_{\beta,0}$, with $\beta > \pi/2$, $y \in X$, and consider the resolvent equation
$$\lambda x - (-1)^{n+1}A^n x = y, \tag{2.4.4}$$
which is equivalent to
$$(-1)^{n+1}\prod_{i=1}^{n}(\alpha_i \lambda^{1/n}I - A)x = y,$$
where the complex numbers α_i, $i = 1, \ldots, n$, are the n-order roots of $(-1)^{n+1}$. We can solve (2.4.4) and estimate the solution provided $\alpha_i \lambda^{1/n} \in S_{\theta,0}$ for every i. This is true if
$$\theta \geq \pi(1 - 1/n) + \beta/n,$$
in which case the unique solution of (2.4.4) is $x = (-1)^{n+1}\prod_{i=1}^{n} R(\alpha_i \lambda^{1/n}, A)y$, and it satisfies
$$\|x\| \leq \frac{M^n}{|\lambda|}\|y\|.$$
So, $(-1)^{n+1}A^n$ is sectorial if $\theta > \pi(1 - 1/n) + \pi/2n = \pi(1 - 1/2n)$, and it satisfies (2.0.1) with θ replaced by $\beta = n[\theta - \pi(1 - 1/n)]$ and M replaced by M^n. ∎

2.5 Bibliographical remarks

For the classical theory of analytic semigroups we refer to E. HILLE, R.S. PHILLIPS [100, §12.8], K. YOSIDA [213, §IX.10]. The non dense domain case was considered by E. SINESTRARI in [177].

The definition (2.2.2) of the spaces $D_A(\alpha,p)$ and their equivalence with the spaces $(X, D(A))_{\alpha,p}$ is due to H. BERENS – P.L. BUTZER [33]. The characterizations of Propositions 2.2.3 and 2.2.4 are due respectively to J.-L. LIONS [127, (I)], P. GRISVARD [93] in the case of the spaces $D_A(\alpha,p)$, and to P.L. BUTZER – H. BERENS [42, Prop. 3.5.8], G. DA PRATO – P. GRISVARD [63] in the case of the spaces $D_A(\alpha)$.

The fractional powers of sectorial operators are treated systematically in K. YOSIDA [213, §X.11], H. KOMATSU [112], H. TRIEBEL [200, §1.15].

A reference for Corollary 2.3.6 is R. TRIGGIANI [199]. Concerning the Spectral Mapping Theorem and related results we refer to R. NAGEL ET AL. [163].

Propositions 2.4.2, 2.4.2 are well known. A proof of Proposition 2.4.3 may be found in W. DESCH – W. SCHAPPACHER [71].

Chapter 3

Generation of analytic semigroups by elliptic operators

Let Ω be an open set in $\mathbb{R}^n$ with sufficiently smooth (possibly empty) boundary $\partial\Omega$, and let $\mathcal{A}(x,D)$ be a strongly elliptic operator of order $2m$ with uniformly continuous and bounded coefficients[1] defined in $\overline{\Omega}$,

$$\mathcal{A}(x,D) = \sum_{|\alpha|\leq 2m} a_\alpha(x) D^\alpha. \tag{3.0.1}$$

If $\Omega \neq \mathbb{R}^n$, let $\mathcal{B}_j(x,D)$, $j = 1,..,m$, be differential operators of order m_j ($0 \leq m_1 < m_2 < ... < m_m \leq 2m - 1$) with sufficiently regular coefficients,

$$\mathcal{B}_j(x,D) = \sum_{|\beta|\leq m_j} b_\beta(x) D^\beta, \tag{3.0.2}$$

and satisfying the usual root, complementing, and normality conditions (for precise definitions and assumptions, see Section 3.2).

We consider the realizations of $\mathcal{A}(\cdot, D)$, with homogeneous boundary conditions $\mathcal{B}_j(\cdot, D) = 0$, in several Banach spaces of functions defined in Ω. The aim of this chapter is to prove that if the coefficients and $\partial\Omega$ are smooth enough, then such realizations are sectorial operators, so that they generate analytic semigroups. As a byproduct of the generation results, we shall characterize some of the relevant real interpolation spaces.

---

[1] In fact, in much of what follows one could assume that the coefficients of the derivatives of order $< 2m$ be measurable and bounded. However, some proofs would become more heavy than the present ones.

For a linear operator be sectorial one needs two things:
(i) the resolvent set contains a sector

$$S = \{\lambda \in \mathbb{C} : \lambda \neq \omega, \ |\arg(\lambda - \omega)| < \theta\},$$

with $\omega \in \mathbb{R}$ and $\theta > \pi/2$;
(ii) there is $M > 0$ such that $\|R(\lambda, A)\|_{L(X)} \leq M/|\lambda - \omega|$ for $\lambda \in S$.

In order to prove (i), we need some existence and uniqueness results for elliptic boundary value problems of the type

$$\begin{cases} \lambda u(x) - \mathcal{A}(x, D)u(x) = f(x), & x \in \Omega, \\ \mathcal{B}_j(x, D)u(x) = g_j, & j = 1, ..., m, \ x \in \partial\Omega. \end{cases} \quad (3.0.3)$$

In fact, although the functions in the domains of the realizations of $\mathcal{A}$ satisfy homogeneous boundary conditions, in many proofs we need to solve auxiliary nonhomogeneous boundary value problems such as (3.0.3).

Concerning point (ii), the main tool will be the Agmon-Douglis-Nirenberg *a priori* estimates. They will be used to show generation results in $L^p(\Omega)$, and then in $C(\overline{\Omega})$. Then we shall see that for noninteger $\theta \in \,]0, 2m[$, the space $C^\theta_\mathcal{B}(\overline{\Omega})$ (consisting of all C^θ functions u such that $\mathcal{B}_j u$ vanishes on $\partial\Omega$ for $m_j \leq [\theta]$) is a real interpolation space. So we will obtain generation results in such spaces from interpolation theory. Moreover, the characterization of Hölder spaces as interpolation spaces will let us prove (at least, in the case $m = 1$) in a simple way the classical Schauder theorems on the Hölder regularity of solutions of elliptic equations in $\mathbb{R}^n$. In their turn, Schauder type theorems will let us characterize the domain of the realization of $\mathcal{A}$ in the Hölder spaces.

Generation results in $C^k_\mathcal{B}(\overline{\Omega})$, $k \in \mathbb{N}$ are also important in the applications to partial differential equations. They do not follow by interpolation, since $C^k_\mathcal{B}(\overline{\Omega})$ is not an interpolation space. The proofs rely on the generation theorem in $C(\overline{\Omega})$ and on *a priori* estimates for the solutions of nonhomogeneous problems such as (3.0.3).

The main ideas of the proofs in the case $m > 1$ are the same of the case $m = 1$. For the sake of simplicity, we give detailed proofs only in the case of second order operators. Second order elliptic operators, with Dirichlet or first order non tangential boundary condition will be considered in Section 3.1. The main results on $2m$-th order operators will be stated without proofs in Section 3.2, together with bibliographical references.

3.1 Second order operators

Let Ω be either $\mathbb{R}^n$, or an open bounded subset of $\mathbb{R}^n$ with uniformly C^2 boundary $\partial\Omega$. We denote by $\nu(x)$ the exterior unit normal vector to $\partial\Omega$ at $x \in \partial\Omega$.

1. Second order operators

We consider a second order differential operator

$$\mathcal{A}(x,D) = \sum_{i,j=1}^{n} a_{ij}(x)D_{ij} + \sum_{i=1}^{n} b_i(x)D_i + c(x)I \tag{3.1.1}$$

with real uniformly continuous and bounded coefficients a_{ij}, b_i, c.

We assume that the matrix $[a_{ij}]$ is symmetric, and that it satisfies the uniform ellipticity condition

$$\sum_{i,j=1}^{n} a_{ij}(x)\xi_i\xi_j \geq \nu|\xi|^2, \quad x \in \overline{\Omega}, \ \xi \in \mathbb{R}^n, \tag{3.1.2}$$

for some $\nu > 0$. Moreover, if $\Omega \neq \mathbb{R}^n$, we consider a first order differential operator acting on the boundary

$$\mathcal{B}(x,D) = \sum_{i=1}^{n} \beta_i(x)D_i + \gamma(x)I \tag{3.1.3}$$

with real differentiable coefficients. We assume that β_i, γ, belong to $UC^1(\overline{\Omega})$, and that the uniform nontangentiality condition

$$\inf_{x\in\partial\Omega} \left|\sum_{i=1}^{n} \beta_i(x)\nu_i(x)\right| > 0 \tag{3.1.4}$$

holds.

Our main purpose is to prove that the realizations of $\mathcal{A}(\cdot,D)$ (with homogeneous boundary conditions, if $\Omega \neq \mathbb{R}^n$) in several Banach spaces of functions defined in Ω are sectorial. Fundamental tools will be the Agmon-Douglis-Nirenberg a priori estimates for elliptic problems in the whole space $\mathbb{R}^n$ and in regular domains, which we recall below. They hold for operators with complex coefficients, under weaker ellipticity assumptions than (3.1.2), namely

$$\left|\sum_{i,j=1}^{n} a_{ij}(x)\xi_i\xi_j\right| \geq \mu|\xi|^2, \quad x \in \overline{\Omega}, \ \xi \in \mathbb{R}^n, \tag{3.1.5}$$

$$\begin{cases} \text{if } \xi, \eta \in \mathbb{R}^n \text{ are linearly independent, then for each } x \in \overline{\Omega} \text{ the} \\ \text{polynomial } \tau \mapsto P(\tau) = \sum_{i,j=1}^{n} a_{ij}(x)(\xi_i + \tau\eta_i)(\xi_j + \tau\eta_j) \\ \text{has a unique root with positive imaginary part.} \end{cases} \tag{3.1.6}$$

Theorem 3.1.1 *(i) Let a_{ij}, b_i, $c : \mathbb{R}^n \mapsto \mathbb{C}$ be uniformly continuous and bounded functions, satisfying (3.1.5), (3.1.6). Let $\mathcal{A}(x,D)$ be defined by (3.1.1). Then for every $p \in\ ]1,+\infty[$ there is $c_p > 0$ such that for every $u \in W^{2,p}(\mathbb{R}^n)$*

$$\|u\|_{W^{2,p}(\mathbb{R}^n)} \leq c_p(\|u\|_{L^p(\mathbb{R}^n)} + \|\mathcal{A}(\cdot,D)u\|_{L^p(\mathbb{R}^n)}). \tag{3.1.7}$$

(ii) Let Ω be an open set in $\mathbb{R}^n$ with uniformly C^2 boundary, and let a_{ij}, b_i, $c : \overline{\Omega} \mapsto \mathbb{C}$ be uniformly continuous and bounded functions, satisfying (3.1.5), (3.1.6). Let $\mathcal{A}(x, D)$ be defined by (3.1.1), and for every $u \in W^{2,p}(\Omega)$, with $1 < p < \infty$, set $f = \mathcal{A}(\cdot, D)u$, $g = u_{|\partial\Omega}$. Then there is $c_p > 0$ such that

$$\|u\|_{W^{2,p}(\Omega)} \leq c_p(\|u\|_{L^p(\Omega)} + \|f\|_{L^p(\Omega)} + \|g_0\|_{W^{2,p}(\Omega)}), \tag{3.1.8}$$

where g_0 is any $W^{2,p}$ extension of g to the whole Ω.

(iii) Under the assumptions of statement (ii), let in addition β_i, γ be uniformly continuous and bounded, together with their first order derivatives, in $\overline{\Omega}$. For every $u \in W^{2,p}(\Omega)$, with $1 < p < \infty$, set $f = \mathcal{A}(\cdot, D)u$, $g = \mathcal{B}(\cdot, D)u_{|\partial\Omega}$. Then there is $c_p > 0$ such that

$$\|u\|_{W^{2,p}(\Omega)} \leq c_p(\|u\|_{L^p(\Omega)} + \|f\|_{L^p(\Omega)} + \|g_1\|_{W^{1,p}(\Omega)}), \tag{3.1.9}$$

where g_1 is any $W^{1,p}$ extension of g to the whole Ω.

One sees immediately that if the coefficients a_{ij} are real and satisfy (3.1.2), then the roots condition (3.1.6) holds. If the coefficients a_{ij} are complex, and they satisfy (3.1.5), then (3.1.6) holds as well provided $n \geq 3$. It can be shown that it holds also for $n = 2$, if the coefficients satisfy the *strong ellipticity* assumption

$$\text{Re} \sum_{i,j=1}^{n} a_{ij}(x)\xi_i\xi_j \geq \nu|\xi|^2, \quad x \in \overline{\Omega}, \ \xi \in \mathbb{R}^n. \tag{3.1.10}$$

The reason why we have to consider complex valued coefficients and to introduce assumption (3.1.6) is the fact that in Subsection 3.1.1 we shall use estimates (3.1.7), (3.1.8), (3.1.9) with $\mathcal{A}(x, D)$ replaced by the operator $\mathcal{A}(x, D) + e^{i\theta}D_{tt}$, in $n+1$ variables (x, t), with $\theta \in [-\pi/2, \pi/2]$, which is not necessarily strongly elliptic, but satisfies (3.1.6).

3.1.1 Generation in $L^p(\Omega)$, $1 < p < \infty$

Let $X = L^p(\Omega)$, $1 < p < \infty$, be endowed with its usual norm $\|\cdot\|_p$. If $u \in W^{1,p}(\Omega)$, we set $\|Du\|_p = \sum_{i=1}^{n} \|D_i u\|_p$; if $u \in W^{2,p}(\Omega)$, we set $\|D^2 u\|_p = \sum_{i,j=1}^{n} \|D_{ij} u\|_p$.

If $\Omega = \mathbb{R}^n$ we set

$$D(A) = W^{2,p}(\mathbb{R}^n), \quad Au = \mathcal{A}(\cdot, D)u \text{ for } u \in D(A),$$

where $\mathcal{A}(\cdot, D)$ is the operator defined in (3.1.1). A is said to be the realization of $\mathcal{A}(\cdot, D)$ in $L^p(\mathbb{R}^n)$. If $\Omega \neq \mathbb{R}^n$, we set

$$D(A_0) = W^{2,p}(\Omega) \cap W^{1,p}_0(\Omega), \quad A_0 u = \mathcal{A}(\cdot, D)u \text{ for } u \in D(A_0),$$

$$D(A_1) = \{u \in W^{2,p}(\Omega) : \mathcal{B}(\cdot, D)u = 0 \text{ in } \partial\Omega\}, \quad A_1 u = \mathcal{A}(\cdot, D)u \text{ for } u \in D(A_1).$$

1. SECOND ORDER OPERATORS

Here $\mathcal{B}$ is the operator defined in (3.1.3). A_0 (respectively, A_1) is said to be the realization of $\mathcal{A}(\cdot, D)$ in $L^p(\mathbb{R}^n)$ with homogeneous Dirichlet (respectively, oblique) boundary condition in $L^p(\Omega)$. The assumptions on the coefficients of $\mathcal{A}$ and $\mathcal{B}$ are those stated at the beginning of the section, and will not be mentioned again.

The resolvent sets of A, A_0, A_1 contain complex half planes, as the next theorem states.

Theorem 3.1.2 *Fix $p \in \,]1, +\infty[$.*

(i) *There exists $\omega \in \mathbb{R}$ such that $\rho(A) \supset \{\lambda \in \mathbb{C} : \operatorname{Re} \lambda \geq \omega\}$.*

(ii) *Let Ω be an open set in $\mathbb{R}^n$ with uniformly C^2 boundary. There exists $\omega_0 \in \mathbb{R}$ such that if $\operatorname{Re} \lambda \geq \omega_0$, then for every $f \in L^p(\Omega)$ and $g \in W^{2,p}(\Omega)$ the problem*
$$\lambda u - \mathcal{A}u = f \text{ in } \Omega, \quad u = g \text{ in } \partial\Omega,$$
has a unique solution $u \in W^{2,p}(\Omega)$, depending continuously on f and g. Taking $g = 0$, it follows that $\rho(A_0) \supset \{\lambda \in \mathbb{C} : \operatorname{Re} \lambda \geq \omega_0\}$.

(iii) *Let Ω be an open set in $\mathbb{R}^n$ with uniformly C^2 boundary. There exists $\omega_1 \in \mathbb{R}$ such that if $\operatorname{Re} \lambda \geq \omega_1$, then for every $f \in L^p(\Omega)$ and $g \in W^{1,p}(\Omega)$, the problem*
$$\lambda u - \mathcal{A}u = f \text{ in } \Omega, \quad \mathcal{B}u = g \text{ in } \partial\Omega,$$
has a unique solution $u \in W^{2,p}(\Omega)$, depending continuously on f and g. Taking $g = 0$, it follows that $\rho(A_1) \supset \{\lambda \in \mathbb{C} : \operatorname{Re} \lambda \geq \omega_1\}$.

If Ω is bounded, then ω_0, ω_1 do not depend on p.

In view of Proposition 2.1.11, to prove that A, A_0, A_1 are sectorial, we need a bound of the type (2.1.12) on the norms of their resolvent operators. In following theorem we prove more general estimates, which will be used in the next subsections.

Theorem 3.1.3 *Fix $p \in \,]1, +\infty[$.*

(i) *There exist $\omega_p \geq \omega$, $M_p > 0$ such that if $\operatorname{Re} \lambda \geq \omega_p$, then for every $u \in W^{2,p}(\mathbb{R}^n)$ we have*
$$|\lambda| \|u\|_p + |\lambda|^{1/2} \|Du\|_p + \|D^2 u\|_p \leq M_p \|\lambda u - \mathcal{A}(\cdot, D)u\|_p. \quad (3.1.11)$$

(ii) *Let Ω be an open set with uniformly C^2 boundary. Then there exist $\omega_p \geq \omega_0$, $M_p > 0$ such that if $\operatorname{Re} \lambda \geq \omega_p$, then for every $u \in W^{2,p}(\Omega)$ we have, setting $g = u_{|\partial\Omega}$,*
$$|\lambda| \|u\|_p + |\lambda|^{1/2} \|Du\|_p + \|D^2 u\|_p \leq \\ M_p(\|\lambda u - \mathcal{A}(\cdot, D)u\|_p + |\lambda| \|g_0\|_p + |\lambda|^{1/2} \|Dg_0\|_p + \|D^2 g_0\|_p), \quad (3.1.12)$$
where g_0 is any extension of g belonging to $W^{2,p}(\Omega)$.

(iii) Let Ω be an open set with uniformly C^2 boundary. Then there exist $\omega_p \geq \omega_1$, $M_p > 0$ such that if $\operatorname{Re} \lambda \geq \omega_p$, then for every $u \in W^{2,p}(\Omega)$ we have, setting $g = \mathcal{B}(\cdot, D)u_{|\partial\Omega}$,

$$|\lambda|\,\|u\|_p + |\lambda|^{1/2}\|Du\|_p + \|D^2 u\|_p \leq M_p(\|\lambda u - \mathcal{A}(\cdot, D)u\|_p + |\lambda|^{1/2}\|g_1\|_p + \|Dg_1\|_p), \qquad (3.1.13)$$

where g_1 is any extension of g belonging to $W^{1,p}(\Omega)$.

It follows that the operators A, A_0, A_1 defined above are sectorial.

Proof — For $-\pi/2 \leq \theta \leq \pi/2$, consider the operator in $n+1$ variables

$$\mathcal{A}_\theta = \mathcal{A}(x, D) + e^{i\theta} D_{tt}, \quad x \in \overline{\Omega},\ t \in \mathbb{R},$$

which satisfies the ellipticity condition (3.1.5) with constant $\mu = (\nu \wedge 1)/\sqrt{2}$. If $n \geq 2$, then $n+1 \geq 3$, so that $\mathcal{A}_\theta$ satisfies the roots condition (3.1.6). If $n = 1$, one checks that (3.1.6) holds by computing explicitly the roots of $P(\tau)$.

Let $\zeta \in C^\infty(\mathbb{R})$ be such that $\zeta(t) \equiv 1$ for $|t| \leq 1/2$, $\zeta(t) \equiv 0$ for $|t| \geq 1$. For every $u \in W^{2,p}(\Omega)$ and $r > 0$ set

$$v(x, t) = \zeta(t)e^{irt}u(x), \quad x \in \Omega,\ t \in \mathbb{R}.$$

Then

$$\mathcal{A}_\theta v = \zeta(t)e^{irt}(\mathcal{A}(\cdot, D) - r^2 e^{i\theta})u + e^{i(\theta + rt)}(\zeta''(t) + 2ir\zeta'(t))u.$$

Now we can prove statements (i), (ii), (iii).

(i) Estimate (3.1.7), applied to the function v, gives

$$\begin{aligned}
\|v\|_{W^{2,p}(\mathbb{R}^{n+1})} &\leq c_p(\|v\|_{L^p(\mathbb{R}^{n+1})} + \|\mathcal{A}_\theta v\|_{L^p(\mathbb{R}^{n+1})}) \\
&\leq c_p\Big[\|u\|_{L^p(\mathbb{R}^n)}(\|\zeta\|_{L^p(\mathbb{R})} + 2r\|\zeta'\|_{L^p(\mathbb{R})} + \|\zeta''\|_{L^p(\mathbb{R})}) \\
&\quad + \|(\mathcal{A}(\cdot, D) - r^2 e^{i\theta})u\|_{L^p(\mathbb{R}^n)}\Big] \\
&\leq c_p'\Big(\|u\|_{L^p(\mathbb{R}^n)}(1 + r) + \|(\mathcal{A}(\cdot, D) - r^2 e^{i\theta})u\|_{L^p(\mathbb{R}^n)}\Big),
\end{aligned} \qquad (3.1.14)$$

with

$$c_p' = c_p \max\{\|\zeta\|_{L^p(\mathbb{R})}, 2\|\zeta'\|_{L^p(\mathbb{R})}\}.$$

On the other hand, since $\zeta \equiv 1$ in $[-1/2, 1/2]$, then

$$\begin{aligned}
\|v\|_{W^{2,p}(\mathbb{R}^n \times]-1/2, 1/2[)}^p &= \int_{\mathbb{R}^n \times]-1/2, 1/2[} \sum_{|\alpha| \leq 2} |D^\alpha(ue^{irt})|^p dx\,dt \\
&= \int_{\mathbb{R}^n} \Big[(1 + r^p + r^{2p})|u|^p + (1 + 2r^p)\sum_{j=1}^n |D_j u|^p + \sum_{j,k=1}^n |D_{jk}u|^p\Big] dx \\
&\geq r^{2p}\|u\|_{L^p(\mathbb{R}^n)}^p + r^p\|Du\|_{L^p(\mathbb{R}^n)}^p + \|D^2 u\|_{L^p(\mathbb{R}^n)}^p.
\end{aligned}$$

1. Second order operators 75

Taking into account (3.1.14), it follows

$$\begin{aligned}r^2\|u\|_{L^p(\mathbb{R}^n)} &+ r\|Du\|_{L^p(\mathbb{R}^n)} + \|D^2 u\|_{L^p(\mathbb{R}^n)}\\ &\leq 3\|v\|_{W^{2,p}(\mathbb{R}^n\times]-1/2,1/2[)} \leq 3\|v\|_{W^{2,p}(\mathbb{R}^{n+1})} \\ &\leq 3c'_p\left(\|u\|_{L^p(\mathbb{R}^n)}(1+r) + \|(\mathcal{A}(\cdot,D) - r^2 e^{i\theta})u\|_{L^p(\mathbb{R}^n)}\right).\end{aligned} \quad (3.1.15)$$

Taking $\lambda = r^{1/2}e^{i\theta}$, with r so large that $3c'_p(1+r) \leq r^2/2$, statement (i) follows, with $M_p = 6c'_p$.

(ii) The procedure is the same as above, using estimate (3.1.8) instead of (3.1.7) in $\Omega \times \mathbb{R}$. Let g_0 be any regular extension to Ω of the trace $u_{|\partial\Omega}$. Then (3.1.14) has to be replaced by

$$\begin{aligned}\|v\|_{W^{2,p}(\Omega\times\mathbb{R})} &\leq c_p\left(\|v\|_{L^p(\Omega\times\mathbb{R})} + \|\mathcal{A}_\theta v\|_{L^p(\Omega\times\mathbb{R})} + \|g_0 \zeta e^{irt}\|_{W^{2,p}(\Omega\times\mathbb{R})}\right)\\ &\leq c''_p\big((r+1)\|u\|_{L^p(\Omega)} + \|(\mathcal{A}(\cdot,D) - r^2 e^{i\theta})u\|_{L^p(\Omega)}\\ &+ (r^2 + r + 1)\|g_0\|_{L^p(\Omega)} + (r+1)\|Dg_0\|_{L^p(\Omega)} + \|D^2 g_0\|_{L^p(\Omega)}\big),\end{aligned} \quad (3.1.16)$$

Accordingly, (3.1.15) has to be replaced by

$$\begin{aligned}r^2\|u\|_{L^p(\Omega)} &+ r\|Du\|_{L^p(\Omega)} + \|D^2 u\|_{L^p(\Omega)} \leq 3\|v\|_{W^{2,p}(\Omega\times]-1/2,1/2[)}\\ &\leq 3\|v\|_{W^{2,p}(\Omega\times\mathbb{R})} \leq 3c''_p\big(\|u\|_{L^p(\Omega)}(1+r) + \|(\mathcal{A}(\cdot,D) - r^2 e^{i\theta})u\|_{L^p(\Omega)}\\ &+ (r^2+r+1)\|g_0\|_{L^p(\Omega)} + (r+1)\|Dg_0\|_{L^p(\Omega)} + \|D^2 g_0\|_{L^p(\Omega)}\big).\end{aligned} \quad (3.1.17)$$

As before, taking $\lambda = r^{1/2}e^{i\theta}$, with r so large that $3c''_p(1+r) \leq r^2/2$, statement (ii) follows.

(iii) Again, the procedure is the same as above, using now estimate (3.1.9) in $\Omega \times \mathbb{R}$. Let g_1 be any regular extension to Ω of the trace $(\mathcal{B}(\cdot,D)u)_{|\partial\Omega}$. Then (3.1.16) has to be replaced by

$$\begin{aligned}\|v\|_{W^{2,p}(\Omega\times\mathbb{R})} &\leq c_p\big(\|v\|_{L^p(\Omega\times\mathbb{R})} + \|\mathcal{A}_\theta v\|_{L^p(\Omega\times\mathbb{R})} + \|g_1 \zeta e^{irt}\|_{W^{1,p}(\Omega\times\mathbb{R})}\big)\\ &\leq c'''_p\big((1+r)\|u\|_{L^p(\Omega)} + \|(\mathcal{A}(\cdot,D) - r^2 e^{i\theta})u\|_{L^p(\Omega)}\\ &+ (r+1)\|g_1\|_{L^p(\Omega)} + \|Dg_1\|_{L^p(\Omega)}\big),\end{aligned} \quad (3.1.18)$$

and, accordingly, (3.1.17) has to be replaced by

$$\begin{aligned}r^2\|u\|_{L^p(\Omega)} &+ r\|Du\|_{L^p(\Omega)} + \|D^2 u\|_{L^p(\Omega)} \leq 3\|v\|_{W^{2,p}(\Omega\times]-1/2,1/2[)}\\ &\leq 3\|v\|_{W^{2,p}(\Omega\times\mathbb{R})} \leq 3c'''_p\big((r+1)\|u\|_{L^p(\Omega)} + \|(\mathcal{A}(\cdot,D) - r^2 e^{i\theta})u\|_{L^p(\Omega)}\\ &+ (r+1)\|g_1\|_{L^p(\Omega)} + \|Dg_1\|_{L^p(\Omega)}\big).\end{aligned} \quad (3.1.19)$$

Statement (iii) follows as before, taking $\lambda = r^{1/2}e^{i\theta}$, with r so large that $3c'''_p(r+1) \leq r^2/2$. ∎

3.1.2 Generation in $L^\infty(\mathbb{R}^n)$ and in spaces of continuous functions in $\mathbb{R}^n$

Throughout the subsection, the ess sup norm is denoted by $\|\cdot\|_\infty$. If a function u is differentiable, we set $\|Du\|_\infty = \sum_{i=1}^n \|D_i u\|_\infty$; if u is twice differentiable, we set $\|D^2 u\|_\infty = \sum_{i,j=1}^n \|D_{ij} u\|_\infty$. Moreover, for each $k \in \mathbb{N}$ and $p \in [1, \infty]$ we denote by $W_{loc}^{k,p}(\mathbb{R}^n)$ the space of the functions $u : \mathbb{R}^n \mapsto \mathbb{C}$ such that for every $r > 0$ the restriction $u_{|B(0,r)}$ belongs to $W^{k,p}(B(0,r))$.

In the present and in the next subsection we shall use the well known Sobolev embeddings, stated in the following lemma.

Lemma 3.1.4 *Let Ω be either $\mathbb{R}^n$, or an open set in $\mathbb{R}^n$ with uniformly C^1 boundary. Let $p > n$ and set $\alpha = 1 - n/p$. Then $W^{1,p}(\Omega) \subset C^\alpha(\overline{\Omega})$. Moreover, there exists $C > 0$ such that for every $\varphi \in W_{loc}^{1,p}(\Omega)$ and for every $x_0 \in \Omega$ we have*

$$\begin{aligned}(i) \quad & \|\varphi\|_{L^\infty(\Omega \cap B(x_0,r))} \leq C r^{-n/p} \left(\|\varphi\|_{L^p(\Omega \cap B(x_0,r))} + r \|D\varphi\|_{L^p(\Omega \cap B(x_0,r))} \right), \\ (ii) \quad & [\varphi]_{C^\alpha(\Omega \cap B(x_0,r))} \leq C \|D\varphi\|_{L^p(\Omega \cap B(x_0,r))}.\end{aligned}$$
(3.1.20)

The proof of the resolvent estimate (and, consequently, of the generation theorem) in $L^\infty(\mathbb{R}^n)$ relies on the first part of Lemma 3.1.4 and on a Caccioppoli type inequality in the L^p norm, which we prove in the next lemma.

Lemma 3.1.5 *Let $p > 1$, and let $u \in W_{loc}^{2,p}(\mathbb{R}^n)$. For $\operatorname{Re} \lambda \geq \omega_p$ (ω_p is given in Theorem 3.1.3(i)), set $f = \lambda u - \mathcal{A}u$. Then there is $C_p > 0$, not depending on u and λ, such that for every $x_0 \subset \mathbb{R}^n$, $r \leq 1$, $\alpha \geq 1$,*

$$\begin{aligned}&|\lambda| \|u\|_{L^p(B(x_0,r))} + |\lambda|^{1/2} \|Du\|_{L^p(B(x_0,r))} + \|D^2 u\|_{L^p(B(x_0,r))} \\ &\leq C_p \big(\|f\|_{L^p(B(x_0,(\alpha+1)r))} \\ &\quad + \tfrac{1}{\alpha}(r^{-2} \|u\|_{L^p(B(x_0,(\alpha+1)r))} + r^{-1} \|Du\|_{L^p(B(x_0,(\alpha+1)r))})\big).\end{aligned}$$
(3.1.21)

Proof — Let $\theta_0 : \mathbb{R}^n \mapsto \mathbb{R}$ be any smooth function such that $\theta_0 \equiv 1$ in $B(0,r)$, $\theta_0 \equiv 0$ outside $B(0,(\alpha+1)r)$, and

$$\|\theta_0\|_\infty + \alpha r \|D\theta_0\|_\infty + \alpha^2 r^2 \|D^2 \theta_0\|_\infty \leq K,$$

with K independent on α and r. Fixed any $x_0 \in \mathbb{R}^n$, set $\theta(x) = \theta_0(x - x_0)$, and set

$$v(x) = \theta(x) u(x), \quad x \in \mathbb{R}^n.$$

Since u and v coincide in $B(x_0, r)$, we have of course

$$\begin{aligned}&|\lambda| \|u\|_{L^p(B(x_0,r))} + |\lambda|^{1/2} \|Du\|_{L^p(B(x_0,r))} + \|D^2 u\|_{L^p(B(x_0,r))} \\ &\leq |\lambda| \|v\|_{L^p(\mathbb{R}^n)} + |\lambda|^{1/2} \|Dv\|_{L^p(\mathbb{R}^n)} + \|D^2 v\|_{L^p(\mathbb{R}^n)}.\end{aligned}$$

1. Second order operators

The function v satisfies

$$\lambda v - \mathcal{A}(\cdot, D)v = \theta f - u \sum_{i,j=1}^n a_{ij} D_{ij}\theta - 2\sum_{i,j=1}^n a_{ij} D_i u D_j \theta - u \sum_{i=1}^n b_i D_i \theta = F. \quad (3.1.22)$$

Since Re $\lambda \geq \omega_p$, we may use (3.1.11), getting

$$|\lambda| \|v\|_{L^p(\mathbb{R}^n)} + |\lambda|^{1/2} \|Dv\|_{L^p(\mathbb{R}^n)} + \|D^2 v\|_{L^p(\mathbb{R}^n)} \leq M_p \|F\|_{L^p(\mathbb{R}^n)}. \quad (3.1.23)$$

Set $C_1 = \max_{i,j} \|a_{ij}\|_\infty + \max_i \|b_i\|_\infty$. Then

$$\|F\|_{L^p(\mathbb{R}^n)} \leq \|f\|_{L^p(B(x_0,(\alpha+1)r))} + C_1 K \left(\frac{\|u\|_{L^p(B(x_0,(\alpha+1)r))}}{\alpha^2 r^2} \right.$$
$$\left. + \frac{2\|Du\|_{L^p(B(x_0,(\alpha+1)r))}}{\alpha r} + \frac{\|u\|_{L^p(B(x_0,(\alpha+1)r))}}{\alpha r} \right). \quad (3.1.24)$$

Replacing in (3.1.23) and recalling that $r \leq 1$, $\alpha \geq 1$, the statement follows, with $C_p - 2C_1 K M_p$. ∎

Now we are ready to prove resolvent estimates in the sup norm. We recall that in order that the realization A of $\mathcal{A}(\cdot, \mathcal{D})$ in $L^\infty(\mathbb{R}^n)$ be a sectorial operator, one needs an estimate like $|\lambda| \|u\|_\infty \leq C \|\lambda u - Au\|_\infty$, for Re λ large. But in the following — precisely, to prove that $R(\lambda, A)$ exists for Re λ large — we will need more refined estimates, such as the next (3.1.25) and (3.1.26).

Theorem 3.1.6 *Fix $p > n$. Then there is $K_p > 0$ such that for every $\lambda \in \mathbb{C}$ with Re $\lambda \geq \Lambda_p = \omega_p \vee 1$ and for every $u \in C^1(\mathbb{R}^n) \cap W^{2,p}_{loc}(\mathbb{R}^n)$ it holds*

$$|\lambda| \|u\|_\infty + |\lambda|^{1/2} \|Du\|_\infty + |\lambda|^{n/2p} \sup_{x_0 \in \mathbb{R}^n} \|D^2 u\|_{L^p(B(x_0,|\lambda|^{-1/2}))}$$
$$\leq K_p |\lambda|^{n/2p} \sup_{x_0 \in \mathbb{R}^n} \|\lambda u - \mathcal{A}u\|_{L^p(B(x_0,|\lambda|^{-1/2}))}. \quad (3.1.25)$$

It follows that if $\mathcal{A}u \in L^\infty(\mathbb{R}^n)$, then

$$|\lambda| \|u\|_\infty + |\lambda|^{1/2} \|Du\|_\infty + |\lambda|^{n/2p} \sup_{x_0 \in \mathbb{R}^n} \|D^2 u\|_{L^p(B(x_0,|\lambda|^{-1/2}))}$$
$$\leq K_p \gamma_n^{1/p} \|\lambda u - \mathcal{A}(\cdot, D)u\|_\infty, \quad (3.1.26)$$

where γ_n is the measure of the unit ball in $\mathbb{R}^n$.

Proof — To get (3.1.25), we shall estimate

$$\mathcal{Q} = |\lambda| \|u\|_{L^\infty(B(x_0,r))} + |\lambda|^{1/2} \|Du\|_{L^\infty(B(x_0,r))} + |\lambda|^{n/2p} \|D^2 u\|_{L^p(B(x_0,r))},$$

where $x_0 \in \mathbb{R}^n$ is arbitrary, $|\lambda| \geq 1$, Re $\lambda \geq \omega_p$, and $r = |\lambda|^{-1/2}$. Using the Sobolev inequality (3.1.20)(i) to estimate the first two addenda, we get

$$\mathcal{Q} \leq (2C+1)|\lambda|^{n/2p} \big[|\lambda| \|u\|_{L^p(B(x_0,r))} + |\lambda|^{1/2} \|Du\|_{L^p(B(x_0,r))}$$
$$+ \|D^2 u\|_{L^p(B(x_0,r))} \big] \quad (3.1.27)$$

To estimate the right hand side of (3.1.27), we use Lemma 3.1.5, getting, for every $\alpha > 0$,

$$\mathcal{Q} \leq C_p(2C+1)|\lambda|^{n/2p}\left(\|f\|_{L^p(B_\alpha)} + \frac{|\lambda|}{\alpha}\|u\|_{L^p(B_\alpha)} + \frac{|\lambda|^{1/2}}{\alpha}\|Du\|_{L^p(B_\alpha)}\right)$$

$$\leq C_p(2C+1)\left(|\lambda|^{n/2p}\|f\|_{L^p(B_\alpha)} + \gamma_n^{1/p}\frac{(\alpha+1)^{n/p}}{\alpha}(|\lambda|\,\|u\|_\infty + |\lambda|^{1/2}\|Du\|_\infty)\right),$$

where $B_\alpha = B(x_0, (\alpha+1)|\lambda|^{-1/2})$. Taking the supremum over $x_0 \in \mathbb{R}^n$ of the three addenda in $\mathcal{Q}$ and summing up we get

$$|\lambda|\,\|u\|_\infty + |\lambda|^{1/2}\|Du\|_\infty + |\lambda|^{n/2p}\sup_{x_0 \in \mathbb{R}^n}\|D^2u\|_{L^p(B(x_0,|\lambda|^{-1/2}))}$$

$$\leq 3(2C+1)C_p\Big(|\lambda|^{n/2p}\sup_{x_0 \in \mathbb{R}^n}\|f\|_{L^p(B(x_0,(\alpha+1)|\lambda|^{-1/2}))}$$

$$+\gamma_n^{1/p}\frac{(\alpha+1)^{n/p}}{\alpha}(|\lambda|\,\|u\|_\infty + |\lambda|^{1/2}\|Du\|_\infty)\Big).$$

Taking α sufficiently large, in such a way that

$$3(2C+1)C_p\gamma_n^{1/p}\frac{(\alpha+1)^{n/p}}{\alpha} \leq 1/2,$$

one obtains

$$|\lambda|\,\|u\|_\infty + |\lambda|^{1/2}\|Du\|_\infty + |\lambda|^{n/2p}\sup_{x_0 \in \mathbb{R}^n}\|D^2u\|_{L^p(B(x_0,|\lambda|^{-1/2}))}$$

$$\leq 6(2C+1)C_p\sup_{x_0 \in \mathbb{R}^n}|\lambda|^{n/2p}\|f\|_{L^p(B(x_0,(\alpha+1)|\lambda|^{-1/2}))},$$

and the statement follows, recalling that every ball with radius $(\alpha+1)|\lambda|^{-1/2}$ may be covered by a finite number (not depending on λ) of balls with radius $|\lambda|^{-1/2}$. ∎

The *a priori* estimate provided by Theorem 3.1.6, together with the existence result of Theorem 3.1.2(i), is the basic tool to prove the generation theorem in $L^\infty(\mathbb{R}^n)$ and in several spaces of continuous functions in $\mathbb{R}^n$. We begin with the generation theorem in $L^\infty(\mathbb{R}^n)$.

Set

$$\begin{cases} D(A_\infty) = \{u \in \bigcap_{p\geq 1} W^{2,p}_{loc}(\mathbb{R}^n) \,:\, u,\,\mathcal{A}(\cdot,D)u \in L^\infty(\mathbb{R}^n)\}, \\ A_\infty u = \mathcal{A}(\cdot,D)u. \end{cases}$$

Theorem 3.1.7 *The operator $A_\infty : D(A_\infty) \mapsto L^\infty(\mathbb{R}^n)$ defined above is sectorial. Its resolvent set contains the halfplane $\{\lambda \in \mathbb{C} : \operatorname{Re}\lambda > \Lambda_0\}$, where*

$$\Lambda_0 = \inf_{p>n}\Lambda_p,$$

and Λ_p is given by Theorem 3.1.6. Moreover, $\overline{D(A_\infty)} = UC(\mathbb{R}^n)$, and $D(A_\infty)$ is continuously embedded in $C^{1+\alpha}(\mathbb{R}^n)$ for every $\alpha \in\,]0,1[$.

1. Second order operators

Proof — Fix $p > n$. We are going to prove that $\rho(A_\infty)$ contains the halfplane $\operatorname{Re} \lambda \geq \Lambda_p\}$.

For any $f \in L^\infty(\mathbb{R}^n)$ and $k \in \mathbb{N}$, set $f_k = \theta_k f$, where θ_k is any smooth cutoff function such that

$$0 \leq \theta_k \leq 1, \quad \theta_k \equiv 1 \text{ in } B(0,k), \quad \theta_k \equiv 0 \text{ outside } B(0,2k).$$

Let $\operatorname{Re} \lambda \geq \Lambda_p$. By Theorems 3.1.2(i) and 3.1.3(i), the problem

$$\lambda u_k - \mathcal{A}(\cdot, D)u_k = f_k \tag{3.1.28}$$

has a unique solution $u_k \in W^{2,p}(\mathbb{R}^n)$, and $\|u_k\|_{W^{2,p}(\mathbb{R}^n)} \leq M_p \|f_k\|_{L^p(\mathbb{R}^n)}$. Thanks to the Sobolev embedding lemma 3.1.4, u_k belongs to $C^1(\mathbb{R}^n)$, so that we may apply estimate (3.1.26), which gives

$$\|u_k\|_{C^1(\mathbb{R}^n)} + \sup_{x \in \mathbb{R}^n} \|D^2 u_k\|_{L^p(B(x,|\lambda|^{-1/2}))} \leq C(\lambda)\|f_k\|_\infty \leq C(\lambda)\|f\|_\infty. \tag{3.1.29}$$

Therefore, the sequence $\{u_k\}_{k \in \mathbb{N}}$ is bounded in $C^1(\mathbb{R}^n)$, so that there exists a subsequence converging uniformly on each compact subset to a function $u \in C(\mathbb{R}^n) \cap Lip(\mathbb{R}^n)$, which satisfies

$$\|u\|_\infty + [u]_{Lip(\mathbb{R}^n)} \leq C(\lambda)\|f\|_\infty. \tag{3.1.30}$$

We are going to show that $u \in W^{2,p}_{loc}(\mathbb{R}^n)$, and that $\lambda u - \mathcal{A}u = f$ in $\mathbb{R}^n$.

Fix any closed ball $B(0,R)$, with $R \geq 4|\lambda|^{-1/2}$. Estimate (3.1.29) implies that the sequence $\{u_k\}$ is bounded in $W^{2,p}(B(0,R))$, so that the limit function u belongs to $W^{2,p}(B(0,R))$. Since R is arbitrary, then $u \in W^{2,p}_{loc}(\mathbb{R}^n)$. Moreover, a subsequence $\{u_{\varphi(k)}\}_{k \in \mathbb{N}}$ converges to u in $W^{1,p}(B(0,R))$, and for h, k sufficiently large we have

$$\lambda(u_{\varphi(h)} - u_{\varphi(k)}) - \mathcal{A}(\cdot, D)(u_{\varphi(h)} - u_{\varphi(k)}) = 0 \text{ in } B(0,R).$$

Lemma 3.1.5 implies that for every $v \in W^{2,p}_{loc}(\mathbb{R}^n)$ and $x_0 \in \mathbb{R}^n$

$$\begin{aligned}\|v\|_{W^{2,p}(B(x_0,|\lambda|^{-1/2}))} &\leq C_1(\lambda)\big(\|\lambda v - \mathcal{A}v\|_{L^p(B(x_0,2|\lambda|^{-1/2}))} \\ &+ \|v\|_{W^{1,p}(B(x_0,2|\lambda|^{-1/2}))}\big).\end{aligned}$$

Applying the above estimate to the function $v = u_{\varphi(h)} - u_{\varphi(k)}$, with any $x_0 \in B(0, R/2)$ (so that $B(x_0, |\lambda|^{-1/2}) \subset B(0,R)$), we get

$$\|u_{\varphi(h)} - u_{\varphi(k)}\|_{W^{2,p}(B(x_0,|\lambda|^{-1/2}))} \leq C_1(\lambda)\|u_{\varphi(h)} - u_{\varphi(k)}\|_{W^{1,p}(B(x_0,2|\lambda|^{-1/2}))}$$
$$\leq C_1(\lambda)\|u_{\varphi(h)} - u_{\varphi(k)}\|_{W^{1,p}(B(0,R))} \to 0 \text{ as } h, k \to \infty.$$

Covering $B(0, R/2)$ by a finite number of balls with radius $|\lambda|^{-1/2}$ we see that $\{u_{\varphi(k)}\}_{k \in \mathbb{N}}$ converges in $W^{2,p}(B(0, R/2))$, so that, letting $k \to \infty$ in (3.1.28), we get $\lambda u - \mathcal{A}u = f$ in $B(0, R/2)$. Since R is arbitrary, then

$$\lambda u - \mathcal{A}u = f \text{ in } \mathbb{R}^n.$$

Fixed any $q > n$, equation (3.1.28) is equivalent to

$$\Lambda_q u_k - \mathcal{A} u_k = (\Lambda_q - \lambda) u_k + f_k,$$

where the right hand side belongs to $L^\infty(\mathbb{R}^n)$, and its sup norm is bounded by a constant independent of k. The above arguments prove that $u \in W^{2,q}_{loc}(\mathbb{R}^n)$. Since q is arbitrary, $u \in D(A_\infty)$. Therefore, the resolvent set of A_∞ contains $\{\operatorname{Re} \lambda > \Lambda_p\}$, for every $p > n$. This proves the first part of the statement. Now, from estimate (3.1.26) and Proposition 2.1.11 it follows that A_∞ is sectorial.

Let us prove that $D(A_\infty) \subset C^{1+\alpha}(\mathbb{R}^n)$ for every $\alpha \in]0,1[$. Due to the Sobolev embedding, every $u \in D(A_\infty)$ is continuously differentiable. To show that its gradient is bounded, fix any $p > n$ and set $f = \Lambda_p u - \mathcal{A} u$. Estimate (3.1.30) implies that

$$\|Du\|_\infty \leq const. \, (\|u\|_\infty + \|\mathcal{A} u\|_\infty).$$

Moreover, choosing $p = n/(1-\alpha)$ and using the Sobolev inequality (3.1.20)(ii) and estimate (3.1.26) with $\lambda = \Lambda_p$, one finds, for $i = 1, \ldots, n$,

$$|D_i u(x) - D_i u(y)| \leq const. \, |x-y|^\alpha (\|u\|_\infty + \|\mathcal{A} u\|_\infty),$$

for all $x, y \in \mathbb{R}^n$ such that $|x-y| \leq \Lambda_p^{-1/2}$. On the other hand, if $|x-y| \geq \Lambda_p^{-1/2}$, then

$$\frac{|D_i u(x) - D_i u(y)|}{|x-y|^\alpha} \leq 2 \|D_i u\|_\infty \Lambda_p^{\alpha/2} \leq const. \, (\|u\|_\infty + \|\mathcal{A} u\|_\infty).$$

Therefore, $D(A_\infty) \subset C^{1+\alpha}(\mathbb{R}^n)$.

Let us prove that $\overline{D(A_\infty)} = UC(\mathbb{R}^n)$. If a function u belongs to the closure of $D(A_\infty)$, then it is the uniform limit of a sequence of uniformly continuous functions (because $D(A) \subset UC(\mathbb{R}^n)$), so that it is uniformly continuous. Conversely, any uniformly continuous function is the uniform limit of a sequence of functions belonging to $C^2(\mathbb{R}^n)$ (for instance, the standard mollifiers). Therefore, it belongs to $\overline{D(A_\infty)}$. ∎

Remark 3.1.8 In the proof of Proposition 3.1.7 we have proved a more refined result than the one stated: if $\operatorname{Re} \lambda \geq \Lambda_p$, then for every $f \in L^p_{loc}(\mathbb{R}^n)$ such that $\sup_{x \in \mathbb{R}^n} \|f\|_{L^p(B(x,1))} < \infty$, the problem $\lambda u - \mathcal{A} u = f$ has a unique solution $u \in W^{2,p}_{loc}(\mathbb{R}^n) \cap C^1(\mathbb{R}^n)$. Indeed, estimate (3.1.29) may be replaced, using (3.1.25) instead of (3.1.26), by

$$\|u_k\|_{C^1(\mathbb{R}^n)} + \sup_{x \in \mathbb{R}^n} \|D^2 u_k\|_{L^p(B(x,|\lambda|^{-1/2}))}$$
$$\leq C(\lambda) \sup_{x \in \mathbb{R}^n} \|f_k\|_{L^p(B(x,|\lambda|^{-1/2}))} \leq C(\lambda) \sup_{x \in \mathbb{R}^n} \|f\|_{L^p(B(x,|\lambda|^{-1/2}))}.$$

The rest of the proof follows as above.

This existence result will be used later, to get a generation result in $C^1(\mathbb{R}^n)$.

1. SECOND ORDER OPERATORS

From Theorems 3.1.6 and 3.1.7 several generation results in spaces of continuous functions follow easily.

Corollary 3.1.9 *Let Λ_0 be defined as in Theorem 3.1.7.*
(i) Set

$$\begin{cases} D(A) = \{u \in \bigcap_{p \geq 1} W^{2,p}_{loc}(\mathbb{R}^n) : u, \mathcal{A}(\cdot, D)u \in C(\mathbb{R}^n)\}, \\ A : D(A) \mapsto C(\mathbb{R}^n), \quad Au = \mathcal{A}(\cdot, D)u. \end{cases} \qquad (3.1.31)$$

Then the resolvent set of A contains the halfplane $\{\lambda \in \mathbb{C} : \operatorname{Re} \lambda > \Lambda_0\}$, and A is sectorial. Moreover, $D(A) \subset C^{1+\alpha}(\mathbb{R}^n)$ for every $\alpha \in {]}0,1{[}$, and $\overline{D(A)} = UC(\mathbb{R}^n)$. If $n = 1$, then $D(A) = C^2(\mathbb{R})$.
(ii) Set

$$\begin{cases} D(A_{UC}) = \{u \in \bigcap_{p \geq 1} W^{2,p}_{loc}(\mathbb{R}^n) : u, \mathcal{A}(\cdot, D)u \in UC(\mathbb{R}^n)\}, \\ A_{UC} : D(A) \mapsto UC(\mathbb{R}^n), \quad A_{UC}u = \mathcal{A}(\cdot, D)u. \end{cases} \qquad (3.1.32)$$

Then $\rho(A_{UC})$ contains $\{\lambda \in \mathbb{C} : \operatorname{Re} \lambda > \Lambda_0\}$, $D(A_{UC})$ is dense in $UC(\mathbb{R}^n)$, and A_{UC} is sectorial. Moreover, $UC^2(\mathbb{R}^n)$ (and hence $UC^k(\mathbb{R}^n)$ for every k) is dense in $D(A_{UC})$. If $n = 1$, then $D(A_{UC}) = UC^2(\mathbb{R})$.
(iii) Set $C_0(\mathbb{R}^n) = \{u \in C(\mathbb{R}^n) : \lim_{|x| \to \infty} u(x) = 0\}$, and

$$\begin{cases} D(A_0) = \{u \in \bigcap_{p \geq 1} W^{2,p}_{loc}(\mathbb{R}^n) : u, \mathcal{A}(\cdot, D)u \in C_0(\mathbb{R}^n)\}, \\ A_0 : D(A_0) \mapsto C_0(\mathbb{R}^n), \quad A_0 u = \mathcal{A}(\cdot, D)u. \end{cases} \qquad (3.1.33)$$

Then $\rho(A_0)$ contains $\{\lambda \in \mathbb{C} : \operatorname{Re} \lambda > \Lambda_0\}$, $D(A_0)$ is dense in $C_0(\mathbb{R}^n)$, and A_0 is sectorial. If $n = 1$, then $D(A_0) = \{u \in C^2(\mathbb{R}) : u, \mathcal{A}(\cdot, D)u \in C_0(\mathbb{R})\}$.

Proof — (i) Since $D(A_\infty) \subset C(\mathbb{R}^n)$, then $\rho(A_\infty) \subset \rho(A)$. Therefore, $\rho(A)$ contains the halfplane $\{\operatorname{Re} \lambda > \Lambda_0\}$. Estimate (3.1.26) and Proposition 2.1.11 prove that A is sectorial. The embedding $D(A) \subset D(A_\infty)$ yields $D(A) \subset C^{1+\alpha}(\mathbb{R}^n)$. The equality $\overline{D(A)} = UC(\mathbb{R}^n)$ follows from the argument used in the proof of Theorem 3.1.7. The last statement is obvious.

The proof of statement (ii) (with the exception of the density of $UC^2(\mathbb{R}^n)$ in $D(A)$) is similar to the one of statement (i); it is sufficient to remark that $D(A_\infty)$ is continuously embedded in $UC(\mathbb{R}^n)$.

Let us prove that $UC^2(\mathbb{R}^n)$ is dense in $D(A_{UC})$. Let $u \in D(A_{UC})$. Let $\lambda > \Lambda_0$ and set

$$f = \lambda u - \mathcal{A}u.$$

Since a_{ij}, b_i, c, f are uniformly continuous and bounded, they may be approximated in the sup norm by sequences of smooth functions with bounded derivatives

a_{ij}^n, b_i^n, etc., with modulus of continuity independent of n (for instance, the usual mollifiers enjoy such a property). For every $n \in \mathbb{N}$ consider the problem

$$\lambda u_n - \mathcal{A}_n u_n = f_n,$$

where $\mathcal{A}_n = \sum_{i,j=1}^n a_{ij}^n D_{ij} + \sum_{i=1}^n b_i^n D_i + c$. Since Λ_0 does not depend on n, for n large enough the above problem is uniquely solvable in $C(\mathbb{R}^n)$, and the solution u_n belongs to $UC^2(\mathbb{R}^n)$ (indeed, it belongs to $C^{2+\alpha}(\mathbb{R}^n)$ for every $\alpha \in]0,1[$, thanks to next Corollary 3.1.16).

The difference $u - u_n$ satisfies

$$\lambda(u - u_n) - \mathcal{A}(u - u_n) = (\mathcal{A} - \mathcal{A}_n)u_n + f - f_n.$$

Fix any $p > 1$. Thanks to estimate (3.1.25) we have

$$\|u - u_n\|_\infty + \|\mathcal{A}(u - u_n)\|_\infty$$
$$\leq K(\lambda, p) \sup_{x_0 \in \mathbb{R}^n} \|(\mathcal{A} - \mathcal{A}_n)u_n + f - f_n\|_{L^p(B(x_0, |\lambda|^{-1/2p}))}$$
$$\leq K_1 \big[\sup_{i,j}(\|a_{ij} - a_{ij}^n\|_\infty + \sup_i \|b_i - b_i^n\|_\infty + \|c - c^n\|_\infty) \cdot$$
$$\cdot \sup_{x_0 \in \mathbb{R}^n} \|u_n\|_{W^{2,p}(B(x_0, |\lambda|^{-1/2p}))} + \|f - f_n\|_\infty \big].$$

If we show that $\sup_{x_0 \in \mathbb{R}^n} \|u_n\|_{W^{2,p}(B(x_0, |\lambda|^{-1/2p}))}$ is bounded by a constant independent of n, then it follows that $u_n \to u$ in $D(\mathcal{A}_{UC})$. The Caccioppoli type Lemma 3.1.5 gives

$$\|u_n\|_{W^{2,p}(B(x_0, |\lambda|^{-1/2p}))} \leq K_2(\|u_n\|_{W^{1,p}(B(x_0, (\alpha+1)|\lambda|^{-1/2p}))}$$
$$+ \|f_n\|_{L^p(B(x_0, (\alpha+1)|\lambda|^{-1/2p}))}),$$

for every $x_0 \in \mathbb{R}^n$, and estimate (3.1.25) implies

$$\|u_n\|_{W^{1,p}(B(x_0, (\alpha+1)|\lambda|^{-1/2p}))} \leq K_3 \|f_n\|_\infty,$$

where K_2, K_3 depend on n through the ellipticity constant of $\mathcal{A}_n$, the sup norm and the modulus of continuity of the coefficients of $\mathcal{A}_n$. Therefore there is $K > 0$ such that

$$\sup_{x_0 \in \mathbb{R}^n} \|u_n\|_{W^{2,p}(B(x_0, |\lambda|^{-1/2p}))} \leq K, \quad \forall n \geq n_0,$$

and statement (ii) follows.

Let us prove that statement (iii) holds. Fix $f \in C_0(\mathbb{R}^n)$. For every $k \in \mathbb{N}$, let θ_k be a smooth cutoff function such that

$$\theta_k \equiv 1 \text{ in } B(0,k), ;\ \theta_k \equiv 0 \text{ outside } B(0, k+1), \ 0 \leq \theta_k(x) \leq 1 \ \forall x.$$

Then $f_k = \theta_k f$ converges to f in $C_0(\mathbb{R}^n)$ as $k \to \infty$. Fix any $p > n$, and for $\operatorname{Re} \lambda > \Lambda_p$ consider the problem

$$\lambda u_k - \mathcal{A} u_k = f_k.$$

1. Second order operators

Since $f_k \in L^p(\mathbb{R}^n)$, by Theorem 3.1.2(i) the above problem has a unique solution $u_k \in W^{2,p}(\mathbb{R}^n)$, and $\lim_{|x|\to\infty} u_k(x) = 0$. Moreover, since $\Lambda_p \geq \Lambda_0$, the problem

$$\lambda u - \mathcal{A}u = f$$

has a unique solution $u \in D(A_\infty)$, thanks to Theorem 3.1.7. By estimate (3.1.26), the sequence $\{u_k\}_{k\in\mathbb{N}}$ converges to u in $L^\infty(\mathbb{R}^n)$, and consequently the sequence $\{\mathcal{A}u_k\}_{k\in\mathbb{N}}$ converges to $\mathcal{A}u$ in $L^\infty(\mathbb{R}^n)$. Since every u_k belongs to $D(A_0)$, then $u \in D(A_0)$. Since p is arbitrary, $\rho(A)$ contains $\{\operatorname{Re}\lambda > \Lambda_0\}$. The resolvent estimate (3.1.26) and Proposition 2.1.11 imply that A_0 is sectorial.

To prove that $D(A)$ is dense in $C_0(\mathbb{R}^n)$, it is sufficient to remark that if $f \in C_0(\mathbb{R}^n)$ then the usual mollifiers belong to $C_0(\mathbb{R}^n) \cap C^2(\mathbb{R}^n) \subset D(A)$, and converge to f uniformly. Therefore, $f \in \overline{D(A)}$. ∎

A useful property of the functions belonging to $D(A)$ is given by the following proposition. It will enable us to use the maximum principle in second order parabolic equations.

Proposition 3.1.10 *Assume that the coefficient c of the differential operator $\mathcal{A}$ vanishes. Let u be a real function defined in an open neighborhood $\mathcal{U}$ of $x_0 \in \mathbb{R}^n$. Assume that $u \in W^{2,p}(\Omega)$ for every $p \in [1, +\infty[$, that $\mathcal{A}u$ is continuous in $\mathcal{U}$, and that u has a relative maximum (respectively, minimum) value at x_0. Then $\mathcal{A}u(x_0) \leq 0$ (respectively, $\mathcal{A}u(x_0) \geq 0$).*

Proof — Let $r > 0$ be such that $u(x_0) \geq u(x)$ for $|x - x_0| \leq r$. Possibly replacing u by $u + c$ we may assume without loss of generality that $u(x) \geq 0$ for $|x - x_0| \leq r$. Let $\theta : \mathbb{R}^n \mapsto \mathbb{R}$ be a smooth function with support contained in $B(0,1)$, such that $0 \leq \theta(x) \leq 1$ for every x, and $\theta(0) > \theta(x)$ for every $x \neq 0$. Define

$$\tilde{u}(x) \begin{cases} = u(x)\theta((x-x_0)/r), & x \in \mathcal{U}, \\ = 0, & x \in \mathbb{R}^n \setminus \mathcal{U}. \end{cases}$$

Then $\tilde{u}(x_0)$ is the maximum value of $\tilde{u}$, and it is attained uniquely at $x = x_0$. Moreover, $\tilde{u}$ and $\mathcal{A}\tilde{u}$ are uniformly continuous and bounded, so that, by Corollary 3.1.9(ii), there is a sequence $\{\tilde{u}_n\}_{n\in\mathbb{N}} \subset C^2(\mathbb{R}^n)$ converging to $\tilde{u}$ in $D(A)$. Since x_0 is the unique point at which $\tilde{u}$ attains its maximum, there is a sequence x_n converging to x_0 such that for each n, u_n has a relative maximum point at x_n. Since u_n is twice continuously differentiable, then $\mathcal{A}u_n(x_n) \leq 0$. Letting $n \to \infty$, it follows $\mathcal{A}u(x_0) \leq 0$.

The proof of the statement about relative minimum values is similar. ∎

3.1.3 Characterization of interpolation spaces and generation results in Hölder spaces in $\mathbb{R}^n$

In this section the space X is $C(\mathbb{R}^n)$, and the operator A is defined by (3.1.31). It is clear however that next statements may be adapted as well to the cases $X = L^\infty(\mathbb{R}^n)$, $X = UC(\mathbb{R}^n)$, $X = C_0(\mathbb{R}^n)$.

We know already that $C^1(\mathbb{R}^n)$ (respectively, $C^{2\alpha}(\mathbb{R}^n)$ with $\alpha \neq 1/2$), is an intermediate space belonging to the class $J_{1/2}$ (respectively, J_α) between $C(\mathbb{R}^n)$ and $C^2(\mathbb{R}^n)$: see Proposition 1.1.2. Estimate (3.1.26) allows to prove that the same is true if $C^2(\mathbb{R}^n)$ is replaced by $D(A)$, even if $D(A)$ is not contained in $C^2(\mathbb{R}^n)$ for $n > 1$. Such a property is important not only in itself, but also because it allows to characterize the interpolation spaces $D_A(\theta, \infty)$, for $0 < \theta < 1$.

Proposition 3.1.11 *(i) The space $C^1(\mathbb{R}^n)$ belongs to the class $J_{1/2}$ between $C(\mathbb{R}^n)$ and $D(A)$.*
(ii) For $0 < \alpha < 1$, the space $C^{1+\alpha}(\mathbb{R}^n)$ belongs to the class $J_{(1+\alpha)/2}$ between $C(\mathbb{R}^n)$ and $D(A)$.

Proof — If $n = 1$, $D(A) = C^2(\mathbb{R})$, and the statement follows from Proposition 1.1.2(i). Let $n > 1$. Since $D(A) = D(A - \omega I)$ for every $\omega \in \mathbb{R}$, then a Banach space Y belongs to the class J_θ between X and $D(A)$ if and only if it belongs to the class J_θ between X and $D(A - \omega I)$. Therefore we may assume, without loss of generality, that $\sigma(A) \subset \{\lambda \in \mathbb{C} : \operatorname{Re} \lambda < 0\}$.

(i) Let $u \in D(A)$, and fix any $p > n$. Estimate (3.1.26) implies that for real $\lambda \geq \Lambda_p$ it holds

$$\sum_{i=1}^n \|D_i u\|_\infty \leq \frac{K_p \gamma_n^{1/p}}{\lambda^{1/2}} (\lambda \|u\|_\infty + \|Au\|_\infty).$$

Moreover, for every $i = 1, \ldots, n$, the function $\lambda \mapsto \lambda^{1/2} D_i R(\lambda, A)$ is continuous with values in $L(C(\mathbb{R}^n))$ in the interval $[0, \Lambda_p]$, which is contained in $\rho(A)$ by assumption. Therefore, it is bounded in such interval. It follows that there is $M > 0$ such that

$$\sum_{i=1}^n \|D_i u\|_\infty \leq \frac{M}{\lambda^{1/2}} (\lambda \|u\|_\infty + \|Au\|_\infty), \quad \lambda > 0.$$

Taking the minimum over $]0, +\infty[$, we get

$$\sum_{i=1}^n \|D_i u\|_\infty \leq 2M \|u\|_\infty^{1/2} \|Au\|_\infty^{1/2}, \quad u \in D(A),$$

so that

$$\|u\|_{C^1(\mathbb{R}^n)} = \|u\|_\infty + \sum_{i=1}^n \|D_i u\|_\infty \leq 2\sqrt{2} M \|u\|_\infty^{1/2} \|u\|_{D(A)}^{1/2}, \quad u \in D(A), \quad (3.1.34)$$

which proves statement (i).

1. Second order operators

(ii) Let $p = n/(1-\alpha)$. Estimate (3.1.26), together with the Sobolev inequality (3.1.20)(ii) implies that for every $u \in D(A)$, $x_0 \in \mathbb{R}^n$ and for every real $\lambda \geq \Lambda_p$ it holds
$$\sup_{x_0 \in \mathbb{R}^n} [D_i u]_{C^\alpha(B(x_0, |\lambda|^{-1/2}))} \leq \frac{C_1}{\lambda^{n/2p}} \|\lambda u - Au\|_\infty, \quad i = 1, \ldots, n,$$

with $C_1 = CK_p \gamma_n^{1/p}$. If $|x-y| \leq \lambda^{-1/2}$ we get
$$\frac{|D_i u(x) - D_i u(y)|}{|x-y|^\alpha} \leq \frac{C_1}{\lambda^{n/2p}} (\lambda \|u\|_\infty + \|Au\|_\infty).$$

On the other hand, if $|x-y| \geq \lambda^{-1/2}$ we have, due again to (3.1.26),
$$\frac{|D_i u(x) - D_i u(y)|}{|x-y|^\alpha} \leq \frac{2\|D_i u\|_\infty}{\lambda^{-1/2+n/2p}} \leq \frac{C_2}{\lambda^{n/2p}} (\lambda \|u\|_\infty + \|Au\|_\infty),$$

with $C_2 = 2K_p \gamma_n^{1/p}$. Therefore, for every $\lambda \geq \Lambda_p$ and for $i = 1, \ldots, n$ it holds
$$[D_i u]_{C^\alpha(\mathbb{R}^n)} \leq \max\{C_1, C_2\} \lambda^{-n/2p} (\lambda \|u\|_\infty + \|Au\|_\infty).$$

Moreover, since $D(A)$ is continuously embedded in $C^{1+\alpha}(\mathbb{R}^n)$, the function $\lambda \mapsto \lambda^{n/2p} R(\lambda, A)$ is continuous with values in $L(C(\mathbb{R}^n), C^{1+\alpha}(\mathbb{R}^n))$ in the interval $[0, \Lambda_p]$, which is contained in $\rho(A)$ by assumption. Therefore, it is bounded in such interval. It follows that there is $M > 0$ such that
$$\sum_{i=1}^n [D_i u]_{C^\alpha(\mathbb{R}^n)} \leq M \lambda^{-n/2p} (\lambda \|u\|_\infty + \|Au\|_\infty), \quad \lambda > 0.$$

Taking the minimum for $\lambda > 0$, we get
$$\sum_{i=1}^n [D_i u]_{C^\alpha(\mathbb{R}^n)} \leq c(\alpha) M \|u\|_\infty^{(1-\alpha)/2} \|Au\|_\infty^{(1+\alpha)/2}, \quad u \in D(A), \qquad (3.1.35)$$

so that
$$\begin{aligned} \|u\|_{C^{1+\alpha}(\mathbb{R}^n)} &= \|u\|_\infty + \sum_{i=1}^n \|D_i u\|_\infty + \sum_{i=1}^n [D_i u]_{C^\alpha} \\ &\leq \|u\|_\infty + C_3 \|u\|_\infty^{1/2} \|Au\|_\infty^{1/2} + C_4 \|u\|_\infty^{(1-\alpha)/2} \|Au\|_\infty^{(1+\alpha)/2} \\ &\leq C_5 \|u\|_\infty^{(1-\alpha)/2} \|u\|_{D(A)}^{(1+\alpha)/2}, \quad u \in D(A), \end{aligned}$$
$$\qquad (3.1.36)$$

and statement (ii) follows. ∎

We recall that $\mathcal{C}^1(\mathbb{R}^n)$ is the subspace of $C(\mathbb{R}^n)$ consisting of the uniformly continuous and bounded functions f such that
$$[f]_{\mathcal{C}^1} = \sup_{x,y \in \mathbb{R}^n;\, x \neq y} \frac{|f(x) - 2f(\frac{x+y}{2}) + f(y)|}{|x-y|} < +\infty.$$

It is endowed with the norm $\|f\|_{\mathcal{C}^1} = \|f\|_\infty + [f]_{\mathcal{C}^1}$.

Theorem 3.1.12 *Let $0 < \theta < 1$. Then*

$$D_A(\theta, \infty) = \begin{cases} C^{2\theta}(\mathbb{R}^n), & \text{if } \theta \neq 1/2, \\ C^1(\mathbb{R}^n), & \text{if } \theta = 1/2. \end{cases}$$

with equivalence of the respective norms. Moreover, $D_A(\theta) = h^{2\theta}(\mathbb{R}^n)$ for every $\theta \neq 1/2$.

Proof — We recall that, due to Theorem 1.2.17, we have

$$(C(\mathbb{R}^n), C^2(\mathbb{R}^n))_{\theta, \infty} = \begin{cases} C^{2\theta}(\mathbb{R}^n), & \text{if } \theta \neq 1/2, \\ C^1(\mathbb{R}^n), & \text{if } \theta = 1/2, \end{cases} \quad (3.1.37)$$

with equivalence of the respective norms. If $n = 1$, the statement follows. If $n > 1$, since $C^2(\mathbb{R}^n)$ is continuously embedded in $D(A)$, then (3.1.37) implies

$$C^{2\theta}(\mathbb{R}^n) \subset D_A(\theta, \infty) \text{ if } \theta \neq 1/2, \ C^1(\mathbb{R}^n) \subset D_A(1/2, \infty).$$

To prove the opposite inclusions, we use Proposition 3.1.11(ii). Fix any $\alpha \in]0,1[$ such that $1 + \alpha > 2\theta$. Then $C^{1+\alpha}(\mathbb{R}^n) \in J_{(1+\alpha)/2}(C(\mathbb{R}^n), C^2(\mathbb{R}^n))$. Part (ii) of the Reiteration Theorem 1.2.15 implies

$$(X, D(A))_{\theta, \infty} \subset (C(\mathbb{R}^n), C^{1+\alpha}(\mathbb{R}^n))_{2\theta/(1+\alpha), \infty}.$$

On the other hand, according to (3.1.37), $C^{1+\alpha}(\mathbb{R}^n) = (C(\mathbb{R}^n), C^2(\mathbb{R}^n))_{(1+\alpha)/2, \infty}$, so that, due again to the Reiteration Theorem and to (3.1.37), the space $(C(\mathbb{R}^n), C^{1+\alpha}(\mathbb{R}^n))_{2\theta/(1+\alpha), \infty}$ coincides with $C^{2\theta}(\mathbb{R}^n)$ if $\theta \neq 1/2$ and it coincides with $C^1(\mathbb{R}^n)$, if $\theta = 1/2$. The first statement is so proved.

Concerning the characterization of the continuous interpolation spaces, it is sufficient to follow the above procedure replacing $D_A(\theta, \infty)$ by $D_A(\theta)$, and Hölder spaces by little-Hölder spaces. An alternative proof is the following: $h^{2\theta}(\mathbb{R}^n)$ is the closure of $D(A)$ in $C^{2\theta}(\mathbb{R}^n)$, since it is the closure of $C^\alpha(\mathbb{R}^n)$ in $C^{2\theta}(\mathbb{R}^n)$, for every $\alpha > 2\theta$, and $C^2(\mathbb{R}^n) \subset D(A) \subset C^{2\theta+\varepsilon}(\mathbb{R}^n)$ for $0 < \varepsilon < 2-2\theta$. Recalling that $D_A(\theta)$ is the closure of $D(A)$ in $D_A(\theta, \infty)$, It follows that $D_A(\theta) = h^{2\theta}(\mathbb{R}^n)$ for $\theta \neq 1/2$. ∎

Remark 3.1.13 The perturbation results of Subsection 3.4 may be used to obtain other generation theorems. In particular, if $B \in L(C^\theta(\mathbb{R}^n), C(\mathbb{R}^n))$ or $B \in L(D(A), C^\theta(\mathbb{R}^n))$ for some $\theta \in]0,2[$, then by Proposition 2.4.1 the operator $A+B : D(A) \mapsto C(\mathbb{R}^n)$ is sectorial in $C(\mathbb{R}^n)$. For instance, for any $x_0, \ldots, x_n \in \mathbb{R}^n$ and $\beta_0, \ldots, \beta_n \in C(\mathbb{R}^n)$, the realization of the operator

$$u \mapsto \mathcal{A}(x, D)u(x) + \beta_0(x)u(x_0) + \sum_{i=1}^{n} \beta_i(x) D_i u(x_i),$$

1. Second order operators

in $C(\mathbb{R}^n)$ is sectorial. Similarly, by Proposition 2.4.3, the realization of the operator

$$u \mapsto \mathcal{A}(x,D)u(x) + \mathcal{A}(x_0,D)u(x_0)\beta_0(x)$$

in $UC(\mathbb{R}^n)$ is sectorial.

We know from Proposition 2.2.7 that the parts of A in $D_A(\theta, \infty)$ and in $D_A(\theta)$, defined by

$$A_\theta : D_A(\theta+1, \infty) \mapsto D_A(\theta, \infty), \quad A_\theta u = Au,$$

$$\tilde{A}_\theta : D_A(\theta+1) \mapsto D_A(\theta), \quad \tilde{A}_\theta u = Au,$$

are sectorial operators in $D_A(\theta, \infty)$ and in $D_A(\theta)$, respectively. Since $D_A(\theta, \infty) = C^{2\theta}(\mathbb{R}^n)$ and $D_A(\theta) = h^{2\theta}(\mathbb{R}^n)$ for $\theta \neq 1/2$, a generation result in the Hölder spaces follows immediately.

Theorem 3.1.14 *Let $0 < \theta < 1$, $\theta \neq 1/2$. Then the realization A_θ (respectively, $\tilde{A}_\theta$) of $\mathcal{A}(\cdot, D)$ in $C^{2\theta}(\mathbb{R}^n)$ (respectively, in $h^{2\theta}(\mathbb{R}^n)$) defined above is sectorial.*

For the characterization of $D_A(\theta+1, \infty)$ we will prove the well known Schauder estimates.

Theorem 3.1.15 *Let a_{ij}, b_i, $c \in C^{2\theta}(\mathbb{R}^n)$, with $\theta \neq 1/2$. If a function $u \in D(A)$ is such that $\mathcal{A}u \in C^{2\theta}(\mathbb{R}^n)$, then $u \in C^{2\theta+2}(\mathbb{R}^n)$, and there is $C > 0$ independent of u such that*

$$\|u\|_{C^{2\theta+2}(\mathbb{R}^n)} \leq C(\|u\|_\infty + \|\mathcal{A}u\|_{C^{2\theta}(\mathbb{R}^n)}). \tag{3.1.38}$$

Proof — The proof is in two steps. First we show that the statement holds when the coefficients of $\mathcal{A}$ are constant. Then we consider the case of variable coefficients by a localizing procedure.

Step 1. Let the coefficients of $\mathcal{A}$ be constant. We may assume, without loss of generality, that the coefficient c is negative. We shall see that there is $C > 0$ such that

$$\|e^{tA}f\|_{C^2(\mathbb{R}^n)} \leq \frac{C}{t^{1-\theta}}\|f\|_{C^{2\theta}(\mathbb{R}^n)}, \quad i,j = 1, \ldots, n, \ t > 0, \tag{3.1.39}$$

for every $f \in C^{2\theta}(\mathbb{R}^n)$. This can be proved writing down the explicit formula for $e^{tA}f$,

$$e^{tA}f(x) = \frac{e^{ct}(\det[\alpha_{ij}])^{1/2}}{(4\pi t)^{n/2}} \int_{\mathbb{R}^n} \exp\left(-\frac{1}{4t}\sum_{i,j=1}^n \alpha_{ij}(x_i - y_i)(x_j - y_j)\right) f(y+tb)dy, \tag{3.1.40}$$

where $[\alpha_{ij}] = [a_{ij}]^{-1}$, $b = (b_1, \ldots, b_n)$, and then estimating the second order derivatives. Another way to prove (3.1.39) is the following. For $0 \leq \alpha < 1$ we have $(C^\alpha(\mathbb{R}^n), D(A))_{\frac{1-\alpha}{2-\alpha}, 1} = (C(\mathbb{R}^n), D(A))_{1/2, 1} \subset C^1(\mathbb{R}^n)$, so that $C^1(\mathbb{R}^n)$ belongs to the class $J_{(1-\alpha)/(2-\alpha)}$ between $C^\alpha(\mathbb{R}^n)$ and $D(A)$. Moreover, $\|e^{tA}\|_{L(C^\alpha(\mathbb{R}^n))}$, $\|t^{1-\alpha/2}Ae^{tA}\|_{L(C^\alpha(\mathbb{R}^n),C(\mathbb{R}^n))}$ are bounded, so that

$$\|e^{tA}\varphi\|_{C^1(\mathbb{R}^n)} \leq k\|e^{tA}\varphi\|_{C^\alpha(\mathbb{R}^n)}^{1-\frac{1-\alpha}{2-\alpha}}\|e^{tA}\varphi\|_{D(A)}^{\frac{1-\alpha}{2-\alpha}} \leq \frac{k'}{t^{(1-\alpha)/2}}\|\varphi\|_{C^\alpha(\mathbb{R}^n)}, \quad \forall \varphi \in C^\alpha(\mathbb{R}^n),$$

from which (3.1.39) follows. Indeed, for $\theta < 1/2$ and $f \in C^{2\theta}(\mathbb{R}^n)$, we get $\|D_{ij}e^{tA}f\|_\infty \leq \|D_i e^{tA/2}\|_{L(C(\mathbb{R}^n))}\|D_j e^{tA/2}f\|_\infty \leq ct^{-1/2}ct^{-1/2+\theta}\|f\|_{C^{2\theta}(\mathbb{R}^n)}$. Similarly, for $\theta > 1/2$ and $f \in C^{2\theta}(\mathbb{R}^n)$, we get $\|D_{ij}e^{tA}f\|_\infty = \|D_i e^{tA}D_j f\|_\infty \leq ct^{1-\theta}\|D_j f\|_{C^{2\theta-1}(\mathbb{R}^n)} \leq ct^{1-\theta}\|f\|_{C^{2\theta}(\mathbb{R}^n)}$, and (3.1.39) is proved.

Let $u \in D(A)$ be such that $\mathcal{A}u \in C^{2\theta}(\mathbb{R}^n)$. Let $\lambda \in \mathbb{R} \cap \rho(A)$, and set $f = \lambda u - \mathcal{A}u$. Formula (2.1.10) gives

$$u = R(\lambda,A)f = \int_0^{+\infty} e^{-\lambda t}e^{tA}f\, dt.$$

From (3.1.39) it follows that the integral has values in $C^2(\mathbb{R}^n)$, and

$$\|u\|_{C^2(\mathbb{R}^n)} \leq C\lambda^{-\theta}\Gamma(\theta)\|f\|_{C^{2\theta}(\mathbb{R}^n)}, \qquad (3.1.41)$$

where the Euler Γ function is defined by (2.2.24). Let us prove now that the second order derivatives of u belong to $C^{2\theta}(\mathbb{R}^n)$. Since $C^{2\theta}(\mathbb{R}^n) = D_A(\theta,\infty)$, it is sufficient to prove that

$$\sup_{0 < \xi \leq 1} \|\xi^{1-\theta}Ae^{\xi A}D_{ij}u\|_\infty < \infty, \quad i,j = 1,\ldots,n.$$

For $0 < \xi \leq 1$ it holds

$$\|\xi^{1-\theta}Ae^{\xi A}D_{ij}u\|_\infty = \left\|\int_0^{+\infty}\xi^{1-\theta}e^{-\lambda t}Ae^{(\xi+t/2)A}D_{ij}e^{tA/2}f\, dt\right\|_\infty$$
$$\leq \int_0^{+\infty}\xi^{1-\theta}\frac{M_1 C}{(\xi+t/2)(t/2)^{1-\theta}}dt\,\|f\|_{C^{2\theta}(\mathbb{R}^n)} \leq \int_0^{+\infty}\frac{2M_1 C}{(1+s)s^{1-\theta}}ds\,\|f\|_{C^{2\theta}(\mathbb{R}^n)}, \qquad (3.1.42)$$

where $M_1 = \sup_{t>0}\|tAe^{tA}\|_{L(C(\mathbb{R}^n))}$.

We know already that $\|u\|_{C^1} \leq C(\lambda)\|f\|_\infty$. Taking $\lambda = 1$, we get

$$\|u\|_{C^{2\theta+2}} \leq K(\|u\|_{C^{2\theta}} + \|\mathcal{A}u\|_{C^{2\theta}}). \qquad (3.1.43)$$

Step 2. The main point is to show that for every $u \in C^{2+2\theta}(\mathbb{R}^n)$ the *a priori* estimate (3.1.38) holds. Then the statement will follow by the standard method of continuity.

Let $\theta < 1/2$. By estimate (3.1.43), for every $x_0 \in \mathbb{R}^n$ and $u \in C^{2+2\theta}(\mathbb{R}^n)$, it holds

$$\|u\|_{C^{2\theta+2}(\mathbb{R}^n)} \leq K\left(\|u\|_{C^{2\theta}(\mathbb{R}^n)} + \left\|\sum_{i,j=1}^n a_{ij}(x_0)D_{ij}u\right\|_{C^{2\theta}(\mathbb{R}^n)}\right).$$

1. SECOND ORDER OPERATORS

Since the coefficients a_{ij} are Hölder continuous, there is $r > 0$ such that

$$|x-y| \leq 2r \Rightarrow |a_{ij}(x) - a_{ij}(y)| \leq \frac{1}{2K}, \quad i,j = 1,\ldots,n.$$

Let η_0 be a smooth cutoff function such that $0 \leq \eta_0(x) \leq 1$ for every x, $\eta_0 \equiv 1$ in $B(0,r)$, $\eta_0 \equiv 0$ outside $B(0,2r)$, and set $\eta(x) = \eta_0(x - x_0)$. For every $u \in C^{2+2\theta}(\mathbb{R}^n)$, the function $v = \eta u$ satisfies

$$\sum_{i,j=1}^n a_{ij}(x_0) D_{ij}(\eta u) = \eta \mathcal{A} u + \eta \sum_{i,j=1}^n (a_{ij}(x_0) - a_{ij}(x)) D_{ij} u$$

$$+ u \sum_{i,j=1}^n a_{ij}(x_0) D_{ij}\eta + \sum_{i,j=1}^n a_{ij}(x_0) D_i \eta D_j u - \eta \sum_{i=1}^n b_i D_i u - \eta c u$$

$$= F,$$

so that, thanks to step (i),

$$\|\eta u\|_{C^{2+2\theta}} \leq K(\|F\|_{C^{2\theta}} + \|u\|_{C^{2\theta}}) \leq K\bigg(\|\eta\|_{C^{2\theta}}\|f\|_{C^{2\theta}}$$

$$+ \sum_{i,j=1}^n \bigg(\frac{1}{2K}[D_{ij}u]_{C^{2\theta}} + [a_{ij}]_{C^{2\theta}}\|D_{ij}u\|_\infty + \frac{[\eta]_{C^{2\theta}}}{2K}\|D_{ij}u\|_\infty\bigg)$$

$$+ \max_{i,j}\|a_{ij}\|_\infty(\|u\|_{C^{2\theta}}\|\eta\|_{C^{2\theta+2}} + \|u\|_{C^{2\theta+1}}\|\eta\|_{C^{2\theta+1}})$$

$$+ \|\eta\|_{C^{2\theta}} \max_i \|b_i\|_{C^{2\theta}}\|u\|_{C^{2\theta+1}} + \|\eta c\|_{C^{2\theta}}\|u\|_{C^{2\theta}}\bigg).$$

This implies that

$$\|u\|_{C^{2+2\theta}(B(x_0,r))} = \|\eta u\|_{C^{2+2\theta}(B(x_0,r))}$$
$$\leq \tfrac{1}{2}\|u\|_{C^{2+2\theta}(\mathbb{R}^n)} + C_1(\|u\|_{C^2(\mathbb{R}^n)} + \|\mathcal{A}u\|_{C^{2\theta}(\mathbb{R}^n)}),$$

where C_1 does not depend on u. Using the inequality

$$\|u\|_{C^{2+2\theta}(\mathbb{R}^n)} \leq \sup_{x_0 \in \mathbb{R}^n} \|u\|_{C^{2+2\theta}(B(x_0,r))} + \frac{2\|D^2 u\|_\infty}{r^{2\theta}},$$

which holds for every function in $C^{2+2\theta}(\mathbb{R}^n)$, we get

$$\|u\|_{C^{2+2\theta}(\mathbb{R}^n)} \leq 2C_1(\|u\|_{C^2(\mathbb{R}^n)} + \|\mathcal{A}u\|_{C^{2\theta}(\mathbb{R}^n)}) + \frac{4\|D^2 u\|_\infty}{r^{2\theta}}$$
$$\leq C_3(\|u\|_{C^2(\mathbb{R}^n)} + \|\mathcal{A}u\|_{C^{2\theta}(\mathbb{R}^n)}),$$

with C_3 independent of u. Thanks to Proposition 1.1.3, the space $C^2(\mathbb{R}^n)$ belongs to the class $J_{1/(1+\theta)}$ between $C(\mathbb{R}^n)$ and $C^{2+2\theta}(\mathbb{R}^n)$, so that

$$\|u\|_{C^2(\mathbb{R}^n)} \leq c\|u\|_{C^{2+2\theta}(\mathbb{R}^n)}^{1/(1+\theta)}\|u\|_\infty^{\theta/(1+\theta)}.$$

We recall that for every positive a, b, ε it holds $ab \leq \varepsilon^p a^p/p + \varepsilon^{-q}b^q/q$ whenever $1/p + 1/q = 1$. Choosing $p = 1 + \theta$, $q = (1+\theta)/\theta$, we get

$$\|u\|_{C^{2+2\theta}(\mathbb{R}^n)} \leq \frac{C_3 c}{1+\theta} \left(\varepsilon^{1+\theta} \|u\|_{C^{2+2\theta}(\mathbb{R}^n)} + \theta \varepsilon^{-(1+\theta)/\theta} \|u\|_\infty \right) + C_3 \|\mathcal{A}u\|_{C^{2\theta}(\mathbb{R}^n)},$$

and (3.1.38) follows, taking $\varepsilon = [(1+\theta)/(2C_3 c)]^{1/(1+\theta)}$.

The case where $\theta > 1/2$ is similar, and it is left to the reader. ∎

Corollary 3.1.16 *Let a_{ij}, b_i, $c \in C^{2\theta}(\mathbb{R}^n)$, with $\theta \neq 1/2$. Then*

$$D_\mathcal{A}(\theta+1, \infty) = C^{2\theta+2}(\mathbb{R}^n),$$

with equivalence of the respective norms. If in addition a_{ij}, b_i, $c \in h^{2\theta}(\mathbb{R}^n)$, then

$$D_\mathcal{A}(\theta+1) = h^{2\theta+2}(\mathbb{R}^n).$$

3.1.4 Generation in $C^1(\mathbb{R}^n)$

Even if $C^1(\mathbb{R}^n)$ is not an interpolation space (see the bibliographical notes in the next section), we will prove in the next proposition the realization of $\mathcal{A}$ in $C^1(\mathbb{R}^n)$ is sectorial. Proposition 3.1.17 is devoted to the case where the coefficients are continuously differentiable. Proposition 3.1.18 deals with the case of coefficients in $C(\mathbb{R}^n)$, and its proof relies on Proposition 3.1.17.

Define the realization of $\mathcal{A}(\cdot, D)$ in $C^1(\mathbb{R}^n)$ by

$$\begin{cases} D_1(A) = \{u \in W^{2,p}_{loc}(\mathbb{R}^n) \: \forall p \geq 1 : u, \mathcal{A}(\cdot, D)u \in C^1(\mathbb{R}^n)\}, \\ A : D_1(A) \mapsto C^1(\mathbb{R}^n), Au = \mathcal{A}(\cdot, D)u. \end{cases} \quad (3.1.44)$$

Proposition 3.1.17 *Let a_{ij}, b_i, $c \in UC^1(\mathbb{R}^n)$. Then $A : D_1(A) \mapsto C^1(\mathbb{R}^n)$ is sectorial, and $D_1(A) \subset W^{3,p}_{loc}(\mathbb{R}^n) \cap C^{2+\alpha}(\mathbb{R}^n)$ for every $p \geq 1$ and $\alpha \in \,]0,1[$.*

Proof — Let Λ_0 be the number given by Proposition 3.1.7. If $\operatorname{Re}\lambda > \Lambda_0$, for every $f \in C^1(\mathbb{R}^n)$ the equation

$$\lambda u - \mathcal{A}(x, D)u = f$$

has a unique solution $u \in D(A)$, and $\|u\|_{C^1(\mathbb{R}^n)} \leq C(\lambda)\|f\|_\infty$. Therefore, the resolvent set of A contains the half plane $\{\operatorname{Re}\lambda > \Lambda_0\}$.

Next, we show that $D_1(A) \subset W^{3,p}_{loc}(\mathbb{R}^n)$ for every p. This will be done by using the results of Theorem 3.1.6, Proposition 3.1.7 and a standard regularization procedure.

1. Second order operators

Fix $p > n$ and $\lambda \in \mathbb{C}$ such that $\operatorname{Re}\lambda > \Lambda_p$. For every $u \in D_1(A)$ set $f = \lambda u - \mathcal{A}u$. For $k = 1, \ldots, n$, consider the equation formally satisfied by $v = D_k u$,

$$\lambda v - \mathcal{A}v = D_k f + \sum_{i,j=1}^{n}(D_k a_{ij})D_{ij}u + \sum_{i=1}^{n}(D_k b_i)D_i u + (D_k c)u. \tag{3.1.45}$$

The right hand side g_k belongs to $L^p_{loc}(\mathbb{R}^n)$, and thanks to estimate (3.1.25) we have

$$\sup_{x \in \mathbb{R}^n} \|g_k\|_{L^p(B(x,1))} < \infty,$$

so that, due to Remark 3.1.8, problem (3.1.45) has a unique solution $v \in C^1(\mathbb{R}^n) \cap W^{2,p}_{loc}(\mathbb{R}^n)$. For $k = 1, \ldots, n$, let $e_k = (\delta_{1k}, \ldots, \delta_{nk})$ be the k-th vector of the basis of $\mathbb{R}^n$. The incremental ratio $\tau_{k,h}u(x) = h^{-1}(u(x + he_k) - u(x))$ satisfies

$$\lambda \tau_{k,h}u(x) - \mathcal{A}(x,D)\tau_{k,h}u(x) = \tau_{k,h}f(x) + (\tau_{k,h}c)(x)u(x + he_k)$$
$$+ \sum_{i=1}^{n}(\tau_{k,h}b_i)(x)D_i u(x + he_k) + \sum_{i,j=1}^{n}(\tau_{k,h}a_{ij})(x)D_{ij}u(x + he_k).$$

Therefore, $v - \tau_{k,h}u = u_1 + u_2$, where u_1 satisfies

$$\lambda u_1(x) - \mathcal{A}(x,D)u_1(x) = (D_k c(x) - \tau_{k,h}c(x))u(x + he_k) + D_k c(x) \cdot$$
$$\cdot (u(x + he_k) - u(x)) + \sum_{i=1}^{n}(D_k b_i(x) - \tau_{k,h}b_i(x))D_i u(x + he_k)$$
$$+ \sum_{i=1}^{n} D_k b_i(x)(D_i u(x + he_k) - D_i u(x)),$$

and u_2 satisfies

$$\lambda u_2(x) - \mathcal{A}(x,D)u_2(x) = \sum_{i,j=1}^{n}(D_k a_{ij}(x) - \tau_{k,h}a_{ij})D_{ij}u(x + he_k)$$
$$+ \sum_{i,j=1}^{n}(D_k a_{ij}(x))(D_{ij}u(x + he_k) - D_{ij}u(x))D_k f(x) - \tau_{k,h}f(x).$$

By estimate (3.1.26),

$$\|u_1\|_\infty \leq |\lambda|^{-1}K_p \gamma_n^{1/p}(\|(D_k c - \tau_{k,h}c)u(\cdot + he_k)\|_\infty$$
$$+ \|D_k c(D_i u(\cdot + he_k) - u)\|_\infty + \|\sum_{i=1}^{n}(D_k b_i - \tau_{k,h}b_i)D_i u(\cdot + he_k)\|_\infty$$
$$+ \|\sum_{i=1}^{n} D_k b_i(D_i u(\cdot + he_k) - D_i u)\|_\infty),$$

whereas by (3.1.25)

$$\|u_2\|_\infty \leq |\lambda|^{-1}K_p \sup_{x_0 \in \mathbb{R}^n}(\|D_k f - \tau_{h,k}f\|_{L^p(B(x_0,|\lambda|^{-1/2}))}$$
$$+ \|\sum_{i,j=1}^{n}(D_k a_{ij} - \tau_{k,h}a_{ij})D_{ij}u(\cdot + he_k)\|_{L^p(B(x_0,|\lambda|^{-1/2}))}$$
$$+ \|\sum_{i,j=1}^{n}(D_k a_{ij})(D_{ij}u(\cdot + he_k) - D_{ij}u)\|_{L^p(B(x_0,|\lambda|^{-1/2}))}).$$

All the terms in the right hand sides of the above estimates go to 0 as h goes to 0, so that $v = D_k u$. Hence, $u \in W^{3,p}_{loc}(\mathbb{R}^n)$.

Now we can show that A is sectorial in $C^1(\mathbb{R}^n)$. By applying estimates (3.1.25) and (3.1.26) to (3.1.45) we get, for $k = 1, \ldots, n$,

$$|\lambda| \|D_k u\|_\infty + |\lambda|^{1/2} \|D(D_k u)\|_\infty + |\lambda|^{n/2p} \sup_{x_0 \in \mathbb{R}^n} \|D^2(D_k u)\|_{L^p(B(x_0, |\lambda|^{-1/2}))}$$
$$\leq K_p \gamma_n^{1/p} \|D_k f + \sum_{i=1}^n (D_k b_i) D_i u + (D_k c) u\|_\infty$$

$$+ K_p |\lambda|^{n/2p} \sup_{x_0 \in \mathbb{R}^n} \|\sum_{i,j=1}^n (D_k a_{ij}) D_{ij} u\|_{L^p(B(x_0, |\lambda|^{-1/2}))}$$
$$\leq K_p \gamma_n^{1/p} (\|D_k f\|_\infty + K_p \gamma_n^{1/p} (\max_{i=1,\ldots,n} \|D_k b_i\|_\infty |\lambda|^{-1/2} \|f\|_\infty$$
$$+ \|D_k c\|_\infty |\lambda|^{-1} \|f\|_\infty)) + K_p \max_{i,j=1,\ldots,n} \|D_k a_{ij}\|_\infty K_p \gamma_n^{1/p} \|f\|_\infty$$
$$\leq C \|f\|_{C^1(\mathbb{R}^n)}.$$
(3.1.46)

It follows that

$$\|u\|_{C^1(\mathbb{R}^n)} = \|u\|_\infty + \sum_{k=1}^n \|D_k u\|_\infty \leq |\lambda|^{-1} (K_p \gamma_n^{1/p} + nC) \|f\|_{C^1(\mathbb{R}^n)}, \quad (3.1.47)$$

which implies that A is sectorial in $C^1(\mathbb{R}^n)$.

Finally, using estimates (3.1.46), the embedding $D_1(A) \subset C^{2+\alpha}(\mathbb{R}^n)$ for $0 < \alpha < 1$ follows as in the proof of Theorem 3.1.7. ■

Now we consider the case where the coefficients are not necessarily differentiable. Then the domain of the realization of $\mathcal{A}$ in $C^1(\mathbb{R}^n)$ is not as nice as in the case of differentiable coefficients, and the inclusion $D_1(A) \subset W^{3,p}_{loc}(\mathbb{R}^n)$ is not true. However, the following generation result holds.

Proposition 3.1.18 *Let a_{ij}, b_i, c belong to $UC(\mathbb{R}^n)$, and let $D_1(A)$ be defined by (3.1.44). Then $A : D_1(A) \mapsto C^1(\mathbb{R}^n)$ is sectorial. Moreover, $\overline{D(A)} = UC^1(\mathbb{R}^n)$.*

Proof — From Theorem 3.1.7 one deduces, as in Proposition 3.1.17, that $\rho(A)$ contains the halfplane $\{\text{Re}\,\lambda > \Lambda_0\}$. We shall prove that there exists $N > 0$ such that for $\text{Re}\,\lambda$ sufficiently large

$$\|\lambda R(\lambda, A)\|_{L(C^1(\mathbb{R}^n))} \leq N. \quad (3.1.48)$$

So, the statement will follow from Proposition 2.1.11.

For every $x_0 \in \mathbb{R}^n$, the realizations of the operator with constant coefficients $\mathcal{A}(x_0, D)$ in $C(\mathbb{R}^n)$ and in $C^1(\mathbb{R}^n)$ are sectorial, thanks to Theorem 3.1.7 and to Proposition 3.1.17. Fix any $p > n$. By Theorem 3.1.6 there exists a constant K_1, not depending on x_0, such that for $\text{Re}\,\lambda$ sufficiently large and $f \in L^p_{loc}(\mathbb{R}^n)$, the solution v of

$$\lambda v - \sum_{i,j=1}^n a_{ij}(x_0) D_{ij} v = f$$

1. Second order operators

satisfies

$$|\lambda|^{1/2}\|v\|_{C^1(\mathbb{R}^n)} + |\lambda|^{n/2p} \sup_{x\in\mathbb{R}^n} \|D^2v\|_{L^p(B(x,|\lambda|^{-1/2}))} + \|\mathcal{A}v\|_\infty \\ \leq K_1|\lambda|^{n/2p} \sup_{x\in\mathbb{R}^n} \|f\|_{L^p(B(x,|\lambda|^{-1/2}))}, \quad (3.1.49)$$

and, consequently,

$$|\lambda|^{1/2}\|v\|_{C^1(\mathbb{R}^n)} + |\lambda|^{n/2p} \sup_{x\in\mathbb{R}^n} \|D^2v\|_{L^p(B(x,|\lambda|^{-1/2}))} + \|\mathcal{A}v\|_\infty \leq K_1\gamma_n^{1/p}\|f\|_\infty, \quad (3.1.50)$$

if f belongs to $C(\mathbb{R}^n)$. Moreover, by estimate (3.1.46), there is K_2 such that

$$|\lambda|\,\|v\|_{C^1(\mathbb{R}^n)} + |\lambda|^{1/2+n/2p} \sup_{x\in\mathbb{R}^n} \|D^2v\|_{L^p(B(x,|\lambda|^{-1/2})} + |\lambda|^{1/2}\|\mathcal{A}v\|_\infty \\ \leq K_2\|f\|_{C^1(\mathbb{R}^n)}, \quad (3.1.51)$$

if f belongs to $C^1(\mathbb{R}^n)$.

Let $r > 0$ be so small that $|a_{ij}(x) - a_{ij}(y)| \leq 1/(2K_1)$ for $|x-y| \leq r$. For every $x_0 \in \mathbb{R}^n$ let θ be a smooth cutoff function such that $0 \leq \theta \leq 1$, $\theta \equiv 1$ in $B(x_0, r)$, $\theta \equiv 0$ outside $B(x_0, r)$, and such that $\|\theta\|_{C^2(\mathbb{R}^n)} \leq C_1$, with C_1 independent of x_0.

Let in addition $\operatorname{Re}\lambda$ be so large that λ belongs to the resolvent set of the realization of $\mathcal{A}$ in $C(\mathbb{R}^n)$. For $f \in C^1(\mathbb{R}^n)$ let u be the solution of

$$\lambda u - \mathcal{A}u = f.$$

Then θu satisfies

$$\lambda\theta u - \sum_{i,j=1}^n a_{ij}(x_0)D_{ij}(\theta u) = \theta f + \theta \sum_{i,j=1}^n (a_{ij}(x) - a_{ij}(x_0))D_{ij}u \\ + u\left(c\theta - \sum_{i,j=1}^n a_{ij}(x_0)D_{ij}\theta\right) - 2\sum_{i,j=1}^n a_{ij}(x_0)D_i\theta D_j u + \theta\sum_{i=1}^n b_i D_i u,$$

so that $\theta u = v_1 + v_2 + v_3$, where

$$\lambda v_1 - \sum_{i,j=1}^n a_{ij}(x_0)D_{ij}v_1 = \theta \sum_{i,j=1}^n (a_{ij}(x_0) - a_{ij}(x))D_{ij}u,$$

$$\lambda v_2 - \sum_{i,j=1}^n a_{ij}(x_0)D_{ij}v_2 = \\ = u\left(c\theta - \sum_{i,j=1}^n a_{ij}(x_0)D_{ij}\theta\right) - 2\sum_{i,j=1}^n a_{ij}(x_0)D_i\theta D_j u + \theta\sum_{i=1}^n b_i D_i u,$$

$$\lambda v_3 - \mathcal{A}(x_0, D)v_3 = \theta f.$$

Set

$$C_2 = \sum_{i,j=1}^{n} \|a_{ij}\|_\infty + \sum_{i=1}^{n} \|b_i\|_\infty + \|c\|_\infty.$$

By estimate (3.1.49) we get

$$|\lambda|\,\|v_1\|_{C^1(\mathbb{R}^n)} + |\lambda|^{1/2+n/2p}\sup_{x\in\mathbb{R}^n}\|D^2 v_1\|_{L^p(B(x,|\lambda|^{-1/2}))} + |\lambda|^{1/2}\|Av_1\|_\infty$$
$$\leq |\lambda|^{1/2}K_1\left(\frac{1}{2K_1}|\lambda|^{n/2p}\sup_{x\in\mathbb{R}^n}\|D^2 u\|_{L^p(B(x,|\lambda|^{-1/2}))}\right)$$
$$= \tfrac{1}{2}|\lambda|^{1/2+n/2p}\sup_{x\in\mathbb{R}^n}\|D^2 u\|_{L^p(B(x,|\lambda|^{-1/2}))}.$$

By estimate (3.1.50),

$$|\lambda|\,\|v_2\|_{C^1(\mathbb{R}^n)} + |\lambda|^{1/2+n/2p}\sup_{x\in\mathbb{R}^n}\|D^2 v_2\|_{L^p(B(x,|\lambda|^{-1/2}))} + |\lambda|^{1/2}\|Av_2\|_\infty$$
$$\leq |\lambda|^{1/2}\gamma_n^{1/p}K_1(C_1 C_2\|u\|_\infty + 2C_1 C_2\|Du\|_\infty)$$
$$\leq \gamma_n^{2/p}K_1 K_p C_1 C_2(|\lambda|^{-1/2} + 2)\|f\|_\infty \leq C_3\|f\|_\infty,$$

where K_p is the constant of estimate (3.1.26). By estimate (3.1.51),

$$|\lambda|\,\|v_3\|_{C^1(\mathbb{R}^n)} + |\lambda|^{1/2+n/2p}\sup_{x\in\mathbb{R}^n}\|D^2 v_3\|_{L^p(B(x,|\lambda|^{-1/2}))}$$
$$+|\lambda|^{1/2}\|Av_3\|_\infty \leq K_2 C_1\|f\|_{C^1(\mathbb{R}^n)}.$$

Summing up, we find

$$|\lambda|\,\|u\|_{C^1(\mathbb{R}^n)} + |\lambda|^{1/2+n/2p}\sup_{x\in\mathbb{R}^n}\|D^2 u\|_{L^p(B(x,|\lambda|^{-1/2}))}|\lambda|^{1/2}\|Au\|_\infty$$
$$\leq \tfrac{1}{2}|\lambda|^{1/2+n/2p}\sup_{x\in\mathbb{R}^n}\|D^2 u\|_{L^p(B(x,|\lambda|^{-1/2}))} + C_3\|f\|_\infty + K_2 C_1\|f\|_{C^1(\mathbb{R}^n)},$$
(3.1.52)

and estimate (3.1.48) follows. ∎

3.1.5 Generation in $L^\infty(\Omega)$ and in spaces of continuous functions in $\overline{\Omega}$

The case where $\mathbb{R}^n$ is replaced by a domain $\overline{\Omega} \neq \mathbb{R}^n$ with uniformly C^2 boundary can be treated as in Subsection 3.1.2, with suitable modifications due to the boundary conditions.

Through the whole subsection we set, for $x_0 \in \mathbb{R}^n$ and $r > 0$,

$$\Omega_{x_0,r} = \Omega \cap B(x_0, r).$$

1. Second order operators

To begin with, we consider the Dirichlet boundary condition. The first step, as in the case $\Omega = \mathbb{R}^n$, is an *a priori* estimate similar to (3.1.25).

Theorem 3.1.19 *For every $p > n$ there are $K_p, \Lambda_p > 0$, such that for $\lambda \in \mathbb{C}$ with $\operatorname{Re} \lambda \geq \Lambda_p$, for every $u \in W^{2,p}_{loc}(\Omega) \cap C^1(\overline{\Omega})$*

$$|\lambda| \|u\|_\infty + |\lambda|^{1/2} \|Du\|_\infty + |\lambda|^{n/2p} \sup_{x_0 \in \overline{\Omega}} \|D^2 u\|_{L^p(\Omega_{x_0, |\lambda|^{-1/2}})}$$
$$\leq K_p (|\lambda|^{n/2p} \sup_{x_0 \in \overline{\Omega}} \|\lambda u - \mathcal{A}(\cdot, D) u\|_{L^p(\Omega_{x_0, |\lambda|^{-1/2}})} \quad (3.1.53)$$
$$+ |\lambda| \|g_0\|_\infty + |\lambda|^{1/2} \|Dg_0\|_\infty + |\lambda|^{n/2p} \sup_{x_0 \in \overline{\Omega}} \|D^2 g_0\|_{L^p(\Omega_{x_0, |\lambda|^{-1/2}})}),$$

where g_0 is any $W^{2,p}_{loc}$ extension of $g = u_{|\partial\Omega}$ to the whole Ω. Consequently, there is $\tilde{K}_p > 0$ such that if $\mathcal{A}(\cdot, D) u \in C(\overline{\Omega})$ and $u_{|\partial\Omega} \in C^2(\partial\Omega)$, then

$$|\lambda| \|u\|_\infty + |\lambda|^{1/2} \|Du\|_\infty + |\lambda|^{n/2p} \sup_{x_0 \in \overline{\Omega}} \|D^2 u\|_{L^p(\Omega_{x_0, |\lambda|^{-1/2}})}$$
$$\leq \tilde{K}_p (\|\lambda u - \mathcal{A}(\cdot, D) u\|_\infty + |\lambda| \|u\|_{C(\partial\Omega)} + |\lambda|^{1/2} \|u\|_{C^1(\partial\Omega)} + \|u\|_{C^2(\partial\Omega)}). \quad (3.1.54)$$

Proof — We adapt the proof of Theorem 3.1.6 to the present situation. The main point is the modification of Lemma 3.1.5. We shall show that for every $u \in W^{2,p}_{loc}(\Omega)$, $x_0 \in \overline{\Omega}$, $r \leq 1$, $\alpha > 0$, and $\operatorname{Re} \lambda \geq \omega_p$ (ω_p is the constant given by Theorem 3.1.3(ii)) it holds

$$|\lambda| \|u\|_{L^p(\Omega_{x_0,r})} + |\lambda|^{1/2} \|Du\|_{L^p(\Omega_{x_0,r})} + \|D^2 u\|_{L^p(\Omega_{x_0,r})}$$
$$\leq C_p \Big[\|f\|_{L^p(\Omega_{x_0,(\alpha+1)r})} + (|\lambda| + \frac{|\lambda|^{1/2}}{\alpha r} + \frac{1}{\alpha^2 r^2}) \|g_0\|_{L^p(\Omega_{x_0,(\alpha+1)r})}$$
$$+ (|\lambda|^{1/2} + \frac{1}{\alpha r}) \|Dg_0\|_{L^p(\Omega_{x_0,(\alpha+1)r})} + \|D^2 g_0\|_{L^p(\Omega_{x_0,(\alpha+1)r})} \quad (3.1.55)$$
$$+ \frac{1}{\alpha} (r^{-2} \|u\|_{L^p(\Omega_{x_0,(\alpha+1)r})} + r^{-1} \|Du\|_{L^p(\Omega_{x_0,(\alpha+1)r})}) \Big]$$

where g_0 is any extension of $u_{|\partial\Omega}$ belonging to $W^{2,p}_{loc}(\Omega)$, and $f = \lambda u - \mathcal{A} u$. The proof of (3.1.55) is similar to the one of (3.1.21). Let $u \in W^{2,p}_{loc}(\Omega)$, and let θ be the cutoff function used in the proof of Lemma 3.1.5. Then the function $v = \theta u$ satisfies

$$\lambda v - \mathcal{A}(\cdot, D) v = F \text{ in } \Omega, \quad v = \theta g \text{ in } \partial\Omega,$$

where F is defined in (3.1.22). Since $\operatorname{Re} \lambda \geq \Lambda_p \geq \omega_p$, by (3.1.12) we get

$$|\lambda| \|v\|_{L^p(\Omega)} + |\lambda|^{1/2} \|Dv\|_{L^p(\Omega)} + \|D^2 v\|_{L^p(\Omega)}$$
$$\leq M_p (\|F\|_{L^p(\Omega)} + |\lambda| \|\theta g_0\|_{L^p(\Omega)} + |\lambda|^{1/2} \|D(\theta g_0)\|_{L^p(\Omega)} + \|D^2(\theta g_0)\|_{L^p(\Omega)}). \quad (3.1.56)$$

An estimate for $\|F\|_{L^p(\Omega)}$ is provided by (3.1.24), with $\mathbb{R}^n$ replaced of course by Ω and $B(x_0, (\alpha+1)r)$ replaced by $\Omega_{x_0,(\alpha+1)r}$.

Moreover, we have

$$|\lambda|\,\|\theta g_0\|_{L^p(\Omega)} + |\lambda|^{1/2}\|D(\theta g_0)\|_{L^p(\Omega)} + \|D^2(\theta g_0)\|_{L^p(\Omega)}$$

$$\leq |\lambda|\,\|g_0\|_{L^p(\Omega_{x_0,(\alpha+1)r})} + |\lambda|^{1/2}\big[\|D\theta\|_\infty \|g_0\|_{L^p(\Omega_{x_0,(\alpha+1)r})}$$

$$+\|Dg_0\|_{L^p(\Omega_{x_0,(\alpha+1)r})}\big] + \|D^2\theta\|_\infty \|g_0\|_{L^p(\Omega_{x_0,(\alpha+1)r})}$$

$$+2\|D\theta\|_\infty \|Dg_0\|_{L^p(\Omega_{x_0,(\alpha+1)r})} + \|D^2 g_0\|_{L^p(\Omega_{x_0,(\alpha+1)r})}$$

$$\leq \left(|\lambda| + \frac{|\lambda|^{1/2}K}{\alpha r} + \frac{K}{\alpha^2 r^2}\right) \|g_0\|_{L^p(\Omega_{x_0,(\alpha+1)r})} +$$

$$+ \left(|\lambda|^{1/2} + \frac{2K}{\alpha r}\right) \|Dg_0\|_{L^p(\Omega_{x_0,(\alpha+1)r})} + \|D^2 g_0\|_{L^p(\Omega_{x_0,(\alpha+1)r})}.$$

Replacing the above estimate and (3.1.24) in (3.1.56), we get (3.1.55).

Once (3.1.55) is established, the *a priori* estimate (3.1.53) follows arguing as in the proof of Theorem 3.1.6. To prove (3.1.54), we use (3.1.53), which implies

$$|\lambda|\,\|u\|_\infty + |\lambda|^{1/2}\|Du\|_\infty + |\lambda|^{n/2p}\sup_{x_0\in\overline{\Omega}}\|D^2 u\|_{L^p(\Omega_{x_0,|\lambda|^{-1/2}})}$$

$$\leq K_p\big[\gamma_n^{1/p}(\|\lambda u - \mathcal{A}(\cdot,D)u\|_\infty + \|D^2 g_0\|_\infty) + |\lambda|\,\|g_0\|_\infty + |\lambda|^{1/2}\|Dg_0\|_\infty\big].$$

Choosing now $g_0 = \mathcal{E}(u_{|\partial\Omega})$, where $\mathcal{E}$ is any extension operator belonging to $L(C(\partial\Omega), C(\overline{\Omega})) \cap L(C^1(\partial\Omega), C^1(\overline{\Omega})) \cap L(C^2(\partial\Omega), C^2(\overline{\Omega}))$ (see e.g. Theorem 0.3.2(i)), estimate (3.1.54) follows. ∎

Theorem 3.1.19 has several consequences. The first one is concerned with the solvability, for Re λ large, of the boundary value problem

$$\lambda u - \mathcal{A}u = f \text{ in } \Omega, \quad u = g \text{ in } \partial\Omega, \qquad (3.1.57)$$

with $f \in L^p_{loc}(\Omega)$, $g \in W^{2,p}_{loc}(\Omega) \cap C^1(\overline{\Omega})$. If Ω is bounded, Theorem 3.1.3(ii) guarantees that (3.1.57) has a unique solution. If Ω is unbounded, arguing exactly as in the proof of Theorem 3.1.7 and Remark 3.1.8, we get

Proposition 3.1.20 *Fix $p \in\,]1,+\infty[$. Let $f \in L^p_{loc}(\Omega)$, $g \in W^{2,p}_{loc}(\Omega) \cap C^1(\overline{\Omega})$ be such that for some $r > 0$*

$$\sup_{x\in\overline{\Omega}}\|f\|_{L^p(\Omega_{x,r})} + \sup_{x\in\overline{\Omega}}\|D^2 g\|_{L^p(\Omega_{x,r})} < \infty.$$

Then for Re $\lambda \geq \Lambda_p$ problem (3.1.57) has a unique solution $u \in W^{2,p}_{loc}(\Omega) \cap C^1(\overline{\Omega})$.

From Theorem 3.1.19 and Proposition 3.1.20, arguing as in the case $\Omega = \mathbb{R}^n$, we get a number of generation results.

1. Second order operators

Corollary 3.1.21 *Let Λ_p be the constant given by Theorem 3.1.19, and set $\Lambda_0 = \inf_{p>n} \Lambda_p$.*
(i) Set

$$\begin{cases} D(A_\infty) = \{u \in \bigcap_{p\geq 1} W^{2,p}_{loc}(\Omega) \,:\, u, \mathcal{A}(\cdot,D)u \in L^\infty(\Omega),\, u_{|\partial\Omega} = 0\}, \\ A_\infty : D(A_\infty) \mapsto L^\infty(\Omega),\; A_\infty u = \mathcal{A}(\cdot,D)u. \end{cases}$$

Then the resolvent set of A_∞ contains the halfplane $\{\lambda \in \mathbb{C} : \operatorname{Re}\lambda > \Lambda_0\}$, and A_∞ is sectorial. Moreover, $D(A_\infty) \subset C^{1+\alpha}(\overline{\Omega})$, for every $\alpha \in\,]0,1[$, and $\overline{D(A_\infty)} = \{u \in UC(\overline{\Omega}) : u_{|\partial\Omega} = 0\}$.
(ii) Set

$$\begin{cases} D(A_0) = \{u \in \bigcap_{p\geq 1} W^{2,p}_{loc}(\Omega) \,:\, u, \mathcal{A}(\cdot,D)u \in C(\overline{\Omega}),\, u_{|\partial\Omega} = 0\}, \\ A_0 : D(A_0) \mapsto C(\overline{\Omega}),\; A_0 u = \mathcal{A}(\cdot,D)u. \end{cases} \qquad (3.1.58)$$

Then the resolvent set of A_0 contains the halfplane $\{\lambda \in \mathbb{C} : \operatorname{Re}\lambda > \Lambda_0\}$, and A_0 is sectorial. Moreover, $\overline{D(A_0)} = \{u \in UC(\overline{\Omega}) : u_{|\partial\Omega} = 0\}$. If $n = 1$, then $D(A_0) = \{u \in C^2(\overline{\Omega}),\, u_{|\partial\Omega} = 0\}$.
(iii) If Ω is unbounded, set

$$\begin{cases} D(A_{UC}) = \{u \in \bigcap_{p\geq 1} W^{2,p}_{loc}(\Omega) \,:\, u, \mathcal{A}(\cdot,D)u \in UC(\overline{\Omega}),\, u_{|\partial\Omega} = 0\}, \\ A_{UC} : D(A_{UC}) \mapsto UC(\overline{\Omega}),\; A_{UC} u = \mathcal{A}(\cdot,D)u. \end{cases}$$

Then the resolvent set of A_{UC} contains the halfplane $\{\lambda \in \mathbb{C} : \operatorname{Re}\lambda > \Lambda_0\}$, and A_{UC} is sectorial. Moreover, $\overline{D(A_{UC})} = \{u \in UC(\overline{\Omega}) : u_{|\partial\Omega} = 0\}$, and the set $\{u \in UC^2(\overline{\Omega}) : u_{|\partial\Omega} = 0\}$ is dense in $D(A_{UC})$.
  If $n = 1$, then $D(A_{UC}) = \{u \in UC^2(\overline{\Omega}),\, u_{|\partial\Omega} = 0\}$.
(iv) If Ω is unbounded, set $C_*(\overline{\Omega}) = \{u \in C(\overline{\Omega}) : \lim_{|x|\to\infty} u(x) = 0\}$, and

$$\begin{cases} D(A_*) = \{u \in \bigcap_{p\geq 1} W^{2,p}_{loc}(\Omega) \,:\, u, \mathcal{A}(\cdot,D)u \in C_*(\overline{\Omega}),\, u_{|\partial\Omega} = 0\} \\ A_* : D(A_*) \mapsto C_*(\overline{\Omega}),\; A_* u = \mathcal{A}(\cdot,D)u. \end{cases}$$

Then the resolvent set of A_* contains the halfplane $\{\lambda \in \mathbb{C} : \operatorname{Re}\lambda > \Lambda_0\}$, and A_* is sectorial. Moreover, $\overline{D(A_*)} = \{u \in C_*(\overline{\Omega}) : u_{\partial\Omega} = 0\}$. If $n = 1$, then $D(A_*) = C^2(\overline{\Omega}) \cap C_*(\overline{\Omega})$.

Now we consider the first order oblique boundary condition. Also in this case, the starting point is an *a priori* estimate similar to (3.1.25). We recall that $\mathcal{A} = \mathcal{A}(\cdot, D)$ is the operator defined in (3.1.1), and $\mathcal{B} = \mathcal{B}(\cdot, D)$ is the operator defined in (3.1.3).

Theorem 3.1.22 *For every $p > n$ there are $K_p, \Lambda_p > 0$, such that for $\lambda \in \mathbb{C}$ with $\operatorname{Re} \lambda \geq \Lambda_p$, for every $u \in W^{2,p}_{loc}(\Omega) \cap C^1(\overline{\Omega})$*

$$\begin{aligned}
&|\lambda| \|u\|_\infty + |\lambda|^{1/2} \|Du\|_\infty + |\lambda|^{n/2p} \sup\nolimits_{x_0 \in \overline{\Omega}} \|D^2 u\|_{L^p(\Omega_{x_0, |\lambda|^{-1/2}})} \\
&\leq K_p(|\lambda|^{n/2p} \sup\nolimits_{x_0 \in \overline{\Omega}} \|\lambda u - \mathcal{A}(\cdot, D)u\|_{L^p(\Omega_{x_0, |\lambda|^{-1/2}})} \\
&+ |\lambda|^{1/2} \|g_1\|_\infty + |\lambda|^{n/2p} \sup\nolimits_{x_0 \in \overline{\Omega}} \|Dg_1\|_{L^p(\Omega_{x_0, |\lambda|^{-1/2}})})
\end{aligned} \quad (3.1.59)$$

where g_1 is any extension of $g = \mathcal{B}u_{|\partial\Omega}$ belonging to $W^{1,p}_{loc}(\Omega)$. In addition, there is $\widetilde{K}_p > 0$ such that if $\mathcal{A}(\cdot, D)u \in C(\overline{\Omega})$ and $\mathcal{B}u_{|\partial\Omega} \in C^1(\partial\Omega)$, then

$$\begin{aligned}
&|\lambda| \|u\|_\infty + |\lambda|^{1/2} \|Du\|_\infty + |\lambda|^{n/2p} \sup\nolimits_{x_0 \in \overline{\Omega}} \|D^2 u\|_{L^p(\Omega_{x_0, |\lambda|^{-1/2}})}) \\
&\leq \widetilde{K}_p(\|\lambda u - \mathcal{A}(\cdot, D)u\|_\infty + |\lambda|^{1/2} \|\mathcal{B}u\|_{C(\partial\Omega)} + \|\mathcal{B}u\|_{C^1(\partial\Omega)}).
\end{aligned} \quad (3.1.60)$$

Proof — Again, the proof is similar to the proof of Theorem 3.1.6. We have only to modify estimate (3.1.21): we shall show that for $u \in W^{2,p}_{loc}(\Omega)$, $x_0 \in \overline{\Omega}$, $r \leq 1$, $\alpha \geq 1$, and $\operatorname{Re} \lambda \geq \omega_p$ (ω_p is given by Theorem 3.1.3(iii)),

$$\begin{aligned}
&|\lambda| \|u\|_{L^p(\Omega_{x_0, r})} + |\lambda|^{1/2} \|Du\|_{L^p(\Omega_{x_0, r})} + \|D^2 u\|_{L^p(\Omega_{x_0, r})} \\
&\leq C_p \big[\|f\|_{L^p(\Omega_{x_0, (\alpha+1)r})} + \tfrac{1}{\alpha}(r^{-2} \|u\|_{L^p(\Omega_{x_0, (\alpha+1)r})} + r^{-1} \|Du\|_{L^p(\Omega_{x_0, (\alpha+1)r})}) \\
&+ (|\lambda|^{1/2} + \tfrac{1}{\alpha r}) \|g_1\|_{L^p(\Omega_{x_0, (\alpha+1)r})} + \|Dg_1\|_{L^p(\Omega_{x_0, (\alpha+1)r})}\big]
\end{aligned} \quad (3.1.61)$$

where $f = \lambda u - \mathcal{A}u$, and g_1 is any extension to $\overline{\Omega}$ of $\mathcal{B}u_{|\partial\Omega}$. To prove that (3.1.61) holds, we introduce the function $v = \theta u$ (where θ is the cutoff function used in the proof of Theorem 3.1.6), which satisfies

$$\lambda v - \mathcal{A}(\cdot, D)v = F \ \text{ in } \Omega, \quad \mathcal{B}v = \theta g + u \sum_{i=1}^n \beta_i D_i \theta \ \text{ in } \partial\Omega,$$

F being defined in (3.1.22). Since $\operatorname{Re} \lambda \geq \omega_p$, estimate (3.1.13) gives

$$\begin{aligned}
&|\lambda| \|v\|_{L^p(\Omega)} + |\lambda|^{1/2} \|Dv\|_{L^p(\Omega)} + \|D^2 v\|_{L^p(\Omega)} \\
&\leq M_p(\|F\|_{L^p(\Omega)} + |\lambda|^{1/2} \|\theta g_1 + u \sum_{i=1}^n \beta_i D_i \theta\|_{L^p(\Omega)} \\
&+ \|D(\theta g_1) + D(u \sum_{i=1}^n \beta_i D_i \theta)\|_{L^p(\Omega)}).
\end{aligned} \quad (3.1.62)$$

1. Second order operators

The estimate of $\|F\|_{L^p(\Omega)}$ is similar to (3.1.24); it is sufficient to replace $\mathbb{R}^n$ by Ω and $B(x_0, (\alpha+1)r)$ by $\Omega_{x_0,(\alpha+1)r}$. Moreover,

$$|\lambda|^{1/2}\|u\sum_{i=1}^n \beta_i D_i\theta\|_{L^p(\Omega)} + \|D(u\sum_{i=1}^n \beta_i D_i\theta)\|_{L^p(\Omega)}$$
$$\leq |\lambda|^{1/2}\sum_{i=1}^n \|\beta_i\|_\infty \frac{K}{\alpha r}\|u\|_{L^p(\Omega_{x_0,(\alpha+1)r})}$$
$$+ \sum_{i=1}^n \left(\|D\beta_i\|_\infty \frac{K}{\alpha r} + \|\beta_i\|_\infty \frac{K}{\alpha^2 r^2}\right)\|u\|_{L^p(\Omega_{x_0,(\alpha+1)r})}$$
$$+ \sum_{i=1}^n \|\beta_i\|_\infty \frac{K}{\alpha r}\|Du\|_{L^p(\Omega_{x_0,(\alpha+1)r})}$$
$$\leq \frac{C_0 K}{\alpha}\left[\left(\frac{|\lambda|^{1/2}}{r} + \frac{2}{r^2}\right)\|u\|_{L^p(\Omega_{x_0,(\alpha+1)r})} + \frac{1}{r}\|Du\|_{L^p(\Omega_{x_0,(\alpha+1)r})}\right],$$

with $C_0 = \sum_{i=1}^n \|\beta_i\|_{C^1(\overline{\Omega})}$, and

$$|\lambda|^{1/2}\|\theta g_1\|_{L^p(\Omega)} + \|D(\theta g_1)\|_{L^p(\Omega)}$$
$$\leq |\lambda|^{1/2}\|g_1\|_{L^p(\Omega_{x_0,(\alpha+1)r})} + \frac{K}{\alpha r}\|g_1\|_{L^p(\Omega_{x_0,(\alpha+1)r})} + \|Dg_1\|_{L^p(\Omega_{x_0,(\alpha+1)r})}.$$

Replacing the above estimates and (3.1.24) in (3.1.62), (3.1.61) follows.

Now the same procedure of Theorem 3.1.6 gives estimate (3.1.59). To prove (3.1.60), we proceed as in Theorem 3.1.19: we use (3.1.59), getting

$$|\lambda|\,\|u\|_\infty + |\lambda|^{1/2}\|Du\|_\infty + |\lambda|^{n/2p}\sup_{x_0\in\overline{\Omega}}\|D^2 u\|_{L^p(\Omega_{x_0,|\lambda|^{-1/2}})}$$
$$\leq K_p\left[\gamma_n^{1/p}(\|\lambda u - \mathcal{A}(\cdot, D)u\|_\infty + \|g_1\|_\infty) + \|Dg_1\|_\infty\right].$$

Choosing now $g_1 = \mathcal{E}(\mathcal{B}u_{|\partial\Omega})$, where $\mathcal{E}$ is any extension operator belonging to $L(C(\partial\Omega), C(\overline{\Omega})) \cap L(C^1(\partial\Omega), C^1(\overline{\Omega}))$ (see e.g. Theorem 0.3.2(i)), estimate (3.1.60) follows. ∎

In the case where Ω is unbounded, Theorem 3.1.22 and Theorem 3.1.3(iii) yield an existence result for problem

$$\lambda u - \mathcal{A}u = f \text{ in } \Omega, \quad \mathcal{B}u = g \text{ in } \partial\Omega, \qquad (3.1.63)$$

with $f \in L^p_{loc}(\Omega)$, $g \in W^{1,p}_{loc}(\Omega) \cap C(\overline{\Omega})$.

Proposition 3.1.23 *Let $p > n$, and $\operatorname{Re}\lambda \geq \Lambda_p$ (Λ_p is given by Theorem 3.1.22). If $f \in L^p_{loc}(\Omega)$ and $g \in W^{1,p}_{loc}(\Omega) \cap C(\overline{\Omega})$ are such that for some $r > 0$*

$$\sup_{x\in\overline{\Omega}}\|f\|_{L^p(\Omega_{x,r})} + \sup_{x\in\overline{\Omega}}\|Dg\|_{L^p(\Omega_{x,r})} < \infty$$

then problem (3.1.63) has a unique solution $u \in W^{2,p}_{loc}(\Omega) \cap C^1(\overline{\Omega})$.

If Ω is bounded, Proposition 3.1.23 reduces to statement (iii) of Theorem 3.1.2. If Ω is unbounded, the proof is similar to the proofs of Theorem 3.1.7 and Remark 3.1.8, and it is omitted.

Using Theorem 3.1.22 and Proposition 3.1.23, arguing as in the case $\Omega = \mathbb{R}^n$, several generation results follow.

Corollary 3.1.24 *Let Λ_p be the constant given by Theorem 3.1.22, and set $\Lambda_1 = \inf_{p>n} \Lambda_p$.*
(i) Set

$$\begin{cases} D(A_{1\infty}) = \{u \in \bigcap_{p\geq 1} W^{2,p}_{loc}(\Omega) \, : \, u, \, \mathcal{A}u \in L^\infty(\Omega), \, \mathcal{B}u_{|\partial\Omega} = 0\}, \\ A_{1\infty} : D(A_\infty) \mapsto L^\infty(\Omega), \, A_{1\infty}u = \mathcal{A}u. \end{cases}$$

Then the resolvent set of $A_{1\infty}$ contains the halfplane $\{\lambda \in \mathbb{C} \, : \, \operatorname{Re}\lambda > \Lambda_1\}$, and $A_{1\infty}$ is sectorial. Moreover, $D(A_\infty) \subset C^{1+\alpha}(\overline{\Omega})$ for every $\alpha \in]0,1[$, and $\overline{D(A_{1\infty})} = UC(\overline{\Omega})$.
(ii) Set

$$\begin{cases} D(A_1) = \{u \in \bigcap_{p\geq 1} W^{2,p}_{loc}(\Omega) \, : \, u, \, \mathcal{A}u \in C(\overline{\Omega}), \, \mathcal{B}u_{|\partial\Omega} = 0\}, \\ A_1 : D(A_1) \mapsto C(\overline{\Omega}), \, A_1 u = \mathcal{A}u. \end{cases} \qquad (3.1.64)$$

Then the resolvent set of A_1 contains the halfplane $\{\lambda \in \mathbb{C} : \operatorname{Re}\lambda > \Lambda_1\}$, and A_1 is sectorial. Moreover, $D(A_1) \subset C^{1+\alpha}(\overline{\Omega})$, for every $\alpha \in]0,1[$, and $\overline{D(A_1)} = UC(\overline{\Omega})$. If Ω is bounded then $D(A_1)$ is dense in $C(\overline{\Omega})$.
(iii) If Ω is unbounded, set

$$\begin{cases} D(A_{1UC}) = \{u \in \bigcap_{p\geq 1} W^{2,p}_{loc}(\Omega) \, : \, u, \, \mathcal{A}(\cdot,D)u \in UC(\overline{\Omega}), \, \mathcal{B}u_{|\partial\Omega} = 0\} \\ A_{1UC} : D(A_{1UC}) \mapsto UC(\overline{\Omega}), \, A_{1UC}u = \mathcal{A}(\cdot,D)u. \end{cases}$$

Then the resolvent set of A_{1UC} contains the halfplane $\{\lambda \in \mathbb{C} : \operatorname{Re}\lambda > \Lambda_1\}$, A_{1UC} is sectorial, and $D(A_{1UC})$ is dense in $UC(\overline{\Omega})$.
(iv) If Ω is unbounded, let $C_*(\overline{\Omega})$ be defined as in Corollary 3.1.21(iv), and set

$$\begin{cases} D(A_{1*}) = \{u \in \bigcap_{p\geq 1} W^{2,p}_{loc}(\Omega) \, : \, u, \, \mathcal{A}u \in C_*(\overline{\Omega}), \, \mathcal{B}u_{|\partial\Omega} = 0\} \\ A_{1*} : D(A_{1*}) \mapsto C_*(\overline{\Omega}), \, A_{1*}u = \mathcal{A}(\cdot,D)u. \end{cases}$$

Then the resolvent set of A_{1*} contains the halfplane $\{\lambda \in \mathbb{C} : \operatorname{Re}\lambda > \Lambda_1\}$, A_{1*} is sectorial, and $D(A_{1*})$ is dense in $C_*(\overline{\Omega})$.

1. Second order operators

Generation in C^1 spaces

As in the case $\Omega = \mathbb{R}^n$, generation results in spaces of C^1 functions may be deduced from the generation results in spaces of continuous functions. We set

$$C_0^1(\overline{\Omega}) = \{u \in C^1(\overline{\Omega}) : u_{|\partial\Omega} = 0\}.$$

Theorem 3.1.25 Let A_0 be defined by (3.1.58), and set

$$\begin{cases} A : D(A) = \{u \in D(A_0) : \mathcal{A}u \in C_0^1(\overline{\Omega})\} \mapsto C_0^1(\overline{\Omega}), \\ Au = \mathcal{A}(\cdot, D)u. \end{cases} \quad (3.1.65)$$

Then A is sectorial in $C_0^1(\overline{\Omega})$. More precisely, there exist $\overline{\lambda}$, $K > 0$ such that $\rho(A) \supset \{\lambda \in \mathbb{C} : \operatorname{Re}\lambda \geq \overline{\lambda}\}$, and for every $u \in C_0^1(\overline{\Omega}) \cap W_{loc}^{2,p}(\Omega)$ it holds

$$|\lambda|\,\|u\|_{C^1(\overline{\Omega})} + |\lambda|^{1/2}\|Au\|_\infty + |\lambda|^{1/2+n/2p} \sup_{x \in \overline{\Omega}} \|D^2 u\|_{L^p(\Omega_{x,|\lambda|^{-1/2}})} \\ \leq C\|\lambda u - Au\|_{C^1(\overline{\Omega})} \quad (3.1.66)$$

with $C > 0$ independent of λ and u.

If in addition the boundary $\partial\Omega$ is uniformly C^3 and the coefficients of $\mathcal{A}$ belong to $UC^1(\overline{\Omega})$, then $D(A)$ is contained in $W_{loc}^{3,p}(\Omega) \cap C^{2+\alpha}(\overline{\Omega})$ for every $p \geq 1$ and $\alpha \in \,]0,1[$.

Proof — We first consider the case where $\overline{\Omega}$ is the half space $\mathbb{R}_+^n$, and the coefficients of $\mathcal{A}$ belong to $UC^1(\mathbb{R}_+^n)$. We follow as far as possible the proof of Propositions 3.1.17 and 3.1.18.

The resolvent set of A contains the half plane $\{\operatorname{Re}\lambda \geq \Lambda_0\}$, where Λ_0 is given by Corollary 3.1.21: indeed, from Theorem 3.1.19, we know that for every $f \in C(\overline{\Omega})$ (and hence, for every $f \in C^1(\overline{\Omega})$), the problem

$$\lambda u - \mathcal{A}u = f \text{ in } \mathbb{R}_+^n, \quad u_{|\partial\mathbb{R}_+^n} = 0,$$

has a unique solution $u \in D(A_0)$, and $\|u\|_{C^1(\overline{\Omega})} \leq C(\lambda)\|f\|_\infty$.

Let us prove that $D(A)$ is continuously embedded in $W_{loc}^{3,p}(\mathbb{R}^n)$ for every p. This is also a preliminary step to prove the decay estimate of the resolvent. Fixed any $p > n$, let Λ_p, K_p be the constants given by Theorem 3.1.19. For $u \in D(A)$ and $\operatorname{Re}\lambda \geq \Lambda_p$ set $f = \lambda u - \mathcal{A}(\cdot, D)u$, and consider the problems formally satisfied by the derivative $v = D_k u$:

$$\begin{cases} \lambda v - \mathcal{A}v = D_k f + \sum_{i,j=1}^n (D_k a_{ij}) D_{ij} u + \sum_{i=1}^n (D_k b_i) D_i u + (D_k c) u, \\ v_{|\partial\mathbb{R}_+^n} = 0, \end{cases} \quad (3.1.67)$$

for $k = 1, \ldots, n-1$, and

$$\begin{cases} \lambda v - \mathcal{A}v = D_n f + \sum_{i,j=1}^{n}(D_n a_{ij})D_{ij}u + \sum_{i=1}^{n}(D_n b_i)D_i u + (D_n c)u, \\ (a_{nn}D_n v + b_n v)_{|\partial \mathbb{R}_+^n} = 0, \end{cases}$$
(3.1.68)

for $k = n$. The right hand sides g_k of (3.1.67) and (3.1.68) belong to $L^p_{loc}(\mathbb{R}_+^n)$, and thanks to (3.1.53) we have

$$\sup_{x \in \mathbb{R}_+^n} \|g_k\|_{L^p(B(x,r) \cap \mathbb{R}_+^n)} < \infty, \quad k = 1, \ldots, n$$

for every $r > 0$. Due to Propositions 3.1.20 and 3.1.23, problems (3.1.67) and (3.1.68) have unique solutions $v_k \in C^1(\mathbb{R}_+^n) \cap W^{2,p}_{loc}(\mathbb{R}_+^n)$, for $k = 1, \ldots, n$. Arguing as in the proof of Theorem 3.1.7, one sees that the incremental ratios $\tau_{h,k}u(x) = h^{-1}(u(x+he_k) - u(x))$ converge to v_k in $C(\mathbb{R}_+^n)$ as $h \to 0$, for $k = 1, \ldots, n-1$. Therefore, $v = D_k u$. Moreover, from the equality $\lambda u - \mathcal{A}u = f$ we get

$$D_{nn}u = \frac{1}{a_{nn}}\left(\lambda u - \sum_{(i,j) \neq (n,n)} a_{ij}D_{ij}u - \sum_{i=1}^{n} b_i D_i u - cu - f\right).$$

All the addenda in the right hand side belong to $C(\mathbb{R}_+^n) \cap W^{1,p}_{loc}(\mathbb{R}_+^n)$, so that also $D_{nn}u$ does. Therefore, $u \in W^{3,p}_{loc}(\mathbb{R}_+^n) \cap C^2(\mathbb{R}_+^n)$, and $D_n u$ is the solution of (3.1.68).

By applying estimates (3.1.54) and (3.1.59) to equations (3.1.67) and (3.1.68) respectively, and arguing as in estimate (3.1.46), we find

$$|\lambda|\,\|D_k u\|_\infty + |\lambda|^{1/2}\|D(D_k u)\|_\infty$$
$$+ \sup_{x \in \mathbb{R}_+^n} \|D^2(D_k u)\|_{L^p(B(x,|\lambda|^{-1/2}) \cap \mathbb{R}_+^n)} \leq C\|f\|_{C^1}, \quad k = 1, \ldots, n.$$

Moreover, the same argument used in the proof of Theorem 3.1.7 gives that $D_k u \in C^{1+\alpha}(\mathbb{R}_+^n)$ for every $k = 1, \ldots, n$, $\alpha \in\,]0,1[$.

Using once again (3.1.54) we get

$$\|u\|_{C^1} = \|u\|_\infty + \sum_{k=1}^{n} \|D_k u\|_\infty \leq |\lambda|^{-1}(K_p \gamma_n^{1/p} + nC)\|f\|_{C^1},$$

so that $\mathcal{A}$ is sectorial in $C^1(\overline{\Omega})$.

Let us consider now the case where Ω is any open set with uniformly C^2 boundary, and the coefficients of $\mathcal{A}$ belong to $UC^1(\overline{\Omega})$. As in the case $\Omega = \mathbb{R}_+^n$, one deduces from Theorem 3.1.19 that the resolvent set of $\mathcal{A}$ contains the half plane $\{\mathrm{Re}\,\lambda > \Lambda_0\}$. The resolvent estimate (3.1.66) for $\mathrm{Re}\,\lambda$ large, and the inclusion $D(A) \subset C^{2+\alpha}(\overline{\Omega}) \cap W^{3,p}_{loc}(\Omega)$ may be proved flattening the boundary and using the corresponding results in the cases $\Omega = \mathbb{R}^n$ and $\Omega = \mathbb{R}_+^n$.

The proof of the statement in the case of uniformly continuous and bounded coefficients is identical to the proof of Proposition 3.1.18. ∎

1. Second order operators

Theorem 3.1.26 *Let A_1 be defined by (3.1.64), and set*

$$\begin{cases} A: D(A) = \{u \in D(A_1): \mathcal{A}u \in C^1(\overline{\Omega})\} \mapsto C^1(\overline{\Omega}), \\ Au = \mathcal{A}(\cdot, D)u. \end{cases}$$

Then A is sectorial in $C^1(\overline{\Omega})$. More precisely, there exist $\overline{\lambda}, K > 0$ such that $\rho(A) \supset \{\lambda \in \mathbb{C}: \operatorname{Re}\lambda \geq \overline{\lambda}\}$, and for every $u \in C^1(\overline{\Omega}) \cap W^{2,p}_{loc}(\Omega)$ estimate (3.1.66) holds. If in addition $\partial\Omega$ is uniformly C^3, the coefficients of $\mathcal{A}$ belong to $UC^1(\overline{\Omega})$, then $D(A)$ is continuously embedded in $W^{3,p}_{loc}(\Omega) \cap C^{2+\alpha}(\overline{\Omega})$ for every $p \geq 1, \alpha \in]0,1[$.

Proof — Theorem 3.1.22 implies easily that the resolvent set of A contains the half plane $\{\operatorname{Re}\lambda > \Lambda_1\}$, where Λ_1 is the constant given by Corollary 3.1.24. To prove the decay estimate of the resolvent, we follow the procedure of Theorem 3.1.25. However, there is an additional difficulty arising from the fact that, if $u \in D(A)$, then the derivatives $D_k u$ do not necessarily satisfy homogeneous boundary conditions.

As a first step, we consider the case where $\Omega = \mathbb{R}^n_+$, the coefficients of the operator $\mathcal{A}$ belong to $UC^1(\mathbb{R}^n_+)$, and the coefficients of the operator $\mathcal{B}$ belong to $UC^2(\mathbb{R}^n_+)$. We fix $p > n$, $\lambda \in \mathbb{C}$ with $\operatorname{Re}\lambda \geq \Lambda_p$ (Λ_p is given by Theorem 3.1.22). For every $u \in D(A)$ we set $f = \lambda u - \mathcal{A}u$. Then, for $k = 1, \ldots, n-1$, we consider the problem formally satisfied by $v = D_k u$,

$$\begin{cases} \lambda v - \mathcal{A}v = D_k f + \sum_{i,j=1}^n (D_k a_{ij}) D_{ij} u + \sum_{i=1}^n (D_k b_i) D_i u + (D_k c) u \\ \qquad = f_k, \\ \mathcal{B}v_{|\partial\mathbb{R}^n_+} = (-u D_k \gamma - \sum_{i=1}^n (D_k \beta_i) D_i u)_{|\partial\mathbb{R}^n_+}, \end{cases} \quad (3.1.69)$$

and the problem formally satisfied by $v = D_n u$,

$$\begin{cases} \lambda v - \mathcal{A}v = D_n f + \sum_{i,j=1}^n (D_n a_{ij}) D_{ij} u + \sum_{i=1}^n (D_n b_i) D_i u + (D_n c) u \\ \qquad = f_n, \\ v_{|\partial\mathbb{R}^n_+} = -\frac{1}{\beta_n}(\gamma u + \sum_{i=1}^{n-1} \beta_i D_i u)_{|\partial\mathbb{R}^n_+}. \end{cases} \quad (3.1.70)$$

First, we consider problem (3.1.69). For $k = 1, \ldots, n-1$ the function f_k belongs to $L^p_{loc}(\mathbb{R}^n_+)$; the function

$$g_k = -u D_k \gamma - \sum_{i=1}^n (D_k \beta_i) D_i u$$

belongs to $W^{1,p}_{loc}(\mathbb{R}^n_+)$, and thanks to estimate (3.1.59) we have

$$\sup_{x \in \mathbb{R}^n_+} \|f_k\|_{L^p(B(x,r) \cap \mathbb{R}^n_+)} + \|g\|_\infty + \sup_{x \in \mathbb{R}^n_+} \|Dg_k\|_{L^p(B(x,r) \cap \mathbb{R}^n_+)} < \infty,$$

for every $r > 0$. So, by Proposition 3.1.23, problem (3.1.69) has a unique solution $v_k \in C^1(\mathbb{R}^n_+) \cap W^{1,p}_{loc}(\mathbb{R}^n_+)$. Arguing as in the proof of Proposition 3.1.17, one sees that the incremental ratios $\tau_{h,k} u(x) = h^{-1}(u(x+he_k) - u(x))$ converge to v_k in $C(\mathbb{R}^n_+)$ as $h \to 0$. Therefore, $v_k = D_k u$. From the equality $\lambda u - \mathcal{A}u = f$ we get

$$D_{nn}u = \frac{1}{a_{nn}}\left(\lambda u - \sum_{(i,j) \neq (n,n)} a_{ij} D_{ij} u - \sum_{i=1}^n b_i D_i u - cu - f\right).$$

All the addenda in the right hand side belong to $C(\mathbb{R}^n_+) \cap W^{1,p}_{loc}(\mathbb{R}^n_+)$, so that also $D_{nn}u$ does. Therefore, $u \in W^{3,p}_{loc}(\mathbb{R}^n_+) \cap C^2(\mathbb{R}^n_+)$, and the function $v = D_n u$ satisfies (3.1.70).

To prove that A is sectorial, we apply estimates (3.1.59) and (3.1.54) to problems (3.1.69) and (3.1.70), respectively. For $k = 1, \ldots, n-1$, one finds, setting $U = \mathbb{R}^n_+ \cap B(x, |\lambda|^{-1/2})$,

$$\begin{aligned}&|\lambda| \|D_k u\|_\infty + |\lambda|^{1/2} \|D(D_k u)\|_\infty + |\lambda|^{n/2p} \sup_{x \in \mathbb{R}^n_+} \|D^2(D_k u)\|_{L^p(U)} \\ &\leq K_p \big(|\lambda|^{n/2p} \sup_{x \in \mathbb{R}^n_+} \|f_k\|_{L^p(U)} \\ &+ |\lambda|^{n/2p} \sup_{x \in \mathbb{R}^n_+} \|Dg_k\|_{L^p(U)} + |\lambda|^{1/2} \|g_k\|_\infty\big),\end{aligned} \qquad (3.1.71)$$

The estimation of f_k, $k = 1, \ldots, n$ can be obtained as in the proof of Proposition 3.1.17: one gets

$$|\lambda|^{n/2p} \sup_{x \in \mathbb{R}^n_+} \|f_k\|_{L^p(U)} \leq C \|f\|_{C^1(\mathbb{R}^n_+)}$$

(see (3.1.46)). Let us estimate g_k. Setting

$$C_1 = \|\gamma\|_{C^2(\mathbb{R}^n_+)} + \sum_{i=1}^n \|\beta_i\|_{C^2(\mathbb{R}^n_+)},$$

we have

$$\|g_k\|_\infty \leq C_1(\|u\|_\infty + \|Du\|_\infty) \leq C_1 K_p \gamma_n^{1/p}(|\lambda|^{-1} + |\lambda|^{-1/2})\|f\|_\infty,$$

and

$$\begin{aligned}\|Dg_k\|_{L^p(U)} &\leq C_1(\|u\|_{L^p(U)} + 2\|Du\|_{L^p(U)} + \|D^2 u\|_{L^p(U)}) \\ &\leq C_1(\gamma_n^{1/p}|\lambda|^{-n/2p}(\|u\|_\infty + 2\|Du\|_\infty) + \|D^2 u\|_{L^p(U)}) \\ &\leq C_1 K_p(\gamma_n^{1/p}(|\lambda|^{-1-n/2p} + 2|\lambda|^{-1/2-n/2p}) + |\lambda|^{-n/2p})\|f\|_\infty.\end{aligned}$$

Replacing in (3.1.71), we get, for $k = 1, \ldots, n-1$,

$$\begin{aligned}&|\lambda| \|D_k u\|_\infty + |\lambda|^{1/2} \|D(D_k u)\|_\infty + |\lambda|^{n/2p} \sup_{x \in \mathbb{R}^n_+} \|D^2(D_k u)\|_{L^p(U)} \\ &\leq const. \|f\|_{C^1}.\end{aligned} \qquad (3.1.72)$$

1. Second order operators

For $k = n$ we use estimate (3.1.53). Recalling that $D_n u$ satisfies (3.1.70) we get

$$|\lambda|\,\|D_n u\|_\infty + |\lambda|^{1/2}\|D(D_n u)\|_\infty + \sup_{x \in \mathbb{R}^n_+} \|D^2(D_n u)\|_{L^p(U)}$$
$$\leq K_p(|\lambda|^{n/2p} \sup_{x \in \mathbb{R}^n_+} \|f_n\|_{L^p(U)} \qquad (3.1.73)$$
$$+ |\lambda|^{n/2p} \sup_{x \in \mathbb{R}^n_+} \|D^2 g_n\|_{L^p(U)} + |\lambda|^{1/2}\|D g_n\|_\infty + |\lambda|\,\|g_n\|_\infty),$$

where

$$g_n = \frac{1}{\beta_n}\left(\gamma u + \sum_{i=1}^{n-1} \beta_i D_i u\right).$$

Let us estimate g_n. Since $1/\beta_n \in C^2(\mathbb{R}^n_+)$, it is sufficient to estimate $\varphi = \gamma u + \sum_{i=1}^{n-1} \beta_i D_i u$. It holds

$$\|\varphi\|_\infty \leq C_1\left(\|u\|_\infty + \sum_{k=1}^{n-1}\|D_k u\|_\infty\right) \leq C_1\left(\frac{K_p}{|\lambda|}\|f\|_\infty + \frac{(n-1)C_p}{|\lambda|}\|f\|_{C^1}\right),$$

$$\|D\varphi\|_\infty \leq \|\gamma\|_{C^1}\|u\|_{C^1} + \sum_{k=1}^{n-1}\|\beta_k\|_{C^1}\|D_k u\|_{C^1}$$
$$\leq C_1\left(\frac{K_p}{|\lambda|^{1/2}}\|f\|_\infty + (n-1)\frac{C_p}{|\lambda|^{1/2}}\|f\|_{C^1}\right),$$

and

$$\|D\varphi\|_{L^p(U)} \leq \|\gamma\|_{C^2}\|u\|_{W^{2,p}(U)} + \sum_{k=1}^{n-1}\|\beta_k\|_{C^2}\|D_k u\|_{W^{2,p}(U)}$$
$$\leq C_1(\gamma_n^{1/p}|\lambda|^{-n/2p}(\|u\|_\infty + \|Du\|_\infty) + \|D^2 u\|_{L^p(U)}$$
$$+ \gamma_n^{1/p}|\lambda|^{-n/2p}\sum_{k=1}^{n-1}(\|D_k u\|_\infty + \|D(D_k u)\|_\infty) + \|D^2(D^k u)\|_{L^p(U)})$$
$$\leq C_1[K_p(\gamma_n^{1/p}(|\lambda|^{-1-n/2p} + |\lambda|^{-1/2-n/2p}) + |\lambda|^{-n/2p}]\|f\|_\infty$$
$$+ (n-1)C_p[\gamma_n^{1/p}(|\lambda|^{-1-n/2p} + |\lambda|^{-1/2-n/2p}) + |\lambda|^{-n/2p})\|f\|_{C^1}].$$

Replacing in (3.1.73), we get

$$|\lambda|\,\|D_n u\|_\infty + |\lambda|^{1/2}\|D(D_n u)\|_\infty + \sup_{x \in \mathbb{R}^n_+}\|D^2(D_n u)\|_{L^p(U)}$$
$$\leq \text{const.}\,\|f\|_{C^1(\mathbb{R}^n_+)},$$

which, added to (3.1.72), gives

$$|\lambda|\,\|Du\|_\infty + |\lambda|^{1/2}\|D^2 u\|_\infty$$
$$+ |\lambda|^{n/2p}\sup_{x \in \mathbb{R}^n_+}\|D^3 u\|_{L^p(B(x,|\lambda|^{-1/2})\cap \mathbb{R}^n_+)} \leq \text{const.}\,\|f\|_{C^1(\mathbb{R}^n_+)}. \qquad (3.1.74)$$

Recalling estimate (3.1.59) and Proposition 2.1.11, we get that A is sectorial in $C^1(\mathbb{R}^n_+)$.

In the case where the coefficients of $\mathcal{A}$ are uniformly continuous and bounded, and the coefficients of $\mathcal{B}$ belong to $C^1(\overline{\mathbb{R}^n_+})$, the statement may be proved as in Proposition 3.1.18. A remark is worth to be made: since $\beta_n(x) \geq \varepsilon$ for $x \in \partial\mathbb{R}^n_+$, we may assume without loss of generality that $\beta_n(x) \geq \varepsilon$ for every $x \in \overline{\mathbb{R}^n_+}$, so that for every $x_0 \in \overline{\mathbb{R}^n_+}$ the operator with constant coefficients $\sum_{i=1}^n \beta_i(x_0)D_i$ satisfies the uniform nontangentiality assumption (3.1.4). Referring to Proposition 3.1.18 for notation, fixed any $x_0 \in \overline{\mathbb{R}^n_+}$ it is convenient to split θu in the sum $\theta u = v_1 + v_2 + v_3$, where

$$\begin{cases} \lambda v_1 - \sum_{i,j=1}^n a_{ij}(x_0)D_{ij}v_1 = \theta \sum_{i,j=1}^n (a_{ij}(x_0) - a_{ij}(\cdot))D_{ij}u & \text{in } \mathbb{R}^n_+, \\ \sum_{i=1}^n \beta_i(x_0)D_i v_1 = \theta \sum_{i=1}^n (\beta_i(x_0) - \beta_i(\cdot))D_i u & \text{in } \partial\mathbb{R}^n_+, \end{cases}$$

$$\begin{cases} \lambda v_2 - \sum_{i,j=1}^n a_{ij}(x_0)D_{ij}v_2 = -u\left(\sum_{i,j=1}^n a_{ij}(x_0)D_{ij}\theta + c\theta\right) \\ \qquad -2\sum_{i,j=1}^n a_{ij}(x_0)D_i\theta D_j u - \theta\sum_{i=1}^n b_i D_i u & \text{in } \mathbb{R}^n_+, \\ \sum_{i=1}^n \beta_i(x_0)D_i v_2 = u\left(-\theta\gamma + \sum_{i=1}^n \beta_i(x_0)D_i\theta\right) & \text{in } \partial\mathbb{R}^n_+, \end{cases}$$

$$\begin{cases} \lambda v_3 - \mathcal{A}(x_0, D)v_3 = \theta f & \text{in } \mathbb{R}^n_+, \\ \sum_{i=1}^n \beta_i(x_0)D_i v_3 = 0 & \text{in } \partial\mathbb{R}^n_+. \end{cases}$$

From now on it is sufficient to follow the proof of Proposition 3.1.18, with the obvious modifications due to the boundary conditions. As usual, the case of a general Ω may be reduced to the cases $\Omega = \mathbb{R}^n$, $\Omega = \mathbb{R}^n_+$. ∎

Interpolatory inclusions and characterization of interpolation spaces

From now on we work in the space $X = C(\overline{\Omega})$, and we consider the operators A_0 and A_1 defined respectively by (3.1.58) and (3.1.64). The results which follow may be easily extended to the spaces $L^\infty(\Omega)$, $UC(\overline{\Omega})$, $C_0(\overline{\Omega})$, $C_*(\overline{\Omega})$, and to the realizations of $\mathcal{A}(\cdot, D)$ in such spaces, with Dirichlet or oblique boundary condition. See Corollaries 3.1.21 and 3.1.24.

We are going to prove some interpolatory inclusions, which will be used in Chapter 7 in the applications to semilinear parabolic problems. Moreover, they will let us characterize the interpolation spaces $D_{A_0}(\theta, \infty)$ and $D_{A_1}(\theta, \infty)$. See next theorems 3.1.29, 3.1.30.

1. Second order operators   107

Proposition 3.1.27 *The space $C_0^1(\overline{\Omega})$ (respectively, $C^1(\overline{\Omega})$) belongs to the class $K_{1/2}$ between $C(\overline{\Omega})$ and $D(A_0)$ (respectively, $D(A_1)$). In other words,*

$$C_0^1(\overline{\Omega}) \subset D_{A_0}(1/2, \infty), \quad C^1(\overline{\Omega}) \subset D_{A_1}(1/2, \infty).$$

Proof — Proposition 2.2.6 yields that if A is any sectorial operator, then $f \in D_A(1/2, \infty)$ if and only if $\limsup_{\lambda \mapsto +\infty} \|\lambda^{1/2} AR(\lambda, A)f\| < \infty$. For $f \in C_0^1(\overline{\Omega})$ (respectively, $f \in C^1(\overline{\Omega})$) such an estimate is provided by (3.1.66). ∎

Proposition 3.1.28 *Let $0 \leq \alpha < 1$. Then the space $C^{1+\alpha}(\overline{\Omega})$ belongs to the class $J_{(1+\alpha)/2}$ between $C(\overline{\Omega})$ and $D(A_0)$, and between $C(\overline{\Omega})$ and $D(A_1)$. In other words,*

$$D_{A_0}((1+\alpha)/2, 1) \subset C^{1+\alpha}(\overline{\Omega}), \quad D_{A_1}((1+\alpha)/2, 1) \subset C^{1+\alpha}(\overline{\Omega}).$$

The proof is the same as in Proposition 3.1.11, and it is omitted.

We consider now some spaces of functions defined in $\overline{\Omega}$, satisfying homogeneous boundary conditions. If Y is any of the symbols C^α, h^α ($\alpha > 0$), C^1, then $Y_0(\overline{\Omega})$ denotes the subset of $Y(\overline{\Omega})$ consisting of all the functions in $Y(\overline{\Omega})$ which vanish on the boundary. Moreover, for $\alpha \geq 0$ we set

$$C_{\mathcal{B}}^{1+\alpha}(\overline{\Omega}) = \{u \in C^{1+\alpha}(\overline{\Omega}) : \mathcal{B}(x,D)u_{|\partial\Omega} = 0\},$$

$$h_{\mathcal{B}}^{1+\alpha}(\overline{\Omega}) = \{u \in h^{1+\alpha}(\overline{\Omega}) : \mathcal{B}(x,D)u_{|\partial\Omega} = 0\}.$$

Theorem 3.1.29 *Let $0 < \theta < 1$. Then*

$$D_{A_0}(\theta, \infty) = \begin{cases} C_0^{2\theta}(\overline{\Omega}), & \text{if } \theta \neq 1/2, \\ \mathcal{C}_0^1(\overline{\Omega}), & \text{if } \theta = 1/2. \end{cases}$$

with equivalence of the respective norms. Moreover, $D_{A_0}(\theta) = h_0^{2\theta}(\overline{\Omega})$ if $\theta \neq 1/2$.

Proof — Due to Corollary 1.2.18(ii) we have

$$(C(\overline{\Omega}), C^2(\overline{\Omega}))_{\theta,\infty} = \begin{cases} C^{2\theta}(\overline{\Omega}), & \text{if } \theta \neq 1/2, \\ \mathcal{C}^1(\overline{\Omega}), & \text{if } \theta = 1/2, \end{cases} \quad (3.1.75)$$

with equivalence of the respective norms. It follows easily that

$$(C(\overline{\Omega}), C_0^2(\overline{\Omega}))_{\theta,\infty} = \begin{cases} C_0^{2\theta}(\overline{\Omega}), & \text{if } \theta \neq 1/2, \\ \mathcal{C}_0^1(\overline{\Omega}), & \text{if } \theta = 1/2. \end{cases} \quad (3.1.76)$$

Indeed, if $f \in (C(\overline{\Omega}), C_0^2(\overline{\Omega}))_{\theta,\infty}$, then it belongs to $C^{2\theta}(\overline{\Omega})$ (respectively, to $\mathcal{C}^1(\overline{\Omega})$) if $\theta = 1/2$, and it vanishes on $\partial\Omega$ because $(C(\overline{\Omega}), C_0^2(\overline{\Omega}))_{\theta,\infty}$ is contained in the closure of $C_0^2(\overline{\Omega})$ in $C(\overline{\Omega})$. Conversely, if $f \in C_0^{2\theta}(\overline{\Omega})$ (respectively, $f \in \mathcal{C}_0^1(\overline{\Omega})$ if

$\theta = 1/2$), then it is the trace at $t = 0$ of a function $u \in W(1-\theta, \infty, C^2(\overline{\Omega}), C(\overline{\Omega}))$. Setting $\widetilde{u} = u - \mathcal{D}u_{|\partial\Omega}$, where $\mathcal{D}$ is the extension operator provided by Theorem 0.3.2(i), then $\widetilde{u}$ belongs to $W(1-\theta, \infty, C_0^2(\overline{\Omega}), C(\overline{\Omega}))$, and $\widetilde{u}(0) = f - \mathcal{D}f_{|\partial\Omega} = f$. Therefore, $f \in (C(\overline{\Omega}), C_0^2(\overline{\Omega}))_{\theta,\infty}$, and (3.1.76) is proved.

The characterizations of $D_{A_0}(\theta, \infty)$ and $D_A(\theta)$ follow now as in the proof of Theorem 3.1.12, using Lemma 3.1.28 instead of Proposition 3.1.11. ∎

The characterizations of $D_{A_1}(\theta, \infty)$ and $D_{A_1}(\theta)$ are more complicated in the case $\theta > 1/2$, due to the first order boundary condition[2].

Theorem 3.1.30 *Let $0 < \theta < 1$. Then*

$$D_{A_1}(\theta, \infty) = \begin{cases} C^{2\theta}(\overline{\Omega}), & \text{if } \theta < 1/2, \\ C_{\mathcal{B}}^{2\theta}(\overline{\Omega}), & \text{if } \theta > 1/2, \end{cases}$$

$$D_{A_1}(\theta) = \begin{cases} h^{2\theta}(\overline{\Omega}), & \text{if } \theta < 1/2, \\ h_{\mathcal{B}}^{2\theta}(\overline{\Omega}), & \text{if } \theta > 1/2, \end{cases}$$

with equivalence of the respective norms.

Proof — We characterize here $D_{A_1}(\theta, \infty)$, the characterization of $D_{A_1}(\theta)$ being similar. From Proposition 3.1.27 and Proposition 3.1.28 we know that $C^1(\overline{\Omega})$ belongs to the class $J_{1/2} \cap K_{1/2}$ between $C(\overline{\Omega})$ and $D(A_1)$. So, we may use the Reiteration Theorem, which gives

$$(C(\overline{\Omega}), D(A_1))_{\theta,\infty} = (C(\overline{\Omega}), C^1(\overline{\Omega}))_{2\theta,\infty}, \quad 0 < \theta < 1/2.$$

On the other hand, from Proposition 2.2.2 we get $D_{A_1}(\theta, \infty) = (C(\overline{\Omega}), D(A_1))_{\theta,\infty}$, and from Corollary 1.2.18(ii) we get $(C(\overline{\Omega}), C^1(\overline{\Omega}))_{2\theta,\infty} = C^{2\theta}(\overline{\Omega})$. The statement is proved for $0 < \theta < 1/2$.

Let us consider now the case $1/2 < \theta < 1$. We recall preliminarily that for $0 \leq \alpha < 1$, the space $D_{A_1}((1+\alpha)/2, 1)$ is continuously embedded in $C^{1+\alpha}(\overline{\Omega})$, thanks again to Proposition 3.1.28. Since $D(A_1)$ is dense in $D_{A_1}((1+\alpha)/2, 1)$, then $D_{A_1}((1+\alpha)/2, 1)$ is contained in $C_{\mathcal{B}}^{1+\alpha}(\overline{\Omega})$. Therefore, the space $C_{\mathcal{B}}^{1+\alpha}(\overline{\Omega})$ belongs to the class $J_{(1+\alpha)/2}$ between $C(\overline{\Omega})$ and $D(A_1)$.

Taking $\alpha = 0$, $C_{\mathcal{B}}^1(\overline{\Omega})$ belongs to the class $J_{1/2} \cap K_{1/2}$. From the Reiteration Theorem it follows that

$$(C(\overline{\Omega}), D(A_1))_{\alpha,\infty} = (C_{\mathcal{B}}^1(\overline{\Omega}), D(A))_{2\alpha-1,\infty} \supset (C_{\mathcal{B}}^1(\overline{\Omega}), C_{\mathcal{B}}^2(\overline{\Omega}))_{2\alpha-1,\infty}. \quad (3.1.77)$$

It is easy now to see that

$$(C_{\mathcal{B}}^1(\overline{\Omega}), C_{\mathcal{B}}^2(\overline{\Omega}))_{2\alpha-1,\infty} = C_{\mathcal{B}}^{2\alpha}(\overline{\Omega}), \quad (3.1.78)$$

---

[2]One could in fact follow the proof of Theorem 3.1.29 provided one could prove that $(C(\overline{\Omega}), C_{\mathcal{B}}^2(\overline{\Omega}))_{\alpha,\infty} = C^{2\alpha}(\overline{\Omega})$ for $0 < \alpha < 1/2$, $(C(\overline{\Omega}), C_{\mathcal{B}}^2(\overline{\Omega}))_{\alpha,\infty} = C_{\mathcal{B}}^{2\alpha}(\overline{\Omega})$ for $1/2 < \alpha < 1$. The presently available proofs of such equivalences seem to work only in the case where the coefficients β_i, γ belong to $UC^2(\overline{\Omega})$.

1. SECOND ORDER OPERATORS

with equivalence of the respective norms. Indeed, from Corollary 1.2.19 it follows that $(C^1(\overline{\Omega}), C^2(\overline{\Omega}))_{2\alpha-1,\infty} = C^{2\alpha}(\overline{\Omega})$, with equivalence of the norms. Therefore, $(C_{\mathcal{B}}^1(\overline{\Omega}), C_{\mathcal{B}}^2(\overline{\Omega}))_{2\alpha-1,\infty}$ is obviously contained in $C_{\mathcal{B}}^{2\alpha}(\overline{\Omega})$. Conversely, if $f \in C_{\mathcal{B}}^{2\alpha}(\overline{\Omega})$, then it is the trace at $t = 0$ of a function $u \in W(2-2\alpha, \infty, C^2(\overline{\Omega}), C^1(\overline{\Omega}))$. Setting $\tilde{u} = u - \mathcal{N}\mathcal{B}u_{|\partial\Omega}$, where $\mathcal{N}$ is the operator provided by Theorem 0.3.2(ii), then $\tilde{u}$ belongs to $W(2-2\alpha, \infty, C_{\mathcal{B}}^2(\overline{\Omega}), C_{\mathcal{B}}^1(\overline{\Omega}))$, and $\tilde{u}(0) = f - \mathcal{N}\mathcal{B}f_{|\partial\Omega} = f$. Therefore, $f \in (C_{\mathcal{B}}^1(\overline{\Omega}), C_{\mathcal{B}}^2(\overline{\Omega}))_{2\alpha-1,\infty}$. So, (3.1.78) follows.

From (3.1.78) and the Reiteration Theorem it follows that

$$(C_{\mathcal{B}}^1(\overline{\Omega}), C_{\mathcal{B}}^{2\alpha}(\overline{\Omega}))_{\sigma,\infty} = C_{\mathcal{B}}^{1+(2\alpha-1)\sigma}(\overline{\Omega}), \ 0 < \sigma < 1. \quad (3.1.79)$$

Moreover, Proposition 2.2.2 and (3.1.77), (3.1.78) imply that

$$D_{A_1}(\alpha, \infty) \supset C_{\mathcal{B}}^{2\alpha}(\overline{\Omega}),$$

which means that $C_{\mathcal{B}}^{2\alpha}(\overline{\Omega})$ belongs to the class K_α between $C(\Omega)$ and $D(A_1)$. We have seen above that it belongs to the class J_α between $C(\Omega)$ and $D(A_1)$. We know already that $C_{\mathcal{B}}^1(\overline{\Omega})$ belongs to the class $J_{1/2} \cap K_{1/2}$ between $C(\Omega)$ and $D(A_1)$. Let now $1/2 < \theta < \alpha < 1$. Using once again the Reiteration Theorem, and (3.1.79), we get

$$C(\overline{\Omega}, D(A_1))_{\alpha,\infty} = (C_{\mathcal{B}}^1(\overline{\Omega}), C_{\mathcal{B}}^{2\alpha}(\overline{\Omega}))_{\frac{2\theta-1}{2\alpha-1},\infty} = C_{\mathcal{B}}^{2\theta}(\overline{\Omega}),$$

and the statement follows. ∎

Concerning the critical case $\theta = 1/2$, in view of Theorem 3.1.12 one expects that $D_{A_1}(1/2, \infty)$ be a subspace of $\mathcal{C}^1(\overline{\Omega})$. Since the functions in $\mathcal{C}^1(\overline{\Omega})$ are not necessarily differentiable, the boundary condition $\mathcal{B}(x, D)u_{|\partial\Omega} = 0$ does not make sense for every $u \in \mathcal{C}^1(\overline{\Omega})$. However, we define $\beta(x) = (\beta_1(x), \ldots, \beta_n(x))$,

$$[[u]]_{1,\beta} = \sup_{x \in \partial\Omega,\, h \in \mathbb{R},\, x-h\beta(x) \in \overline{\Omega}} \frac{|u(x - h\beta(x)) - u(x)|}{h},$$

which is meaningful for every $u \in C(\overline{\Omega})$, and we set

$$\mathcal{C}_{\mathcal{B}}^1(\overline{\Omega}) = \{u \in \mathcal{C}^1(\overline{\Omega}) : [[u]]_{1,\beta} < \infty\},$$

$$h_{\mathcal{B}}^1(\overline{\Omega}) = \left\{ u \in \mathcal{C}^1(\overline{\Omega}) : \lim_{h \to 0} \sup_{\substack{x,y \in \overline{\Omega} \\ |x-y| \le h}} \frac{|u(x) + u(y) - 2u(\frac{x+y}{2})|}{h} = 0, \right.$$

$$\left. \lim_{h \to 0} \sup_{x \in \partial\Omega,\, x-h\beta(x) \in \overline{\Omega}} \frac{|u(x - h\beta(x)) - u(x)|}{h} = 0 \right\}.$$

$\mathcal{C}_{\mathcal{B}}^1(\overline{\Omega})$ is endowed with the norm of $\mathcal{C}^1(\overline{\Omega})$. The following characterization holds.

Theorem 3.1.31 *We have*

$$D_{A_1}(1/2, \infty) = \mathcal{C}_{\mathcal{B}}^1(\overline{\Omega}), \ D_{A_1}(1/2) = h_{\mathcal{B}}^1(\overline{\Omega}),$$

with equivalence of the respective norms.

The proof of Theorem 3.1.31 is omitted. See next section for bibliographical references.

Corollary 3.1.32 *Let $0 < \theta < 1$, $\theta \neq 1/2$. Then:*
(i) The operator

$$A_{0|C_0^{2\theta}(\overline{\Omega})} : D_{A_0}(\theta+1,\infty) = \{u \in D(A_0) : \mathcal{A}u \in C_0^{2\theta}(\overline{\Omega})\} \mapsto C_0^{2\theta}(\overline{\Omega})$$

is sectorial in $C_0^{2\theta}(\overline{\Omega})$.
(ii) If $\theta < 1/2$, the operator

$$A_{1|C^{2\theta}(\overline{\Omega})} : D_{A_1}(\theta+1,\infty) = \{u \in D(A_1) : \mathcal{A}u \in C^{2\theta}(\overline{\Omega})\} \mapsto C^{2\theta}(\overline{\Omega})$$

is sectorial in $C^{2\theta}(\overline{\Omega})$.
(iii) If $\theta > 1/2$, the operator

$$A_{1|C_{\mathcal{B}}^{2\theta}(\overline{\Omega})} : D_{A_1}(\theta+1,\infty) = \{u \in D(A_1) : \mathcal{A}u \in C_{\mathcal{B}}^{2\theta}(\overline{\Omega})\} \mapsto C_{\mathcal{B}}^{2\theta}(\overline{\Omega})$$

is sectorial in $C_{\mathcal{B}}^{2\theta}(\overline{\Omega})$.
The same conclusions hold if $C^{2\theta}(\overline{\Omega})$, $C_0^{2\theta}(\overline{\Omega})$, $C_{\mathcal{B}}^{2\theta}(\overline{\Omega})$ are replaced respectively by $h^{2\theta}(\overline{\Omega})$, $h_0^{2\theta}(\overline{\Omega})$, $h_{\mathcal{B}}^{2\theta}(\overline{\Omega})$, and $D_{A_0}(\theta+1,\infty)$, $D_{A_1}(\theta+1,\infty)$ are replaced by $D_{A_0}(\theta+1)$, $D_{A_1}(\theta+1)$.

The boundary conditions in the Hölder spaces above arise naturally from interpolation. However, the spaces $C_0^{2\theta}(\overline{\Omega})$ and $C_{\mathcal{B}}^{2\theta}(\overline{\Omega})$ cannot be replaced by $C^{2\theta}(\overline{\Omega})$, as the following counterexamples show.

Example 3.1.33 *Let $0 < \alpha \leq 1$. Set*

$$\begin{cases} A : D(A) = \{u \in C^{2+\alpha}([0,\pi]) : u(0) = u(\pi) = 0\} \mapsto C^{\alpha}([0,\pi]), \\ Au = u''; \end{cases}$$

$$\begin{cases} B : D(B) = \{u \in C^{3+\alpha}([0,\pi]) : u'(0) = u'(\pi) = 0\} \mapsto C^{1+\alpha}([0,\pi]), \\ Bu = u''. \end{cases}$$

Then nor A neither B is a sectorial operator.

Proof — Let $f \equiv 1$. Then for $\lambda > 0$

$$(R(\lambda,A)f)(x) = \frac{1}{\lambda}\left(1 - \frac{\sinh(\sqrt{\lambda}x) + \sinh(\sqrt{\lambda}(\pi-x))}{\sinh(\sqrt{\lambda}\pi)}\right).$$

For $\lambda > \pi^{-2}$ set $u = R(\lambda,A)f$. Then, if $\alpha < 1$,

$$\lambda[u]_{C^{\alpha}([0,\pi])} \geq \lambda \frac{u(\lambda^{-1/2})}{\lambda^{-\alpha/2}}$$

$$= \frac{\lambda^{\alpha/2}}{e^{\sqrt{\lambda}\pi} - e^{-\sqrt{\lambda}\pi}}\left(e^{\sqrt{\lambda}\pi}(1-e^{-1}) + e^{-\sqrt{\lambda}\pi}(e-1) - e + e^{-1}\right),$$

1. SECOND ORDER OPERATORS

which is not bounded as $\lambda \to +\infty$. If $\alpha = 1$,

$$\lambda \|u'\|_{C([0,\pi])} \geq \lambda |u'(0)| = \sqrt{\lambda}\, \frac{e^{\sqrt{\lambda}\pi} + e^{-\sqrt{\lambda}\pi} - 2}{e^{\sqrt{\lambda}\pi} - e^{-\sqrt{\lambda}\pi}},$$

which is not bounded as $\lambda \to +\infty$. Therefore, A is not sectorial.

Let now $f(x) = x$, $\lambda > 0$, and set $u = R(\lambda, B)f$. Then $v = u'$ satisfies $\lambda v - v'' = 1$ in $[0, \pi]$, $v(0) = v(1) = 0$. By the above considerations, for every $C > 0$ the inequality $\lambda [u']_{C^\alpha([0,\pi])} \leq C$ fails to be true for λ large, so that $\|\lambda R(\lambda, B)\|_{L(C^\alpha([0,\pi]))}$ is not bounded as $\lambda \to +\infty$, and B is not sectorial. ∎

We end the section with the Schauder estimates, which let us characterize the spaces $D_{A_0}(\theta + 1, \infty)$ and $D_{A_1}(\theta + 1, \infty)$ when the coefficients are Hölder continuous.

Theorem 3.1.34 *Let $a_{ij}, b_i, c \in C^{2\theta}(\overline{\Omega})$ (respectively, $C^{2\theta}(\overline{\Omega})$), with $\theta \neq 1/2$, and let $\partial\Omega$ be uniformly $C^{2\theta+2}$ (respectively, $h^{2\theta+2}$). Then:*

(i) If a function $u \in D(A_0)$ is such that $\mathcal{A}u \in C^{2\theta}(\overline{\Omega})$ (respectively, $h^{2\theta}(\overline{\Omega})$), then $u \in C^{2\theta+2}(\overline{\Omega})$ (respectively, $u \in h^{2\theta+2}(\overline{\Omega})$). There is $C > 0$, independent of u, such that

$$\|u\|_{C^{2\theta+2}(\overline{\Omega})} \leq C(\|u\|_\infty + \|\mathcal{A}u\|_{C^{2\theta}(\overline{\Omega})}). \tag{3.1.80}$$

(ii) If $\beta_i, \gamma \in C^{2\theta+1}(\overline{\Omega})$ (respectively, $h^{2\theta+1}(\overline{\Omega})$), and a function $u \in D(A_1)$ is such that $\mathcal{A}u \in C^{2\theta}(\overline{\Omega})$ (respectively, $h^{2\theta}(\overline{\Omega})$), then $u \in C^{2\theta+2}(\overline{\Omega})$ (respectively, $h^{2\theta+2}(\overline{\Omega})$), and there is $C > 0$, independent of u, such that (3.1.80) holds.

See next section for bibliographical remarks.

Corollary 3.1.35 *Let $a_{ij}, b_i, c \in C^{2\theta}(\overline{\Omega})$, and let $\partial\Omega$ be uniformly $C^{2\theta+2}$, with $\theta \neq 1/2$. Then*

$$D_{A_0}(\theta + 1, \infty) = \{u \in C^{2\theta+2}(\overline{\Omega}) : u_{|\partial\Omega} = \mathcal{A}u_{|\partial\Omega} = 0\}.$$

If in addition $\beta_i, \gamma \in C^{2\theta+1}(\overline{\Omega})$, then

$$D_{A_1}(\theta + 1, \infty) \begin{cases} = C_\mathcal{B}^{2\theta+2}(\overline{\Omega}), & \text{if } \theta < 1/2, \\ = \{u \in C_\mathcal{B}^{2\theta+2}(\overline{\Omega}) : \mathcal{B}\mathcal{A}u_{|\partial\Omega} = 0\}, & \text{if } \theta > 1/2. \end{cases}$$

3.2 Higher order operators and bibliographical remarks

This section is devoted to give information about the extension of the results of Section 3.1 to a wide class of differential operators in $\overline{\Omega}$, where Ω is either $\mathbb{R}^n$ or a domain with uniformly C^{2m} boundary, $m \in \mathbb{N}$.

We consider a differential operator of order $2m$ with uniformly continuous and bounded coefficients,

$$\mathcal{A}(x, D) = \sum_{|\alpha| \leq 2m} a_\alpha(x) D^\alpha \tag{3.2.1}$$

under the following ellipticity assumption: there exists $\nu > 0$ such that for each $x \in \overline{\Omega}$, $\theta \in [-\pi/2, \pi/2]$, $\xi \in \mathbb{R}^n$, $r \geq 0$ with $|\xi|^2 + r^2 \neq 0$ we have

$$\left| \sum_{|\gamma|=2m} a_\gamma(x) \xi^\gamma - (-1)^m r^{2m} e^{i\theta} \right| \geq \nu(|\xi|^{2m} + r^{2m}). \tag{3.2.2}$$

If $\partial \Omega \neq \emptyset$ we assume that the following extended roots condition holds.

$$\begin{cases} \text{For each } x \in \partial\Omega, \, \theta \in [-\pi/2, \pi/2], \, \xi \in \mathbb{R}^n, \, r > 0 \\ \text{with } |\xi|^2 + r^2 \neq 0, \text{ and } \langle \xi, \nu(x) \rangle = 0, \text{ the polynomial} \\ p(z) = \sum_{|\gamma|=2m} a_\gamma(x) [\xi + z\nu(x)]^\gamma - (-1)^m r^{2m} e^{i\theta} \\ \text{has exactly } m \text{ roots } z_j^+(x, \xi, r, \eta), \, j = 1, \cdots, m, \text{ with} \\ \text{positive imaginary part.} \end{cases} \tag{3.2.3}$$

Taking $r = 0$ in (3.2.2) and in (3.2.3), we see that $\mathcal{A}$ is uniformly elliptic and satisfies the roots condition in the usual sense. The full assumptions (3.2.2) and (3.2.3) are made in order that for every $\theta \in [-\pi/2, \pi/2]$, the operator in $n+1$ variables (x, t), $\mathcal{A}_\theta = \mathcal{A} - (-1)^m e^{i\theta} D_t^{2m}$, is elliptic and satisfies the roots condition. The operators $\mathcal{A}_\theta$ play an important role in the generation theorem in L^p spaces: see Theorem 3.1.2.

If $\Omega \neq \mathbb{R}^n$, we consider a family $\{\mathcal{B}_j\}_{j=1,\ldots,m}$ of differential operators acting on the boundary,

$$\mathcal{B}_j(x, D) = \sum_{|\beta| \leq m_j} b_{j\beta}(x) D^\beta \tag{3.2.4}$$

The operators $\mathcal{B}_j$ are assumed to satisfy a uniform normality condition,

$$\begin{cases} 0 \leq m_1 < m_2 < \cdots < m_m \leq 2m - 1, \\ \left| \sum_{|\beta|=m_j} b_{j\beta}(x) (\nu(x))^\beta \right| \geq \varepsilon > 0, \, x \in \partial\Omega, \, j = 1, \ldots, m. \end{cases} \tag{3.2.5}$$

Therefore, the operator $\mathcal{B}_j$ has order m_j. Concerning the regularity of the coefficients, we assume that

$$b_{j\beta} \in UC^{2m-m_j}(\overline{\Omega}), \, j = 1, \ldots, m. \tag{3.2.6}$$

Finally, we assume that a complementing condition holds:

$$\begin{cases} \text{for each } x \in \partial\Omega,\ \xi \in \mathbb{R}^n,\ \theta \in [-\pi/2, \pi/2],\ r \geq 0 \text{ with} \\ |\xi|^2 + r^2 \neq 0 \text{ and } \langle \xi, \nu(x) \rangle = 0,\ \text{the polynomials} \\ P_j(z) = \sum_{|\beta|=m_j} b_{j\beta}(x)(\xi + z\nu(x))^\beta,\ j = 1, \cdots, m \\ \text{are linearly independent modulo the polynomial} \\ Q(z) = \prod_{j=1}^{m}(z - z_j^+(x, \xi, r, \theta)), \\ \text{where } z_j^+(x, \xi, r, \theta) \text{ are defined in } (3.2.3). \end{cases} \quad (3.2.7)$$

As in the case of second order operators, the first step consists in the *a priori* estimates in the $W^{2m,p}$ norm due to S. AGMON – A. DOUGLIS – L. NIRENBERG ([13, Thm. 15.2]). Using such estimates, one shows generation results in L^p, $1 < p < \infty$. Then, using the generation results in L^p, one shows generation results in L^∞ and in spaces of continuous functions. Generation results in $W^{k+\alpha,p}$ and in Hölder spaces follow by interpolation.

Theorem 3.2.1 *(i) For every $p \in\ ]1, +\infty[$ there is $c_p > 0$ such that for every $u \in W^{2m,p}(\mathbb{R}^n)$ we have*

$$\|u\|_{W^{2m,p}(\mathbb{R}^n)} \leq c_p(\|u\|_{L^p(\mathbb{R}^n)} + \|\mathcal{A}(\cdot, D)u\|_{L^p(\mathbb{R}^n)}). \quad (3.2.8)$$

(ii) Let Ω be an open set in $\mathbb{R}^n$ with uniformly C^{2m} boundary. For $u \in W^{2m,p}(\Omega)$, with $1 < p < \infty$, set $f = \mathcal{A}(\cdot, D)u$, $g_j = \mathcal{B}_j(\cdot, D)u_{|\partial\Omega}$. Then there is $c_p > 0$ such that

$$\|u\|_{W^{2m,p}(\Omega)} \leq c_p \left(\|u\|_{L^p(\Omega)} + \|f\|_{L^p(\Omega)} + \sum_{j=1}^{m} \|\tilde{g}_j\|_{W^{2m-m_j,p}(\Omega)} \right), \quad (3.2.9)$$

where $\tilde{g}_j$ is any extension of g_j belonging to $W^{2m-m_j,p}(\Omega)$.

To be precise, in the paper [13] the above estimates are stated in the cases where Ω is bounded. However, they may be extended to the case of unbounded Ω. See e.g. [17, Thm. 12.1].

If $\Omega = \mathbb{R}^n$, the realization of $\mathcal{A}(\cdot, D)$ in $L^p(\mathbb{R}^n)$ is defined by

$$D(A_p) = W^{2m,p}(\mathbb{R}^n),\ A_p u = \mathcal{A}(\cdot, D)u.$$

If $\Omega \neq \mathbb{R}^n$, we define some subsets of the Sobolev spaces satisfying suitable homogeneous boundary conditions. For $s > 0$ we set

$$W_\mathcal{B}^{s,p}(\Omega) = \{u \in W^{s,p}(\Omega) : \mathcal{B}_j u_{|\partial\Omega} = 0 \text{ if } m_j < s - 1/p\}.$$

If $\Omega \neq \mathbb{R}^n$, the realization of $\mathcal{A}(\cdot, D)$ with homogeneous boundary conditions in $L^p(\Omega)$ is defined by

$$D(A_p) = W_\mathcal{B}^{2m,p}(\Omega),\ A_p u = \mathcal{A}(\cdot, D)u.$$

Theorem 3.2.2 *(i) Let $\Omega = \mathbb{R}^n$, and fix $p \in \,]1, +\infty[$. Then there exist $\omega_p \in \mathbb{R}$, $M_p > 0$ such that $\rho(A_p) \supset \{\lambda \in \mathbb{C} : \operatorname{Re}\lambda \geq \omega_p\}$, and*

$$\sum_{k=0}^{2m} |\lambda|^{1-\frac{k}{2m}} \|D^k u\|_p \leq M_p \|\lambda u - \mathcal{A}u\|_p \qquad (3.2.10)$$

for every λ with $\operatorname{Re}\lambda \geq \omega_p$ and $u \in W^{2m,p}(\mathbb{R}^n)$.

(ii) Let Ω be an open set in $\mathbb{R}^n$ with uniformly C^{2m} boundary, and fix $p \in \,]1, +\infty[$. Then there exist $\omega_p \in \mathbb{R}$, $M_p > 0$ such that for every $u \in W^{2m,p}(\Omega)$ it holds

$$\sum_{k=0}^{2m} |\lambda|^{1-\frac{k}{2m}} \|D^k u\|_p \leq M_p \left(\|\lambda u - \mathcal{A}u\|_p + \sum_{j=1}^{m} \sum_{k=0}^{2m-m_j} |\lambda|^{1-\frac{m_j+k}{2m}} \|D^k \tilde{g}_j\|_p \right), \quad (3.2.11)$$

$\tilde{g}_j$ being any extension of $g_j = \mathcal{B}_j u_{|\partial\Omega}$ belonging to $W^{2m-m_j,p}(\Omega)$. In addition, $\rho(A_p) \supset \{\lambda \in \mathbb{C} : \operatorname{Re}\lambda \geq \omega_p\}$.

The proof of the fact that the resolvent set contains a halfplane may be found in G. GEYMONAT – P. GRISVARD [89], where also systems are considered. See also the exposition in the book of H. TANABE [193, Section 3.8] for the case of a single operator. The proof of estimates (3.2.10) and (3.2.11) is quite similar to the corresponding proof in the case of second order operators (Theorem 3.1.2), the main tools being the *a priori* estimates (3.2.8) and (3.2.9). For details, see the paper by S. AGMON [12], and the book [193, Section 3.8]. For operators with variational structure, estimates (3.2.10)-(3.2.11) may be obtained by a direct method, without using the *a priori* estimates (3.2.8) and (3.2.9): see A. PAZY [166, Section 7.3]. However, (3.2.8) and (3.2.9) are necessary to characterize the domain $D(A_p)$.

Further generalization to certain systems of order $2m$ — roughly, to systems such that estimates of the type (3.2.8)-(3.2.9) are available — may be found in G. GEYMONAT – P. GRISVARD [89] and in the papers of H. AMANN [17], [18].

The characterization of the interpolation spaces $D_{A_p}(\theta, p)$ is provided by the next theorem.

Theorem 3.2.3 *(i) Let $\Omega = \mathbb{R}^n$, and let $p \in \,]1, +\infty[$, $0 < \theta < 1$ be such that $\theta p \notin \mathbb{N}$. Then*

$$D_{A_p}(\theta, p) = W^{\theta p, p}(\mathbb{R}^n),$$

with equivalence of the respective norms. Consequently, the part of A_p in $W^{\theta p, p}(\mathbb{R}^n)$ is sectorial.

(ii) Let Ω be an open set with uniformly C^{2m} boundary, and let the coefficients $b_{j\beta}$ belong to $C^{2m}(\overline{\Omega})$. Fix $p \in \,]1, +\infty[$ and $0 < \theta < 1$ such that θp and $\theta p - 1/p$ are not integers. Then

$$D_{A_p}(\theta, p) = W^{\theta p, p}_{\mathcal{B}}(\Omega).$$

Consequently, the part of A_p in $W^{\theta p, p}_{\mathcal{B}}(\Omega)$ is sectorial.

2. HIGHER ORDER OPERATORS AND BIBLIOGRAPHICAL REMARKS

We recall that for $0 < \alpha < 1$, $W^{\alpha,p}(\Omega)$ is the subspace of $L^p(\Omega)$ consisting of those functions f such that

$$[f]_{W^{\alpha,p}} = \int_{\Omega \times \Omega} \frac{|f(x) - f(y)|}{|x-y|^{n+\alpha p}} \, dx \, dy < \infty.$$

It is endowed with the norm $\|f\|_{\alpha,p} = \|f\|_{L^p} + [f]_{W^{\alpha,p}}$. If $\alpha > 1$, $\alpha = k + \sigma$, with $k \in \mathbb{N}$ and $0 < \sigma < 1$, $W^{\alpha,p}(\Omega)$ is the subspace of $W^{k,p}(\Omega)$ consisting of the functions f such that $D^\beta f \in W^{\sigma,p}(\Omega)$ for every multi-index β with $|\beta| = k$. It is endowed with the norm $\|f\|_{W^{\alpha,p}} = \|f\|_{W^{k,p}} + \sum_{|\beta|=k} [D^\beta f]_{W^{\sigma,p}}$.

Part (i) of Theorem 3.2.3 can be proved by a procedure similar to the one of Theorem 1.2.17, or by other classical interpolation methods. See e.g. H. TRIEBEL [200, Chapter 2]. Part (ii) was proved by P. GRISVARD in [93], [94]. See also [95] for the case $p = 2$. In these papers, also the critical cases $\theta p \in \mathbb{N}$ and $\theta p - 1/p \in \mathbb{N} \cup \{0\}$ have been considered.

The case $p = 1$ is much more complicated. Results similar to Theorem 3.2.1 do not hold, and the domain of the realization of $\mathcal{A}(\cdot, D)$ in $L^1(\Omega)$ is not a subset of $W^{2m,1}(\Omega)$ if $n > 1$. Generation results in L^1 spaces for operators with smooth coefficients were stated by H. TANABE [192, 194]. Publications [192, 194] are not easily available. A simple proof, which makes use of duality arguments, may be found in the book of A. PAZY [166, Sect. 7.3]. See also H. AMANN [16]. The case of second order operators with C^α coefficients was considered by V. VESPRI [202].

Generation results in $C(\overline{\Omega})$ may be deduced from Theorem 3.2.2, as in the case of second order operators, arguing as in the proof of Theorem 3.1.6 in the case $\Omega = \mathbb{R}^n$ and as in the proof of Theorem 3.1.22 in the case of general Ω.

Theorem 3.2.4 *Let $\Omega = \mathbb{R}^n$. Set*

$$\begin{cases} D(A) = \{u \in \bigcap_{p \geq 1} W^{2m,p}_{loc}(\mathbb{R}^n) : u, \mathcal{A}(\cdot, D)u \in C(\mathbb{R}^n)\}, \\ A : D(A) \mapsto C(\mathbb{R}^n), \quad Au = \mathcal{A}(\cdot, D)u. \end{cases} \quad (3.2.12)$$

Then the resolvent set of A contains a halfplane $\{\lambda \in \mathbb{C} : \operatorname{Re}\lambda \geq \omega\}$. Moreover, for every $p > n$ there exists $K_p > 0$ such that for every $u \in D(A)$ and for $\operatorname{Re}\lambda \geq \omega$ it holds

$$\sum_{k=1}^{2m-1} |\lambda|^{1-\frac{k}{2m}} \|D^k u\|_\infty + |\lambda|^{n/2p} \sup_{x \in \mathbb{R}^n} \|D^{2m} u\|_{L^p(B(x, |\lambda|^{-1/2m}))} \leq K_p \|\lambda u - Au\|_\infty. \quad (3.2.13)$$

It follows that A is sectorial in $C(\mathbb{R}^n)$.

Theorem 3.2.5 *Let Ω be an open set with uniformly C^{2m} boundary. Set*

$$\begin{cases} D(A) = \{u \in \bigcap_{p \geq 1} W^{2m,p}_{loc}(\Omega) : u, \mathcal{A}(\cdot, D)u \in C(\overline{\Omega}), \\ \qquad\qquad \mathcal{B}_j(\cdot, D)u_{|\partial \Omega} = 0, \; j = 1, \ldots, m\}, \\ A : D(A) \mapsto C(\overline{\Omega}), \quad Au = \mathcal{A}(\cdot, D)u. \end{cases} \quad (3.2.14)$$

Then the resolvent set of A contains a halfplane $\{\lambda \in \mathbb{C} : \operatorname{Re}\lambda \geq \omega\}$. Moreover, for every $p > n$ there exists $K_p > 0$ such that for every $u \in W^{2m,p}_{loc}(\Omega)$ and for $\operatorname{Re}\lambda$ sufficiently large it holds, setting $U = B(x, |\lambda|^{-1/2m}) \cap \Omega$,

$$\begin{aligned}
&\sum_{k=1}^{2m-1} |\lambda|^{1-\frac{k}{2m}} \|D^k u\|_\infty + |\lambda|^{n/2mp} \sup_{x \in \overline{\Omega}} \|D^{2m} u\|_{L^p(U)} \\
&\leq K_p \bigg(|\lambda|^{n/2mp} \|\lambda u - Au\|_{L^p(U)} + \sum_{j=1}^{m} \Big(|\lambda|^{n/2mp} \sup_{x \in \overline{\Omega}} \|D^{2m-m_j} \tilde{g}_j\|_{L^p(U)} \\
&\quad + \sum_{k=0}^{2m-m_j-1} |\lambda|^{1-\frac{m_j+k}{2m}} \|D^k \tilde{g}_j\|_\infty \Big) \bigg),
\end{aligned} \quad (3.2.15)$$

$\tilde{g}_j$ being any extension of $g_j = \mathcal{B}_j u_{|\partial \Omega}$ belonging to $W^{2m-m_j}_{loc}(\Omega)$.
It follows that A is sectorial.

The above generation results go back to H.B. STEWART [187], [188]. In the proof of the resolvent *a priori* estimates of Theorems 3.1.6, 3.1.19 and 3.1.22 we have followed the approach of P. ACQUISTAPACE – B. TERRENI [7, Appendix], to which we refer for the case of operators of order $2m$. In the proof of the existence of $R(\lambda, A)$ for $\operatorname{Re}\lambda$ large in the case of unbounded domain, we have used a technique similar to the one of P. CANNARSA – V. VESPRI [48, Section 5]. Such a paper, together with the subsequent paper [49], is devoted to generation results in $L^p(\mathbb{R}^n)$, $1 \leq p \leq \infty$, and in $C(\mathbb{R}^n)$, in the case of elliptic operators with possibly unbounded coefficients.

A proof of Lemma 3.1.4 may be found in R.A. ADAMS [10, Lemmas 5.15, 5.17] in the case $r = 1$. The general case follows by standard dilation arguments.

An alternative approach to generation results in several Banach spaces, which does not make use of the Agmon-Douglis-Nirenberg estimates but uses techniques of Morrey spaces, may be found in P. CANNARSA – B. TERRENI – V. VESPRI [50], for second order systems with Dirichlet boundary condition.

Also the characterization of the spaces $D_A(\theta, \infty)$ is similar to the one in the second order case. Indeed, the following result holds.

Theorem 3.2.6 *(i) Let $\Omega = \mathbb{R}^n$, and let A be the operator defined in (3.2.12). Then for every $\theta \in \,]0, 1[$ we have*

$$D_A(\theta, \infty) = \mathcal{C}^{2m\theta}(\mathbb{R}^n),$$

2. Higher order operators and bibliographical remarks

with equivalence of the respective norms. In particular, if $2m\theta$ is not integer, then

$$D_A(\theta, \infty) = C^{2m\theta}(\mathbb{R}^n).$$

(ii) Let Ω be an open set with uniformly C^{2m} boundary, and let A be the operator defined in (3.2.14). Then for every $\theta \in \,]0,1[$ such that $2m\theta \notin \mathbb{N}$ we have

$$D_A(\theta, \infty) = C^{2m\theta}_\mathcal{B}(\overline{\Omega}),$$

with equivalence of the respective norms.

The proof of statement (i) is quite similar to the proof of Theorem 3.1.12. Statement (ii) in its generality was proved by P. ACQUISTAPACE – B. TERRENI [7]. For the case of second order operators with the Dirichlet boundary condition see A. LUNARDI [137]. In Section 3.1 (Theorems 3.1.29 and 3.1.30) we have followed a simpler method, the main tools being the Reiteration Theorem and the interpolatory estimates obtained as byproducts from the generation theorems. The idea of using the Reiteration Theorem for characterizing interpolation spaces is not new, see e.g. J.O. ADEYEYE [11]. However, the present proof of Theorem 3.1.30 is new.

The critical cases $2m\theta \in \mathbb{N}$ have been considered by P. ACQUISTAPACE in [1]. The spaces $D_A(\theta, \infty)$ and $D_A(\theta)$ turn out to be subspaces of $C^{2m\theta}(\overline{\Omega})$ consisting of functions which satisfy conditions near the boundary similar to the ones of Theorem 3.1.31. The space $C^{2m\theta}_\mathcal{B}(\overline{\Omega})$ is continuously embedded in $D_A(\theta, \infty)$.

From the above characterizations, generation results in Hölder spaces follow through Proposition 2.2.7. Specifically, the part of A in $C^{2m\theta}(\mathbb{R}^n)$, defined by

$$\begin{cases} \mathcal{A}_{2m\theta} : D_A(\theta+1, \infty) = \{u \in D(A) : Au \in C^{2m\theta}(\mathbb{R}^n)\} \mapsto C^{2m\theta}(\mathbb{R}^n), \\ \mathcal{A}_{2m\theta}u = Au, \end{cases}$$

is sectorial in $C^{2m\theta}(\mathbb{R}^n)$. If the coefficients of $\mathcal{A}$ belong to $C^{2m\theta}(\mathbb{R}^n)$, then it can be proved, using the method of Subsection 2.1.3 and the Hölder *a priori* estimates of S. AGMON – A. DOUGLIS – L. NIRENBERG [13], that $D_A(\theta+1, \infty) = C^{2m+2m\theta}(\mathbb{R}^n)$.

Similarly, if Ω is an open set with uniformly C^{2m} boundary, and A is the operator defined by (3.2.14), then the part of A in $C^{2m\theta}_\mathcal{B}(\overline{\Omega})$, defined by

$$\begin{cases} \mathcal{A}_{2m\theta} : D_A(\theta+1, \infty) = \{u \in D(A) : Au \in C^{2m\theta}_\mathcal{B}(\overline{\Omega})\} \mapsto C^{2m\theta}_\mathcal{B}(\overline{\Omega}), \\ \mathcal{A}_{2m\theta}u = Au, \end{cases}$$

is sectorial in $C^{2m\theta}_\mathcal{B}(\overline{\Omega})$.

If the coefficients of $\mathcal{A}$ belong to $C^{2m\theta}(\overline{\Omega})$, the coefficients of $\mathcal{B}_j$ belong to $C^{2m-m_j+2m\theta}(\overline{\Omega})$, and the boundary $\partial\Omega$ is uniformly $C^{2m+2m\theta}$, then by the regularity results of AGMON – DOUGLIS – NIRENBERG [13, Thm. 7.3], we have

$$D_A(\theta+1, \infty) = \{u \in C^{2m+2m\theta}(\overline{\Omega}) : \mathcal{B}_j u_{|\partial\Omega} = 0 \,\forall j, \; \mathcal{B}_j Au_{|\partial\Omega} = 0 \text{ for } m_j < 2m\theta\}.$$

In the case $m = 1$, the proof of the Schauder estimates of Theorem 3.1.34 may be found in several books about classical results in elliptic equations. See e.g. D. GILBARG – N.S. TRUDINGER [91, Chapter 6], which contains also a clear exposition of the method of continuity. Results similar to the Schauder Theorem in the case of little-Hölder functions may be easily proved by density arguments. See e.g. A. LUNARDI [134].

Generation results in Hölder spaces may be shown also without making use of the above characterization of the interpolation spaces, but using techniques of Morrey spaces. See S. CAMPANATO [45], [46] for second order operators in divergence form, P. CANNARSA – V. VESPRI [47] and P. CANNARSA – B. TERRENI – V. VESPRI [50] for non divergence form second order systems with Dirichlet boundary condition.

The counterexample 3.1.33 is due to W. VON WAHL [203]. Such counterexample was the first motivation of the theory of analytic semigroups not bounded near $t = 0$, developed by W. VON WAHL [204]. In such a theory, the assumption that $\|\lambda R(\lambda, A)\|_{L(X)}$ is bounded for λ in a sector is replaced by the assumption that there exists $\alpha \in \,]0,1[$ such that $|\lambda|^\alpha \|R(\lambda, A)\|_{L(X)}$ is bounded for λ in a sector.

The C^k spaces, with k integer, are not in general interpolation spaces. See e.g. B.S. MITJAGIN – E.M. SEMENOV [161]. However, a proof similar to the one of Proposition 3.1.11(i) yields an interpolatory result.

Proposition 3.2.7 *Let Ω be either $\mathbb{R}^n$, or an open set in $\mathbb{R}^n$ with uniformly C^{2m} boundary. Let $D(A)$ be defined by (3.2.12) in the case $\Omega = \mathbb{R}^n$, by (3.2.14) in the case $\Omega \neq \mathbb{R}^n$. Then, for $1 \leq k \leq 2m - 1$, the space $C^k(\overline{\Omega})$ belongs to the class $J_{k/2m}$ between $C(\overline{\Omega})$ and $D(A)$. More precisely, there is C such that*

$$\|u\|_{C^k(\overline{\Omega})} \leq C \|u\|_\infty^{1-k/2m} (\|\mathcal{A}u\|_\infty + \|u\|_\infty)^{k/2m}, \quad \forall u \in D(A).$$

Generation results in C^k spaces with suitable boundary conditions, and $k = 1, \ldots, 2m$, may be found in a recent paper by G. COLOMBO – V. VESPRI [55]. In the proofs of Theorems 3.1.25, 3.1.26 we have followed their method. A completely different approach has been followed by X. MORA in [160]. He uses results of the Russian school about parabolic differential initial boundary value problems to prove generation results in C^k spaces with boundary conditions, with $k \in \mathbb{N} \cup \{0\}$. His technique needs C^α coefficients of the operator $\mathcal{A}$ and $C^{2m-m_j+\alpha}$ coefficients of the operators $\mathcal{B}_j$, with $\alpha > 0$, while the direct methods of Section 3.1 and of [55] allow us to take $\alpha = 0$. This improvement, however, is not essential in many of the applications that will be given in the next chapters. What is more important is the fact that in the next chapters we are going to prove existence and regularity results for parabolic problems, using the generation results of the present chapter. So, we cannot follow the approach of Mora.

Through the whole chapter we have dealt essentially with nondivergence type operators. Second order operators of divergence type, with uniformly continuous

2. Higher order operators and bibliographical remarks

and bounded coefficients,

$$\mathcal{A}u = \sum_{i,j=1}^{n} D_i(a_{ij}(x)D_j u(x)) + \sum_{i=1}^{n} D_i(b_i(x)u(x)) + c(x)u(x),$$

associated with suitable boundary conditions, generate analytic semigroups even in dual spaces and subspaces of dual spaces, such as $W^{-1,p}(\Omega)$, $C^{-1,\alpha}(\Omega)$ ($0 < \alpha < 1$), and others. See H. TANABE [195], V. VESPRI [201, 202], A. LUNARDI – V. VESPRI [157]. Generation in Hölder spaces has been considered by S. CAMPANATO [45, 46]. Concerning higher order operators, the method followed by A. PAZY in [166, Sect. 7.3] to prove estimate (3.2.11) in the case of smooth coefficients and Dirichlet boundary conditions works as well for the realization in $L^p(\Omega)$ of operators with variational structure.

Chapter 4

Nonhomogeneous equations

We study the solvability of the initial value problem

$$u'(t) = Au(t) + f(t), \quad t > 0; \quad u(0) = u_0, \qquad (4.0.1)$$

where A is a linear sectorial operator in general Banach space X. The function f is defined and continuous in a (possibly unbounded) interval I such that $\inf I = 0$. We prove several properties of the solution, in a bounded interval $[0, T]$ and in the half line $[0, +\infty[$. We consider also backward solutions, solutions defined in the whole real line, and periodic solutions.

This chapter is the heart of the book. Together with the obvious applications to linear parabolic PDE's, we shall use the results of this chapter in the second part of the book, where nonlinear problems will be studied by linearization techniques, and in Chapter 6, where the parabolic evolution operator for nonautonomous equations will be constructed by "freezing" the principal part at fixed time $t = t_0$.

Several notions of solution are commonly considered in the literature. Comparison between different types of solutions are made in Section 4.1, where we also show that, under reasonable assumptions on the data, any solution of (4.0.1) in an arbitrary interval $[0, T]$ is given by the variation of constants formula

$$u(t) = e^{tA} u_0 + \int_0^t e^{(t-s)A} f(s) ds, \quad 0 \le t \le T. \qquad (4.0.2)$$

So, the study of the solution of (4.0.1) is reduced to the study of the representation formula (4.0.2).

Section 4.2 is devoted to the properties of the function u defined by (4.0.2) when f is measurable and bounded in every interval $[\varepsilon, T]$ with $0 < \varepsilon < T$ and integrable in $[0, T]$. Such assumptions do not guarantee that u has values in $D(A)$

and that it is differentiable with values in X. It solves (4.0.1) in the weak sense precised in Section 4.1.

Sufficient conditions on f in order that u has values in $D(A)$ and it is differentiable with values in X, at least for $t > 0$, are given in Section 4.3. Subsection 4.3.1 deals with the case where f enjoys some further regularity properties (*time regularity*), and Subsection 4.3.2 deals with the case where f has values in some intermediate space between X and $D(A)$ (*space regularity*). In both cases, particular attention is paid to optimal regularity results, that is to the cases where u' and Au enjoy the same regularity properties of f. To be specific, let $\mathcal{B}$ be a Banach space of functions defined on $[0,T]$, with values in X. We say that the space $\mathcal{B}$ enjoys the optimal regularity property (or the maximal regularity property) if for every $f \in \mathcal{B}$ the solution u of (4.0.1) is such that both u' and Au belong to $\mathcal{B}$ (of course, provided u_0 satisfy the necessary compatibility conditions). Such a property is crucial in the applications to nonautonomous and to fully nonlinear problems, which will be treated in the next chapters, and also in other applications such as integrodifferential and delay problems, which are not considered in this book but may be found in the papers [176], [144], [155], [27], [74], [75], [73], [178].

Next example 4.1.7 shows that, in general, the space $C([0,T];X)$ does not enjoy the optimal regularity property. But many other spaces do, such as, for instance, the Hölder spaces $C^\alpha([0,T];X)$, with $0 < \alpha < 1$, and $C([0,T]; D_A(\theta))$, with $0 < \theta < 1$.

In Section 4.4 we study asymptotic behavior: we treat bounded solutions, solutions with exponential decay or exponential growth, periodic solutions, backward solutions. The results depend heavily on the spectral properties of A, and rely upon the estimates of Section 2.3.

Through the whole chapter, we fix $T > 0$ and we set

$$M_k = \sup_{0 < t \leq T+1} \|t^k A^k e^{tA}\|_{L(X)}, \quad k \in \mathbb{N} \cup \{0\}, \qquad (4.0.3)$$

and, for $\alpha \in \,]0,1]$,

$$M_{k,\alpha} = \sup_{0 < t \leq T+1} \|t^{k-\alpha} A^k e^{tA}\|_{L(D_A(\alpha,\infty),X)}, \quad k \in \mathbb{N}, \qquad (4.0.4)$$

$$K_{k,\alpha} = \sup_{0 < t \leq T+1} t^{k+\alpha} \|A^k e^{tA}\|_{L(X,D_A(\alpha,1))}, \quad k \in \mathbb{N} \cup \{0\}. \qquad (4.0.5)$$

Due to estimates (2.2.20), we have M_k, $M_{k,\alpha}$, $K_{k,\alpha} < \infty$ for every k.

Due to the singular behavior near $t = 0$ of the derivatives of $t \mapsto e^{tA}x$, it will be sometimes convenient to work with weighted functional spaces. In particular, we shall consider spaces of functions defined in a bounded interval $]a,b]$, which are bounded or Hölder continuous in each interval $[a+\varepsilon, b]$ but that are not necessarily bounded (respectively, Hölder continuous) up to $t = a$. Let $\mu \in \mathbb{R}$, and set

$$B_\mu(]a,b];X) = \{f : \,]a,b] \mapsto X : \|f\|_{B_\mu(]a,b];X)} = \sup_{a < t \leq b} (t-a)^\mu \|f(t)\| < \infty\}, \qquad (4.0.6)$$

1. SOLUTIONS OF LINEAR PROBLEMS

$$C_\mu(]a,b];X) = C(]a,b];X) \cap B_\mu(]a,b];X), \quad \|f\|_{C_\mu(]a,b];X)} = \|f\|_{C_\mu(]a,b];X)}. \quad (4.0.7)$$

For $0 < \alpha < 1$, $\beta > 0$, set

$$\begin{aligned}C_\beta^\alpha(]a,b];X) &= \{f \in B_{\beta-\alpha}(]a,b];X) \cap C^\alpha([a+\varepsilon,b];X) \ \forall \varepsilon \in]0, b-a[:\\ [f]_{C_\beta^\alpha(]a,b];X)} &= \sup_{0<\varepsilon<b-a} \varepsilon^\beta [f]_{C^\alpha([a+\varepsilon,b];X)} < +\infty\};\\ \|f\|_{C_\beta^\alpha(]a,b];X)} &= \|f\|_{B_{\beta-\alpha}(]a,b];X)} + [f]_{C_\beta^\alpha(]a,b];X)}.\end{aligned} \quad (4.0.8)$$

In particular, for $\beta = \alpha$, the space $C_\alpha^\alpha(]a,b];X)$ is the set of the bounded functions $f :]a,b] \mapsto X$ such that

$$\sup_{0<\varepsilon<b-a} \varepsilon^\alpha [f]_{C^\alpha([a+\varepsilon,b];X)} < +\infty.$$

It may be easily seen that it coincides with the set of the bounded functions $f :]a, b] \mapsto X$ such that $t \mapsto g(t) = (t - a)^\alpha f(t)$ belongs to $C^\alpha(]a, b]; X)$, and that the norm $[[f]]_\alpha = \|f\|_\infty + [g]_{C^\alpha(]a,b];X)}$ is equivalent to the C_α^α norm. More generally, one can show that $C_\beta^\alpha(]a,b];X)$ is the space of the functions $f :]a,b] \mapsto X$ such that $h(t) = (t-a)^{\beta-\alpha} f(t)$ is bounded and $g(t) = (t-a)^\beta f(t)$ is α-Hölder continuous in $]a, b]$ with values in X, and that the norms $f \mapsto \|f\|_{C_\beta^\alpha(]a,b];X)}$ and

$$f \mapsto \|h\|_{B(]a,b];X)} + [g]_{C^\alpha(]a,b];X)}$$

are equivalent. However, in what follows it is more convenient to use the norm defined in (4.0.8).

4.1 Solutions of linear problems

We consider several types of solutions of problem (4.0.1).

Definition 4.1.1 *Let $T > 0$, let $f : [0,T] \mapsto X$ be a continuous function, and let $u_0 \in X$. Then:*

(i) *A function $u \in C^1([0,T];X) \cap C([0,T];D(A))$ is said to be a strict solution of (4.0.1) in the interval $[0,T]$ if $u'(t) = Au(t) + f(t)$ for each $t \in [0,T]$, and $u(0) = u_0$.*

(ii) *A function $u \in C([0,T];X)$ is said to be a strong solution of (4.0.1) in the interval $[0,T]$ if there is a sequence $\{u_n\}_{n\in\mathbb{N}} \subset C^1([0,T];X) \cap C([0,T];D(A))$ such that*

$$u_n \to u, \quad u_n' - Au_n \to f \text{ in } C([0,T];X) \text{ as } n \to +\infty.$$

Let now $f :]0, T] \mapsto X$ be continuous. Then

(iii) A function $u \in C^1(]0,T];X) \cap C(]0,T];D(A)) \cap C([0,T];X)$ is said to be a *classical solution* of (4.0.1) in the interval $[0,T]$ if $u'(t) = Au(t) + f(t)$ for each $t \in\,]0,T]$, and $u(0) = u_0$.

If f is defined in $[0,+\infty[$, a function $u : [0,+\infty[\, \mapsto X$ is said to be a strict (respectively classical, strong) solution of (4.0.1) in $[0,+\infty[$ if for every $T > 0$ the restriction $u_{|[0,T]}$ is a strict (respectively classical, strong) solution of (4.0.1) in $[0,T]$.

From Definition 4.1.1 it follows easily that if problem (4.0.1) has a strict solution then

$$u_0 \in D(A), \quad Au_0 + f(0) \in \overline{D(A)}, \qquad (4.1.1)$$

whereas if problem (4.0.1) has a classical or strong solution, then

$$x \in \overline{D(A)}. \qquad (4.1.2)$$

Moreover, any strict solution is also strong and classical.

We are going to show that if $f \in L^1(0,T;X)$ then any type of solution of (4.0.1) may be represented by the variation of constants formula (4.0.2). We begin with the classical solution.

Proposition 4.1.2 *Let $f \in L^1(0,T;X) \cap C(]0,T],X)$, and let $u_0 \in \overline{D(A)}$. If u is a classical solution of (4.0.1), then*

$$u(t) = e^{tA}u_0 + \int_0^t e^{(t-s)A}f(s)ds, \ \ 0 \leq t \leq T.$$

Proof — Let u be a classical solution of (4.0.1) in $[0,T]$, and let $t \in\,]0,T]$. Since $u \in C^1(]0,T];X) \cap C([0,T];X) \cap C(]0,T];D(A))$, then $u(t)$ belongs to $D(A)$ for $0 < t \leq T$, so that the function

$$v(s) = e^{(t-s)A}u(s), \ \ 0 \leq s \leq t,$$

belongs to $C([0,t];X) \cap C^1(]0,t[,X)$, and

$$v(0) = e^{tA}u_0, \ \ v(t) = u(t),$$

$$v'(s) = -Ae^{(t-s)A}u(s) + e^{(t-s)A}u'(s) = e^{(t-s)A}f(s), \ \ 0 < s < t.$$

Then, for $0 < 2\varepsilon < t$,

$$v(t-\varepsilon) - v(\varepsilon) = \int_\varepsilon^{t-\varepsilon} e^{(t-s)A}f(s)ds,$$

so that, letting $\varepsilon \to 0$, we get

$$v(t) - v(0) = \int_0^t e^{(t-s)A}f(s)ds,$$

and the statement follows. ∎

1. Solutions of linear problems

Proposition 4.1.2 implies that if $f \in L^1(0,T;X) \cap C(]0,T],X)$ then the classical solution of (4.0.1) is unique. In particular if $f \in C([0,T],X)$ the strict solution of (4.0.1) is unique and it is given by (4.0.2). As a consequence, also the strong solution of (4.0.1) is unique, and it is given by (4.0.2).

In the applications we shall find situations in which $D(A)$ is not dense in X, and the initial datum u_0 does not belong to $\overline{D(A)}$. Then the initial condition $u(0) = u_0$ has to be understood in a weak sense, such as

$$\lim_{t \to 0} R(\lambda, A)u(t) = R(\lambda, A)u_0$$

for some $\lambda \in \rho(A)$ — and hence, for every $\lambda \in \rho(A)$. Also in this case the representation formula (4.0.2) holds, as the next corollary states.

Corollary 4.1.3 *Let $f \in L^1(0,T;X) \cap C(]0,T],X)$, and let $u_0 \in X$. If $u \in C^1(]0,T],X) \cap C(]0,T],D(A))$ satisfies $u'(t) = Au(t) + f(t)$ for $t > 0$, and $\lim_{t \to 0} R(\lambda, A)u(t) = R(\lambda, A)u_0$ for some $\lambda \in \rho(A)$, then u is given by (4.0.2).*

Proof — The function $v(t) = R(\lambda, A)u(t)$ is a classical solution of

$$v'(t) = Av(t) + R(\lambda, A)f(t), \quad 0 < t \leq T, \quad v(0) = R(\lambda, A)u_0.$$

By Proposition 4.1.2,

$$\begin{aligned}
v(t) &= e^{tA} R(\lambda, A) u_0 + \int_0^t e^{(t-s)A} R(\lambda, A) f(s) ds \\
&= R(\lambda, A) \left(e^{tA} u_0 + \int_0^t e^{(t-s)A} f(s) ds \right), \quad 0 \leq t \leq T.
\end{aligned}$$

By applying $\lambda I - A$ to both sides, (4.0.2) follows. ∎

It is easy to see that (4.0.2) makes sense whenever $f \in L^1(0,T;X)$, $u_0 \in X$. Therefore we give the following definition.

Definition 4.1.4 *Let $f \in L^1(0,T;X)$, and let $x \in X$. The function u defined in (4.0.2) is called the mild solution of (4.0.1).*

Due to estimate (4.0.3), with $k = 0$, the mild solution satisfies

$$\|u(t)\| \leq M_0 \left(\|u_0\| + \int_0^t \|f(s)\| ds \right), \quad 0 \leq t \leq T. \tag{4.1.3}$$

It is easy to see that if $u_0 \in \overline{D(A)}$ then u belongs to $C([0,T];X)$.

The mild solution is also an "integral solution", in the sense specified by the next proposition.

Chapter 4. Nonhomogeneous equation with continuous data

Proposition 4.1.5 *Let $f \in L^1(0,T;X)$, and let $u_0 \in X$. If u is defined by (4.0.2), then for every $t \in [0,T]$ the integral $\int_0^t u(s)ds$ belongs to $D(A)$, and*

$$u(t) = u_0 + A \int_0^t u(s)ds + \int_0^t f(s)ds, \quad 0 \leq t \leq T. \tag{4.1.4}$$

Proof — For every $t \in [0,T]$ we have

$$\begin{aligned}\int_0^t u(s)ds &= \int_0^t e^{sA} u_0 ds + \int_0^t ds \int_0^s e^{(s-\sigma)A} f(\sigma)d\sigma \\ &= \int_0^t e^{sA} u_0 ds + \int_0^t d\sigma \int_\sigma^t e^{(s-\sigma)A} f(\sigma)ds.\end{aligned}$$

By Proposition 2.1.4(ii), the integral $\int_0^t u(s)ds$ belongs to $D(A)$, and

$$A \int_0^t u(s)ds = e^{tA} u_0 - u_0 + \int_0^t (e^{(t-\sigma)A} - 1) f(\sigma) d\sigma, \quad 0 \leq t \leq T,$$

so that (4.1.4) holds. ■

The result of Proposition 4.1.5 is used in the next lemma, where we give sufficient conditions in order that a mild solution be classical or strict.

Lemma 4.1.6 *Let $f \in L^1(0,T;X) \cap C(]0,T];X)$, let $u_0 \in \overline{D(A)}$, and let u be the mild solution of (4.0.1). The following conditions are equivalent.*

(a) $u \in C(]0,T];D(A))$,

(b) $u \in C^1(]0,T];X)$,

(c) u *is a classical solution of (4.0.1).*

If in addition $f \in C([0,T];X)$, then the following conditions are equivalent.

(a') $u \in C([0,T];D(A))$,

(b') $u \in C^1([0,T];X)$,

(c') u *is a strict solution of (4.0.1).*

Proof — Of course, (c) is stronger than (a) and (b). Let us show that if either (a) or (b) holds, then u is a classical solution. By Proposition 2.1.4(i), $t \mapsto e^{tA} u_0$ belongs to $C([0,T];X)$. From estimates (4.0.3) it follows that $t \mapsto \int_0^t e^{(t-s)A} f(s)ds$ belongs to $C([0,T];X)$, so that u belongs to $C([0,T];X)$ (see also next Proposition 4.2.1). Moreover, by Proposition 4.1.5, u satisfies (4.1.4). Therefore, for every t, h such that $t, t+h \in]0,T]$,

$$\frac{u(t+h) - u(t)}{h} = \frac{1}{h} A \int_t^{t+h} u(s)ds + \frac{1}{h} \int_t^{t+h} f(s)ds. \tag{4.1.5}$$

1. SOLUTIONS OF LINEAR PROBLEMS

Since f is continuous at t, then

$$\lim_{h \to 0} \frac{1}{h} \int_t^{t+h} f(s)ds = f(t). \tag{4.1.6}$$

Let (a) hold. Then Au is continuous at t, so that

$$\lim_{h \to 0} \frac{1}{h} A \int_t^{t+h} u(s)ds = \lim_{h \to 0} \frac{1}{h} \int_t^{t+h} Au(s)ds = Au(t).$$

By (4.1.5) and (4.1.6) we get now that u is differentiable at the point t, with $u'(t) = Au(t) + f(t)$. Since both Au and f are continuous in $]0, T]$, then u' too is continuous, and u is a classical solution.

Let now (b) hold. Since u is continuous at t, then

$$\lim_{h \to 0} \frac{1}{h} \int_t^{t+h} u(s)ds = u(t).$$

On the other hand, by (4.1.5) and (4.1.6), there exists the limit

$$\lim_{h \to 0} A \left(\frac{1}{h} \int_t^{t+h} u(s)ds \right) = u'(t) - f(t).$$

Since A is a closed operator, then $u(t)$ belongs to $D(A)$, and $Au(t) = u'(t) - f(t)$. Since both u' and f are continuous in $]0, T]$, then also Au is continuous in $]0, T]$, so that u is a classical solution.

The equivalence of (a'), (b'), (c') may be proved in the same way. ∎

Lemma 4.1.6 is very useful, because it lets one get down half of the job in proving existence of strict or classical solutions. Note that, if f is an arbitrary function in $C([0, T]; X)$, then the mild solution of (4.0.1) is not necessarily strict, as the following counterexample shows.

Example 4.1.7 Let $X = L^2(0, \pi)$, $D(A) = H^2(0, \pi) \cap H_0^1(0, \pi)$, $A\varphi(\xi) = \varphi''(\xi)$. Then A generates an analytic semigroup; moreover $\sigma(A) = \{-n^2 : n \in \mathbb{N}\}$, and setting $e_n(\xi) = (2/\pi)^{1/2} \sin n\xi$ we have $\|e_n\| = 1$ and $e^{tA} e_n = e^{-tn^2} e_n$.

Choose $T = 1$, and assume by contradiction that for every $f \in C([0, 1]; X)$ there is a strict solution u of problem

$$u'(t) = Au(t) + f(t), \quad 0 \leq t \leq 1; \quad u(0) = 0.$$

Then the mapping $C([0, 1]; D(A)) \cap C^1([0, 1], X) \mapsto C([0, 1], X)$, $u \mapsto u' - Au$, is continuous and onto, so that there exists $C > 0$ such that

$$\|u\|_{C([0,1];D(A))} \leq C \|f\|_{C([0,1];X)}. \tag{4.1.7}$$

For $k \in \mathbb{N}$, let $a_k = 1 - 2^{-2k}$, $b_k = a_k + 2^{-2k-1}$, and let $\varphi_k \in C^\infty(\mathbb{R})$ be such that

$$0 \le \varphi_k(\xi) \le 1, \quad \varphi_k(\xi) = 1 \text{ for } \xi \in [a_k, b_k],$$

$$\varphi_k(0) = 0, \quad \text{supp } \varphi_k \cap \text{supp } \varphi_h = \emptyset \text{ for } h \ne k.$$

For every $n \in \mathbb{N}$ set

$$f_n(t) = \sum_{k=1}^n \varphi_k(t) e_{2^k}, \quad 0 \le t \le 1.$$

Since every φ_k is continuous and $\text{supp } \varphi_k \cap \text{supp } \varphi_h = \emptyset$ for $h \ne k$, then f_n is continuous and

$$\sup_{0 \le t \le 1} \|f_n(t)\| = 1, \quad \forall n \in \mathbb{N}.$$

Moreover, it is easy to check that $u_n(t) = \int_0^t e^{(t-s)A} f_n(s) ds$ belongs to $D(A)$ for every $t \in [0, 1]$, and

$$A u_n(1) = \sum_{k=1}^n \int_0^1 \varphi_k(s) e^{(1-s)A} A e_{2^k} ds = -\sum_{k=1}^n \int_0^1 2^{2k} \varphi_k(s) e^{-2^{2k}(1-s)} ds \, e_{2^k},$$

so that

$$\|A u_n(1)\|^2 \ge \sum_{k=1}^n \left(\int_{a_k}^{b_k} 2^{2k} e^{-2^{2k}(1-s)} ds \right)^2$$

$$= \sum_{k=1}^n \left(e^{-2^{2k}(1-b_k)} - e^{-2^{2k}(1-a_k)} \right)^2 \ge \sum_{k=1}^n (e^{-1/2} - e^{-1})^2.$$

Therefore, there is no number $C > 0$ such that $\|Au_n\|_{C([0,1];X)} \le C \|f_n\|_{C([0,1];X)}$ for every $n \in \mathbb{N}$, so that (4.1.7) cannot hold. ∎

Now we discuss the relationship between mild and strong solutions. It is clear from the definition that if problem (4.0.1) has a strong solution, then $f \in C([0, T]; X)$ and $x \in \overline{D(A)}$. Indeed, these conditions are also sufficient for the mild solution be strong, as the following proposition shows.

Proposition 4.1.8 *Let $f \in C([0, T]; X)$, $x \in \overline{D(A)}$. Then the mild solution of problem (4.0.1) is strong.*

Proof — Fix $\lambda \in \rho(A)$. Then $u_0 + (A - \lambda I)^{-1} f(0)$ belongs to $\overline{D(A)}$. Since $D(A^2)$ is dense in $\overline{D(A)}$, there exists a sequence $\{y_n\} \subset D(A^2)$ such that

$$\lim_{n \to \infty} y_n = u_0 + (A - \lambda I)^{-1} f(0).$$

Moreover, there exists a sequence $\{f_n\}_{n \in \mathbb{N}}$ of functions in $C^1([0, T]; X)$ such that

$$f_n(0) = f(0), \quad \lim_{n \to \infty} \|f_n - f\|_{C([0,T];X)} = 0.$$

2. MILD SOLUTIONS

For instance, f_n could be the n-th Bernstein polynomial of f, see Subsection 0.1.1. Set
$$x_n = y_n - (A - \lambda I)^{-1} f(0), \quad n \in \mathbb{N}.$$
Then $x_n \in D(A)$, $Ax_n + f_n(0) = Ay_n - \lambda(A - \lambda I)^{-1} f(0) \in \overline{D(A)}$. We shall show later (Theorem 4.3.1(ii)) that the problem
$$u'_n(t) = Au_n(t) + f_n(t), \quad t > 0; \quad u_n(0) = x_n,$$
has a unique strict solution u_n. By estimate (4.0.3), with $k = 0$, we have
$$\|u_n - u\|_{C([0,T];X)} \leq M_0(\|x_n - x\| + T\|f_n - f\|_{C([0,T];X)}).$$
Since $x_n \to u_0$ and $f_n \to f$ as $n \to \infty$, the statement follows. ∎

4.2 Mild solutions

This section is devoted to the properties of the mild solution of (4.0.1) in the case where f belongs either to $L^\infty(0,T;X)$ or to $C(]0,T];X) \cap L^1(0,T;X)$. Since the properties of the function $t \mapsto e^{tA}x$ were studied in Chapter 2, we focus our attention on the function
$$v(t) = (e^{tA} * f)(t) = \int_0^t e^{(t-s)A} f(s) ds, \quad 0 \leq t \leq T. \tag{4.2.1}$$

Proposition 4.2.1 *Let $f \in L^\infty(0,T;X)$. Then, for every $\alpha \in]0,1[$, $v \in C^\alpha([0,T]; X) \cap C([0,T]; D_A(\alpha,1))$. Precisely, it belongs to $C^{1-\alpha}([0,T]; D_A(\alpha,1))$, and there is C independent of f such that*
$$\|v\|_{C^{1-\alpha}([0,T];D_A(\alpha,1))} \leq C\|f\|_{L^\infty(0,T;X)}. \tag{4.2.2}$$
It follows that v belongs to $C^{1-\alpha}([0,T]; X_\alpha)$ for every space $X_\alpha \in J_\alpha(X, D(A))$.

Proof — Since $s \mapsto \|e^{(t-s)A}\|_{L(X, D_A(\alpha,1))}$ belongs to $L^1(0,t)$ for every $t \in]0,T]$, then $v(t)$ belongs to $D_A(\alpha,1)$ for every $\alpha \in]0,1[$, and
$$\|v(t)\|_{D_A(\alpha,1)} \leq K_{0,\alpha}(1-\alpha)^{-1} T^{1-\alpha} \|f\|_{L^\infty(0,T;X)}, \tag{4.2.3}$$
Moreover, for $0 \leq s \leq t \leq T$,
$$\begin{aligned} v(t) - v(s) &= \int_0^s \left(e^{(t-\sigma)A} - e^{(s-\sigma)A} \right) f(\sigma) d\sigma + \int_s^t e^{(t-\sigma)A} f(\sigma) d\sigma \\ &= \int_0^s d\sigma \int_{s-\sigma}^{t-\sigma} Ae^{\tau A} f(\sigma) d\tau + \int_s^t e^{(t-\sigma)A} f(\sigma) d\sigma, \end{aligned}$$

which implies

$$\|v(t) - v(s)\|_{D_A(\alpha,1)} \leq K_{1,\alpha} \int_0^s d\sigma \int_{s-\sigma}^{t-\sigma} \tau^{-1-\alpha} d\tau \, \|f\|_\infty$$
$$+ K_{0,\alpha} \int_s^t (t-\sigma)^{-\alpha} d\sigma \, \|f\|_\infty \leq \left(\frac{K_{1,\alpha}}{\alpha(1-\alpha)} + \frac{K_{0,\alpha}}{1-\alpha} \right) (t-s)^{1-\alpha} \|f\|_\infty, \quad (4.2.4)$$

so that v is $(1-\alpha)$-Hölder continuous with values in $D_A(\alpha,1)$. Estimate (4.2.2) follows now from (4.2.3) and (4.2.4). ∎

Corollary 4.2.2 *Let $f \in L^\infty(0,T;X)$, $u_0 \in X$, and let u be the mild solution of (4.0.3). Then $u \in L^\infty(0,T;X)$, and*

$$\|u\|_{L^\infty(0,T;X)} \leq M_0(\|u_0\| + T\|f\|_{L^\infty(0,T;X)}). \quad (4.2.5)$$

For every $\alpha \in \,]0,1[$ and $\varepsilon \in \,]0,T[$, $u \in C^{1-\alpha}([\varepsilon,T]; D_A(\alpha,1))$. Consequently, u belongs to $C^{1-\alpha}([\varepsilon,T]; X_\alpha)$ for every space $X_\alpha \in J_\alpha(X, D(A))$. There is C independent of f such that

$$\|u\|_{C^\alpha([\varepsilon,T];X)} + \|u\|_{B([\varepsilon,T];D_A(\alpha,1))} \leq C(\varepsilon^{-\alpha}\|u_0\| + \|f\|_{L^\infty(0,T;X)}),$$
$$\|u\|_{C^{1-\alpha}([\varepsilon,T];D_A(\alpha,1))} \leq C(\varepsilon^{-1}\|u_0\| + \|f\|_{L^\infty(0,T;X)}). \quad (4.2.6)$$

Moreover, $u \in C([0,T];X) \iff u_0 \in \overline{D(A)}$. For $0 < \alpha < 1$, $u \in C^\alpha([0,T];X) \iff x \in D_A(\alpha,\infty)$.

In the applications to nonlinear problems we shall consider frequently functions f which are bounded in every interval $[\varepsilon,T]$ with $\varepsilon \in \,]0,T[$, and blow up at $t=0$. Next proposition deals with functions belonging to the weighted spaces defined in (4.0.6), (4.0.8).

Proposition 4.2.3 *Let $f \in L^1(0,T;X) \cap L^\infty(\varepsilon,T;X)$ for every $\varepsilon \in \,]0,T[$. Then $v = e^{tA} * f$ belongs to $C([0,T];X) \cap C^{1-\alpha}([\varepsilon,T]; D_A(\alpha,1))$ for every $\alpha \in \,]0,1[$, $\varepsilon \in \,]0,T[$.*

If in addition $t \mapsto t^\theta f(t)$ is bounded and measurable in $]0,T]$, for some $\theta \in \,]0,1[$ (so that $f \in B_\theta(]0,T];X)$), then the following statements hold.

(i) $v \in C^{1-\theta}([0,T];X)$, and

$$\|v\|_{C^{1-\theta}([0,T];X)} \leq C\|f\|_{B_\theta(]0,T];X)}.$$

(ii) *For every $\alpha \in \,]0,1[$, $v \in C_\theta^{1-\alpha}(]0,T]; D_A(\alpha,1))$, and*

$$\|v\|_{C_\theta^{1-\alpha}(]0,T];D_A(\alpha,1))} \leq C\|f\|_{B_\theta(]0,T];X)}.$$

In particular, v is bounded with values in $D_A(1-\theta,1)$, and $t \mapsto t^{\theta+\alpha-1}v(t) \in B([0,T]; D_A(\alpha,1))$, $t \mapsto t^\theta v(t) \in C^{1-\alpha}([0,T]; D_A(\alpha,1))$ for every $\alpha \in \,]0,1[$. If in addition $\lim_{t \to 0} \|t^\theta f(t)\| = 0$, then $v \in C([0,T]; D_A(1-\theta,1))$.

2. Mild solutions

(iii) If $\alpha + \theta < 1$, then $v \in C^{1-\alpha-\theta}([0,T]; D_A(\alpha,1))$, and

$$\|v\|_{C^{1-\alpha-\theta}([0,T];D_A(\alpha,1))} \leq C\|f\|_{B_\theta(]0,T];X)}.$$

Proof — v is obviously continuous with values in X up to $t = 0$. For $0 < \varepsilon \leq t \leq T$ we have

$$v(t) = e^{(t-\varepsilon/2)A} \int_0^{\varepsilon/2} e^{(\varepsilon/2-s)A} f(s) ds + \int_{\varepsilon/2}^t e^{(t-s)A} f(s) ds = v_1(t) + v_2(t).$$

By Proposition 4.2.1, v_2 belongs to $C^{1-\alpha}([\varepsilon/2,T]; D_A(\alpha,1))$ for every α. Moreover, v_1 belongs to $C^\infty([\varepsilon,T]; D(A^n))$ for every $n \in \mathbb{N}$, due to Proposition 2.1.1(i)-(iv). Therefore, $v \in C([0,T];X) \cap C^{1-\alpha}([\varepsilon,T]; D_A(\alpha,1))$.

Let now $t \mapsto t^\theta f(t)$ be measurable and bounded. Then for $0 \leq r \leq t \leq T$

$$\|v(t) - v(r)\| \leq \left\| \int_0^r (e^{(t-s)A} - e^{(r-s)A}) f(s) ds \right\| + \left\| \int_r^t e^{(t-s)A} f(s) ds \right\|$$

$$\leq \left(M_1 \int_0^r \frac{ds}{s^\theta} \int_{r-s}^{t-s} \frac{d\sigma}{\sigma} + M_0 \int_r^t \frac{ds}{s^\theta} \right) \|f\|_{B_\theta(]0,T];X)}$$

$$\leq \left(M_1 \int_0^r \frac{ds}{(r-s)^{1-\theta} s^\theta} \int_{r-s}^{t-s} \frac{d\sigma}{\sigma^\theta} + \frac{M_0}{1-\theta}(t-r)^{1-\theta} \right) \|f\|_{B_\theta(]0,T];X)}$$

$$\leq \frac{1}{1-\theta} \left(M_1 \int_0^1 \frac{d\sigma}{(1-\sigma)^{1-\theta}\sigma^\theta} + M_0 \right) (t-r)^{1-\theta} \|f\|_{B_\theta(]0,T];X)},$$

and statement (i) follows.

Let us prove that statement (ii) holds. For $0 < t \leq T$ we have

$$\|v(t)\|_{D_A(\alpha,1)} \leq K_{0,\alpha} \int_0^t \frac{ds}{(t-s)^\alpha s^\theta} \|f\|_{B_\theta(]0,T];X)}$$

$$= K_{0,\alpha} t^{1-\alpha-\theta} \int_0^1 \frac{d\sigma}{(1-\sigma)^\alpha \sigma^\theta} \|f\|_{B_\theta(]0,T];X)}, \qquad (4.2.7)$$

so that $t \mapsto t^{\theta+\alpha-1} v(t)$ is bounded in $]0,T]$ with values in $D_A(\alpha,1)$. Moreover, if N is such that $\|tAe^{tA}\|_{L(D_A(\alpha,1))} \leq N$ for $0 < t \leq T$, then for $0 < \varepsilon \leq r \leq t \leq T$

$$\|v_1(t) - v_1(r)\|_{D_A(\alpha,1)} = \left\| \int_{r-\varepsilon/2}^{t-\varepsilon/2} Ae^{\sigma A} d\sigma \int_0^{\varepsilon/2} e^{(\varepsilon/2-s)A} f(s) ds \right\|_{D_A(\alpha,1)}$$

$$\leq \frac{NK_{0,\alpha}}{(r-\varepsilon/2)^{1-\alpha}} \int_{r-\varepsilon/2}^{t-\varepsilon/2} \frac{d\sigma}{\sigma^\alpha} \int_0^1 \frac{d\tau}{(1-\tau)^\alpha \tau^\theta} \left(\frac{\varepsilon}{2}\right)^{1-\alpha-\theta} \|f\|_{B_\theta(]0,T];X)}$$

$$\leq \frac{NK_{0,\alpha}}{1-\alpha} \int_0^1 \frac{d\sigma}{(1-\sigma)^\alpha \sigma^\theta} \left(\frac{\varepsilon}{2}\right)^{-\theta} (t-r)^{1-\alpha} \|f\|_{B_\theta(]0,T];X)},$$

and by estimate (4.2.4) there is $C > 0$ independent of ε such that

$$\|v_2(t) - v_2(r)\|_{D_A(\alpha,1)}$$
$$\leq C(t-r)^{1-\alpha}\|f\|_{B([\varepsilon/2,T];X)} \leq C(t-r)^{1-\alpha}\left(\frac{\varepsilon}{2}\right)^{-\theta}\|f\|_{B_\theta(]0,T];X)},$$

so that $v \in C_\theta^{1-\alpha}(]0,T]; D_A(\alpha,1))$. Choosing $\alpha = 1 - \theta$, we see that v is bounded up to $t = 0$ and continuous for $t > 0$ with values in $D_A(1-\theta,1)$. If in addition

$$\lim_{t \to 0} \|t^\theta f(t)\| = 0,$$

then v is continuous up to $t = 0$ with values in $D_A(1-\theta,1)$: indeed, for every $\varepsilon > 0$ there is $\delta > 0$ such that for $0 < s \leq \delta$ we have $\|s^\theta f(s)\| \leq \varepsilon$, so that for $0 < t \leq \delta$ estimate (4.2.7) can be replaced by

$$\|v(t)\|_{D_A(1-\theta,1)} \leq K_{0,1-\theta}\,\varepsilon \int_0^1 \frac{d\sigma}{(1-\sigma)^{1-\theta}\sigma^\theta},$$

which implies that $\lim_{t \to 0} \|v(t)\|_{D_A(1-\theta,1)} = 0$.

The proof of statement (iii) is similar to the one of statement (i) and is left to the reader. ∎

In the next Corollary we combine the above results with the results of Chapter 2 about the function $t \mapsto e^{tA}u_0$.

Corollary 4.2.4 *Let $u_0 \in X$, $f \in L^1(0,T;X)$, and let u be the mild solution of (4.0.1). Then u belongs to $C(]0,T];X)$ and*

$$\sup_{0 \leq t \leq T} \|u(t)\| \leq C(\|u_0\| + \|f\|_{L^1(0,T;X)}).$$

Moreover, if $u_0 \in D_A(\theta,\infty)$ and $t \mapsto t^{1-\theta}f(t)$ is measurable and bounded, for some $\theta \in\,]0,1[$, then $u \in C^\theta([0,T];X) \cap C_{1-\theta}^{1-\alpha}(]0,T]; D_A(\alpha,1))$ for every $\alpha \in\,]0,1[$ (so that in particular $t \mapsto t^{1-\theta}u(t) \in C^{1-\alpha}([0,T]; D_A(\alpha,1))$), and

$$\|u\|_{C^\theta([0,T];X)} \leq C(\|u_0\|_{D_A(\theta,\infty)} + \|f\|_{B_\theta(]0,T];X)}),$$

$$\|u\|_{C_{1-\theta}^{1-\alpha}(]0,T];D_A(\alpha,1))} \leq C(\|u_0\|_{D_A(\theta,\infty)} + \|f\|_{B_\theta(]0,T];X)}).$$

For $\alpha > \theta$, $t \mapsto t^{\alpha-\theta}u(t) \in B([0,T]; D_A(\alpha,1))$, and

$$\sup_{0 \leq t \leq T} \|t^{\alpha-\theta}u(t)\|_{D_A(\alpha,1)} \leq C(\|u_0\|_{D_A(\theta,\infty)} + \|f\|_{B_\theta(]0,T];X)}).$$

For $\alpha < \theta$, $u \in C^{\theta-\alpha}([0,T]; D_A(\alpha,1))$, and

$$\|u\|_{C^{\theta-\alpha}([0,T];D_A(\alpha,1))} \leq C(\|u_0\|_{D_A(\theta,\infty)} + \|f\|_{B_\theta(]0,T];X)}).$$

Note that, if u_0 does not belong to the closure of $D(A)$, then u is not continuous up to $t = 0$. However, a mild continuity result at $t = 0$ holds. See next proposition.

Proposition 4.2.5 *Let $f \in L^1(0, T; X)$, $u_0 \in X$, and let u be the mild solution of (4.0.1). For every $\lambda \in \rho(A)$ it holds*

$$\lim_{t \to 0} \|R(\lambda, A)(u(t) - u_0)\|_{D_A(\theta, p)} = 0,$$

for each $\theta \in {]}0, 1[$, $p \geq 1$.

Proof — The function $v = e^{tA} * f$ is continuous up to $t = 0$ with values in X, so that $\lim_{t \to 0} \|R(\lambda, A)v(t)\|_{D_A(\theta,p)} = 0$; moreover, since $R(\lambda, A)u_0 \in D(A)$, then $\lim_{t \to 0} \|R(\lambda, A)(e^{tA}u_0 - u_0)\|_{D_A(\theta,p)} = \lim_{t \to 0} \|(e^{tA} - 1)R(\lambda, A)u_0\|_{D_A(\theta,p)} = 0$. ∎

4.3 Strict and classical solutions, and optimal regularity

In this section we state and prove several existence theorems for classical and strict solutions. Example 4.1.7 shows that if the datum f is merely continuous with values in X, then problem (4.0.1) has not necessarily a strict solution. So, we have to make further assumptions on f to get strict or classical solutions. We begin with the case where f is Hölder continuous, either up to $t = 0$ or in each interval $[\varepsilon, T]$ with $\varepsilon > 0$ (what we call *time regularity*). In Subsection 4.3.2 we shall consider the case where f has values in some intermediate space between X and $D(A)$ (what we call *space regularity*). As in Section 4.2, we use the constants M_k, $M_{k,\alpha}$, $K_{k,\alpha}$ defined in (4.0.3), (4.0.4), (4.0.5).

Even if not stated explicitly, the constants C appearing in the estimates on the solution proved in the next theorems are nondecreasing with respect to T. In particular, they remain bounded when the interval $[0, T]$ is replaced by a smaller interval $[a, b] \subset [0, T]$. This property is important in the applications to nonlinear problems, where small intervals are considered.

4.3.1 Time regularity

The results of this subsection can be roughly grouped in two parts: the first part, concerning the case where f is Hölder continuous up to $t = 0$ (Theorem 4.3.1, Corollary 4.2.4, Remark 4.3.3), and the second part, concerning the case where f is Hölder continuous in each interval $[\varepsilon, T]$, with $\varepsilon > 0$ (Theorems 4.3.5, 4.3.7, and Corollary 4.3.6). At a first reading, the second part may be skipped, since it deals with more complicated results, obtained however by similar techniques.

The case where f is Hölder continuous up to $t = 0$

Let u be the mild solution of (4.0.1), and set $u = u_1 + u_2$, where

$$\begin{cases} u_1(t) = \int_0^t e^{(t-s)A}(f(s) - f(t))ds, & 0 \leq t \leq T, \\ u_2(t) = e^{tA}u_0 + \int_0^t e^{(t-s)A}f(t)ds, & 0 \leq t \leq T. \end{cases} \quad (4.3.1)$$

Theorem 4.3.1 *Let $0 < \alpha < 1$, $f \in C^\alpha([0,T],X)$, $u_0 \in X$, and let u be the mild solution of (4.0.1). Then u belongs to $C^\alpha([\varepsilon,T],D(A)) \cap C^{1+\alpha}([\varepsilon,T],X)$ for every $\varepsilon \in]0,T[$, and*

(i) *if $u_0 \in \overline{D(A)}$, then u is a classical solution of (4.0.1);*

(ii) *if $u_0 \in D(A)$ and $Au_0 + f(0) \in \overline{D(A)}$, then u is a strict solution of (4.0.1), and there is C such that*

$$\|u\|_{C^1([0,T],X)} + \|u\|_{C([0,T],D(A))} \leq C(\|f\|_{C^\alpha([0,T],X)} + \|u_0\|_{D(A)}); \quad (4.3.2)$$

(iii) *if $u_0 \in D(A)$ and $Au_0 + f(0) \in D_A(\alpha,\infty)$, then both u' and Au belong to $C^\alpha([0,T],X)$, u' belongs to $B([0,T];D_A(\alpha,\infty))$, and there is C such that*

$$\|u\|_{C^{1+\alpha}([0,T],X)} + \|Au\|_{C^\alpha([0,T],X)} + \|u'\|_{B([0,T],D_A(\alpha,\infty))} \\ \leq C(\|f\|_{C^\alpha([0,T],X)} + \|u_0\|_{D(A)} + \|Au_0 + f(0)\|_{D_A(\alpha,\infty)}). \quad (4.3.3)$$

Proof — Thanks to Lemma 4.1.6, to prove statements (i) and (ii) it is sufficient to show that u belongs to $C(]0,T];D(A))$ in the case where $u_0 \in \overline{D(A)}$, and to $C([0,T];D(A))$ in the case where $u_0 \in D(A)$ and $Au_0 + f(0) \in \overline{D(A)}$.

Let u_1 and u_2 be defined by (4.3.1). Then $u_1(t) \in D(A)$ for $t \geq 0$, $u_2(t) \in D(A)$ for $t > 0$, and

$$\begin{cases} (i) & Au_1(t) = \int_0^t Ae^{(t-s)A}(f(s) - f(t))ds, \quad 0 \leq t \leq T, \\ (ii) & Au_2(t) = Ae^{tA}u_0 + (e^{tA} - 1)f(t), \quad 0 < t \leq T. \end{cases} \quad (4.3.4)$$

If $u_0 \in D(A)$, then (4.3.4)(ii) holds also for $t = 0$.

Let us show that Au_1 is Hölder continuous in $[0,T]$. For $0 \leq s \leq t \leq T$

$$Au_1(t) - Au_1(s) = \int_0^s A\left(e^{(t-\sigma)A} - e^{(s-\sigma)A}\right)(f(\sigma) - f(s))d\sigma \\ + (e^{tA} - e^{(t-s)A})(f(s) - f(t)) + \int_s^t Ae^{(t-\sigma)A}(f(\sigma) - f(t))d\sigma, \quad (4.3.5)$$

3. STRICT AND CLASSICAL SOLUTIONS, AND OPTIMAL REGULARITY

so that

$$\|Au_1(t) - Au_1(s)\| \leq M_2 \int_0^s (s-\sigma)^\alpha \int_{s-\sigma}^{t-\sigma} \tau^{-2} d\tau \, d\sigma \, [f]_{C^\alpha}$$

$$+ 2M_0(t-s)^\alpha [f]_{C^\alpha} + M_1 \int_s^t (t-\sigma)^{\alpha-1} d\sigma \, [f]_{C^\alpha} \quad (4.3.6)$$

$$\leq M_2 \int_0^s d\sigma \int_{s-\sigma}^{t-\sigma} \tau^{\alpha-2} d\tau \, [f]_{C^\alpha} + (2M_0 + M_1 \alpha^{-1})(t-s)^\alpha [f]_{C^\alpha}$$

$$\leq \left(\frac{M_2}{\alpha(1-\alpha)} + 2M_0 + \frac{M_1}{\alpha} \right) (t-s)^\alpha [f]_{C^\alpha}.$$

Therefore, Au_1 is α-Hölder continuous in $[0,T]$. Moreover, Au_2 is obviously continuous in $]0,T]$: hence, if $u_0 \in \overline{D(A)}$, then $u \in C([0,T],X)$, and $Au \in C(]0,T];X)$, so that, by Lemma 4.1.6, u is a classical solution of (4.0.1), and statement (i) is proved.

If $u_0 \in D(A)$ we have

$$Au_2(t) = e^{tA}(Au_0 + f(0)) + e^{tA}(f(t) - f(0)) - f(t), \quad 0 \leq t \leq T, \quad (4.3.7)$$

so that if $Au_0 + f(0) \in \overline{D(A)}$ then Au_2 is continuous also at $t=0$, and statement (ii) follows.

In the case where $Ax + f(0) \in D_A(\alpha, \infty)$, from (4.3.7) we get, for $0 \leq s \leq t \leq T$,

$$\|Au_2(t) - Au_2(s)\| \leq \|(e^{tA} - e^{sA})(Au_0 + f(0))\|$$
$$+ \|(e^{tA} - e^{sA})(f(s) - f(0))\| + \|(e^{tA} - 1)(f(t) - f(s))\|$$

$$\leq \int_s^t \|Ae^{\sigma A}\|_{L(D_A(\alpha,\infty),X)} d\sigma \, \|Au_0 + f(0)\|_{D_A(\alpha,\infty)}$$

$$+ s^\alpha \left\| A \int_s^t e^{\sigma A} d\sigma \right\|_{L(X)} [f]_{C^\alpha} + (M_0 + 1)(t-s)^\alpha [f]_{C^\alpha}$$

$$\leq \frac{M_{1,\alpha}}{\alpha} \|Au_0 + f(0)\|_{D_A(\alpha,\infty)} (t-s)^\alpha + \left(\frac{M_1}{\alpha} + M_0 + 1 \right)(t-s)^\alpha [f]_{C^\alpha},$$
$$(4.3.8)$$

so that also Au_2 is Hölder continuous, and the estimate

$$\|u\|_{C^{1+\alpha}([0,T];X)} + \|Au\|_{C^\alpha([0,T];X)}$$
$$\leq C(\|f\|_{C^\alpha([0,T],X)} + \|u_0\|_{D(A)} + \|Au_0 + f(0)\|_{D_A(\alpha,\infty)})$$

follows easily.

Since u' and Au are Hölder continuous, from Proposition 2.2.12(i) it follows that u' is bounded with values in $D_A(\alpha, \infty)$. However, the embedding constant given by Proposition 2.2.12(i) depends on the lenght T of the interval in such a way that it blows up as $T \to 0$. So, we estimate $[u'(t)]_{D_A(\alpha,\infty)}$ directly. For $0 \leq t \leq T$

we have, by (4.3.4),

$$u'(t) = \int_0^t Ae^{(t-s)A}(f(s) - f(t))ds + e^{tA}(Au_0 + f(0)) + e^{tA}(f(t) - f(0)),$$

so that for $0 < \xi \leq 1$

$$\|\xi^{1-\alpha} Ae^{\xi A} u'(t)\| \leq \left\| \xi^{1-\alpha} \int_0^t A^2 e^{(t+\xi-s)A}(f(s) - f(t))ds \right\|$$
$$+ \|\xi^{1-\alpha} Ae^{(t+\xi)A}(Au_0 + f(0))\| + \|\xi^{1-\alpha} Ae^{(t+\xi)A}(f(t) - f(0))\|$$
$$\leq M_2 \xi^{1-\alpha} \int_0^t (t-s)^\alpha (t+\xi-s)^{-2} ds \, [f]_{C^\alpha} \qquad (4.3.9)$$
$$+ M_0 [Au_0 + f(0)]_{D_A(\alpha,\infty)} + M_1 \xi^{1-\alpha} (t+\xi)^{-1} t^\alpha \, [f]_{C^\alpha}$$
$$\leq M_2 \int_0^\infty \sigma^\alpha (\sigma+1)^{-2} d\sigma \, [f]_{C^\alpha} + M_0 [Au_0 + f(0)]_{D_A(\alpha,\infty)} + M_1 [f]_{C^\alpha}.$$

Therefore, $[u'(t)]_{D_A(\alpha,\infty)}$ is bounded in $[0, T]$, and the proof is complete. ∎

Corollary 4.3.2 *Let $0 < \alpha < 1$, and let $f \in h^\alpha([0,T];X)$, $u_0 \in D(A)$. Assume moreover that*

$$Au_0 + f(0) \in D_A(\alpha). \qquad (4.3.10)$$

Then the solution u of problem (4.0.1) belongs to $h^{1+\alpha}([0,T],X) \cap h^\alpha([0,T], D(A))$, and u' belongs to $C([0,T]; D_A(\alpha))$.

Proof — Let $\{f_n\}_{n \in \mathbb{N}} \subset C^\infty([0,T],X)$ be such that $f_n \to f$ in $C^\alpha([0,T],X)$. Let $\lambda \in \rho(A)$ be fixed. Since $D(A^2)$ is dense in $D_A(\alpha+1)$, and $u_0 + (A - \lambda I)^{-1} f(0) = (A - \lambda I)^{-1}(Ax + f(0) - \lambda u_0) \in D_A(\alpha+1)$ thanks to (4.3.10), there is a sequence $\{y_n\}_{n \in \mathbb{N}} \subset D(A^2)$ such that $y_n \to u_0 + (A - \lambda I)^{-1} f(0)$ in $D_A(\alpha+1, \infty)$. Set

$$x_n = y_n - (A - \lambda I)^{-1} f_n(0), \quad n \in \mathbb{N}.$$

Then $x_n \in D(A)$, and $Ax_n + f_n(0) = Ay_n - \lambda(A-\lambda)^{-1} f_n(0) \in D(A)$. By Theorem 4.3.1, the solution u_n of problem

$$u_n'(t) = Au_n(t) + f_n(t), \ 0 \leq t \leq T; \quad u_n(0) = x_n$$

belongs to $C^{1+\alpha+\varepsilon}([0,T],X) \cap C^{\alpha+\varepsilon}([0,T], D(A))$, and $u_n' \in B([0,T]; D_A(\alpha+\varepsilon,\infty))$ for every $\varepsilon \in \,]0, 1-\alpha[$, so that $u_n \in h^{1+\alpha}([0,T],X) \cap h^\alpha([0,T], D(A))$ and $u_n' \in C([0,T]; D_A(\alpha))$ thanks to Proposition 2.2.12(ii). Moreover,

$$f_n \to f \text{ in } C^\alpha([0,T],X), \quad x_n \to u_0 \text{ in } D(A),$$
$$Ax_n + f_n(0) \to Au_0 + f(0) \text{ in } D_A(\alpha, \infty),$$

so that, due again to Theorem 4.3.1, $u_n \to u$ in $C^{1+\alpha}([0,T],X) \cap C^\alpha([0,T], D(A))$ as $n \to \infty$, and $\{u_n'\}_{n \in \mathbb{N}}$ is a Cauchy sequence in $C([0,T]; D_A(\alpha))$. The statement follows now easily. ∎

3. STRICT AND CLASSICAL SOLUTIONS, AND OPTIMAL REGULARITY

Remark 4.3.3 From Proposition 2.2.12 (or else, looking at the proofs of Theorem 4.3.1 and of Corollary 4.2.4) it follows that if $f \in C^\alpha([0,T];X)$ then the condition $Au_0 + f(0) \in D_A(\alpha, \infty)$ is necessary to get u', $Au \in C^\alpha([0,T];X)$, and if $f \in h^\alpha([0,T];X)$, then the condition $Au_0 + f(0) \in D_A(\alpha)$ is necessary to get u', $Au \in h^\alpha([0,T];X)$.

The case where f is not Hölder continuous up to $t = 0$

In many applications to nonlinear or nonautonomous problems we have to deal with functions f that are not necessarily Hölder continuous up to $t = 0$. See Chapters 6, 8, 9. Next result will be useful.

Theorem 4.3.4 *Let $f \in L^1(0,T;X) \cap C^\alpha([\varepsilon,T];X)$ for every $\varepsilon \in \,]0,T[$, and let $u_0 \in X$. Then the mild solution u of problem (4.0.1) belongs to $C(]0,T];D(A)) \cap C^1(]0,T];X) \cap C^\alpha([\varepsilon,T];D(A)) \cap C^{1+\alpha}([\varepsilon,T];X)$, for every $\varepsilon \in \,]0,T[$. If in addition $u_0 \in \overline{D(A)}$, then u is a classical solution.*

Proof — Let us split u as $u(t) = e^{tA}u_0 + v(t)$, where $v = e^{tA} * f$. The first term is obviously continuous in $]0,T]$ with values in $D(A)$, and it belongs to $C([0,T];X)$ if and only if $u_0 \in \overline{D(A)}$. So, we consider only the function v.

We showed in Proposition 4.2.3 that v is continuous in $[0,T]$ with values in X. Moreover, for $0 < \varepsilon \leq t \leq T$,

$$v(t) = e^{(t-\varepsilon/2)A}v(\varepsilon/2) + \int_{\varepsilon/2}^t e^{(t-s)A}f(s)ds = v_1(t) + v_2(t),$$

where $v_1 \in C^\infty([\varepsilon,T];D(A^n))$ for every n, $v_2 \in C^\alpha([\varepsilon,T];D(A)) \cap C^{\alpha+1}([\varepsilon,T];X)$ thanks to Theorem 4.3.1. Since ε is arbitrary, $u \in C(]0,T];D(A)) \cap C^1(]0,T];X)$.

If in addition $u_0 \in \overline{D(A)}$, then $u \in C([0,T];X)$ and it is a classical solution thanks to Lemma 4.1.6. The statement follows. ∎

If we have more precise information on the behavior of the Hölder seminorm of f near $t = 0$, we can study the behavior of u near $t = 0$. In the next theorem we assume that f belongs to the weighted Hölder space $C_\alpha^\alpha(]0,T];X)$, or to $C([0,T];X) \cap C_\alpha^\alpha(]0,T];X)$, with $0 < \alpha < 1$. We recall that $f \in C_\alpha^\alpha(]0,T];X)$ means that f is bounded and that $t \mapsto t^\alpha f(t)$ is α-Hölder continuous in $]0,T]$. Weighted Hölder spaces naturally arise in the study of parabolic evolution equations: for instance, the function

$$t \mapsto e^{tA}x, \;\; 0 \leq t \leq T,$$

belongs to $C_\alpha^\alpha(]0,T];X)$ for all $x \in X$, and to $C([0,T];X) \cap C_\alpha^\alpha(]0,T];X)$ for all $x \in \overline{D(A)}$.

Theorem 4.3.5 *Let $0 < \alpha < 1$, $f \in C_\alpha^\alpha(]0,T];X)$, and set $v = e^{tA} * f$. Then $v \in C_\alpha^\alpha(]0,T];D(A))$, it is differentiable in $]0,T]$ with values in X, and $v' \in C_\alpha^\alpha(]0,T];X) \cap B_\alpha(]0,T];D_A(\alpha,\infty))$. There is $C > 0$ such that*

$$\|v'\|_{C_\alpha^\alpha(X)} + \|Av\|_{C_\alpha^\alpha(X)} + \|v'\|_{B_\alpha(D_A(\alpha,\infty))} \leq C\|f\|_{C_\alpha^\alpha(X)}. \tag{4.3.11}$$

138   CHAPTER 4. NONHOMOGENEOUS EQUATIONS WITH CONTINUOUS DATA

Proof — By Theorem 4.3.4, v belongs to $C(]0,T]; D(A)) \cap C^1(]0,T]; X)$. Let us show that it is bounded with values in $D(A)$. For $0 \le t \le T$ we have

$$\|v(t)\| \le M_0 T \|f\|_{L^\infty(0,T;X)}.$$

Moreover, using the decomposition $v(t) = \int_0^t e^{(t-s)A}(f(s) - f(t))ds + \int_0^t e^{\sigma A} f(t)d\sigma$, we get

$$\begin{aligned}\|Av(t)\| &\le M_1 \int_0^t (t-s)^{\alpha-1} s^{-\alpha} ds\, [f]_{C_\alpha^\alpha(X)} + \|(e^{tA} - I)f(t)\| \\ &= M_1 \int_0^1 (1-s)^{\alpha-1} s^{-\alpha} ds\, [f]_{C_\alpha^\alpha(X)} + (M_0 + 1)\|f\|_\infty.\end{aligned}$$

Therefore,

$$\|v(t)\|_{D(A)} \le C_1 \|f\|_{C_\alpha^\alpha(]0,T];X)}, \ \ 0 < t \le T.$$

As a second step, we prove that $t^\alpha v'(t)$ is bounded with values in $D_A(\alpha, \infty)$, which means that $v' \in B_\alpha(]0,T]; D_A(\alpha, \infty))$. Since

$$v'(t) = Av(t) + f(t) = \int_0^t A e^{(t-s)A}(f(s) - f(t))ds + e^{tA} f(t),$$

then for each $t \in]0,T]$ and $\xi \in]0,1]$ we have

$$t^\alpha \xi^{1-\alpha} \|A e^{\xi A} v'(t)\| \le t^\alpha \left\| \xi^{1-\alpha} \int_0^{t/2} A^2 e^{(t+\xi-s)A}(f(s) - f(t))ds \right\|$$
$$+ t^\alpha \xi^{1-\alpha} \left(\left\| \int_{t/2}^t A^2 e^{(t+\xi-s)A}(f(s) - f(t))ds \right\| + \|A e^{(t+\xi)A} f(t)\| \right)$$
$$\le M_2 t^\alpha \xi^{1-\alpha} \int_0^{t/2} (t+\xi-s)^{-2} ds\, 2\|f\|_\infty$$
$$+ M_2 t^\alpha \xi^{1-\alpha} \int_{t/2}^t \frac{(t-s)^\alpha}{s^\alpha (t+\xi-s)^2} ds\, [f]_{C_\alpha^\alpha(X)} + M_1 t^\alpha \xi^{1-\alpha}(t+\xi)^{-1} \|f(t)\|$$
$$\le 2^{\alpha+1} M_2 \|f\|_\infty + 2^\alpha M_2 \int_0^{t/2} \sigma^\alpha (\sigma+1)^{-2} d\sigma\, [f]_{C_\alpha^\alpha(X)} + M_1 \|f\|_\infty.$$

Therefore, $t^\alpha v'(t)$ is bounded in $]0,T]$ with values in $D_A(\alpha, \infty)$, which means that v' belongs to $B_\alpha(]0,T]; D_A(\alpha, \infty))$. Moreover, there is C_2 such that

$$\|v'\|_{B_\alpha(]0,T];D_A(\alpha,\infty))} \le C_2 \|f\|_{C_\alpha^\alpha(]0,T];X)}.$$

Let us prove that $v \in C_\alpha^\alpha(]0,T]; D(A))$. For every $\varepsilon \in]0,T[$ it holds

$$v(t) = e^{(t-\varepsilon)A} v(\varepsilon) + \int_\varepsilon^t e^{(t-s)A} f(s) ds, \ \ \varepsilon \le t \le T.$$

3. Strict and classical solutions, and optimal regularity

Since $f \in C^\alpha([\varepsilon, T]; X)$ and $v(\varepsilon) \in D(A)$, $Av(\varepsilon) + f(\varepsilon) = v'(\varepsilon) \in D_A(\alpha, \infty)$, then from Theorem 4.3.1(iii), applied in the interval $[\varepsilon, T]$ instead of $[0, T]$, it follows that $v \in C^\alpha([\varepsilon, T]; D(A)) \cap C^{\alpha+1}([\varepsilon, T]; X)$, and that $v' \in B([\varepsilon, T]; D_A(\alpha, \infty))$. From estimate (4.3.3) we get

$$\|v\|_{C^\alpha([\varepsilon,T];D(A))} + \|v'\|_{C^\alpha([\varepsilon,T];X)} + \|v'\|_{B([\varepsilon,T];D_A(\alpha,\infty))}$$
$$\leq C(\|v(\varepsilon)\|_{D(A)} + \|v'(\varepsilon)\|_{D_A(\alpha,\infty)} + \|f\|_{C^\alpha([\varepsilon,T];X)})$$
$$\leq C\left(C_1 \|f\|_{C^\alpha_\alpha(]0,T];X)} + \frac{C_2}{\varepsilon^\alpha}\|f\|_{C^\alpha_\alpha(]0,T];X)} + \frac{1}{\varepsilon^\alpha}\|f\|_{C^\alpha_\alpha(]0,T];X)}\right),$$

so that $v \in C^\alpha_\alpha(]\varepsilon, T]; D(A))$, and estimate (4.3.11) follows. ∎

Theorem 4.3.5 could be shown also without using the results of Theorem 4.3.1. One could estimate $[Av]_{C^\alpha([\varepsilon,T];X)}$ by splitting Au into the sum $Au_1 + Au_2$, as in the proof of Theorem 4.3.1. However, the direct proof would be longer than the present one.

Corollary 4.3.6 *Let $0 < \alpha < 1$, $f \in C^\alpha_\alpha(]0, T]; X)$, $u_0 \in X$, and let u be the mild solution of (4.0.1). The following statements hold.*

(i) *If $u_0 \in \overline{D(A)}$, then u is a classical solution of (4.0.1);*

(ii) *if $u_0 \in D_A(1, \infty)$, then u' and Au belong to $C^\alpha_\alpha(]0, T]; X)$, u' belongs to $B_\alpha(]0, T]; D_A(\alpha, \infty))$, and there is $C > 0$ such that*

$$\|u'\|_{C^\alpha_\alpha(]0,T];X)} + \|Au\|_{C^\alpha_\alpha(]0,T];X)} + \|u'\|_{B_\alpha(]0,T];D_A(\alpha,\infty))} \quad (4.3.12)$$
$$\leq C(\|f\|_{C^\alpha_\alpha(]0,T];X)} + \|u_0\|_{D_A(1,\infty)});$$

(iii) *if $f \in C([0,T]; X)$, $u_0 \in D(A)$, and $Au_0 + f(0) \in \overline{D(A)}$, then u', $Au \in C([0,T]; X)$, and u is a strict solution of problem (4.0.1).*

Proof — Statement (i) follows obviously from Proposition 2.1.4(i) and Theorem 4.3.5.

To prove statement (ii) we have to check that for every $x \in D_A(1, \infty)$ the function $t \mapsto e^{tA}x$ belongs to $C^\alpha_\alpha(]0, T]; D(A))$, and that $\|t^\alpha Ae^{tA}x\|_{D_A(\alpha,\infty)}$ is bounded. For $0 < \varepsilon \leq r \leq t \leq T$ it holds

$$\|A(e^{tA}x - e^{rA}x)\| = \left\|\int_r^t A^2 e^{\sigma A} x \, d\sigma\right\|$$
$$\leq \frac{M_{2,1}}{\varepsilon^\alpha} \int_r^t \frac{d\sigma}{\sigma^{1-\alpha}} \|x\|_{D_A(1,\infty)} \leq \frac{M_{2,1}}{\alpha \varepsilon^\alpha}(t-r)^\alpha \|x\|_{D_A(1,\infty)}.$$

Moreover, estimates (4.0.4) yield

$$\|t^\alpha A e^{tA} x\|_{D_A(\alpha,\infty)} = \|t^\alpha A e^{tA} x\| + \sup_{0 < \xi \leq 1} \|t^\alpha \xi^{1-\alpha} A^2 e^{(t+\xi)A} x\|$$
$$\leq T^\alpha M_{1,1} \|x\|_{D_A(1,\infty)} + M_{2,1} \|x\|_{D_A(1,\infty)}, \quad 0 < t \leq T.$$

Statement (ii) follows now easily.

To prove statement (iii) we will show that Au is continuous up to $t = 0$, provided f is continuous up to $t = 0$ and $Au_0 + f(0) \in \overline{D(A)}$. Then the statement will follow from Lemma 4.1.6.

We know already that Au is continuous in $]0, T]$, so that we have only to prove that $Au(t) \to Au_0$ as $t \to 0$. Fix $\varepsilon > 0$, and let $\delta \in]0, 1[$ be such that

$$\int_{1-\delta}^{1} (1-\sigma)^{\alpha-1} \sigma^{-\alpha} d\sigma \leq \varepsilon.$$

Split again $u(t)$ as $u(t) = u_1(t) + u_2(t)$, where u_1 and u_2 are defined by (4.3.1). If t is so small that

$$\sup_{0 \leq s \leq t} \|f(s) - f(t)\| \leq \varepsilon |\log \delta|^{-1},$$

then

$$\|Au_1(t)\| \leq \left\| \int_0^{t(1-\delta)} Ae^{(t-s)A}(f(s) - f(t)) ds \right\|$$

$$\left\| \int_{t(1-\delta)}^{t} Ae^{(t-s)A}(f(s) - f(t)) ds \right\| \leq M_1 |\log \delta| \sup_{0 \leq s \leq t} \|f(s) - f(t)\|$$

$$+ M_1 \int_{1-\delta}^{1} (1-\sigma)^{\alpha-1} \sigma^{-\alpha} d\sigma [f]_{C_\alpha^\alpha} \leq M_1 \varepsilon + M_1 \varepsilon [f]_{C_\alpha^\alpha}.$$

Therefore, $Au_1(t) \to 0$ as $t \to 0$. Moreover, by (4.3.7) it follows easily that $Au_2(t) \to Au_0$ as $t \to 0$. Hence, Au is continuous up to $t = 0$. ∎

Now we consider the case where f is unbounded near $t = 0$. The introduction of the weighted Hölder spaces $C_{\alpha+\mu}^\alpha(]0, T]; X)$ and $C_{\alpha+\mu}^\alpha(]0, T]; D(A))$ is motivated again by the behavior of $e^{tA}x$ as $t \to 0$: indeed, it is easy to see that if $x \in D_A(1-\mu, \infty)$, $0 < \mu < 1$, then $t \mapsto e^{tA}x$ belongs to $C_{\alpha+\mu}^\alpha(]0, T]; D(A))$ for each $\alpha \in]0, 1[$. We recall that $f \in C_{\alpha+\mu}^\alpha(]0, T]; X)$ means that $t \mapsto t^\mu f(t)$ is bounded and $t \mapsto t^{\alpha+\mu} f(t)$ is α-Hölder continuous in $]0, T]$ with values in X. Of course, since f is possibly unbounded, we will not get a strict solution but only a classical one.

Theorem 4.3.7 Let $0 < \alpha$, $\mu < 1$, $f \in C_{\alpha+\mu}^\alpha(]0, T]; X)$. Then $v = e^{tA} * f$ belongs to $C_{\alpha+\mu}^\alpha(]0, T]; D(A))$, it is differentiable in $]0, T]$ with values in X, $v' \in C_{\alpha+\mu}^\alpha(]0, T]; X) \cap B_{\alpha+\mu}(]0, T]; D_A(\alpha, \infty))$, and there is C such that

$$\|v\|_{C_{\alpha+\mu}^\alpha(D(A))} + \|v'\|_{C_{\alpha+\mu}^\alpha(X)} + \|v'\|_{B_{\alpha+\mu}(D_A(\alpha,\infty))} \leq C \|f\|_{C_{\alpha+\mu}^\alpha(X)}. \quad (4.3.13)$$

3. STRICT AND CLASSICAL SOLUTIONS, AND OPTIMAL REGULARITY   141

Consequently,

(i) *if $u_0 \in \overline{D(A)}$, then the mild solution u of (4.0.1) is classical;*

(ii) *if $u_0 \in D_A(1-\mu, \infty)$, then u' and Au belong to $C^\alpha_{\alpha+\mu}(0,T;X)$, u' belongs to $B_{\alpha+\mu}(D_A(\alpha, \infty))$, and there is C such that*

$$\|u\|_{C^\alpha_{\alpha+\mu}(D(A))} + \|u'\|_{C^\alpha_{\alpha+\mu}(X)} + \|u'\|_{B_{\alpha+\mu}(D_A(\alpha,\infty))} \qquad (4.3.14)$$
$$\leq C(\|u_0\|_{D_A(1-\mu,\infty)} + \|f\|_{C^\alpha_{\alpha+\mu}(X)}).$$

Proof — v belongs to $C(]0,T];D(A)) \cap C^1(]0,T];X)$ thanks to Theorem 4.3.4. To estimate $Av(t)$, we split it in three addenda, setting

$$Av(t) = \int_0^{t/2} Ae^{(t-s)A} f(s)ds + \int_{t/2}^t Ae^{(t-s)A}(f(s) - f(t))ds + (e^{tA/2} - I)f(t),$$

and we get

$$t^\mu \|Av(t)\| \leq M_1 t^\mu \int_0^{t/2} (t-s)^{-1} s^{-\mu} ds \, \|f\|_{B_\mu(X)} + (M_0+1)\|f\|_{B_\mu(X)}$$
$$+ M_1 t^\mu \int_{t/2}^t (t-s)^{\alpha-1} s^{-\alpha-\mu} ds \, [f]_{C^\alpha_{\alpha+\mu}(X)}$$
$$\leq M_1 \int_0^{1/2} (1-\sigma)^{-1} \sigma^{-\mu} d\sigma \, \|f\|_{B_\mu(X)} + (M_0+1)\|f\|_{B_\mu(X)}$$
$$+ M_1 \int_{1/2}^1 (1-\sigma)^{\alpha-1} \sigma^{-\alpha-\mu} ds \, [f]_{C^\alpha_{\alpha+\mu}(X)}.$$

So, Av belongs to $B_\mu(]0,T];X)$. Now we estimate $[v'(t)]_{D_A(\alpha,\infty)}$. For $0 < t \leq T$,

$$t^{\alpha+\mu} \|\xi^{1-\alpha} Ae^{\xi A} v'(t)\| \leq t^{\alpha+\mu} \xi^{1-\alpha} \left\| \int_0^{t/2} A^2 e^{(t+\xi-s)A}(f(s) - f(t))ds \right\|$$
$$+ t^{\alpha+\mu} \xi^{1-\alpha} \left\| \int_{t/2}^t A^2 e^{(t+\xi-s)A}(f(s) - f(t))ds \right\|$$
$$+ t^{\alpha+\mu} \|\xi^{1-\alpha} Ae^{(t+\xi)A} f(t)\| = I_1 + I_2 + I_3,$$

where

$$I_1 \leq t^{\alpha+\mu} \xi^{1-\alpha} \left\| \int_0^{t/2} A^2 e^{(t+\xi-s)A} f(s)ds \right\|$$
$$+ t^{\alpha+\mu} \xi^{1-\alpha} \|(Ae^{(t+\xi)A} - Ae^{(t/2+\xi)A})f(t)\|$$
$$\leq M_2 t^{\alpha+\mu} \xi^{1-\alpha} \int_0^{t/2} (t+\xi-s)^{-2} s^{-\mu} ds \, \|f\|_{B_\mu(X)}$$
$$+ t^{\alpha+\mu} \xi^{1-\alpha} \|(Ae^{(t+\xi)A} - Ae^{(t/2+\xi)A})f(t)\|$$
$$\leq M_2 \int_0^{1/2} (1-\sigma)^{-2} \sigma^{-\mu} d\sigma \, \|f\|_{B_\mu(X)} + M_2(1+2^\alpha)\|f\|_{B_\mu(X)};$$

$$I_2 \leq M_2 t^{\alpha+\mu} \xi^{1-\alpha} \int_{t/2}^{t} \frac{(t-s)^\alpha}{s^{\alpha+\mu}(t+\xi-s)^2} ds\, [f]_{C^\alpha_{\alpha+\mu}(X)}$$

$$\leq 2^{\alpha+\mu} M_2 \int_0^{+\infty} \sigma^\alpha (\sigma+1)^{-2} d\sigma\, [f]_{C^\alpha_{\alpha+\mu}(X)};$$

$$I_3 \leq M_1 t^{\alpha+\mu} \xi^{1-\alpha}(t+\xi)^{-1} \|f(t)\| \leq M_1 \|f\|_{B_\mu(X)}.$$

To conclude the proof it is sufficient to argue as we did in the proof of Theorem 4.3.5 and of Corollary 4.3.6(i)(ii). ∎

4.3.2 Space regularity

We consider several regularity assumptions on f, which are suggested by the behavior of $t \mapsto e^{tA} u_0$ and of $t \mapsto d/dt\, e^{tA} u_0 = A e^{tA} u_0$ as functions with values in $D_A(\alpha, \infty)$. To begin with, we consider the case where $u_0 \in D_A(\alpha+1, \infty)$: then $A e^{tA} u_0$ is continuous for $t > 0$ and bounded near $t = 0$ with values in $D_A(\alpha, \infty)$. If in addition $u_0 \in D_A(\alpha+1)$, then $t \mapsto A e^{tA} u_0$ is continuous up to $t = 0$ with values in $D_A(\alpha)$. So, first we consider the case where f is bounded with values in $D_A(\alpha, \infty)$, and it has some continuity property, at least for $t > 0$. Then we consider the case of unbounded f.

The case where f is bounded with values in $D_A(\alpha, \infty)$

Theorem 4.3.8 *Let $0 < \alpha < 1$, and let $f \in C(]0,T]; X) \cap B(]0,T]; D_A(\alpha, \infty))$. Then $v = e^{tA} * f$ has values in $D(A)$, it is differentiable for $t > 0$ with values in X, and it is the classical solution of*

$$v'(t) = Av(t) + f(t),\quad 0 < t \leq T,\quad v(0) = 0. \tag{4.3.15}$$

Moreover, v' and Av belong to $C(]0,T]; X) \cap B(]0,T]; D_A(\alpha, \infty))$, Av belongs to $C^\alpha([0,T]; X)$, and there is C such that

$$\|v'\|_{B(D_A(\alpha,\infty))} + \|Av\|_{B(D_A(\alpha,\infty))} + \|Av\|_{C^\alpha(X)} \leq C\|f\|_{B(D_A(\alpha,\infty))}. \tag{4.3.16}$$

In addition,

(i) *if $f \in C([0,T]; X) \cap B([0,T]; D_A(\alpha, \infty))$, then v' and Av are continuous with values in X up to $t = 0$, and v is a strict solution of (4.3.15);*

(ii) *if $f \in C(]0,T]; D_A(\alpha, \infty)) \cap B([0,T]; D_A(\alpha, \infty))$, then $v \in C(]0,T]; D_A(\alpha+1, \infty)) \cap B([0,T]; D_A(\alpha+1, \infty)) \cap C^1(]0,T]; D_A(\alpha, \infty))$.*

Proof — Let us show that v is a classical solution of (4.3.15), and that (4.3.16) holds. For $0 \leq t \leq T$, $v(t)$ belongs to $D(A)$, and

$$\|Av(t)\| \leq M_{1,\alpha} \int_0^t (t-s)^{\alpha-1} ds\, \|f\|_{B(D_A(\alpha,\infty))} = \frac{T^\alpha M_{1,\alpha}}{\alpha} \|f\|_{B(D_A(\alpha,\infty))}. \tag{4.3.17}$$

3. Strict and classical solutions, and optimal regularity 143

Moreover, for $0 < \xi \leq 1$,

$$\|\xi^{1-\alpha} Ae^{\xi A} Av(t)\| = \xi^{1-\alpha} \left\| \int_0^t A^2 e^{(t+\xi-s)A} f(s) ds \right\|$$
$$\leq M_{2,\alpha} \xi^{1-\alpha} \int_0^t (t+\xi-s)^{\alpha-2} ds \|f\|_{B(D_A(\alpha,\infty))} \leq \frac{M_{2,\alpha}}{1-\alpha} \|f\|_{B(D_A(\alpha,\infty))}, \quad (4.3.18)$$

so that Av is bounded with values in $D_A(\alpha, \infty)$.

Let us show that Av is Hölder continuous with values in X: for $0 \leq s \leq t \leq T$ we have

$$\|Av(t) - Av(s)\| \leq \left\| A \int_0^s \left(e^{(t-\sigma)A} - e^{(s-\sigma)A} \right) f(\sigma) d\sigma \right\|$$
$$+ \left\| A \int_s^t e^{(t-\sigma)A} f(\sigma) d\sigma \right\| \leq M_{2,\alpha} \int_0^s d\sigma \int_{s-\sigma}^{t-\sigma} \tau^{\alpha-2} d\tau \|f\|_{B(D_A(\alpha,\infty))} \quad (4.3.19)$$
$$+ M_{1,\alpha} \int_s^t (t-\sigma)^{\alpha-1} d\sigma \|f\|_{B(D_A(\alpha,\infty))}$$
$$\leq \left(\frac{M_{2,\alpha}}{\alpha(1-\alpha)} + \frac{M_{1,\alpha}}{\alpha} \right) (t-s)^\alpha \|f\|_{B(D_A(\alpha,\infty))},$$

so that Av is α-Hölder continuous in $[0,T]$. Estimate (4.3.16) follows now from (4.3.17), (4.3.18), (4.3.19). Moreover, thanks to Lemma 4.1.6, v is a classical solution of (4.3.15) if $f \in C(]0,T]; X)$, and it is a strict solution if $f \in C([0,T]; X)$.

Let us prove statement (ii). Let $f \in C(]0,T]; D_A(\alpha, \infty)) \cap B([0,T]; D_A(\alpha, \infty))$. We are going to show that Av belongs to $C([a,T]; D_A(\alpha, \infty))$ for every $a \in]0,T[$. For $\varepsilon > 0$, let $\delta \in]0, \varepsilon \wedge a]$ be such that for $t_1, t_2 \in [a,T]$, $|t_1 - t_2| \leq \delta$, we have $\|f(t_1) - f(t_2)\|_{D_A(\alpha,\infty)} \leq \varepsilon$. Then for $a \leq r < t \leq T$, $t-r \leq \delta$, and for $0 < \xi \leq 1$

$$\|\xi^{1-\alpha} Ae^{\xi A} (Av(t) - Av(r))\| \leq \left\| \xi^{1-\alpha} A^2 \int_0^{r-\delta} e^{(\xi+s)A} [f(t-s) - f(r-s)] ds \right\|$$
$$+ \left\| \xi^{1-\alpha} A^2 \int_{r-\delta}^r e^{(\xi+s)A} [f(t-s) - f(r-s)] ds \right\|$$
$$+ \left\| \xi^{1-\alpha} A^2 \int_r^t e^{(\xi+s)A} f(t-s) ds \right\| \leq \frac{M_{2,\alpha}}{1-\alpha} \xi^{1-\alpha} (\xi^{-1+\alpha} - (\xi+r-\delta)^{-1+\alpha}) \varepsilon$$

$$+ \frac{M_{2,\alpha}}{1-\alpha} \xi^{1-\alpha} ((\xi+r-\delta)^{-1+\alpha} - (\xi+r)^{-1+\alpha}) 2\|f\|_{B([0,T]; D_A(\alpha,\infty))}$$
$$+ \frac{M_{2,\alpha}}{1-\alpha} \xi^{1-\alpha} ((\xi+r)^{-1+\alpha} - (\xi+t)^{-1+\alpha}) \|f\|_{B([0,T]; D_A(\alpha,\infty))}$$
$$\leq \frac{M_{2,\alpha}}{1-\alpha} \left(\varepsilon + 3\|f\|_{B([0,T]; D_A(\alpha,\infty))} \frac{\varepsilon^{1-\alpha}}{a^{1-\alpha}} \right).$$

Taking the supremum over $\xi \in \,]0,1]$, we see that $[Av(t) - Av(r)]_{D_A(\alpha,\infty)}$ goes to 0 as $t - r$ goes to 0. Since Av is continuous with values in X, then $v \in C(]0,T]; D_A(\alpha+1,\infty))$. From the equality $v' = Av + f$ it follows that v' is continuous in $]0,T]$ with values in $D_A(\alpha,\infty)$, and hence that v is continuously differentiable in $]0,T]$ with values in $D_A(\alpha,\infty)$. ∎

Corollary 4.3.9 *Let $0 < \alpha < 1$, $u_0 \in X$, $f \in C(]0,T]; X) \cap B(]0,T]; D_A(\alpha,\infty))$, and let u be the mild solution of (4.0.1). Then $u \in C^1(]0,T]; X) \cap C(]0,T]; D(A))$, and $u \in B([\varepsilon,T]; D_A(\alpha+1,\infty))$ for every $\varepsilon \in \,]0,T[$. Moreover, the following statements hold.*

(i) if $u_0 \in \overline{D(A)}$, then u is a classical solution;

(ii) if $u_0 \in D(A)$, $Au_0 \in \overline{D(A)}$, and $f \in C([0,T]; X) \cap B([0,T]; D_A(\alpha,\infty))$, then u is a strict solution;

(iii) if $u_0 \in D_A(\alpha+1,\infty)$ and $f \in C([0,T]; X) \cap B([0,T]; D_A(\alpha,\infty))$, then u' and Au belong to $C([0,T]; X) \cap B([0,T]; D_A(\alpha,\infty))$, Au belongs to $C^\alpha([0,T]; X)$, and there is C such that

$$\|u'\|_{B(D_A(\alpha,\infty))} + \|Au\|_{B(D_A(\alpha,\infty))} + \|Au\|_{C^\alpha([0,T];X)} \\ \leq C(\|f\|_{B(D_A(\alpha,\infty))} + \|u_0\|_{D_A(\alpha,\infty)}). \qquad (4.3.20)$$

(iv) if $u_0 \in D_A(\alpha+1,\infty)$ and $f \in C(]0,T]; D_A(\alpha,\infty))$, then $u \in C^1(]0,T]; D_A(\alpha,\infty)) \cap C(]0,T]; D_A(\alpha+1,\infty))$.

Proof — We have $u(t) = e^{tA}u_0 + (e^{tA} * f)(t)$. If $u_0 \in \overline{D(A)}$, the function $t \mapsto e^{tA}u_0$ is the classical solution of $w' = Aw$, $t > 0$, $w(0) = u_0$, by Propositions 2.1.1(iv) and 2.1.4(i). If $u_0 \in D(A)$ and $Au_0 \in \overline{D(A)}$ it is a strict solution, by Proposition 2.1.4(iii). If $x \in D_A(\alpha+1,\infty)$, it is a strict solution, and moreover it belongs to $C^1([0,T]; X) \cap C^1(]0,T]; D_A(\alpha,\infty)) \cap B([0,T]; D_A(\alpha+1,\infty)) \cap C(]0,T]; D_A(\alpha+1,\infty))$, by Proposition 2.2.2. The statement follows now from Theorem 4.3.8. ∎

Corollary 4.3.10 *Let $0 < \alpha < 1$, and let $f \in C([0,T], D_A(\alpha))$, $u_0 \in D_A(\alpha+1)$. Then the solution u of problem (4.0.1) belongs to $C([0,T], D_A(\alpha+1)) \cap C^1([0,T], D_A(\alpha))$, and u' belongs to $h^\alpha([0,T]; X)$.*

Proof — Let $\{f_n\}_{n \in \mathbb{N}} \subset C([0,T], D(A))$ and $\{x_n\}_{n \in \mathbb{N}} \subset D(A^2)$ be such that

$$f_n \to f \text{ in } C([0,T]; D_A(\alpha,\infty)), \quad x_n \to u_0 \text{ in } X.$$

Then, by Corollary 4.3.9(iii), the solution u_n of problem

$$u'_n(t) = Au_n(t) + f_n(t), \quad 0 \leq t \leq T, \quad u_n(0) = x_n,$$

3. STRICT AND CLASSICAL SOLUTIONS, AND OPTIMAL REGULARITY   145

belongs to $B([0,T]; D_A(\alpha+\varepsilon+1,\infty)) \cap C([0,T], D(A))$, and u'_n belongs to $B([0,T]; D_A(\alpha+\varepsilon,\infty)) \cap C^{\alpha+\varepsilon}([0,T], X)$ for every $\varepsilon \in \,]0, 1-\alpha[$, so that $u_n \in C([0,T]; D_A(\alpha+1,\infty)) \cap C^1([0,T]; D_A(\alpha,\infty))$, and $u'_n \in h^{\alpha}([0,T], X)$. Due to estimate (4.3.20), $u_n \to u$ in $C([0,T]; D_A(\alpha+1,\infty)) \cap C^1([0,T]; D_A(\alpha,\infty))$, and $u'_n \to u'$ in $C^{\alpha}([0,T], X)$. Therefore $u \in C([0,T]; D_A(\alpha+1)) \cap C^1([0,T]; D_A(\alpha))$, and $u' \in h^{\alpha}([0,T]; X)$. ∎

The case where f is unbounded with values in $D_A(\alpha,\infty)$

In the next theorem we show that problem (4.0.1) may have a classical solution even if f is unbounded with values in $D_A(\alpha,\infty)$ near $t=0$.

Theorem 4.3.11 Let $0 < \alpha < 1$, $f \in L^1(0,T;X) \cap C(]0,T];X) \cap B([\varepsilon,T]; D_A(\alpha,\infty))$ for every $\varepsilon \in \,]0,T[$, and let $u_0 \in X$. Then the mild solution u of problem (4.0.1) belongs to $C(]0,T]; D(A)) \cap C^1(]0,T]; X) \cap B([\varepsilon,T]; D_A(\alpha+1,\infty))$, for every $\varepsilon \in \,]0,T[$. If in addition $u_0 \in \overline{D(A)}$, then u is a classical solution.

If f belongs also to $C(]0,T]; D_A(\alpha,\infty))$ (respectively, to $C(]0,T]; D_A(\alpha))$), then v' and Av belong to $C(]0,T]; D_A(\alpha,\infty))$ (respectively, to $C(]0,T]; D_A(\alpha))$).

Proof — The proof is similar to the proof of Theorem 4.3.4. We split u as $u(t) = e^{tA}u_0 + v(t)$, where $v = e^{tA} * f$. The first addendum is obviously continuous in $]0,T]$ with values in $D(A)$, and it belongs to $C([0,T]; X)$ if and only if $u_0 \in \overline{D(A)}$.

Let us consider the function v. By Proposition 4.2.3, v is continuous in $[0,T]$ with values in X. Moreover, for $0 < \varepsilon \le t \le T$,

$$v(t) = e^{(t-\varepsilon)A}v(\varepsilon) + \int_\varepsilon^t e^{(t-s)A} f(s)ds = v_1(t) + v_2(t),$$

where v_1 belongs to $C^{\infty}(]\varepsilon,T]; D(A^n))$ for every n, and v_2 belongs to $C([\varepsilon,T]; D(A)) \cap C^1([\varepsilon,T]; X) \cap B([\varepsilon,T]; D_A(\alpha+1,\infty))$ thanks to Theorem 4.3.8. Since ε is arbitrary, then $u \in C(]0,T]; D(A)) \cap C^1(]0,T]; X)$.

If in addition $u_0 \in \overline{D(A)}$, then $u \in C([0,T]; X)$, so that it is a classical solution thanks to Lemma 4.1.6.

Let now f be continuous in $]0,T]$ with values in $D_A(\alpha,\infty)$ (respectively, with values in $D_A(\alpha)$). For $0 < \varepsilon \le t \le T$, split $v(t) = v_1(t) + v_2(t)$ as above. Then v_1 belongs to $C(]\varepsilon,T]; D(A^n))$ for every n, and v_2 belongs to $C(]\varepsilon,T]; D_A(\alpha+1,\infty))$ (respectively, to $C([\varepsilon,T]; D_A(\alpha+1))$) thanks to Theorem 4.3.8(ii) (respectively, to Corollary 4.3.10) applied in the interval $[\varepsilon,T]$. Since ε is arbitrary, then v' and Av belong to $C(]0,T]; D_A(\alpha,\infty))$ (respectively, to $C(]0,T]; D_A(\alpha))$). ∎

If we know how $\|f(t)\|_{D_A(\alpha,\infty)}$ blows up as $t \to 0$, we may give precise information on the behavior of $u(t)$ as $t \to 0$, getting also optimal regularity results. In the next theorem we consider the case where f belongs to $B_\theta(]0,T]; D_A(\alpha,\infty))$, that is $\|t^\theta f(t)\|_{D_A(\alpha,\infty)}$ is bounded, for some $\theta \in \,]0,1[$.

Theorem 4.3.12 Let $0<\alpha, \theta<1$, and let $f \in C(]0,T]; X) \cap B_\theta(]0,T]; D_A(\alpha, \infty))$. Then $v = e^{tA} * f$ has values in $D(A)$, it is differentiable with values in X for $t > 0$, and it is a classical solution of (4.3.15). Moreover, v' and Av belong to $C(]0,T]; X) \cap B_\theta(]0,T]; D_A(\alpha, \infty))$, Av belongs to $C^\alpha_\theta(]0,T]; X)$, and there is C such that

$$\|v'\|_{B_\theta(D_A(\alpha,\infty))} + \|Av\|_{B_\theta(D_A(\alpha,\infty))} + \|Av\|_{C^\alpha_\theta(X)} \leq C\|f\|_{B_\theta(D_A(\alpha,\infty))}. \quad (4.3.21)$$

In particular, if $\theta = \alpha$ then v is bounded with values in $D(A)$. If $\theta < \alpha$, then Av belongs to $C^{\alpha-\theta}([0,T]; X)$, and there is C such that

$$\|Av\|_{C^{\alpha-\theta}(X)} \leq C\|f\|_{B_\theta(D_A(\alpha,\infty))}. \quad (4.3.22)$$

If f belongs also to $C(]0,T]; D_A(\alpha, \infty))$ (respectively, to $C(]0,T]; D_A(\alpha))$), then v' and Av belong to $C(]0,T]; D_A(\alpha, \infty))$ (respectively, to $C(]0,T]; D_A(\alpha))$).

Proof — By Theorem 4.3.11, v is continuous in $]0,T]$ with values in $D(A)$, and continuously differentiable in $]0,T]$ with values in X. Moreover,

$$\|Av(t)\| \leq M_{1,\alpha} \int_0^t (t-s)^{\alpha-1} s^{-\theta} ds \, \|f\|_{B_\theta(D_A(\alpha,\infty))}$$
$$= M_{1,\alpha} t^{\alpha-\theta} \int_0^1 (1-\sigma)^{\alpha-1} \sigma^\theta d\sigma \, \|f\|_{B_\theta(D_A(\alpha,\infty))} = K t^{\alpha-\theta} \|f\|_{B_\theta(D_A(\alpha,\infty))}. \quad (4.3.23)$$

Fix now any $r \in]0,T[$. For $t \geq r$ it holds

$$Av(t) = Ae^{(t-r/2)A} v(r/2) + \int_{r/2}^t Ae^{(t-s)A} f(s) ds = Av_1(t) + Av_2(t).$$

Denoting by C_α the norm of the embedding $D_A(\alpha, \infty) \subset D_A(\alpha, 1)$ and using (4.0.5) we get

$$\|Av_1(t)\|_{D_A(\alpha,\infty)} \leq \frac{K_{0,\alpha} C_\alpha}{(t-r/2)^\alpha} \|Av(r/2)\|$$
$$\leq \frac{K_{0,\alpha} C_\alpha}{(r/2)^\alpha} K(r/2)^{\alpha-\theta} \|f\|_{B_\theta(D_A(\alpha,\infty))} = \frac{K_{0,\alpha} C_\alpha K}{(r/2)^\theta} \|f\|_{B_\theta(D_A(\alpha,\infty))}.$$

Moreover, from Theorem 4.3.8 applied in the interval $[r/2, T]$, we get for every $t \in [r/2, T]$

$$\|Av_2(t)\|_{D_A(\alpha,\infty)} \leq C \sup_{r/2 \leq t \leq T} \|f(s)\|_{D_A(\alpha,\infty)} \leq \frac{C}{(r/2)^\theta} \|f\|_{B_\theta(D_A(\alpha,\infty))}.$$

Summing up and taking $t = r$ we see that

$$\|r^\theta Av(r)\|_{D_A(\alpha,\infty)} \leq (2^\theta M_{0,\alpha} K + 2^\theta C) \|f\|_{B_\theta(D_A(\alpha,\infty))},$$

which implies that $Av \in B_\theta(]0,T]; D_A(\alpha,\infty))$.

3. STRICT AND CLASSICAL SOLUTIONS, AND OPTIMAL REGULARITY

The proof of the statements about the Hölder continuity of Av is similar to the proof of statements (ii)-(iii) of Proposition 4.2.3. We have seen above that $v(r/2)$ belongs to $D(A)$. Moreover, we have seen in the proof of Corollary 4.3.6 that for every $x \in D(A)$ the function $t \mapsto Ae^{(t-r/2)A}x$ belongs to $C^\alpha_\alpha(]r/2,T];X)$, and its C^α_α-norm is less or equal to $C\|Ax\|$, for some $C > 0$. Using estimate (4.3.23) we get

$$\|Av_1\|_{C^\alpha([r,T];X)} \leq C(r/2)^{-\alpha}\|Av(r/2)\| \leq CK(r/2)^{-\theta}\|f\|_{B_\theta(D_A(\alpha,\infty))}.$$

Due again to Theorem 4.3.8, Av_2 belongs to $C^\alpha([r/2,T];X)$, and that

$$\|Av_2\|_{C^\alpha([r/2,T];X)} \leq C\|f\|_{B([r/2,T];D_A(\alpha,\infty))} \leq C(r/2)^{-\theta}\|f\|_{B_\theta(D_A(\alpha,\infty))}.$$

Summing up we get

$$\|Av_2\|_{C^\alpha([r,T];X)} \leq \text{const.}\, r^{-\theta}\|f\|_{B_\theta(D_A(\alpha,\infty))},$$

which implies that Av belongs to $C^\alpha_\theta(]0,T];X)$. Estimate (4.3.21) follows now easily.

Let us prove that if $\theta < \alpha$ then $Av \in C^{\alpha-\theta}([0,T];X)$. For $0 \leq r \leq t \leq T$ we have

$$\|Av(t) - Av(r)\| \leq \left\|\int_0^r \left(\int_{r-s}^{t-s} A^2 e^{\sigma A} d\sigma\right) f(s)ds\right\| + \left\|\int_r^t Ae^{(t-s)A}f(s)ds\right\|$$

$$\leq \left(M_{2,\alpha}\int_0^r s^{-\theta}\int_{r-s}^{t-s}\frac{1}{\sigma^{2-\alpha}}d\sigma ds + M_{1,\alpha}\int_r^t \frac{1}{(t-s)^{1-\alpha}s^\theta}ds\right)\|f\|_{B_\theta(D_A(\alpha,\infty))}.$$

Following the proof of statement (iii) of Proposition 4.2.3, we find

$$\|Av(t) - Av(r)\| \leq \frac{1}{\alpha-\theta}\left(M_{2,\alpha}\int_0^1 \frac{d\sigma}{(1-\sigma)^{1-\theta}\sigma^\theta}\right.$$

$$\left.+M_{1,\alpha}\int_0^1 \frac{d\sigma}{(1-\sigma)^{1-\alpha}\sigma^\theta}\right)(t-r)^{\alpha-\theta}\|f\|_{B_\theta(D_A(\alpha,\infty))},$$

and estimate (4.3.22) follows.

The last statement follows from Theorem 4.3.11. ∎

Following the method employed in the proof of Theorem 4.3.8, it is possible to prove Theorem 4.3.12 also without using the results of Theorem 4.3.8. However, the direct proof would be much longer.

Corollary 4.3.13 *Let $0 < \alpha < 1$, $u_0 \in X$, $f \in C(]0,T];X) \cap B_\theta(]0,T];D_A(\alpha,\infty))$, and let u be the mild solution of problem (4.0.1). The following statements hold.*

(i) If $u_0 \in \overline{D(A)}$, then u is a classical solution;

(ii) If $u_0 \in D_A(\alpha + 1 - \theta, \infty)$, then u' and Au belong to $B_\theta(]0,T]; D_A(\alpha, \infty))$, Au belongs to $C_\theta^\alpha(]0,T];X)$, and there is C such that

$$\|u'\|_{B_\theta(D_A(\alpha,\infty))} + \|Au\|_{B_\theta(D_A(\alpha,\infty))} + \|Au\|_{C_\theta^\alpha(X)} \leq C(\|f\|_{B_\theta(D_A(\alpha,\infty))} + \|u_0\|_{D_A(\alpha+1-\theta,\infty)}). \quad (4.3.24)$$

Moreover, in the case where $\theta < \alpha$, then $u \in C^{\alpha-\theta}([0,T]; D(A))$, and there is C such that

$$\|Au\|_{C^{\alpha-\theta}(X)} \leq C(\|f\|_{B_\alpha(D_A(\alpha,\infty))} + \|u_0\|_{D_A(\alpha+1-\theta,\infty)}). \quad (4.3.25)$$

(iii) If in addition $f \in C(]0,T]; D_A(\alpha, \infty))$ (respectively, $f \in C(]0,T]; D_A(\alpha))$), then u', Au belong to $C(]0,T]; D_A(\alpha, \infty))$ (respectively, to $C(]0,T]; D_A(\alpha))$).

(iv) In the case $\theta = \alpha$, if $f \in C([0,T];X)$, $u_0 \in D(A)$, $Au_0 \in \overline{D(A)}$, and $f \in C([0,T];X) \cap B_\alpha(]0,T]; D_A(\alpha, \infty))$, $\lim_{t \to 0} \|t^\alpha f(t)\|_{D_A(\alpha,\infty)} = 0$, then u is a strict solution and $\lim_{t \to 0} \|t^\alpha Au(t)\|_{D_A(\alpha,\infty)} = \lim_{t \to 0} \|t^\alpha u'(t)\|_{D_A(\alpha,\infty)} = 0$.

Proof — (i) We split as usual $u = e^{tA}u_0 + v$. By Theorem 4.3.11, v is a classical solution of (4.3.15). Concerning $t \mapsto Ae^{tA}u_0$, we know from Proposition 2.1.4(i)-(iii) that it belongs to $C([0,T];X) \cap C^\infty(]0,T]; D(A))$. Therefore, due to Lemma 4.1.6, u is a classical solution of (4.0.1).

(ii) Thanks to Theorem 4.3.12, v belongs to $B_\theta(]0,T]; D_A(\alpha+1, \infty))$, while Av belongs to $C_\theta^\alpha(]0,T];X)$. Moreover, from estimates (4.0.4) it follows that $t \mapsto e^{tA}u_0$ belongs to $C_\theta^\alpha(]0,T]; D(A)) \cap B_\theta(]0,T]; D_A(\alpha+1, \infty))$.

In the case $\theta < \alpha$, we know from Theorem 4.3.12 that $v \in C^{\alpha-\theta}([0,T]; D(A))$. The fact that $t \mapsto e^{tA}u_0$ belongs to $C^{\alpha-\theta}([0,T]; D(A))$ may be proved using once again estimates (4.0.4).

(iii) The statement is an obvious consequence of Theorem 4.3.12, since the function $t \mapsto e^{tA}u_0$ belongs to $C^\infty(]0,T]; D(A^n))$ for every n.

(iv) We know already that $u \in C(]0,T]; D(A))$. Thanks to Lemma 4.1.6, to verify that u is a strict solution it is sufficient to show that $u \in C([0,T]; D(A))$.

Note that, since f is continuous with values in X up to $t = 0$, and $f(t) \in D_A(\alpha, \infty) \subset \overline{D(A)}$ for $t > 0$, then $f(0) \in \overline{D(A)}$, so that $Au_0 + f(0) \in \overline{D(A)}$. Fix any $\varepsilon > 0$, and choose $\delta > 0$ so small that $\|f(s) - f(0)\| \leq \varepsilon$ and $\|s^\alpha f(s)\|_{D_A(\alpha,\infty)} \leq \varepsilon$ for $0 \leq s \leq \delta$.

3. STRICT AND CLASSICAL SOLUTIONS, AND OPTIMAL REGULARITY

For $0 \leq t \leq \delta$,

$$\|Au(t) - Au_0\| \leq \|A(e^{tA}u_0 - u_0) + (e^{tA} - I)f(t)\|$$
$$+ \left\|A\left(\int_0^{t/2} + \int_{t/2}^t\right)e^{(t-s)A}(f(s) - f(t))ds\right\|$$
$$\leq \|(e^{tA} - I)(Au_0 + f(0))\| + \|(e^{tA} - I)(f(t) - f(0))\|$$
$$+ M_1 \int_0^{t/2} \frac{\varepsilon}{t-s}ds + M_{1,\alpha}\int_{t/2}^t \frac{\varepsilon}{(t-s)^{1-\alpha}}\left(\frac{1}{s^\alpha} + \frac{1}{t^\alpha}\right)ds$$
$$\leq \|(e^{tA} - I)(Au_0 + f(0))\| + (M_0 + 1)\varepsilon$$
$$+ M_1\varepsilon \log 2 + M_{1,\alpha}\left(\int_{1/2}^1 \frac{1}{(1-\sigma)^{1-\alpha}\sigma^\alpha}d\sigma + (2\alpha)^{\alpha-1}\right)\varepsilon.$$

Therefore, $\lim_{t\to 0}\|Au(t) - Au_0\| = 0$, which implies that u is continuous at $t = 0$ with values in $D(A)$, and it is a strict solution.

From estimate (4.3.21), applied in every interval $[0, \varepsilon]$, it follows that $t^\alpha v(t)$ goes to 0 in $D_A(\alpha + 1, \infty)$ as $t \to 0$. Moreover, it is easy to check that, since $Ax \in \overline{D(A)}$, then $\lim_{t\to 0}\|t^\alpha e^{tA}u_0\|_{D_A(\alpha+1,\infty)} = 0$. Statement (iv) follows. ∎

It is also possible to study the case where f is highly singular (with singularity of order greater than 1) at $t = 0$ with values in $D_A(\alpha, \infty)$, provided it is mildly singular with values in X (with singularity of order less than 1). The order of the singularities are again suggested by the behavior of $Ae^{tA}u_0$ as $t \to 0$, when u_0 belongs to suitable interpolation spaces.

Theorem 4.3.14 *Let $0 < \alpha, \mu < 1$, and let $f \in C(]0, T]; X) \cap B_\mu(]0, T]; X) \cap B_{\alpha+\mu}(]0, T]; D_A(\alpha, \infty))$. Then $v = e^{tA} * f$ is a classical solution of (4.3.15), v', Av belong to $B_\mu(]0, T]; X) \cap B_{\alpha+\mu}(]0, T]; D_A(\alpha, \infty))$, Av belongs to $C^\alpha_{\alpha+\mu}(]0, T]; X)$, and there is C such that*

$$\|v'\|_{B_\mu(X)} + \|v'\|_{B_{\alpha+\mu}(D_A(\alpha,\infty))} + \|Av\|_{B_{\alpha+\mu}(D_A(\alpha,\infty))} \tag{4.3.26}$$
$$+ \|Av\|_{C^\alpha_{\alpha+\mu}(X)} \leq C(\|f\|_{B_\mu(X)} + \|f\|_{B_{\alpha+\mu}(D_A(\alpha,\infty))}).$$

If f belongs also to $C(]0, T]; D_A(\alpha, \infty))$ (respectively, to $C(]0, T]; D_A(\alpha))$), then v' and Av belong to $C(]0, T]; D_A(\alpha, \infty))$ (respectively, to $C(]0, T]; D_A(\alpha))$).

Proof — By Theorem 4.3.11, v is continuous in $]0, T]$ with values in $D(A)$. Let us prove that Av belongs to $B_\mu(]0, T]; X)$. For $0 \leq t \leq T$ we have

$$\|t^\mu Av(t)\| \leq t^\mu \left\|\int_0^{t/2} Ae^{(t-s)A}f(s)ds\right\| + t^\mu \left\|\int_{t/2}^t Ae^{(t-s)A}f(s)ds\right\|,$$

so that

$$\begin{aligned}
\|t^\mu Av(t)\| &\leq M_1 t^\mu \int_0^{t/2} (t-s)^{-1} s^{-\mu} ds\, \|f\|_{B_\mu(X)} \\
&\quad + 2^\mu M_{1,\alpha} \int_{t/2}^t (t-s)^{\alpha-1} s^{-\alpha} ds\, \|f\|_{B_{\alpha+\mu}(D_A(\alpha,\infty))} \\
&\leq M_1 \int_0^{1/2} (1-\sigma)^{-1} \sigma^{-\mu} d\sigma\, \|f\|_{B_\mu(X)} \\
&\quad + 2^\mu M_{1,\alpha} \int_{1/2}^1 (1-\sigma)^{\alpha-1} \sigma^{-\alpha} d\sigma\, \|f\|_{B_{\alpha+\mu}(D_A(\alpha,\infty))}.
\end{aligned}$$

From now on, the proof is similar to the proof of Theorem 4.3.12, and it is left to the reader. ∎

Corollary 4.3.15 *Let $0 < \alpha$, $\mu < 1$, $u_0 \in X$, $f \in C(]0,T]; X) \cap B_\mu(]0,T];X) \cap B_{\alpha+\mu}(]0,T]; D_A(\alpha,\infty))$, and let u be the mild solution of problem (4.0.1). Then*

(i) *if $u_0 \in \overline{D(A)}$, u is a classical solution;*

(ii) *if $u_0 \in D_A(1-\mu,\infty)$, then u' and Au belong to $B_\mu(0,T;X) \cap B_\alpha(]0,T]; D_A(\alpha,\infty))$, Au belongs to $C^\alpha_{\alpha+\mu}(]0,T]; X)$, and there is $C > 0$ such that*

$$\|u'\|_{B_\mu(X)} + \|u'\|_{B_{\alpha+\mu}(D_A(\alpha,\infty))} + \|Au\|_{B_{\alpha+\mu}(D_A(\alpha,\infty))} + \|Au\|_{C^\alpha_{\alpha+\mu}(X)}$$
$$\leq C(\|f\|_{B_\mu(X)} + \|f\|_{B_{\alpha+\mu}(D_A(\alpha,\infty))} + \|u_0\|_{D_A(1-\mu,\infty)}). \qquad (4.3.27)$$

Proof — We split u into the sum $u(t) = e^{tA} u_0 + v(t)$. If $u_0 \in \overline{D(A)}$, the function $w(t) = e^{tA} u_0$ is the classical solution of $w' = Aw$, $t > 0$, $w(0) = u_0$. By Theorem 4.3.14, the function $v = e^{tA} * f$ is the classical solution of (4.3.15). Statement (i) is so proved.

To prove statement (ii), one has to use Theorem 4.3.14, and then to show that if $u_0 \in D_A(1-\mu,\infty)$, then $t \mapsto Ae^{tA} u_0$ belongs to $B_\mu(]0,T]; X) \cap C^\alpha_{\alpha+\mu}(]0,T]; X) \cap B_{\alpha+\mu}(]0,T]; D_A(\alpha,\infty))$. The proof of these properties is similar to the proof of statement (ii) of Corollary 4.3.13, and it is left to the reader. ∎

4.3.3 A further regularity result

Following the methods of the previous subsections, it is possible to give several further maximal regularity results. For the sake of brevity we state only one further regularity result, which will be used in the sequel, and leave the (infinitely many) others to the interested reader.

3. Strict and classical solutions, and optimal regularity

Theorem 4.3.16 *Let $0 < \theta, \beta < 1$, with $\theta + \beta \neq 1$. Let*

$$f \in C^\theta([0,T]; D_A(\beta, \infty)), \quad u_0 \in D(A), \quad Au_0 + f(0) \in D_A(\theta + \beta, \infty).$$

Then the mild solution u of problem (4.0.1) is strict, u' and Au belong to $C^\theta([0,T]; D_A(\beta, \infty))$, $u' \in B([0,T]; D_A(\theta + \beta, \infty))$, $Au \in C^{\theta+\beta}([0,T]; X)$, and there exists $C > 0$ such that

$$\|u\|_{C^{\theta+1}(D_A(\beta,\infty))} + \|Au\|_{C^\theta(D_A(\beta,\infty))} + \|u'\|_{B(D_A(\theta+\beta,\infty))}$$
$$+ \|Au\|_{C^{\theta+\beta}(X)} \leq C(\|f\|_{C^\theta(D_A(\beta,\infty))} + \|u_0\|_{D(A)} + \|Au_0 + f(0)\|_{D_A(\theta+\beta,\infty)}). \tag{4.3.28}$$

Proof — Let us consider problem (4.0.1) as an evolution equation in the space $Y = D_A(\beta, \infty)$. The domain of the part of A in Y is $D(A_Y) = D_A(\beta + 1, \infty)$, and the Reiteration Theorem 1.2.15 yields

$$(Y, D(A_Y))_{\theta, \infty} = D_A(\theta + \beta, \infty).$$

The function f is Hölder continuous with values in Y, moreover u_0 belongs to $D(A_Y)$ and $Au_0 + f(0)$ belongs to $D_A(\theta + \beta, \infty) = (Y, D(A_Y))_{\theta, \infty}$. By Theorem 4.3.1(iii), the mild solution in Y is strict and belongs to $C^{\theta+1}([0,T]; D_A(\beta, \infty))$, moreover $u' \in B([0,T]; D_A(\theta + \beta, \infty))$.

It remains to show that $Au \in C^{\theta+\beta}([0,T]; X)$. Thanks to (4.3.4),

$$Au(t) = \int_0^t Ae^{(t-s)A}(f(s) - f(t))ds + (e^{tA} - 1)f(t) + Ae^{tA}u_0.$$

In the case $\theta + \beta < 1$, $Au(t) - Au(r)$ may be splitted for $0 \leq r \leq t \leq T$ as

$$Au(t) - Au(r) = \int_0^r A(e^{(t-s)A} - e^{(r-s)A})(f(s) - f(r))ds$$
$$+ \int_r^t Ae^{(t-s)A}(f(s) - f(t))ds + \left[(e^{(t-r)A} - I)(f(t) - f(r))\right]$$
$$+ \left[(e^{tA} - e^{rA})(f(r) - f(0))\right] + \left[(e^{tA} - e^{rA})(Au_0 + f(0))\right] = \sum_{k=1}^5 I_k.$$

Each addendum I_k may be estimated using the methods of Subsection 4.3.2, getting finally

$$\|Au(t) - Au(r)\| \leq \left(\frac{M_{2,\beta}}{(1-\theta-\beta)(\theta+\beta)} + \frac{2M_{1,\beta}}{\theta+\beta} + \frac{M_{1,\beta}}{\beta} + \frac{M_{1,\beta}}{\theta+\beta}\right).$$
$$\cdot (t-r)^{\theta+\beta}[f]_{C^\theta(D_A(\beta,\infty))} + \frac{M_{1,\theta+\beta}}{\theta+\beta}(t-r)^{\theta+\beta}\|Au_0 + f(0)\|_{D_A(\theta+\beta,\infty)}.$$

Therefore, Au is $(\theta + \beta)$-Hölder continuous with values in X.

Let us consider now the case $\theta + \beta > 1$. To prove that Au belongs to $C^{\theta+\beta}([0,T];X)$ it is sufficient to show that u' belongs to $C^{\theta+\beta-1}([0,T];D(A))$, because in this case Au' is continuous, so that

$$\left\| \frac{Au(t+h) - Au(t)}{h} - Au'(t) \right\| = \left\| \frac{1}{h} \int_t^{t+h} (Au'(s) - Au'(t)) ds \right\|.$$

Letting $h \to 0$, we find that Au is differentiable and

$$(Au)' = Au' \in C^{\theta+\beta}([0,T];X).$$

For $0 \le t \le T$ we have

$$u'(t) = \int_0^t A e^{(t-s)A}(f(s) - f(t)) ds + e^{tA}(f(t) - f(0)) + e^{tA}(f(0) + Au_0),$$

from which it follows that $u'(t) \in D(A)$, and

$$\|Au'(t)\| \le \left(\frac{M_{2,\beta}}{\theta + \beta - 1} + M_{1,\beta} \right) T^{\theta+\beta-1} [f]_{C^\theta(D_A(\beta,\infty))} + M_0 \|A(f(0) + Au_0)\|.$$

Moreover, for $0 \le r \le t \le T$

$$Au'(t) - Au'(r) = \int_0^r A^2 (e^{(t-s)A} - e^{(r-s)A})(f(s) - f(r)) ds$$

$$+ \int_r^t A^2 e^{(t-s)A}(f(s) - f(t)) ds + A e^{tA}(f(r) - f(0))$$

$$+ A e^{(t-r)A}(f(t) - f(r)) + A(e^{tA} - e^{rA})(Au_0 + f(0)) = \sum_{k=1}^5 I_k.$$

Again, each I_k may be estimated arguing as in Subsection 4.3.2, getting

$$\|Au'(t) - Au'(r)\| \le \left(\frac{M_{3,\beta}}{(2-\beta-\theta)(\theta+\beta-1)} + \frac{2M_{2,\beta}}{\theta+\beta-1} + M_{1,\beta} \right) \cdot$$
$$\cdot (t-r)^{\theta+\beta-1} [f]_{C^\theta(D_A(\beta,\infty))} + M_{1,\beta}(t-r)^{\theta+\beta-1} \|A(Au_0 + f(0))\|_{D_A(\theta+\beta-1,\infty)}.$$

Therefore, Au' is $(\theta+\beta-1)$-Hölder continuous, and the statement follows. ∎

4.4 The nonhomogeneous problem in unbounded time intervals

Given a function $f : [0, +\infty[\mapsto X$ and $u_0 \in X$, we study the asymptotic behavior as $t \to +\infty$ of the (strong, classical, strict) solution u of (4.0.1) in $[0, +\infty[$. We have seen in the previous section that the behavior of u near $t = 0$ is determined by the regularity properties of f near $t = 0$ and the compatibility conditions between x and $f(0)$. Therefore it is sufficient to estimate the restriction of u to half lines $[a, +\infty[$, for some $a > 0$. Due to estimates of Section 2.3, it is convenient to work in functional spaces with exponential weights at $+\infty$, giving conditions in order that the function

$$t \mapsto e^{\omega t}u(t), \quad t \geq a, \ \omega \in \mathbb{R},$$

be bounded (or Hölder continuous) in $[a, +\infty[$, with values in some Banach spaces, such as X, $D_A(\alpha, p)$, $D(A)$, and so on.

Special attention will be paid to the case $\omega = 0$, since the general case may be reduced to this one by setting $w(t) = e^{\omega t}u(t)$.

We shall also treat backward solutions (that is, solutions of problem (4.0.1) in the half line $]-\infty, 0]$) and solutions defined in the whole real line, in particular periodic solutions.

We define the spaces of functions with exponential decay or (not more than) exponential growth that we shall use in the sequel.

Let I be either a half line or the whole real line, let $\omega \in \mathbb{R}$, let Y be a Banach space, and let H be any of the symbols B, C, C^α, $C^{k+\alpha}$ ($0 < \alpha < 1$, $k \in \mathbb{N}$). We denote by $H(I; Y, \omega)$ the space of the functions $f : I \mapsto Y$ such that $t \mapsto e^{-\omega t}f(t)$ belongs to $H(I, Y)$. $H(I; Y, \omega)$ is endowed with the norm

$$\|f\|_{H(I;Y,\omega)} = \|e^{-\omega \cdot}f(\cdot)\|_{H(I;Y)}.$$

4.4.1 Bounded solutions in $[0, +\infty[$

The case where the spectrum of A does not intersect the imaginary axis

We assume here that

$$\sigma(A) \cap i\mathbb{R} = \emptyset. \tag{4.4.1}$$

Therefore, setting $\sigma_-(A) = \{\lambda \in \sigma(A) : \operatorname{Re} \lambda < 0\}$, and $\sigma_+(A) = \{\lambda \in \sigma(A) : \operatorname{Re} \lambda > 0\}$, both $\sigma_-(A)$ and $\sigma_+(A)$ are spectral sets, and

$$-\omega_- = \sup\{\operatorname{Re} \lambda : \lambda \in \sigma_-(A)\} < 0 < \omega_+ = \inf\{\operatorname{Re} \lambda : \lambda \in \sigma_+(A)\}. \tag{4.4.2}$$

$\sigma_-(A)$ and $\sigma_+(A)$ may be possibly empty: in these cases we set respectively $\omega_- = +\infty$, $\omega_+ = +\infty$.

We use notation and results from Section 2.3. In particular, P is the projection associated to σ_+, defined in (2.3.7). Moreover, we fix once and for all a positive number ω such that

$$-\omega_- < -\omega < \omega < \omega_+,$$

and we set
$$M_k^- = \sup_{t>0} \|t^k e^{\omega t} A^k e^{tA}(I-P)\|_{L(X)}, \quad k \in \mathbb{N} \cup \{0\}, \qquad (4.4.3)$$

$$M_{k,\alpha}^- = \sup_{t>0} \|t^{k-\alpha} e^{\omega t} A^k e^{tA}(I-P)\|_{L(D_A(\alpha,\infty),X)}, \quad k \in \mathbb{N}, \qquad (4.4.4)$$

$$C_{k,\alpha}^- = \sup_{t>0} \|t^{k+\alpha} e^{\omega t} A^k e^{tA}(I-P)\|_{L(X,D_A(\alpha,1))}, \quad k \in \mathbb{N} \cup \{0\}, \qquad (4.4.5)$$

$$M_k^+ = \sup_{t<0} \|e^{-\omega t} A^k e^{tA} P\|_{L(X)}, \quad k \in \mathbb{N} \cup \{0\}. \qquad (4.4.6)$$

Due to the estimates of Section 2.3, the quantities defined above are finite.

To begin with, we study two linear operators Φ, Ψ, which will play an important role in what follows. Let $f :]0,+\infty[\mapsto X$ be such that for some $a > 0$ its restrictions to $]0,a[$ and to $]a,+\infty[$ belong respectively to $L^1(0,a;X)$ and to $L^\infty(a,\infty;X)$. We shall write simply $f \in L^1(0,a;X) \cap L^\infty(a,\infty;X)$. We define

$$\Phi f(t) = (e^{tA} * (I-P)f)(t) = \int_0^t e^{(t-s)A}(I-P)f(s)ds, \quad t \geq 0, \qquad (4.4.7)$$

$$\Psi f(t) = -\int_t^{+\infty} e^{(t-s)A} Pf(s)ds, \quad t \geq 0. \qquad (4.4.8)$$

Proposition 4.4.1 *Let $f :]0,+\infty[\mapsto X$ belong to $L^1(0,a;X) \cap L^\infty(a,\infty;X)$ for some $a > 0$, and let $0 < \alpha < 1$. The following statements hold.*

(i) *Φf belongs to $C^{1-\alpha}([a+\varepsilon,+\infty[;D_A(\alpha,1))$ for every $\varepsilon > 0$, and there is $C = C(\alpha,\varepsilon) > 0$ such that*

$$\|\Phi f\|_{C^{1-\alpha}([a+\varepsilon,+\infty[;D_A(\alpha,1))} \leq C(\|f\|_{L^1(0,a;X)} + \|f\|_{L^\infty(a,+\infty;X)}).$$

In particular, if $E \in J_\alpha(X,D(A))$, then Φf belongs to $C^{1-\alpha}([a+\varepsilon,+\infty[;E)$.

(ii) *If in addition $f \in C^\alpha([a,+\infty[;X)$, then Φf belongs to $C^{1+\alpha}([a+\varepsilon,+\infty[;X) \cap C^\alpha([a+\varepsilon,+\infty[;D(A))$, $(\Phi f)'$ belongs to $B([a+\varepsilon,+\infty[;D_A(\alpha,\infty))$ for every $\varepsilon > 0$, and there is $C = C(\alpha,\varepsilon) > 0$ such that*

$$\|\Phi f\|_{C^{1+\alpha}([a+\varepsilon,+\infty[;X)} + \|A\Phi f\|_{C^\alpha([a+\varepsilon,+\infty[;X)}$$
$$+\|(\Phi f)'\|_{B([a+\varepsilon,+\infty[;D_A(\alpha,\infty))} \leq C(\|f\|_{L^1(0,a;X)} + \|f\|_{C^\alpha([a,+\infty[;X)}).$$

(iii) *If in addition $f \in B([a,+\infty[;D_A(\alpha,\infty))$, then Φf is continuously differentiable for $t > a$, $(\Phi f)'$ and $A\Phi f$ belong to $B([a+\varepsilon,+\infty[;D_A(\alpha,\infty))$, $A\Phi f$ belongs to $C^\alpha([a+\varepsilon,+\infty[;X)$, and there is $C = C(\alpha,\varepsilon) > 0$ such that*

$$\|(\Phi f)'\|_{B([a+\varepsilon,+\infty[;D_A(\alpha,\infty))} + \|A\Phi f\|_{B([a+\varepsilon,+\infty[;D_A(\alpha,\infty))}$$
$$+\|A\Phi f\|_{C^\alpha([a+\varepsilon,+\infty[;X)} \leq C(\|f\|_{L^1(0,a;X)} + \|f\|_{B([a,+\infty[;D_A(\alpha,\infty))}).$$

4. The nonhomogeneous problem in unbounded time intervals

Proof — For $t \geq a$ we set $\Phi f = v + z$, where

$$z(t) = \int_0^a e^{(t-s)A}(1-P)f(s)ds = e^{(t-a)A}(1-P)\int_0^a e^{(a-s)A}f(s)ds,$$

$$v(t) = \int_a^t e^{(t-s)A}(1-P)f(s)ds.$$

Then $z \in C^\infty(]a, +\infty[; D(A^k))$ for each k, and using (4.4.3) we get, for $t \geq 1$, $k \in \mathbb{N} \cup \{0\}$,

$$\left\|\frac{d^k}{dt^k}z(t)\right\| = \|A^k z(t)\| \leq \frac{M_k^-}{(t-a)^k}e^{-\omega(t-a)}M_0^-\|f\|_{L^1(0,a;X)}.$$

Therefore it is sufficient to study the function v. For $t \geq a$ we have

$$\|v(t)\|_{D_A(\alpha,1)} \leq C_{0,\alpha}^- \int_a^t \frac{e^{-\omega(t-s)}}{(t-s)^{-\alpha}}ds \, \|f\|_{L^\infty(a,\infty;X)} \leq \frac{C_{0,\alpha}^- \Gamma(1-\alpha)}{\omega^{1-\alpha}}\|f\|_{L^\infty(a,\infty;X)},$$

where the Γ denotes the Euler function. Concerning Hölder regularity, $[v(t) - v(s)]_{D_A(\alpha,1)}$ can be estimated as in formula (4.2.4) of Proposition 4.2.1, getting, for $t \geq s \geq a$,

$$[v(t) - v(s)]_{D_A(\alpha,1)} \leq \left(\frac{C_{1,\alpha}^-}{\alpha(1-\alpha)} + \frac{C_{0,\alpha}^-}{1-\alpha}\right)(t-s)^{1-\alpha}\|f\|_{L^\infty(a,\infty;X)}.$$

Therefore v belongs to $C^{1-\alpha}([a, +\infty[; D_A(\alpha, 1))$ for every $\alpha \in]0, 1[$, and statement (i) follows.

Let us prove (ii). If f is Hölder continuous in $[a, \infty[$, then $v(t)$ belongs to $D(A)$ for each $t \geq a$, and

$$\|Av(t)\| \leq \left\|\int_a^t Ae^{(t-s)A}(1-P)[f(s) - f(t)]ds\right\| + \|(e^{(t-a)A} - I)(1-P)f(t)\|$$

$$\leq M_1^- \int_a^t e^{-\omega(t-s)}(t-s)^{\alpha-1}ds[f]_{C^\alpha([a,\infty[;X)} + (M_0^- + 1)[f]_{L^\infty(a,\infty;X)}$$

$$\leq (M_1^- \omega^{-\alpha}\Gamma(\alpha) + M_0^- + 1)\|f\|_{C^\alpha([a,\infty[;X)}, \quad t \geq a.$$

Moreover, for $t \geq s > a + \varepsilon$,

$$\|Av(t) - Av(s)\| \leq$$

$$\leq \left\|\int_a^t Ae^{(t-\sigma)A}(1-P)[f(\sigma) - f(t)]d\sigma - \int_a^s Ae^{(s-\sigma)A}(1-P)[f(\sigma) - f(s)]d\sigma\right\|$$

$$+ \|(e^{(t-a)A} - I)(1-P)(f(t) - f(s))\| + \|(e^{(t-a)A} - e^{(s-a)A})(1-P)f(s)\|$$

$$= I_1 + I_2 + I_3.$$

The first addendum I_1 can be estimated as in formula (4.3.6) of Theorem 4.3.1, getting

$$I_1 \leq \left(\frac{M_2^-}{\alpha(1-\alpha)} + 2M_0^- + \frac{M_1^-}{\alpha}\right)(t-s)^\alpha [f]_{C^\alpha([a,\infty[;X)}, \quad t \geq s \geq a.$$

Moreover we get easily

$$I_2 \leq (M_0^- + \|I - P\|_{L(X)})(t-s)^\alpha [f]_{C^\alpha([a,\infty[;X)}, \quad t \geq s \geq a,$$

and, for $t \geq s \geq a + \varepsilon$,

$$\begin{aligned} I_3 &= \left\|\int_{s-a}^{t-a} Ae^{\tau A}(1-P)d\tau f(s)\right\| \leq M_1^- \int_{s-a}^{t-a} \frac{e^{-\omega\tau}}{\tau} d\tau \, \|f(s)\| \\ &\leq \frac{M_1^-}{(s-a)^\alpha} \int_{s-a}^{t-a} \frac{1}{\tau^{\alpha-1}} d\tau \, \|f(s)\| \leq \frac{M_1^-}{\alpha\varepsilon^\alpha}(t-s)^\alpha \|f\|_{L^\infty(a,\infty;X)}. \end{aligned}$$

Therefore Av is α-Hölder continuous in $[a+\varepsilon, +\infty[$. Since $v' = Av + f$, then also v' is α-Hölder continuous in $[a+\varepsilon, +\infty[$. Boundedness of $[v'(t)]_{D_A(\alpha,\infty)}$ in $[a+\varepsilon, +\infty[$ follows now from Proposition 2.2.12(i).

Let us prove statement (iii). For $t \geq a$ we have, by (4.4.4),

$$\|Av(t)\| \leq M_{1,\alpha}^- \int_a^t \frac{e^{-\omega(t-s)}}{(t-s)^{1-\alpha}} ds \, \|f\|_{B(D_A(\alpha,\infty))} \leq M_{1,\alpha}^- \Gamma(\alpha)\omega^{-\alpha} \|f\|_{B(D_A(\alpha,\infty))}.$$

The other estimates can be obtained following step by step estimates (4.3.18) and (4.3.19). One finds, for $t \geq s \geq a$,

$$[Av(t)]_{D_A(\alpha,\infty)} \leq M_{2,\alpha}^-(1-\alpha)^{-1}\|f\|_{B(D_A(\alpha,\infty))};$$

$$\|Av(t) - Av(s)\| \leq \left(\frac{M_{2,\alpha}^-}{\alpha(1-\alpha)} + \frac{M_{1,\alpha}^-}{\alpha}\right)(t-s)^\alpha \|f\|_{B(D_A(\alpha,\infty))},$$

and statement (iii) follows. ■

Let us consider now the operator Ψ defined in (4.4.8).

Proposition 4.4.2 *Let f belong to $L^\infty(a, +\infty; X)$. The following statements hold:*

(i) Ψf belongs to $Lip([a, +\infty[; D(A^k))$ for each $k \in \mathbb{N}$, and there is $C > 0$ such that

$$\|A^k \Psi f\|_{Lip([a,+\infty[;X)} \leq C\|f\|_{L^\infty([a,+\infty[;X)}.$$

If in addition $f \in C([a, +\infty[; X)$, then Ψf belongs to $C^1([a, +\infty[; D(A^k))$ for each $k \in \mathbb{N}$.

4. THE NONHOMOGENEOUS PROBLEM IN UNBOUNDED TIME INTERVALS   157

(ii) *If* $f \in C^\alpha([a, +\infty[; X)$ *for some* $\alpha \in \,]0, 1[$, *then* Ψf *belongs to* $C^{1+\alpha}([a, +\infty[; D(A^k))$ *for each* $k \in \mathbb{N}$, *and there is* $C = C(k) > 0$ *such that*

$$\|A^k \Phi f\|_{C^{1+\alpha}([a,+\infty[;X)} \leq C \|f\|_{C^\alpha([a,+\infty[;X)}.$$

The proof is a simple consequence of estimates (4.4.6), and it is left to the reader.

Now we give necessary and sufficient conditions in order that the solution of (4.0.1) remains bounded as $t \to +\infty$.

Theorem 4.4.3 *Let* $f : \,]0, +\infty[\,\mapsto\, X$ *belong to* $L^1(0, a; X) \cap L^\infty(a, +\infty; X)$ *for some* $a > 0$, *let* $u_0 \in X$ *and let* u *be the function defined by (4.0.2). Then* u *is bounded with values in* X *in* $[0, +\infty[$ *if and only if*

$$P u_0 = - \int_0^{+\infty} e^{-sA} P f(s) ds. \tag{4.4.9}$$

In this case, u *is given by*

$$u(t) = e^{tA}(1-P)u_0 + \int_0^t e^{(t-s)A}(1-P)f(s)ds - \int_t^{+\infty} e^{(t-s)A} P f(s) ds. \tag{4.4.10}$$

Moreover,

(i) *if* $f \in C^\alpha([a, +\infty[; X)$, $\alpha \in \,]0, 1[$, *then* u *belongs to* $C^{1+\alpha}([a + \varepsilon, +\infty[; X) \cap C^\alpha([a + \varepsilon, +\infty[; D(A))$, u' *belongs to* $B([a + \varepsilon, +\infty[; D_A(\alpha, \infty))$ *for every* $\varepsilon > 0$, *and there is* $C = C(\varepsilon) > 0$ *such that*

$$\|u\|_{C^{1+\alpha}([a+\varepsilon,+\infty[;X)} + \|Au\|_{C^\alpha([a+\varepsilon,+\infty[;X)} + \|u'\|_{B([a+\varepsilon,+\infty[;D_A(\alpha,\infty))}$$
$$\leq C \left(\|u_0\| + \|f\|_{L^1(0,a;X)} + \|f\|_{C^\alpha([a,+\infty[;X)} \right);$$

(ii) *if* $f \in C([a, +\infty[; X) \cap B([a, +\infty[; D_A(\alpha, \infty))$ *for some* $\alpha \in \,]0, 1[$, *then* u *is continuously differentiable for* $t > a$, u' *and* Au *belong to* $B([a + \varepsilon, +\infty[; D_A(\alpha, \infty))$, Au *belongs to* $C^\alpha([a + \varepsilon, +\infty[; X)$ *for every* $\varepsilon > 0$, *and there is* $C = C(\varepsilon) > 0$ *such that*

$$\|u'\|_{B([a+\varepsilon,+\infty[;D_A(\alpha,\infty))} + \|Au\|_{B([a+\varepsilon,+\infty[;D_A(\alpha,\infty))} + \|Au\|_{C^\alpha([a+\varepsilon,+\infty[;X)}$$
$$\leq C \left(\|u_0\| + \|f\|_{L^1(0,a;X)} + \|f\|_{B([a,+\infty[;D_A(\alpha,\infty))} \right).$$

Proof — Set $u = u_1 + u_2$, with

$$u_1(t) = e^{tA}(1-P)u_0 + \int_0^t e^{(t-s)A}(1-P)f(s)ds - \int_t^{+\infty} e^{(t-s)A} P f(s) ds$$

$$u_2(t) = e^{tA} P u_0 + \int_0^{+\infty} e^{(t-s)A} P f(s) ds, \quad t \geq 0.$$

By estimate (4.4.3) with $k = 0$ and Propositions 4.4.1 and 4.4.2, u_1 is bounded in $[0, +\infty[$. Moreover

$$u_2(t) = e^{tA}\left(Pu_0 + \int_0^{+\infty} e^{-sA}Pf(s)ds\right), \quad t \geq 0.$$

By (4.4.6) we get $\|e^{tA}y\| \geq e^{\omega t}(M_0^+)^{-1}\|y\|$ for each $y \in P(X)$ and $t > 0$. Therefore, u_2 is bounded in $[0, +\infty[$ if and only if (4.4.9) holds.

Statements (i) and (ii) follow now easily from estimates (4.4.3) and Propositions 4.4.1 and 4.4.2. ∎

Note that condition (4.4.9) is satisfied whenever $\sigma_+ = \emptyset$, in which case for every initial datum and for every bounded f the solution of (4.0.1) is bounded. Moreover, we shall see in Subsection 4.4.4 that if f decays exponentially, then the solution decays exponentially.

The case where A generates a bounded semigroup

We assume now that there are M_0, $M_1 > 0$ such that

$$\|e^{tA}\|_{L(X)} \leq M_0, \quad \|Ae^{tA}\|_{L(X)} \leq \frac{M_1}{t}, \quad \forall t > 0. \tag{4.4.11}$$

Assumption (4.4.11) is independent of (4.4.1). Due to Remark 2.1.10, A satisfies (4.4.11) if and only if it satisfies (2.0.1) with $\omega = 0$. In particular, if (4.4.11) holds, then the type ω_A of A is nonpositive. In the noncritical case $\omega_A < 0$, the norms of e^{tA} and of its derivatives decay exponentially as $t \to +\infty$, and the results of next subsection 4.3.4 are stonger than the present ones. In the critical case $\omega_A = 0$, A is not invertible, and 0 needs not to be isolated in $\sigma(A)$, but it is the unique element of $\sigma(A)$ with zero imaginary part. Examples of operators A satisfying (4.4.11) are frequently found in partial differential equations in unbounded domains, or else in bounded domains with certain boundary conditions (for instance, when A is the realization of the Laplace operator Δ with Neumann boundary condition in $L^p(\Omega)$ or in $C(\overline{\Omega})$).

Assumption (4.4.11) implies that there are constants M_n such that $\|A^n e^{tA}\|_{L(X)} \leq M_n t^{-n}$ for every $t > 0$. It follows that for every $\theta \in]0,1[$ the norms $x \mapsto \|x\|_{D_A(\theta,\infty)}$ and

$$x \mapsto \|x\| + |x|_{D_A(\theta,\infty)} = \|x\| + \sup_{t>0} \|t^{1-\theta}Ae^{tA}x\| \tag{4.4.12}$$

are equivalent, even if the seminorms $[\cdot]_{D_A(\theta,\infty)}$ and $|\cdot|_{D_A(\theta,\infty)}$ are not necessarily equivalent. In what follows, it is convenient to work with the seminorm $|\cdot|_{D_A(\theta,\infty)}$. So, throughout the subsection we replace the norm of $D_A(\theta,\infty)$ by the norm defined in (4.4.12).

4. THE NONHOMOGENEOUS PROBLEM IN UNBOUNDED TIME INTERVALS

For $\theta \in \,]0,1]$ and $n \in \mathbb{N}$ there are constants $M_{n,\theta}$ such that
$$\|t^{n-\theta}A^n e^{tA}\|_{L(D_A(\theta,\infty),X)} \leq M_{n,\theta}, \quad t > 0. \tag{4.4.13}$$

Similarly, for $\theta \in \,]0,1[$ and $n \in \mathbb{N} \cup \{0\}$ there are constants $C_{n,\theta}$ such that
$$|t^{n+\theta}A^n e^{tA}x|_{D_A(\theta,\infty)} \leq C_{n,\theta}\|x\|, \quad t > 0, \ x \in X. \tag{4.4.14}$$

In the applications to nonlinear problems we shall also need estimates on u in a bounded interval $[0,T]$, with constants independent of T. The theorems of Section 4.3 provide estimates with constants possibly depending on T; in particular, the constants may blow up when $T \to \infty$. However, in Section 4.3 we gave uniform estimates for small values of T, say $T \leq 2$. So, it is sufficient to consider the case $T > 2$.

It is clear that for $T \leq +\infty$, $f \in L^1(0,T;X)$ and $u_0 \in X$, then
$$\|u(t)\| \leq M_0(\|u_0\| + \|f\|_{L^1(0,T;X)}), \quad 0 \leq t \leq T. \tag{4.4.15}$$

Proposition 4.4.4 *Let $2 < T \leq +\infty$, and let $f \in L^\infty(0,T;X)$. If $T = +\infty$, assume moreover that $f \in L^1(0,+\infty;X)$. Let u be the mild solution of (4.0.1), and set $\varphi(t) = \max\{1,t\}$. Then:*

(i) If $u_0 \in D_A(\theta,\infty)$ for some $\theta \in \,]0,1[$, then
$$\sup_{0<t<T} \varphi(t)^\theta |u(t)|_{D_A(\theta,\infty)}$$
$$\leq C(\|u_0\|_{D_A(\theta,\infty)} + \|f\|_{L^1(0,T;X)} + \sup_{0<t<T} \varphi(t)^\theta \|f(t)\|), \tag{4.4.16}$$

with C independent of T;

(ii) if in addition $f \in C(]0,T];X) \cap B([0,T];D_A(\theta,\infty))$ and $u_0 \in D_A(\theta+1,\infty)$ for some $\theta \in \,]0,1[$, then
$$\sup_{0<t<T} \varphi(t)\|Au(t)\| + \sup_{0<t<T} \varphi(t)^{1+\theta}|Au(t)|_{D_A(\theta,\infty)}$$
$$\leq C(\|u_0\|_{D_A(\theta+1,\infty)} + \|f\|_{L^1(0,T;X)} + \sup_{0<t<T} \varphi(t)^{1+\theta}|f(t)|_{D_A(\theta,\infty)}), \tag{4.4.17}$$

with C independent of T.

Proof — (i) By Proposition 4.2.1 there is K such that
$$\|u(t)\|_{D_A(\theta,\infty)} \leq K(\|u_0\|_{D_A(\theta,\infty)} + \|f\|_{L^\infty(0,2;X)}), \quad 0 \leq t \leq 2.$$

For $t > 2$ it holds
$$|t^\theta u(t)|_{D_A(\theta,\infty)} \leq C_{0,\theta}\|u_0\| + t^\theta \left(\int_0^{t/2} + \int_{t/2}^t\right) \frac{C_{0,\theta}}{(t-s)^\theta}\|f(s)\|ds$$
$$\leq C_{0,\theta}\left(\|u_0\| + 2^\theta \|f\|_{L^1(0,T;X)} + \varphi(t)\int_{1/2}^1 \frac{1}{(1-\sigma)^\theta \sigma}d\sigma \sup_{s\geq 1}\|sf(s)\|\right).$$

Taking into account (4.4.15), statement (i) holds.

Let us prove (ii). Theorem 4.3.8 and Corollary 4.3.9 give, for $0 \leq t \leq 2$,
$$\|u(t)\| + \varphi(t)\|Au(t)\| + \varphi(t)^{1+\theta}|Au(t)|_{D_A(\theta,\infty)} \\ \leq K(\|u_0\|_{D_A(\theta+1,\infty)} + \sup_{0 \leq s \leq T} \|f(s)\|_{D_A(\theta,\infty)}). \tag{4.4.18}$$

For $t \geq 2$ we have
$$\|Av(t)\| \leq \frac{M_1}{t}\|u_0\| + \left\|\left(\int_0^{t/2} + \int_{t/2}^t\right) A e^{(t-s)A} f(s) ds\right\|$$
$$\leq \frac{M_1}{t}\|u_0\| + M_1 \int_0^{t/2} \frac{\|f(s)\|}{t-s} ds + M_{1,\theta} \int_{t/2}^t \frac{|f(s)|_{D_A(\theta,\infty)}}{(t-s)^{1-\theta}} ds$$
$$\leq \frac{M_1}{t}\|u_0\| + \frac{2M_1}{t}\|f\|_{L^1(0,T;X)} + \frac{2M_{1,\theta}}{\theta t} \sup_{1 \leq s \leq T} |s^{\theta+1} f(s)|_{D_A(\theta,\infty)}.$$

Moreover, for $t \geq 2$ and for every $\xi > 0$,
$$\|\xi^{1-\theta} A^2 e^{\xi A} v(t)\| \leq \frac{\xi^{1-\theta} M_2}{(t+\xi)^2}\|u_0\| + \left\|\xi^{1-\theta}\left(\int_0^{t/2} + \int_{t/2}^t\right) A^2 e^{(\xi+t-s)A} f(s) ds\right\|$$
$$\leq \frac{M_2}{t^{\theta+1}}\|u_0\| + M_2 \xi^{1-\theta} \int_0^{t/2} \frac{\|f(s)\|}{(\xi+t-s)^2} ds + C_{2,\theta} \xi^{1-\theta} \int_{t/2}^t \frac{|f(s)|_{D_A(\theta,\infty)}}{(\xi+t-s)^{2-\theta}} ds$$
$$\leq \frac{M_2}{t^{\theta+1}}\|u_0\| + \frac{2^{\theta+1} M_2}{t^{\theta+1}}\|f\|_{L^1(0,T;X)} + \frac{2^{\theta+1} C_{2,\theta}}{(1-\theta)t^{\theta+1}} \sup_{1 \leq s \leq T} |s^{\theta+1} f(s)|_{D_A(\theta,\infty)},$$

which implies
$$[Av(t)]_{D_A(\theta,\infty)} \leq \frac{M_2}{t^{\theta+1}}\|u_0\|$$
$$+ \frac{2^{\theta+1}}{t^{\theta+1}}\left(M_2\|f\|_{L^1(0,T;X)} + \frac{C_{2,\theta}}{1-\theta} \sup_{1 \leq s \leq T} |s^{\theta+1} f(s)|_{D_A(\theta,\infty)}\right).$$

Therefore,
$$\varphi(t)\|Av(t)\| + \varphi(t)^{1+\theta}|Av(t)|_{D_A(\theta,\infty)} \\ \leq C(\|u_0\| + \|f\|_{L^1(0,T;X)} + \sup_{1 \leq s \leq T} |\varphi(s)^{\theta+1} f(s)|_{D_A(\theta,\infty)}), \quad 2 \leq t \leq T,$$

which, added to (4.4.18), proves statement (ii). ∎

4.4.2 Bounded solutions in $]-\infty, 0]$

Throughout the subsection we assume again that (4.4.1) holds. To study the bounded backward solutions we introduce the operator Λ defined on $L^\infty(-\infty, 0; X)$ by
$$(\Lambda g)(t) = \int_{-\infty}^t e^{(t-s)A}(1-P)g(s) ds, \quad t \leq 0. \tag{4.4.19}$$

4. The nonhomogeneous problem in unbounded time intervals   161

Note that, for every $a > 0$, Λg satisfies

$$\Lambda g(t) = e^{(t+a)A}(\Lambda g)(-a) + \int_{-a}^{t} e^{(t-s)A}(I - P)g(s)ds, \quad -a \leq t \leq 0,$$

so that it is a mild solution of $u'(t) = Au(t)$, $u(-a) = (\Lambda g)(-a)$ in the interval $[-a, 0]$.

Proposition 4.4.5 *The operator Λ defined in (4.4.19) is continuous:*

(i) *from $L^\infty(-\infty, 0; X)$ to $C^{1-\alpha}(]-\infty, 0]; D_A(\alpha, 1))$, for each $\alpha \in \,]0, 1[$;*

(ii) *from $C^\alpha(]-\infty, 0]; X)$ to $C^\alpha(]-\infty, 0]; D(A))$, for each $\alpha \in \,]0, 1[$;*

(iii) *from $C(]-\infty, 0]; X) \cap B(]-\infty, 0]; D_A(\alpha, \infty))$ to $C_b(]-\infty, 0]; D(A)) \cap B(]-\infty, 0]; D_A(\alpha+1, \infty))$, from $C_b(]-\infty, 0]; D_A(\alpha, \infty))$ to $C_b(]-\infty, 0]; D_A(\alpha+1, \infty))$, and from $C_b(]-\infty, 0]; D_A(\alpha))$ to $C_b(]-\infty, 0]; D_A(\alpha+1))$.*

Moreover, if g belongs either to $C^\alpha(]-\infty, 0]; X)$ or to $C(]-\infty, 0]; X) \cap B(]-\infty, 0]; D_A(\alpha, \infty))$, then Λg is continuously differentiable in $]-\infty, 0]$, and

$$(\Lambda g)'(t) = A\Lambda g(t) + g(t), \quad t \leq 0.$$

Proof — Let us prove statement (i). For $s < t \leq 0$ we have

$$\|(\Lambda g)(t) - (\Lambda g)(s)\| \leq$$

$$\leq \left\| \int_{-\infty}^{s} d\sigma \int_{s-\sigma}^{t-\sigma} Ae^{\tau A} d\tau (1 - P)g(\sigma) \right\| + \left\| \int_{s}^{t} e^{(t-\sigma)A}(1 - P)g(\sigma) d\sigma \right\|$$

$$\leq M_1^- \int_{-\infty}^{s} \frac{e^{-\omega(s-\sigma)}}{(s-\sigma)^\alpha} d\sigma \int_{s-\sigma}^{t-\sigma} \tau^{\alpha-1} d\tau \, \|g\|_{L^\infty(X)} + M_0^- \int_{s}^{t} e^{-\omega(t-\sigma)} d\sigma \, \|g\|_{L^\infty(X)}$$

$$\leq \left(\frac{M_1^- \Gamma(1-\alpha)}{\omega^{1-\alpha}} + M_0^- \left(\frac{1-\alpha}{\omega} \right)^{1-\alpha} \right) \|g\|_{L^\infty(X)} (t-s)^\alpha,$$

so that Λg belongs to $C^\alpha(]-\infty, 0]; X)$. In addition, $\|(\Lambda g)(t) - (\Lambda g)(s)\|_{D_A(\alpha, 1)}$ can be estimated as in formula (4.2.4) of Proposition 4.2.1, getting, for $s < t \leq 0$,

$$\|(\Lambda g)(t) - (\Lambda g)(s)\|_{D_A(\alpha, 1)} \leq \left(\frac{M_2^-}{\alpha(1-\alpha)} + \frac{M_1^-}{\alpha} \right) (t-s)^{1-\alpha} \|g\|_\infty.$$

Therefore Λg belongs to $C^{1-\alpha}(]-\infty, 0]; D_A(\alpha, 1))$ for every $\alpha \in \,]0, 1[$, and statement (i) follows.

The proof of (ii) is similar to the proof of (ii) of Proposition 4.4.2, and even simpler. Of course, integrals over $[0, t[$ must be replaced by integrals over $]-\infty, t]$. We get

$$\|A(\Lambda g)(t)\| \leq (M_1^- \omega^{-\alpha} \Gamma(\alpha) + 2M_0^-) \|g\|_{C^\alpha}, \quad t \leq 0.$$

Moreover, for $s \leq t \leq 0$,

$$\|A(\Lambda g)(t) - A(\Lambda g)(s)\| \leq \|(1-P)(g(t) - g(s))\|$$
$$+ \left\|A(I-P)\left(\int_{-\infty}^{t} e^{(t-\sigma)A}[g(\sigma) - g(t)]d\sigma - \int_{-\infty}^{s} e^{(s-\sigma)A}[g(\sigma) - g(s)]d\sigma\right)\right\|$$
$$= I_1 + I_2.$$

The first addendum I_1 is obviously less or equal to $c[g]_{C^\alpha}(t-s)^\alpha$. The second one can be estimated as in formula (4.3.6) of Theorem 4.3.1, getting

$$I_2 \leq \left(\frac{M_1^-}{\alpha(1-\alpha)} + 2M_0^- + \frac{M_1^-}{\alpha}\right)(t-s)^\alpha[g]_{C^\alpha}.$$

Therefore, Λg is Hölder continuous with values in $D(A)$. By Lemma 4.1.6, Λg belongs to $C^1([-a,0];X)$ for each $a > 0$, and $(\Lambda g)'(t) = (\Lambda g)(t) + g(t)$ for every $t \leq 0$. Moreover, by Proposition 4.4.9, $(\Lambda g)'$ is bounded with values in $D_A(\alpha, \infty)$.

The proof of statement (iii) is similar to the proof of Proposition 4.4.1(iii) and of estimates (4.3.18), (4.3.19). For every $g \in C(]-\infty,0];X) \cap B(]-\infty,0]; D_A(\alpha, \infty))$ we get

$$\|A(\Lambda g)\|_{B(]-\infty,0];D_A(\alpha,\infty))} \leq \left(\frac{M_{1,\alpha}^- \Gamma(\alpha)}{\omega^\alpha} + \frac{M_{2,\alpha}^-}{1-\alpha}\right)\|g\|_{B(]-\infty,0];D_A(\alpha,\infty))}, \ t \leq 0;$$

$$[A(\Lambda g)]_{C^\alpha(]-\infty,0];X)} \leq \left(\frac{M_{2,\alpha}^-}{\alpha(1-\alpha)} + \frac{M_{1,\alpha}^-}{\alpha}\right)\|g\|_{B(]-\infty,0];D_A(\alpha,\infty))}.$$

Moreover, since $A(\Lambda g)$ is continuous, then, thanks to Lemma 4.1.6, Λg belongs to $C^1([-a,0];X)$ for each $a > 0$, and $(\Lambda g)'(t) = (\Lambda g)(t) + g(t)$ for every $t \leq 0$.

If in addition $f \in C(]-\infty,0]; D_A(\alpha,\infty))$ (respectively, $f \in C(]-\infty,0]; D_A(\alpha))$), then $(I-P)f \in C([-a,0]; D_A(\alpha,\infty))$ (respectively, $(I-P)f \in C([-a,0]; D_A(\alpha))$) for every $a > 0$, therefore $A\Lambda g$ and $(\Lambda g)'$ belong to $C(]-a,0]; D_A(\alpha,\infty))$ thanks to Corollary 4.3.9(iv) (respectively, to $C([-a,0]; D_A(\alpha))$ thanks to Corollary 4.3.10). ∎

Now we can study the backward problem

$$v'(t) = Av(t) + g(t), \ t \leq 0; \ v(0) = v_0, \quad (4.4.20)$$

giving necessary and sufficient conditions for the existence of a bounded solution, and describing the properties of the bounded solutions.

Theorem 4.4.6 *Let $g : \,]-\infty,0] \mapsto X$ be continuous and bounded, and let $v_0 \in X$. Then problem (4.4.20) has a bounded strong solution v if and only if*

$$(1-P)v_0 = \int_{-\infty}^{0} e^{-sA}(1-P)g(s)ds \quad (4.4.21)$$

4. THE NONHOMOGENEOUS PROBLEM IN UNBOUNDED TIME INTERVALS

In this case, v is given by

$$v(t) = e^{tA}Pv_0 + \int_0^t e^{(t-s)A}Pg(s)ds + \int_{-\infty}^t e^{(t-s)A}(1-P)g(s)ds, \quad t \leq 0, \quad (4.4.22)$$

and it belongs to $C^{1-\alpha}(]-\infty,0]; D_A(\alpha,1))$ for every $\alpha \in]0,1[$. If, in addition, g belongs either to $C^\alpha(]-\infty,0]; X)$ or to $B(]-\infty,0]; D_A(\alpha,\infty))$ for some $\alpha \in]0,1[$, then v is a strict solution of (4.4.20). More precisely,

(i) *if $g \in C^\alpha(]-\infty,0]; X)$, then $v \in C^{1+\alpha}(]-\infty,0]; X) \cap C^\alpha(]-\infty,0]; D(A))$, $v' \in B(]-\infty,0]; D_A(\alpha,\infty))$, and there is $C > 0$ such that*

$$\|v\|_{C^{1+\alpha}(]-\infty,0];X)} + \|Av\|_{C^\alpha(]-\infty,0];X)} + \|v'\|_{B(]-\infty,0];D_A(\alpha,\infty))}$$
$$\leq C(\|v_0\| + \|g\|_{C^\alpha(]-\infty,0];X)});$$

(ii) *if $g \in B(]-\infty,0]; D_A(\alpha,\infty))$, then Av belongs to $C^\alpha(]-\infty,0]; X)$, v' and Av belong to $B(]-\infty,0]; D_A(\alpha,\infty))$, and there is $C > 0$ such that*

$$\|v'\|_{B(]-\infty,0];D_A(\alpha,\infty))} + \|Av\|_{B(]-\infty,0];D_A(\alpha,\infty))}$$
$$+ \|Av\|_{C^\alpha(]-\infty,0];X)} \leq C(\|v_0\| + \|g\|_{B(]-\infty,0];D_A(\alpha,\infty))}).$$

If in addition $g \in C(]-\infty,0]; D_A(\alpha,\infty))$ (respectively, $g \in C(]-\infty,0]; D_A(\alpha))$), then v' and Av belong to $C(]-\infty,0]; D_A(\alpha,\infty))$ (respectively, to $C(]-\infty,0]; D_A(\alpha))$).

Proof — First we show that if problem (4.4.20) has a bounded strong solution then (4.4.21) holds. If v is any strong solution of (4.4.20), for every $a < 0$ and $t \in [a,0]$ we have

$$\begin{aligned} v(t) &= e^{(t-a)A}v(a) + \int_a^t e^{(t-s)A}g(s)ds \\ &= e^{(t-a)A}(1-P)\left(v(a) - \int_{-\infty}^a e^{(a-s)A}(1-P)g(s)ds\right) \quad (4.4.23) \\ &\quad + \int_{-\infty}^t e^{(t-s)A}(1-P)g(s)ds + Pv(t). \end{aligned}$$

Using Proposition 4.4.5(i) we get

$$\sup_{a \leq 0}\left\|\int_{-\infty}^a e^{(a-s)A}(1-P)g(s)ds\right\| < +\infty.$$

If v is bounded, then $\sup_{a\leq 0}\|(1-P)v(a)\| < +\infty$. Therefore, letting $a \to -\infty$ in (4.4.23), and using estimate (4.4.3), we find

$$v(t) = \int_{-\infty}^t e^{(t-s)A}(1-P)g(s)ds + Pv(t), \quad t \leq 0, \quad (4.4.24)$$

which implies that
$$(1-P)v(t) = \int_{-\infty}^{t} e^{(t-s)A}(1-P)g(s)ds, \ t \leq 0.$$

Taking $t = 0$ we find that (4.4.21) holds. Moreover, Pv is the mild solution of
$$w'(t) = A_+ w(t) + Pg(t), \ t \leq 0; \ w(0) = Pv_0,$$
where $A_+ = A_{|P(X)}$ (see Proposition 2.3.3), so that, since $A_+ \in L(P(X))$,
$$Pv(t) = e^{tA_+} Pv_0 + \int_0^t e^{(t-s)A_+} Pg(s)ds, \ t \leq 0,$$
and (4.4.22) follows from (4.4.24). The rest of the statement follows easily from Proposition 4.4.5. ∎

4.4.3 Bounded solutions in $\mathbb{R}$

Here we apply the results of Subsections 4.4.1 and 4.4.2 to study the problem
$$z'(t) = Az(t) + h(t), \ t \in \mathbb{R}, \qquad (4.4.25)$$
where $h : \mathbb{R} \mapsto X$ is continuous and bounded.

Theorem 4.4.7 *Let (4.4.1) hold, and let $h \in C_b(\mathbb{R}; X)$. Then problem (4.4.25) has a unique bounded strong solution z, given by*
$$z(t) = \int_{-\infty}^{t} e^{(t-s)A}(1-P)h(s)ds - \int_t^{+\infty} e^{(t-s)A} Ph(s)ds, \ t \in \mathbb{R}, \qquad (4.4.26)$$
and it belongs to $C^{1-\alpha}(\mathbb{R}; D_A(\alpha, 1))$ for every $\alpha \in]0, 1[$. If in addition h belongs either to $C^\alpha(\mathbb{R}; X)$ or to $B(\mathbb{R}; D_A(\alpha, \infty))$ for some $\alpha \in]0, 1[$, then z is a strict solution of (4.4.25). Moreover,

(i) *if $h \in C^\alpha(\mathbb{R}; X)$, then z belongs to $C^{1+\alpha}(\mathbb{R}; X) \cap C^\alpha(\mathbb{R}; D(A))$, $z' \in B(\mathbb{R}; D_A(\alpha, \infty))$, and there is $C > 0$ such that*
$$\|z\|_{C^{1+\alpha}(\mathbb{R};X)} + \|Az\|_{C^\alpha(\mathbb{R};X)} + \|z'\|_{B(\mathbb{R};D_A(\alpha,\infty))} \leq C\|h\|_{C^\alpha(\mathbb{R};X)};$$

(ii) *if $h \in B(\mathbb{R}; D_A(\alpha, \infty))$, then Az belongs to $C^\alpha(\mathbb{R}; X)$, z' and Az belong to $B(\mathbb{R}; D_A(\alpha, \infty))$, and there is $C > 0$ such that*
$$\|z'\|_{B(\mathbb{R};D_A(\alpha,\infty))} + \|Az\|_{B(\mathbb{R};D_A(\alpha,\infty))} + \|Az\|_{C^\alpha(\mathbb{R};X)} \leq C\|h\|_{B(\mathbb{R};D_A(\alpha,\infty))}.$$

If in addition $h \in C_b(\mathbb{R}; D_A(\alpha, \infty))$ (respectively, $h \in C_b(\mathbb{R}; D_A(\alpha))$) then z' and Az belong to $C_b(\mathbb{R}; D_A(\alpha, \infty))$ (respectively, to $C_b(\mathbb{R}; D_A(\alpha))$).

4. The nonhomogeneous problem in unbounded time intervals

Proof — Let $h \in C_b(\mathbb{R}; X)$, and let z be defined by (4.4.26). By Propositions 4.4.2 (with $[1, +\infty[$ replaced by $\mathbb{R}$) and 4.4.5 (with $]-\infty, 0]$ replaced by $\mathbb{R}$), z is bounded in $\mathbb{R}$. Moreover, it is easy to see that for every $a \in \mathbb{R}$ we have

$$z(t) = e^{(t-a)A} z(a) + \int_a^t e^{(t-s)A} h(s) ds, \ t \geq a,$$

so that z is a strong solution of (4.4.25). Now we have only to show that z is the unique bounded strong solution of (4.4.25), since the regularity properties stated in (i) and (ii) follow from Theorems 4.3.1 and 4.3.8.

Let z be a bounded strong solution of (4.4.25) in $\mathbb{R}$. Then, for every $a \in \mathbb{R}$, z is a bounded strong solution of both

$$z' = Az + h, \ t \geq a,$$

and

$$z' = Az + h, \ t \leq a,$$

so that, by Theorem 4.4.3,

$$Pz(t) = -\int_t^{+\infty} e^{(t-s)A_+} Ph(s) ds, \ t \geq a,$$

and by Theorem 4.4.6,

$$(1-P)z(t) = -\int_{-\infty}^t e^{(t-s)A}(1-P)h(s) ds, \ t \leq a.$$

Since a is arbitrary and $z(t) = Pz(t) + (1-P)z(t)$, (4.4.26) follows. ∎

It is not difficult to see that if $h : \mathbb{R} \mapsto X$ is continuous and T-periodic, then the function z defined in (4.4.26) is T-periodic; moreover if h is constant, then z is constant. However, for the existence of periodic solutions to (4.4.25), assumption (4.4.2) is too restrictive. In fact, if $h \in C(\mathbb{R}; X)$ is T-periodic, a strong solution z of (4.4.25) is T-periodic if and only if

$$z(t) = z(t+T) = e^{TA} z(t) + \int_t^{t+T} e^{(t+T-s)A} h(s) ds, \ t \in \mathbb{R}. \quad (4.4.27)$$

Therefore, the operator $I - e^{TA}$ plays a crucial role in existence and uniqueness of periodic solutions. In particular, if $I - e^{TA}$ is invertible, then we get a representation formula for the (unique) strong periodic solution of (4.4.25),

$$\begin{aligned} z(t) &= (I - e^{TA})^{-1} \int_t^{t+T} e^{(t+T-s)A} Ph(s) ds, \\ &= (I - e^{TA})^{-1} \int_0^T e^{sA} Ph(t-s) ds, \ t \in \mathbb{R}. \end{aligned} \quad (4.4.28)$$

Proposition 2.3.6 allows us to give simple conditions for the existence of periodic solutions of (4.4.25), and to give also a representation formula.

Proposition 4.4.8 Let $h : \mathbb{R} \mapsto X$ be continuous and T-periodic. Then:

(i) If $1 \in \rho(e^{TA})^{(1)}$, then problem (4.4.25) has a unique T-periodic strong solution, given by the representation formula (4.4.28).

(ii) If $\rho(A) \supset \{2k\pi i/T : k \in \mathbb{Z}, k \neq k_1, ..., k_N\}$, and for every $j = 1, .., N$, $2k_j\pi i/T$ is a semisimple eigenvalue of A, then problem (4.4.25) has a T-periodic strong solution if and only if

$$\int_0^T e^{-2k_j\pi i s/T} P_j h(s) ds = 0, \; j = 1, ..., N, \quad (4.4.29)$$

where P_j, $j = 1, .., N$, are the projections defined in (2.3.20). If (4.4.29) holds, all the T-periodic strong solutions of (4.4.25) are given by

$$\begin{aligned} z(t) &= \sum_{j=1}^N e^{2k_j\pi i t/T}\left(x_j + \int_0^t e^{-2k_j\pi i(t-s)/T} P_j h(s) ds\right) \\ &+ \left(I - e^{TA}\Big|_{(I-P)(X)}\right)^{-1}\int_0^T e^{sA}(I-P)h(t-s)ds, \end{aligned} \quad (4.4.30)$$

where for $j = 1, .., N$, x_j are arbitrary elements of $P_j(X)$, and the inverse of $(I - e^{TA})\big|_{(I-P)(X)}$ is given by formula (2.3.21).

Proof — Concerning part (i) of the statement, it is not difficult to see that the function z defined in (4.4.28) satisfies the variation of constants formula

$$z(t) = e^{(t-a)A}z(a) + \int_a^t e^{(t-s)A}h(s)ds, \; t \geq a, \quad (4.4.31)$$

for each $a \in \mathbb{R}$: therefore, z is a strong solution of (4.4.25). Moreover, it is T-periodic. We already know that the T-periodic strong solution of (4.4.25) is unique, so that z is the unique T-periodic strong solution of (4.4.25).

To show part (ii), we recall that a strong solution z of (4.4.25) is T-periodic if and only if (4.4.27) holds. On the other hand, (4.4.27) is equivalent to

$$\begin{cases} 0 = P_j(I - e^{TA})z(t) = P_j \int_t^{t+T} e^{(t+T-s)A}h(s)ds, \; t \in \mathbb{R}, j = 1, ..., N; \\ (I-P)(I - e^{TA})z(t) = (I-P)\int_t^{t+T} e^{(t+T-s)A}h(s)ds, \; t \in \mathbb{R}, \end{cases}$$

---
[1]We recall that if $\rho(A) \supset \{2k\pi i/T : k \in \mathbb{Z}\}$, then $1 \in \rho(e^{TA})$, and $(I - e^{TA})^{-1}$ is given by formula (2.3.17).

4. THE NONHOMOGENEOUS PROBLEM IN UNBOUNDED TIME INTERVALS

which are equivalent, due to (2.3.15) and to Proposition 2.3.9, to

$$\begin{cases} (a) & \int_t^{t+T} e^{-2\pi k_j is/T} P_j h(s) ds = 0, \quad t \in \mathbb{R}, \; j = 1, ..., N; \\ (b) & (I - P)z(t) = \\ & \left(I - e^{TA}\Big|_{(I-P)(X)}\right)^{-1} \int_t^{t+T} e^{(t+T-s)A}(I - P)h(s) ds, \quad t \in \mathbb{R}. \end{cases}$$
(4.4.32)

As easily seen, (4.4.32)(a) is equivalent to (4.4.29). If (4.4.29) holds, we get the representation formula

$$z(t) = e^{tA} Pz(0) + \int_0^t e^{(t-s)A} Ph(s) ds + (I - P)z(t), \quad t \in \mathbb{R},$$

which, thanks to (2.3.15) and to (4.4.32)(b), gives (4.4.30). Therefore, if problem (4.4.25) has a T-periodic strong solution z, then (4.4.29) holds, and z is one of the functions defined by formula (4.4.30). Conversely, it is not difficult to see that if (4.4.29) holds, then the functions defined in (4.4.30) are T-periodic, and satisfy the variation of constants formula (4.4.31), so that they are strong solutions to problem (4.4.25). ■

Let us state some further regularity results for the periodic solutions of (4.4.25). They are immediate consequences of the regularity results of Sections 4.2 and 4.3.

Proposition 4.4.9 *Let $h \in C(\mathbb{R}; X)$ be a T-periodic function, and let $z : \mathbb{R} \mapsto X$ be a T-periodic strong solution of (4.4.25). Then*

(i) *$z \in C^{1-\alpha}(\mathbb{R}; D_A(\alpha, 1))$ for each $\alpha \in \,]0,1[$, and there is $C > 0$ such that*

$$\|z\|_{C^{1-\alpha}(\mathbb{R}; D_A(\alpha,1))} \leq C(\|h\|_{C(\mathbb{R};X)} + \|z\|_{C(\mathbb{R};X)});$$

(ii) *if $h \in C^\alpha(\mathbb{R}; X)$ for some $\alpha \in \,]0,1[$, then z is a strict solution to (4.4.25), it belongs to $C^\alpha(\mathbb{R}; D(A)) \cap C^{1+\alpha}(\mathbb{R}; X)$, and there is $C > 0$ such that*

$$\|z\|_{C^\alpha(\mathbb{R}; D(A))} + \|z\|_{C^{1+\alpha}(\mathbb{R};X)} \leq C(\|h\|_{C^\alpha(\mathbb{R};X)} + \|z\|_{C(\mathbb{R};X)});$$

(iii) *if $h \in C(\mathbb{R}; X) \cap B(\mathbb{R}; D_A(\alpha, \infty))$ for some $\alpha \in \,]0,1[$, then z is a strict solution to (4.4.25), moreover Az and z' belong to $B(\mathbb{R}; D_A(\alpha, \infty))$. There is $C > 0$ such that*

$$\|z\|_{B(\mathbb{R}; D_A(\alpha+1,\infty))} + \|z'\|_{B(\mathbb{R}; D_A(\alpha,\infty))} \leq C(\|h\|_{B(\mathbb{R}; D_A(\alpha,\infty))} + \|z\|_{C(\mathbb{R};X)}).$$

If in addition $h \in C(\mathbb{R}; D_A(\alpha, \infty))$ (respectively, $h \in C(\mathbb{R}; D_A(\alpha))$) then Az and z' belong to $C(\mathbb{R}; D_A(\alpha, \infty))$ (respectively, to $C(\mathbb{R}; D_A(\alpha)))$.

If $1 \in \rho(e^{TA})$ then the term $\|z\|_{C(\mathbb{R};X)}$ may be dropped from the above estimates.

4.4.4 Exponentially decaying and exponentially growing solutions

The results of subsections 4.3.1, 4.3.2, 4.3.3 can be applied to study existence and properties of exponentially decaying solutions of problems (4.0.1), (4.4.20) and (4.4.25).

The spaces of functions which will be considered here are those defined at the beginning of the Section.

Assumption (4.4.1) is now replaced by

$$\sigma(A) \cap \{\lambda \in \mathbb{C} : \operatorname{Re}\lambda = \omega\} = \emptyset, \tag{4.4.33}$$

where $\omega \in \mathbb{R}$. Let $\sigma_+^\omega = \{\lambda \in \sigma(A) : \operatorname{Re}\lambda > \omega\}$, $\sigma_-^\omega = \{\lambda \in \sigma(A) : \operatorname{Re}\lambda < \omega\}$,, let γ be a regular curve around σ_+^ω, with support in the halfplane $\operatorname{Re}\lambda > \omega$, and let

$$P = \frac{1}{2\pi i} \int_\gamma R(\lambda, A) d\lambda$$

be the projection associated to the spectral set σ_+^ω.

Proposition 4.4.10 *Let $\omega \in \mathbb{R}$ be such that (4.4.33) holds, and let $u_0 \in X$, $f \in L^1(0, a; X) \cap L^\infty(a, +\infty; X, \omega)$, for some $a > 0$. Then the mild solution u of problem (4.0.1) belongs to $L^\infty([0, +\infty[; X, \omega)$ if and only if (4.4.9) holds. In addition,*

(i) *if $f \in C^\alpha([a, +\infty[; X, \omega)$ with $\alpha \in\,]0, 1[$, then for every $\varepsilon > 0$ u' and Au belong to $C^\alpha([a + \varepsilon, +\infty[; X, \omega)$, u' belongs to $B([a + \varepsilon, +\infty[; D_A(\alpha, \infty), \omega)$, and there is $C = C(\varepsilon) > 0$ such that*

$$\|u'\|_{C^\alpha([a+\varepsilon,+\infty[;X,\omega)} + \|Au\|_{C^\alpha([a+\varepsilon,+\infty[;X,\omega)} + \|u'\|_{B([a+\varepsilon,+\infty[;D_A(\alpha,\infty),\omega)}$$
$$\leq C(\|u_0\| + \|f\|_{L^1(0,a;X)} + \|f\|_{C^\alpha([a,+\infty[;X,\omega)}); \tag{4.4.34}$$

(ii) *if $f \in B([a, +\infty[; D_A(\alpha, \infty), \omega)$ with $\alpha \in\,]0, 1[$, then for every $\varepsilon > 0$, u' and Au belong to $B([a + \varepsilon, +\infty[; D_A(\alpha, \infty), \omega)$, Au belongs to $C^\alpha([a + \varepsilon, +\infty[; X, \omega)$, and there is $C = C(\varepsilon) > 0$ such that*

$$\|u'\|_{B([a+\varepsilon,+\infty[;D_A(\alpha,\infty),\omega)} + \|Au\|_{B([a+\varepsilon,+\infty[;D_A(\alpha,\infty),\omega)}$$
$$+ \|Au\|_{C^\alpha([a+\varepsilon,+\infty[,X,\omega)} \tag{4.4.35}$$
$$\leq C(\|u_0\| + \|f\|_{L^1(0,a;X)} + \|f\|_{B([a,+\infty[;D_A(\alpha,\infty),\omega)}).$$

Proof — Let u be the mild solution of (4.0.1). Set $g(t) = e^{-\omega t} f(t)$, and $v(t) = e^{-\omega t} u(t)$. Due to the representation formula (4.0.2), v is the mild solution of problem

$$v'(t) = (A - \omega I)v(t) + g(t), \quad t \geq 0, \quad v(0) = u_0. \tag{4.4.36}$$

The operator $A - \omega I : D(A) \mapsto X$ is obviously sectorial, and it satisfies assumption (4.4.1). Therefore Theorem 4.4.3 may be applied to problem (4.4.36), and the statements follow. ∎

4. THE NONHOMOGENEOUS PROBLEM IN UNBOUNDED TIME INTERVALS   169

Remark 4.4.11 Condition (4.4.33) is obviously satisfied by every $\omega > \omega_A$, ω_A being the type of A. In the case where $\omega_A < 0$, choosing $\omega \in]\omega_A, 0[$ we get $P = 0$, so that (4.4.9) holds for every $u_0 \in X$ and $f \in L^1(0, a; X) \cap L^\infty(a, +\infty; X, \omega)$. Then the mild solution u of problem (4.0.1) decays exponentially as $t \to \infty$.

Proposition 4.4.12 Let $\omega \in \mathbb{R}$ be such that (4.4.33) holds, and let $v_0 \in X$, $g \in C(]-\infty, 0]; X, \omega)$. Then problem (4.4.20) has a mild solution v belonging to $C(]-\infty, 0]; X, \omega)$ if and only if (4.4.21) holds. In addition,

(i) if $g \in C^\alpha(]-\infty, 0]; X, \omega)$ for some $\alpha \in]0, 1[$, then v' and Av belong to $C^\alpha(]-\infty, 0]; X, \omega)$, v' belongs to $B(]-\infty, 0]; D_A(\alpha, \infty), \omega)$, and there is $C > 0$ such that

$$\|v'\|_{C^\alpha(]-\infty,0];X,\omega)} + \|Av\|_{C^\alpha(]-\infty,0];X,\omega)} \\ + \|v'\|_{B(]-\infty,0];D_A(\alpha,\infty),\omega)} \leq C(\|v_0\| + \|g\|_{C^\alpha(]-\infty,0];X,\omega)}); \quad (4.4.37)$$

(ii) if $g \in C(]-\infty, 0]; X) \cap B(]-\infty, 0]; D_A(\alpha, \infty), \omega)$ (respectively, $g \in C(]-\infty, 0]; D_A(\alpha, \infty), \omega)$, $g \in C(]-\infty, 0]; D_A(\alpha), \omega))$ for some $\alpha \in]0, 1[$, then v' and Av belong to $B(]-\infty, 0]; D_A(\alpha, \infty), \omega)$ (respectively, to $C(]-\infty, 0]; D_A(\alpha, \infty), \omega)$, $C(]-\infty, 0]; D_A(\alpha), \omega))$, Av belongs to $C^\alpha(]-\infty, 0]; X, \omega)$, and there is $C > 0$ such that

$$\|v'\|_{B(]-\infty,0];D_A(\alpha,\infty),\omega)} + \|Av\|_{B(]-\infty,0];D_A(\alpha,\infty),\omega)} \\ + \|Av\|_{C^\alpha(]-\infty,0];X,\omega)} \leq C(\|v_0\| + \|g\|_{B(]-\infty,0];D_A(\alpha,\infty),\omega)}). \quad (4.4.38)$$

Proof — It is sufficient to argue as in the proof of Proposition 4.4.10, applying Theorem 4.4.6 instead of Theorem 4.4.3. ∎

Proposition 4.4.13 Let $\omega \in \mathbb{R}$ be such that (4.4.33) holds, and let $h \in C(\mathbb{R}; X, \omega)$. Then problem (4.4.25) has a unique mild solution z belonging to $C(\mathbb{R}; X, \omega)$, given by the representation formula (4.4.26). In addition,

(i) if $h \in C^\alpha(\mathbb{R}; X, \omega)$ for some $\alpha \in]0, 1[$, then u' and Au belong to $C^\alpha(\mathbb{R}; X, \omega)$;

(ii) if $h \in C(\mathbb{R}; X, \omega) \cap B(\mathbb{R}; D_A(\alpha, \infty), \omega)$ (respectively, $h \in C(\mathbb{R}; D_A(\alpha, \infty), \omega)$, $h \in C(\mathbb{R}; D_A(\alpha, \omega))$ for some $\alpha \in]0, 1[$, then u' and Au belong to $B(\mathbb{R}; D_A(\alpha, \infty), \omega)$ (respectively, to $C(\mathbb{R}; D_A(\alpha, \infty), \omega)$, $C(\mathbb{R}; D_A(\alpha), \omega))$.

Proof — Again, it is sufficient to argue as in the proof of Proposition 4.4.10, applying Theorem 4.4.7 instead of Theorem 4.4.3. ∎

4.5 Bibliographical remarks

The definitions of strict, classical, strong, and mild solutions given in Section 4.1 are not used by all authors. Somebody calls "strong solution" our strict solution ([166]), or "solution" our classical solution ([99]), and so on.

The old literature (e.g., [107, 213, 166]) deals only with the case of dense domain $D(A)$. However, the extension to the case of nondense domain looks quite natural.

Proposition 4.1.2 is well known, while Corollary 4.1.3 is new. Proposition 4.1.5 may be found in G. DA PRATO – E. SINESTRARI [68]. Lemma 4.1.6 is known, at least in the case where $D(A)$ is dense, but the present proof seems to be new.

The counterexample 4.1.7 is adapted from G. DA PRATO – P. GRISVARD [63]. Other counterexamples are left as exercises in the book of P. CLÉMENT ET AL. [54, pages 157 – 158]. Concerning optimal regularity in $C([0,T]; X)$, it was proved by J.B. BAILLON [28] that if $C([0,T]; X)$ enjoys the optimal regularity property, then X contains a subspace isomorphic to c_0 (the space of all sequences of real numbers $\{x_n\}$ going to 0 as $n \to \infty$, endowed with the sup norm). In particular, X cannot be reflexive. An extended proof of Baillon's result may be found in [78]. See also [198].

Most of the results of Section 4.2 about mild solutions are old and well known, with the exception perhaps of Proposition 4.2.3(ii)(iii) and of Proposition 4.2.5.

On the contrary, most of the results of Section 4.3 about classical and strict solutions, and optimal regularity, have been proved rather recently. Theorem 4.3.1(i), under the assumption that $D(A)$ is dense in X, may be found in the book of T. KATO [107, Ch. IX, Sect. 1.7]. The extension to the nondense domain case is trivial. Statements (ii) and (iii) of Theorem 4.3.1 were proved by E. SINESTRARI in [177]. Moreover, in the case where $u_0 = 0$, $f(0) = 0$, the fact that $u \in C^{\alpha}([0,T]; D(A))$ was proved by different methods by G. DA PRATO – P. GRISVARD [62], and is implicit in Kato's proof of statement (i). Corollary 4.2.4 is due to E. SINESTRARI [177]. Theorem 4.3.4 is classical in the case where $D(A)$ is dense, see e.g. R. H. MARTIN [158, Prop. 4.4].

Part (ii) of Corollary 4.3.6, which will be used in the study of fully nonlinear problems, was stated without proof by P. E. SOBOLEVSKIĬ [182], then it was forgotten. Statements (ii) and (iii), as well as statement (ii) of Theorem 4.3.7, were essentially proved by P. ACQUISTAPACE – B. TERRENI [8], in the more general context of nonautonomous equations of the type $u'(t) = A(t)u(t) + f(t)$. The spaces $C^{\alpha}_{\alpha+\mu}$ are called $Z_{\mu,\alpha}$ in [8], and an equivalent norm — namely, $|f| = \sup_{0<t\leq T} \|t^{\mu} f(t)\| + \sup_{0<\varepsilon<T} \varepsilon^{\alpha+\mu} [f]_{C^{\alpha}([\varepsilon/2,\varepsilon];X)}$ — is considered. A variant of statement (iii) was shown by A. LUNARDI [141] and it was applied to fully nonlinear problems. However, we will not need it in Chapters 9 and 10.

Concerning the space regularity results, Theorem 4.3.8 and the subsequent Corollary 4.3.9 were proved by E. SINESTRARI in [177], with the exception of statements (ii) of Theorem 4.3.8 and (iv) of Corollary 4.3.9, which may be found

5. BIBLIOGRAPHICAL REMARKS

in A. LUNARDI [152]. Corollary 4.3.10 was first shown by a different method in G. DA PRATO – P. GRISVARD [63]. We have followed the simpler approach of E. SINESTRARI [177]. It must be said, however, that the paper [63] was the starting point of all subsequent investigations about optimal space regularity. The method of [63] was followed by S. ANGENENT [25] who showed a part of Theorem 4.3.12 (precisely, he considered the case where f is continuous for $t > 0$ with values in $D_A(\alpha)$, $\lim_{t \to 0} \|t^\alpha f(t)\|_{D_A(\alpha,\infty)} = 0$, and $u_0 \in D(A)$, when $D(A)$ is dense in X). Our proof is simpler.

Statements (iii) of Corollary 4.3.13 and (ii) of Corollary 4.3.15 were proved in the above mentioned paper [8].

The remaining space regularity results of Subsection 4.3.2 (e.g. Theorem 4.3.11, the most part of Theorem 4.3.12 and of Corollary 4.3.13, and statement (i) of Corollary 4.3.15) are new.

Finally, the further regularity result of Theorem 4.3.16 was proved in A. LUNARDI – E. SINESTRARI – W. VON WAHL [156].

Many of the asymptotic behavior results of Section 4.4 are implicit in the book of D. HENRY [99]. With the exception of Proposition 4.4.4, they were proved in the present form by A. LUNARDI [143] for nonautonomous equations with constant domains. The case of time depending domains was studied by M. FUHRMAN [87]. A part of Proposition 4.4.4 was shown in [152].

Chapter 5

Linear parabolic problems

This chapter contains several applications of the abstract results of Chapter 4 to parabolic initial (boundary) value problems in cylindrical domains $[0,T] \times \mathbb{R}^n$, $[0,T] \times \overline{\Omega}$, where Ω is an open set in $\mathbb{R}^n$ with regular boundary. Applying abstract results to "concrete" parabolic problems is possible thanks to the generation results and to the characterizations of interpolation spaces of Chapter 3.

Instead of giving a systematic treatment we select some of the results which, in our opinion, are more interesting and useful. They may serve as a guideline in other applications. In particular, we choose $X = C(\overline{\Omega})$, the space of all continuous and bounded functions in $\overline{\Omega}$. To save room, we do not consider the case $T = +\infty$.

We focus our attention on continuity and Hölder continuity results, with particular emphasis to regularity up to $t = 0$. Let us explain what we mean by a simple example,

$$\begin{cases} u_t(t,x) = \Delta u(t,x) + f(t,x), & 0 < t \leq T, \ x \in \mathbb{R}^n, \\ u(0,x) = u_0(x), & x \in \mathbb{R}^n. \end{cases} \quad (5.0.1)$$

Here Δ is the Laplace operator: $\Delta u = \sum_{i=1}^{n} u_{x_i x_i}$. The function f is uniformly continuous and bounded with respect to (t,x), and u_0 is continuous and bounded. These assumptions imply that problem (5.0.1) has a solution in a weak sense. In general, such a solution is not differentiable with respect to time, nor twice differentiable with respect to the space variables x. However, if f is uniformly Hölder continuous either with respect to t or with respect to x, then u_t and Δu exist continuous for $t > 0$.

We give conditions on u_0 in order that the solution enjoys certain regularity properties up to $t = 0$. For instance, we show that

(i) if u_0 is continuous and bounded, then u is continuous and bounded, and $u(t,\cdot)$ converges to u_0 uniformly on every compact set $K \subset \mathbb{R}^n$ as $t \to 0$;

(ii) if u_0 is uniformly continuous and bounded, then $u(t,\cdot)$ converges to u_0 uniformly on $\mathbb{R}^n$ as $t \to 0$;

(iii) if $u_0 \in C^{2\theta}(\mathbb{R}^n)$, with $0 < \theta < 1$, $\theta \neq 1/2$, then $u \in C^{\theta,2\theta}([0,T] \times \mathbb{R}^n)$[1] ;

(iv) if u_0 is differentiable, with continuous and bounded first order derivatives, then the space derivatives u_{x_i} are continuous and bounded in $[0,T] \times \mathbb{R}^n$.

Special attention is paid to the optimal regularity results, i.e. to the situations where u_t and Δu enjoy the same regularity properties as f. Three types of optimal regularity are considered: time Hölder regularity, space Hölder regularity, and time-space Hölder regularity. Precisely, we show that

(v) if f is α-Hölder continuous with respect to t, uniformly with respect to x, with $0 < \alpha < 1$, and $u_0 \in D(\Delta)$ (that is, $u_0 \in W^{2,p}_{loc}(\mathbb{R}^n)$ for every $p \geq 1$, u_0, Δu_0 are continuous and bounded), and $\Delta u_0 + f(0,\cdot)$ belongs to $C^{2\alpha}(\mathbb{R}^n)$, then u_t and Δu are α-Hölder continuous with respect to t, uniformly with respect to x;

(vi) if f is α-Hölder continuous with respect to x, uniformly with respect to t, with $0 < \alpha < 1$, and and $u_0 \in C^{2+\alpha}(\mathbb{R}^n)$, then u_t and Δu are α-Hölder continuous with respect to x, uniformly with respect to t; in addition all the derivatives $u_{x_i x_j}$, $i,j = 1, \ldots, n$, are α-Hölder continuous with respect to x, uniformly with respect to t;

(vii) if $f \in C^{\alpha/2,\alpha}([0,T] \times \mathbb{R}^n)$, with $0 < \alpha < 1$, and $u_0 \in C^{2+\alpha}(\mathbb{R}^n)$, then $u \in C^{1+\alpha/2, 2+\alpha}([0,T] \times \mathbb{R}^n)$[2]. In particular, u_t and $u_{x_i x_j}$, $i, j = 1, \ldots, n$, belong to $C^{\alpha/2,\alpha}([0,T] \times \mathbb{R}^n)$.

Most of the results (i) through (vii) are extended to the case where the Laplace operator Δ is replaced by any uniformly elliptic operator with uniformly continuous and bounded coefficients, and to the case of initial boundary value problems in $[0,T] \times \overline{\Omega}$, where Ω is an open set in $\mathbb{R}^n$ with regular boundary $\partial\Omega$,

$$\begin{cases} u_t(t,x) = \sum_{i,j=1}^n a_{ij}(t,x) u_{x_i x_j}(t,x) + \sum_{i=1}^n b_i(t,x) u_{x_i}(t,x) \\ \qquad\qquad + c(t,x) u(t,x), \quad 0 \leq t \leq T, \ x \in \overline{\Omega}, \\ u(0,x) = u_0(x), \quad x \in \overline{\Omega}, \end{cases}$$

supported either with Dirichlet boundary condition

$$u(t,x) = g(t,x), \quad 0 \leq t \leq T, \ x \in \partial\Omega,$$

---

[1] $C^{\theta,2\theta}([0,T] \times \mathbb{R}^n)$ is the usual parabolic Hölder space. See the definition at the beginning of Section 5.1.

[2] Also the space $C^{1+\alpha/2, 2+\alpha}([0,T] \times \mathbb{R}^n)$ is very common in the study of parabolic equations; it is defined at the beginning of Section 5.1.

1. Second order equations

or with oblique derivative boundary condition

$$\sum_{i=1}^{n} \beta_i(t,x)u_{x_i}(t,x) + \gamma(t,x)u(t,x) = g(t,x), \ 0 \leq t \leq T, \ x \in \partial\Omega.$$

Section 5.1 deals with second order problems. We consider only single equations, without nonlocal terms. However, we avoid as far as possible the use of the maximum principle and of other techniques which work only in the context of second order equations, so that many results may be generalized without any difficulty to higher order equations, or to nonlocal equations and systems.

In Section 5.2 bibliographical references are given, also for higher order problems.

5.1 Second order equations

Throughout the section, we use the notation of Section 3.1. Then Ω is either $\mathbb{R}^n$ or an open set in $\mathbb{R}^n$ with uniformly C^2 boundary, $\mathcal{A} = \mathcal{A}(x,D) = \sum_{i,j=1}^{n} a_{ij}(x)D_{ij} + \sum_{i=1}^{n} b_i(x)D_i + c(x)$ is a second order elliptic operator with uniformly continuous and bounded coefficients in $\overline{\Omega}$, and, if $\Omega \neq \mathbb{R}^n$, $\mathcal{B} = \mathcal{B}(x,D) = \sum_{i=1}^{n} \beta_i(x)D_i + \gamma(\dot{x})$ is a first order differential operator acting on the boundary $\partial\Omega$, satisfying a nontangentiality condition. The precise assumptions on $\mathcal{A}$ and $\mathcal{B}$, unless otherwise specified, are those stated at the beginning of Section 3.1. We are going to apply some of the results of Chapter 4 to parabolic initial (boundary) value problems in $[0,T] \times \Omega$, which is possible thanks to the generation results of Chapter 3.

We deal with functions u depending on time $t \in [a,b] \subset [0,T]$ and space variables $x = (x_1, \ldots, x_n) \in \overline{\Omega}$. We denote by $D_i u$ or u_{x_i} (respectively, $D_{ij}u$ or $u_{x_i x_j}$) the first (respectively, second) order derivatives of u with respect to space variables, and by $D_t u$ or u_t, the derivative with respect to time. Du denotes the gradient with respect to the space variables, $D^2 u$ denotes the matrix of the second order derivatives of u with respect to the space variables.

We recall that $C([a,b] \times \overline{\Omega})$ (respectively, $UC([a,b] \times \overline{\Omega})$) is the space of all the continuous (respectively, uniformly continuous) and bounded functions in $[a,b] \times \overline{\Omega}$, endowed with the sup norm $\|\cdot\|_\infty$. We introduce spaces of continuous functions that are Hölder continuous or continuously differentiable either with respect to time or with respect to the space variables. For $\alpha > 0$ and $a < b$ we set

$$C^{\alpha,0}([a,b] \times \overline{\Omega}) = \{f \in C([a,b] \times \overline{\Omega}) : f(\cdot,x) \in C^\alpha([a,b]) \ \forall x \in \overline{\Omega},$$
$$\|f\|_{C^{\alpha,0}} = \sup_{x \in \overline{\Omega}} \|f(\cdot,x)\|_{C^\alpha([a,b])} < \infty\},$$

and, similarly,

$$C^{0,\alpha}([a,b] \times \overline{\Omega}) = \{f \in C([a,b] \times \overline{\Omega}) : f(t,\cdot) \in C^\alpha(\overline{\Omega}) \ \forall t \in [a,b],$$
$$\|f\|_{C^{0,\alpha}} = \sup_{a \leq t \leq b} \|f(t,\cdot)\|_{C^\alpha(\overline{\Omega})} < \infty\}.$$

The classical space where one looks for solutions is $C^{1,2}([a,b] \times \overline{\Omega})$, defined by

$$C^{1,2}([a,b] \times \overline{\Omega}) = \{f \in C([a,b] \times \overline{\Omega}) : \exists D_t f, \ D_{ij} f \in C([a,b] \times \overline{\Omega}), \ i,j = 1,\ldots,n\},$$

$$\|f\|_{C^{1,2}([a,b] \times \overline{\Omega})} = \|f\|_\infty + \sum_{i=1}^n \|D_i f\|_\infty + \|f_t\|_\infty + \sum_{i,j=1}^n \|D_{ij} f\|_\infty.$$

We shall consider also more regular solutions. For $0 < \alpha < 1$ we set

$$C^{1,2+\alpha}([a,b] \times \overline{\Omega}) = \{f \in C^{1,2}([a,b] \times \overline{\Omega}) : D_t f, \ D_{ij} f \in C^{0,\alpha}([a,b] \times \overline{\Omega}), \ \forall i,j\},$$

$$\|f\|_{C^{1,2+\alpha}([a,b] \times \overline{\Omega})} = \|f\|_\infty + \sum_{i=1}^n \|D_i f\|_\infty + \|f_t\|_{C^{0,\alpha}} + \sum_{i,j=1}^n \|D_{ij} f\|_{C^{0,\alpha}}.$$

The following interpolatory inclusions will be useful to point out further regularity properties of the solutions.

Lemma 5.1.1 *Let $0 < \alpha < 2$. If f belongs to $C^{1,2+\alpha}([a,b] \times \overline{\Omega})$, then the function $t \mapsto \tilde{f}(t) = f(t,\cdot)$ belongs to $C^{1/2}([a,b]; C^{1+\alpha}(\overline{\Omega}))$ and to $Lip([a,b]; C^\alpha(\overline{\Omega}))$, and there is $K_\alpha > 0$ not depending on $b-a$, f, such that*

$$\|\tilde{f}\|_{C^{1/2}([a,b];C^{1+\alpha}(\overline{\Omega}))} + \|\tilde{f}\|_{Lip([a,b];C^\alpha(\overline{\Omega}))} \leq K_\alpha \|f\|_{C^{1,2+\alpha}([a,b] \times \overline{\Omega})}. \tag{5.1.1}$$

Moreover, the second order space derivatives $D_{ij}f$ belong to $C^{\alpha/2,0}([a,b] \times \overline{\Omega})$, and there is $C_\alpha > 0$ (not depending on $b-a$, f), such that

$$\sum_{i,j=1}^n \|D_{ij}f\|_{C^{\alpha/2,0}([a,b] \times \overline{\Omega})} \leq C_\alpha \|f\|_{C^{1,2+\alpha}([a,b] \times \overline{\Omega})}. \tag{5.1.2}$$

If $0 < \alpha < 1$, then the first order space derivatives $D_i f$ belong to $C^{(1+\alpha)/2,0}([a,b] \times \overline{\Omega})$, and there is $C'_\alpha > 0$, not depending on $b-a$, f, such that

$$\sum_{i=1}^n \|D_i f\|_{C^{(1+\alpha)/2,0}([a,b] \times \overline{\Omega})} \leq C'_\alpha \|f\|_{C^{1,2+\alpha}([a,b] \times \overline{\Omega})}. \tag{5.1.3}$$

Proof — Let $f \in C^{1,2+\alpha}([a,b] \times \overline{\Omega})$. Then $f_t(t,\cdot)$ belongs to $C^\alpha(\overline{\Omega})$ for every t, and its C^α norm is bounded by a constant independent of t, so that the function $\tilde{f}$ defined above belongs to $Lip([a,b]; C^\alpha(\overline{\Omega}))$. Moreover, $\tilde{f}$ is obviously bounded with values in $C^{2+\alpha}(\overline{\Omega})$. Due to Proposition 1.1.3(ii)(iii), the spaces $C^{1+\alpha}(\overline{\Omega})$, $C^2(\overline{\Omega})$, belong respectively to the classes $J_{1/2}$, $J_{(2-\alpha)/2}$ between $C^\alpha(\overline{\Omega})$ and $C^{2+\alpha}(\overline{\Omega})$. If $0 < \alpha < 1$, then the space $C^1(\overline{\Omega})$ belongs to the class $J_{(1-\alpha)/2}$ between $C^\alpha(\overline{\Omega})$ and $C^{2+\alpha}(\overline{\Omega})$. From Proposition 1.1.4, it follows that $\tilde{f}$ belongs to $C^{1/2}([a,b]; C^{1+\alpha}(\overline{\Omega})) \cap C^{\alpha/2}([a,b]; C^2(\overline{\Omega}))$, with

$$\|\tilde{f}\|_{C^{1/2}([a,b];C^{1+\alpha}(\overline{\Omega}))} + \|\tilde{f}\|_{C^{\alpha/2}([a,b];C^2(\overline{\Omega}))}$$
$$\leq c(\|\tilde{f}\|_{B([a,b];C^{2+\alpha}(\overline{\Omega}))} + \|\tilde{f}\|_{Lip([a,b];C^\alpha(\overline{\Omega}))}),$$

1. SECOND ORDER EQUATIONS

where c does not depend on $\tilde{f}$ and $b - a$. This implies that (5.1.1) and (5.1.2) hold. If $0 < \alpha < 1$, again from Proposition 1.1.4, it follows that $\tilde{f}$ belongs to $C^{(1+\alpha)/2}([a,b]; C^1(\overline{\Omega}))$, and

$$\|\tilde{f}\|_{C^{(1+\alpha)/2}([a,b];C^1(\overline{\Omega}))} \leq c(\|\tilde{f}\|_{B([a,b];C^{2+\alpha}(\overline{\Omega}))} + \|\tilde{f}\|_{Lip([a,b];C^\alpha(\overline{\Omega}))}),$$

with c independent of $\tilde{f}$ and $b - a$. So, (5.1.3) follows. ∎

Now we define the familiar "parabolic" Hölder spaces. For $0 < \alpha < 2$ we set

$$C^{\alpha/2,\alpha}([a,b] \times \overline{\Omega}) = C^{\alpha/2,0}([a,b] \times \overline{\Omega}) \cap C^{0,\alpha}([a,b] \times \overline{\Omega}),$$

$$\|f\|_{C^{\alpha/2,\alpha}([a,b]\times\overline{\Omega})} = \|f\|_{C^{\alpha/2,0}([a,b]\times\overline{\Omega})} + \|f\|_{C^{0,\alpha}([a,b]\times\overline{\Omega})},$$

and

$$C^{1+\alpha/2,2+\alpha}([a,b]\times\overline{\Omega}) = \{f \in C^{1,2}([a,b]\times\overline{\Omega}) : D_t f, D_{ij} f \in C^{\alpha/2,\alpha}([a,b]\times\overline{\Omega}), \forall i,j\}$$

$$\|f\|_{C^{1+\alpha/2,2+\alpha}([a,b]\times\overline{\Omega})} = \|f\|_\infty + \sum_{i=1}^n \|D_i f\|_\infty$$
$$+ \|D_t f\|_{C^{\alpha/2,\alpha}} + \sum_{i,j=1}^n \|D_{ij} f\|_{C^{\alpha/2,\alpha}([a,b]\times\overline{\Omega})}.$$

Lemma 5.1.1 implies that for $0 < \alpha < 1$ the space $C^{1,2+\alpha}([a,b] \times \overline{\Omega})$ is continuously embedded in $C^{(1+\alpha)/2,1+\alpha}([a,b] \times \overline{\Omega})$, with embedding constant independent of $b - a$. Since the space $C^{1+\alpha/2,2+\alpha}([a,b] \times \overline{\Omega})$ is continuously embedded in $C^{1,2+\alpha}([a,b] \times \overline{\Omega})$, then it is continuously embedded in $C^{(1+\alpha)/2,1+\alpha}([a,b] \times \overline{\Omega})$, with embedding constant independent of $b - a$. These facts will be used later.

Where no danger of confusion will arise, we shall drop the symbol $([a,b] \times \overline{\Omega})$. So, we shall write $C^{\alpha/2,\alpha}$ instead of $C^{\alpha/2,\alpha}([a,b] \times \overline{\Omega})$, $C^{1,2+\alpha}$ instead of $C^{1,2+\alpha}([a,b] \times \overline{\Omega})$, and so on.

Similar definitions may be given if $\overline{\Omega}$ is replaced by its boundary $\partial\Omega$. If Y is any of the symbols $C^{0,\alpha}$, $C^{\alpha,0}$, $C^{\alpha/2,\alpha}$, $C^{1,2}$, $C^{1,2+\alpha}$, $C^{1+\alpha/2,2+\alpha}$, we denote by $Y([a,b] \times \partial\Omega)$ the space of the functions f defined in $[a,b] \times \partial\Omega$ such that for every $j \in \mathbb{N}$ the composition $f \circ \varphi_j^{-1}$ belongs to $Y([a,b] \times B(0,1))$ and its norm is bounded by a constant independent of j. Here the functions φ_j are the diffeomorphisms which flatten the boundary, defined at the beginning of Chapter 0. The space $Y([a,b] \times \partial\Omega)$ is endowed with the norm

$$\|f\|_{Y([a,b]\times\partial\Omega)} = \sup_{j\in\mathbb{N}} \|f \circ \varphi_j^{-1}\|_{Y([a,b]\times B(0,1))}.$$

5.1.1 Initial value problems in $[0,T] \times \mathbb{R}^n$

To begin with, we consider the case of time independent coefficients,

$$\begin{cases} u_t(t,x) = \mathcal{A}u(t,x) + f(t,x), & 0 < t \leq T, \; x \in \mathbb{R}^n, \\ u(0,x) = u_0(x), & x \in \mathbb{R}^n, \end{cases} \quad (5.1.4)$$

where $f : [0,T] \times \mathbb{R}^n \mapsto \mathbb{R}$ is a continuous function.

We write problem (5.1.4) as an evolution equation in the space

$$X = C(\mathbb{R}^n),$$

by setting

$$u(t) = u(t,\cdot), \quad f(t) = f(t,\cdot),$$

and

$$\begin{cases} D(A) = \{\varphi \in \bigcap_{p \geq 1} W^{2,p}_{loc}(\mathbb{R}^n) \; : \; \varphi, \; \mathcal{A}\varphi \in C(\mathbb{R}^n)\}, \\ A : D(A) \mapsto C(\mathbb{R}^n), \; A\varphi = \mathcal{A}\varphi. \end{cases}$$

We recall that, due to Corollary 3.1.9(i), the realization A of $\mathcal{A}$ in $C(\mathbb{R}^n)$ is a sectorial operator, and that $D(A) = C^2(\mathbb{R})$ if $n = 1$. Moreover,

$$\overline{D(A)} = UC(\mathbb{R}^n),$$

and thanks to Theorem 3.1.12,

$$D_A(\theta,\infty) = \begin{cases} C^{2\theta}(\mathbb{R}^n), & \text{if } \theta \neq 1/2, \\ C^1(\mathbb{R}^n), & \text{if } \theta = 1/2. \end{cases}$$

A number of regularity results for the solution of (5.1.4) will be obtained by applying the results of Chapter 4 to problem

$$u'(t) = Au(t) + f(t), \quad 0 < t \leq T; \quad u(0) = u_0, \quad (5.1.5)$$

with A, u, f as above. Some of them follow immediately from the abstract results of Chapter 4 and the characterization of the interpolation spaces. Other results require more complicated proofs. For instance, we shall show in Theorem 5.1.2 that u is continuous in $[0,T] \times \mathbb{R}^n$ if u_0 is merely continuous and bounded; since u_0 does not belong necessarily to $\overline{D(A)}$ we cannot prove the continuity of u near $t = 0$ using only Corollary 4.2.4, but we have to use a suitable localization procedure.

Mild solutions

To begin with, we consider data with low regularity properties, which do not imply the existence of a classical solution. Indeed, the function u given by the variation of constants formula below solves (5.1.5) only in a weak sense; it satisfies

1. SECOND ORDER EQUATIONS

pointwise (5.1.5) when the data are more regular. However, the estimates in low norms, which hold even when the data are not very regular, are as important as the estimates in high norms which may be found only in the case of regular data.

Let $f : [0,T] \times \mathbb{R}^n \mapsto \mathbb{R}$ be a continuous function such that $t \mapsto f(t, \cdot)$ belongs to $C([0,T]; C(\mathbb{R}^n))$[3], and let $u_0 \in C(\mathbb{R}^n)$. Set

$$u(t,x) = (e^{tA}u_0)(x) + \int_0^t e^{(t-s)A}f(s,\cdot)ds(x), \quad 0 \le t \le T, \; x \in \mathbb{R}^n. \tag{5.1.6}$$

Theorem 5.1.2 *Let $f : [0,T] \times \mathbb{R}^n \mapsto \mathbb{R}$ be a continuous function such that $t \mapsto f(t,\cdot)$ belongs to $C([0,T]; C(\mathbb{R}^n))$, and let $u_0 \in C(\mathbb{R}^n)$. Then the function u defined by (5.1.6) belongs to $C([0,T] \times \mathbb{R}^n) \cap C^{\theta,2\theta}([\varepsilon,T] \times \mathbb{R}^n)$ for every $\varepsilon \in \,]0,T[$ and $\theta \in \,]0,1[$, and there are $C > 0$, $C(\varepsilon,\theta) > 0$ such that*

$$\|u\|_\infty \le C(\|u_0\|_\infty + \|f\|_\infty), \tag{5.1.7}$$

$$\|u\|_{C^{\theta,2\theta}([\varepsilon,T]\times\mathbb{R}^n)} \le C(\varepsilon,\theta)(\varepsilon^{-\theta}\|u_0\|_\infty + \|f\|_\infty). \tag{5.1.8}$$

In particular, $\|t^\theta u(t,\cdot)\|_{C^{2\theta}(\mathbb{R}^n)}$ is bounded, and taking $\theta = 1/2$, $\|t^{1/2}Du\|_\infty$ is bounded.

Moreover, for every $\varepsilon \in \,]0,T[$ there exists a sequence $\{u_n\}_{n\in\mathbb{N}} : [\varepsilon,T] \times \mathbb{R}^n \mapsto \mathbb{R}$ such that $D_t u_n$, $A u_n$ exist continuous, and $u_n \to u$, $D_t u_n - A u_n - f \to 0$ in $L^\infty([\varepsilon,T] \times \mathbb{R}^n)$. In addition:

(i) *If $u_0 \in UC(\mathbb{R}^n)$, then $u(t,\cdot)$ converges to u_0 uniformly in $\mathbb{R}^n$ as $t \to 0$. Moreover there exists a sequence $\{u_n\}_{n\in\mathbb{N}} \subset UC([0,T] \times \mathbb{R}^n)$ such that $\partial u_n/\partial t$, $A u_n$ exist continuous in $[0,T] \times \mathbb{R}^n$, and $\lim_{n\to\infty}\|u_n - u\|_\infty = 0$, $\lim_{n\to\infty}\|D_t u_n - A u_n - f\|_\infty = 0$.*

(ii) *If $u_0 \in C^{2\theta}(\mathbb{R}^n)$, with $0 < \theta < 1$, then u belongs to $C^{\theta,2\theta}([0,T] \times \mathbb{R}^n)$, and*

$$\|u\|_{C^{\theta,2\theta}([0,T]\times\mathbb{R}^n)} \le C(\|u_0\|_{C^{2\theta}(\mathbb{R}^n)} + \|f\|_\infty).$$

Proof — First we show that the function u defined by (5.1.6) enjoys the regularity properties stated above.

Since $t \mapsto f(t,\cdot)$ is continuous in $[0,T]$ with values in X, then problem (5.1.5) has a mild solution $u : [0,T] \mapsto X$, given by the variation of constants formula (4.0.2). The function $u(t,x)$ defined in (5.1.6) is nothing but $u(t)(x)$.

Corollary 4.2.2 states that $t \mapsto u(t)$ is continuous and bounded in $\,]0,T]$ with values in $X = C(\mathbb{R}^n)$. This implies that $u \in C(]0,T] \times \mathbb{R}^n)$, and estimate (4.2.5) implies that (5.1.7) holds. Corollary 4.2.2 states also that $t \mapsto u(t)$ belongs to $C^\theta([\varepsilon,T]; X) \cap B([\varepsilon,T]; D_A(\theta,1))$ for every $\theta \in \,]0,1[$ and $\varepsilon \in \,]0,T[$, with norm bounded by $c\varepsilon^{-\theta}(\|u_0\|_X + \|f\|_{C([0,T];X)})$. Since $C^{2\theta}(\mathbb{R}^n) = D_A(\theta,\infty) \supset D_A(\theta,1)$ for $\theta \neq 1/2$, and $C^1(\mathbb{R}^n) \supset D_A(1/2,1)$ for $\theta = 1/2$ (see Proposition 3.1.11(i)), then u belongs to $C^{\theta,2\theta}([\varepsilon,T] \times \mathbb{R}^n)$, and estimate (5.1.8) follows.

---
[3] Note that the assumption $t \mapsto f(t,\cdot) \in C([0,T]; C(\overline{\Omega}))$ is stronger than $f \in C([0,T] \times \overline{\Omega})$ whenever Ω is unbounded.

Now we prove that $u(t,x)$ is continuous in $[0,T] \times \mathbb{R}^n$. Note that, in general, $t \mapsto u(t,\cdot)$ is not continuous up to $t=0$ with values in $C(\mathbb{R}^n)$, because u_0 does not belong necessarily to $\overline{D(A)}$. However, we shall show below that $(t,x) \mapsto e^{tA}u_0(x)$ belongs to $C([0,T] \times B(x_0,r))$ for every $x_0 \in \mathbb{R}^n$ and $r > 0$. The other addendum $v(t,x) = \int_0^t e^{(t-s)A}f(s,\cdot)ds(x)$ is more regular, since it belongs to $C^{\theta,2\theta}([0,T] \times \mathbb{R}^n)$ for every $\theta \in \,]0,1[$. Indeed, Proposition 4.2.1 implies that $t \mapsto v(t,\cdot)$ belongs to $C^\theta([0,T];X) \cap B([0,T]; D_A(\theta,\infty))$ for every $\theta \in \,]0,1[$, which means that $v \in C^{\theta,2\theta}([0,T] \times \mathbb{R}^n)$ for every $\theta \in \,]0,1[$.

Thanks to Proposition 4.2.5, for every $\lambda \in \rho(A)$

$$\lim_{t \to 0} \|R(\lambda,A)e^{tA}u_0 - R(\lambda,A)u_0\|_{D_A(\alpha,1)} = 0, \quad 0 < \alpha < 1.$$

In particular, since $D_A(1/2,1)$ is continuously embedded in $C^1(\mathbb{R}^n)$, then

$$\lim_{t \to 0} \|R(\lambda,A)(e^{tA}u_0 - u_0)\|_{C^1(\mathbb{R}^n)} = 0. \tag{5.1.9}$$

For every $x_0 \in \mathbb{R}^n$ and $r > 0$ let θ be a cutoff function such that

$$\theta \in C^\infty(\mathbb{R}^n), \quad \theta \equiv 1 \text{ in } B(x_0,r), \quad \theta \equiv 0 \text{ outside } B(x_0,2r). \tag{5.1.10}$$

Then $\theta e^{tA}u_0$ satisfies

$$D_t(\theta e^{tA}u_0) = A(\theta e^{tA}u_0) - 2\sum_{i,j=1}^n a_{ij}D_i\theta D_j u$$
$$-e^{tA}u_0 \left(\sum_{i,j=1}^n a_{ij}D_{ij}\theta + \sum_{i=1}^n b_i D_i\theta\right) = A(\theta e^{tA}u_0) + \psi, \quad 0 < t \le T, \tag{5.1.11}$$

where $\psi \in L^1(0,T;X)$ thanks to (5.1.8). Moreover, for every $v \in D(A)$ it holds

$$R(\lambda,A)(\theta v) = \theta R(\lambda,A)v + 2R(\lambda,A)\sum_{i,j=1}^n a_{ij}D_i\theta D_j(R(\lambda,A)v)$$
$$+R(\lambda,A)\left[\left(\sum_{i,j=1}^n a_{ij}D_{ij}\theta + \sum_{i=1}^n b_i D_i\theta\right) \cdot R(\lambda,A)v\right]. \tag{5.1.12}$$

(5.1.12) follows easily writing down the equation satisfied by $\theta R(\lambda,A)v$. Taking $v = e^{tA}u_0$, and then letting $t \to 0$ and using (5.1.9), we get

$$\lim_{t \to 0} \|R(\lambda,A)(\theta e^{tA}u_0 - \theta u_0)\|_X = 0.$$

By Corollary 4.1.3 we have

$$\theta e^{tA}u_0 = e^{tA}(\theta u_0) + \int_0^t e^{(t-s)A}\psi(s)ds, \quad 0 \le t \le T. \tag{5.1.13}$$

Since θu_0 has compact support, then it is uniformly continuous. Therefore it belongs to the closure of $D(A)$. From Proposition 2.1.4(i) it follows that $t \mapsto \theta e^{tA}u_0$ is continuous with values in X in $[0,T]$, and in particular $\lim_{t \to 0} \|\theta u(t,\cdot) - \theta u_0\|_\infty = 0$. Since $\theta \equiv 1$ in $B(x_0,r)$, then $(t,x) \mapsto e^{tA}u_0(x)$ is continuous in

1. SECOND ORDER EQUATIONS

$[0,T] \times B(x_0, r)$, and $\lim_{t\to 0} e^{tA}u_0(x) = u_0(x)$ uniformly on $B(x_0, r)$. Since x_0 is arbitrary, it follows that $(t,x) \mapsto e^{tA}u_0(x) \in C([0,T] \times \mathbb{R}^n)$. The first part of the theorem is so proved.

To prove the second part we note that for every $\varepsilon \in \,]0,T[$, $u(\varepsilon, \cdot)$ belongs to $\overline{D(A)}$. Moreover, u satisfies

$$u(t,x) = e^{(t-\varepsilon)A}u(\varepsilon, \cdot)(x) + \int_\varepsilon^t e^{(t-s)A}f(s,\cdot)ds(x), \quad \varepsilon \leq t \leq T,\ x \in \mathbb{R}^n.$$

Proposition 4.1.8, applied in the interval $[\varepsilon, T]$ instead of $[0,T]$, gives the second part of the statement.

Let us prove now (i) and (ii).

Statement (i) follows from Corollary 4.2.4(i) and Proposition 4.1.8, since $u_0 \in UC(\mathbb{R}^n) = \overline{D(A)}$.

If $\theta \neq 1/2$, statement (ii) follows from Corollary 4.2.4. Indeed, since $u_0 \in D_A(\theta, \infty)$, then $t \mapsto u(t, \cdot)$ belongs to $C^\theta([0,T];X) \cap B([0,T]; D_A(\theta, \infty))$, which means that $u \in C^{\theta, 2\theta}([0,T] \times \mathbb{R}^n)$.

If $\theta = 1/2$ and $u_0 \in C^1(\mathbb{R}^n)$, we consider again the representation formula (5.1.6). We know already that the second addendum $(t,x) \mapsto \int_0^t e^{(t-s)A}f(s)ds(x)$ belongs to $C^{\theta, 2\theta}([0,T] \times \mathbb{R}^n)$ for every $\theta \in \,]0,1[$. In particular, its first order space derivatives are continuous and bounded in $[0,T] \times \mathbb{R}^n$. Let us consider now the function $(t,x) \mapsto e^{tA}u_0(x)$. Due to Proposition 3.1.18, the realization of $\mathcal{A}$ in $C^1(\mathbb{R}^n)$, with domain $D_1(A)$ defined in (3.1.44), is sectorial in $C^1(\mathbb{R}^n)$. It follows that $e^{tA}u_0$ is bounded with values in $C^1(\mathbb{R}^n)$, and $\|e^{tA}u_0\|_{C^1(\mathbb{R}^n)} \leq c\|u_0\|_{C^1(\mathbb{R}^n)}$ for $0 \leq t \leq T$. Summing up we find that the derivatives $D_i u$ are bounded in $[0,T] \times \mathbb{R}^n$, and

$$\|D_i u\|_\infty \leq C(\|u_0\|_{C^1(\mathbb{R}^n)} + \|f\|_\infty), \quad i=1,\ldots,n.$$

Let us prove that the derivatives $D_i(e^{tA}u_0)$ are continuous in $[0,T] \times \mathbb{R}^n$. It is sufficient to show that they are continuous in $[0,T] \times B(x_0, r)$, for every $x_0 \in \mathbb{R}^n$ and $r > 0$. Let θ, ψ be the functions defined in (5.1.10), (5.1.11), respectively. Since θu_0 is continuously differentiable and has compact support, then it belongs to $UC^1(\mathbb{R}^n)$, which is the closure of $D_1(A)$ in $C^1(\mathbb{R}^n)$ (see Proposition 3.1.18). It follows that $t \mapsto e^{tA}(\theta u_0)$ belongs to $C([0,T];C^1(\mathbb{R}^n))$. Moreover, since $e^{tA}u_0$ and $D_i(e^{tA}u_0)$ are bounded in $[0,T] \times \mathbb{R}^n$, then $\psi \in L^\infty(0,T;X)$, and from Proposition 4.2.1 we get that $t \mapsto \int_0^t e^{(t-s)A}\psi(s)ds$ is continuous in $[0,T]$ with values in $D_A(\beta, \infty)$ for every β, so that it is continuous with values in $C^1(\mathbb{R}^n)$. Summing up, we see that $t \mapsto \theta e^{tA}u_0$ belongs to $C([0,T];C^1(\mathbb{R}^n))$. Since $\theta e^{tA}u_0(x) \equiv e^{tA}u_0(x)$ in $[0,T] \times B(x_0, r)$, it follows that $t \mapsto e^{tA}u_{0|B(x_0,r)}$ belongs to $C([0,T];C^1(B(x_0,r)))$. Since x_0 is arbitrary, then $(t,x) \mapsto D_i e^{tA}u_0(x)$ is continuous and bounded in $[0,T] \times \mathbb{R}^n$ for every i, which implies that $(t,x) \mapsto e^{tA}u_0(x)$ belongs to $C^{0,1}([0,T] \times \mathbb{R}^n)$.

It remains to show that $(t,x) \mapsto (e^{tA}u_0)(x)$ belongs to $C^{1/2,0}([0,T] \times \mathbb{R}^n)$. To this aim, we recall that $C^1(\mathbb{R}^n)$ is continuously embedded in $D_A(1/2, \infty)$, so that u_0 belongs to $D_A(1/2, \infty)$. Then, by Remark 2.2.5, $t \mapsto e^{tA}u_0$ belongs to $C^{1/2}([0,T];X)$, which means that $(t,x) \mapsto e^{tA}u_0(x) \in C^{1/2,0}([0,T] \times \mathbb{R}^n)$. Therefore, $(t,x) \mapsto e^{tA}u_0(x) \in C^{1/2,1}([0,T] \times \mathbb{R}^n)$, and statement (ii) is proved also for $\theta = 1/2$. ∎

Note that the regularity assumptions on the initial datum u_0 in statements (i) and (ii) are necessary for the solution enjoy the regularity properties stated. Specifically,

(i) If $u(t,\cdot)$ converges uniformly to u_0 as $t \to 0$, then u_0 belongs to the closure of $D(A)$, thanks to Proposition 2.1.4(i).

(ii) If $0 < \theta < 1$ and $u \in C^{\theta,0}([0,T] \times \mathbb{R}^n)$, then $t \mapsto u(t,\cdot)$ belongs to $C^\theta([0,T];X)$. It follows that also $t \mapsto e^{tA}u_0$ belongs to $C^\theta([0,T];X)$, so that, by Remark 2.2.5, u_0 belongs to $D_A(\theta,\infty) = C^{2\theta}(\mathbb{R}^n)$ for $\theta \neq 1/2$.

(iii) If $u \in C^{0,2\theta}([0,T] \times \mathbb{R}^n)$, then $t \mapsto u(t,\cdot)$ belongs to $B([0,T]; C^{2\theta}(\mathbb{R}^n))$, so that u_0 belongs to $C^{2\theta}(\mathbb{R}^n)$.

Regular solutions and optimal regularity

Theorem 5.1.3 *Let $f \in C^{\alpha,0}([0,T] \times \mathbb{R}^n)$, with $\alpha \in\,]0,1[$, and let $u_0 \in C(\mathbb{R}^n)$. Then the mild solution u of problem (5.1.4) is differentiable with respect to t in $]0,T] \times \mathbb{R}^n$, $u(t,\cdot)$ belongs to $W^{2,p}_{loc}(\mathbb{R}^n)$ for every $p \geq 1$, and $u_t, \mathcal{A}u \in C^{\alpha,0}([\varepsilon,T] \times \mathbb{R}^n)$ for $0 < \varepsilon < T$. Moreover, u satisfies (5.1.4), and it is the unique solution of (5.1.4) belonging to $C([0,T] \times \mathbb{R}^n)$ and enjoying the above regularity properties. There is $C > 0$ such that*

$$\sup_{0 < t \leq T} t(\|u_t(t,\cdot)\|_\infty + \|\mathcal{A}u\|_\infty) \leq C(\|u_0\|_\infty + \|f\|_{C^{\alpha,0}([0,T] \times \mathbb{R}^n)}). \tag{5.1.14}$$

Moreover, $u_t(t,\cdot)$ belongs to $C^{2\alpha}(\mathbb{R}^n)$ for every $t \in\,]0,T]$. In particular, for $\alpha \neq 1/2$, $u_t \in C^{\alpha,2\alpha}([\varepsilon,T] \times \mathbb{R}^n)$ for every $\varepsilon \in\,]0,T[$; for $\alpha = 1/2$, $u_t(t,\cdot)$ belongs to $C^1(\mathbb{R}^n)$ for every $t \in\,]0,T]$.

In addition:

(i) *If $u_0 \in D(\mathcal{A})$, then u_t and $\mathcal{A}u$ belong to $C([0,T] \times \mathbb{R}^n)$, and*

$$\begin{aligned}\|u\|_\infty + \|Du\|_\infty + \|u_t\|_\infty + \|\mathcal{A}u\|_\infty \\ \leq C(\|u_0\|_\infty + \|\mathcal{A}u_0\|_\infty + \|f\|_{C^{\alpha,0}([0,T] \times \mathbb{R}^n)});\end{aligned} \tag{5.1.15}$$

(ii) *If $u_0 \in D(\mathcal{A})$ and $\mathcal{A}u_0 + f(0,\cdot) \in UC(\mathbb{R}^n)$, then $\lim_{t \to 0} \mathcal{A}u(t,x) = \mathcal{A}u_0(x)$, uniformly for $x \in \mathbb{R}^n$;*

(iii) *If $\alpha \neq 1/2$, $u_0 \in D(\mathcal{A})$ and $\mathcal{A}u_0 + f(0,\cdot) \in C^{2\alpha}(\mathbb{R}^n)$, then u_t belongs to $C^{\alpha,2\alpha}([0,T] \times \mathbb{R}^n)$, $\mathcal{A}u$ belongs to $C^{\alpha,0}([0,T] \times \mathbb{R}^n)$, and*

$$\begin{aligned}\|u_t\|_{C^{\alpha,2\alpha}([0,T] \times \mathbb{R}^n)} + \|\mathcal{A}u\|_{C^{\alpha,0}([0,T] \times \mathbb{R}^n)} \leq C(\|u_0\|_\infty \\ + \|\mathcal{A}u_0\|_\infty + \|\mathcal{A}u_0 + f(0,\cdot)\|_{C^{2\alpha}(\mathbb{R}^n)} + \|f\|_{C^{\alpha,0}([0,T] \times \mathbb{R}^n)}).\end{aligned} \tag{5.1.16}$$

1. Second order equations

If $u_0 \in D(\mathcal{A})$ and $\mathcal{A}u_0 + f(0,\cdot) \in \mathcal{C}^1(\mathbb{R}^n)$, then u_t and $\mathcal{A}u$ belong to $C^{1/2,0}([0,T] \times \mathbb{R}^n)$, $u_t(t,\cdot)$ belongs to $\mathcal{C}^1(\mathbb{R}^n)$ for every $t \in [0,T]$, and

$$\|u_t\|_{C^{1/2,0}([0,T]\times\mathbb{R}^n)} + \|\mathcal{A}u\|_{C^{1/2,0}([0,T]\times\mathbb{R}^n)} + \sup_{0\leq t\leq T}\|u_t(t,\cdot)\|_{\mathcal{C}^1(\mathbb{R}^n)}$$
$$\leq C(\|u_0\|_\infty + \|\mathcal{A}u_0\|_\infty + \|\mathcal{A}u_0 + f(0,\cdot)\|_{\mathcal{C}^1(\mathbb{R}^n)} + \|f\|_{C^{1/2,0}([0,T]\times\mathbb{R}^n)}).$$

Proof — The function $t \mapsto f(t,\cdot)$ belongs to $C^\alpha([0,T];X)$. Let $u: [0,T] \mapsto X$ be the function defined in (4.0.2). Thanks to Theorem 4.3.1, u is differentiable with values in X, u' and Au belong to $C^\alpha([\varepsilon,T];X)$, u' belongs to $B([\varepsilon,T];D_A(\alpha,\infty))$ for every $\varepsilon \in {]}0,T[$, and $u' = Au + f(t,\cdot)$ for $0 < t \leq T$. This implies that, setting $u(t,x) = u(t)(x)$ for $0 \leq t \leq T$ and $x \in \mathbb{R}^n$, u is differentiable with respect to time at any point $(t,x) \in {]}0,T] \times \mathbb{R}^n$, u_t and $\mathcal{A}u$ belong to $C^{\alpha,0}([\varepsilon,T] \times \mathbb{R}^n)$, and u satisfies the differential equation in (5.1.4). Moreover, for every $t \in {]}0,T]$, $u_t(t,\cdot) \in \mathcal{C}^{2\alpha}(\mathbb{R}^n)$ if $\alpha \neq 1/2$, $u_t(t,\cdot) \in \mathcal{C}^1(\mathbb{R}^n)$ if $\alpha = 1/2$.

Let us prove that estimate (5.1.14) holds. By (2.1.3), for $0 < t \leq T$ we have $\|d/dt\, e^{tA}u_0\|_X \leq ct^{-1}\|u_0\|_X$, whereas by (4.3.11) we get $\|d/dt \int_0^t e^{(t-s)A} f(s)ds\|_X \leq c\|f\|_{C^\alpha([0,T];X)}$ for $0 \leq t \leq T$. Summing up, (5.1.14) follows.

Uniqueness of a solution of (5.1.4) enjoying the regularity properties of u can be proved by adapting the maximum principle to our situation. See Lemma 5.1.6.

Let us prove statements (i), (ii), (iii).

Since $u_0 \in D(A)$, then $Ae^{tA}u_0 = d/dt\, e^{tA}u_0$ is bounded in $[0,T]$ with values in X. Moreover, due to Theorem 4.3.5, $A(e^{tA} * f)$ and $d/dt\,(e^{tA} * f)$ are bounded in $[0,T]$ with values in X. Therefore, $t \mapsto u(t,\cdot)$ is bounded with values in $D(A)$, $t \mapsto u_t(t,\cdot)$ is bounded with values in X, and estimate (5.1.15) follows from (2.1.3), (4.3.11).

To prove the continuity of u_t and $\mathcal{A}u$ we argue as in the proof of Theorem 5.1.2, showing that u_t and $\mathcal{A}u$ are continuous in $[0,T] \times B(x_0,r)$ for every $x_0 \in \mathbb{R}^n$ and $r > 0$. Let θ and ψ be the functions defined in (5.1.10), (5.1.11), respectively. The function $(t,x) \mapsto \theta(x)u(t,x)$ satisfies

$$D_t(\theta u) = \mathcal{A}(\theta u) + \widetilde{\psi},$$

where $\widetilde{\psi} = \psi + \theta f$. Then $\widetilde{\psi}$ belongs to $C^\beta([0,T];X)$, with $\beta = \min\{\alpha, 1/2\}$. Indeed, from Proposition 1.1.4 it follows that $t \mapsto u(t,\cdot)$ belongs to the space $C^{1/2}([0,T];Z)$ for every Banach space $Z \in J_{1/2}(X, D(A))$. In particular, since $\mathcal{C}^1(\mathbb{R}^n)$ belongs to $J_{1/2}(X, D(A))$, then $t \mapsto u(t,\cdot) \in C^{1/2}([0,T]; \mathcal{C}^1(\mathbb{R}^n))$. Moreover θu_0 belongs to $D(A)$, and $A(\theta u_0) + \widetilde{\psi}(0,\cdot)$ belongs to $UC(\mathbb{R}^n)$ because it is continuous and has compact support. By Theorem 4.3.1(ii), $t \mapsto \theta u(t,\cdot)$ belongs to $C([0,T]; D(A)) \cap C^1([0,T];X)$. Since $\theta \equiv 1$ in $B(x_0,r)$, then u_t and $\mathcal{A}u$ are continuous in $[0,T] \times B(x_0,r)$, and statement (i) follows.

Statement (ii) is an immediate consequence of Theorem 4.3.1(iii). Indeed, the condition $\mathcal{A}u_0 + f(0,\cdot) \in UC(\mathbb{R}^n)$ means that $Au_0 + f(0) \in \overline{D(A)}$. By Theorem 4.3.1(ii), $u', Au \in C([0,T];X)$. In particular, $\lim_{t\to 0} u_t(t,\cdot) = \mathcal{A}u_0 + f(0,\cdot)$ in X.

Statement (iii) is a consequence of Theorem 4.3.1(iii). The condition $\mathcal{A}u_0 + f(0,\cdot) \in C^{2\alpha}(\mathbb{R}^n)$ if $\alpha \neq 1/2$, $\mathcal{A}u_0 + f(0,\cdot) \in C^1(\mathbb{R}^n)$ if $\alpha = 1/2$ means that $\mathcal{A}u_0 + f(0) \in D_A(\alpha, \infty)$. Theorem 4.3.1(iii) implies that u', $\mathcal{A}u \in C^\alpha([0,T];X)$ and $u' \in B([0,T]; D_A(\alpha, \infty))$. Note that, for $\alpha \neq 1/2$, $u' \in C^\alpha([0,T];X) \cap B([0,T]; D_A(\alpha, \infty))$ means that $u_t \in C^{\alpha, 2\alpha}([0,T] \times \mathbb{R}^n)$. The statement follows. ∎

Note that the assumption $u_0 \in D(A)$, $\mathcal{A}u_0 + f(0,\cdot) \in C^{2\alpha}(\mathbb{R}^n)$ is necessary for statement (iii) hold. Indeed, if u is uniformly continuous and u_t, $\mathcal{A}u$ belong to $C^{\alpha, 0}([0,T] \times \mathbb{R}^n)$, then $u(t, \cdot)$ belongs to $D(A)$ for every t, and in particular $u_0 \in D(A)$; moreover $t \mapsto u(t, \cdot)$ belongs to $C^{1+\alpha}([0,T]; X) \cap C^\alpha([0,T]; D(A))$. From Remark 4.3.3 we get $\mathcal{A}u_0 + f(0,\cdot) \in D_A(\alpha, \infty)$, which means that $\mathcal{A}u_0 + f(0,\cdot) \in C^{2\alpha}(\mathbb{R}^n)$.

The next theorem deals with the case where f is Hölder continuous with respect to the space variables x.

Theorem 5.1.4 *Let $f \in UC([0,T] \times \mathbb{R}^n) \cap C^{0,\alpha}([0,T] \times \mathbb{R}^n)$, with $\alpha \in {]0,2[}$, $\alpha \neq 1$, and let $u_0 \in C(\mathbb{R}^n)$. Then the mild solution u of problem (5.1.4) is differentiable with respect to t in ${]0,T]} \times \mathbb{R}^n$, $u(t,\cdot)$ belongs to $W^{2,p}_{loc}(\mathbb{R}^n)$ for every $p \geq 1$, and u_t, $\mathcal{A}u \in UC([\varepsilon,T] \times \mathbb{R}^n) \cap C^{0,\alpha}([\varepsilon,T] \times \mathbb{R}^n)$ for $0 < \varepsilon < T$. Moreover, u satisfies pointwise (5.1.4), and it is the unique solution of (5.1.4) belonging to $C([0,T] \times \mathbb{R}^n)$ and enjoying the above regularity properties. There is $C > 0$ such that*

$$\sup_{0 < t \leq T} t(\|u_t(t,\cdot)\|_\infty + \|\mathcal{A}u\|_\infty) \leq C(\|u_0\|_\infty + \|f\|_{C^{0,\alpha}([0,T] \times \mathbb{R}^n)}). \quad (5.1.17)$$

Moreover, $\mathcal{A}u$ belongs to $C^{\alpha/2, \alpha}([\varepsilon, T] \times \mathbb{R}^n)$ for every $\varepsilon \in {]0,T]}$.
In addition:

(i) *If $u_0 \in D(A)$, then u_t and $\mathcal{A}u$ belong to $C([0,T] \times \mathbb{R}^n)$, and*

$$\begin{aligned}\|u\|_\infty + \|Du\|_\infty + \|u_t\|_\infty + \|\mathcal{A}u\|_\infty \\ \leq C(\|u_0\|_\infty + \|\mathcal{A}u_0\|_\infty + \|f\|_{C^{0,\alpha}([0,T] \times \mathbb{R}^n)});\end{aligned} \quad (5.1.18)$$

(ii) *If $u_0 \in D(A)$ and $\mathcal{A}u_0 \in UC(\mathbb{R}^n)$, then $\lim_{t \to 0} \mathcal{A}u(t,x) = \mathcal{A}u_0(x)$, uniformly for $x \in \mathbb{R}^n$;*

(iii) *If $u_0 \in D(A)$ and $\mathcal{A}u_0 \in C^\alpha(\mathbb{R}^n)$, then $\mathcal{A}u$ belongs to $C^{\alpha/2, \alpha}([0,T] \times \mathbb{R}^n)$, u_t belongs to $C^{0,\alpha}([0,T] \times \mathbb{R}^n)$, and*

$$\|u_t\|_{C^{0,\alpha}} + \|\mathcal{A}u\|_{C^{\alpha/2,\alpha}} \leq C(\|u_0\|_\infty + \|\mathcal{A}u_0\|_{C^\alpha(\mathbb{R}^n)} + \|f\|_{C^{0,\alpha}}); \quad (5.1.19)$$

(iv) *If the coefficients a_{ij}, b_i, c belong to $C^\alpha(\mathbb{R}^n)$, then $u \in C^{1,2+\alpha}([\varepsilon, T] \times \mathbb{R}^n)$ for every $\varepsilon \in {]0,T[}$, and*

$$\|u\|_{C^{1,2+\alpha}([\varepsilon,T] \times \mathbb{R}^n)} \leq \frac{C}{\varepsilon^{\alpha/2+1}} \left(\|u_0\|_\infty + \|f\|_{C^{0,\alpha}([0,T] \times \mathbb{R}^n)} \right).$$

1. SECOND ORDER EQUATIONS   185

If also $u_0 \in C^{2+\alpha}(\mathbb{R}^n)$, then $u \in C^{1,2+\alpha}([0,T] \times \mathbb{R}^n)$, and

$$\|u\|_{C^{1,2+\alpha}([0,T]\times\mathbb{R}^n)} \leq C(\|u_0\|_{C^{2+\alpha}(\mathbb{R}^n)} + \|f\|_{C^{0,\alpha}([0,T]\times\mathbb{R}^n)}). \qquad (5.1.20)$$

Proof — Due to the characterization of the interpolation spaces $D_A(\theta,\infty)$, a function φ defined in $[0,T] \times \mathbb{R}^n$ belongs to $UC([0,T] \times \mathbb{R}^n) \cap C^{0,\alpha}([0,T] \times \mathbb{R}^n)$, with $0 < \alpha < 2$, $\alpha \neq 1$, if and only if $t \mapsto \varphi(t,\cdot)$ belongs to $C([0,T];X) \cap B([0,T];D_A(\alpha/2,\infty))$. Therefore, $t \mapsto f(t,\cdot)$ belongs to $C([0,T];X) \cap B([0,T];D_A(\alpha/2,\infty))$. By Corollary 4.3.9 we get that u' and Au belong to $C([\varepsilon,T];X) \cap B([\varepsilon,T];D_A(\alpha/2,\infty))$ for every $\varepsilon \in \,]0,T[$, Au belongs to $C^{\alpha/2}([\varepsilon,T];X)$ and $u'(t) = Au(t) + f(t,\cdot)$ for $0 < t \leq T$. So, setting again $u(t,x) = u(t)(x)$ for $0 \leq t \leq T$ and $x \in \mathbb{R}^n$, then u_t belongs to $UC([\varepsilon,T] \times \mathbb{R}^n) \cap C^{0,\alpha}([\varepsilon,T] \times \mathbb{R}^n)$, Au belongs to $C^{\alpha/2,\alpha}([\varepsilon,T] \times \mathbb{R}^n)$ and u satisfies the differential equation in (5.1.4).

Let us prove that the estimates (5.1.18) hold. By (2.1.3) for $0 < t \leq T$ we have $\|d/dt\, e^{tA}u_0\|_X \leq ct^{-1}\|u_0\|_X$, whereas by (4.3.16) we get $\|d/dt \int_0^t e^{(t-s)A}f(s)ds\|_X \leq c\|f\|_{B([0,T];C^{\alpha}(\mathbb{R}^n))}$ for $0 \leq t \leq T$. Summing up, (5.1.18) follows.

The fact that $u(t,x)$ is continuous in $[0,T] \times \mathbb{R}^n$ and that $u(0,x) = u_0(x)$ follows from Theorem 5.1.2.

The statement about uniqueness follows from the version of the maximum principle stated in next Lemma 5.1.6.

The proof of statement (i) is quite similar to the proof of statement (i) of Theorem 5.1.3, and is omitted. Statements (ii) and (iii) follow from Corollary 4.3.9(ii)(iii), since $UC(\mathbb{R}^n) = \overline{D(A)}$ and $C^{\alpha}(\mathbb{R}^n) = D_A(\alpha/2,\infty)$. Note that Corollary 4.3.9(iii) gives $Au \in C^{\alpha/2}([0,T];X) \cap B([0,T];D_A(\alpha/2,\infty))$, which implies that $Au \in C^{\alpha/2,\alpha}([0,T] \times \mathbb{R}^n)$.

Let us consider the case where the coefficients a_{ij}, b_i, c belong to $C^{\alpha}(\mathbb{R}^n)$. Since u belongs to $C([\varepsilon,T];D(A)) \cap B([\varepsilon,T];D_A(\alpha/2+1,\infty))$ for every $\varepsilon \in \,]0,T[$, then by Proposition 1.1.4(iii) it belongs to $C([\varepsilon,T];D_A(\beta/2+1,\infty))$ for $0 < \beta < \alpha$. Since the coefficients of $\mathcal{A}$ are α-Hölder continuous, by Corollary 3.1.16 we have $D_A(\alpha/2+1,\infty) = C^{2+\alpha}(\mathbb{R}^n)$ and $D_A(\beta/2+1,\infty) = C^{2+\beta}(\mathbb{R}^n)$, so that $t \mapsto u(t,\cdot)$ belongs to $B([\varepsilon,T];C^{2+\alpha}(\mathbb{R}^n)) \cap C([\varepsilon,T];C^{2+\beta}(\mathbb{R}^n))$. Therefore the derivatives $D_{ij}u$ belong to $C^{0,\alpha}([\varepsilon,T] \times \mathbb{R}^n)$, and the first part of statement (iv) follows. Concerning the second part, if $u_0 \in C^{2+\alpha}(\mathbb{R}^n)$ then $u_0 \in D_A(\alpha/2+1,\infty)$, so that the above arguments work as well taking $\varepsilon = 0$, thanks to Corollary 4.3.9(iii). ∎

In the limiting case $\alpha = 1$, if $f \in C^{0,1}([0,T] \times \mathbb{R}^n)$ we do not expect that u_t and Au belong to $C^{0,1}([\varepsilon,T] \times \mathbb{R}^n)$ for any $\varepsilon \geq 0$. However, a proof analogous to the one of Theorem 5.1.4 yields a similar result, provided $C^1(\mathbb{R}^n)$ is replaced by $\mathcal{C}^1(\mathbb{R}^n)$.

Proposition 5.1.5 *Let $f \in UC([0,T] \times \mathbb{R}^n)$ be such that $f(t,\cdot) \in \mathcal{C}^1(\mathbb{R}^n)$ for every $t \in [0,T]$, and $\sup_{0 \leq t \leq T} \|f(t,\cdot)\|_{\mathcal{C}^1(\mathbb{R}^n)} < \infty$. Then u_t belongs to $UC([\varepsilon,T] \times \mathbb{R}^n)$, Au belongs to $C^{1/2,0}([\varepsilon,T] \times \mathbb{R}^n)$ for every $\varepsilon \in \,]0,T[$, moreover for every $t \in \,]0,T]$,*

$u_t(t,\cdot)$ and $\mathcal{A}u(t,\cdot)$ belong to $\mathcal{C}^1(\mathbb{R}^n)$, and

$$\|u\|_\infty + \|Du\|_\infty + \sup_{0<t\leq T}\|tu_t(t,\cdot)\|_\infty + \sup_{0<t\leq T}[t^{3/2}u_t(t,\cdot)]_{\mathcal{C}^1}$$
$$\leq C(\|u_0\|_\infty + \sup_{0\leq t\leq T}\|f(t,\cdot)\|_{\mathcal{C}^1}).$$

If in addition $u_0 \in D(\mathcal{A})$ is such that $\mathcal{A}u_0 \in \mathcal{C}^1(\mathbb{R}^n)$, then the above statements hold also for $\varepsilon = 0$, and

$$\|u\|_\infty + \|Du\|_\infty + \sup_{0\leq t\leq T}\|u_t(t,\cdot)\|_{\mathcal{C}^1} + \|\mathcal{A}u\|_{C^{1/2,0}([0,T]\times\mathbb{R}^n)}$$
$$\leq C(\|u_0\|_\infty + \|\mathcal{A}u_0\|_{\mathcal{C}^1} + \sup_{0\leq t\leq T}\|f(t,\cdot)\|_{\mathcal{C}^1}).$$

We have called "regular" the solutions provided by Theorems 5.1.3, 5.1.4, and Proposition 5.1.5, because they are continuously differentiable with respect to time and $\mathcal{A}u$ is continuous. We remark that the continuity of $\mathcal{A}u$ does not imply, in general, that the second order derivatives $D_{ij}u$ are continuous. We can only say that for each $i,j = 1,\ldots,n$, and $R > 0$, the function $t \mapsto u(t,\cdot)_{|B(0,R)}$ is continuous (in the case of Theorem 5.1.3, α-Hölder continuous) with values in $W^{2,p}(B(0,R))$ for every $p \in [1,\infty[$. In particular, the differential equation in (5.1.4) is satisfied for every $t > 0$, almost everywhere in x. If $n = 1$, however, $u(t,\cdot)$ belongs to $C^2(\mathbb{R})$, and, consequently, the differential equation in (5.1.4) is satisfied pointwise. Moreover, in the case of statement (iv) of Theorem 5.1.4, the second order derivatives $D_{ij}u$ belong to $C^{0,\alpha}([\varepsilon,T]\times\mathbb{R}^n)$, so that they are continuous for $t > 0$ and the differential equation in (5.1.4) is satisfied pointwise.

The statements about uniqueness in Theorems 5.1.3 and 5.1.4 follow from next Lemma.

Lemma 5.1.6 *Let $u \in C([0,T]\times\mathbb{R}^n)$ be such that u_t exists continuous in $]0,T]\times\mathbb{R}^n$, $u(t,\cdot) \in W^{2,p}_{loc}(\mathbb{R}^n)$ for every $p \geq 1$, and*

$$u_t = \mathcal{A}u, \quad 0 < t \leq T, \ x \in \mathbb{R}^n; \quad u(0,x) = 0, \ x \in \mathbb{R}^n.$$

Then $u \equiv 0$.

Proof — Fix any $\omega > \|c\|_\infty$ (recall that $\mathcal{A} = \sum_{i,j=1}^n a_{ij}D_{ij} + \sum_{i=1}^n b_i D_i + c$), and consider the function $v(t,x) = u(t,x)e^{-\omega t}$. It has the same regularity properties of u and satisfies, for $0 < t \leq T$, $x \in \mathbb{R}^n$,

$$\begin{cases} v_t = \sum_{i,j=1}^n a_{ij}D_{ij}v + \sum_{i=1}^n b_i D_i v + (c-\omega)v + f(t,x)e^{-\omega t}, \\ v(0,x) = 0. \end{cases} \quad (5.1.21)$$

Let us prove that $\sup v \leq 0$.

Assume by contradiction that $\sup v > 0$. At any relative maximum point (t_0,x_0) for v we have $D_i v(t_0,x_0) = 0$ for each i, and by Proposition 3.1.10, $\sum_{i,j=1}^n a_{ij}D_{ij}v(t_0,x_0) \leq 0$. Since $v_t(t_0,x_0) \geq 0$ unless $t_0 = 0$, from (5.1.21) it

1. SECOND ORDER EQUATIONS

follows that v has not positive maximum, so that $\sup v$ is not attained at any point. Choose a sequence $(\bar{t}_n, \bar{x}_n)$ such that $v(\bar{t}_n, \bar{x}_n) \geq \sup v - 1/n$, and set

$$v_n(t,x) = v(t,x) + \frac{2}{n}\theta(x - \bar{x}_n),$$

where θ is any smooth nonnegative function such that

$$0 \leq \theta(x) \leq 1, \quad \theta(0) = 1, \quad \theta(x) \equiv 0 \text{ for } |x| \geq 1.$$

Then $\lim_{n\to\infty} \sup v_n = \sup v$, and $\sup v_n = \max v_n = v_n(t_n, x_n)$ for some (t_n, x_n). Then either (i) $t_n = 0$ for infinitely many values of n, or (ii) $t_n > 0$ for every $n \in \mathbb{N}$.

In case (i), since $v_n(0, x_n) \leq \sup v_0(\cdot) + 2/n$, then $\sup v = \lim_{n\to\infty} v_n(0, x_n) = 0$, a contradiction.

Let us consider case (ii). If (ii) holds, then $D_t v_n(t_n, x_n) \geq 0$, $D_i v_n(t_n, x_n) = 0$ for every i, and $\sum_{i,j=1}^{n} a_{ij} D_{ij} v_n(t_n, x_n) \leq 0$ thanks to Proposition 3.1.10. From the equalities $D_t v(t_n, x_n) = D_t v_n(t_n, x_n)$, $D_i v(t_n, x_n) = D_i v_n(t_n, x_n) - \frac{2}{n} D_i \theta(x_n - \bar{x}_n)$, $D_{ij} v(t_n, x_n) = D_{ij} v_n(t_n, x_n) - \frac{2}{n} D_{ij}\theta(x_n - \bar{x}_n)$ we get

$$\liminf_{n\to\infty} D_t v(t_n, x_n) - \sum_{i,j=1}^{n} a_{ij}(x_n) D_{ij} v(t_n, x_n) - \sum_{i=1}^{n} b_i(x_n) D_i v(t_n, x_n) \geq 0.$$

Writing (5.1.21) at $t = t_n$, $x = x_n$ and letting $n \to \infty$ we get

$$\liminf_{n\to\infty} (c(t_n, x_n) - \omega) v(t_n, x_n) \geq 0.$$

This is a contradiction, because $c(t_n, x_n) - \omega \leq -\varepsilon < 0$ for every n, whereas $\lim_{n\to\infty} v_n(t_n, x_n) = \sup v > 0$. So, neither (i) nor (ii) can hold. It follows that $\sup v \leq 0$. Arguing similarly, one proves that $\inf v \geq 0$. Therefore, $v \equiv 0$, and the statement follows. ∎

Note that Lemma 5.1.6 is not extendable to general higher order problems nor to systems, because it is based on the Maximum Principle. So, we give below another uniqueness lemma, which works under slightly more restrictive assumptions, but can be easily extended to other situations.

Lemma 5.1.7 *Let $u \in C([0,T] \times \mathbb{R}^n)$ be such that u_t exists continuous in $]0,T] \times \mathbb{R}^n$, $u(t,\cdot) \in W^{2,p}_{loc}(\mathbb{R}^n)$ for every $p \geq 1$, $t \mapsto u(t,\cdot)$, $t \mapsto \mathcal{A}u(t,\cdot)$ belong to $C([\varepsilon, T]; C(\mathbb{R}^n))$ for every $\varepsilon \in]0,T[$, and*

$$u_t = \mathcal{A}u, \quad 0 < t \leq T, \; x \in \mathbb{R}^n, \quad u(0,x) = 0, \; x \in \mathbb{R}^n.$$

Assume moreover that

$$\sup_{0 < t \leq T, \, x \in \mathbb{R}^n} |t^\alpha Du(t,x)| < \infty, \tag{5.1.22}$$

for some $\alpha \in]0,1[$. Then $u \equiv 0$.

Proof — Thanks to Corollary 4.1.3, it is sufficient to show that if u is any solution of (5.1.4) satisfying (5.1.22) and enjoying the regularity properties stated, then

$$X - \lim_{t \to 0} R(\lambda, A)u(t, \cdot) = R(\lambda, A)u_0,$$

for some $\lambda \in \rho(A)$.

Fix any $x_0 \in \mathbb{R}^n$, $r > 0$, and let θ be a cutoff function satisfying (5.1.10). Then $\theta u(t, \cdot) \to \theta u_0$ in X as $t \to 0$, so that $R(\lambda, A)(\theta u(t, \cdot)) \to R(\lambda, A)(\theta u_0)$, for every $\lambda \in \rho(A)$. On the other hand,

$$\frac{d}{dt} R(\lambda, A)u(t, \cdot) = \lambda R(\lambda, A)u(t, \cdot) - u(t, \cdot)$$

belongs to $C(]0, T]; X) \cap L^1(0, T; C^1(\mathbb{R}^n))$, thanks to the regularity assumptions made on u and to (5.1.22). It follows that there exists the limit

$$\lim_{t \to 0} R(\lambda, A)u(t, \cdot) = v_0 \text{ in } C^1(\mathbb{R}^n).$$

We have to show that $v_0 = R(\lambda, A)u_0$. From the equality (5.1.12), with $v = u(t, \cdot)$, letting $t \to 0$ we get

$$R(\lambda, A)(\theta u_0) = \theta v_0 + 2R(\lambda, A) \sum_{i,j=1}^n a_{ij} D_i \theta D_j v_0$$
$$+ R(\lambda, A) \left[\left(\sum_{i,j=1}^n a_{ij} D_{ij}\theta + \sum_{i=1}^n b_i D_i \theta \right) v_0 \right],$$

which implies that $\theta v_0 \in D(A)$. In particular, $v_0 \in W^{2,p}(B(x_0, r))$ for every $p \geq 1$. Applying $\lambda I - A$ to both sides we get

$$\theta u_0 = \lambda \theta v_0 - \mathcal{A}(\theta v_0) + 2 \sum_{i,j=1}^n a_{ij} D_i \theta D_j v_0$$
$$+ \left(\sum_{i,j=1}^n a_{ij} D_{ij}\theta + \sum_{i=1}^n b_i D_i \theta \right) v_0 = \lambda \theta v_0 - \theta \mathcal{A} v_0,$$

so that $u_{0|B(x_0,r)} = (\lambda v_0 - \mathcal{A} v_0)_{|B(x_0,r)}$. Since x_0 and r are arbitrary, then $u_0 = \lambda v_0 - \mathcal{A} v_0$, that is $v_0 = R(\lambda, A)u_0$. Thus, $X - \lim_{t \to 0} R(\lambda, A)u(t, \cdot) = R(\lambda, A)u_0$, and the statement follows. ∎

The results of Theorems 5.1.3 and 5.1.4 may be easily extended to nonlocal equations, such as for instance

$$\begin{cases} u_t(t, x) = \mathcal{A}u(t, x) + \sum_{i=1}^n \beta_i(x) u(t, x_i) + \gamma(x) u(t, x_0) + f(t, x), \\ \qquad 0 < t \leq T, \ x \in \mathbb{R}^n, \\ u(0, x) = u_0(x), \ x \in \mathbb{R}^n, \end{cases}$$

where x_i, $i = 0, \ldots, n$, are given points in $\mathbb{R}^n$. Indeed, the operator $B : D(B) = D(A) \mapsto C(\mathbb{R}^n)$, defined by $B\varphi(x) = \mathcal{A}\varphi(x) + \sum_{i=1}^n \beta_i(x)\varphi(x_i) + \gamma(x)\varphi(x_0)$, is sectorial in $C(\mathbb{R}^n)$, thanks to Remark 3.1.13, and $D_B(\theta, \infty) = D_A(\theta, \infty) = \mathcal{C}^{2\theta}(\mathbb{R}^n)$ for every $\theta \in]0, 1[$.

As a corollary of Theorem 5.1.4 we get a classical result.

1. Second order equations

Theorem 5.1.8 *Let the coefficients of $\mathcal{A}$ belong to $C^\alpha(\mathbb{R}^n)$, with $0 < \alpha < 2$, $\alpha \neq 1$. Let $f \in C^{\alpha/2,\alpha}([0,T] \times \mathbb{R}^n)$, and $u_0 \in C(\mathbb{R}^n)$. Then the solution u of (5.1.4) belongs to $C^{1+\alpha/2,2+\alpha}([\varepsilon,T] \times \mathbb{R}^n)$ for every $\varepsilon \in]0,T[$. If in addition $u_0 \in C^{2+\alpha}(\mathbb{R}^n)$, then u belongs to $C^{1+\alpha/2,2+\alpha}([0,T] \times \mathbb{R}^n)$, and*

$$\|u\|_{C^{1+\alpha/2,2+\alpha}([0,T]\times\mathbb{R}^n)} \leq C(\|u_0\|_{C^{2+\alpha}(\mathbb{R}^n)} + \|f\|_{C^{\alpha/2,\alpha}([0,T]\times\mathbb{R}^n)}). \quad (5.1.23)$$

Proof — Due to Theorem 5.1.4(iv), u belongs to $C^{1,2+\alpha}([\varepsilon,T] \times \mathbb{R}^n)$, for every $\varepsilon \in]0,T[$. From Lemma 5.1.1 it follows that the derivatives $D_{ij}u$ belong to $C^{\alpha/2,\alpha}([\varepsilon,T]\times\mathbb{R}^n)$. From the equality $u_t = \mathcal{A}u + f$ we get $u_t \in C^{\alpha/2,\alpha}([\varepsilon,T]\times\mathbb{R}^n)$, and the first part of the statement follows. Concerning the second part, due again to Theorem 5.1.4(iv), the above arguments hold with ε replaced by 0. ∎

The case of coefficients depending on time

We assume that $a_{ij}, b_i, c : [0,T] \times \mathbb{R}^n \mapsto \mathbb{R}$ are uniformly continuous in $[0,T] \times \mathbb{R}^n$, and satisfy the uniform ellipticity condition

$$\sum_{i,j=1}^n a_{ij}(t,x)\xi_i\xi_j \geq \nu|\xi|^2, \quad 0 \leq t \leq T, \; x, \; \xi \in \mathbb{R}^n, \quad (5.1.24)$$

for some $\nu > 0$. For $0 \leq t \leq T$, $x \in \mathbb{R}^n$, we set

$$\mathcal{A}(t,x) = \sum_{i,j=1}^n a_{ij}(t,x)D_{ij} + \sum_{i=1}^n b_i(t,x)D_i + c(t,x),$$

and we consider the problem

$$\begin{cases} u_t(t,x) = \mathcal{A}(t,x)u(t,x) + f(t,x), & 0 < t \leq T, \; x \in \mathbb{R}^n, \\ u(0,x) = u_0(x), & x \in \mathbb{R}^n. \end{cases} \quad (5.1.25)$$

We shall show several existence and regularity results using the previous results together with perturbation arguments.

Theorem 5.1.9 *Let a_{ij}, b_i, c, f be uniformly continuous functions belonging to $C^{0,\alpha}([0,T] \times \mathbb{R}^n)$, with $0 < \alpha < 1$, and let $u_0 \in C^{2+\alpha}(\mathbb{R}^n)$. Assume moreover that the ellipticity condition (5.1.24) is satisfied. Then problem (5.1.25) has a unique solution $u \in C^{1,2+\alpha}([0,T] \times \mathbb{R}^n)$, and*

$$\|u\|_{C^{1,2+\alpha}([0,T]\times\mathbb{R}^n)} \leq C(\|u_0\|_{C^{2+\alpha}(\mathbb{R}^n)} + \|f\|_{C^{0,\alpha}([0,T]\times\mathbb{R}^n)}).$$

Proof — It is sufficient to show that there exists $\delta > 0$ such that if $0 \leq a < b \leq T$ and $b - a \leq \delta$, then for every $v_0 \in C^{2+\alpha}(\mathbb{R}^n)$ the problem

$$\begin{cases} v_t(t,x) = \mathcal{A}(t,x)v(t,x) + f(t,x), & a \leq t \leq b, \ x \in \mathbb{R}^n, \\ v(a,x) = v_0(x), & x \in \mathbb{R}^n, \end{cases} \quad (5.1.26)$$

has a unique solution $v \in C^{1,2+\alpha}([a,b] \times \mathbb{R}^n)$, and

$$\|v\|_{C^{1,2+\alpha}([a,b]\times\mathbb{R}^n)} \leq K(\|v_0\|_{C^{2+\alpha}(\mathbb{R}^n)} + \|f\|_{C^{0,\alpha}([a,b]\times\mathbb{R}^n)}).$$

Define

$$Y = \{v \in C^{1,2+\alpha}([a,b] \times \mathbb{R}^n) : v(a) = v_0\},$$

and for every $v \in Y$ set $\Gamma v = u$, where u is the solution of

$$\begin{cases} u_t(t,x) = \mathcal{A}(a,x)u(t,x) + [\mathcal{A}(t,x) - \mathcal{A}(a,x)]v(t,x) + f(t,x), \\ \qquad a \leq t \leq b, \ x \in \mathbb{R}^n, \\ u(a,x) = v_0(x), \quad x \in \mathbb{R}^n. \end{cases} \quad (5.1.27)$$

Clearly, every fixed point of Γ in Y is a solution of (5.1.26).

For every $v \in Y$, the function $(t,x) \mapsto [\mathcal{A}(t,x) - \mathcal{A}(a,x)]v(t,x) + f(t,x)$ belongs to $C^{0,\alpha}([a,b] \times \mathbb{R}^n)$. So, by Theorem 5.1.4(iv), Γv belongs to $C^{1,2+\alpha}([a,b] \times \mathbb{R}^n)$. Therefore, Γ maps Y into itself. Moreover, by estimate (5.1.20), there is C, independent of $b - a$, such that for every couple of functions $v, w \in Y$ it holds

$$\|\Gamma v - \Gamma w\|_{C^{1,2+\alpha}([a,b]\times\mathbb{R}^n)} \leq C\|\varphi\|_{C^{0,\alpha}([a,b]\times\mathbb{R}^n)},$$

where

$$\varphi(t,x) = (\mathcal{A}(t,x) - \mathcal{A}(a,x))(v(t,x) - w(t,x)).$$

Let us estimate $\|\varphi\|_{C^{0,\alpha}}$. Let C_α, C'_α be the embedding constants given by Lemma 5.1.1, and set

$$\begin{cases} K_1 = \max_{i,j} \|a_{ij}\|_\infty + \max_i \|b_i\|_\infty + \|c\|_\infty, \\ K_2 = \max_{i,j}[a_{ij}]_{C^{0,\alpha}} + \max_i[b_i]_{C^{0,\alpha}} + [c]_{C^{0,\alpha}}. \end{cases} \quad (5.1.28)$$

Then we have, recalling that $(v - w)(a, \cdot) \equiv 0$,

$$\begin{aligned} \|\varphi\|_\infty &\leq \sum_{i,j=1}^n 2\|a_{ij}\|_\infty \|D_{ij}(v-w)\|_\infty \\ &\quad + \sum_{i=1}^n 2\|b_i\|_\infty \|D_i(v-w)\|_\infty + 2\|c\|_\infty \|v-w\|_\infty \\ &\leq 2K_1\Big(\sum_{i,j=1}^n \delta^{\alpha/2}[D_{ij}(v-w)]_{C^{\alpha/2,0}} \\ &\quad + \sum_{i=1}^n \delta^{(1+\alpha)/2}[D_i(v-w)]_{C^{(1+\alpha)/2,0}} + \delta\|D_t(v-w)\|_\infty\Big) \\ &\leq 2K_1(C_\alpha \delta^{\alpha/2} + C'_\alpha \delta^{(1+\alpha)/2} + \delta)\|v-w\|_{C^{1,2+\alpha}}. \end{aligned} \quad (5.1.29)$$

1. SECOND ORDER EQUATIONS 191

Set moreover

$$\varepsilon(\delta) = \sup_{|t-s|\leq\delta,\, i,j=1,\ldots,n} \|a_{ij}(t,\cdot) - a_{ij}(s,\cdot)\|_\infty$$
$$+ \sup_{|t-s|\leq\delta,\, i=1,\ldots,n} \|b_i(t,\cdot) - b_i(s,\cdot)\|_\infty + \sup_{|t-s|\leq\delta} \|c(t,\cdot) - c(s,\cdot)\|_\infty. \tag{5.1.30}$$

Since the coefficients a_{ij}, b_i, c are uniformly continuous, then

$$\lim_{\delta\to 0} \varepsilon(\delta) = 0.$$

Recalling that $\|\xi\|_{C^\alpha(\mathbb{R}^n)} \leq 2\|\xi\|_{C^1(\mathbb{R}^n)}$ for every $\xi \in C^1(\mathbb{R}^n)$, we get

$$[\varphi]_{C^{0,\alpha}} \leq \sum_{i,j=1}^n (2[a_{ij}]_{C^{0,\alpha}} \|D_{ij}(v-w)\|_\infty$$
$$+ \|a_{ij}(t,\cdot) - a_{ij}(a,\cdot)\|_\infty [D_{ij}(v-w)]_{C^{0,\alpha}})$$
$$+ \sum_{i=1}^n (2[b_i]_{C^{0,\alpha}} \|D_i(v-w)\|_\infty + \|b_i(t,\cdot) - b_i(a,\cdot)\|_\infty [D_i(v-w)]_{C^{0,\alpha}})$$
$$+ 2[c]_{C^{0,\alpha}} \|v-w\|_\infty + \|c(t,\cdot) - c(a,\cdot)\|_\infty [v-w]_{C^{0,\alpha}}$$

so that

$$[\varphi]_{C^{0,\alpha}} \leq \sum_{i,j=1}^n 2K_2 \delta^{\alpha/2}[D_{ij}(v-w)]_{C^{\alpha/2,0}}$$
$$+ \varepsilon(\delta)[D_{ij}(v-w)]_{C^{0,\alpha}} + \sum_{i=1}^n 2K_2 \delta^{(1+\alpha)/2}[D_i(v-w)]_{C^{1/2+\alpha/2,0}} \tag{5.1.31}$$
$$+ \varepsilon(\delta)[D_i(v-w)]_{C^{0,\alpha}} + 2K_2\delta\|D_t(v-w)\|_\infty + \varepsilon(\delta)[v-w]_{C^{0,\alpha}}$$
$$\leq (2K_2(C_\alpha \delta^{\alpha/2} + C'_\alpha \delta^{(1+\alpha)/2} + \delta) + 3\varepsilon(\delta))\|v-w\|_{C^{1,2+\alpha}}.$$

Summing up, we get

$$\|\varphi\|_{C^{0,\alpha}} \leq [2(K_1+K_2)(C_\alpha \delta^{\alpha/2} + C'_\alpha \delta^{(1+\alpha)/2} + \delta) + 3\varepsilon(\delta)]\|v-w\|_{C^{1,2+\alpha}}.$$

Since $\lim_{\delta\to 0} \varepsilon(\delta) = 0$, if δ is sufficiently small Γ is a contraction with constant $1/2$, so that there exists a unique fixed point v of Γ in Y, which is the unique solution of (5.1.26) belonging to $C^{1,2+\alpha}([a,b]\times\mathbb{R}^n)$.

To estimate the norm of v, we use again Theorem 5.1.4(iv). Let $\tilde v_0$ denote the function constant in time $\tilde v_0(t,x) = v_0(x)$. The Contraction Theorem gives $\|v-\tilde v_0\|_{C^{1,2+\alpha}} \leq 2\|\Gamma\tilde v_0 - \tilde v_0\|_{C^{1,2+\alpha}}$, and from estimate (5.1.20) we get

$$\|\Gamma\tilde v_0 - \tilde v_0\|_{C^{1,2+\alpha}} \leq \|\Gamma\tilde v_0\|_{C^{1,2+\alpha}} + \|v_0\|_{C^{2+\alpha}(\mathbb{R}^n)}$$
$$\leq C(\|\psi\|_{C^{0,\alpha}} + \|v_0\|_{C^{2+\alpha}(\mathbb{R}^n)}) + \|v_0\|_{C^{2+\alpha}(\mathbb{R}^n)},$$

where $\psi(t,x) = [\mathcal{A}(t,x) - \mathcal{A}(a,x)]v_0(x) + f(t,x)$. Estimates similar to (5.1.29) and (5.1.31) give

$$\|\psi\|_{C^{0,\alpha}([a,b]\times\mathbb{R}^n)} \leq (4K_1+2K_2)\|v_0\|_{C^{2+\alpha}(\mathbb{R}^n)} + \|f\|_{C^{0,\alpha}([a,b]\times\mathbb{R}^n)},$$

so that

$$\|v\|_{C^{1,2+\alpha}} \leq [C(4K_1+2K_2+1)+1]\|v_0\|_{C^{2+\alpha}(\mathbb{R}^n)} + C\|f\|_{C^{0,\alpha}},$$

and the statement follows. ∎

A similar result holds also for $1 < \alpha < 2$. The technique of the proof is the same. We leave the details to the reader.

We are now in position of proving a classical result.

Theorem 5.1.10 *Let a_{ij}, b_i, c, f belong to $C^{\alpha/2,\alpha}([0,T] \times \mathbb{R}^n)$, with $0 < \alpha < 1$, and let $u_0 \in C^{2+\alpha}(\mathbb{R}^n)$. Assume moreover that the uniform ellipticity condition (5.1.24) holds. Then problem (5.1.25) has a unique solution $u \in C^{1+\alpha/2, 2+\alpha}([0,T] \times \mathbb{R}^n)$, and*

$$\|u\|_{C^{1+\alpha/2,2+2\alpha}([0,T]\times\mathbb{R}^n)} \leq C(\|u_0\|_{C^{2+\alpha}(\mathbb{R}^n)} + \|f\|_{C^{\alpha/2,\alpha}([0,T]\times\mathbb{R}^n)}).$$

Proof — By Theorem 5.1.9, problem (5.1.25) has a unique solution u belonging to $C^{1,2+\alpha}([0,T] \times \mathbb{R}^n)$. Lemma 5.1.1 yields that the derivatives $D_{ij}u$ belong to $C^{\alpha/2,\alpha}([0,T] \times \mathbb{R}^n)$. Since the coefficients of $\mathcal{A}(t,x)$ and the function f belong to $C^{\alpha/2,\alpha}([0,T] \times \mathbb{R}^n)$, from the equality $u_t = \mathcal{A}u + f$ we deduce that also u_t belongs to $C^{\alpha/2,\alpha}([0,T] \times \mathbb{R}^n)$, and the statement follows. ∎

5.1.2 Initial boundary value problems in $[0,T] \times \overline{\Omega}$

The Dirichlet boundary condition

We begin with the case of coefficients independent of time, and homogeneous Dirichlet boundary condition.

$$\begin{cases} u_t(t,x) = \mathcal{A}u(t,x) + f(t,x), & 0 < t \leq T, \ x \in \overline{\Omega}, \\ u(0,x) = u_0(x), & x \in \overline{\Omega}, \\ u(t,x) = 0, & 0 < t \leq T, \ x \in \partial\Omega. \end{cases} \quad (5.1.32)$$

We recall that $\mathcal{A}$ is a uniformly elliptic operator with uniformly continuous and bounded coefficients. As in the case $\Omega = \mathbb{R}^n$, we consider problem (5.1.32) as an evolution equation in the space $X = C(\overline{\Omega})$, by setting, according to (3.1.56),

$$D(A_0) = \{u \in \bigcap_{p \geq 1} W^{2,p}_{loc}(\Omega) \ : \ u, \mathcal{A}u \in C(\overline{\Omega}), \ u_{|\partial\Omega} = 0\}, \quad A_0 u = \mathcal{A}u.$$

We recall that if $n = 1$, then $D(A_0) = \{u \in C^2(\overline{\Omega}), \ u_{|\partial\Omega} = 0\}$.

Then we apply some of the results of Chapter 4 to the problem

$$u'(t) = A_0 u(t) + f(t), \quad 0 < t \leq T; \quad u(0) = u_0, \quad (5.1.33)$$

with

$$u(t) = u(t, \cdot), \quad f(t) = f(t, \cdot),$$

and we go back to problem (5.1.32), getting a number of regularity results.

Due to the homogeneous Dirichlet boundary condition, we shall often consider functions vanishing at $\partial\Omega$. As in Section 3.1, if Y is any subspace of $C(\overline{\Omega})$, we denote by Y_0 the subset consisting of the functions $\varphi \in Y$ which vanish on $\partial\Omega$.

Next theorem deals with mild solutions.

1. SECOND ORDER EQUATIONS

Theorem 5.1.11 *Let $f : [0,T] \times \overline{\Omega} \mapsto \mathbb{R}$ be a continuous function such that $t \mapsto f(t,\cdot)$ belongs to $C([0,T]; C(\overline{\Omega}))$, and let $u_0 \in C(\overline{\Omega})$. Then the function*

$$u(t,x) = e^{tA_0}u_0(x) + \int_0^t e^{(t-s)A_0} f(s,\cdot)ds(x), \ \ 0 \le t \le T, \ x \in \overline{\Omega}, \quad (5.1.34)$$

belongs to $C([0,T] \times \Omega) \cap C^{\theta,2\theta}([\varepsilon,T] \times \overline{\Omega})$ for every $\varepsilon \in \,]0,T[$ and $\theta \in \,]0,1[$, and there are $C > 0$, $C(\varepsilon,\theta) > 0$ such that

$$\|u\|_\infty \le C(\|u_0\|_\infty + \|f\|_\infty), \quad (5.1.35)$$

$$\|u\|_{C^{\theta,2\theta}([\varepsilon,T]\times\overline{\Omega})} \le C(\varepsilon,\theta)(\varepsilon^{-\theta}\|u_0\|_\infty + \|f\|_\infty). \quad (5.1.36)$$

In particular, for every $\theta \in \,]0,1[$, $\|t^\theta u(t,\cdot)\|_{C^{2\theta}(\overline{\Omega})}$ is bounded, and taking $\theta = 1/2$, $\|t^{1/2}Du\|_\infty$ is bounded.

Moreover, for every $\varepsilon \in \,]0,T[$ there exists a sequence $\{u_n\}_{n\in\mathbb{N}} \subset UC([\varepsilon,T] \times \overline{\Omega})$ such that $\partial u_n/\partial t$, $\mathcal{A}u_n$ exist continuous, and $\lim_{n\to\infty} \|u_n - u\|_{L^\infty([\varepsilon,T]\times\overline{\Omega})} = 0$, $\lim_{n\to\infty} \|D_t u_n - \mathcal{A}u_n - f\|_{L^\infty([\varepsilon,T]\times\overline{\Omega})} = 0$. In addition:

(i) *If $u_0 \in UC_0(\overline{\Omega})$, then $u \in C([0,T] \times \overline{\Omega})$, and $u(t,\cdot)$ converges to u_0 uniformly in $\overline{\Omega}$ as $t \to 0$. Moreover there exists a sequence $\{u_n\}_{n\in\mathbb{N}} \subset UC([0,T] \times \overline{\Omega})$ such that $\partial u_n/\partial t$, $\mathcal{A}u_n$ exist continuous in $[0,T] \times \overline{\Omega}$, and $\lim_{n\to\infty} \|u_n - u\|_\infty = 0$, $\lim_{n\to\infty} \|D_t u_n - \mathcal{A}u_n - f\|_\infty = 0$.*

(ii) *If $u_0 \in C_0^{2\theta}(\overline{\Omega})$, with $0 < \theta < 1$, then u belongs to $C^{\theta,2\theta}([0,T] \times \overline{\Omega})$, and*

$$\|u\|_{C^{\theta,2\theta}([0,T]\times\overline{\Omega})} \le C(\|u_0\|_{C^{2\theta}(\overline{\Omega})} + \|f\|_\infty).$$

Proof — The proof is quite similar to the proof of Theorem 5.1.2. We point out only the differences. In the present case we have, by Corollary 3.1.21(ii), $\overline{D(A_0)} = UC_0(\overline{\Omega})$, and by Theorem 3.1.29, $D_{A_0}(\theta,\infty) = C_0^{2\theta}(\overline{\Omega})$, if $\theta \ne 1/2$, $D_{A_0}(\theta,\infty) = \mathcal{C}_0^1(\overline{\Omega})$, if $\theta = 1/2$.

Now all the statements may be shown following the proof of Theorem 5.1.2, with obvious modifications. ∎

Next theorems deal with regular solutions.

Theorem 5.1.12 *Let $f \in C^{\alpha,0}([0,T]\times\overline{\Omega})$, with $\alpha \in \,]0,1[$, and let $u_0 \in C_0(\overline{\Omega})$. Then the function u defined in (5.1.34) is differentiable with respect to t in $\,]0,T] \times \overline{\Omega}$, $u(t,\cdot)$ belongs to $W^{2,p}_{loc}(\Omega)$ for every $p \ge 1$, and u_t, $\mathcal{A}u \in C^{\alpha,0}([\varepsilon,T] \times \overline{\Omega})$ for $0 < \varepsilon < T$. Moreover, u satisfies (5.1.32), and it is the unique solution of (5.1.32) belonging to $C([0,T] \times \overline{\Omega})$ and enjoying the above regularity properties. There is $C > 0$ such that*

$$\sup_{0 < t \le T} t(\|u_t(t,\cdot)\|_\infty + \|\mathcal{A}u(t,\cdot)\|_\infty) \le C(\|u_0\|_\infty + \|f\|_{C^{\alpha,0}([0,T]\times\overline{\Omega})}). \quad (5.1.37)$$

Moreover, $u_t(t,\cdot)$ belongs to $\mathcal{C}^{2\alpha}(\overline{\Omega})$ for every $t \in \,]0,T]$. In particular, for $\alpha \neq 1/2$, $u_t \in C^{\alpha,2\alpha}([\varepsilon,T]\times\overline{\Omega})$ for every $\varepsilon \in \,]0,T[$; for $\alpha = 1/2$, $u_t(t,\cdot)$ belongs to $\mathcal{C}^1(\overline{\Omega})$ for every $t \in \,]0,T]$.

In addition:

(i) If $u_0 \in D(A_0)$, then u_t and $\mathcal{A}u$ belong to $C([0,T]\times\Omega)$, and
$$\|u\|_\infty + \|Du\|_\infty + \|u_t\|_\infty + \|\mathcal{A}u\|_\infty \leq C(\|u_0\|_\infty + \|\mathcal{A}u_0\|_\infty + \|f\|_{C^{\alpha,0}}); \quad (5.1.38)$$

(ii) If $u_0 \in D(A_0)$ and $\mathcal{A}u_0 + f(0,\cdot) \in UC_0(\overline{\Omega})$, then u_t and $\mathcal{A}u$ belong to $C([0,T]\times\overline{\Omega})$, and $\lim_{t\to 0}\mathcal{A}u(t,x) = \mathcal{A}u_0(x)$, uniformly for $x \in \overline{\Omega}$;

(iii) If $\alpha \neq 1/2$, $u_0 \in D(A_0)$ and $\mathcal{A}u_0 + f(0,\cdot) \in \mathcal{C}_0^{2\alpha}(\overline{\Omega})$, then u_t belongs to $C^{\alpha,2\alpha}([0,T]\times\overline{\Omega})$, $\mathcal{A}u$ belongs to $C^{\alpha,0}([0,T]\times\overline{\Omega})$, and
$$\begin{aligned}\|u_t\|_{C^{\alpha,2\alpha}} + \|\mathcal{A}u\|_{C^{\alpha,0}} &\leq C(\|u_0\|_\infty + \|\mathcal{A}u_0\|_\infty \\ &\quad + \|\mathcal{A}u_0 + f(0,\cdot)\|_{C^{2\alpha}(\overline{\Omega})} + \|f\|_{C^{\alpha,0}}).\end{aligned} \quad (5.1.39)$$

If $\alpha = 1/2$, $u_0 \in D(A)$ and $\mathcal{A}u_0 + f(0,\cdot) \in \mathcal{C}_0^1(\overline{\Omega})$, then u_t and $\mathcal{A}u$ belong to $C^{1/2,0}([0,T]\times\overline{\Omega})$, $u_t(t,\cdot)$ belongs to $\mathcal{C}^1(\overline{\Omega})$ for every $t \in [0,T]$, and
$$\begin{aligned}&\|u_t\|_{C^{1/2,0}} + \|\mathcal{A}u\|_{C^{1/2,0}} + \sup_{0\leq t\leq T}\|u_t(t,\cdot)\|_{\mathcal{C}^1(\overline{\Omega})} \\ &\leq C(\|u_0\|_\infty + \|\mathcal{A}u_0\|_\infty + \|\mathcal{A}u_0 + f(0,\cdot)\|_{\mathcal{C}^1(\overline{\Omega})} + \|f\|_{C^{1/2,0}}).\end{aligned}$$

Also the proof of Theorem 5.1.12 is quite similar to that of Theorem 5.1.3, and it is omitted.

Now we consider the case where the data are Hölder continuous with respect to x.

Theorem 5.1.13 *Let $\alpha \in \,]0,2[$, $\alpha \neq 1$, and let $f \in UC([0,T]\times\overline{\Omega}) \cap C^{0,\alpha}([0,T]\times\overline{\Omega})$ be such that $f(t,x) = 0$ for every $t \in [0,T]$ and $x \in \partial\Omega$. Let moreover $u_0 \in C(\overline{\Omega})$. Then the mild solution u of problem (5.1.32) is differentiable with respect to t in $\,]0,T]\times\overline{\Omega}$, $u(t,\cdot)$ belongs to $W^{2,p}_{loc}(\Omega)$ for every $p \geq 1$, and u_t, $\mathcal{A}u \in UC([\varepsilon,T]\times\overline{\Omega}) \cap C^{0,\alpha}([\varepsilon,T]\times\overline{\Omega})$ for $0 < \varepsilon < T$. Moreover, u satisfies (5.1.32), and it is the unique solution of (5.1.32) belonging to $C([0,T]\times\Omega)$ and enjoying the above properties. There is $C > 0$ such that*
$$\sup_{0<t\leq T} t(\|u_t(t,\cdot)\|_\infty + \|\mathcal{A}u\|_\infty) \leq C(\|u_0\|_\infty + \|f\|_{C^{0,\alpha}([0,T]\times\overline{\Omega})}). \quad (5.1.40)$$

Moreover, $\mathcal{A}u$ belongs to $C^{\alpha/2,\alpha}([\varepsilon,T]\times\overline{\Omega})$ for every $\varepsilon \in \,]0,T]$.

In addition:

(i) If $u_0 \in D(A_0)$, then u_t, $\mathcal{A}u$ belong to $C([0,T]\times\Omega)$, and
$$\|u\|_\infty + \|Du\|_\infty + \|u_t\|_\infty + \|\mathcal{A}u\|_\infty \leq C(\|u_0\|_\infty + \|\mathcal{A}u_0\|_\infty + \|f\|_{C^{0,\alpha}}); \quad (5.1.41)$$

1. Second order equations

(ii) If $u_0 \in D(A_0)$ and $\mathcal{A}u_0 \in UC_0(\overline{\Omega})$, then u_t, $\mathcal{A}u$ belong to $C([0,T] \times \overline{\Omega})$, and $\lim_{t \to 0} \mathcal{A}u(t,x) = \mathcal{A}u_0(x)$, uniformly for $x \in \overline{\Omega}$;

(iii) If $u_0 \in D(A)$ and $\mathcal{A}u_0 \in C_0^\alpha(\overline{\Omega})$, with $0 < \alpha < 2$, $\alpha \neq 1$, then $\mathcal{A}u$ belongs to $C^{\alpha/2,\alpha}([0,T] \times \overline{\Omega})$, u_t belongs to $C^{0,\alpha}([0,T] \times \overline{\Omega})$, and

$$\|u_t\|_{C^{0,\alpha}} + \|\mathcal{A}u\|_{C^{\alpha/2,\alpha}} \leq C(\|u_0\|_\infty + \|\mathcal{A}u_0\|_{C^\alpha(\overline{\Omega})} + \|f\|_{C^{0,\alpha}}); \quad (5.1.42)$$

(iv) If $\partial\Omega$ is uniformly $C^{2+\alpha}$ and the coefficients a_{ij}, b_i, c belong to $C^\alpha(\overline{\Omega})$, with $0 < \alpha < 1$, then $u \in C^{1,2+\alpha}([\varepsilon,T] \times \overline{\Omega})$ for every $\varepsilon \in]0,T[$, and

$$\|u\|_{C^{1,2+\alpha}([\varepsilon,T] \times \overline{\Omega})} \leq \frac{C}{\varepsilon^{\alpha/2+1}} \left(\|u_0\|_\infty + \|f\|_{C^{0,\alpha}([0,T] \times \overline{\Omega})} \right).$$

If also $u_0 \in C^{2+\alpha}(\overline{\Omega})$, then $u \in C^{1,2+\alpha}([0,T] \times \overline{\Omega})$, and there is C such that

$$\|u\|_{C^{1,2+\alpha}([0,T] \times \overline{\Omega})} \leq C(\|u_0\|_{C^{2+\alpha}(\overline{\Omega})} + \|f\|_{C^{0,\alpha}([0,T] \times \overline{\Omega})}). \quad (5.1.43)$$

The proof of Theorem 5.1.13 is similar to the one of Theorem 5.1.4, and it is omitted. The assumption that $f(t,x) = 0$ for every $t \in [0,T]$ and $x \in \partial\Omega$ is rather restrictive. It implies that $f(t,\cdot)$ belongs to $D_{A_0}(\alpha/2,\infty)$, so that the regularity results of Subsection 4.3.2 may be applied. However, some results about regular solutions of (5.1.32) are available also in the case where f does not vanish on $\partial\Omega$. See the bibliographical remarks in the next Section.

Remark 5.1.14 Results similar to the ones of Theorems 5.1.12 and 5.1.13 hold for the problem

$$\begin{cases} u_t(t,x) = \mathcal{A}u(t,x) + f(t,x), & 0 < t \leq T, \; x \in \overline{\Omega}, \\ u(0,x) = u_0(x), & x \in \overline{\Omega}, \\ u(t,x) = g(t,x), & 0 < t \leq T, \; x \in \partial\Omega, \end{cases} \quad (5.1.44)$$

provided g is the trace on $[0,T] \times \partial\Omega$ of a function $G \in C([0,T] \times \overline{\Omega})$ such that $G_t \in C^{\alpha,0}([0,T] \times \overline{\Omega})$ (respectively, $G_t \in UC([0,T] \times \overline{\Omega}) \cap C^{0,\alpha}([0,T] \times \overline{\Omega})$), $G(t,\cdot) \in W^{2,p}_{loc}(\Omega)$ for every $p > 1$, and $\mathcal{A}G \in C^{\alpha,0}([0,T] \times \overline{\Omega})$ (respectively, $\mathcal{A}G \in C^{0,\alpha}([0,T] \times \overline{\Omega})$). Indeed, consider the problem which should be satisfied by $v = u - G$:

$$\begin{cases} v_t(t,x) = \mathcal{A}v(t,x) + f(t,x) + \mathcal{A}G(t,x) - G_t(t,x), & 0 < t \leq T, \; x \in \overline{\Omega}, \\ v(0,x) = u_0(x) - G(0,x), & x \in \overline{\Omega}, \\ v(t,x) = 0, & 0 < t \leq T, \; x \in \partial\Omega. \end{cases} \quad (5.1.45)$$

The function $(t,x) \mapsto \mathcal{A}G(t,x) - G_t(t,x)$ belongs to $C^{\alpha,0}([0,T] \times \overline{\Omega})$ (respectively, to $UC([0,T] \times \overline{\Omega}) \cap C^{0,\alpha}([0,T] \times \overline{\Omega})$), and $u_0 - G(0,\cdot) \in C(\overline{\Omega})$. So, the results of Theorems 5.1.12 and 5.1.13 may be applied to problem (5.1.45). We leave the details to the reader.

Now we prove a classical optimal regularity result.

Theorem 5.1.15 *Let $\partial\Omega$ be uniformly $C^{2+\alpha}$, with $0 < \alpha < 1$, and let a_{ij}, b_i, c $\in C^\alpha(\overline{\Omega})$, $f \in C^{\alpha/2,\alpha}([0,T]\times\overline{\Omega})$, $g \in C^{1+\alpha/2,2+\alpha}([0,T]\times\partial\Omega)$, $u_0 \in C^{2+\alpha}(\overline{\Omega})$ be such that*

$$(i)\ g(0,x) = u_0(x), \quad (ii)\ g_t(0,x) = \mathcal{A}u_0(x) + f(0,x), \quad x \in \partial\Omega. \tag{5.1.46}$$

Then the solution u of (5.1.44) belongs to $C^{1+\alpha/2,2+\alpha}([0,T]\times\overline{\Omega})$, and

$$\|u\|_{C^{1+\alpha/2,2+\alpha}} \le C(\|u_0\|_{C^{2+\alpha}(\overline{\Omega})} + \|f\|_{C^{\alpha/2,\alpha}} + \|g\|_{C^{1+\alpha/2,2+\alpha}}).$$

Proof — Let us extend g to $[0,T]\times\overline{\Omega}$ by means of the extension operator $\mathcal{D}$ given by Theorem 0.3.2(i). The function $G(t,\cdot) = \mathcal{D}g(t,\cdot)$ belongs to $C^{1+\alpha/2,2+\alpha}([0,T]\times\overline{\Omega})$, and

$$\|G\|_{C^{1+\alpha/2,2+\alpha}([0,T]\times\overline{\Omega})} \le C\|g\|_{C^{1+\alpha/2,2+\alpha}([0,T]\times\partial\Omega)}.$$

The function $v = u - G$ should satisfy problem (5.1.45). We are going to prove that problem (5.1.45) has indeed a solution $v \in C^{1+\alpha/2,2+\alpha}([0,T]\times\overline{\Omega})$.

The function $t \mapsto \varphi(t) = f(t,\cdot) + \mathcal{A}G(t,\cdot) - G_t(t,\cdot)$ belongs to $C^{\alpha/2}([0,T]; C(\overline{\Omega}))$, and

$$\|\varphi\|_{C^{\alpha/2}([0,T];C(\overline{\Omega}))} \le C(\|f\|_{C^{\alpha/2,0}([0,T]\times\overline{\Omega})} + \|g\|_{C^{1+\alpha/2,2+\alpha}([0,T]\times\partial\Omega)}).$$

Moreover, $u_0 - G(0,\cdot)$ belongs to $D(A_0)$, due to the compatibility condition (5.1.46)(i). $A_0(u_0 - G(0,\cdot)) + \varphi(0) = \mathcal{A}u_0 - G_t(0,\cdot)$ belongs to $C^\alpha(\overline{\Omega})$ due to the regularity assumptions, and vanishes on $\partial\Omega$ due to the compatibility condition (5.1.46)(ii): therefore, it belongs to $D_{A_0}(\alpha/2,\infty)$. By Theorem 4.3.1(iii), the problem

$$v' = A_0 v + \varphi, \ 0 \le t \le T; \ v(0) = u_0 - G(0,\cdot),$$

has a unique solution $v \in C^{1+\alpha/2}([0,T]; C(\overline{\Omega})) \cap C^{\alpha/2}([0,T]; D(A_0))$, such that $v'(t) \in D_{A_0}(\alpha/2,\infty)$ for every t, and

$$\|v\|_{C^{1+\alpha/2}([0,T];C(\overline{\Omega}))} + \sup_{0\le t\le T} \|v'(t)\|_{D_{A_0}(\alpha/2,\infty)} \le C(\|u_0 - G(0,\cdot)\|_{D(A_0)}$$
$$+ \|A_0(u_0 - G(0,\cdot)) + \varphi(0)\|_{D_{A_0}(\alpha/2,\infty)} + \|\varphi\|_{C^{\alpha/2}([0,T];C(\overline{\Omega}))}).$$

Set $v(t,x) = v(t)(x)$. Then, v is differentiable with respect to t and twice differentiable with respect to the space variables, with $D_{ij}v \in L^p_{loc}(\Omega)$ for every $p \ge 1$, and v_t, $\mathcal{A}v$ belong to $C^{\alpha/2,0}([0,T]\times\overline{\Omega})$. Recalling that $D_{A_0}(\alpha/2,\infty) = C^\alpha_0(\overline{\Omega})$, it follows that $t \mapsto v_t(t,\cdot)$ is bounded with values in $C^\alpha(\overline{\Omega})$. Then, v_t belongs to $C^{\alpha/2,\alpha}([0,T]\times\overline{\Omega})$, and

$$\|v_t\|_{C^{\alpha/2,\alpha}} \le \text{const.}\ (\|u_0\|_{C^{2+\alpha}} + \|f\|_{C^{\alpha/2,0}} + \|g\|_{C^{1+\alpha/2,2+\alpha}}).$$

Moreover, from the equality $v_t = \mathcal{A}v + f + \mathcal{A}G - G_t$, it follows that $\mathcal{A}v(t,\cdot) \in C^\alpha(\overline{\Omega})$ for every t. Due to the Schauder Theorem 3.1.34(i), $v(t,\cdot)$ belongs to $C^{2+\alpha}(\overline{\Omega})$

1. SECOND ORDER EQUATIONS

for every t. Since $t \mapsto \mathcal{A}v(t,\cdot)$ belongs to $C([0,T]; C(\overline{\Omega})) \cap B([0,T]; C^\alpha(\overline{\Omega}))$, then it belongs to $C([0,T]; C^\beta(\overline{\Omega}))$, for every $\beta < \alpha$, due to Propositions 1.1.3(iii) and 1.1.4(iii). Due again to the Schauder Theorem, the function $t \mapsto v(t,\cdot)$ belongs to $C([0,T]; C^{2+\beta}(\overline{\Omega}))$. In particular, the second order space derivatives $D_{ij}v$ are continuous in $[0,T] \times \overline{\Omega}$. It follows that for every $i,j = 1,\ldots,n$, $D_{ij}v$ belongs to $C^{\alpha/2,\alpha}([0,T] \times \overline{\Omega})$, and

$$\|D_{ij}v\|_{C^{\alpha/2,\alpha}} \leq \text{const.} \,(\|u_0\|_{C^{2+\alpha}} + \|f\|_{C^{\alpha/2,0}} + \|g\|_{C^{1+\alpha/2,2+\alpha}}).$$

Therefore, $v \in C^{1+\alpha/2,2+\alpha}([0,T] \times \overline{\Omega})$. Since $u = v + G$, the statement follows. ∎

The case of coefficients depending on time

In the case where the coefficients a_{ij}, b_i, c depend also on time, a result similar to the one of Theorem 5.1.15 holds. We set, for $0 \leq t \leq T$, $x \in \overline{\Omega}$,

$$\mathcal{A}(t,x)\varphi = \sum_{i,j=1}^n a_{ij}(t,x)D_{ij}\varphi + \sum_{i=1}^n b_i(t,x)D_i\varphi + c(t,x)\varphi.$$

Theorem 5.1.16 *Let $\partial\Omega$ be uniformly $C^{2+\alpha}$, with $0 < \alpha < 1$, and let a_{ij}, b_i, c, $f \in C^{\alpha/2,\alpha}([0,T] \times \overline{\Omega})$, $g \in C^{1+\alpha/2,2+\alpha}([0,T] \times \partial\Omega)$, $u_0 \in C^{2+\alpha}(\overline{\Omega})$ be such that*

$$g(0,x) = u_0(x), \quad g_t(0,x) = \mathcal{A}(0,x)u_0(x) + f(0,x), \quad x \in \partial\Omega. \tag{5.1.47}$$

Let moreover the ellipticity condition (5.1.24) be satisfied for $0 \leq t \leq T$, $x \in \overline{\Omega}$, $\xi \in \mathbb{R}^n$. Then the problem

$$\begin{cases} u_t(t,x) = \mathcal{A}(t,x)u(t,x) + f(t,x), & 0 < t \leq T, \; x \in \overline{\Omega}, \\ u(0,x) = u_0(x), & x \in \overline{\Omega}, \\ u(t,x) = g(t,x), & 0 < t \leq T, \; x \in \partial\Omega, \end{cases}$$

has a unique solution u belonging to $C^{1+\alpha/2,2+\alpha}([0,T] \times \overline{\Omega})$, and

$$\|u\|_{C^{1+\alpha/2,2+\alpha}} \leq C(\|u_0\|_{C^{2+\alpha}(\overline{\Omega})} + \|f\|_{C^{\alpha/2,\alpha}} + \|g\|_{C^{1+\alpha/2,2+\alpha}}).$$

Proof — We follow the method used in the proof of Theorem 5.1.9. It is sufficient to show that there exists $\delta > 0$ such that if $0 \leq a < b \leq T$ and $b - a \leq \delta$, then for every $v_0 \in C^{2+\alpha}(\overline{\Omega})$ such that $v_0(x) = g(a,x)$ and $\mathcal{A}(a,x)v_0(x) + f(a,x) = g_t(a,x)$ for $x \in \partial\Omega$, the problem

$$\begin{cases} v_t(t,x) = \mathcal{A}(t,x)v(t,x) + f(t,x), & a \leq t \leq b, \; x \in \overline{\Omega}, \\ v(a,x) = v_0(x), & x \in \overline{\Omega}, \\ v(t,x) = g(t,x), & a \leq t \leq b, \; x \in \partial\Omega, \end{cases} \tag{5.1.48}$$

has a unique solution $v \in C^{1+\alpha/2, 2+\alpha}([a,b] \times \overline{\Omega})$, and

$$\|v\|_{C^{1+\alpha/2,2+\alpha}} \leq K(\|v_0\|_{C^{2+\alpha}(\overline{\Omega})} + \|f\|_{C^{\alpha/2,\alpha}} + \|g\|_{C^{1+\alpha/2,2+\alpha}}),$$

with K independent of $b - a$ and v_0. Define

$$Y = \{v \in C^{1+\alpha/2, 2+\alpha}([a,b] \times \overline{\Omega}) : v(a) = v_0\},$$

and for every $v \in Y$ set $\Gamma v = u$, where u is the solution of

$$\begin{cases} u_t(t,x) = \mathcal{A}(a,x)u(t,x) + [\mathcal{A}(t,x) - \mathcal{A}(a,x)]v(t,x) + f(t,x), \\ \qquad a \leq t \leq b,\ x \in \overline{\Omega}, \\ u(a,x) = v_0(x),\ x \in \overline{\Omega}, \\ u(t,x) = g(t,x),\ a \leq t \leq b,\ x \in \partial\Omega. \end{cases} \qquad (5.1.49)$$

Clearly, every fixed point of Γ in Y is a solution of (5.1.48).

For every $v \in Y$, the function $(t,x) \mapsto [\mathcal{A}(t,x) - \mathcal{A}(a,x)]v(t,x) + f(t,x)$ belongs to $C^{\alpha/2,\alpha}([a,b] \times \overline{\Omega})$. By Theorem 5.1.15, Γv belongs to $C^{1+\alpha/2,2+\alpha}([a,b] \times \overline{\Omega})$. In particular, Γ maps Y into itself. Moreover, due again to Theorem 5.1.15, there is C, independent on $b-a$, such that for every $v, w \in Y$ it holds

$$\|\Gamma v - \Gamma w\|_{C^{1+\alpha/2,2+\alpha}([a,b]\times\overline{\Omega})} \leq C\|\varphi\|_{C^{\alpha/2,\alpha}([a,b]\times\overline{\Omega})},$$

where

$$\varphi(t,x) = (\mathcal{A}(t,x) - \mathcal{A}(a,x))(v(t,x) - w(t,x)).$$

The estimation of $\|\varphi\|_{C^{0,\alpha}}$ has been done in the proof of Theorem 5.1.9: from the formula after (5.1.31) we get

$$\|\varphi\|_{C^{0,\alpha}} \leq [2(K_1 + K_2)(C_\alpha \delta^{\alpha/2} + C'_\alpha \delta^{(1+\alpha)/2} + \delta) + 3\varepsilon(\delta)]\|v-w\|_{C^{1,2+\alpha}},$$

where K_1, K_2 are defined in (5.1.28), C_α and C'_α are the constants given by Lemma 5.1.1, and $\varepsilon(\delta)$ is defined in (5.1.30), with $\mathbb{R}^n$ replaced by $\overline{\Omega}$. In the present case, setting

$$K_3 = \max_{i,j} \|a_{ij}\|_{C^{\alpha/2,0}} + \max_i \|b_i\|_{C^{\alpha/2,0}} + \|c\|_{C^{\alpha/2,0}},$$

we get

$$\varepsilon(\delta) \leq K_3 \delta^{\alpha/2}.$$

It remains to estimate $\|\varphi\|_{C^{\alpha/2,0}}$. For every $x \in \overline{\Omega}$ it holds

$$[\varphi(\cdot,x)]_{C^{\alpha/2,0}} \leq K_3 \sum_{i,j=1}^n (\|D_{ij}(v-w)\|_\infty + \delta^{\alpha/2}[D_{ij}(v-w)]_{C^{\alpha/2,0}})$$
$$+ K_3 \sum_{i=1}^n (\|D_i(v-w)\|_\infty + \delta^{\alpha/2}[D_i(v-w)]_{C^{\alpha/2,0}})$$
$$+ K_3(\|v-w\|_\infty + \delta^{\alpha/2}[v-w]_{C^{\alpha/2,0}})$$
$$\leq 2K_3(\delta^{\alpha/2}[D_{ij}(v-w)]_{C^{\alpha/2,0}} + \delta^{(1+\alpha)/2}[D_i(v-w)]_{C^{(1+\alpha)/2,0}} + \delta\|D_t(v-w)\|_\infty)$$
$$\leq 2K_3(C_\alpha \delta^{\alpha/2} + C'_\alpha \delta^{(1+\alpha)/2} + \delta)\|v-w\|_{C^{1+\alpha/2,2+\alpha}}.$$

1. SECOND ORDER EQUATIONS

Summing up, we find
$$\|\varphi\|_{C^{\alpha/2,\alpha}} \leq const.\, \delta^{\alpha/2}\|v-w\|_{C^{1+\alpha/2,2+\alpha}},$$
so that, if δ is small enough, Γ is a contraction with constant $1/2$, in which case there exists a unique fixed point v of Γ in Y. v is the unique solution of (5.1.48) belonging to $C^{1+\alpha/2,2+\alpha}([a,b]\times\overline{\Omega})$.

The estimate for $\|v\|_{C^{1+\alpha/2,2+\alpha}([a,b]\times\overline{\Omega})}$ follows now as in the proof of Theorem 5.1.9. ■

The oblique boundary condition

We consider an oblique derivative initial boundary value problem,
$$\begin{cases} u_t(t,x) = \mathcal{A}u(t,x) + f(t,x), & 0 < t \leq T,\ x \in \overline{\Omega}, \\ u(0,x) = u_0(x),\ x \in \overline{\Omega}, \\ \mathcal{B}u(t,x) = g(t,x),\ 0 < t \leq T,\ x \in \partial\Omega, \end{cases} \qquad (5.1.50)$$

where $\mathcal{A} = \sum_{i,j=1}^n a_{ij}(x)D_{ij} + \sum_{i=1}^n b_i(x)D_i + c(x)$, $\mathcal{B} = \sum_{i=1}^n \beta_i(x)D_i + \gamma(x)$ satisfy the assumptions stated at the beginning of the section. We define the realization of $\mathcal{A}$ in $X = C(\overline{\Omega})$ with homogeneous first order boundary condition,
$$D(A_1) = \{u \in \bigcap_{p\geq 1} W_{loc}^{2,p}(\Omega)\ :\ u,\, \mathcal{A}u \in C(\overline{\Omega}),\, \mathcal{B}u_{|\partial\Omega} = 0\},\quad A_1 u = \mathcal{A}u.$$

We recall that if $n = 1$ then $D(A_1) = \{u \in C^2(\overline{\Omega}),\, \mathcal{B}u_{|\partial\Omega} = 0\}$. Due to Corollary 3.1.24(ii) and to Theorems 3.1.30, 3.1.31,
$$\overline{D(A_1)} = UC(\overline{\Omega}),\qquad D_{A_1}(\theta,\infty) = \begin{cases} C^{2\theta}(\overline{\Omega}), & \text{if } \theta \leq 1/2, \\ C_\mathcal{B}^1(\overline{\Omega}), & \text{if } \theta = 1/2, \\ C_\mathcal{B}^{2\theta}(\overline{\Omega}), & \text{if } \theta \geq 1/2. \end{cases}$$

The subscript $\mathcal{B}$ has the same meaning as in Chapter 3: if Y is any Banach space contained in $C^1(\overline{\Omega})$, we denote by $Y_\mathcal{B}$ the subspace of Y consisting of those functions φ such that $\mathcal{B}\varphi$ vanishes on $\partial\Omega$. The definition of $C_\mathcal{B}^1(\overline{\Omega})$ may be found before Theorem 3.1.31.

Note that problem (5.1.50) is not homogeneous at the boundary, so that the abstract theory may be applied directly only in the case where $g \equiv 0$, because $D(A_1)$ is endowed with homogeneous boundary conditions. If g is the trace at $[0,T] \times \partial\Omega$ of $\mathcal{B}G$, where G is a sufficiently smooth function defined in $[0,T] \times \overline{\Omega}$, then problem (5.1.50) may be reduced to a homogeneous problem by means of the procedure of Remark 5.1.14. Setting $v = u - G$, v should satisfy
$$\begin{cases} v_t(t,x) = \mathcal{A}v(t,x) + f(t,x) + \mathcal{A}G(t,x) - G_t(t,x), & 0 < t \leq T,\ x \in \overline{\Omega}, \\ v(0,x) = u_0(x) - G(0,x),\ x \in \overline{\Omega}, \\ \mathcal{B}v(t,x) = 0,\ 0 < t \leq T,\ x \in \partial\Omega, \end{cases} \qquad (5.1.51)$$

and problem (5.1.51) could be treated as an evolution equation in X by the methods used above in the case $\Omega = \mathbb{R}^n$ and in the case of the Dirichlet boundary condition. However, for the right hand side of the differential equation in (5.1.51) to make sense, we need that $\mathcal{A}G - G_t$ exists, while such assumption is clearly redundant for the solvability of problem (5.1.50). So, we use a trick which lets us consider also less regular functions g.

If g is smooth, problem (5.1.51) has a unique solution v, given by the variation of constants formula

$$v(t,\cdot) = e^{tA}(u_0 - G(0,\cdot)) + \int_0^t e^{(t-s)A}[f(s,\cdot) + \mathcal{A}G(s,\cdot)]ds - \int_0^t e^{(t-s)A}G_s(s,\cdot)ds.$$

Integrating formally by parts the last integral, we get

$$\begin{aligned} v(t,\cdot) &= e^{tA}(u_0 - G(0,\cdot)) + \int_0^t e^{(t-s)A}[f(s,\cdot) + \mathcal{A}G(s,\cdot)]ds \\ &\quad - A\int_0^t e^{(t-s)A}G(s,\cdot)ds + e^{tA}G(0,\cdot) - G(t,\cdot), \end{aligned}$$

so that

$$\begin{aligned} u(t,\cdot) &= e^{tA}(u_0 - G(0,\cdot)) + \int_0^t e^{(t-s)A}[f(s,\cdot) + \mathcal{A}G(s,\cdot)]ds \\ &\quad - A\int_0^t e^{(t-s)A}[G(s,\cdot) - G(0,\cdot)]ds + G(0,\cdot), \quad 0 \le t \le T. \end{aligned} \quad (5.1.52)$$

We will not try to justify the computations leading to (5.1.52): they are purely formal. But we will see that, under reasonable assumptions on the data, formula (5.1.52) makes sense and it gives in fact the unique solution of (5.1.50).

In the following, G will be chosen as

$$G(t,\cdot) = \mathcal{N}g(t,\cdot), \quad 0 \le t \le T, \quad (5.1.53)$$

where $\mathcal{N}$ is the operator given by Theorem 0.3.2(ii). We recall that $\mathcal{N}$ is linear and continuous from $C^\theta(\partial\Omega)$ to $C^{\theta+1}(\overline{\Omega})$ for $0 \le \theta \le 1$ (if $\partial\Omega$ is uniformly $C^{2+\alpha}$, also for $1 < \theta \le \alpha + 1$), from $UC(\partial\Omega)$ to $UC^1(\overline{\Omega})$, and from $UC^1(\partial\Omega)$ to $UC^2(\overline{\Omega})$. Moreover,

$$\mathcal{B}\mathcal{N}g = g \text{ in } [0,T] \times \partial\Omega.$$

Theorem 5.1.17 *Let $f : [0,T] \times \overline{\Omega} \mapsto \mathbb{R}$ be a continuous function such that $t \mapsto f(t,\cdot)$ belongs to $C([0,T]; C(\overline{\Omega}))$, and let $u_0 \in C(\overline{\Omega})$, $g \in C^{0,1}([0,T]; \times\partial\Omega)$ be such that $t \mapsto g(t,\cdot)$ belongs to $C([0,T]; C^1(\partial\Omega))$. Then the function u defined in (5.1.52) (with G given by (5.1.53)) is continuous in $[0,T] \times \overline{\Omega}$, and it belongs to $C^{\theta/2,\theta}([\varepsilon,T] \times \overline{\Omega})$ for every $\varepsilon \in]0,T[$ and $\theta \in]0,1[$, and there are $C > 0$, $C(\varepsilon,\theta) > 0$ such that*

$$\|u\|_\infty \le C(\|u_0\|_\infty + \|f\|_\infty + \|g\|_{C^{0,1}}), \quad (5.1.54)$$

1. SECOND ORDER EQUATIONS

$$\|u\|_{C^{\theta/2,\theta}([\varepsilon,T]\times\overline{\Omega})} \leq C(\varepsilon,\theta)(\varepsilon^{-\theta/2}\|u_0\|_\infty + \|f\|_\infty + \|g\|_{C^{0,1}}). \quad (5.1.55)$$

In particular, $\|t^{\theta/2}u(t,\cdot)\|_{C^\theta(\overline{\Omega})}$ is bounded for every $\theta \in \,]0,1[$.

Moreover, for every $\varepsilon \in \,]0,T[$ there exists a sequence $\{u_n\}_{n\in\mathbb{N}} \subset UC([\varepsilon,T]\times\overline{\Omega})$ such that $\partial u_n/\partial t$, $\mathcal{A}u_n$ exist continuous, and $\lim_{n\to\infty}\|u_n - u\|_{L^\infty([\varepsilon,T]\times\overline{\Omega})} = 0$, $\lim_{n\to\infty}\|\mathcal{B}u_n - g\|_{L^\infty([\varepsilon,T]\times\partial\Omega)} = 0$, $\lim_{n\to\infty}\|D_t u_n - \mathcal{A}u_n - f\|_{L^\infty([\varepsilon,T]\times\overline{\Omega})} = 0$.
In addition:

(i) If $u_0 \in UC(\overline{\Omega})$, then $u(t,\cdot)$ converges to u_0 uniformly in $\overline{\Omega}$ as $t \to 0$. Moreover there exists a sequence $\{u_n\}_{n\in\mathbb{N}} \subset UC([0,T]\times\overline{\Omega})$ such that $\partial u_n/\partial t$, $\mathcal{A}u_n$ exist continuous in $[0,T]\times\overline{\Omega}$, and $\lim_{n\to\infty}\|u_n - u\|_\infty = 0$, $\lim_{n\to\infty}\|\mathcal{B}u_n - g\|_\infty = 0$, $\lim_{n\to\infty}\|D_t u_n - \mathcal{A}u_n - f\|_\infty = 0$.

(ii) If $u_0 \in C^{2\theta}(\overline{\Omega})$, with $0 < \theta < 1/2$, then u belongs to $C^{\theta,2\theta}([0,T]\times\overline{\Omega})$, and

$$\|u\|_{C^{\theta,2\theta}([0,T]\times\overline{\Omega})} \leq C(\|u_0\|_{C^{2\theta}(\overline{\Omega})} + \|f\|_\infty).$$

Proof — We split u into the sum $u = u_1 + u_2$, where, for $0 \leq t \leq T$,

$$u_1(t,\cdot) = -A\int_0^t e^{(t-s)A}[G(s,\cdot) - G(0,\cdot)]ds + G(0,\cdot), \quad (5.1.56)$$

$$u_2(t,\cdot) = e^{tA}(u_0 - G(0,\cdot)) + \int_0^t e^{(t-s)A}[f(s,\cdot) + \mathcal{A}G(s,\cdot)]ds. \quad (5.1.57)$$

First we consider the function u_1. Since $t \mapsto g(t,\cdot)$ belongs to $C([0,T];C^1(\partial\Omega))$, then $t \mapsto G(t,\cdot)$ belongs to $C([0,T];C^2(\overline{\Omega}))$. Since $C^2(\overline{\Omega})$ is continuously embedded in $D_{A_1}(1/2,\infty)$, then $t \mapsto G(t,\cdot)$ belongs to $C([0,T];D_{A_1}(1/2,\infty))$, and its norm is less than $\mathrm{const.}\,\|g\|_{C^{0,1}([0,T]\times\partial\Omega)}$. Set

$$z(t) = \int_0^t e^{(t-s)A_1}[G(s,\cdot) - G(0,\cdot)]ds, \quad 0 \leq t \leq T. \quad (5.1.58)$$

By Theorem 4.3.8(ii), with $\alpha = 1/2$, $f(t) = G(t,\cdot) - G(0,\cdot)$, we get that z is continuously differentiable in $]0,T]$ with values in $D_{A_1}(1/2,\infty)$, z' and $A_1 z$ are bounded with values in $D_{A_1}(1/2,\infty)$, $A_1 z$ belongs to $C^{1/2}([0,T];X)$, and

$$\|z'\|_{B([0,T];D_{A_1}(1/2,\infty))} + \|A_1 z\|_{C^{1/2}([0,T];X)} \leq \mathrm{const.}\,\|g\|_{C^{0,1}([0,T]\times\partial\Omega)}.$$

Since $u_1(t,\cdot) = -A_1 z(t) + G(0,\cdot)$, it follows that $u_1(t,\cdot) - G(t,\cdot) = -z'(t)$ belongs to $D_{A_1}(1/2,\infty)$ for every t, so that it belongs to $C^\theta(\overline{\Omega})$ for every $\theta \in \,]0,1[$. Moreover, $t \mapsto u_1(t,\cdot) - G(0,\cdot) = -A_1 z(t)$ belongs to $C^{1/2}([0,T];X)$, so that u_1 belongs to $C^{\theta,2\theta}([0,T]\times\overline{\Omega})$ for every $\theta \in \,]0,1/2[$, and

$$\|u_1\|_{C^{\theta,2\theta}([0,T]\times\overline{\Omega})} \leq C(\theta)\|g\|_{C^{0,1}([0,T]\times\partial\Omega)}.$$

Concerning the function u_2, since $t \mapsto \mathcal{A}G(t,\cdot)$ belongs to $C([0,T];C(\overline{\Omega}))$, one can follow the procedure of Theorem 5.1.2, with obvious modifications, proving that statements (i), (ii) hold with u replaced by u_2. Summing up and arguing as in the proof of Theorem 5.1.2, the result follows. ∎

Let us consider now regular solutions.

Theorem 5.1.18 *Let $f \in C^{\alpha,0}([0,T] \times \overline{\Omega})$, $g \in C^{(1+\alpha)/2, 1+\alpha}([0,T] \times \partial\Omega)$, with $\alpha \in {]0,1[}$, and let $u_0 \in C(\overline{\Omega})$. Then the function u given by (5.1.52) is differentiable with respect to t in $]0,T] \times \overline{\Omega}$, $u(t,\cdot)$ belongs to $W^{2,p}_{loc}(\Omega)$ for every $p \geq 1$, and u_t, $\mathcal{A}u \in C^{\alpha,0}([\varepsilon,T] \times \overline{\Omega})$ for $0 < \varepsilon < T$. There is $C > 0$ such that*

$$\sup_{0 < t \leq T} t(\|u_t(t,\cdot)\|_\infty + \|\mathcal{A}u(t,\cdot)\|_\infty)$$
$$\leq C(\|u_0\|_\infty + \|f\|_{C^{\alpha,0}([0,T]\times\overline{\Omega})} + \|g\|_{C^{(1+\alpha)/2,1+\alpha}([0,T]\times\partial\Omega)}). \tag{5.1.59}$$

u satisfies pointwise (5.1.50), and it is the unique solution of (5.1.50) belonging to $C([0,T] \times \overline{\Omega})$ enjoying the above properties. Moreover, $u_t(t,\cdot)$ belongs to $\mathcal{C}^{2\alpha}(\overline{\Omega})$ for every $t \in {]0,T]}$. It follows that for $\alpha \neq 1/2$, $u_t \in C^{\alpha,2\alpha}([\varepsilon,T] \times \overline{\Omega})$ for every $\varepsilon \in {]0,T[}$; for $\alpha = 1/2$, $u_t(t,\cdot)$ belongs to $\mathcal{C}^1(\overline{\Omega})$ for every $t \in {]0,T]}$.
In addition:

(i) *If $u_0 \in C^{2\theta}(\overline{\Omega})$, with $1/2 \leq \theta < 1$, and*

$$\mathcal{B}u_0(x) = g(0,x), \quad x \in \partial\Omega, \tag{5.1.60}$$

then u belongs to $C^{\theta,2\theta}([0,T] \times \overline{\Omega})$, and

$$\|u\|_{C^{\theta,2\theta}([0,T]\times\overline{\Omega})} \leq C(\|u_0\|_{C^{2\theta}(\overline{\Omega})} + \|f\|_\infty + \|g\|_{C^{(1+\alpha)/2,1+\alpha}}).$$

(ii) *If $u_0 \in \cap_{p\geq 1} W^{2,p}_{loc}(\Omega)$ satisfies (5.1.60), and $\mathcal{A}u_0$ is uniformly continuous and bounded in Ω, then u_t and $\mathcal{A}u$ belong to $C([0,T] \times \overline{\Omega})$, and*

$$\|u\|_\infty + \|Du\|_\infty + \|u_t\|_\infty + \|\mathcal{A}u\|_\infty \leq C(\|u_0\|_\infty + \|\mathcal{A}u_0\|_\infty$$
$$+ \|f\|_{C^{\alpha,0}([0,T]\times\overline{\Omega})} + \|g\|_{C^{(1+\alpha)/2,1+\alpha}([0,T]\times\partial\Omega)}).$$

(iii) *If $\alpha \neq 1/2$, $u_0 \in \cap_{p\geq 1} W^{2,p}_{loc}(\Omega)$ is such that $\mathcal{A}u_0 + f(0,\cdot) \in \mathcal{C}^{2\alpha}(\overline{\Omega})$, (if $0 < \alpha < 1/2$), $\mathcal{A}u_0 + f(0,\cdot) \in \mathcal{C}^{2\alpha}_\mathcal{B}(\overline{\Omega})$ (if $1/2 < \alpha < 1$), and (5.1.60) holds, then u_t belongs to $C^{\alpha,2\alpha}([0,T] \times \overline{\Omega})$, $\mathcal{A}u$ belongs to $C^{\alpha,0}([0,T] \times \overline{\Omega})$, and*

$$\|u_t\|_{C^{\alpha,2\alpha}} + \|\mathcal{A}u\|_{C^{\alpha,0}} \leq C(\|u_0\|_\infty + \|\mathcal{A}u_0\|_\infty$$
$$+ \|\mathcal{A}u_0 + f(0,\cdot)\|_{\mathcal{C}^{2\alpha}(\overline{\Omega})} + \|f\|_{C^{\alpha,0}} + \|g\|_{C^{(1+\alpha)/2,1+\alpha}}). \tag{5.1.61}$$

Proof — We split again u into the sum $u = u_1 + u_2$, where u_1 and u_2 are defined by (5.1.56) and (5.1.57), respectively. u_1 and u_2 will be shown to satisfy

$$\begin{cases} D_t u_1(t,x) = \mathcal{A}u_1(t,x) - \mathcal{A}G(t,x), & 0 < t \leq T, \ x \in \overline{\Omega}, \\ u_1(0,x) = G(0,x), & x \in \overline{\Omega}, \\ \mathcal{B}u_1(t,x) = g(t,x), & 0 < t \leq T, \ x \in \partial\Omega, \end{cases} \tag{5.1.62}$$

1. SECOND ORDER EQUATIONS

$$\begin{cases} D_t u_2(t,x) = \mathcal{A} u_2(t,x) + f(t,x) + \mathcal{A} G(t,x), & 0 < t \leq T, \; x \in \overline{\Omega}, \\ u_2(0,x) = u_0(x) - G(0,x), & x \in \overline{\Omega}, \\ \mathcal{B} u_2(t,x) = 0, & 0 < t \leq T, \; x \in \partial\Omega. \end{cases} \quad (5.1.63)$$

First we consider the function u_1, with G defined by (5.1.53). Since $t \mapsto g(t,\cdot)$ belongs to $C^{(1+\alpha)/2}([0,T]; C(\partial\Omega))$, then $t \mapsto G(t,\cdot)$ belongs to $C^{(1+\alpha)/2}([0,T]; C^1(\overline{\Omega}))$. Since $C^1(\overline{\Omega})$ is continuously embedded in $D_{A_1}(1/2, \infty)$, then $t \mapsto G(t,\cdot)$ belongs to $C^{(1+\alpha)/2}([0,T]; D_{A_1}(1/2, \infty))$, and its norm is less than $c\|g\|_{C^{(1+\alpha)/2, 1+\alpha}}$. Let z be the function defined in (5.1.58). By Theorem 4.3.16, applied with $\theta = (1+\alpha)/2$, $\beta = 1/2$, $f(t) = G(t,\cdot) - G(0,\cdot)$, and $u_0 = 0$, we get that z is continuously differentiable with values in $D(A_1)$, $d/dt(A_1 z) = A_1 z'$ belongs to $B([0,T]; D_{A_1}(\alpha/2, \infty)) \cap C^{\alpha/2}([0,T]; X)$, and

$$\|z\|_{C^{1+\alpha/2}([0,T]; D(A_1))} + \|A_1 z'\|_{B([0,T]; D_{A_1}(\alpha/2, \infty))} \leq c\|g\|_{C^{(1+\alpha)/2, 1+\alpha}([0,T] \times \partial\Omega)}.$$

Moreover, z satisfies

$$z'(t) = A_1 z(t) + G(t,\cdot) - G(0,\cdot), \quad 0 \leq t \leq T.$$

Since $u_1(t,\cdot) = -A_1 z(t) + G(0,\cdot)$, it follows that $u_1(t,\cdot) - G(t,\cdot) = -A_1 z(t) - G(t,\cdot) + G(0,\cdot) = -z'(t)$ belongs to $D(A_1)$ for every t, so that

$$\mathcal{B} u_1(t,x) = g(t,x), \quad 0 \leq t \leq T, \; x \in \partial\Omega,$$

and $u_1(t,\cdot)$ belongs to $W^{2,p}_{loc}(\Omega)$ for every $p \geq 1$. Moreover, $t \mapsto u_1(t,\cdot) - G(0,\cdot) = -A_1 z(t)$ belongs to $C^{1+\alpha/2}([0,T]; X)$, so that u is differentiable with respect to time, and $D_t u_1 \in C^{\alpha/2, 0}([0,T] \times \overline{\Omega})$. Since $(A_1 z)'$ is bounded with values in $D_{A_1}(\alpha/2, \infty)$, it follows that $D_t u_1(t,\cdot) = -(A_1 z)'(t)$ is bounded with values in $C^\alpha(\overline{\Omega})$. Therefore, $D_t u_1$ belongs to $C^{\alpha/2, \alpha}([0,T] \times \overline{\Omega})$, and

$$\|D_t u_1\|_{C^{\alpha/2,\alpha}([0,T] \times \overline{\Omega})} \leq \text{const.} \; \|g\|_{C^{(1+\alpha)/2, 1+\alpha}([0,T] \times \partial\Omega)}.$$

From the equality $D_t u_1(t,\cdot) = -A_1 z'(t) = A_1(u_1(t,\cdot) - G(t,\cdot))$ we get

$$D_t u_1(t,x) = \mathcal{A} u_1(t,x) - \mathcal{A} G(t,x).$$

In particular, u_1 is a solution of (5.1.62), and $\mathcal{A} u_1$ is continuous in $[0,T] \times \overline{\Omega}$.

Concerning the function u_2 and the problem (5.1.63), we remark that $t \mapsto f(t,\cdot) + \mathcal{A} G(t,\cdot)$ belongs to $C^{\alpha/2}([0,T]; C(\overline{\Omega}))$. Indeed, since $g \in C^{(1+\alpha)/2, 1+\alpha}([0,T] \times \partial\Omega)$, then $t \mapsto g(t,\cdot)$ belongs to $C^{(1+\alpha)/2}([0,T]; C(\partial\Omega)) \cap B([0,T]; C^{1+\alpha}(\partial\Omega))$. Then $t \mapsto G(t,\cdot)$ belongs to $C^{(1+\alpha)/2}([0,T]; C^1(\overline{\Omega})) \cap B([0,T]; C^{2+\alpha}(\overline{\Omega}))$. Since $C^2(\overline{\Omega})$ belongs to the class $J_{1/(1+\alpha)}(C^1(\overline{\Omega}), C^{2+\alpha}(\overline{\Omega}))$ thanks to Proposition 1.1.3, from Proposition 1.1.4(ii) it follows that $t \mapsto G(t,\cdot)$ belongs to $C^{\alpha/2}([0,T]; C^2(\overline{\Omega}))$. Therefore, $t \mapsto \mathcal{A} G(t,\cdot)$ belongs to $C^{\alpha/2}([0,T]; C(\overline{\Omega}))$. From now on, one can follow the procedure of Theorem 5.1.3, with obvious modifications, proving that statements (i), (ii), (iii) hold with u replaced by u_2. Summing up and arguing as in the proof of Theorem 5.1.3, the result follows. ∎

Some remarks on the assumptions of Theorem 5.1.18 are worth to be made. For u_t and $\mathcal{A}u$ belong to $C^{\alpha/2,0}([0,T]\times\overline{\Omega})$, the conditions $f \in C^{\alpha/2,0}([0,T]\times\overline{\Omega})$ and $\mathcal{A}u_0 + f(0,\cdot) \in C^\alpha(\overline{\Omega})$ are necessary (the former is obviously necessary, for the latter see Remark 4.3.3). But the condition $g \in C^{(1+\alpha)/2,1+\alpha}([0,T]\times\partial\Omega)$ is not necessary. In fact, there is not a simple characterization of the traces on $[0,T]\times\partial\Omega$ of the derivatives of the functions u such that u_t and $\mathcal{A}u$ belong to $C^{\alpha/2,0}([0,T]\times\overline{\Omega})$.

Results similar to the ones of Theorem 5.1.18 hold when f belongs to $UC([0,T]\times\overline{\Omega}) \cap C^{0,\alpha}([0,T]\times\overline{\Omega})$.

Theorem 5.1.19 *Let $\alpha \in\,]0,1[$, let $f \in UC([0,T]\times\overline{\Omega}) \cap C^{0,\alpha}([0,T]\times\overline{\Omega})$, $g \in C^{(1+\alpha)/2,1+\alpha}([0,T]\times\partial\Omega)$, and let $u_0 \in C(\overline{\Omega})$. Then the function u given by (5.1.52) is differentiable with respect to t in $]0,T]\times\overline{\Omega}$, $u(t,\cdot)$ belongs to $W^{2,p}_{loc}(\Omega)$ for every $p \geq 1$, and $u_t \in C^{0,\alpha}([\varepsilon,T]\times\overline{\Omega})$, $\mathcal{A}u \in C^{\alpha/2,\alpha}([\varepsilon,T]\times\overline{\Omega})$ for $0 < \varepsilon < T$. There is $C > 0$ such that*

$$\sup_{0<t\leq T} t(\|u_t(t,\cdot)\|_\infty + \|\mathcal{A}u(t,\cdot)\|_\infty)$$
$$\leq C(\|u_0\|_\infty + \|f\|_{C^{0,\alpha}([0,T]\times\overline{\Omega})} + \|g\|_{C^{(1+\alpha)/2,1+\alpha}([0,T]\times\partial\Omega)}).$$

Moreover, u satisfies (5.1.50), and it is the unique solution of (5.1.50) belonging to $C([0,T]\times\Omega)$ enjoying the above properties.

Statements (i) and (ii) of Theorem 5.1.18 hold (in the estimates, the norm $\|f\|_{C^{\alpha,0}([0,T]\times\overline{\Omega})}$ has to be replaced by $\|f\|_{C^{0,\alpha}([0,T]\times\overline{\Omega})}$).

In addition, if $\partial\Omega$ is uniformly $C^{2+\alpha}$, and $a_{ij}, b_i, c \in C^\alpha(\overline{\Omega})$, $u_0 \in C^{2+\alpha}(\overline{\Omega})$ and the compatibility condition (5.1.60) holds, then u belongs to $C^{1,2+\alpha}([0,T]\times\overline{\Omega})$, and (5.1.50) is satisfied pointwise. Moreover,

$$\|u\|_{C^{1,2+\alpha}([0,T]\times\overline{\Omega})} \leq C(\|u_0\|_{C^{2+\alpha}(\overline{\Omega})} + \|f\|_{C^{0,\alpha}} + \|g\|_{C^{(1+\alpha)/2,1+\alpha}}).$$

Proof — The first part of the statement may be shown as in the proof of Theorem 5.1.18, with obvious modifications. So, we give only the proof of the last statement.

We split again $u = u_1 + u_2$, with u_1 and u_2 defined by (5.1.56), (5.1.57). We know from the proof of Theorem 5.1.18 that $D_t u_1$ belongs to $C^{\alpha/2,\alpha}([0,T]\times\overline{\Omega})$. Since $D_t u_1 = \mathcal{A}u_1 - \mathcal{A}G$, it follows that $\mathcal{A}u_1(t,\cdot)$ belongs to $C^\alpha(\overline{\Omega})$ for every $t \in [0,T]$. By Theorem 3.1.34(ii), $u_1(t,\cdot)$ belongs to $C^{2+\alpha}(\overline{\Omega})$ for every t, and the differential equation in (5.1.62) is satisfied pointwise. Moreover, $\mathcal{A}u_1(t,\cdot)$ belongs to $B([0,T]; C^\alpha(\overline{\Omega})) \cap C([0,T]; C(\overline{\Omega}))$, so that it belongs to $C([0,T]; C^\beta(\overline{\Omega}))$ for every $\beta < \alpha$ thanks to Proposition 1.1.4(iii). Due again to Theorem 3.1.34(ii), $t \mapsto u_1(t,\cdot)$ belongs to $C([0,T]; C^{2+\beta}(\overline{\Omega}))$. In particular, the derivatives $D_{ij}u_1$ are uniformly continuous in $[0,T]\times\overline{\Omega}$. It follows that u_1 belongs to $C^{1,2+\alpha}([0,T]\times\overline{\Omega})$, and

$$\|u_1\|_{C^{1,2+\alpha}([0,T]\times\overline{\Omega})} + \|D_t u_1\|_{C^{\alpha/2,\alpha}([0,T]\times\overline{\Omega})} \leq c\|g\|_{C^{(1+\alpha)/2,1+\alpha}([0,T]\times\partial\Omega)}. \quad (5.1.64)$$

1. SECOND ORDER EQUATIONS

As far as u_2 is concerned, the function $t \mapsto f(t,\cdot) + \mathcal{A}G(t,\cdot)$ belongs to $C([0,T]; C(\overline{\Omega})) \cap B([0,T]; C^\alpha(\overline{\Omega}))$. The initial datum $u_0 - G(0,\cdot)$ belongs to $C^{2+\alpha}(\overline{\Omega})$, and $\mathcal{B}(u_0 - G(0,\cdot))$ vanishes on $\partial\Omega$. Therefore, $u_0 - G(0,\cdot)$ belongs to $D_{A_1}(\alpha/2+1,\infty)$. Corollary 4.3.9(iii) implies that $t \mapsto D_t u_2(t,\cdot)$ and $t \mapsto \mathcal{A}u_2(t,\cdot)$ belong to $C([0,T]; C(\overline{\Omega})) \cap B([0,T]; C^\alpha(\overline{\Omega}))$. This means that $D_t u_2$ and $\mathcal{A}u_2$ belong to the space $UC([0,T] \times \overline{\Omega}) \cap C^{0,\alpha}([0,T] \times \overline{\Omega})$. The same argument used for the regularity of u_1 shows now that $u_2 \in C^{1,2+\alpha}([0,T] \times \overline{\Omega})$, and

$$\|u_2\|_{C^{1,2+\alpha}} + \|D_t u_2\|_{C^{0,\alpha}} \leq \text{const.} \, (\|u_0\|_{C^{2+\alpha}(\overline{\Omega})} + \|f\|_{C^{0,\alpha}} + \|g\|_{C^{(1+\alpha)/,1+\alpha}}). \tag{5.1.65}$$

Recalling that $u = u_1 + u_2$, the statement follows. ∎

Note that the assumptions on the data f, g, u_0 in the last statement of Theorem 5.1.19 are necessary in order that the solution u belong to $C^{1,2+\alpha}([0,T] \times \overline{\Omega})$. Indeed, if $u \in C^{1,2+\alpha}([0,T] \times \overline{\Omega})$ and u_t, $D_{ij}u$ are uniformly continuous, then $u_0 \in C^{2+\alpha}(\overline{\Omega})$ and $u_t - \mathcal{A}u = f \in UC([0,T] \times \overline{\Omega}) \cap C^{0,\alpha}([0,T] \times \overline{\Omega})$. Moreover, from Lemma 5.1.1 we know that the derivatives $D_i u$, $i = 1,\ldots,n$, belong to $C^{(1+\alpha)/2,1+\alpha}([0,T] \times \overline{\Omega})$. Then $\mathcal{B}u = g \in C^{(1+\alpha)/2,1+\alpha}([0,T] \times \partial\Omega)$.

As a corollary of Theorem 5.1.19 we get an optimal regularity result.

Theorem 5.1.20 *Let $\partial\Omega$ be uniformly $C^{2+\alpha}$, with $0 < \alpha < 1$, and let a_{ij}, b_i, $c \in C^\alpha(\overline{\Omega})$, $f \in C^{\alpha/2,\alpha}([0,T] \times \overline{\Omega})$, β_i, $\gamma \in C^{1+\alpha}(\overline{\Omega})$, $g \in C^{(1+\alpha)/2,1+\alpha}([0,T] \times \partial\Omega)$, $u_0 \in C^{2+\alpha}(\overline{\Omega})$. Assume moreover that the compatibility condition (5.1.60) holds. Then the solution u of problem (5.1.50), given by formula (5.1.52), belongs to $C^{1+\alpha/2,2+\alpha}([0,T] \times \overline{\Omega})$. Moreover,*

$$\|u\|_{C^{1+\alpha/2,2+\alpha}} \leq C(\|u_0\|_{C^{2+\alpha}(\overline{\Omega})} + \|f\|_{C^{\alpha/2,\alpha}} + \|g\|_{C^{(1+\alpha)/2,1+\alpha}}).$$

Proof — Due to Theorem 5.1.19, u belongs to $C^{1,2+\alpha}([0,T] \times \overline{\Omega})$. Lemma 5.1.1 yields that the second order space derivatives $D_{ij}u$ belong to $C^{\alpha/2,\alpha}([0,T] \times \overline{\Omega})$. From the equality $u_t = \mathcal{A}u + f$ we deduce that u_t belongs to $C^{\alpha/2,\alpha}([0,T] \times \overline{\Omega})$, and the statement follows. ∎

We end this section with a nonautonomous problem,

$$\begin{cases} u_t(t,x) = \mathcal{A}(t,x)u(t,x) + f(t,x), & 0 < t \leq T, \ x \in \overline{\Omega}, \\ u(0,x) = u_0(x), & x \in \overline{\Omega}, \\ \mathcal{B}(t,x)u(t,x) = g(t,x), & 0 < t \leq T, \ x \in \partial\Omega, \end{cases} \tag{5.1.66}$$

where, as usual,

$$\mathcal{A}(t,x)\varphi = \sum_{i,j=1}^n a_{ij}(t,x)D_{ij}\varphi + \sum_{i=1}^n b_i(t,x)D_i\varphi + c(t,x)\varphi,$$
$$\mathcal{B}(t,x)\varphi = \sum_{i=1}^n \beta_i(t,x)D_i\varphi + \gamma(t,x)\varphi.$$

Theorem 5.1.21 *Let $\partial\Omega$ be uniformly $C^{2+\alpha}$, with $0 < \alpha < 1$, and let a_{ij}, b_i, c, $f \in UC([0,T] \times \overline{\Omega}) \cap C^{0,\alpha}([0,T] \times \overline{\Omega})$, β_i, γ, $g \in C^{(1+\alpha)/2,1+\alpha}([0,T] \times \partial\Omega)$, $u_0 \in C^{2+\alpha}(\overline{\Omega})$. Moreover, assume that*

$$\mathcal{B}(0,x)u_0(x) = g(0,x), \quad x \in \partial\Omega. \tag{5.1.67}$$

Then problem (5.1.66) has a unique solution solution $u \in C^{1,2+\alpha}([0,T] \times \overline{\Omega})$, and

$$\|u\|_{C^{1,2+\alpha}} \leq C(\|u_0\|_{C^{2+\alpha}(\overline{\Omega})} + \|f\|_{C^{0,\alpha}} + \|g\|_{C^{(1+\alpha)/2,1+\alpha}}).$$

Proof — The proof is similar to that of Theorem 5.1.9. It is sufficient to show that there exists $\delta > 0$ such that if $0 \leq a < b \leq T$ and $b - a \leq \delta$, then for every $v_0 \in C^{2+\alpha}(\overline{\Omega})$ such that $\mathcal{B}(a,x)v_0(x) = g(a,x)$ for $x \in \partial\Omega$, the problem

$$\begin{cases} v_t(t,x) = \mathcal{A}(t,x)v(t,x) + f(t,x), & a \leq t \leq b, \ x \in \overline{\Omega}, \\ v(a,x) = v_0(x), & x \in \overline{\Omega}, \\ \mathcal{B}(t,x)v(t,x) = g(t,x), & a \leq t \leq b, \ x \in \partial\Omega, \end{cases} \tag{5.1.68}$$

has a unique solution $v \in C^{1,2+\alpha}([a,b] \times \overline{\Omega})$, and

$$\|v\|_{C^{1,2+\alpha}} \leq K(\|v_0\|_{C^{2+\alpha}(\overline{\Omega})} + \|f\|_{C^{0,\alpha}([a,b]\times\overline{\Omega})} + \|g\|_{C^{(1+\alpha)/2,1+\alpha}([a,b]\times\partial\Omega)}),$$

with K independent of a and b. Define

$$Y = \{v \in C^{1,2+\alpha}([a,b] \times \overline{\Omega}) : v(a) = v_0\},$$

and for every $v \in Y$ set $\Gamma v = u$, where u is the solution of

$$\begin{cases} u_t(t,x) = \mathcal{A}(a,x)u(t,x) + [\mathcal{A}(t,x) - \mathcal{A}(a,x)]v(t,x) + f(t,x), & a \leq t \leq b, \ x \in \overline{\Omega}, \\ u(a,x) = v_0(x), & x \in \overline{\Omega}, \\ \mathcal{B}(a,x)u(t,x) = [\mathcal{B}(a,x) - \mathcal{B}(t,x)]u(t,x) + g(t,x), & a \leq t \leq b, \ x \in \partial\Omega. \end{cases} \tag{5.1.69}$$

Clearly, every fixed point of Γ in Y is a solution of (5.1.68).

For every $v \in Y$, the second order derivatives $D_{ij}v$ are uniformly continuous, and the first order derivatives D_iv belong to $C^{(1+\alpha)/2,1+\alpha}([a,b] \times \overline{\Omega})$, thanks to Lemma 5.1.1. Therefore, the function $(t,x) \mapsto [\mathcal{A}(t,x) - \mathcal{A}(a,x)]v(t,x) + f(t,x)$ belongs to $UC([a,b] \times \overline{\Omega}) \cap C^{0,\alpha}([a,b] \times \overline{\Omega})$, and the function $(t,x) \mapsto [\mathcal{B}(a,x) - \mathcal{B}(t,x)]v(t,x) + g(t,x)$ belongs to $C^{(1+\alpha)/2,1+\alpha}([a,b]\times\partial\Omega)$. So, by Theorem 5.1.19, Γv belongs to $C^{1,2+\alpha}([a,b] \times \overline{\Omega})$. In particular, Γ maps Y into itself. Moreover, by Theorem 5.1.19, there is C, independent on $b - a$, such that for every couple of functions $v, w \in Y$ it holds

$$\|\Gamma v - \Gamma w\|_{C^{1,2+\alpha}} \leq C(\|\varphi\|_{C^{0,\alpha}([a,b]\times\overline{\Omega})} + \|\psi\|_{C^{(1+\alpha)/2,1+\alpha}([a,b]\times\partial\Omega)}),$$

where

$$\varphi(t,x) = (\mathcal{A}(t,x) - \mathcal{A}(a,x))(v(t,x) - w(t,x)), \quad a \leq t \leq b, \ x \in \overline{\Omega},$$
$$\psi(t,x) = (\mathcal{B}(a,x) - \mathcal{B}(t,x))(v(t,x) - w(t,x)), \quad a \leq t \leq b, \ x \in \partial\Omega.$$

1. Second order equations

The estimation of $\|\varphi\|_{C^{0,\alpha}([a,b]\times\overline{\Omega})}$ is identical to the corresponding one in the proof of Theorem 5.1.9: if C_α, C'_α are the embedding constants given by Lemma 5.1.1, and K_1, K_2, $\varepsilon(\delta)$ are defined respectively by (5.1.28) and (5.1.30), with $\mathbb{R}^n$ replaced by $\overline{\Omega}$, then

$$\|\varphi\|_{C^{0,\alpha}} \leq [2(K_1+K_2)(C_\alpha \delta^{\alpha/2} + C'_\alpha \delta^{1/2+\alpha/2} + \delta) + 2\varepsilon(\delta)]\,\|v-w\|_{C^{1,2+\alpha}}. \quad (5.1.70)$$

Now we estimate $\|\psi\|_{C^{(1+\alpha)/2,0}([a,b]\times\overline{\Omega})}$. Setting

$$K_3 = \max_i \|\beta_i\|_{C^{(1+\alpha)/2,1+\alpha}([0,T]\times\overline{\Omega})} + \|\gamma\|_{C^{(1+\alpha)/2,1+\alpha}([0,T]\times\overline{\Omega})},$$

we get

$$[\psi]_{C^{(1+\alpha)/2,0}} \leq \sum_{i=1}^n ([\beta_i]_{C^{(1+\alpha)/2,0}} \|D_i(v-w)\|_\infty$$
$$+ \sup_{a\leq t\leq b} \|\beta_i(a,\cdot) - \beta_i(t,\cdot)\|_\infty [D_i(v-w)]_{C^{(1+\alpha)/2,0}}) + [\gamma]_{C^{(1+\alpha)/2,0}} \|(v-w)\|_\infty$$
$$+ \sup_{a\leq t\leq b} \|\gamma(a,\cdot) - \gamma(t,\cdot)\|_\infty [(v-w)]_{C^{(1+\alpha)/2,0}}$$
$$\leq 2K_3(\delta^{(1+\alpha)/2} \sum_{i=1}^n [D_i(v-w)]_{C^{(1+\alpha)/2,0}} + \delta\|D_t(v-w)\|_\infty).$$

It follows that

$$\|\psi\|_{C^{(1+\alpha)/2,0}} \leq 2K_3(\delta^{(1+\alpha)/2} + 1)(C_\alpha \delta^{(1+\alpha)/2} + \delta)\|u-v\|_{C^{1,2+\alpha}}. \quad (5.1.71)$$

Moreover, for $k = 1,\ldots,n$

$$\|D_k\psi\|_\infty \leq \sum_{i=1}^n (2\|D_k\beta_i\|_\infty \|D_i(v-w)\|_\infty$$
$$+ \delta^{(1+\alpha)/2}[\beta_i]_{C^{(1+\alpha)/2,0}} \|D_{ik}(v-w)\|_\infty)$$
$$+ 2\|D_k\gamma\|_\infty \|(v-w)\|_\infty + \delta^{(1+\alpha)/2}[\gamma]_{C^{(1+\alpha)/2,0}} \|D_k(v-w)\|_\infty$$
$$\leq 2K_3(\delta^{(1+\alpha)/2}C_\alpha + \delta^{(1+\alpha)/2} + \delta)\|v-w\|_{C^{1,2+\alpha}},$$

and for every $t \in [a,b]$

$$[D_k\psi(t,\cdot)]_{C^\alpha(\overline{\Omega})} \leq \sum_{i=1}^n (2[D_k\beta_i]_{C^{0,\alpha}} \|D_i(v-w)\|_\infty$$
$$+ 2\|D_k\beta_i\|_\infty [D_i(v-w)]_{C^{0,\alpha}} + 2[\beta_i]_{C^{0,\alpha}} \|D_{ik}(v-w)\|_\infty +$$
$$+ \delta^{(1+\alpha)/2}[\beta_i]_{C^{(1+\alpha)/2,0}} [D_{ik}(v-w)]_{C^{0,\alpha}})$$
$$+ 2[D_k\gamma]_{C^{0,\alpha}} \|v-w\|_\infty + 2\|D_k\gamma\|_\infty [v-w]_{C^{0,\alpha}}$$
$$+ 2[\gamma]_{C^{0,\alpha}} \|D_k(v-w)\|_\infty + \delta^{(1+\alpha)/2}[\gamma]_{C^{(1+\alpha)/2,0}} [D_k(v-w)]_{C^{0,\alpha}},$$

so that

$$[D_k\psi(t,\cdot)]_{C^\alpha(\overline{\Omega})} \leq K_3 \sum_{i=1}^n (2\delta^{(1+\alpha)/2}[D_i(v-w)]_{C^{(1+\alpha)/2,0}}$$
$$+ 2\delta^{1/2}\|v-w\|_{C^{1/2}([a,b];C^{1+\alpha}(\overline{\Omega}))} + 2\delta^{\alpha/2}[D_{ik}(v-w)]_{C^{\alpha/2,0}}$$
$$+ \delta^{(1+\alpha)/2}[D_{ik}(v-w)]_{C^{0,\alpha}})$$
$$+ K_3(2\delta\|D_t(v-w)\|_\infty + 2\delta\|v-w\|_{Lip([a,b];C^{1+\alpha}(\overline{\Omega}))}$$
$$+ 2\delta^{(1+\alpha)/2}[D_k(v-w)]_{C^{(1+\alpha)/2,0}} + \delta^{(1+\alpha)/2}[\gamma]_{C^{(1+\alpha)/2,0}} [D_k(v-w)]_{C^\alpha(\overline{\Omega})}).$$
$$(5.1.72)$$

Due to Lemma 5.1.1,

$$\|v-w\|_{C^{1/2}([a,b];C^{1+\alpha}(\overline{\Omega}))} + \|v-w\|_{Lip([a,b];C^{\alpha}(\overline{\Omega}))} \leq K_{\alpha}\|v-w\|_{C^{1,2+\alpha}([a,b]\times\overline{\Omega})},$$

with K_{α} independent of $b-a$. So, from (5.1.70), (5.1.71), and (5.1.72) it follows that

$$\|\psi\|_{C^{(1+\alpha)/2,1+\alpha}([a,b]\times\overline{\Omega})} \leq const.\ \delta^{\alpha/2}\|v-w\|_{C^{1,2+\alpha}([a,b]\times\overline{\Omega})}.$$

Since $\lim_{\delta\to 0}\varepsilon(\delta) = 0$, if δ is sufficiently small then Γ is a contraction with constant $1/2$, so that there exists a unique fixed point v of Γ in Y, which is the unique solution of (5.1.68) belonging to $C^{1,2+\alpha}([a,b]\times\mathbb{R}^n)$.

The norm of $\|v\|_{C^{1,2+\alpha}([a,b]\times\overline{\Omega})}$ may be estimated as in the proof of Theorem 5.1.9. ∎

As usual, from the optimal regularity result in $UC([0,T]\times\overline{\Omega}) \cap C^{0,\alpha}([0,T]\times\overline{\Omega})$ we get an optimal regularity result in $C^{\alpha/2,\alpha}([0,T]\times\overline{\Omega})$.

Corollary 5.1.22 *Let $\partial\Omega$ be uniformly $C^{2+\alpha}$, with $0 < \alpha < 1$, and let a_{ij}, b_i, c, $f \in C^{\alpha/2,\alpha}([0,T]\times\overline{\Omega})$, β_i, γ, $g \in C^{(1+\alpha)/2,1+\alpha}([0,T]\times\partial\Omega)$, $u_0 \in C^{2+\alpha}(\overline{\Omega})$, and assume that (5.1.67) holds. Then the solution u of problem (5.1.66) belongs to $C^{1+\alpha/2,2+\alpha}([0,T]\times\overline{\Omega})$, and*

$$\|u\|_{C^{1+\alpha/2,2+\alpha}} \leq C(\|u_0\|_{C^{2+\alpha}(\overline{\Omega})} + \|f\|_{C^{\alpha/2,\alpha}} + \|g\|_{C^{(1+\alpha)/2,1+\alpha}}).$$

Proof — By Theorem 5.1.21, problem (5.1.66) has a unique solution u, which belongs to $C^{1,2+\alpha}([0,T]\times\overline{\Omega})$. By Lemma 5.1.1 the derivatives $D_{ij}u$ belong to $C^{\alpha/2,\alpha}([0,T]\times\overline{\Omega})$. Since the coefficients a_{ij}, b_i, c and the function f belong to $C^{\alpha/2,\alpha}([0,T]\times\overline{\Omega})$, from the equality $u_t = \mathcal{A}u + f$ we get $u_t \in C^{\alpha/2,\alpha}([0,T]\times\overline{\Omega})$, and the statement follows. ∎

5.2 Bibliographical remarks

Existence and regularity results for linear second order and higher order parabolic problems with $C^{\alpha/2,\alpha}$ coefficients are well known. The classical approach may be found in O.A. LADYZHENSKAJA – V.A. SOLONNIKOV – N.N. URAL'CEVA [124, Chapter 4], V.A. SOLONNIKOV [184] and in A.M. ILIN – A.S. KALASHNIKOV – O.A. OLEINIK [103]. [124] and [184] are good references also for L^p regularity, which is not treated here. See also A. FRIEDMAN [84], and M.S. AGRANOVICH – M.I. VISHIK [14], J.-L. LIONS – E. MAGENES [128] for the Hilbert space approach.

Continuous and Hölder continuous regularity results in the case of less regular coefficients are less known. S.N. KRUZHKOV – A. CASTRO – M. LOPES [116, 117] considered the case of $C^{0,\alpha}$ coefficients for second order equations in $[0,T]\times\mathbb{R}^n$. Most of the L^p regularity theory is developed in the case of uniformly continuous

2. Bibliographical remarks

and bounded coefficients. Of course, some regularity results may be deduced from the L^p theory by Sobolev embeddings, but they are not optimal.

The abstract evolution equation approach presented here has not systematically developed up to now. There is a number of results, spread in many papers. Just to mention a few, we quote E. SINESTRARI [177], E. SINESTRARI – W. VON WAHL [179], A. LUNARDI [146], A. LUNARDI – E. SINESTRARI – W. VON WAHL [156] for continuous and Hölder regularity results, and I. LASIECKA [125], B. TERRENI [197], H. AMANN [24] for L^p regularity results. We have followed the method of [146, 156] in the study of the nonhomogeneous at the boundary problems (5.1.44), (5.1.50).

In this chapter we have considered three types of optimal regularity results: $C^{\alpha,0}$ (Theorems 5.1.3(iii), 5.1.12(iii), 5.1.18(iii)), $C^{0,\alpha}$ (Theorems 5.1.4(iv), 5.1.9, 5.1.13, 5.1.19, 5.1.21), and $C^{\alpha/2,\alpha}$ (Theorems 5.1.8, 5.1.10, 5.1.15, 5.1.16, 5.1.20).

The $C^{0,\alpha}$ optimal regularity results hold in the case where Ω is the whole space $\mathbb{R}^n$, or else Ω is a regular open set and the boundary condition is of the first order. They have been shown in [116, 117] by the potential theoretical approach when $\Omega = \mathbb{R}^n$, and in [146] by the semigroup approach when Ω is bounded. In the case of the Dirichlet boundary condition, they fail to be true, as a counterexample in [179] shows. To be more precise, the counterexample shows that there are bounded C^∞ domains Ω, functions $f \in C^{0,\alpha}([0,T] \times \overline{\Omega})$, with $f(0,\cdot) = 0$, such that the solution of $u_t = \Delta u + f$ in $[0,T] \times \overline{\Omega}$, $u(0,x) = 0$ in $\overline{\Omega}$, $u(t,x) = 0$ for $x \in \partial\Omega$, does not belong to $C^{1,2+\alpha}([0,T] \times \overline{\Omega})$. However, u belongs to $C^{1,2+\alpha}([0,T] \times \overline{\Omega}')$, for every Ω' with closure contained in Ω. See [179].

The idea of integrating by parts in the case of the oblique boundary condition to get the representation formula (5.1.52) goes back to A.V. BALAKRISHNAN [29], and was used in [125] and in [146].

The $C^{\alpha/2,\alpha}$ optimal regularity results are the most familiar to people working in PDE's. As we mentioned above, the classical treatment may be found in [124, Chapter 4] for second order equations, and in [184] for higher order equations and systems. The semigroup approach may be found in [156].

Chapter 6

Linear nonautonomous equations

In this chapter we treat equations of the type

$$u'(t) = A(t)u(t) + f(t), \quad 0 < t \leq T, \quad u(0) = x, \qquad (6.0.1)$$

where, for every $t \in [0, T]$, the linear operator $A(t) : D(A(t)) \subset X \mapsto X$ is sectorial, and the function $t \mapsto A(t)$ has a certain degree of smoothness.

The literature about linear nonautonomous parabolic equations is rather wide, and giving an overview of all the existing results is out of the aim of this book. We shall consider only the simplest case, when the operators $A(t)$ have common domains $D(A(t)) \equiv D$, and $t \mapsto A(t)$ is Hölder continuous with values in $L(D, X)$.

Strict, classical, and strong solutions of (6.0.1) are defined as in the autonomous case, see Definition 4.1.1. The role of the analytic semigroup e^{tA} is played now by the evolution operator $G(t, s)$.

Definition 6.0.1 *A family of linear operators $\{G(t, s) : 0 \leq s \leq t \leq T\} \subset L(X)$ is said to be an evolution operator for problem (6.0.1) if*

$$\begin{cases} (a) & G(t, s)G(s, r) = G(t, r), \quad G(s, s) = I, \quad 0 \leq r \leq s \leq t \leq T, \\ (b) & G(t, s) \in L(X, D) \text{ for } 0 \leq s < t \leq T, \\ (c) & t \mapsto G(t, s) \text{ is differentiable in }]s, T] \text{ with values in } L(X), \text{ and} \\ & \frac{\partial}{\partial t} G(t, s) = A(t)G(t, s), \, 0 \leq s < t \leq T. \end{cases} \qquad (6.0.2)$$

The construction of the evolution operator is the subject of Section 6.1. Several properties of $G(t, s)$ are similar to the ones of $e^{(t-s)A}$, where A is any sectorial operator: for instance, we shall see that there is a constant C such that

$$\|G(t, s)\|_{L(X)} \leq C, \quad \|A(t)G(t, s)\|_{L(X)} \leq \frac{C}{t - s}, \quad 0 \leq s < t \leq T.$$

Once existence and properties of the evolution operator are established, we show that, under minimal assumption on $x \in X$ and $f : [s,T] \subset [0,T] \mapsto X$, any solution (strict, classical, strong) of problem

$$u'(t) = A(t)u(t) + f(t), \ s < t \leq T; \quad u(s) = x, \quad (6.0.3)$$

may be represented by the variation of constants formula,

$$u(t) = G(t,s)x + \int_s^t G(t,\sigma)f(\sigma)d\sigma, \ s \leq t \leq T. \quad (6.0.4)$$

Then, using the properties of the evolution operator shown before, we are able to state a number of regularity results, quite similar to the ones obtained in the autonomous case. Also the techniques are similar, and some proofs will be only sketched.

As in the autonomous case, for every $f \in L^1(0,T;X)$ and $x \in X$ the function u defined by (6.0.4) is called the *mild solution* of (6.0.3).

The asymptotic behavior as $t - s \to +\infty$ is considered in Section 6.3, in the periodic case $A(t) = A(t+T)$. The results will be used in Section 9.3, where periodic solutions of nonlinear problems will be studied.

6.1 Construction and properties of the evolution operator

Let D be a Banach space, continuously embedded in X, and let $T > 0$, $0 < \alpha < 1$. For $0 \leq t \leq T$, let $A(t) : D(A(t)) \mapsto X$ be a linear operator such that

$$\begin{cases} (a) & \forall t \in [0,T], \ A(t) \text{ is sectorial and } D(A(t)) \simeq D, \\ (b) & t \mapsto A(t) \in C^\alpha([0,T]; L(D,X)). \end{cases} \quad (6.1.1)$$

Then the operators $e^{sA(t)}$ satisfy some estimates which will be used throughout.

Lemma 6.1.1 *Let (6.1.1) hold. Then*

(i) *There are $\omega \in \mathbb{R}$, $\theta \in \ ]\pi/2, \pi[$, $M > 0$ such that for every $t \in [0,T]$ the operator $A(t)$ satisfies assumption (2.0.1) with constants ω, θ, M;*

(ii) *There are constants γ, $\gamma_{\alpha,p} \geq 1$, such that for $0 \leq t \leq T$ we have*

(a) $\gamma^{-1}\|y\|_D \leq \|y\| + \|A(t)y\| \leq \gamma\|y\|_D, \ \forall y \in D,$

(b) $\gamma_{\alpha,p}^{-1}\|y\|_{(X,D)_{\alpha,p}} \leq \|y\|_{D_{A(t)}(\alpha,p)} \leq \gamma_{\alpha,p}\|y\|_{(X,D)_{\alpha,p}}, \ \forall y \in (X,D)_{\alpha,p};$

$$(6.1.2)$$

1. Construction and properties of the evolution operator

(iii) For $k \in \mathbb{N} \cup \{0\}$ there are M_k, C_k, such that for $0 < s \leq T$, $0 \leq r$, $t \leq T$ we have

$$\begin{cases} (a) & \|s^k A(t)^k e^{sA(t)}\|_{L(X)} \leq M_k, \\ (b) & \|s^k (A(t)^k e^{sA(t)} - A(r)^k e^{sA(r)})\|_{L(X)} \leq C_k (t-r)^\alpha. \end{cases} \quad (6.1.3)$$

(iv) For $k = 1, 2$ there are constants N_k, c_k such that

$$(a) \quad \|s e^{sA(t)}\|_{L(X,D)} \leq N_1, \quad \|s^2 A(t) e^{sA(t)}\|_{L(X,D)} \leq N_2,$$

$$(b) \begin{cases} \|s(e^{sA(t)} - e^{sA(r)})\|_{L(X,D)} \leq c_1 (t-r)^\alpha, \\ \|s^2 (A(t) e^{sA(t)} - A(r) e^{sA(r)})\|_{L(X,D)} \leq c_2 (t-r)^\alpha. \end{cases} \quad (6.1.4)$$

Sketch of the proof — Statement (i) follows from perturbation arguments similar to the ones of Section 2.4. Estimate (6.1.2)(a) is an easy consequence of assumption (6.1.1)(a) and of the continuity of $A(\cdot)$. Estimate (6.1.2)(b) follows from (6.1.2)(a) through Corollary 2.2.3(i). Statement (iii) follows from (i), through the procedure of Proposition 2.1.1(iii), recalling that for $0 \leq t$, $r \leq T$ and $\lambda \in \rho(A(t)) \cap \rho(A(s))$,

$$(\lambda - A(t))^{-1} - (\lambda - A(r))^{-1} = (\lambda - A(t))^{-1}(A(t) - A(r))(\lambda - A(r))^{-1}.$$

Estimates (6.1.4) follow from (6.1.2)(a) and (6.1.3). ∎

For $0 \leq s < T$ we consider problem (6.0.3).

A first uniqueness lemma for the strict solution of (6.0.3) can be easily shown. Precisely, one shows uniqueness in the set of the classical solutions u such that $\|(t-s)^\beta u(t)\|_D$ is bounded for some $\beta \in \,]0,1[$. Since every strict solution is bounded with values in D, then the strict solution of (6.0.3) is unique. Stronger uniqueness results will be proved later.

Lemma 6.1.2 *Let $0 < \beta < 1$, and let $u \in C(]s, T]; D)$ be a classical solution of (6.0.3) with $f \equiv 0$, $x = 0$, such that $\|(t-s)^\beta u(t)\|_D$ is bounded. Then $u \equiv 0$.*

Proof — Set $t_0 = \sup\{t \in [s,T] : u_{|[s,t]} \equiv 0\}$.
Since u is continuous, then $u(t_0) = 0$. If $t_0 = T$, there is nothing to show. If not, for every $\delta \in \,]0, T - t_0[$ we have

$$K(\delta) = \sup\{(t - t_0)^\beta \|u(t)\|_D : t_0 < t \leq t_0 + \delta\} < \infty.$$

From the equality

$$\frac{\partial}{\partial s} e^{(t-s)A(t)} u(s) = e^{(t-s)A(t)}(A(s) - A(t))u(s), \quad t_0 < s < t,$$

we get

$$u(t) = \int_{t_0}^{t} e^{(t-s)A(t)}(A(s) - A(t))u(\sigma)ds, \quad t_0 \leq t \leq T,$$

so that for $t_0 < t \leq t_0 + \delta \leq T$

$$(t-t_0)^\beta \|u(t)\|_D \leq M_1[A]_{C^\alpha} K(\delta) \int_{t_0}^t (t-s)^{\alpha-1}(s-t_0)^{-\beta} ds.$$

It follows that

$$K(\delta) \leq M_1[A]_{C^\alpha} \int_0^1 (1-\sigma)^{\alpha-1} \sigma^\beta d\sigma \; \delta^\alpha K(\delta),$$

which is impossible if δ is small. Therefore, $t_0 = T$, and the statement is proved. ∎

To begin with, we solve (6.0.3) in the simplest case, where f is Hölder continuous, $x \in D$, and f, x satisfy a suitable compatibility condition.

Proposition 6.1.3 *Let $f \in C^\alpha([s,T];X)$, $x \in D$ be such that*

$$A(s)x + f(s) \in (X,D)_{\alpha,\infty}. \tag{6.1.5}$$

Then problem (6.0.3) has a unique solution $u \in C^\alpha([s,T];D) \cap C^{1+\alpha}([s,T];X)$, and there is $C > 0$, independent of f, x, s, such that

$$\begin{aligned}&\|u\|_{C^\alpha([s,T];D)} + \|u'\|_{C^\alpha([s,T];X)} + \|u'\|_{B([s,T];(X,D)_{\alpha,\infty})} \\ &\leq C(\|x\|_D + \|A(s)x + f(s)\|_{(X,D)_{\alpha,\infty}} + \|f\|_{C^\alpha([s,T];X)}).\end{aligned} \tag{6.1.6}$$

Proof — We first show that problem (6.0.3) can be uniquely solved in $C^\alpha([s,s+\delta];D)$, where $\delta \in {]}0,T-s]$ is sufficiently small.

Any solution $u \in C^\alpha([s,s+\delta];D)$ of problem (6.0.3) is a fixed point of the operator Γ, defined on $C^\alpha([s,s+\delta];D)$ by

$$\Gamma v(t) = e^{(t-s)A(s)}x + \int_s^t e^{(t-\sigma)A(s)}[(A(\sigma)-A(s))v(\sigma) + f(\sigma)]d\sigma. \tag{6.1.7}$$

We shall prove that Γ maps the closed set

$$Y = \{v \in C^\alpha([s,s+\delta];D) \; : \; v(s) = x\}$$

into itself and it is a contraction in Y, provided δ is sufficiently small.

For each $v \in C^\alpha([s,s+\delta];D)$ the function $t \mapsto g(t) = (A(t)-A(s))v(t) + f(t)$ belongs obviously to $C^\alpha([s,s+\delta];X)$. Moreover, $x \in D$, and by assumption (6.1.5)

$$A(s)x + g(s) = A(s)x + f(s) \in (X,D)_{\alpha,\infty} = D_{A(s)}(\alpha,\infty).$$

By Theorem 4.3.1(iii), Γv belongs to $C^\alpha([s,s+\delta];D)$. Let now C be the constant given by Theorem 4.3.1(iii), and let γ, $\gamma_{\alpha,\infty}$ be the constants given by formula (6.1.2). For v_1, v_2 in Y we have

$$\begin{aligned}\|\Gamma v_1 - \Gamma v_2\|_{C^\alpha([s,s+\delta];D)} &\leq \gamma \|\Gamma v_1 - \Gamma v_2\|_{C^\alpha([s,s+\delta];D(A(s)))} \\ &\leq \gamma C \|(A(\cdot) - A(s))(v_1 - v_2)\|_{C^\alpha([s,s+\delta];X)}.\end{aligned}$$

1. CONSTRUCTION AND PROPERTIES OF THE EVOLUTION OPERATOR 215

Let us estimate the Hölder norm of $(A(\cdot) - A(s))(v_1 - v_2)$. For $s \leq r \leq t \leq s+\delta$ we have
$$\|(A(t) - A(s))(v_1(t) - v_2(t))\| \leq [A]_{C^\alpha} \delta^\alpha \|v_1 - v_2\|_\infty,$$
and
$$\|(A(t) - A(s))(v_1(t) - v_2(t)) - (A(r) - A(s))(v_1(r) - v_2(r))\|$$
$$\leq \|(A(t) - A(r))(v_1(t) - v_2(t))\|$$
$$+ \|(A(r) - A(s))(v_1(t) - v_2(t) - v_1(r) + v_2(r))\|$$
$$\leq [A]_{C^\alpha}(t-r)^\alpha [v_1 - v_2]_{C^\alpha(D)} \delta^\alpha + [A]_{C^\alpha} \delta^\alpha [v_1 - v_2]_{C^\alpha(D)}(t-r)^\alpha,$$
where we have used the equality $v_1(s) - v_2(s) = 0$. Then,
$$\|(A(\cdot) - A(s))(v_1 - v_2)\|_{C^\alpha(X)} \leq 2[A]_{C^\alpha} \delta^\alpha \|v_1 - v_2\|_{C^\alpha(D)}, \quad (6.1.8)$$
so that Γ is a 1/2-contraction provided
$$\delta = \min\{T - s, (4\gamma C[A]_{C^\alpha})^{-1/\alpha}\}.$$

In that case there is a unique fixed point u of Γ in Y, and u is the unique solution of (6.0.3) belonging to $C^\alpha([s, s+\delta]; D)$.

The Contraction Theorem gives $\|u\|_{C^\alpha(D)} \leq 2\|\Gamma \overline{u}\|_{C^\alpha(D)} + \|\overline{u}\|_{C^\alpha(D)}$, where $\overline{u}$ is the constant function $\overline{u}(t) \equiv x$, which obviously belongs to Y. $\Gamma \overline{u}$ is the solution w of
$$w'(t) = A(s)w(t) + (A(t) - A(s))x + f(t), \quad s < t \leq s+\delta, \quad w(s) = x.$$

An estimate similar to (6.1.6) (with $[s, T]$ replaced by $[s, s+\delta]$) follows from Theorem 4.3.1(iii) and estimates (6.1.3). If $s + \delta = T$, the proof is finished. If not, since $u(s+\delta) \in D$ and $u'(s+\delta) = A(s+\delta)u(s+\delta) + f(s+\delta) \in (X, D)_{\alpha,\infty}$, we apply the above procedure to problem (6.0.3), with s replaced by $s+\delta$ and x replaced by $u(s+\delta)$. It is clear that, iterating the procedure, in a finite number of steps we construct a solution u of (6.0.3) satisfying estimate (6.1.6). ∎

Now we come back to the general case, when $x \in X$. To overcome the difficulty caused by the singularity at $t = s$ of $t \mapsto e^{(t-s)A(s)}x$ as a function with values in D, we introduce the new unknown
$$w(t) = u(t) - e^{(t-s)A(s)}x.$$

Then w should satisfy
$$w'(t) = A(t)w(t) + (A(t) - A(s))e^{(t-s)A(s)}x + f(t), \quad s < t \leq T, \quad w(s) = 0. \quad (6.1.9)$$

As it is easy to check, the function
$$t \mapsto (A(t) - A(s))e^{(t-s)A(s)}x, \quad s < t \leq T,$$

belongs to $C_1^\alpha(]s,T];X)^{(1)}$. This space enjoys the optimal regularity property (see Theorem 4.3.7(ii)). Therefore, we try to solve problem (6.1.9) by a fixed point theorem as in the proof of Proposition 6.1.3.

Theorem 6.1.4 *Let $x \in X$ and $f \in C_1^\alpha(]s,T];X)$. Then problem (6.1.9) has a unique solution $w \in C_1^\alpha(]s,T];D)$. Moreover, w belongs to $C^\alpha([s,T];X) \cap C_{1-\alpha}^{1-\alpha}(]s,T];(X,D)_{\alpha,1})$ (in particular, it is bounded with values in $(X,D)_{\alpha,1}$), and there is $C > 0$, independent of s, f, and x, such that*

$$\|w\|_{C_1^\alpha(]s,T];D)} + \|w'\|_{C_1^\alpha(]s,T];X)} + \|w'\|_{B_1(]s,T];(X,D)_{\alpha,\infty})} + \|w\|_{C^\alpha([s,T];X)}$$
$$+ \|w\|_{C_{1-\alpha}^{1-\alpha}(]s,T];(X,D)_{\alpha,1})} \leq C(\|x\| + \|f\|_{C_1^\alpha(]s,T];X)}).$$
(6.1.10)

Proof — As in the proof of Proposition 6.1.3, first we solve (6.1.9) in a small interval $[s, s+\delta] \subset [0,T]$. In the present case we look for a fixed point in $C_1^\alpha(]s, s+\delta]; D)$ of the operator Λ defined by $\Lambda v = w$, where w is the mild solution of

$$\begin{cases} w'(t) = A(s)w(t) + (A(t) - A(s))(v(t) + e^{(t-s)A(s)}x) + f(t), & s < t \leq s+\delta, \\ w(s) = 0. \end{cases}$$

Let us show that Λ maps $C_1^\alpha(]s, s+\delta]; D)$ into itself. For every $v \in C_1^\alpha(]s, s+\delta]; D)$ the function $(A(\cdot) - A(s))(v + e^{(\cdot-s)A(s)}x)$ belongs to $C_1^\alpha(]s, s+\delta]; X)$. Indeed

$$(t-s)^{1-\alpha} \|(A(t) - A(s))v(t)\| \leq [A]_{C^\alpha} \delta^\alpha \|v\|_{B_{1-\alpha}(D)},$$

and for $s < s+\varepsilon \leq r \leq t \leq s+\delta$

$$\varepsilon \|(A(t) - A(s))v(t) - (A(r) - A(s))v(r)\|$$
$$\leq \varepsilon \|(A(t) - A(r))v(t)\| + \varepsilon \|(A(r) - A(s))(v(t) - v(r))\|$$
$$\leq [A]_{C^\alpha}(t-r)^\alpha \varepsilon^\alpha \|v\|_{B_{1-\alpha}(D)} + [A]_{C^\alpha} \delta^\alpha [v]_{C_1^\alpha(D)}(t-r)^\alpha,$$

so that $(A(\cdot) - A(s))v$ belongs to $C_1^\alpha(]s, s+\delta]; X)$, and

$$\|(A(\cdot) - A(s))v\|_{C_1^\alpha(X)} \leq 2[A]_{C^\alpha} \delta^\alpha \|v\|_{C_1^\alpha(D)}.$$
(6.1.11)

Moreover, for $s < t \leq s+\delta$ we have

$$(t-s)^{1-\alpha} \|(A(t) - A(s))e^{(t-s)A(s)}x\| \leq [A]_{C^\alpha} N_1 \|x\|,$$
(6.1.12)

and, for $s < s+\varepsilon \leq r \leq t \leq s+\delta$,

$$\varepsilon \|(A(r) - A(s))e^{(r-s)A(s)}x\| \leq \varepsilon \|(A(t) - A(r))e^{(t-s)A(s)}x\|$$
$$+ \varepsilon \|(A(r) - A(s))(e^{(t-s)A(s)}x - e^{(r-s)A(s)}x)\|$$
$$\leq [A]_{C^\alpha}(t-r)^\alpha N_1 \|x\| + \varepsilon [A]_{C^\alpha}(r-s)^\alpha N_2 \int_{r-s}^{t-s} \sigma^{-2} d\sigma \|x\|$$
$$\leq [A]_{C^\alpha}(t-r)^\alpha N_1 \|x\| + [A]_{C^\alpha}(t-r)^\alpha \alpha^{-1} N_2 \|x\|,$$
(6.1.13)

---
[1] The definitions of the weighted spaces which are used here may be found in the introduction of Chapter 4.

1. Construction and properties of the evolution operator

so that also $t \mapsto (A(t) - A(s))e^{(t-s)A(s)}x$ belongs to $C_1^\alpha(]s, s+\delta]; X)$, and

$$\|(A(\cdot) - A(s))e^{(\cdot - s)A(s)}x\|_{C_1^\alpha(X)} \leq K\|x\|, \qquad (6.1.14)$$

with $K = [A]_{C^\alpha}(2N_1 + \gamma\alpha^{-1}N_2)$. By Theorem 4.3.7(ii), Λv belongs to $C_1^\alpha(s, s+\delta; D) \cap C^{1-\alpha}([s, s+\delta]; X) \cap B([s, s+\delta]; (X, D)_{1-\alpha,\infty})$, $\Lambda v'$ belongs to $C_1^\alpha(]s, s+\delta]; X)$; moreover, if C is the constant given by (4.3.13), for v_1, v_2 in $C_1^\alpha(]s, s+\delta]; D)$ we have, due to (6.1.11),

$$\|\Lambda v_1 - \Lambda v_2\|_{C_1^\alpha(D)} \leq 2\gamma C[A]_{C^\alpha}\delta^\alpha\|v_1 - v_2\|_{C_1^\alpha(D)},$$

so that Λ is a 1/2-contraction if

$$\delta = \min\{T - s, (4\gamma C[A]_{C^\alpha})^{-1/\alpha}\}.$$

In this case there is a unique fixed point w of Λ in $C_1^\alpha(]s, s+\delta]; D)$. Again by Theorem 4.3.7(ii), we have, using estimates (6.1.11), (6.1.14),

$$\|w\|_{C_1^\alpha(D(A(s)))} + \|w'\|_{C_1^\alpha(X)} + \|w\|_{C^{1-\alpha}(X)} + \|w'\|_{B_1(D_{A(s)}(\alpha,\infty))}$$
$$+ \|w\|_{C^{1-\alpha}(X)} + \|w\|_{C_{1-\alpha}^{1-\alpha}(D_{A(s)}(\alpha,1))}$$
$$\leq C([A]_{C^\alpha}\delta^\alpha\|w\|_{C_1^\alpha(D)} + K\|x\| + \|f\|_{C_1^\alpha(X)}),$$

so that

$$\|w\|_{C_1^\alpha(D)} + \|w'\|_{C_1^\alpha(X)} + \|w'\|_{B_1((X,D)_{\alpha,\infty})} + \|w\|_{C^{1-\alpha}(X)} + \|w\|_{C_{1-\alpha}^{1-\alpha}((X,D)_{\alpha,1})}$$
$$\leq \max\{1, \gamma_{\alpha,1}/2\gamma\}2\gamma CK(\|x\| + \|f\|_{C_1^\alpha(X)}).$$

If $s + \delta = T$, the proof is finished. If $s + \delta < T$, since $w(s+\delta) \in D$ and

$$A(s+\delta)w(s+\delta) + (A(s+\delta) - A(s))e^{\delta A}x + f(s+\delta) = w'(s+\delta) \in (X, D)_{\alpha,\infty},$$

we apply Proposition 6.1.3 to problem (6.1.9), with s replaced by $s + \delta$ and x replaced by $w(s+\delta)$, and we continue w to the whole $[s, T]$, in such a way that the continuation satisfies estimate (6.1.10). ∎

The regularity and the behavior at $t = s$ of w depend of course on the regularity of x and f, as the following proposition states.

Proposition 6.1.5 *Let $x \in X$, $f \in C_1^\alpha(]s, T]; X)$, and let w be the solution of (6.1.9) given by Theorem 6.1.4. Then:*

(i) *If $0 < \theta < 1 - \alpha$, $x \in (X, D)_{\theta,\infty}$ and $f \in C_{1-\theta}^\alpha(]s, T]; X)$, then $w \in C_{1-\theta}^\alpha(]s, T]; D) \cap C^{\alpha+\theta}([s, T]; X) \cap C_{1-\alpha-\theta}^{1-\alpha-\theta}(]s, T]; (X, D)_{\alpha+\theta,1})$, and $w' \in C_{1-\theta}^\alpha(]s, T]; X) \cap B_{1-\theta}(]s, T]; (X, D)_{\alpha,\infty})$;*

(ii) *If $x \in (X, D)_{1-\alpha,\infty}$ and $f \in C_\alpha^\alpha(]s, T]; X)$, then $w \in C_\alpha^\alpha(]s, T]; D)$ and $w' \in C_\alpha^\alpha(]s, T]; X) \cap B_\alpha(]s, T]; (X, D)_{\alpha,\infty})$;*

(iii) If $1 - \alpha < \theta < 1$, $x \in (X, D)_{\theta, \infty}$ and $f \in C^{\alpha + \theta - 1}(]s, T]; X)$, with $f(s) \in (X, D)_{\alpha + \theta - 1, \infty}$, then $w \in C^{\alpha + \theta - 1}(]s, T]; D)$, and $w' \in C^{\alpha + \theta - 1}(]s, T]; X)$;

(iv) If $x \in D$ and $f \in C^{\alpha}([s, T]; X)$, with $f(s) \in (X, D)_{\alpha, \infty}$, then $w \in C^{\alpha}([s, T]; D)$, and $w' \in C^{\alpha}([s, T]; X)$.

Moreover, the mapping $(x, f) \mapsto (w, w')$ is continuous:

(i) from $(X, D)_{\theta, \infty} \times C^{\alpha}_{1-\theta}(]s, T]; X)$ (with $0 < \theta < 1 - \alpha$) to $(C^{\alpha}_1(]s, T]; D) \cap C^{\alpha + \theta}([s, T]; X) \cap C^{1-\alpha-\theta}_{1-\alpha-\theta}(]s, T]; (X, D)_{\alpha+\theta, 1})) \times C^{\alpha}_{1-\theta}(]s, T]; X)$;

(ii) from $(X, D)_{1-\alpha, \infty} \times C^{\alpha}_{\alpha}(]s, T]; X)$ to $C^{\alpha}_{\alpha}(]s, T]; D) \times C^{\alpha}_{\alpha}(]s, T]; X)$;

(iii) from

$$(X, D)_{\theta, \infty} \times \{ f \in C^{\alpha + \theta - 1}([s, T]; X) : f(s) \in (X, D)_{\alpha + \theta - 1, \infty} \}$$

with $1 - \alpha < \theta < 1$, to

$$C^{\alpha + \theta - 1}([s, T]; D) \times \left(C^{\alpha + \theta - 1}([s, T]; X) \cap B([s, T]; (X, D)_{\alpha + \theta - 1, \infty}) \right);$$

(iv) from $D \times \{ f \in C^{\alpha}([s, T]; X) : f(s) \in (X, D)_{\alpha, \infty} \}$ to $C^{\alpha}([s, T]; D) \times (C^{\alpha}([s, T]; X) \cap B([s, T]; (X, D)_{\alpha, \infty}))$.

Proof — We only give the guidelines, and leave the details to the reader. Set

$$\varphi(t) = (A(t) - A(s))e^{(t-s)A(s)}x, \ s < t \leq T.$$

(i) It is sufficient to follow the proof of Theorem 6.1.4 taking into account that $\varphi \in C^{\alpha}_{1-\theta}(]s, s + \delta]; X)$. One has to replace the space $C^{\alpha}_1(]s, s + \delta]; D)$ by $C^{\alpha}_{1-\theta}(]s, s + \delta]; D)$.

(ii) In this case, φ belongs to $C^{\alpha}_{\alpha}(]s, T]; X)$. Therefore one has to work in the space $C^{\alpha}_{\alpha}(]s, T]; D)$.

(iii), (iv) In the case where $x \in (X, D)_{\theta, \infty}$, with $\theta > 1 - \alpha$, or $x \in D$, the function φ can be extended at $t = s$ in such a way that the extension belongs to $C^{\alpha + \theta - 1}([s, T]; X)$ if $x \in (X, D)_{\theta, \infty}$, to $C^{\alpha}([s, T]; X)$ if $x \in D$. Then one can work in the metric spaces $\{ u \in C^{\alpha + \theta - 1}([s, s+\delta]; D) : u(s) = 0 \}$, $\{ u \in C^{\alpha}([s, s+\delta]; D) : u(s) = 0 \}$ respectively, arguing as in Proposition 6.1.3. ∎

We are already able to give some results on problem (6.0.3), which are easy consequences of Propositions 6.1.3, 6.1.5, and Theorem 6.1.4. For $x \in X$ set

$$u(t) = w(t) + e^{(t-s)A(s)}x, \ s \leq t \leq T,$$

where w is the solution of problem (6.1.9) given by Theorem 6.1.4.

Corollary 6.1.6 *Let $0 < \theta \leq \alpha$. Then:*

(i) *If $f \in C_1^\theta(]s,T];X)$ and $x \in \overline{D}$, u belongs to $C([s,T];X) \cap C^{1+\theta}([s+\varepsilon,T];X) \cap C^\theta([s+\varepsilon,T];D)$ for every $\varepsilon \in]0, T-s[$, and it is a classical solution of problem (6.0.3).*

(ii) *If $f \in C_\theta^\theta(]s,T];X)$ and $x \in D$, u belongs to $C_\theta^\theta(]s,T];D)$, u' belongs to $C_\theta^\theta(]s,T];X) \cap B_\theta(]s,T];(X,D)_{\theta,\infty})$.*

(iii) *If $f \in C^\theta([s,T];X)$ and $x \in D$, $A(s)x + f(s) \in \overline{D}$, u is a strict solution of problem (6.0.3).*

(iv) *If $f \in C^\theta([s,T];X)$ and $x \in D$, $A(s)x + f(s) \in (X,D)_{\theta,\infty}$, u is a strict solution of problem (6.0.3) and belongs to $C^{1+\theta}([s,T];X) \cap C^\theta([s,T];D)$.*

Proof — (i) and (ii) are obvious consequences of Theorem 6.1.4, Proposition 6.1.5(ii) and of the properties of $t \mapsto e^{(t-s)A(s)}x$. (iv) is a part of the statement of Proposition 6.1.3. To show that (iii) holds, we set $u = v + z$, where $v(t) = e^{(t-s)A(s)}x + \int_s^t e^{(t-\sigma)A(s)} f(s) d\sigma$ and $z(t)$ are respectively the solutions of

$$v'(t) = A(s)v(t) + f(s), \quad s \leq t \leq T; \quad v(s) = x,$$

$$z'(t) = A(t)z(t) + (A(t) - A(s))v(t) + f(t) - f(s), \quad s \leq t \leq T, \quad z(s) = 0.$$

Then v belongs to $C^1([s,T];X) \cap C([s,T];D)$, moreover it is easy to see that $(A(\cdot) - A(s))v(\cdot)$ belongs to $C^{\alpha \wedge \theta}([s,T];X)$. By applying Proposition 6.1.3 we get $z \in C^{1+\alpha \wedge \theta}([s,T];X) \cap C^{\alpha \wedge \theta}([s,T];D)$. ■

If $f \equiv 0$, we can solve problem (6.1.9) for every $x \in X$, and the mapping $x \mapsto w$ is linear. So we can define an evolution operator as follows.

Definition 6.1.7 *For $0 \leq s \leq t \leq T$ and for $x \in X$ set*

$$G(t,s)x = W(t,s)x + e^{(t-s)A(s)}x,$$

where $W(t,s)x$ is the solution $w(t)$ of problem (6.1.9) with $f \equiv 0$.

The family $\{G(t,s) : 0 \leq s \leq t \leq T\}$ is an evolution operator for problem (6.0.1): properties (b), (c) of Definition 6.0.1 follow easily from Theorem 6.1.4 and from the properties of $e^{(t-s)A(s)}x$. (a) will be shown in Corollary 6.1.9. From Propositions 6.1.3, 6.1.5 and Theorem 6.1.4, together with the results of Chapter 2, one deduces easily several other properties of $G(t,s)$. Next corollary deals with estimates.

Corollary 6.1.8 *Let $0 \leq s < r < t \leq T$.*

(i) *There is $C > 0$ such that for every $x \in X$*

$$\|G(t,s)x\| \leq C\|x\|, \quad \|G(t,s)x\|_D \leq \frac{C}{t-s}\|x\|, \qquad (6.1.15)$$

$$\|A(t)G(t,s)x\|_{(X,D)_{\alpha,\infty}} \leq \frac{C}{(t-s)^{1+\alpha}}\|x\|, \qquad (6.1.16)$$

$$\|A(t)G(t,s)x - A(r)G(r,s)x\| \leq C\left(\frac{(t-r)^{\alpha}}{r-s} + \frac{t-r}{(r-s)(t-s)}\right)\|x\|. \qquad (6.1.17)$$

For $1 \leq p \leq \infty$, $0 < \beta < 1$ there is $C > 0$ such that

$$\begin{cases} (a) & \|G(t,s)x\|_{(X,D)_{\beta,p}} \leq \dfrac{C}{(t-s)^{\beta}}\|x\|, \\ (b) & \|G(t,s)x - G(r,s)x\|_{(X,D)_{\beta,p}} \leq C\left(\dfrac{(t-r)^{1-\beta}}{(r-s)^{1-\alpha}} \dfrac{1}{(r-s)^{\beta}} - \dfrac{1}{(t-s)^{\beta}}\right)\|x\|, \\ (c) & \|A(t)G(t,s)x\|_{(X,D)_{\beta,p}} \leq \dfrac{C}{(t-s)^{1+\beta}}\|x\|, \quad 0 < \beta < \alpha. \end{cases} \qquad (6.1.18)$$

(ii) Let $1 \leq p \leq \infty$, $0 < \theta < 1$, $0 < \beta < \theta$. Then there is $C > 0$ such that for $x \in (X,D)_{\theta,p}$

$$\begin{cases} (a) & \|G(t,s)x\|_{(X,D)_{\theta,p}} \leq C\|x\|_{(X,D)_{\theta,p}}, \\ (b) & \|G(t,s)x\|_D \leq \dfrac{C}{(t-s)^{1-\theta}}\|x\|_{(X,D)_{\theta,p}}, \end{cases} \qquad (6.1.19)$$

$$\|G(t,s)x - G(r,s)x\|_{(X,D)_{\beta,1}} \leq C(t-r)^{\theta-\beta}\|x\|_{(X,D)_{\theta,p}}, \qquad (6.1.20)$$

$$\|A(t)G(t,s)x\|_{(X,D)_{\alpha,\infty}} \leq \frac{C}{(t-s)^{1+\alpha-\theta}}\|x\|_{(X,D)_{\theta,p}}, \qquad (6.1.21)$$

$$\|A(t)G(t,s)x - A(r)G(r,s)x\|$$
$$\leq C\left(\frac{(t-r)^{\alpha}}{(r-s)^{1-\theta}} + \frac{1}{(r-s)^{1-\theta}} - \frac{1}{(t-s)^{1-\theta}}\right)\|x\|_{(X,D)_{\theta,p}}, \qquad (6.1.22)$$

and consequently, for $1 \leq p \leq \infty$, $0 < \beta < 1$ there is $C > 0$ such that

$$\begin{cases} \|G(t,s)x\|_{(X,D)_{\beta,p}} \leq \dfrac{C}{(t-s)^{\beta-\theta}}\|x\|_{(X,D)_{\theta,p}}, & \theta < \beta < 1, \\ \|A(t)G(t,s)x\|_{(X,D)_{\beta,p}} \leq \dfrac{C}{(t-s)^{1+\beta-\theta}}\|x\|_{(X,D)_{\theta,p}}, & 0 < \beta < \alpha. \end{cases} \qquad (6.1.23)$$

(iii) Let Y be any Banach space in the class $J_\theta \cap K_\theta$ between X and D, with $0 < \theta < 1$, and such that there is $M > 0$ satisfying

$$\|e^{tA(s)}\|_{L(Y)} \leq M, \quad 0 < s, t \leq T.$$

Then there is $C > 0$ such that for every $y \in Y$

$$\|G(t,s)y\|_Y \leq C\|y\|_Y, \qquad (6.1.24)$$

and if $\theta < \alpha$

$$\|A(t)G(t,s)y\|_Y \leq \frac{C}{t-s}\|y\|_Y. \qquad (6.1.25)$$

1. CONSTRUCTION AND PROPERTIES OF THE EVOLUTION OPERATOR 221

(iv) There is $C > 0$ such that for every $x \in D$

$$\begin{cases} \|A(t)G(t,s)x\| \leq C\|x\|_D, \\ \|A(t)G(t,s)x\|_{(X,D)_{\alpha,\infty}} \leq \dfrac{C}{(t-s)^\alpha}\|x\|_D, \\ \|A(t)G(t,s)x - A(r)G(r,s)x\| \leq C\left((t-s)^\alpha + \log \dfrac{t-s}{r-s}\right)\|x\|_D. \end{cases} \quad (6.1.26)$$

(v) Let $x \in D$, $A(s)x \in (X,D)_{\theta,\infty}$, $0 < \theta \leq \alpha$. Then

$$\begin{cases} \|A(t)G(t,s)x\|_{(X,D)_{\theta,\infty}} \leq C(\|x\|_D + \|A(s)x\|_{(X,D)_{\theta,\infty}}), \\ \|A(t)G(t,s)x - A(r)G(r,s)x\| \leq C(t-r)^\theta(\|x\|_D + \|A(s)x\|_{(X,D)_{\theta,\infty}}). \end{cases} \quad (6.1.27)$$

Proof — (i) Let $x \in X$. Theorem 6.1.4 implies that

$$\|W(t,s)x\| \leq C_1\|x\|, \quad \|W(t,s)x\|_D \leq \dfrac{C_1}{(t-s)^{1-\alpha}}\|x\|,$$

$$\|W_t(t,s)x\|_{(X,D)_{\alpha,\infty}} \leq \dfrac{C_1}{t-s}\|x\|,$$

$$\|W_\sigma(\sigma,s)x_{|\sigma=t} - W_\sigma(\sigma,s)x_{|\sigma=r}\| \leq C_1 \dfrac{(t-r)^\alpha}{r-s}\|x\|.$$

On the other hand, from estimates (6.1.2)(b) and (6.1.3) one gets

$$\|e^{(t-s)A(s)}x\| \leq C_2\|x\|, \quad \|e^{(t-s)A(s)}x\|_D \leq \dfrac{C_2}{t-s}\|x\|,$$

$$\|A(s)e^{(t-s)A(s)}x\|_{(X,D)_{\alpha,\infty}} \leq \dfrac{C_2}{(t-s)^{1+\alpha}}\|x\|,$$

$$\|A(s)(e^{(t-s)A(s)}x - e^{(r-s)A(s)}x)\| \leq C_2\|x\| \int_{r-s}^{t-s} \dfrac{d\sigma}{\sigma^2} = C_2\left(\dfrac{1}{r-s} - \dfrac{1}{t-s}\right)\|x\|.$$

Therefore, (6.1.15), (6.1.16), (6.1.17) follow, recalling that

$$A(t)G(t,s)x = \dfrac{\partial}{\partial t}G(t,s)x = W_t(t,s)x + A(s)e^{(t-s)A(s)}x.$$

Estimates (6.1.18)(a) and (6.1.18)(c) follow by interpolation from (6.1.15) and (6.1.16), recalling that $(X,D)_{\beta,p}$ belongs to the class J_β between X and D, and it belongs to the class $J_{\beta/\alpha}$ between X and $(X,D)_{\alpha,\infty}$ if $\beta < \alpha$. To prove (6.1.18)(b) one has to use again the decomposition $G(t,s) = W(t,s) + e^{(t-s)A(s)}$. From (6.1.10) one gets

$$\|W(t,s)x - W(r,s)x\|_{(X,D)_{\beta,p}} \leq C\dfrac{(t-r)^{1-\beta}}{(r-s)^{1-\alpha}}\|x\|,$$

whereas (6.1.2)(b) and (6.1.3) yield

$$\|e^{(t-s)A(s)}x - e^{(r-s)A(s)}x\|_{(X,D)_{\beta,p}} \leq C \int_{r-s}^{t-s} \frac{d\sigma}{\sigma^{1+\beta}} \|x\|.$$

To prove statements (ii) and (iv) it is sufficient to argue as in the proof of statement (i), and applying Proposition 6.1.5 instead of Theorem 6.1.4.

Concerning (iii), we recall that Y belongs to the class $J_\theta \cap K_\theta$ if and only if $(D,X)_{\theta,1} \subset Y \subset (D,X)_{\theta,\infty}$. From Proposition 6.1.5(i)(ii) we get

$$\|W(t,s)y\|_{(X,D)_{\theta,1}} \leq C\|y\|_{(X,D)_{\theta,\infty}},$$

for every $y \in Y$, which implies

$$\|W(t,s)y\|_Y \leq C'\|y\|_Y.$$

On the other hand, by assumption we have $\|e^{(t-s)A(s)}y\|_Y \leq M\|y\|_Y$. Summing up, estimate (6.1.24) follows. Concerning (6.1.25), from (6.1.10) it follows that $\|\partial/\partial t\, W(t,s)y\|_{(X,D)_{\alpha,\infty}} \leq Ct^{-1}\|y\|$, for every y. Since $\alpha > \theta$, then

$$\|\partial/\partial t\, W(t,s)y\|_Y \leq Ct^{-1}\|y\|.$$

Moreover, since Y belongs to the class $J_\theta \cap K_\theta$ between X and $D(A)$, then

$$\|\partial/\partial t\, e^{(t-s)A(s)}y\|_Y \leq c\|A(s)e^{(t-s)A(s)}y\|_{D(A)}^\theta \|A(s)e^{(t-s)A(s)}y\|^{1-\theta}$$

$$\leq c \left(\frac{C\|y\|_{(X,D)_{\theta,\infty}}}{(t-s)^{2-\theta}}\right)^\theta \left(\frac{C\|y\|_{(X,D)_{\theta,\infty}}}{(t-s)^{1-\theta}}\right)^{1-\theta} \leq \frac{C'}{t-s}\|y\|_Y.$$

Estimate (6.1.25) follows now from the equality

$$A(t)G(t,s)y = \partial/\partial t\, (W(t,s) + e^{(t-s)A(s)})y.$$

Statement (v) is an immediate consequence of Proposition 6.1.3. ∎

Other properties of $G(t,s)$ follow from Theorem 6.1.4 and Corollary 6.1.8.

Corollary 6.1.9 (i) $G(t,r)G(r,s) = G(t,s)$, $0 \leq s \leq r \leq t \leq T$.
(ii) $\exists \lim_{t \to s} G(t,s)x \iff x \in \overline{D}$, and in this case $\lim_{t \to s} G(t,s)x = x$.
(iii) For $0 < \theta < 1$, $G(\cdot,s)x \in C^\theta([s,T];X) \iff x \in (X,D)_{\theta,\infty}$.
(iv) $G(\cdot,s)x \in C([s,T];D) \cap C^1([s,T];X) \iff x \in D$, $A(s)x \in \overline{D}$.
(v) For $0 < \theta \leq \alpha$, $G(\cdot,s)x \in C^\theta([s,T];D) \cap C^{1+\theta}([s,T];X) \iff x \in D$, $A(s)x \in (X,D)_{\theta,\infty}$.

1. Construction and properties of the evolution operator

(vi) Let $Y \in J_\theta(X,D) \cap K_\theta(X,D)$ be such that there exists $M > 0$ satisfying $\|e^{tA(s)}\|_{L(Y)} \leq M$ for every $s, t \in [0,T]$. Let $x \in Y$. Then $G(\cdot,s)x \in C([s,T];Y)$ if and only if x belongs to the closure of D in Y. In particular, for $1 \leq p < \infty$, $G(\cdot,s)x \in C([s,T];(X,D)_{\theta,p})$ for every $x \in (X,D)_{\theta,p}$; $G(\cdot,s)x \in C([s,T];(X,D)_\theta)$ for every $x \in (X,D)_\theta$; $G(\cdot,s)x \in C([s,T];(X,D)_{\theta,\infty})$ if and only if $x \in (X,D)_\theta$.

Proof — (i) By definition $G(r,r) = I$, so that the equality is trivial if $t = r$ or $r = s$. If $r > s$, for each $x \in X$ the function $\psi(t) = G(t,r)G(r,s)x$ is a strict solution of
$$u'(t) = A(t)u(t), \quad r \leq t \leq T; \quad u(r) = G(r,s)x,$$
because $G(r,s)x \in D$ and $A(r)G(r,s)x \in \overline{D}$. Since also $G(\cdot,s)x$ is a strict solution of the same problem, statement (i) follows from the uniqueness lemma 6.1.2.

(ii) For every $x \in X$, $\lim_{t \to s} W(t,s)x = 0$, so that there exists $\lim_{t \to s} G(t,s)x$ if and only if there exists $\lim_{t \to s} e^{(t-s)A(s)}x$. Thanks to Proposition 2.1.4(i), such a limit does exist if and only if $x \in \overline{D}$, in which case it is equal to x.

(iii) Let $0 < \theta \leq \alpha$. Then for every $x \in X$ the function $W(\cdot,s)x$ is θ-Hölder continuous with values in X, so that $G(\cdot,s)x$ is θ-Hölder continuous if and only if $t \mapsto e^{(t-s)A(s)}x$ is θ-Hölder continuous. This happens if and only if $x \in (X,D)_{\theta,\infty}$, thanks to Proposition 2.2.4 and to the equivalence $(X,D)_{\theta,\infty} = D_{A(s)}(\theta,\infty)$.

Let now $\alpha < \theta < 1$. If $x \in (X,D)_{\theta,\infty}$, from (6.1.19)(b) we obtain
$$\|G(t,s)x - G(r,s)x\| = \left\|\int_{r-s}^{t-s} A(\sigma)G(\sigma,s)x\,d\sigma\right\| \leq \text{const.}\,(t-r)^\theta,$$
so that $G(\cdot,s)x \in C^\theta([s,T];X)$. Conversely, if $G(\cdot,s)x \in C^\theta([s,T];X)$, since $\theta \geq \alpha$ from the first part of the proof we get $x \in (X,D)_{\alpha,\infty}$. Then, by Proposition 6.1.5, $W(\cdot,s)x$ belongs to $C^{2\alpha}([s,T];X)$ if $2\alpha < 1$, it is Lipschitz continuous if $2\alpha \geq 1$. In both cases, $t \mapsto e^{(t-s)A(s)}x$ belongs to $C^{2\alpha \wedge \theta}([s,T];X)$, so that $x \in D_{A(s)}(2\alpha \wedge \theta, \infty) = (X,D)_{2\alpha \wedge \theta, \infty}$. Iterating this procedure, we get the statement.

(iv), (v) Let $x \in D, A(s)x \in \overline{D}$. By Proposition 6.1.5(iv), $w \in C^{1+\alpha}([s,T];X) \cap C^\alpha([s,T];D)$, and $t \mapsto e^{(t-s)A(s)}x \in C^1([s,T];X) \cap C([s,T];D)$. If in addition $A(s)x \in (X,D)_{\theta,\infty}$ with $\theta \leq \alpha$, then $t \mapsto e^{(t-s)A(s)}x \in C^{1+\theta}([s,T];X) \cap C^\theta([s,T];D)$.

Conversely, if $G(\cdot,s)x$ is continuous with values in D, then x belongs obviously to D. Then, again by Proposition 6.1.5(iv), $w \in C^{1+\alpha}([s,T];X) \cap C^\alpha([s,T];D)$, so that $G(\cdot,s)x$ belongs to $C^1([s,T];X) \cap C([s,T];D)$, or to $C^{1+\theta}([s,T];X) \cap C^\theta([s,T];D)$, with $\theta \leq \alpha$, if and only if $t \mapsto e^{(t-s)A(s)}x$ does. The statements follows now from Proposition 2.1.4(iii)(iv) and Proposition 2.2.4.

(vi) The function $t \to G(t,s)x$ is continuous with values in D in $]s,T]$, so that it is obviously continuous with values in Y in $]s,T]$. From (6.1.24) we know that $G(\cdot,s)$ is bounded with values in $L(Y)$, up to $t = s$. Moreover, for every $x \in D$, $\lim_{t \to s} G(t,s)x = x$ in $(X,D)_{\beta,1}$, for every $\beta \in\,]0,1[$; in particular, since

$(X, D)_{\theta,1} \subset Y$, then $\lim_{t \to s} G(t,s)x = x$ in Y. Therefore, for every x in the closure of D in Y, $\lim_{t \to s} G(t,s)x = x$ in Y, and the function $G(\cdot, s)x$ belongs to $C([s,T]; Y)$.

Conversely, if $G(\cdot, s)x \in C([s,T]; Y)$, then x belongs to the closure of D in Y because $G(t,s)x \in D$ for $t > s$. ∎

Dependence of $G(t,s)$ on s

Several properties of regularity of $G(t,\cdot)x$ are stated below.

Corollary 6.1.10 *Let $t \in \,]0, T]$.*

(i) *If $x \in \overline{D}$, then $G(t, \cdot)x \in C([0,t]; X)$.*

(ii) *If $x \in (X, D)_{\theta, \infty}$, with $0 < \theta < 1$, then $G(t, \cdot)x \in C^{\theta}([0,t]; X)$.*

(iii) *If $x \in D$, then $G(t, \cdot)x \in Lip([0,t]; X)$.*

(iv) *If $x \in D$ and $A(t)x \in \overline{D}$, then $G(t, \cdot)x \in C^1([0,t]; X) \cap C([0,t]; D)$.*

Proof — For $0 \leq s < r \leq t \leq T$ we have, due to Corollary 6.1.9(i)

$$G(t,r)x - G(t,s)x = G(t,r)(I - G(r,s))x.$$

Then, all the statements are consequences of Corollary 6.1.9(i),...,(iv). ∎

Corollary 6.1.10 does not give informations about the regularity of $G(t, \cdot)x$ if $x \in X$. We are going to show that $G(t, \cdot)x$ is continuous with values in D in the half-open interval $[0, t[$, as in the autonomous case. To this aim, we need a preliminary lemma on the behavior of $W(t, \cdot)$.

Lemma 6.1.11 *For every $\theta \in \,]0, \alpha[$ there is $C_\theta > 0$ such that if $x \in X$ and $0 \leq s < s + h < T$ then*

$$\|W(\cdot, s+h)x - W(\cdot, s)x\|_{C_1^\theta(]s+h,T]; D)} \atop + \|W(\cdot, s+h)x - W(\cdot, s)x\|_{C^\theta([s+h,T]; X)} \leq C_\theta h^{\alpha - \theta} \|x\| \quad (6.1.28)$$

It follows that for every $x \in X$

(i) $\|W(t, s+h)x - W(t, s)x\| \leq C_\theta h^{\alpha - \theta} \|x\|, \quad s+h < t \leq T,$

(ii) $\|W(t, s+h)x - W(t, s)x\|_D \leq \dfrac{C_\theta h^{\alpha - \theta}}{(t - s - h)^{1-\theta}} \|x\|, \quad s+h < t \leq T.$

$$(6.1.29)$$

1. CONSTRUCTION AND PROPERTIES OF THE EVOLUTION OPERATOR

Proof — The function $v(t) = W(t, s+h)x - W(t,s)x + G(t, s+h)W(s+h, s)x$ is a classical solution of

$$v'(t) = A(t)v(t) + \varphi(t), \quad s+h < t \leq T; \quad v(s+h) = 0, \tag{6.1.30}$$

where $\varphi(t) = [A(t) - A(s+h)]e^{(t-s-h)A(s+h)}x - [A(t) - A(s)]e^{(t-s)A(s)}x$ is defined for $s + h < t \leq T$.

From the proof of Theorem 6.1.4 we know that $\varphi \in C_1^\alpha(]s+h, T]; X)$. From estimates (6.1.12), (6.1.13) we get

$$(t-s-h)^{1-\alpha}\|\varphi(t)\| \leq 4[A]_{C^\alpha} N_1 \|x\|,$$

and for $s + h < r \leq t \leq T$

$$\|\varphi(t) - \varphi(r)\| \leq 2[A]_{C^\alpha} \gamma(N_1 + \alpha^{-1} N_2) \frac{(t-r)^\alpha}{r-s-h}\|x\| = K_1 \frac{(t-r)^\alpha}{r-s-h}\|x\|. \tag{6.1.31}$$

On the other hand, by splitting φ as

$$\varphi(t) = [A(t) - A(s+h)](e^{(t-s-h)A(s+h)} - e^{(t-s)A(s+h)})x$$
$$+ [A(t) - A(s+h)](e^{(t-s)A(s+h)} - e^{(t-s)A(s)})x + [A(s+h) - A(s)]e^{(t-s)A(s)}x$$

and using estimates (6.1.4) we get

$$\|\varphi(t)\| \leq K_2 \frac{h^\alpha}{t-s-h}\|x\|,$$

so that, since $t - s - h \geq r - s - h$,

$$\|\varphi(t) - \varphi(r)\| \leq 2K_2 \frac{h^\alpha}{r-s-h}\|x\|. \tag{6.1.32}$$

From (6.1.31) and (6.1.32) we get, for every $\theta \in]0, \alpha[$,

$$\|\varphi(t) - \varphi(r)\| \leq K_1^{\theta/\alpha}(2K_2)^{1-\theta/\alpha} \frac{(t-r)^\theta}{r-s-h} h^{\alpha-\theta}\|x\|,$$

which implies that $\varphi \in C_1^\theta(]s+h, T]; X)$ and that $\|\varphi\|_{C_1^\theta} \leq \text{const.} \, h^{\alpha-\theta}\|x\|$.

Applying now the procedure of Theorem 6.1.4, with s replaced by $s + h$ and α replaced by θ, we see that problem (6.1.30) has a unique solution $w \in C_1^\theta(]s+h, T]; D) \cap C^\theta([s+h, T]; X)$, and $\|w\|_{C_1^\theta(]s+h,T];D)} + \|w\|_{C^\theta([s+h,T];X)} \leq \text{const.} \, h^{\alpha-\theta}\|x\|$. The function $t \mapsto G(t, s+h)W(s+h, s)x$ has the same properties. Indeed, $\|W(s+h, s)x\|_{(X,D)_{\theta,\infty}} \leq Ch^{\alpha-\theta}\|x\|$, so that $\|G(\cdot, s+h)W(s+h, s)x\|_{C_1^\theta(]s+h,T];D)} + \|G(\cdot, s+h)W(s+h, s)x\|_{C^\theta([s+h,T];X)} \leq \text{const.} \, h^{\alpha-\theta}\|x\|$ thanks to (6.1.19) and (6.1.23). Since $W(t, s+h)x - W(t, s)x = w(t) - G(t, s+h)W(s+h, s)x$, the statement follows. ∎

Recalling that $G(t, s) = e^{(t-s)A(s)} + W(t, s)$, and using estimates (6.1.4), (6.1.10), (6.1.29), we get the following Corollary.

Corollary 6.1.12 *Let $\Delta = \{(t,s) \in [0,T]^2 : s < t\}$, $\Delta_\varepsilon = \{(t,s) \in [0,T]^2 : s < t - \varepsilon\}$ ($0 < \varepsilon < T$). Then $(t,s) \mapsto G(t,s)$ is continuous in Δ and θ-Hölder continuous in Δ_ε with values in $L(X, D)$, for every $\theta \in]0, \alpha[$.*

In the following lemma we describe another important property of $G(t, s)$. The proof is rather lenghty, but the statement is important for future use. For $x \in X$ and $0 \leq s < T$, set

$$\varphi(t) = \int_s^t G(t, \sigma) x \, d\sigma, \quad s \leq t \leq T.$$

Thanks to estimate (6.1.15), $\|G(t, \sigma)x\|$ is bounded. Moreover, by Corollary 6.1.12, the function $\sigma \mapsto G(t, \sigma)x$ is continuous in $[0, t[$; in particular it is measurable, so that the integral makes sense.

Lemma 6.1.13 *For $0 \leq s < T$ and $x \in X$, the function φ defined above is continuous with values in D and differentiable with values in X in $]s, T]$, and*

$$\varphi'(t) = A(t)\varphi(t) + x, \quad s < t \leq T. \tag{6.1.33}$$

Moreover there is c (independent of s and x) such that

$$\|\varphi(t)\|_D + \|\varphi'(t)\| \leq c\|x\|, \quad s \leq t \leq T. \tag{6.1.34}$$

Proof — We remark that $\varphi(t) \in D$ for $s \leq t \leq T$. Indeed, splitting $\varphi(t)$ as

$$\varphi(t) = \varphi_1(t) + \varphi_2(t) + \varphi_3(t)$$
$$= \int_s^t W(t,\sigma) x \, d\sigma + \int_s^t e^{(t-\sigma)A(t)} x \, d\sigma + \int_s^t [e^{(t-\sigma)A(\sigma)} - e^{(t-\sigma)A(t)}] x \, d\sigma,$$

we see that $\varphi_1(t)$ belongs to D, thanks to the properties of W proved in Theorem 6.1.4, φ_2 takes values in D thanks to Proposition 2.1.4(ii), and φ_3 takes values in D thanks to (6.1.4)(b). The estimate $\|\varphi(t)\|_D \leq \text{const.} \|x\|$ follows easily. The proof of the continuity of φ with values in D is left to the reader; in fact one can show that φ_1 belongs to $C^{\alpha-\varepsilon}([s,T]; D)$ for every $\varepsilon \in]0, \alpha[$, and $\varphi_2 + \varphi_3$ belong to $C^\theta_\theta(]s,T]; D)$ for every $\theta \in]0, \alpha[$.

Let us show now that φ is right differentiable for $t > s$. For $s < t < t + h < T$ we have

$$\frac{\varphi(t+h) - \varphi(t)}{h} = \frac{1}{h} \left(\int_s^{t+h} W(t+h, \sigma) x \, d\sigma - \int_s^t W(t, \sigma) x \, d\sigma \right)$$
$$+ \frac{1}{h} \left(\int_s^{t+h} e^{(t+h-\sigma)A(\sigma)} x \, d\sigma - \int_s^t e^{(t-\sigma)A(\sigma)} x \, d\sigma \right).$$

The first addendum is easily seen to converge to

$$A(t) \int_s^t W(t, \sigma) x \, d\sigma + \int_s^t (A(t) - A(\sigma)) e^{(t-\sigma)A(\sigma)} x \, d\sigma$$

1. Construction and properties of the evolution operator

as $h \to 0$. It is convenient to split the second addendum into the sum $I_{1,h}(t) + I_{2,h}(t)$, where

$$I_{1,h}(t) = \frac{1}{h} \left(\int_s^{t+h} (e^{(t+h-\sigma)A(\sigma)} - e^{(t+h-\sigma)A(t)}) x \, d\sigma \right.$$
$$\left. - \int_s^t (e^{(t-\sigma)A(\sigma)} - e^{(t-\sigma)A(t)}) x \, d\sigma \right),$$

$$I_{2,h}(t) = \frac{1}{h} \left(\int_s^{t+h} e^{(t+h-\sigma)A(t)} x \, d\sigma - \int_s^t e^{(t-\sigma)A(t)} x \, d\sigma \right).$$

Now we have

$$\left\| I_{1,h}(t) - \int_s^t (A(\sigma) e^{(t-\sigma)A(\sigma)} - A(t) e^{(t-\sigma)A(t)}) x \, d\sigma \right\|$$

$$\leq \left\| \frac{1}{h} \int_t^{t+h} (e^{(t+h-\sigma)A(\sigma)} - e^{(t+h-\sigma)A(t)}) x \, d\sigma \right\|$$

$$+ \left\| \frac{1}{h} \int_s^t d\sigma \int_{t-\sigma}^{t+h-\sigma} [(A(\sigma) e^{\tau A(\sigma)} - A(t) e^{\tau A(t)}) x \right.$$
$$\left. - (A(\sigma) e^{(t-\sigma)A(\sigma)} - A(t) e^{(t-\sigma)A(t)}) x] \, d\tau \right\|$$

$$\leq \sup_{t < \sigma < t+h} \|(e^{(t+h-\sigma)A(\sigma)} - e^{(t+h-\sigma)A(t)}) x\|$$

$$+ \left\| \frac{1}{h} \int_s^t d\sigma \int_{t-\sigma}^{t+h-\sigma} d\tau \int_{t-\sigma}^{\tau} (A(\sigma)^2 e^{\xi A(\sigma)} - A(t)^2 e^{\xi A(t)}) x \, d\xi \right\|.$$

Thanks to (6.1.3)(b), with $k = 0$, the first addendum is less than $C_0 h^\alpha \|x\|$, so that it goes to 0 as $h \to 0$. The second one can be estimated using again (6.1.3)(b) with $k = 2$, and it is less or equal to

$$\frac{C_2}{h} \int_s^t (t-\sigma)^\alpha d\sigma \int_{t-\sigma}^{t+h-\sigma} d\tau \int_{t-\sigma}^\tau \xi^{-2} d\xi \|x\| \leq \frac{4C_2}{\alpha^2} (t-s)^{\alpha/2} h^{\alpha/2} \|x\|.$$

Therefore, $I_{1,h}(t)$ goes to $\int_s^t (A(\sigma) e^{(t-\sigma)A(\sigma)} - A(t) e^{(t-\sigma)A(t)}) x \, d\sigma$ as $h \to 0$. Moreover, fixed any $\lambda \in \rho(A(t))$ for each t, we rewrite $I_{2,h}(t)$ as

$$I_{2,h}(t) = -\frac{1}{h} (e^{(t-s+h)A(t)} - e^{(t-s)A(t)}) R(\lambda, A(t)) x$$
$$+ \frac{1}{h} \left(\int_s^{t+h} e^{(t+h-\sigma)A(t)} \lambda R(\lambda, A(t)) x \, d\sigma - \int_s^t e^{(t-\sigma)A(t)} \lambda R(\lambda, A(t)) x \, d\sigma \right).$$

Since $R(\lambda, A(t)) x \in D$, it is not difficult to see that

$$\lim_{h \to 0} I_{2,h}(t) = e^{(t-s)A(t)} x = A(t) \int_s^t e^{(t-\sigma)A(t)} x \, d\sigma + x.$$

Therefore, φ is right differentiable, with $d^+/dt\, \varphi(t) = A(t)\varphi(t) + x$. Since φ is continuous in $]s,T]$ with values in D, then $d^+/dt\, \varphi$ is continuous in $]s,T[$ with values in X. Therefore, φ is differentiable in $]s,T[$ (see e.g. [107, p.492]). Since the derivative has a continuous extension to $]s,T]$, then φ is differentiable in $]s,T]$, and (6.1.33) holds. ∎

6.2 The variation of constants formula

We shall show that under reasonable assumptions on x and f any type of solution to problem (6.0.3) is given by formula (6.0.4). Then, using the properties of $G(t,s)$ showed in Section 6.1, other properties of the solution will be proved. First we consider the case where f is Hölder continuous.

Lemma 6.2.1 *Let $f \in C^\theta([s,T];X)$, $x \in D$. Then the function u defined in (6.0.4) is the unique classical solution of problem (6.0.3) which is bounded with values in D.*

Proof — Existence of a classical solution is guaranteed by Corollary 6.1.6(i), and uniqueness of a D-bounded classical solution by Lemma 6.1.2. Therefore it is sufficient to show that the function u defined in (6.0.4) is a classical solution of (6.0.3) and that it is bounded with values in D. First, $u(t) \in D$ for $s \leq t \leq T$, because

$$u(t) = G(t,s)x + \int_s^t G(t,\sigma)(f(\sigma) - f(t))d\sigma + \int_s^t G(t,\sigma)f(t)d\sigma.$$

$G(\cdot,s)x$ and the first integral are bounded with values in D thanks to the first of estimates (6.1.26). The second integral is bounded with values in D thanks to Lemma 6.1.13. Therefore, u is bounded with values in D.

Let us show that u is right differentiable in $]s,T[$. Since $G(\cdot,s)x$ is differentiable in $]s,T]$, we take care only of $\int_s^t G(t,\sigma)f(\sigma)d\sigma$. We split the incremental ratio into the sum $I_{1,h} + I_{2,h} + I_{3,h}$, where

$$I_{1,h} = \frac{1}{h}\int_s^t (G(t+h,\sigma) - G(t,\sigma))(f(\sigma) - f(t))d\sigma,$$

$$I_{2,h} = \frac{1}{h}\int_t^{t+h} G(t+h,\sigma)(f(\sigma) - f(t))d\sigma,$$

$$I_{3,h} = \frac{1}{h}\left(\int_s^{t+h} G(t+h,\sigma)f(t)d\sigma - \int_s^t G(t,\sigma)f(t)d\sigma\right).$$

2. THE VARIATION OF CONSTANTS FORMULA

The incremental ratio $h^{-1}(G(t+h,\sigma) - G(t,\sigma))$ converges to $A(t)G(t,\sigma)$ in $L(X)$ as $h \to 0$, and thanks to estimate (6.1.13) its norm is less than $const. \, h^{-1} \int_t^{t+h} (r-\sigma)^{-1} dr \leq (t-\sigma)^{-1}$. By the Dominated Convergence Theorem, $I_{1,h}$ converges to

$$\int_s^t A(t)G(t,\sigma)(f(\sigma) - f(t))d\sigma = A(t) \int_s^t G(t,\sigma)(f(\sigma) - f(t))d\sigma.$$

$I_{2,h}$ converges to 0 because $G(t+h,\sigma)$ is bounded in $L(X)$ and f is continuous. $I_{3,h}$ converges to

$$A(t) \int_s^t G(t,\sigma)f(t)d\sigma + f(t)$$

thanks to Lemma 6.1.13. Therefore, u is right differentiable, and $d^+/dt\, u(t) = A(t)u(t) + f(t)$ for $s < t < T$. u is clearly continuous with values in X. To show that $d^+/dt\, u(t)$ is also continuous, it is sufficient to write the difference $A(t)u(t) - A(r)u(r)$ (for $s < r < t \leq T$) as

$$A(t)u(t) - A(r)u(r) = A(t)G(t,s)x - A(r)G(r,s)x$$
$$+ \int_s^r (A(t)G(t,\sigma) - A(r)G(r,\sigma))(f(\sigma) - f(r))d\sigma$$
$$+ \int_r^t A(t)G(t,\sigma)(f(\sigma) - f(t))d\sigma + A(t)\int_s^t G(t,\sigma)(f(r) - f(t))d\sigma$$
$$+ \left(A(t)\int_s^t G(t,\sigma)d\sigma - A(r)\int_s^r G(r,\sigma)d\sigma \right) f(r),$$

and to use respectively estimates (6.1.26), (6.1.17), (6.1.15), (6.1.34), and the continuity of $t \mapsto A(t)\int_s^t G(t,\sigma)d\sigma x$ which follows from Lemma 6.1.13. Now, both u and $d^+/dt\, u$ are continuous in $]s,T[$, and $d^+/dt\, u$ has a continuous extension to $]s,T]$, so that u is differentiable in $]s,T]$ and it solves (6.0.3). ∎

With the aid of Lemma 6.2.1 we are able to show that any strict solution of (6.0.3) is given by (6.0.4).

Proposition 6.2.2 *Let $f \in C([s,T];X)$, $x \in D$, and let u be a strict solution of problem (6.0.3). Then u satisfies (6.0.4).*

Proof — If u is a strict solution, then $u'(s) = A(s)x + f(s)$ belongs to $\overline{D}$. By Corollary 6.1.6 and Lemma 6.1.2, the problem

$$v'(t) = A(t)v(t) + f(s), \quad s \leq t \leq T; \quad v(s) = x,$$

has a unique strict solution, given by

$$v(t) = G(t,s)x + \int_s^t G(t,\sigma)f(s)d\sigma, \quad s \leq t \leq T,$$

thanks to Lemma 6.2.1. Therefore, we have only to show that if w is a strict solution of

$$w'(t) = A(t)w(t) + f(t) - f(s), \quad s \le t \le T; \quad w(s) = 0, \qquad (6.2.1)$$

then

$$w(t) = \int_s^t G(t,\sigma)(f(\sigma) - f(s))d\sigma, \quad s \le t \le T.$$

Let w be a strict solution of (6.2.1), and set

$$w_n(t) = n \int_s^t e^{-n(t-\sigma)} w(\sigma) d\sigma, \quad n \in \mathbb{N}, \ s \le t \le T.$$

Since $w(s) = 0$, then w_n converges to w in $C([s,T];X)$ as $n \to \infty$. Setting $f_n = w_n' - A(\cdot)w_n$, we have

$$f_n(t) = n \int_s^t e^{-n(t-\sigma)}(f(\sigma) - f(s))d\sigma + n \int_0^{t-s} e^{-n\sigma}(A(t-\sigma) - A(t))w(t-\sigma)d\sigma.$$

The first addendum goes uniformly to f as $n \to \infty$; the norm of the second one is bounded by $\int_0^T ne^{-n\sigma}\sigma^\alpha d\sigma \, [A]_{C^\alpha} \|w\|_{C(D)}$, so that it goes uniformly to 0 as $n \to \infty$. Therefore, f_n converges to f in $C([s,T];X)$ as $n \to \infty$. Moreover, it easy to see that $f_n \in C^\alpha([s,T];X)$, so that by Lemma 6.2.1 we have

$$w_n(t) = \int_s^t G(t,\sigma) f_n(\sigma) d\sigma, \quad n \in \mathbb{N}, \ s \le t \le T.$$

Letting $n \to \infty$ the statement follows. ∎

Proposition 6.2.2 has two important consequences: the first concerning existence and uniqueness of the strong solution of (6.0.3), the second concerning the variation of constants formula for classical solutions.

Corollary 6.2.3 *Let $f \in C([s,T];X)$ and $x \in \overline{D}$. Then the function u defined in (6.0.4) is the unique strong solution of problem (6.0.3).*

Proof — Let $f_n \in C^\alpha([s,T];X)$ be such that

$$\lim_{n \to \infty} \|f_n - f\|_{C([s,T];X)} = 0, \quad f_n(s) = f(s), \ \forall n \in \mathbb{N}$$

(for instance, f_n may be the n-th Bernstein polynomial of f: see Section 0.1.1). Fix $\lambda \in \rho(A(s))$. Since $D(A(s)^2)$ is dense in $\overline{D}$, there is a sequence $\{y_n\}_{n \in \mathbb{N}} \subset D(A(s)^2)$ such that

$$\lim_{n \to \infty} y_n = x - (\lambda I - A(s))^{-1} f(s).$$

2. THE VARIATION OF CONSTANTS FORMULA

Set $x_n = y_n + (\lambda I - A(s))^{-1} f(s)$. Then $x_n \in D$, $A(s)x_n + f_n(s) = A(s)y_n + \lambda(\lambda I - A(s))^{-1} f(s) \in D$. By Corollary 6.1.6(ii), the problem

$$u_n'(t) = A(t)u_n(t) + f_n(t), \quad s \leq t \leq T, \quad u_n(s) = x_n,$$

has a unique strict solution, which is given by the variation of constants formula

$$u_n = G(t,s)x_n + \int_s^t G(t,\sigma) f_n(\sigma) d\sigma, \quad s \leq t \leq T,$$

thanks to Proposition 6.2.2. Since $x_n \to x$ in X and $f_n \to f$ in $C([s,T];X)$, the statement follows by letting $n \to \infty$. ∎

Corollary 6.2.4 *Let $f \in C(]s,T];X) \cap L^1(s,T;X)$ and let $x \in \overline{D}$. If problem (6.0.3) has a classical solution, it is given by formula (6.0.4).*

Proof — Let u be a classical solution of (6.0.3). For every $\varepsilon \in]0, T-s[$, the restriction u_ε of u to $]s+\varepsilon, T]$ is the strict solution of

$$u_\varepsilon'(t) = A(t)u_\varepsilon(t) + f(t), \quad s+\varepsilon \leq t \leq T, \quad u_\varepsilon(s+\varepsilon) = u(s+\varepsilon),$$

so that by Proposition 6.2.2

$$u(t) = G(t, s+\varepsilon)u(s+\varepsilon) + \int_{s+\varepsilon}^t G(t,\sigma) f(\sigma) d\sigma, \quad s+\varepsilon \leq t \leq T.$$

The first addendum may be rewritten as $G(t, s+\varepsilon)[u(s+\varepsilon) - x] + [G(t, s+\varepsilon) - G(t,s)]x + G(t,s)x$. Since $u(s+\varepsilon) \to x \in \overline{D}$, then, using also Corollary 6.1.10(i), one sees that the first addendum goes to $G(t,s)x$ as $\varepsilon \to 0$. Since $f \in L^1(s,T;X)$, the second addendum goes to $\int_s^t G(t,\sigma) f(\sigma) d\sigma$ as $\varepsilon \to 0$, and the statement follows. ∎

Proposition 6.2.2 and Corollary 6.2.4 give uniqueness results. Therefore, the Hölder regularity results of Corollary 6.1.6 concern *the* classical or strict solution of problem (6.1.3). Moreover, from the variation of constants formula we get

$$\|u(t)\| \leq C \left(\|x\| + \int_s^t \|f(\sigma)\| d\sigma \right), \quad s \leq t \leq T. \tag{6.2.2}$$

For the study of regularity be completed we still need some "space" regularity results, which we state in the next proposition. Since the regularity of $t \mapsto G(t,s)x$ has been studied in Corollaries 6.1.8, 6.1.9 and in Lemma 6.1.13, we consider only the function

$$\varphi(t) = \int_s^t G(t,\sigma) f(\sigma) d\sigma, \quad s \leq t \leq T.$$

Proposition 6.2.5 *Let φ be defined as above.*

(i) *If $f \in L^\infty(s,T;X)$, then $\varphi \in C^{1-\theta}([s,T];(X,D)_{\theta,1})$ for each $\theta \in {]0,1[}$.*

(ii) *If $f \in C(]s,T];X) \cap B_\theta(]s,T];(X,D)_{\theta,\infty})$, with $0 < \theta \leq \alpha$, then φ is a strict solution of (6.1.3) with $x = 0$; moreover $A(\cdot)\varphi$ belongs to $B_\theta(]s,T];(X,D)_{\theta,\infty}) \cap C_\theta^\theta(]s,T];X)$.*

(iii) *If $f \in C([s,T];X) \cap B([s,T];(X,D)_{\theta,\infty})$, with $0 < \theta \leq \alpha$, then φ is a classical solution of (6.1.3) with $x = 0$; moreover $A(\cdot)\varphi$ belongs to $B([s,T];(X,D)_{\theta,\infty}) \cap C^\theta([s,T];X)$.*

(iv) *If $f \in C([s,T];(X,D)_\theta)$, with $0 < \theta < \alpha$, then φ' belongs to $C([s,T];(X,D)_\theta)$, and $A(\cdot)\varphi$ belongs to $C([s,T];(X,D)_\theta) \cap h^\theta([s,T];X)$.*

Proof — Statement (i) follows from estimates (6.1.18), using the same procedure of the proof of Proposition 4.2.1.

Let us prove statement (ii). To show that $\varphi(t) \in D$ for every t and that $A(\cdot)\varphi$ belongs to $C_\theta^\theta(]s,T];X)$ it is sufficient to follow closely the proof of the analogous property in the autonomous case (see Theorem 4.3.12), using estimates (6.1.19).

Let us show that $(t-s)^\theta A(t)\varphi(t)$ is bounded with values in $(X,D)_{\theta,\infty}$. By Proposition 6.1.5(i)(ii) we have

$$\left\| \frac{d}{dt} W(t,\sigma)x \right\|_{(X,D)_{\theta,\infty}} \leq \text{const.}\ \sigma^{-(1-\theta)} \|x\|_{(X,D)_{\theta,\infty}}, \quad \forall x \in (X,D)_{\theta,\infty}. \quad (6.2.3)$$

Moreover, from (6.1.4) we get easily

$$\|\tau(A(t)e^{\tau A(t)} - A(\sigma)e^{\tau A(\sigma)})\|_{L((X,D)_{\theta,\infty})} \leq \text{const.}\ (t-\sigma)^\alpha. \quad (6.2.4)$$

For $s < t \leq T$

$$\|A(t)\varphi(t)\|_{(X,D)_{\theta,\infty}} \leq \left\| \int_s^t \frac{d}{dt} W(t,\sigma) f(\sigma) d\sigma \right\|_{(X,D)_{\theta,\infty}}$$
$$+ \left\| \int_s^t (A(\sigma)e^{(t-\sigma)A(\sigma)} - A(t)e^{(t-\sigma)A(t)}) f(\sigma) d\sigma \right\|_{(X,D)_{\theta,\infty}}$$
$$+ \left\| A(t) \int_s^t e^{(t-\sigma)A(t)} f(\sigma) d\sigma \right\|_{(X,D)_{\theta,\infty}}.$$

The first and second addenda are bounded thanks to estimates (6.2.3), (6.2.4) respectively. The third one is bounded by $\text{const.}(t-s)^{-\theta}$ thanks to Theorem 4.3.12.

The proof of the fact that $A(\cdot)\varphi \in C_\theta^\theta(]s,T];X)$ follows from estimates (6.1.19)(b), (6.1.22), and is left to the reader.

2. THE VARIATION OF CONSTANTS FORMULA

Let us show that φ is differentiable in $]s,T]$. Since both φ and $A(\cdot)\varphi$ are continuous, it is sufficient to show that φ is right differentiable, and that the right derivative is $A(\cdot)\varphi + f$. For $s < t < t+h \leq T$,

$$\frac{\varphi(t+h) - \varphi(t)}{h} - (A(t)\varphi(t) + f(t)) = \frac{1}{h}\int_t^{t+h}[(G(t+h,\sigma)f(\sigma) - f(t)]d\sigma$$
$$+ \int_s^t \left[\frac{1}{h}(G(t+h,\sigma) - G(t,\sigma)) - A(t)G(t,\sigma)\right]f(\sigma)d\sigma.$$

The first integral goes to 0 as h goes to 0 because f is continuous and it has values in $\overline{D}$. Using the equality

$$(G(t+h,\sigma) - G(t,\sigma))f(\sigma) = \int_t^{t+h} A(\tau)G(\tau,\sigma)f(\sigma)d\tau,$$

and estimate (6.1.22), we find that the norm of the second integral is less than

$$\frac{C}{h}\int_s^t \sigma^{-\theta}d\sigma \int_t^{t+h}\left(\frac{(t-\tau)^\alpha}{(t-\sigma)^{1-\theta}} + \frac{(t-\tau)^{1-\theta}}{(\tau-\sigma)^{1-\theta}(t-\sigma)^{1-\theta}}\right)d\tau \,\|f\|_{B_\theta((X,D)_\theta,\infty)}$$

$$\leq C\int_0^1 \frac{d\sigma}{\sigma^{-\theta}(1-\sigma)^{\theta-1}}\left(h^\alpha + \frac{1}{h}\int_t^{t+h}\frac{(t-\tau)^{1-\theta}}{(\tau-\sigma)^{1-\theta}}d\tau\right)\|f\|_{B_\theta((X,D)_\theta,\infty)},$$

so that also the second integral goes to 0 as $h \to 0$. Statement (ii) is so proved.

The proof of statement (iii) is analogous and it is left to the reader. To prove statement (iv) it is sufficient to approximate f by a sequence $\{f_n\}_{n\in\mathbb{N}} \subset C([s,T];D)$ and then to use (iii). ∎

With the aid of the variation of constants formula one can investigate further about the dependence of $G(t,s)$ on s, improving Corollary 6.1.10(iv).

Proposition 6.2.6 *For every $t \in]0,T]$ the function $s \mapsto G(t,s)$ is continuously differentiable in $[0,t[$ with values in $L(D)$, and*

$$\frac{\partial}{\partial s}G(t,s)x = -G(t,s)A(s)x, \quad \forall\, x \in D. \tag{6.2.5}$$

Proof — For $x \in D$, the function $t \mapsto G(t,s)x - x$ is the classical solution of

$$u'(t) = A(t)u(t) + A(t)x, \quad s < t \leq T, \quad u(s) = 0.$$

By Corollary 6.2.4 we get

$$G(t,s)x - x = \int_s^t G(t,\sigma)A(\sigma)x\,d\sigma, \quad s \leq t \leq T,$$

so that
$$\frac{G(t,s+h)x - G(t,s)x}{h} = \frac{1}{h}\int_{s+h}^{s} G(t,\sigma)A(\sigma)x\, d\sigma, \quad 0 \leq s,\, s+h < t.$$

By Corollary 6.1.12, $G(t,\cdot)A(\cdot)x$ is continuous in $[0,t[$ with values in D. Therefore, the integral in the right hand side converges to $-G(t,s)A(s)x$ as $h \to 0$, and (6.2.5) follows. ∎

6.3 Asymptotic behavior in the periodic case

If $t \mapsto A(t)$ is periodic, many of the results of Section 4.4 about asymptotic behavior in linear equations may be extended to the nonautonomous case. So, we assume that $A(\cdot)$ satisfies (6.1.1), and that $A(\cdot)$ is periodic with minimum period $T > 0$. We denote again by $G(t,s)$ the evolution operator generated by the family $\{A(t) : t \in \mathbb{R}\}$.

6.3.1 The period map

Due to the periodicity of $A(\cdot)$, the evolution operator enjoys some algebraic properties, stated in the next lemma.

Lemma 6.3.1 *For $s \leq t$ it holds*

(i) $G(t+nT, s) = G(t,s)(G(s+T,s))^n$, $n \in \mathbb{N}$,
(ii) $G(t+T, s+T) = G(t,s)$.

Proof — (i) Let $n = 1$. For every $x \in X$ the function $u(t) = G(t+T,s)x$ is continuously differentiable in $[s, +\infty[$, and it satisfies
$$u'(t) = A(t+T)u(t) = A(t)u(t), \quad t \geq s; \quad u(s) = G(s+T,s)x.$$

Then $u(t) = G(t,s)G(s+T,s)x$, and the statement is proved for $n = 1$. For $n > 1$, the statement follows by recurrence.

Statement (ii) may be proved easily if the domain D is dense, since in this case for every $x \in X$ both $G(\cdot + T, s+T)x$ and $G(\cdot, s)x$ are classical solutions of
$$u'(t) = A(t)u(t), \quad t > s; \quad u(s) = x,$$

so that they do coincide. In the nondense domain case, (ii) is a consequence of (i) if $t \geq s + T$. Otherwise, we split $G(t+T, s+T)x$ and $G(t,s)x$ as
$$G(t+T, s+T)x = e^{(t-s)A(s+T)}x + W(t+T, s+T)x,$$
$$G(t,s)x = e^{(t-s)A(s)}x + W(t,s)x.$$

3. Asymptotic behavior in the periodic case

We have of course $e^{(t-s)A(s+T)}x = e^{(t-s)A(s)}x$. Moreover, $v(t) = W(t+T, s+T)x - W(t,s)x$ is the classical solution of

$$v'(t) = A(t)v(t), \quad t > s; \quad v(s) = 0,$$

and hence $v \equiv 0$. ∎

Define the family of operators

$$V(s) = G(s+T, s), \quad s \in \mathbb{R}. \tag{6.3.1}$$

The function $s \mapsto V(s)$ is called *period map* or *Poincaré map*.

From Corollary 6.1.12 it follows that $s \mapsto V(s) \in C^\theta(\mathbb{R}; L(D)) \cap C^\theta(\mathbb{R}; L(X))$ for every $\theta < \alpha$. By interpolation, $V(\cdot) \in C^\theta(\mathbb{R}; L((X,D)_{\beta,p}))$ for every $\beta \in]0,1[$, $p \in [1, +\infty[$. Moreover, from Lemma 6.3.1 we get

$$V(s+T) = V(s), \quad s \in \mathbb{R}, \tag{6.3.2}$$

$$G(t,s)(V(s))^n = (V(t))^n G(t,s), \quad t \geq s, \ n \in \mathbb{N}. \tag{6.3.3}$$

The asymptotic behavior of $(V(s))^n$, as $n \to \infty$, depends on the spectral properties of $V(s)$. Some of them are stated in the next lemma.

Lemma 6.3.2 *For $t, s \in \mathbb{R}$ it holds*

$$\sigma(V(t)) \setminus \{0\} = \sigma(V(s)) \setminus \{0\}.$$

Moreover, the nonzero eigenvalues of $V(s)$ are independent of s.

Proof — Since $V(\cdot)$ is T-periodic, it is sufficient to consider the case $s \leq t < s+T$. Let $\lambda \neq 0$ belong to $\rho(V(s))$, and for every $y \in X$ consider the resolvent equation

$$\lambda x - V(t)x = y.$$

If x is a solution, then

$$\lambda G(s+T, t)x - G(s+T, t-T)x = G(s+T, t)y,$$

so that $w = G(s+T, t)x$ satisfies

$$\lambda w - V(s)w = G(s+T, t)y,$$

and hence

$$G(s+T, t)x = (\lambda I - V(s))^{-1} G(s+T, t)y,$$

so that

$$G(t, s)(\lambda I - V(s))^{-1} G(s+T, t)y = V(t)x,$$

and
$$x = \frac{1}{\lambda}(G(t,s)(\lambda I - V(s))^{-1}G(s+T,t)y + y). \tag{6.3.4}$$

Therefore, the resolvent equation cannot have more than one solution. Moreover, defining x by (6.3.4) one sees that x solves the resolvent equation. So, $\rho(V(s)) \subset \rho(V(t))$ for $s \leq t < s+T$, and, similarly, $\rho(V(t)) \subset \rho(V(\sigma))$ for $t \leq \sigma \leq t+T$. Choosing $\sigma = s+T$, we get $\rho(V(t)) \subset \rho(V(s+T)) = \rho(V(s))$. Then $\rho(V(s)) = \rho(V(t))$, and the first statement follows.

Let $\lambda \neq 0$ be an eigenvalue of $V(s)$, and let x be a corresponding eigenvector. For $s \leq t \leq s+T$ it holds
$$V(t)G(t,s)x = G(t,s)V(s)x = \lambda G(t,s)x,$$
so that $y = G(t,s)x$ is an eigenvector of $V(t)$ with eigenvalue λ. Note that y is not 0 because $G(t,s)x = 0$ implies $G(t+\sigma,s)x = 0$ for $\sigma > 0$, and in particular $G(s+T,s)x = V(s)x = 0$, which is impossible because $V(s)x = \lambda x \neq 0$. ■

The eigenvalues of $V(0)$ are called *characteristic multipliers* or *Floquet multipliers*.

If the embedding $D \subset X$ is compact, then $V(s)$ is a compact operator for every s, because it belongs to $L(X,D)$. Therefore, $\sigma(V(s)) \setminus \{0\}$ consists of eigenvalues.

A nonzero complex number λ is an eigenvalue of $V(0)$ (and hence of $V(s)$ for every s) if and only if the problem
$$v'(t) = A(t)v(t) - \kappa v(t), \quad t \in \mathbb{R}, \tag{6.3.5}$$
has nontrivial T-periodic solutions, with
$$e^{\kappa T} = \lambda.$$

Indeed, if $\lambda \neq 0$ is an eigenvalue of $V(0)$, then for every eigenvector x it holds $G(T,0)e^{-\kappa T}x = x$, so that problem (6.3.5) admits the T-periodic solution $v(t) = G(t,0)e^{-\kappa t}x$. Conversely, if (6.3.5) has a nontrivial T-periodic solution v, then $v(T) = G(T,0)e^{-\kappa T}v(0) = v(0)$, so that $e^{\kappa T} = \lambda$ is an eigenvalue of $V(0)$.

The complex numbers κ such that $e^{\kappa T} \in \sigma_p(V(0))$ are called *Floquet exponents*.

6.3.2 Estimates on the evolution operator

Since $\sigma(V(s)) \setminus \{0\}$ is independent of s, then the spectral radius $\rho(V(s))$ is independent of s. We set
$$\overline{\rho} = \rho(V(0)) = \rho(V(s)) \quad \forall s \in \mathbb{R}.$$

Then for every $\varepsilon > 0$ there is $n(\varepsilon)$ such that $\|(V(s))^n x\| \leq (\overline{\rho}+\varepsilon)^n \|x\|$ for $n \geq n(\varepsilon)$. By Lemma 6.3.1(i) and Corollary 6.1.8(i) it follows that $K(\varepsilon)$ such that
$$\|G(t,s)\|_{L(X)} \leq K(\varepsilon) \exp\left((t-s)\frac{\log(\overline{\rho}+\varepsilon)}{T}\right), \quad t \geq s$$

3. Asymptotic behavior in the periodic case

(for more details see next Proposition 6.3.3). Therefore, $\|G(t,s)\|_{L(X)}$ decays exponentially as $t - s \to +\infty$ if $\bar{\rho} < 1$. However, even if $\bar{\rho} \geq 1$ it is possible to give more precise informations on the behavior of $G(t,s)x$ as $t - s \to +\infty$. We fix a number $\rho > 0$ such that

$$\{\lambda \in \mathbb{C} : |\lambda| = \rho\} \cap \sigma(V(0)) = \emptyset. \tag{6.3.6}$$

Every $\rho > \bar{\rho}$ satisfies obviously (6.3.6). But the more interesting case is when (6.3.6) is satisfied by some $\rho < \bar{\rho}$. Then we set

$$\sigma(V(0)) = \sigma_- \cup \sigma_+,$$

where

$$\sigma_- = \{\lambda \in \sigma(V(0)) : |\lambda| < \rho\}, \quad \sigma_+ = \{\lambda \in \sigma(V(0)) : |\lambda| > \rho\}.$$

Therefore,

$$\sup\{|\lambda| : \lambda \in \sigma_-\} = \rho_- < \rho < \rho_+ = \inf\{|\lambda| : \lambda \in \sigma_+\}, \tag{6.3.7}$$

and σ_-, σ_+ are spectral sets for $V(s)$, for every $s \in \mathbb{R}$. We define the associated projections,

$$P_-(s) = \frac{1}{2\pi i} \int_{C(0,\rho)} (\lambda I - V(s))^{-1} d\lambda, \quad P_+(s) = I - P_-(s), \quad s \in \mathbb{R}, \tag{6.3.8}$$

where $C(0, \rho)$ denotes as usual the circumference centered at 0 with radius ρ, oriented counterclockwise. Then

$$X = X_-(s) \oplus X_+(s), \quad X_\pm(s) = P_\pm(s)(X), \quad s \in \mathbb{R},$$

and thanks to Proposition A.1.2

$$V(s)(X_\pm(s)) \subset X_\pm(s), \quad \sigma(V(s)_{|X_\pm(s)}) = \sigma_\pm.$$

Moreover, since

$$P_+(s) = -\frac{V(s)}{2\pi i} \int_{C(0,\rho)} [\lambda(\lambda I - V(s))]^{-1} d\lambda,$$

then $X_+(s) \subset D$ and $A(s)(X_+(s)) \subset (X, D)_{\alpha,\infty}$ for every $s \in \mathbb{R}$. Formula (6.3.3) yields $G(t,s)(\lambda I - V(s))^{-1} = (\lambda I - V(t))^{-1} G(t,s)$ for every $\lambda \in C(0, \rho)$ and $t \geq s$, so that

$$G(t,s)P_\pm(s) = P_\pm(t)G(t,s), \quad t \geq s.$$

Now we are able to prove the estimates we need.

Proposition 6.3.3 *Let (6.1.1) and (6.3.6) hold. Fix $\varepsilon > 0$. Then there is $C(\varepsilon) > 0$ such that*

(i) For every $x \in X$ and $t \geq s$

$$\begin{cases} (a) & \|G(t,s)P_-(s)x\| \leq C(\varepsilon)e^{(t-s)\frac{\log(\rho_-+\varepsilon)}{T}}\|x\|, \\ (b) & \|G(t,s)P_-(s)x\|_D \leq \frac{C(\varepsilon)}{t-s}e^{(t-s)\frac{\log(\rho_-+\varepsilon)}{T}}\|x\|, \\ (c) & \|A(t)G(t,s)P_-(s)x\|_{(X,D)_{\alpha,\infty}} \leq \frac{C(\varepsilon)}{(t-s)^{1+\alpha}}e^{(t-s)\frac{\log(\rho_-+\varepsilon)}{T}}\|x\|, \\ (d) & \|(G(t,s) - G(r,s))P_-(s)x\|_{(X,D)_{\theta,1}} \\ & \leq C(\varepsilon)C(\theta)\frac{(t-r)^{1-\theta}}{r-s^{1-\alpha}}e^{(r-s)\frac{\log(\rho_-+\varepsilon)}{T}}\|x\|, \quad 0 < \theta < 1. \end{cases}$$
(6.3.9)

$$\|A(t)G(t,s)P_-(s)x - A(r)G(r,s)P_-(s)x\| \leq C(\varepsilon)\left(\frac{(t-r)^{\alpha}}{r-s} + \frac{t-r}{(r-s)(t-s)}\right)e^{(r-s)\frac{\log(\rho_-+\varepsilon)}{T}}\|x\|. \quad (6.3.10)$$

(ii) For every $x \in D$ and $t \geq s$

$$\begin{cases} (a) & \|A(t)G(t,s)P_-(s)x\| \leq C(\varepsilon)e^{(t-s)\frac{\log(\rho_-+\varepsilon)}{T}}\|x\|_D, \\ (b) & \|A(t)G(t,s)P_-(s)x\|_{(X,D)_{\alpha,\infty}} \leq \frac{C(\varepsilon)}{(t-s)^{\alpha}}e^{(t-s)\frac{\log(\rho_-+\varepsilon)}{T}}\|x\|_D, \\ (c) & \|A(t)G(t,s)P_-(s)x - A(r)G(r,s)P_-(s)x\| \\ & \leq C(\varepsilon)\left((t-s)^{\alpha} + \log\frac{t-s}{r-s}\right)e^{(r-s)\frac{\log(\rho_-+\varepsilon)}{T}}\|x\|_D. \end{cases}$$
(6.3.11)

Proof — The main point is the proof of (6.3.9)(a), the others will be consequence of this one and of the estimates of Corollary 6.1.8. Fix $\varepsilon \in]0, \rho - \rho_-[$. Since $s \mapsto V(s)$ is continuous with values in $L(X)$, then

$$M(\varepsilon) = \sup\{\|(\lambda I - V(s))^{-1}\|_{L(X)} : \lambda \in C(0, \rho_- + \varepsilon), s \in \mathbb{R}\} < +\infty.$$

Let $t > s$, $t = s + \sigma + nT$ with $n \in \mathbb{N} \cup \{0\}$, $\sigma \in [0, T[$. By Lemma 6.3.1(i) and estimate (6.1.15),

$$\|G(t,s)P_-(s)x\| = \frac{1}{2\pi}\left\|G(s+\sigma, s)\int_{C(0,\rho_-+\varepsilon)}(V(s))^n(\lambda I - V(s))^{-1}x\, d\lambda\right\|$$

$$= \frac{1}{2\pi}\left\|G(s+\sigma, s)\int_{C(0,\rho_-+\varepsilon)}\lambda^n(\lambda I - V(s))^{-1}x\, d\lambda\right\|$$

$$\leq CM(\varepsilon)(\rho_-+\varepsilon)^{n+1}\|x\| \leq C(\rho_-+\varepsilon)M(\varepsilon)\exp\left((t-s)\frac{\log(\rho_-+\varepsilon)}{T}\right)\|x\|.$$

3. Asymptotic behavior in the periodic case

Estimate (6.3.9)(a) follows. (6.3.9)(b) is a consequence of (6.1.15) if $t - s \leq T$; if $t - s > T$ we have, using again (6.1.15) and (6.3.9)(a),

$$\begin{aligned}\|G(t,s)P_-(s)x\|_D &\leq \|G(t,t-T)\|_{L(X,D)}\|G(t-T,s)P_-(s)x\| \\ &\leq CT^{-1}C(\varepsilon)\exp\left((t-T-s)\frac{\log(\rho_-+\varepsilon)}{T}\right)\|x\|,\end{aligned}$$

and (6.3.9)(b) follows, since ε is arbitrary. The proof of (6.3.9)(c) is similar.

The proofs of (6.3.9)(d) and (6.3.10) are analogous. We prove here (6.3.10), which follows from (6.3.9)(c) if $t - r \geq T$. Assume that $t - r < T$; then there is $n \in \mathbb{N} \cup \{0\}$ such that $s + nT < r < t < s + (n+2)T$. If $n = 0$, the estimate may be deduced from (6.1.17). If $n \geq 1$, using (6.1.17) and (6.3.9)(a) we find

$$\|A(t)G(t,s)P_-(s)x - A(r)G(r,s)P_-(s)x\|$$
$$= \|[A(t)G(t,s+(n-1)t) - A(r)G(r,s+(n-1)T)]G(s+(n-1)T,s)P_-(s)x\|$$
$$\leq C\left(\frac{(t-r)^\alpha}{T} + \frac{t-r}{T^2}\right)C(\varepsilon)\exp\left((n-1)\log(\rho_-+\varepsilon)\right)\|x\|,$$

and (6.3.10) follows, since ε is arbitrary.

The proof of estimates (6.3.11) is similar and it is left to the reader. ∎

We assume now that σ_+ is not void, and we study the behavior of $G(t,s)P_+(s)$. Note that 0 belongs to the resolvent set of $V(s)_{|X_+(s)}$. It follows that the operator $G(t,s)_{|X_+(s)} : X_+(s) \mapsto X_+(t)$ is invertible, with inverse

$$(G(t,s)_{|X_+(s)})^{-1} = (V(s)_{|X_+(s)})^{-n}G(s+nT,t)$$

for $s + (n-1)T \leq t < s + nT$. Moreover, setting

$$G(s,t) : X_+(t) \mapsto X_+(s), \quad G(s,t) = (G(t,s)_{|X_+(s)})^{-1}, \quad t > s,$$

one gets

$$G(t,s)G(s,r)x = G(t,r)x, \quad t,s,r \in \mathbb{R}, \ x \in X_+(r),$$

and

$$\frac{\partial}{\partial t}G(t,s)x = A(t)G(t,s)x, \quad t,s \in \mathbb{R}, \ x \in X_+(s).$$

Since

$$(V(s)_{|X_+(s)})^{-n} = \frac{1}{2\pi i}\int_{C(0,\overline{\rho}+\varepsilon)-C(0,\rho_+-\varepsilon)}\lambda^{-n}(\lambda I - V(s))^{-1}d\lambda$$

for ε small enough, then for $x \in X_+(s)$

$$\|(V(s)_{|X_+(s)})^{-n}x\| \leq (\overline{\rho}+\rho_++2\varepsilon)(\rho_+-\varepsilon)^{-n}\|x\| \sup_{|\lambda|=\rho_+-\varepsilon, |\lambda|=\overline{\rho}+\varepsilon}\|(\lambda I - V(s))^{-1}\|_{L(X)},$$

which implies, for every $x \in X$,

$$\|G(t,s)P_+(s)x\| \leq K(\varepsilon)e^{(t-s)\frac{\log(\rho_+ -\varepsilon)}{T}}\|x\|, \quad t \leq s. \tag{6.3.12}$$

Arguing as in the proof of Proposition 6.3.3 and recalling that $X_+(s)$ is contained in $D_{A(S)}(\alpha+1,\infty)$, we get the following estimates, for $t < s$, $\varepsilon > 0$, and $x \in X$.

$$\|G(t,s)P_+(s)x\|_D \leq K(\varepsilon)e^{(t-s)\frac{\log(\rho_+ -\varepsilon)}{T}}\|x\|, \quad t \leq s, \tag{6.3.13}$$

$$\|A(t)G(t,s)P_+(s)x\|_{(X,D)_{\alpha,\infty}} \leq K(\varepsilon)e^{(t-s)\frac{\log(\rho_+ -\varepsilon)}{T}}\|x\|, \tag{6.3.14}$$

$$\|(A(t)G(t,s) - A(r)G(r,s))P_+(s)x\| \leq K(\varepsilon)(t-r)^\alpha e^{(r-s)\frac{\log(\rho_+ -\varepsilon)}{T}}\|x\|, \tag{6.3.15}$$

with constant $K(\varepsilon)$ independent of x, t, s.

6.3.3 Asymptotic behavior in nonhomogeneous problems

Techniques similar to the ones used in Section 4.4 give several results about bounded or exponentially decaying forward and backward solutions of the nonhomogeneous problem

$$u'(t) = A(t)u(t) + f(t) \tag{6.3.16}$$

in unbounded intervals.

Bounded solutions

To begin with, we consider bounded solutions in a halfline $[t_0, +\infty[$. We assume that (6.3.6) holds, with $\rho = 1$. As in the time independent case, we shall see that if $f \in L^\infty(t_0, +\infty; X)$ then all the bounded mild solutions of (6.3.16) in $[t_0, +\infty[$ may be represented as

$$u(t) = G(t,t_0)x_- + \int_{t_0}^t G(t,s)P_-(s)f(s)ds - \int_t^\infty G(t,s)P_+(s)f(s)ds, \tag{6.3.17}$$

where x_- is any element of $P_-(t_0)(X)$. To this aim, we need to study the operators Φ and Ψ, defined respectively by

$$\Phi f(t) = \int_{t_0}^t G(t,s)P_-(s)f(s)ds, \quad t \geq t_0,$$

$$\Psi f(t) = -\int_t^\infty G(t,s)P_+(s)f(s)ds, \quad t \geq t_0,$$

for $f \in L^\infty(t_0, +\infty; X)$.

3. Asymptotic behavior in the periodic case

Proposition 6.3.4 *Let (6.1.1) and (6.3.6) hold with $\rho = 1$. Then the mapping Φ is continuous:*

(i) *from $L^\infty(t_0, +\infty; X)$ to $L^\infty(t_0, +\infty; X) \cap C^{1-\theta}([t_0, +\infty[; (X, D)_{\theta,1})$, for every $\theta \in\ ]0, 1[$;*

(ii) *from $C^\theta([t_0, +\infty[; X)$ to $B([t_0, +\infty[; D)$, for every $\theta \in\ ]0, 1[$;*

(iii) *from $C([t_0, +\infty[; X) \cap B([t_0, +\infty[; (X, D)_{\theta,\infty})$ to $B([t_0, +\infty[; D)$, for $0 < \theta < 1$. Moreover, $f \mapsto A(\cdot)\Phi f$ is bounded from $L^\infty(t_0, +\infty; X) \cap B([t_0, +\infty[; (X, D)_{\theta,\infty})$ to $B([t_0, +\infty[; (X, D)_{(\theta,\infty)}) \cap C^\theta([t_0, +\infty[; X)$ for $0 < \theta \leq \alpha$.*

Proof — By interpolation (precisely, using Proposition 1.2.13), estimates (6.3.9)(a) (b) imply that for every $\varepsilon > 0$ there is $C(\varepsilon)$ such that for each $x \in X$

$$\|G(t,s)P_-(s)x\|_{(X,D)_{\theta,1}} \leq \frac{C(\varepsilon)}{(t-s)^\theta} e^{T^{-1}(t-s)\log(\rho_- + \varepsilon)} \|x\|, \quad t \geq s. \qquad (6.3.18)$$

Using (6.3.18) and (6.3.9)(d), and arguing as in the proof of Proposition 4.2.1, statement (i) follows.

Let $f \in C^\theta([t_0, +\infty[; X)$. Thanks to Corollary 6.1.6(ii), Φf is bounded in $[t_0, t_0 + T]$ with values in D. For $t \geq t_0 + T$ we have

$$\|\Phi f(t)\|_D \leq \left\|\int_{t_0}^t G(t,s)P_-(s)(f(s) - f(t))ds\right\|$$
$$+ \left\|P_-(t)\int_{t_0}^{t-T} G(t,s)f(t)ds\right\| + \left\|\int_{t-T}^t G(t,s)P_-(s)f(t)ds\right\|.$$

Using twice (6.3.9)(b) and then (6.1.34), we see that Φf is bounded in $[t_0, +\infty[$ with values in D. Statement (ii) is so proved.

Let now $f \in C([t_0, +\infty[; X) \cap B([t_0, +\infty[; (X, D)_{\theta,\infty})$. Then $s \mapsto P_-(s)f(s)$ belongs to $C([t_0, +\infty[; X) \cap B([t_0, +\infty[; (X, D)_{\theta,\infty})$. Estimates (6.3.9)(b) and (6.3.11)(a) yield by interpolation

$$\|G(t,s)P_-(s)x\|_D \leq \frac{C(\varepsilon)^{1-\theta}}{t-s} e^{(t-s)\frac{\log(\rho_- + \varepsilon)}{T}} \|x\|_{(X,D)_{\theta,\infty}}, \quad x \in (X,D)_{\theta,\infty}.$$

It follows that $\|\Phi f(t)\|_D \leq \text{const.} \|f\|_{B([t_0,+\infty[;(X,D)_{(\theta,\infty)})}$. Moreover, for $t > t_0 + T$

$$A(t)\Phi f(t) = A(t)G(t, t-T)\Phi f(t-T) + {}'A(t)\int_{t-T}^t G(t,s)P_-(s)f(s)ds.$$

Since Φf is bounded with values in D, the first addendum is bounded in $[t_0 + T, +\infty[$ with values in $(X, D)_{\alpha,\infty}$, thanks to estimate (6.1.16). The second addendum is bounded with values in $(X, D)_{\theta,\infty}$ and $\theta \wedge \alpha$-Hölder continuous with values in X in $[t_0 + T, +\infty[$ thanks to Proposition 6.2.5(iii). Statement (iii) follows. ∎

Proposition 6.3.5 *Let (6.1.1) and (6.3.6) hold with $\rho = 1$. For every $f \in L^\infty(t_0, +\infty; X)$, Ψf belongs to $Lip([t_0, +\infty[; X) \cap C^\alpha([t_0, +\infty[; D)$. Moreover, $A(\cdot)\Psi f$ is bounded with values in $(X, D)_{\alpha, \infty}$, and there is $C > 0$ such that*

$$\|\Psi f\|_{B([t_0, +\infty[; D)} + \|A(\cdot)\Psi\|_{B([t_0, +\infty[; (X,D)_{\alpha,\infty})}$$
$$+ \|A(\cdot)\Psi f(\cdot)\|_{C^\alpha([t_0, +\infty[; X)} \leq C\|f\|_{L^\infty(t_0, +\infty; X)}. \quad (6.3.19)$$

If in addition f is continuous, then Ψf is continuously differentiable with values in X, and

$$(\Psi f)'(t) = A(t)(\Psi f)(t) + P_+(t)f(t), \ t \geq t_0. \quad (6.3.20)$$

If $f \in C^\theta([t_0, +\infty[; X)$, with $\theta < \alpha$, then Ψf belongs to $C^{1+\theta}([t_0, +\infty[; X) \cap C^\theta([t_0, +\infty[; D)$, and there is C independent of f such that

$$\|\Psi f\|_{C^{1+\theta}([t_0, +\infty[; X)} + \|\Psi f\|_{C^\theta([t_0, +\infty[; D)} \leq C\|f\|_{C^\theta([t_0, +\infty[; X)}.$$

Proof — Since $s \mapsto P_+(s)$ is in $C(\mathbb{R}; L(X, D))$, then for each $f \in L^\infty(t_0, +\infty; X)$ $P_+(\cdot)f(\cdot)$ belongs to $L^\infty(t_0, +\infty; D)$. Due to estimates (6.3.12), (6.3.13), Ψf is bounded with values in D, and due to (6.3.14), $A(\cdot)\Psi f(\cdot)$ is bounded with values in $(X, D)_{\alpha, \infty}$.

Let us prove that Ψf is α-Hölder continuous with values in D. From (6.3.13) and (6.3.15) it follows that for $t_0 \leq r \leq t \leq s$ it holds

$$\|G(r, s)P_+(s)\|_{L(X,D)} \leq Ce^{-\omega(s-r)},$$

$$\|(G(t, s) - G(r, s))P_+(s)\|_{L(X,D)} \leq C(t-r)^\alpha e^{-\omega(s-r)},$$

with $C, \omega > 0$. Therefore, for $t_0 \leq r < t$,

$$\|(\Psi f)(t) - (\Psi f)(r)\|_D \leq \left\|\int_t^\infty (G(t, s) - G(r, s))P_+(s)f(s)ds\right\|_D$$
$$+ \left\|\int_r^t G(r, s)P_+(s)f(s)\right\|_D \leq C\omega^{-1}((t-r)^\alpha + t - r)\|f\|_{L^\infty(t_0, +\infty; X)}.$$

From the equality

$$(\Psi f)(t) = (\Psi f)(t_0) + \int_{t_0}^t G(t, s)P_+(s)f(s)ds, \ t \geq t_0,$$

it follows that Ψf is a mild solution of (6.3.20) in $[t_0, +\infty[$. The initial datum $\Psi f(t_0)$ belongs to $D_{A(t_0)}(\alpha + 1, \infty)$, and if f is continuous with values in X, then $P_+(\cdot)f(\cdot)$ is continuous with values in D. From Proposition 6.2.5(iii) it follows that Ψf is continuously differentable and it is a strict solution of (6.3.20). ∎

Now we are able to study the bounded forward solutions of the initial value problem

$$u'(t) = A(t)u(t) + f(t), \ t > t_0; \ u(t_0) = x. \quad (6.3.21)$$

3. Asymptotic behavior in the periodic case

Proposition 6.3.6 *Let (6.1.1) and (6.3.6) hold with $\rho = 1$. Let $f \in L^\infty(0, +\infty; X)$, and $x \in X$. Then the mild solution u of (6.3.21) is bounded in $[t_0, +\infty[$ if and only if*

$$P_+(t_0)x = -\int_{t_0}^\infty G(t_0, s)P_+(s)f(s)ds, \tag{6.3.22}$$

in which case u is given by

$$u(t) = G(t,t_0)P_-(t_0)x + \int_{t_0}^t G(t,s)P_-(s)f(s)ds - \int_t^\infty G(t,s)P_+(s)f(s)ds. \tag{6.3.23}$$

The linear mapping $(x, f) \mapsto u$, defined for (x, f) satisfying the compatibility condition (6.3.22), is continuous:

(i) *from $X \times L^\infty(t_0, +\infty; X)$ into $L^\infty(t_0, +\infty; X) \cap C^{1-\theta}([t_0 + \varepsilon, +\infty[; (X, D)_{\theta,1})$, for every $\theta \in]0, 1[$, $\varepsilon > 0$;*

(ii) *from $D \times C_\theta^\theta(]t_0, t_0 + \varepsilon]; X) \cap C^\theta([t_0 + \varepsilon, +\infty[; X)$ into $C_\theta^\theta(]t_0, t_0 + \varepsilon]; D) \cap C^\theta([t_0 + \varepsilon, +\infty[; D)$, for every $\theta \in]0, \alpha]$, $\varepsilon > 0$;*

(iii) *from $D_{A(t_0)}(\theta + 1, \infty) \times L^\infty(t_0, +\infty; X) \cap B([t_0, +\infty[; (X, D)_{\theta,\infty})$ to $C^\theta([t_0, +\infty[; D)$, for every $\theta \in]0, \alpha]$; moreover $(x, f) \mapsto A(\cdot)u(\cdot)$ is continuous with values in $B([t_0, +\infty[; (X, D)_{\theta,\infty})$.*

Proof — Split u into the sum $u(t) = u_1(t) + u_2(t)$, where, for $t \geq t_0$,

$$u_1(t) = G(t,t_0)P_-(t_0)x + \int_{t_0}^t G(t,s)P_-(s)f(s)ds - \int_t^{+\infty} G(t,s)P_+(s)f(s)ds,$$

$$u_2(t) = G(t,t_0)\left(P_+(t_0)x + \int_{t_0}^{+\infty} G(t_0,s)P_+(s)f(s)ds\right).$$

Thanks to estimates (6.3.9)(a), (6.3.12) and to Proposition 6.3.5, u_1 and u_2 are well defined, and u_1 belongs to $L^\infty(t_0, +\infty; X)$. If $P_+(t_0) = 0$, then $u_2 \equiv 0$, and (6.3.22), (6.3.23) hold trivially. If $P_+(t_0) \neq 0$, then

$$y = P_+(t_0)x + \int_{t_0}^{+\infty} G(t_0, s)P_+(s)f(s)ds$$

belongs to $X_+(t_0)$. Let $\varepsilon > 0$ be so small that $\mu = T^{-1}\log(\rho_+ - \varepsilon) > 0$. Then (6.3.12) yields $\|u_2(t)\| \geq K(\varepsilon)^{-1}e^{\mu(t-t_0)}\|y\|$ for $t \geq t_0$, so that u_2 is unbounded in $[t_0, +\infty[$ unless $y = 0$. If $y = 0$, then $u \equiv u_1$, and (6.3.22), (6.3.23) hold.

Statements (i) and (iii) are now consequences of Propositions 6.3.4 and 6.3.5. Let us prove that statement (ii) holds. From Propositions 6.3.4(ii) and 6.3.5 we know that u is bounded in $[t_0, +\infty[$ with values in D. Moreover, it belongs to $C_\theta^\theta(]t_0, t_0 + T]; D)$ thanks to Corollary 6.1.6(ii). So, it is sufficient to prove that u is θ-Hölder continuous with values in D in any interval of lenght T contained in

$[t_0+T, \infty[$, with Hölder constant independent of the interval. For $T \leq r \leq t \leq r+T$ write $u(t)$ as

$$u(t) = G(t, r - T/2)u(r - T/2) + \int_{r-T/2}^{t} G(t, s)f(s)ds.$$

Since $\|u(r - T/2)\|_D$ is bounded by a constant independent of r, then u belongs to $C_\theta^\theta(]r - T/2, r + T]; D)$, and hence to $C^\theta([r, r + T]; D)$, with Hölder constant independent of r, thanks to Corollary 6.1.6(ii). ∎

Let us consider now a backward problem,

$$v'(t) = A(t)v(t) + g(t), \quad t < t_0; \quad v(t_0) = y. \tag{6.3.24}$$

To study (6.3.24) we introduce the operator Λ defined on $L^\infty(-\infty, t_0; X)$ by

$$(\Lambda g)(t) = \int_{-\infty}^{t} G(t, s)P_-(s)g(s)ds, \quad t \leq t_0.$$

Proposition 6.3.7 *Let (6.1.1) and (6.3.6) hold with $\rho = 1$. Then the linear operator Λ is continuous:*

(i) from $L^\infty(-\infty, t_0; X)$ to $C^{1-\theta}(]-\infty, t_0]; (X, D)_{\theta,1})$, for every $\theta \in]0, 1[$;

(ii) from $C^\theta(]-\infty, t_0]; X)$ to $B(]-\infty, t_0]; D)$, for every $\theta \in]0, 1[$;

(iii) from $C(]-\infty, t_0]; X) \cap B(]-\infty, t_0]; (X, D)_{\theta,\infty})$ to $C^\theta(]-\infty, t_0]; D)$; moreover $g \mapsto A(\cdot)\Lambda g(\cdot)$ is bounded from $C(]-\infty, t_0]; X) \cap B(]-\infty, t_0]; (X, D)_{\theta,\infty})$ to $B(]-\infty, t_0]; (X, D)_{\theta,\infty})$, for every $\theta \in]0, \alpha]$.

If g belongs to $C(]-\infty, t_0]; X) \cap B(]-\infty, t_0]; (X, D)_{\theta,\infty})$ or to $C^\theta(]-\infty, t_0]; X)$ for some $\theta > 0$, then Λg is differentiable with values in X, and

$$(\Lambda g)'(t) = A(t)\Lambda g(t) + P_-(t)g(t), \quad t \leq t_0.$$

Proof — Let $\varepsilon > 0$ be so small that

$$\omega = -T^{-1}\log(\rho_- + \varepsilon) > 0.$$

Estimate (6.3.9)(ii) implies that for every $g \in L^\infty(-\infty, t_0; X)$

$$\|\Lambda g\|_{L^\infty(-\infty,t_0;X)} \leq \frac{K(\varepsilon)}{\omega} \|g\|_{L^\infty(-\infty,t_0;X)}. \tag{6.3.25}$$

Moreover, using (6.3.18) we get

$$\|(\Lambda g)(t)\|_{(X,D)_{\theta,1}} \leq \frac{C(\varepsilon)\Gamma(1-\theta)}{\omega^{1-\theta}} \|g\|_{L^\infty(-\infty,t_0;X)}, \quad t \leq t_0.$$

3. ASYMPTOTIC BEHAVIOR IN THE PERIODIC CASE

To see that Λg belongs to $C^{1-\theta}(]-\infty, t_0]; (X, D)_{\theta,1})$, recall that for $k \in \mathbb{Z}$ and $kT \leq t < (k+1)T \wedge t_0$

$$(\Lambda g)(t) = G(t, (k-1)T)(\Lambda g)((k-1)T) + \int_{(k-1)T}^{t} G(t, s) P_-(s) g(s) ds. \quad (6.3.26)$$

By (6.3.25), $\|(\Lambda g)((k-1)T)\|$ is bounded by a constant independent of k. Estimate (6.1.18) implies that $G(\cdot, (k-1)T)(\Lambda g)((k-1)T)$ belongs to $C^{1-\theta}([kT, (k+1)T \wedge t_0]; (X, D)_{\theta,1})$ with norm independent of k. Since $P_-(\cdot) g(\cdot)$ belongs to $L^\infty(-\infty, t_0; X)$, by Proposition 6.2.5(i) the function

$$t \mapsto \int_{(k-1)T}^{t} G(t, s) P_-(s) g(s) ds$$

belongs to $C^{1-\theta}([(k-1)T, (k+1)T \wedge t_0]; (X, D)_{\theta,1})$, and hence to $C^{1-\theta}([kT, (k+1)T \wedge t_0]; (X, D)_{\theta,1})$, with norm independent of k. Statement (i) follows.

Taking into account (6.3.26), the proof of statements (ii) and (iii) are analogous to the proofs of statements (ii) and (iii) of Proposition 6.3.6, and are omitted. ∎

Now we are able to study bounded solutions of (6.3.24).

Proposition 6.3.8 *Let (6.1.1) and (6.3.6) hold with $\rho = 1$. Let $y \in X$, $g \in L^\infty(-\infty, t_0; X)$. Then problem (6.3.24) has a bounded strong solution v in $]-\infty, t_0]$ if and only if*

$$P_-(t_0) y = \int_{-\infty}^{t_0} G(t_0, s) P_-(s) g(s) ds, \quad (6.3.27)$$

in which case v is given by

$$v(t) = G(t, t_0) P_+(t_0) y + \int_{t_0}^{t} G(t, s) P_+(s) g(s) ds + \int_{-\infty}^{t} G(t, s) P_-(s) g(s) ds. \quad (6.3.28)$$

The linear mapping $(y, g) \mapsto v$, defined on the couples (y, g) which satisfy (6.3.27), is continuous:

(i) *from $X \times L^\infty(-\infty, t_0; X)$ to $C^{1-\theta}(]-\infty, t_0]; (X, D)_{\theta,1})$ for every $\theta \in]0, 1[$;*
(ii) *from $X \times C^\theta(]-\infty, t_0]; X)$ to $C^\theta(]-\infty, t_0]; D)$, for every $\theta \in]0, \alpha]$;*
(iii) *from $X \times L^\infty(-\infty, t_0; X) \cap B(]-\infty, t_0]; (X, D)_{\theta,\infty})$ to $C^\theta(]-\infty, t_0]; D)$, for every $\theta \in]0, \alpha]$; moreover $(y, g) \mapsto A(\cdot) v(\cdot)$ is continuous from $X \times L^\infty(-\infty, t_0; X) \cap B(]-\infty, t_0]; (X, D)_{\theta,\infty})$ to $B(]-\infty, t_0]; (X, D)_{\theta,\infty})$.*

Moreover, if $f \in C^\theta(]-\infty, t_0]; X)$ or $f \in C(]-\infty, t_0]; X) \cap B(]-\infty, t_0]; (X, D)_{\theta,\infty})$ for some $\theta > 0$, then v is a strict solution of (6.3.27).

Proof — If v is any bounded strong solution of (6.3.27), then for every $a < t_0$

$$\begin{aligned}v(t) &= G(t,s)v(a) + \int_a^t G(t,s)v(s)ds\\ &= G(t,a)\left[P_-(a)v(a) - \int_{-\infty}^a G(a,s)P_-(s)g(s)ds\right]\\ &\quad + \int_{-\infty}^t G(t,s)P_-(s)g(s)ds + P_+(t)v(t), \quad a \leq t \leq t_0.\end{aligned}$$

Since v and g are bounded, then

$$\left\|P_-(a)v(a) - \int_{-\infty}^a G(a,s)P_-(s)g(s)ds\right\|$$

is bounded by a constant independent of a. Letting $a \to -\infty$, and using estimate (6.3.9)(a), we find

$$v(t) = \int_{-\infty}^t G(t,s)P_-(s)g(s)ds + P_+(t)v(t), \quad t \leq t_0,$$

so that v is given by (6.3.28), and (6.3.27) holds.

The rest of the proof is similar to the proof of Proposition 6.3.6; one has to use of course the results of Proposition 6.3.7 instead of those of Propositions 6.3.4, 6.3.5. ∎

Once the characterization of bounded forward and backward solutions are established, it is easy to study bounded solutions of

$$z'(t) = A(t)z(t) + h(t), \quad t \in \mathbb{R}. \tag{6.3.29}$$

Corollary 6.3.9 *Let (6.1.1) and (6.3.6) hold with $\rho = 1$. If $h \in C_b(\mathbb{R}; X)$, then problem (6.3.29) has a unique bounded strong solution z, given by*

$$z(t) = \int_{-\infty}^t G(t,s)P_-(s)h(s)ds - \int_t^{+\infty} G(t,s)P_+(s)h(s)ds, \quad t \in \mathbb{R}. \tag{6.3.30}$$

The linear mapping $h \mapsto z$ is continuous:

(i) *from $L^\infty(\mathbb{R}; X)$ to $C^{1-\theta}(\mathbb{R}; (X,D)_{\theta,1})$, for every $\theta \in {]}0,1[$;*

(ii) *from $C^\theta(\mathbb{R}; X)$ to $C^\theta(\mathbb{R}; D) \cap C^{\theta+1}(\mathbb{R}; X)$, for every $\theta \in {]}0,\alpha]$;*

(iii) *from $L^\infty(\mathbb{R}; X) \cap B(\mathbb{R}; (X,D)_{\theta,\infty})$ to $C^\theta(\mathbb{R}; D)$; moreover $h \mapsto A(\cdot)z$ is continuous from $L^\infty(\mathbb{R}; X) \cap B(\mathbb{R}; (X,D)_{\theta,\infty})$ to $B(\mathbb{R}; (X,D)_{\theta,\infty})$, for every $\theta \in {]}0,\alpha]$.*

3. Asymptotic behavior in the periodic case

Proof — Let z be any bounded strong solution of (6.3.29). Then, for every $t_0 \in \mathbb{R}$, z is a bounded strong solution of (6.3.21) and (6.3.24), with $x = y = z(t_0)$. By Proposition 6.3.6, we have

$$P_+(t_0)z(t_0) = -\int_{t_0}^{+\infty} G(t_0, s)P_+(s)h(s).$$

By Proposition 6.3.8,

$$P_-(t_0)z(t_0) = \int_{-\infty}^{t_0} G(t_0, s)P_-(s)h(s)ds.$$

Since $z(t_0) = P_+(t_0)z(t_0) + P_-(t_0)z(t_0)$, formula (6.3.30) follows. Statements (i), (ii), (iii) follow now from the corresponding statements of Propositions 6.3.6 and 6.3.8. ■

Exponentially decaying and exponentially growing solutions

We assume now that (6.3.6) holds for some $\rho > 0$, and we set

$$\omega = \frac{\log \rho}{T}. \tag{6.3.31}$$

Proposition 6.3.10 *Let (6.3.6) hold, and define ω by (6.3.31). If $x \in X$ and $f \in L^\infty(t_0, +\infty; X, \omega)$[2], then the mild solution u of (6.3.21) belongs to $L^\infty(t_0, +\infty; X, \omega)$ if and only if (6.3.22) holds, in which case u is given by (6.3.23). In formulas (6.3.22), (6.3.23), $P_-(s)$, $P_+(s)$ are the projections defined by (6.3.8).*

The linear mapping $(x, f) \mapsto u$, defined for (x, f) satisfying the compatibility condition (6.3.22), is continuous:

(i) *from $X \times L^\infty(t_0, +\infty; X, \omega)$ into $L^\infty(t_0, +\infty; X, \omega) \cap C^{1-\theta}([t_0 + \varepsilon, +\infty[; (X, D)_{\theta,1}, \omega)$, for every $\theta \in \,]0, 1[$, $\varepsilon > 0$;*

(ii) *from $D \times (C_\theta^\theta(]t_0, t_0 + \varepsilon]; X) \cap C_\theta^\theta([t_0 + \varepsilon, \infty[; X, \omega))$ into $C_\theta^\theta(]t_0, t_0 + 1]; D) \cap C_\theta^\theta([t_0 + 1, \infty[; D, \omega)$, for every $\theta \in \,]0, \alpha]$, $\varepsilon > 0$;*

(iii) *from $D_{A(t_0)}(\theta + 1, \infty) \times L^\infty(t_0, +\infty; X, \omega) \cap B([t_0, +\infty[; (X, D)_{\theta,\infty}, \omega)$ to $C^\theta([t_0, +\infty[; D, \omega)$ for $0 < \theta \leq \alpha$; moreover $(x, f) \mapsto A(\cdot)u$ is continuous with values in $B([t_0, +\infty[; (X, D)_{\theta,\infty}, \omega) \cap C^\theta([t_0, +\infty[; X, \omega)$.*

Proof — Let u be the mild solution of (6.3.21). Then the function $\widetilde{u}(t) = e^{-\omega(t-t_0)}$ is the mild solution of

$$\widetilde{u}'(t) = \widetilde{A}(t)\widetilde{u}(t) + \widetilde{f}(t), \quad t \geq t_0; \quad \widetilde{u}(t_0) = x, \tag{6.3.32}$$

---
[2] If I is an unbounded interval, the weighted spaces $L^\infty(I; X, \omega)$, $C^\beta(I; X, \omega)$, etc., are defined at the beginning of Section 4.4.

with $\widetilde{A}(t) = A(t) - \omega I$, $\widetilde{f}(t) = e^{-\omega(t-t_0)}f(t)$. The evolution operator associated to the family $\{\widetilde{A}(t) : t \in \mathbb{R}\}$ is $\widetilde{G}(t,s) = G(t,s)e^{-\omega(t-s)}$. For every $s \in \mathbb{R}$, the spectrum of $\widetilde{V}(s) = \widetilde{G}(s+T, s)$ does not intersect the circumference centered at 0 with radius 1. So, the results of Proposition 6.3.6 hold for problem (6.3.32). Therefore, $\widetilde{u}$ is bounded in $[t_0, +\infty[$ if and only if

$$\widetilde{P}_+(t_0)x = -\int_{t_0}^{+\infty} \widetilde{G}(t_0, s)\widetilde{P}_+(s)\widetilde{f}(s)ds, \qquad (6.3.33)$$

where

$$\widetilde{P}_+(s) = I - \frac{1}{2\pi i}\int_{C(0,1)}(\lambda - \widetilde{V}(s))^{-1}d\lambda = P_+(s).$$

An easy computation shows now that (6.3.33) is equivalent to (6.3.22). So, all the statements follow by applying Proposition 6.3.6. ∎

The backward problem (6.3.24) may be treated by the same procedure.

Proposition 6.3.11 *Let (6.3.6) hold, and define ω by (6.3.31). If $y \in X$ and $f \in L^\infty(-\infty, t_0; X, \omega)$, then problem (6.3.24) has a bounded mild solution v in $]-\infty, t_0]$ if and only if (6.3.27) holds, in which case v is given by (6.3.28).*

The linear mapping $(y, g) \mapsto v$, defined for the couples (y, g) which satisfy (6.3.27), is continuous:

(i) *from $X \times L^\infty(-\infty, t_0; X, \omega)$ to $C^{1-\theta}(]-\infty, t_0]; (X, D)_{\theta,1}, \omega)$ for every $\theta \in]0, 1[$;*

(ii) *from $X \times C^\theta(]-\infty, t_0]; X, \omega)$ to $C^\theta(]-\infty, t_0]; D, \omega)$, for every $\theta \in]0, \alpha]$;*

(iii) *from $X \times L^\infty(-\infty, t_0; X, \omega) \cap B(]-\infty, t_0]; (X, D)_{\theta, \infty}, \omega)$ to $C^\theta(]-\infty, t_0]; D, \omega)$, for every $\theta \in]0, \alpha]$; moreover the mapping $(y, g) \mapsto A(\cdot)v(\cdot)$ is continuous from $X \times L^\infty(-\infty, t_0; X, \omega) \cap B(]-\infty, t_0]; (X, D)_{\theta, \infty})$ to $B(]-\infty, t_0]; (X, D)_{\theta, \infty})$.*

The proof is similar to the proof of Proposition 6.3.10, and is omitted.

Periodic solutions

Also the study of the periodic solutions is similar to the correponding one in the autonomous case.

3. Asymptotic behavior in the periodic case

Proposition 6.3.12 *Let $f : \mathbb{R} \mapsto X$ be continuous and T-periodic. Then*

(i) If 1 belongs to $\rho(V(0))$, then problem (6.3.29) has a unique T-periodic strong solution, given by

$$z(t) = (I - V(t))^{-1} \int_t^{t+T} G(t+T, s) f(s) ds, \quad t \in \mathbb{R}. \tag{6.3.34}$$

(ii) If 1 is a semisimple eigenvalue of $V(0)$ (and hence of $V(t)$ for every t), then problem (6.3.29) has a T-periodic strong solution if and only if

$$P(0) \int_0^T G(T, s) f(s) ds = 0, \tag{6.3.35}$$

in which case all the T-periodic strong solutions of (6.3.29) are given by

$$\begin{aligned}
z(t) &= (I - V(t))^{-1}_{|(I-P(t))(X)} \int_t^{t+T} G(t+T, s)(I - P(s)) f(s) ds \\
&\quad + G(t, 0) x + \int_{T[t/T]}^t G(t, s) P(s) f(s) ds, \quad t \in \mathbb{R},
\end{aligned} \tag{6.3.36}$$

($[t/T]$ denotes the maximum integer $\leq t/T$) with $x \in \mathrm{Ker}(V(0) - I)$.

Any periodic strong solution z of (6.3.29) belongs to $C^{1-\theta}(\mathbb{R}; (X, D)_{\theta,1})$ for every $\theta \in {]}0, 1[$. If in addition $f \in C^\theta(\mathbb{R}; X)$, with $0 < \theta \leq \alpha$, then z is a strict solution and it belongs to $C^\theta(\mathbb{R}; D)$, whereas z' belongs to $B(\mathbb{R}; (X, D)_{\theta,\infty})$ $\cap C^\theta(\mathbb{R}; X)$. If $f \in C(\mathbb{R}; X) \cap B(\mathbb{R}; (X, D)_{\theta,\infty})$, with $0 < \theta \leq \alpha$, then z is a strict solution, it belongs to $B(\mathbb{R}; (X, D)_{\theta+1,\infty})$, and $A(\cdot) z$ belongs to $C^\theta(\mathbb{R}; X)$.

Proof — If z is any T-periodic strong solution of (6.3.29), then

$$z(T + t) = V(t) z(t) + \int_t^{t+T} G(t+T, s) f(s) ds = z(t), \quad \forall t \in \mathbb{R}. \tag{6.3.37}$$

Let $1 \in \rho(V(0))$. Then, by Lemma 6.3.2, $1 \in \rho(V(t))$ for every $t \in \mathbb{R}$, so that the unique T-periodic strong solution of (6.3.29) is given by (6.3.34).

Let now 1 be a semisimple eigenvalue of $V(0)$. Then it is a semisimple eigenvalue of $V(t)$, for every t. If problem (6.3.29) has a T-periodic strong solution z, applying $P(t)$ and $I - P(t)$ to both members of (6.3.37) one gets, for every $t \in \mathbb{R}$,

(i) $\displaystyle P(t) \int_t^{t+T} G(t+T, s) f(s) ds = \int_t^{t+T} G(t+T, s) P(s) f(s) ds = 0,$

(ii) $\displaystyle (I - V(t))(I - P(t)) z(t) = \int_t^{t+T} G(t+T, s)(I - P(s)) f(s) ds.$

From (ii) it follows that

$$(I - P(t))z(t) = (I - V(t))^{-1}_{|(I-P(t))(X)} \int_t^{t+T} G(t+T,s)(I - P(s))f(s)ds.$$

Moreover, (i) is equivalent to (6.3.35). Indeed, if (i) holds, taking $t = 0$ one gets (6.3.35). Conversely, after easy computation one sees that if $t = kT + \sigma$, with $k = [t/T] \in \mathbb{Z}$ and $\sigma \in [0, T[$, then

$$\int_t^{t+T} G(t+T,s)P(s)f(s)ds = G(T+\sigma,T)\int_0^T G(T,s)P(s)f(s)ds,$$

so that if (6.3.35) holds then (i) holds, and the function $t \mapsto \int_{kT}^t G(t,s)P(s)f(s)ds$ is T-periodic in $[kT, +\infty[$. (6.3.36) follows. The statements about regularity are obvious consequences of Corollary 6.1.6 and Proposition 6.2.5. ∎

6.4 Bibliographical remarks

Nonautonomous abstract parabolic problems have been studied by many authors. The earliest papers about evolution operators are due to P.E. SOBOLEVSKIĬ [181] and H. TANABE [190], who independently constructed the evolution operator for problem (6.0.1) under assumption (6.1.1), and the additional assumption that the domain D is dense in X. They wrote down an integral equation for the evolution operator and were able to solve it by successive approximations. The theory of Sobolevskiĭ was then exposed in the book [84]. Assumption (6.1.1) was considered also by P. ACQUISTAPACE – B. TERRENI in [4, 5], who did not consider the evolution operator but gave other reprentation formulas, different from the variation of constants formula (6.0.4), for the strict, classical, or strong solution to (6.0.1). Such representation formula let them prove several regularity results, among which the optimal regularity results of Propositions 6.1.3 and 6.1.5(iii).

Here we have followed the approach of A. LUNARDI [140, 143, 147], based on optimal regularity results for the autonomous case and perturbation arguments. That method had been used for the first time by G. DA PRATO – P. GRISVARD [63], under different assumptions, to prove the existence of regular solutions without using the evolution operator. Proposition 6.1.3 is due to A. LUNARDI – E. SINESTRARI [155], who improved an earlier result of G. DA PRATO – E. SINESTRARI [67].

In the case of variable domains several types of assumptions can be done to solve problem (6.0.1). They can be roughly divided in two groups: a first group af assumptions, concerning the case where the function $t \mapsto R(\lambda, A(t))$ is very regular — at least, differentiable — for every $\lambda \in \rho(A(t))$; and a second group where, although the domains $D(A(t))$ are allowed to vary, a suitable intermediate space between X and $D(A(t))$ is independent of time.

The first case was considered by T. KATO – H. TANABE [108], H. TANABE [191], P. ACQUISTAPACE – B. TERRENI [4], A. YAGI [210, 211]. The second case was considered by T. KATO [106], P.E. SOBOLEVSKIĬ [183], H. AMANN [20], who assumed that $D((-A(t))^\rho)$ is constant for some $\rho \in\,]0,1[$, and by P. ACQUISTAPACE – B. TERRENI [6], G. DA PRATO – P. GRISVARD [64], who assumed that $D_{A(t)}(\rho, \infty)$ is constant for some $\rho \in\,]0,1[$.

A unified approach was given by P. ACQUISTAPACE – B. TERRENI [8, 2]. They recovered the results of [181], [190], [191], [183], [140], [4],[5], [210], and found new regularity results for the solution, including optimal regularity in the weighted spaces introduced in Chapter 4.

A discussion about the different assumptions which lead to the construction of the evolution operator may be found in P. ACQUISTAPACE [3].

In the case where not only the domain $D(A(t))$ are constant, but also the spaces $D_{A(t)}(\theta+1, \infty)$ or $D_{A(t)}(\theta+1)$ are independent of time for some $\theta \in\,]0,1[$, the regularity assumptions on $t \mapsto A(t)$ may be reduced to the mere continuity. See G. DA PRATO – P. GRISVARD [64], A. BUTTU [40, 41].

Chapter 7

Semilinear equations

This chapter is devoted to semilinear parabolic problems, both abstract and concrete. The abstract equations we consider are of the kind

$$u'(t) = Au(t) + f(t, u(t)), \quad t > 0, \quad u(0) = u_0, \tag{7.0.1}$$

where $A : D(A) \mapsto X$ is a linear sectorial operator, f is a continuous function defined in $[0, T] \times X_\alpha$ (or in $[0, T] \times \mathcal{O}$, $\mathcal{O}$ being an open set in X_α), and takes values in X. Here $X_0 = X$, and for $0 < \alpha < 1$, X_α is any Banach space continuously embedded in X and such that

$$\begin{cases} (i) & D_A(\alpha, 1) \subset X_\alpha \subset D_A(\alpha, \infty), \\ (ii) & \text{the part of } A \text{ in } X_\alpha \text{ is sectorial in } X_\alpha. \end{cases} \tag{7.0.2}$$

We recall that (7.0.2)(i) means that X_α belongs to $J_\alpha(X, D(A)) \cap K_\alpha(X, D(A))$.

If X_α is a real interpolation space $D_A(\alpha, p)$ or $D_A(\alpha)$, with $0 < \alpha < 1$, $1 \leq p \leq \infty$, assumption (7.0.2) is satisfied thanks to Corollary 2.2.3(ii) and Proposition 2.2.7. Assumption (7.0.2) is satisfied as well when X_α is a complex interpolation space $[D(A), X]_\alpha$, or the domain $D(-A)^\alpha$ of a fractional power of $-A$.

The limiting case $\alpha = 1$, with $X_1 = D(A)$, is much more delicate, and will be considered in Chapter 9.

Note that $D(A)$ is not necessarily dense in X_α. We denote by $\overline{D(A)}$ the closure of $D(A)$ in X, and by $\overline{D(A)}^\alpha$ the closure of $D(A)$ in X_α. For every $x \in \overline{D(A)}^\alpha$ we have $\lim_{t \to 0} \|e^{tA}x - x\|_{X_\alpha} = 0$, thanks to 7.0.2(ii). In particular, if $D(A)$ is dense in X_α, then the semigroup e^{tA} is strongly continuous in X_α. In this situation, some proofs and statements are simpler.

Typical situations in which it is convenient to choose $\alpha = 0$ are parabolic initial value problems where the nonlinearity depends only on the unknown u and not on its derivatives, such as

$$\begin{cases} u_t(t,x) = \mathcal{A}u(t,x) + \varphi(u(t,x)), & x \in \mathbb{R}^n, \ t > 0, \\ u(0,x) = u_0(x), & x \in \mathbb{R}^n, \end{cases} \quad (7.0.3)$$

where $\mathcal{A}$ is a uniformly elliptic operator of order $2m$ with uniformly continuous and bounded coefficients, $\varphi : \mathbb{R} \mapsto \mathbb{R}$ is a smooth nonlinear function, and u_0 is continuous and bounded. It is natural to see (7.0.3) as an evolution equation of the type (7.0.1) in the space $X = C(\mathbb{R}^n)$ of all continuous and bounded functions on $\mathbb{R}^n$, by setting $f(t,v) = f(v) = \varphi(v)$ for every $v \in C(\mathbb{R}^n)$. Note that the same choice can be made also when the nonlinearities depend on nonlocal terms, provided they do not involve derivatives of the solution, such as for instance $\varphi = \varphi(t, x, u(t,x), u(t,x_0), \int_{B(x_1,r)} \psi(u(t,y)) dy)$, with arbitrary x_0, x_1 in $\mathbb{R}^n$, $r > 0$, and regular ψ.

As we said, in the case $0 < \alpha < 1$ X_α may be an interpolation space between X and $D(A)$, or the domain of a fractional power of $-A$. However, also other choices are possible, and may give even better results. For instance, consider the initial boundary value problem

$$\begin{cases} u_t = \Delta u + \varphi(Du), & x \in \overline{\Omega}, \ t > 0, \\ u(0,x) = u_0(x), & x \in \overline{\Omega}, \\ \partial u/\partial \nu = 0, & x \in \partial\Omega, \ t > 0, \end{cases} \quad (7.0.4)$$

where Ω is a bounded open set in $\mathbb{R}^n$ with regular boundary $\partial\Omega$, $\varphi : \mathbb{R}^n \mapsto \mathbb{R}$ and $u_0 : \overline{\Omega} \mapsto \mathbb{R}$ are smooth functions. Again, it is natural to consider (7.0.4) as an evolution equation of the type (7.0.1) in the space $X = C(\overline{\Omega})$, with $f(t,u) = f(u) = \varphi(Du)$. The maximal domain of f in X is $C^1(\overline{\Omega})$, which is neither an interpolation space between X and $D(A)$ nor a domain of some fractional power of $-\Delta$. However, $C^1(\overline{\Omega})$ satisfies assumption (7.0.2), with $\alpha = 1/2$, thanks to Proposition 3.1.27, Theorem 3.1.30, and Theorem 3.1.26. Of course, f is well defined also in $D_\Delta(\theta, \infty) = \{u \in C^{2\theta}(\overline{\Omega}) : \partial u/\partial \nu = 0\}$ for $\theta > 1/2$, and rough results on existence and regularity may be found defining f on $D_\Delta(\theta, \infty)$. However, more refined results on the behavior of the solution near $t = 0$ may be found only if f is defined on its maximal domain.

Let us come back to the abstract problem (7.0.1). As in the case of linear equations, we define mild, classical, and strict solutions. It is convenient to consider problem (7.0.1) starting from an arbitrary initial time a,

$$u'(t) = Au(t) + f(t, u(t)), \quad t > a; \quad u(a) = x. \quad (7.0.5)$$

Definition 7.0.1 *Let $0 \leq a < b \leq T$, and let $x \in \mathcal{O}$. Then:*

INTRODUCTION

(i) *A function* $u \in C^1([a,b]; X) \cap C([a,b]; D(A))$ *such that* $u(t) \in \mathcal{O}$ *for every* $t \in [a,b]$ *is said to be a* strict *solution of (7.0.5) in the interval* $[a,b]$ *if* $u'(t) = Au(t) + f(t, u(t))$ *for each* $t \in [a,b]$, *and* $u(a) = x$.

(ii) *A function* $u \in C^1(]a,b]; X) \cap C(]a,b]; D(A)) \cap C([a,b]; X)$, *such that* $u(t) \in \mathcal{O}$ *for every* $t \in [a,b]$, *is said to be a* classical *solution of (7.0.5) in the interval* $[a,b]$ *if* $u'(t) = Au(t) + f(t, u(t))$ *for each* $t \in]a,b]$, *and* $u(a) = x$.

Let u be any classical solution of (7.0.5), such that $f(\cdot, u(\cdot)) \in L^1(a,b;X)$. Then, by Proposition 4.1.2, u satisfies the variation of constants formula

$$u(t) = e^{(t-a)A}x + \int_a^t e^{(t-s)A} f(s, u(s))ds, \ a \le t \le b. \qquad (7.0.6)$$

Such a representation of u motivates the following definition.

Definition 7.0.2 *A continuous function* $u :]a,b] \mapsto X_\alpha$, *such that* $u(t) \in \mathcal{O}$ *for every* $t \in]a,b]$, *and* $f(\cdot, u(\cdot)) \in L^1(a,b;X)$, *is said to be a* mild *solution of (7.0.5) in the interval* $[a,b]$ *if it satisfies (7.0.6)*.

Strict, classical, and mild solutions in the half-open interval $[a,b[$ are similarly defined. Due to the semigroup property of e^{tA}, if u is a mild solution of (7.0.1) in the interval $[0,b]$, then u is a mild solution of (7.0.5) in $[a,b]$, with $x = u(a)$. Moreover, all the statements concerning the problem with initial time 0 can be extended trivially to the problem with initial time $a > 0$.

We shall prove (Theorem 7.1.3(i)) that if $f(t,x)$ is locally Lipschitz continuous with respect to x, then problem (7.0.1) has a unique local mild solution $u \in C([0, \delta]; X_\alpha)$, provided $u_0 \in \mathcal{O} \cap \overline{D(A)}^\alpha$. The condition $u_0 \in \overline{D(A)}^\alpha$ is easily seen to be necessary for the mild solution to be continuous up to $t = 0$ with values in X_α. In the case of example (7.0.3), with the choice $\alpha = 0$, $u_0 = C(\mathbb{R}^n)$, the condition $u_0 \in \overline{D(A)}^\alpha$ means that u_0 is uniformly continuous and bounded. In the case of example (7.0.4), the condition $u_0 \in \overline{D(A)}^\alpha$ means that u_0 is continuously differentiable in $\overline{\Omega}$ and the normal derivative of u_0 vanishes on $\partial\Omega$.

It is possible to treat also the case of initial data not belonging to $\overline{D(A)}^\alpha$, provided slightly more restrictive assumptions on f are fulfilled. If $\mathcal{O} = X_\alpha$ it is sufficient to assume that f is Lipschitz continuous with respect to x on every bounded subset of X_α. In most applications to PDE's, this condition is satisfied when the nonlinearities are reasonably smooth. In Theorem 7.1.2 we prove that for every $u_0 \in X_\alpha$ problem (7.0.1) has a unique local mild solution $u \in C(]0, \delta]; X_\alpha) \cap L^\infty(0, \delta; X_\alpha)$. In particular, in example (7.0.3) we can allow continuous and bounded initial data, and in example (7.0.4) we can allow C^1 initial data with nonvanishing normal derivative.

If $\alpha > 0$ and f satisfies suitable growth conditions it is possible to solve locally (7.0.1) even for initial data in X or in $D_A(\beta, \infty)$, with $\beta < \alpha$ (Theorems 7.1.5, 7.1.6). We find a unique mild solution $u \in C_\alpha(]0, \delta]; X_\alpha)$ (respectively, $u \in$

$C_{\alpha-\beta}(]0,\delta];X_\alpha))$. This means that $u \in C(]0,\delta];X_\alpha)$ and $t \mapsto t^\alpha u(t)$ (respectively, $t \mapsto t^{\alpha-\beta}u(t)$) is bounded up to $t=0$ with values in X_α. In the case of problem (7.0.4), for instance, the growth assumption on φ is simply

$$|\varphi(p_1) - \varphi(p_2)| \leq C(1 + |p_1|^{\gamma-1} + |p_2|^{\gamma-1})|p_1 - p_2|, \quad \forall p_1, p_2 \in \mathbb{R}^n,$$

with $\gamma \geq 1$. Then we may allow initial data in $C^\beta(\overline{\Omega})$, $0 \leq \beta < 1$, provided $(1-\beta)\gamma < 2 - \beta$.

The local existence results are collected in Subsection 7.1.1. In Subsection 7.1.2 we show that the local mild solution can be continued to a maximal time interval $I = I(u_0)$, which is relatively open in $[0,T]$. Then, in Subsection 7.1.3, we study the regularity of u. Of course, if f is merely continuous with respect to t, a mild solution is not necessarily strict or classical. For this we need that either f is Hölder continuous with respect to time, or that it has values in some intermediate space between X and $D(A)$. The regularity of the solution up to $t=0$, according to the regularity of the initial datum u_0, is discussed in Propositions 7.1.10 and 7.1.11. In the applications to parabolic initial value problems the regularity properties of local solutions near $t=0$ are similar to the ones stated in Chapter 5 for the linear case.

Special attention is paid to existence in the large of the solution, treated in Section 7.2. It is easy to show that if the solution u remains far from the boundary of $\mathcal{O}$ and $f(\cdot, u(\cdot))$ is bounded with values in X, then $I(u_0) = [0,T]$. This happens for instance when $\mathcal{O} = X_\alpha$ and f has not more than linear growth with respect to x; in particular, when f is globally Lipschitz continuous. In the case where f is only locally Lipschitz continuous, the main problem is to find an *a priori* estimate in the X_α-norm. In the applications it happens frequently to find easily *a priori* estimates in X, especially when X is endowed with a L^p norm, $1 \leq p \leq \infty$. However, *a priori* estimates in the X-norm are in general not enough to ensure existence in the large, in the case $\alpha > 0$: several examples of quenching (i.e., the X-norm of the solution remains bounded, but the X_α-norm blows up) are known in abstract and concrete parabolic equations. In order to find *a priori* estimates in the X_α-norm we add suitable growth conditions on f: $f(t,x)$ may grow as $(\|x\|_{X_\alpha})^\gamma$, with $\gamma \leq 1/\alpha$. We show that if either $\gamma < 1/\alpha$ and u is bounded with values in X, or $\gamma = 1/\alpha$ and u is Hölder continuous with values in X, then u is bounded with values in X_α. If γ is greater than the critical exponent $1/\alpha$, it is not possible, in general, to find *a priori* estimates in X_α, even if $\|u'\|$ is bounded.

In the case of problem (7.0.4), the condition $\gamma \leq 1/\alpha$ implies that φ grows not more than quadratically, i.e. $|\varphi(p)| \leq C(1 + |p|^2)$ for every $p \in \mathbb{R}^n$. A priori estimates in the sup norm for u and u_t may be found easily through the maximum principle. Applying Theorem 7.2.4 and its corollary, one gets existence in the large.

In Section 7.3 we apply the abstract theory to semilinear parabolic problems. Together with the above mentioned examples, we consider general semilinear equations of order $2m$, reaction-diffusion systems, and we study in detail some particular cases, such as second order equations with nonlinearities in divergence form, and the Cahn-Hilliard equation, for which it is possible to get local and global solutions even if the initial data are not differentiable.

Through the whole chapter we set

$$M_0 = \sup_{0 \le t \le T} \|e^{tA}\|_{L(X)}, \quad M_\alpha = \sup_{0 \le t \le T} \|e^{tA}\|_{L(X_\alpha)} \quad (7.0.7)$$

$$C_\alpha = \sup_{0 < t \le T} \|t^\alpha e^{tA}\|_{L(X, X_\alpha)}. \quad (7.0.8)$$

If Y is any Banach space, $y \in Y$, and $r > 0$, we denote by $B_Y(y, r)$ the closed ball in Y centered at y with radius r. If no danger of confusion may arise, we omit the subscript Y. Moreover, we shall make use of the following well known generalization of the Gronwall's lemma. A proof may be found for instance in [99, p. 188].

Lemma 7.0.3 *Let $0 \le a < b \le T$, and let $u : [a,b] \mapsto \mathbb{R}$ be an integrable nonnegative function such that*

$$u(t) \le k + h \int_a^t (t-s)^{-\alpha} u(s) ds, \quad a \le t \le b,$$

with $0 \le \alpha < 1$, h, $k > 0$. Then there is $C_1 > 0$ (not depending on a, b, k) such that

$$u(t) \le C_1 k, \quad a \le t \le b.$$

7.1 Local existence and regularity

7.1.1 Local existence results

We will look for a local mild solution of problem (7.0.1), that is for a solution of the integral equation

$$u(t) = e^{tA} u_0 + \int_0^t e^{(t-s)A} f(s, u(s)) ds, \quad 0 \le t \le \delta, \quad (7.1.1)$$

belonging to $C(]0, \delta]; X_\alpha)$, and such that $f(\cdot, u(\cdot))$ belongs to $L^1(0, \delta; X)$. Here δ is a small positive number.

To treat the integral in the right hand side of (7.1.1) we need some estimates which are obvious consequences of Propositions 4.2.3, 4.2.3 and of the embedding $D_A(\alpha, 1) \subset X_\alpha$.

Lemma 7.1.1 *Let $\varphi \in L^\infty(0, a; X)$, $0 < a \le T$, and set*

$$v(t) = (e^{tA} * \varphi)(t) = \int_0^t e^{(t-s)A} \varphi(s) ds, \quad 0 \le t \le a. \quad (7.1.2)$$

Then $v \in C^\gamma([0,a];X)$ for every $\gamma \in \,]0,1[$, and there is $C > 0$, not depending on a and φ, such that
$$\|v\|_{C^\gamma([0,a];X)} \leq C\|\varphi\|_{L^\infty(0,a;X)}. \tag{7.1.3}$$
If $0 < \alpha < 1$, then $v \in C^{1-\alpha}([0,a];X_\alpha)$, and there is $C > 0$, not depending on a and φ, such that
$$\|v\|_{C^{1-\alpha}([0,a];X_\alpha)} \leq C\|\varphi\|_{L^\infty(0,a;X)}. \tag{7.1.4}$$

We begin by studying the case where $f(t,\cdot)$ is defined in the whole space X_α. We assume that for every $R > 0$ there exists $L > 0$ such that

$$\|f(t,x) - f(t,y)\| \leq L\|x - y\|_{X_\alpha}, \quad \forall t \in [0,T], \ \forall x,y \in B_{X_\alpha}(0,R). \tag{7.1.5}$$

Theorem 7.1.2 *Let $0 \leq \alpha < 1$, and let $f : [0,T] \times X_\alpha \mapsto X$ be a continuous function satisfying (7.1.5). Then for every $\overline{u} \in X_\alpha$ there are $r, \delta > 0$, $K > 0$ such that for $\|u_0 - \overline{u}\|_{X_\alpha} \leq r$ problem (7.0.1) has a unique mild solution $u = u(\cdot; u_0) \in L^\infty(0,\delta;X_\alpha)$.*

In addition, u belongs to $C^\alpha([0,\delta];X) \cap C([0,\delta];D_A(\beta,1))$ for every $\beta < \alpha$ (if $\alpha > 0$), and to $C^{1-\beta}([\varepsilon,\delta];D_A(\beta,1))$ for every $\varepsilon \in \,]0,\delta[$ and $\beta \in \,]0,1[$. In particular, u belongs to $C(]0,\delta];X_\alpha)$, and if $\alpha > 0$ then u belongs to $C([0,\delta];X)$. It belongs to $C([0,\delta];X_\alpha)$ if and only if $u_0 \in \overline{D(A)}^\alpha$. If $\alpha = 0$, then $\lim_{t \to 0} R(\lambda,A)u(t) = R(\lambda,A)u_0$ in $D_A(\beta,1)$ for every $\lambda \in \rho(A)$ and $\beta \in \,]0,1[$.

Moreover, for $u_0, u_1 \in B(\overline{u},r)$ it holds

$$\|u(t;u_0) - u(t;u_1)\|_{X_\alpha} \leq K\|u_0 - u_1\|_{X_\alpha}, \ 0 \leq t \leq \delta. \tag{7.1.6}$$

Proof — Fix $R > 0$ such that $R \geq 8M_\alpha\|\overline{u}\|_{X_\alpha}$, so that if $\|u_0 - \overline{u}\|_{X_\alpha} \leq r = R/8M_\alpha$

$$\sup_{0 \leq t \leq T} \|e^{tA}u_0\|_{X_\alpha} \leq R/4.$$

Let moreover L be such that

$$\|f(t,v) - f(t,w)\| \leq L\|v - w\|_{X_\alpha} \ 0 \leq t \leq T, \ v,w \in B(0,R).$$

We look for a local mild solution in the metric space Y defined by

$$Y = \{u \in C(]0,\delta];X_\alpha) : \|u(t)\|_{X_\alpha} \leq R, \ \forall t \in [0,\delta]\},$$

where $\delta \in \,]0,T]$ is to be found. Y is closed in $L^\infty(0,\delta;X_\alpha)$, and for every $v \in Y$, the function $t \mapsto f(\cdot,v(\cdot))$, belongs to $L^\infty(0,\delta;X)$. Define a nonlinear operator Γ on Y by

$$\Gamma(v)(t) = e^{tA}u_0 + \int_0^t e^{(t-s)A}f(s,v(s))ds, \ 0 \leq t \leq \delta. \tag{7.1.7}$$

Clearly, a function $v \in Y$ is a mild solution of (7.0.1) in $[0,\delta]$ if and only if it is a fixed point of Γ.

1. Local existence and regularity

We shall show that Γ is a contraction and maps Y into itself, provided δ is sufficiently small.

Let $v_1, v_2 \in Y$. Then, by Lemma 7.1.1, $\Gamma(v_1)$ and $\Gamma(v_2)$ belong to $C^\gamma([0,\delta]; X_\alpha)$, with $\gamma = 1 - \alpha$ if $\alpha > 0$, with any $\gamma \in {]0,1[}$ if $\alpha = 0$. By estimates (7.1.3), (7.1.4) we have

$$[\Gamma(v_1) - \Gamma(v_2)]_{C^\gamma([0,\delta]; X_\alpha)} \leq C\|f(\cdot, v_1(\cdot)) - f(\cdot, v_2(\cdot))\|_{L^\infty(0,\delta;X)} \tag{7.1.8}$$
$$\leq CL\|v_1 - v_2\|_{L^\infty(0,\delta;X_\alpha)}.$$

Moreover, $(\Gamma(v_1) - \Gamma(v_2))(0) = 0$, so that

$$\|\Gamma(v_1) - \Gamma(v_2)\|_{L^\infty(0,\delta;X_\alpha)} \leq \delta^\gamma [\Gamma(v_1) - \Gamma(v_2)]_{C^\gamma([0,\delta];X_\alpha)}$$
$$\leq CL\delta^\gamma\|v_1 - v_2\|_{L^\infty(0,\delta;X_\alpha)}.$$

Therefore, if

$$\delta \leq \delta_0 = (2CL)^{-1/\gamma},$$

Γ is a 1/2-contraction on Y. Moreover, for every $v \in Y$ and $t \in [0,\delta]$, with $\delta \leq \delta_0$, we have, due again to Lemma 7.1.1,

$$\begin{aligned}\|\Gamma(v)\|_{L^\infty(0,\delta;X_\alpha)} &\leq \|\Gamma(v) - \Gamma(0)\|_{L^\infty(0,\delta;X_\alpha)} + \|\Gamma(0)\|_{L^\infty(0,\delta;X_\alpha)}\\ &\leq R/2 + \|e^{tA}u_0\|_{L^\infty(0,\delta;X_\alpha)} + C\delta^\gamma\|f(\cdot,0)\|_{L^\infty(0,\delta;X)}\\ &\leq R/2 + R/4 + C\delta^\gamma\|f(\cdot,0)\|_{L^\infty(0,\delta;X)}.\end{aligned} \tag{7.1.9}$$

Therefore, if $\delta \leq \delta_0$ is so small that

$$C\delta^\gamma\|f(\cdot,0)\|_{L^\infty(0,\delta;X)} \leq R/4,$$

then Γ maps Y into itself, so that it has a unique fixed point in Y.

The function $t \mapsto e^{tA}u_0$ belongs to $C([\varepsilon,\delta]; D_A(\beta,1))$ for every $\varepsilon \in {]0,\delta[}$, $\beta \in {]0,1[}$, and, if $\alpha > 0$, it belongs also to $C([0,\delta]; D_A(\beta,1))$ for every $\beta \in {]0,\alpha[}$. Then by Proposition 4.2.1 the range of Γ is contained in $C^{1-\beta}([\varepsilon,\delta]; D_A(\beta,1))$ for every $\varepsilon \in {]0,\delta[}$, $\beta \in {]0,1[}$ and if $\alpha > 0$ it is contained in $C^\alpha([0,\delta]; X) \cap C([0,\delta]; D_A(\beta,1))$ for every $\beta \in {]0,\alpha[}$. Therefore the fixed point u of Γ belongs to such spaces. If $\alpha = 0$, then $\lim_{t\to 0} R(\lambda,A)u(t) = R(\lambda,A)u_0$ in $D_A(\theta,1)$ for every θ thanks to Proposition 4.2.5.

Concerning the continuity of u with values in X_α near $t = 0$, we recall that, due to Lemma 7.1.1, the function $t \mapsto u(t) - e^{tA}u_0$ belongs to $C([0,\delta]; X_\alpha)$, whereas $t \mapsto e^{tA}u_0$ belongs to $C([0,\delta]; X_\alpha)$ if and only if $u_0 \in \overline{D(A)}^\alpha$. Therefore, $u \in C([0,\delta]; X_\alpha)$ if and only if $u_0 \in \overline{D(A)}^\alpha$.

Let us prove the statement about dependence on the initial value. Since Γ is a 1/2-contraction, if u_0, u_1 belong to $B(\overline{u},r)$ then

$$\|u(\cdot;u_0) - u(\cdot;u_1)\|_{L^\infty(0,\delta;X_\alpha)} \leq 2\|e^{tA}(u_0 - u_1)\|_{L^\infty(0,\delta;X_\alpha)} \leq 2M_\alpha\|u_0 - u_1\|_{X_\alpha},$$

so that (7.1.6) holds, with $K = 2M_\alpha$.

Let us prove uniqueness: if $u_1, u_2 \in L^\infty(0,\delta; X_\alpha)$ are mild solutions of (7.0.1), they belong to $C(]0,\delta]; X_\alpha)$. Define

$$t_0 = \sup\{t \in [0,\delta] : u_1(s) = u_2(s) \text{ for } 0 \leq s \leq t\}, \qquad (7.1.10)$$

and set $y = u_1(t_0) = u_2(t_0)$. If $t_0 < \delta$, the problem

$$v'(t) = Av(t) + f(t, v(t)), \quad t > t_0, \quad v(t_0) = y, \qquad (7.1.11)$$

has a unique solution in a set

$$Y' = \{u \in C(]t_0, t_0 + \varepsilon]; X_\alpha) : \|u(t)\|_{X_\alpha} \leq R'\},$$

provided R' is large enough and ε is small enough. Since u_1 and u_2 are bounded with values in X_α, there exists R' such that $\|u_i(t)\|_{X_\alpha} \leq R'$ for $t_0 \leq t \leq \delta$, $i = 1, 2$. On the other hand, u_1 and u_2 are two different mild solutions of (7.1.11) in $[t_0, t_0 + \varepsilon]$, for every $\varepsilon \in]0, \delta - t_0]$. This is a contradiction; hence, $t_0 = \delta$, and the mild solution of (7.0.1) is unique in $L^\infty(0,\delta; X_\alpha)$. ∎

In the applications to PDE's one often encounters nonlinearities not defined on the whole of X_α. For instance, second order equations of the type $u_t = u_{xx} + \varphi((u)_x)$, $u_t = u_{xx} + (\varphi(u))_x$, with $\varphi :]a,b[\mapsto \mathbb{R}$, are frequently found in the mathematical literature. Choosing X as a space of continuous functions, $\alpha = 1/2$, X_α as a C^1 space, the nonlinearities $f(u) = \varphi(du/dx)$, $g(u) = d/dx(\varphi(u))$ are defined in the open sets of the functions u such that du/dx (respectively, u) has values in $]a,b[$. We have to modify slightly the previous theorem to take account of such situations.

Theorem 7.1.3 *Let $0 \leq \alpha < 1$, and let $\mathcal{O}$ be an open set in X_α.*

(i) *Let $f : [0,T] \times \mathcal{O} \mapsto X$ be a continuous function such that for every $x_0 \in \mathcal{O}$ there are $\rho, L > 0$ satisfying*

$$\|f(t,x) - f(t,y)\| \leq L\|x - y\|_{X_\alpha} \quad \forall t \in [0,T], \ x, y \in B(x_0, \rho). \qquad (7.1.12)$$

Then for every $\overline{u} \in \overline{D(A)}^\alpha \cap \mathcal{O}$ there are $r, \delta > 0$ such that for $u_0 \in \overline{D(A)}^\alpha$, $\|u_0 - \overline{u}\|_{X_\alpha} \leq r$, problem (7.0.1) has a unique mild solution $u = u(\cdot, u_0) \in C([0,\delta]; X_\alpha)$.

(ii) *Let $0 < \alpha < 1$, and $\mathcal{O} = \mathcal{U} \cap X_\alpha$, where $\mathcal{U}$ is an open set in X. Let $f : [0,T] \times \mathcal{O} \mapsto X$ be a continuous function such that for every $x_0 \in \mathcal{O}$ and for every $R > 0$ there are $\rho, L > 0$ satisfying*

$$\|f(t,x) - f(t,y)\| \leq L\|x - y\|_{X_\alpha}, \quad \forall t \in [0,T], \ x,y \in B_X(x_0, \rho) \cap B_{X_\alpha}(0, R). \qquad (7.1.13)$$

Then for every $\overline{u} \in \mathcal{O}$ there are $r, \delta > 0$ such that for $u_0 \in \mathcal{O}$, $\|u_0 - \overline{u}\|_{X_\alpha} \leq r$ problem (7.0.1) has a unique mild solution $u = u(\cdot, u_0) \in C(]0,\delta]; X_\alpha) \cap L^\infty(0,\delta; X_\alpha) \cap C([0,\delta]; X)$.

1. LOCAL EXISTENCE AND REGULARITY

In both cases (i) and (ii), the conclusions of Theorem 7.1.2 about regularity and dependence on the initial data hold.

Proof — We have to modify suitably the proof of Theorem 7.1.2. We replace the metric space Y respectively by

$$Y' = \{u \in C([0,\delta]; X_\alpha) : \|u(t) - \overline{u}\|_{X_\alpha} \leq r, \; \forall t \in [0,\delta]\},$$

and by

$$Y'' = \{u \in C(]0,\delta]; X_\alpha) : \|u(t) - \overline{u}\| \leq r, \; \|u(t)\|_{X_\alpha} \leq R, \; \forall t \in [0,\delta]\},$$

where $\delta \in \,]0,T]$ is to be determined, $R \geq 8M_\alpha \|\overline{u}\|_{X_\alpha}$, and r is a sufficiently small number. Precisely, in case (i) r has to be chosen such that the X_α distance between $\overline{u}$ and $\partial \mathcal{O}$ is less than r, and there is $L > 0$ such that

$$\|f(t,x) - f(t,y)\| \leq L\|x - y\|_{X_\alpha} \tag{7.1.14}$$

for every x, y in the ball $B_{X_\alpha}(\overline{u}, r)$. In case (ii), r has to be chosen such that the X distance between $\overline{u}$ and $\partial \mathcal{U}$ is less than r, and there is $L > 0$ such that (7.1.14) holds for every x, y in $B_{X_\alpha}(0, R) \cap B_X(\overline{u}, r)$.

The local solution is sought again as a fixed point of the operator Γ defined by (7.1.7) in Y' and in Y'', respectively. Using Lemma 7.1.1, one checks easily that Γ is a 1/2-contraction and maps Y' (respectively, Y'') into itself provided δ is small enough. Indeed, estimate (7.1.8) still holds, whereas (7.1.9) has to be replaced, in case (i), by

$$\|\Gamma(v)(t) - \overline{u}\|_{X_\alpha} \leq \|\Gamma(v)(t) - \Gamma(\overline{u})(t)\|_{X_\alpha} + \|\Gamma(\overline{u})(t) - \overline{u}\|_{X_\alpha}$$
$$\leq r/2 + \|e^{tA}(u_0 - \overline{u})\|_{X_\alpha} + \|e^{tA}\overline{u} - \overline{u}\|_{X_\alpha} + C\delta^\gamma \|f(\cdot, \overline{u})\|_{L^\infty(0,\delta;X)},$$
$$\leq r/2 + M_\alpha \|u_0 - \overline{u}\|_{X_\alpha} + \|e^{tA}\overline{u} - \overline{u}\|_{X_\alpha} + C\delta^\gamma \|f(\cdot, \overline{u})\|_{L^\infty_*(0,\delta;X)}.$$

Since $\overline{u} \in \overline{D(A)}^\alpha$, then $\lim_{t \to 0} \|e^{tA}\overline{u} - \overline{u}\|_{X_\alpha} = 0$. Therefore, if $\|u_0 - \overline{u}\|_{X_\alpha}$ and $\delta \leq \delta_0$ are sufficiently small, then Γ maps Y' into itself.

In case (ii), (7.1.8) and (7.1.9) still hold, and moreover

$$\|\Gamma(v)(t) - \overline{u}\| \leq r/2 + \|e^{tA}(u_0 - \overline{u})\| + \|e^{tA}\overline{u} - \overline{u}\| + C\delta^\gamma \|f(\cdot, \overline{u})\|_{L^\infty(0,\delta;X)}$$
$$\leq r/2 + M_0 \|u_0 - \overline{u}\| + \|e^{tA}\overline{u} - \overline{u}\| + C\delta^\gamma \|f(\cdot, u_0)\|_{L^\infty(0,\delta;X)},$$

Since $\alpha > 0$, then $\overline{u} \in \overline{D(A)}$, so that $\lim_{t \to 0} \|e^{tA}\overline{u} - \overline{u}\| = 0$. Hence, if $\|u_0 - \overline{u}\|$ and $\delta \leq \delta_0$ are sufficiently small, Γ maps Y'' into itself.

Therefore, in both cases (i) and (ii), Γ is a 1/2-contraction mapping Y' (respectively, Y'') into itself, so that it has a unique fixed point u in Y' (respectively, in Y'').

The statements about regularity and dependence on the initial datum of the solution may be proved as in the proof of Theorem 7.1.2.

Concerning uniqueness, one has to pay more attention. In case (i), if u_1, $u_2 \in C([0,\delta]; X_\alpha)$ are mild solutions of (7.0.1), define t_0 by (7.1.10) and assume by contradiction that $t_0 < \delta$. Setting $y = u_1(t_0) = u_2(t_0)$, for small $r > 0$ problem (7.1.11) has a unique solution in a set

$$Z = \{u \in C([t_0, t_0+\varepsilon]; X_\alpha) : \|u(t)\|_{X_\alpha} \le r \ \forall t \in [t_0, t_0+\varepsilon]\},$$

provided ε is small enough. Since u_1 and u_2 are continuous with values in X_α, then for every sufficiently small ε, both $u_{1|[t_0,t_0+\varepsilon]}$ and $u_{2|[t_0,t_0+\varepsilon]}$ belong to Z. On the other hand, u_1 and u_2 are two different mild solutions of (7.1.11) in $[t_0, t_0+\varepsilon]$, for every $\varepsilon \in]0, \delta - t_0]$. This is a contradiction; hence, $t_0 = \delta$, and the mild solution of (7.0.1) is unique in $C([0,\delta]; X_\alpha)$.

In case (ii), uniqueness may be proved by putting together the arguments used above and in the proof of Theorem 7.1.2. We leave the details to the reader. ∎

If $0 < \alpha < 1$ and f satisfies suitable growth assumptions it is possible to get a local mild solution even if the initial datum is "less regular", that is it does not belong to X_α but to a bigger subspace of X. Let $\mathcal{U}$ be an open set in X and set $\mathcal{O} = X_\alpha \cap \mathcal{U}$. Assume that $f : [0,T] \times \mathcal{O} \mapsto X$ is a continuous function such that

$$\begin{cases} \|f(t,x) - f(t,y)\| \le C_1(x,y)\big[(1 + \|x\|_{X_\alpha}^{\gamma-1} + \|y\|_{X_\alpha}^{\gamma-1})\|x-y\|_{X_\alpha} \\ + (\|x\|_{X_\alpha}^\gamma + \|y\|_{X_\alpha}^\gamma)\|x-y\|\big], \ \forall x,y \in \mathcal{O}, \end{cases} \quad (7.1.15)$$

with $\gamma \ge 1$, $C_1 : \mathcal{U} \times \mathcal{U} \mapsto [0, +\infty[$ locally bounded. It follows that

$$\|f(t,x)\| \le C_2(x)(1 + \|x\|_{X_\alpha}^\gamma), \ \forall x \in X_\alpha, \quad (7.1.16)$$

with $C_2 : \mathcal{U} \mapsto [0, +\infty[$ locally bounded.

In Proposition 2.2.9 we have shown that if $u_0 \in D_A(\beta, \infty)$, with $0 \le \beta < \alpha^{(1)}$, then $\|e^{tA}u_0\|_{D_A(\alpha,1)} \le \text{const.}\ t^{-\alpha+\beta}\|u_0\|_{D_A(\beta,\infty)}$ for $0 < t \le T$. Since $D_A(\alpha,1)$ is continuously embedded in X_α, there is $C_{\beta,\alpha} > 0$ such that

$$\|e^{tA}u_0\|_{X_\alpha} \le C_{\beta,\alpha} t^{-\alpha+\beta}\|u_0\|_{D_A(\beta,\infty)}, \ \forall u_0 \in D_A(\beta,\infty),\ 0 \le \beta < \alpha. \quad (7.1.17)$$

This suggests to use weighted spaces. We recall that if Y is any Banach space and $\theta > 0$ the space $C_\theta(]0,\delta]; Y)$ is defined by

$$C_\theta(]0,\delta]; Y) = \{u \in C(]0,\delta]; Y) : \|u\|_{C_\theta(]0,\delta];Y)} = \sup_{0 < t \le \delta} \|t^\theta u(t)\|_Y < \infty\}.$$

Thanks to (7.1.17), if $x \in D_A(\beta, \infty)$ with $0 \le \beta < \alpha$, the function $t \mapsto e^{tA}u_0$ belongs to $C_{\alpha-\beta}(]0,\delta]; X_\alpha)$, for every $\delta \in]0,T]$. Therefore it is reasonable to look for a mild solution of (7.0.1) belonging to such a space. To treat the integral term in (7.1.1) we need to know the behavior of the convolution $v = e^{tA} * \varphi$ when φ belongs to any weighted space $C_\theta(]0,\delta]; X)$. The following lemma is a consequence of Proposition 4.2.3 and of the embedding $D_A(\alpha, 1) \subset X_\alpha$.

---
[1]Recall that $D_A(0, \infty) = X$.

Lemma 7.1.4 *Let $0 < a \leq T$, and let $\varphi \in C_\theta(]0,a]; X)$, with $0 < \theta < 1$. Then the function $v = e^{tA} * \varphi$ belongs to $C^{1-\theta}([0,a]; X) \cap C(]0,a]; X_\alpha)$, and there is $C > 0$, depending only on α and θ, such that*

$$\|v\|_{C^{1-\theta}([0,a];X)} + \sup_{0 < t \leq a} \|t^{\alpha+\theta-1} v(t)\|_{X_\alpha} \leq C \|\varphi\|_{C_\theta(]0,a];X)}, \quad 0 < t \leq a.$$

Theorem 7.1.5 *Let $\alpha > 0$, and let f satisfy assumption (7.1.15). Then the following statements hold.*

(i) *If $\alpha\gamma < 1$, then for every $\overline{u} \in \overline{D(A)} \cap \mathcal{U}$ there are $\delta > 0$, $r > 0$, $K > 0$ such that for each $u_0 \in \overline{D(A)}$ with $\|u_0 - \overline{u}\| \leq r$, problem (7.0.1) has a unique mild solution $u = u(\cdot; u_0) \in C_\alpha(]0,\delta]; X_\alpha)$. Moreover, $u \in C([0,\delta]; X)$. For $0 < t \leq \delta$ and $u_i \in B_X(\overline{u}, r)$, $u_i \in \overline{D(A)}$, $i = 0, 1$,*

$$\|u(t; u_0) - u(t; u_1)\| + \|t^\alpha (u(t;u_0) - u(t;u_1))\|_{X_\alpha} \leq K \|u_0 - u_1\|. \quad (7.1.18)$$

(ii) *If $\beta > 0$ and $\gamma(\alpha - \beta) < 1 - \beta$, then for every $\overline{u} \in D_A(\beta, \infty) \cap \mathcal{U}$ there are $\delta > 0$, $r > 0$, $K > 0$ such that for each $u_0 \in D_A(\beta, \infty) \cap \mathcal{U}$ with $\|u_0 - \overline{u}\|_{D_A(\beta,\infty)} \leq r$, problem (7.0.1) has a unique mild solution $u = u(\cdot; u_0)$ belonging to $C_{\alpha-\beta}(]0,\delta]; X_\alpha)$. Moreover, $u \in C([0,\delta]; X) \cap B([0,\delta]; D_A(\beta,\infty))$, and for $0 < t \leq \delta$, $u_i \in B_{D_A(\beta,\infty)}(\overline{u}, r)$, $i = 0,1$,*

$$\|u(t;u_0) - u(t;\overline{u})\|_{D_A(\beta,\infty)} + \|t^{\alpha-\beta}(u(t;u_0) - u(t;\overline{u}))\|_{X_\alpha} \leq K \|u_0 - \overline{u}\|_{D_A(\beta,\infty)}. \quad (7.1.19)$$

(iii) *If $\beta > 0$ and $\gamma(\alpha - \beta) = 1 - \beta$, then for every $\overline{u} \in D_A(\beta) \cap \mathcal{U}$ there are $\delta > 0$, $r > 0$ such that for every $u_0 \in D_A(\beta) \cap \mathcal{U}$ with $\|u_0 - \overline{u}\|_{D_A(\beta,\infty)} \leq r$, problem (7.0.1) has a mild solution $u = u(\cdot; u_0)$ belonging to $C_{\alpha-\beta}(]0,\delta]; X_\alpha) \cap C([0,\delta]; D_A(\beta))$, and such that $\lim_{t \to 0} \|t^{\alpha-\beta} u(t)\|_{X_\alpha} = 0$. u is the unique mild solution of (7.0.1) in the class of the functions $v \in C_{\alpha-\beta}(]0,\delta]; X_\alpha) \cap C([0,\delta]; D_A(\beta))$ such that $\lim_{t \to 0} \|t^{\alpha-\beta} v(t)\|_{X_\alpha} = 0$. For $0 < t \leq \delta$, $u_i \in B_{D_A(\beta)}(\overline{u}, r)$, $i = 0, 1$, (7.1.19) holds.*

Proof — We give only a sketch, since the ideas are the same as in the proofs of Theorems 7.1.2 and 7.1.3. Here we have to use Lemma 7.1.4, with $\theta = (\alpha - \beta)\gamma$, instead of Lemma 7.1.1. We look for a fixed point of the operator Γ defined by (7.1.7) in the set

$$B = \{u \in C_{\alpha-\beta}(]0,\delta]; X_\alpha) \cap C([0,\delta]; X) : \\ \|u(t) - \overline{u}\| \leq r, \ \|t^{\alpha-\beta} u(t)\|_{X_\alpha} \leq R \ \forall t \in]0,\delta]\},$$

where $\beta = 0$ in the case of statement (i), and $r < X\text{-dist}(\overline{u}, \partial \mathcal{U})$ is such that

$$C_1 = \sup\{C_1(x,y) : \|x - \overline{u}\| \leq r, \ \|y - \overline{u}\| \leq r\} < \infty,$$

and, consequently,

$$C_2 = \sup\{C_2(x) : \|x - \overline{u}\| \leq r, \ \|y - \overline{u}\| \leq r\} < \infty.$$

B is a closed set in $C_{\alpha-\beta}(]0,\delta];X_\alpha) \cap C([0,\delta];X)$. Using (7.1.15), (7.1.16), and Lemma 7.1.4, for $u, v \in B$ one finds

$$\|\Gamma u - \overline{u}\|_{L^\infty(0,\delta;X)} + \|\Gamma u\|_{C_{\alpha-\beta}(]0,\delta];X_\alpha)}$$
$$\leq \|e^{\cdot A}(u_0 - \overline{u})\|_{L^\infty(0,\delta;X)} + \|e^{\cdot A}\overline{u} - \overline{u}\|_{L^\infty(0,\delta;X)} + \|e^{\cdot A}u_0\|_{C_{\alpha-\beta}(]0,\delta];X_\alpha)}$$
$$+ C(\delta^{1-\theta} + \delta^{1-\beta-\theta})\|f(\cdot, u(\cdot))\|_{C_\theta(]0,\delta];X)} \leq M_0\|u_0 - \overline{u}\| + \|e^{\cdot A}\overline{u} - \overline{u}\|_{L^\infty(0,\delta;X)}$$
$$+ C_{\beta,\alpha}(\|u_0 - \overline{u}\| + \|\overline{u}\|) + C(\delta^{1-\theta} + \delta^{1-\beta-\theta})C_2(\delta^\theta + R^\gamma),$$

and

$$\|\Gamma u - \Gamma v\|_{L^\infty(0,\delta;X)} + \|\Gamma u - \Gamma v\|_{C_{\alpha-\beta}(]0,\delta];X_\alpha)}$$
$$\leq C(\delta^{1-\theta} + \delta^{1-\beta-\theta})\|f(\cdot, u(\cdot)) - f(\cdot, v(\cdot))\|_{C_\theta(]0,\delta];X)}$$
$$\leq C(\delta^{1-\theta} + \delta^{1-\beta-\theta})C_1\big[(\delta^{(\alpha-\beta)(\gamma-1)} + 2R^{\gamma-1})\|u - v\|_{C_{\alpha-\beta}(]0,\delta];X_\alpha)}$$
$$+ (\delta^{(\alpha-\beta)\gamma} + 2R^\gamma)\|u - v\|_{L^\infty(0,\delta;X)}\big].$$

In cases (i) and (ii), when $1 - \beta - \theta > 0$, R has to be chosen sufficiently large, in such a way that $\|e^{\cdot A}\overline{u} - \overline{u}\|_{L^\infty(0,T;X)} + C_{\beta,\alpha}\|\overline{u}\| < R$. Then, if δ and $\|u_0 - \overline{u}\|_{D_A(\beta,\infty)}$ are sufficiently small, Γ is a contraction and it maps B into itself, so that it has a unique fixed point in B, which is a mild solution of (7.0.1).

Moreover, in case (ii) the range of Γ is contained in $B([0,\delta]; D_A(\beta,\infty))$ thanks to Proposition 4.2.3, so that its fixed point $u = u(t; u_0)$ belongs to $B([0,\delta]; D_A(\beta,\infty))$.

In both cases (i) and (ii) the statements about uniqueness can be shown by putting together the methods used in the proofs of uniqueness in Theorems 7.1.2 and 7.1.3. Note that, since $u_0 \in \overline{D(A)}$, then every mild solution is in $C([0,\delta];X)$.

In case (iii), when $1 - \beta - \theta = 0$, since $D(A)$ is dense in $D_A(\beta)$ and in $\overline{D(A)}$ and estimate (7.1.17) holds, we have

$$\lim_{t \to 0} \|t^{\alpha-\beta}e^{tA}\overline{u}\|_{X_\alpha} = 0, \quad \lim_{t \to 0} \|e^{tA}\overline{u} - \overline{u}\| = 0.$$

Then Γ is a contraction and it maps B into itself, provided r, R is sufficiently small and $\|u_0 - \overline{u}\|_{D_A(\beta,\infty)}$, δ are sufficiently small (depending on r and R). In this case, Γ has a unique fixed point in B. Note that, since $u_0 \in D_A(\beta) \subset \overline{D(A)}$, then the range of Γ is contained in $C([0,\delta];X)$, so that $u \in C([0,\delta];X)$. Uniqueness of the mild solution in the class of functions $v \in C_{\alpha-\beta}(]0,\delta];X_\alpha) \cap C([0,\delta];X)$ such that $\lim_{t \to 0} \|t^{\alpha-\beta}v(t)\|_{X_\alpha} = 0$ can be shown as in the proof of Theorem 7.1.3.

The statements about continuous dependence on the initial datum follow easily, as in the proof of Theorem 7.1.2. ∎

In the case where $f(t, \cdot)$ is defined in the whole X_α, instead of in an open set, it is possible to consider initial data not belonging to the closure $\overline{D(A)}$. We assume that $f : [0,T] \times X_\alpha \mapsto X$ is a continuous function such that

$$\begin{cases} \|f(t,x) - f(t,y)\| \leq C_1(\|x\| + \|y\|)\big[(1 + \|x\|_{X_\alpha}^{\gamma-1} + \|y\|_{X_\alpha}^{\gamma-1})\|x - y\|_{X_\alpha} \\ + (\|x\|_{X_\alpha}^\gamma + \|y\|_{X_\alpha}^\gamma)\|x - y\|\big], \quad \forall x, y \in X_\alpha, \end{cases}$$

(7.1.20)

1. LOCAL EXISTENCE AND REGULARITY

with $\gamma \geq 1$, $C_1 : [0, +\infty[\mapsto [0, +\infty[$ increasing. It follows that

$$\|f(t,x)\| \leq C_2(\|x\|)(1 + \|x\|_{X_\alpha}^\gamma), \quad \forall x \in X_\alpha, \tag{7.1.21}$$

with $C_2 : [0, +\infty[\mapsto [0, +\infty[$ increasing.

Theorem 7.1.6 *Let $\alpha > 0$, and let f satisfy assumption (7.1.20), with $\alpha\gamma < 1$. Then for every $\overline{u} \in X$ there are $r, \delta > 0$ such that for $\|u_0 - \overline{u}\| \leq r$ problem (7.0.1) has a unique mild solution $u = u(\cdot; u_0)$ belonging to $C_\alpha(]0, \delta]; X_\alpha)$.*

Moreover, u belongs to $L^\infty(0, \delta; X) \cap C([\varepsilon, \delta]; D_A(\beta, 1))$ for every $\varepsilon \in]0, \delta[$ and $\beta \in]0, 1[$.

There is $K = K(\overline{u}) > 0$ such that for $\|u_0 - \overline{u}\| \leq r$ it holds

$$\|u(t; u_0) - u(t; \overline{u})\| + \|t^\alpha(u(t; u_0) - u(t; \overline{u}))\|_{X_\alpha} \leq K\|u_0 - \overline{u}\|, \quad 0 < t \leq \delta. \tag{7.1.22}$$

Proof — It is sufficient to modify suitably the proof of Theorem 7.1.5(i), replacing the set B by

$$Z = \{u \in C_\alpha(]0, \delta]; X_\alpha) : \|t^\alpha u(t)\|_{X_\alpha} \leq R, \|u(t)\| \leq R, 0 < t \leq \delta\}.$$

Z is endowed with the metric of $C_\alpha(]0, \delta]; X_\alpha) \cap L^\infty(0, \delta; X)$. Using Lemma 7.1.4, one sees that if R is sufficiently large, δ and $\|u_0 - \overline{u}\|$ are sufficiently small, then Γ is a contraction and it maps Z into itself, so that it has a unique fixed point in Z, which is a mild solution of (7.0.1). Uniqueness of the mild solution in $C_\alpha(]0, \delta]; X_\alpha)$, further regularity, and continuous dependence on the initial datum can be shown as in the proof of Theorem 7.1.2. ∎

7.1.2 The maximally defined solution

The local existence results of the previous subsection allow to continue the local mild solution of problem (7.0.1) to a maximal interval of existence. Indeed, in all the cases considered, $u(\delta)$ belongs to $\overline{D(A)}^\alpha$, and it can be taken as initial value for problem (7.0.5) with $a = \delta$, in such a way that (7.0.5) has a unique mild solution $v \in C([\delta, \delta + \varepsilon]; X_\alpha)$ for some small ε, thanks to Theorem 7.1.3(i). The function w defined by $w(t) = u(t)$ for $0 \leq t \leq \delta$, $w(t) = v(t)$ for $\delta < t \leq \delta + \varepsilon$, is continuous in $]0, \delta + \varepsilon]$ with values in X_α, and it is easy to see that it satisfies (7.1.1) in the whole interval $[0, \delta + \varepsilon]$. Again, $w(\delta + \varepsilon)$ can be taken as initial value for problem (7.0.5), with $a = \delta + \varepsilon$, and the above procedure may be repeated indefinitely, up to construct a noncontinuable solution defined in a maximal time interval $I(u_0)$.

Definition 7.1.7 *Let f, u_0 satisfy the assumptions of any of Theorems 7.1.2, 7.1.3, 7.1.5, 7.1.6, and define $I(u_0)$ as the union of all the intervals $[0, a]$, such that problem (7.0.1) has a mild solution u_a belonging to $\mathcal{H}$, where*

$$\mathcal{H} = L^\infty(0, a; X_\alpha), \text{ if the assumptions of Theorem 7.1.2 hold;}$$

$\mathcal{H} = C([0,a]; X_\alpha)$, if the assumptions of Theorem 7.1.3(i) hold;
$\mathcal{H} = L^\infty(0,a; X_\alpha) \cap C([0,a]; X)$, if the assumptions of Theorem 7.1.3(ii) hold;
$\mathcal{H} = C_{\alpha-\beta}(]0,a]; X_\alpha)$, if the assumptions of Theorem 7.1.5(i) or Theorem 7.1.5(ii) or Theorem 7.1.6 hold;
$\mathcal{H} = \{u \in C_{\alpha-\beta}(]0,a]; X_\alpha) : \lim_{t \to 0} \|t^{\alpha-\beta} u(t)\|_{X_\alpha} = 0\}$, if the assumptions of Theorem 7.1.5(iii) hold.

Set

$$\begin{cases} \tau(u_0) &= \sup I(u_0), \\ u(t; u_0) &= u_a(t) \text{ if } t \in [0,a] \subset I(u_0). \end{cases} \qquad (7.1.23)$$

$u(\cdot; u_0)$ is well defined thanks to the uniqueness parts of the theorems of the previous subsection. It is clear that if $u(t; u_0)$ has a continuous extension at $t = \tau(u_0)$, then either $u(\tau; u_0) \in \partial \mathcal{O}$ or $\tau(u_0) = T$ and $I(u_0) = [0, T]$. The uniform continuity of u near $t = \tau(u_0)$ is a consequence of the boundedness of $f(\cdot, u(\cdot; u_0))$, as the next proposition shows.

Proposition 7.1.8 *Let $I(u_0) \subset [0, T]$, $I(u_0) \neq [0, T]$, and assume that $\inf_{0 < t < \tau(u_0)} dist(u(t), \partial \mathcal{O}) > 0$. Then*

$$\limsup_{t \to \tau(u_0)} \|f(t, u(t))\| = +\infty. \qquad (7.1.24)$$

Proof — Assume by contradiction that (7.1.24) does not hold. Set $\tau = \tau(u_0)$. Then $t \mapsto f(t, u(t; u_0))$ belongs to $L^\infty(0, \tau; X)$. Since u satisfies the variation of constants formula (7.1.1), by Lemma 7.1.1 it may be extended at $t = \tau$ in such a way that the extension is Hölder continuous on each interval $[\varepsilon, \tau]$ ($0 < \varepsilon < \tau$) with values in X_α. Moreover, $u(\tau) \in \mathcal{O} \cap \overline{D(A)}^u$. By Theorem 7.1.3(i), the problem

$$v'(t) = Av(t) + f(t, v(t)), \quad t \geq \tau, \quad v(\tau) = u(\tau),$$

has a unique mild solution $v \in C([\tau, \tau + \delta]; X_\alpha)$ for some $\delta > 0$. The function w defined by $w(t) = u(t)$ for $0 \leq t < \tau$, $w(t) = v(t)$ for $\tau \leq t \leq \tau + \delta$, is a mild solution of (7.0.1) in $[0, \tau + \delta]$. This is in contrast with the definition of τ. Therefore, $f(\cdot, u(\cdot; u_0))$ cannot be bounded near $t = \tau$. ∎

Note that if f maps bounded subsets of $\mathcal{O}$ into bounded subsets of X, then (7.1.24) may be replaced by

$$\limsup_{t \to \tau(u_0)} \|u(t)\|_{X_\alpha} = +\infty.$$

Now we prove a continuous dependence result for the maximally defined solution of (7.0.1).

Proposition 7.1.9 *(i) Let the assumptions of Theorem 7.1.2 (respectively, Theorem 7.1.3(i), Theorem 7.1.3 (ii)) hold, and fix $\bar{u} \in X_\alpha$ (respectively, $\bar{u} \in \mathcal{O} \cap \overline{D(A)}^\alpha$,*

1. LOCAL EXISTENCE AND REGULARITY

$\overline{u} \in \mathcal{O} \cap \overline{D(A)}$). Then for every $b < \tau(\overline{u})$ there are r, $K > 0$ such that if $u_0 \in X_\alpha$ (respectively, $u_0 \in \mathcal{O} \cap \overline{D(A)}^\alpha$, $u_0 \in \mathcal{O} \cap \overline{D(A)}$) and $\|\overline{u} - u_0\|_{X_\alpha} \leq r$, then

$$\tau(u_0) \geq b, \quad \|u(t; u_0) - u(t; \overline{u})\|_{X_\alpha} \leq K\|u_0 - \overline{u}\|_{X_\alpha}, \quad 0 \leq t \leq b.$$

(ii) Let the assumptions of Theorem 7.1.5(i) (respectively, Theorem 7.1.6) hold, and fix $\overline{u} \in \mathcal{U} \cap \overline{D(A)}$ (respectively, $\overline{u} \in X$). Then for every $b < \tau(\overline{u})$ there are r, $K > 0$ such that if $u_0 \in \mathcal{U} \cap \overline{D(A)}$ (respectively, $u_0 \in X$) and $\|\overline{u} - u_0\| \leq r$, then $\tau(u_0) \geq b$ and

$$\|u(t; u_0) - u(t; \overline{u})\| + \|t^\alpha(u(t; u_0) - u(t; \overline{u}))\|_{X_\alpha} \leq K\|u_0 - \overline{u}\|, \quad 0 < t \leq b.$$

(iii) Let the assumptions of Theorem 7.1.5(ii) (respectively, Theorem 7.1.5(iii)) hold, and fix $\overline{u} \in \mathcal{U} \cap D_A(\beta, \infty)$ (respectively, $\overline{u} \in \mathcal{U} \cap D_A(\beta)$). Then for every $b < \tau(\overline{u})$ there are r, $K > 0$ such that if $u_0 \in \mathcal{U} \cap D_A(\beta, \infty)$ (respectively, $u_0 \in \mathcal{U} \cap D_A(\beta)$) and $\|\overline{u} - u_0\|_{D_A(\beta, \infty)} \leq r$, then $\tau(u_0) \geq b$ and

$$\|u(t; u_0) - u(t; \overline{u})\| + \|t^{\alpha-\beta}(u(t; u_0) - u(t; \overline{u}))\|_{X_\alpha} \leq K\|u_0 - \overline{u}\|_{D_A(\beta, \infty)}, \quad 0 < t \leq b.$$

Proof — Fix $\overline{u}$ satisfying the conditions specified above. In case (i), there are $r > 0$, $\delta > 0$, $K > 0$ such that if $\|u_0 - \overline{u}\|_{X_\alpha} \leq r$, then $u(\cdot; u_0)$ is well defined in $[0, \delta]$, and

$$\|u(t; u_0) - u(t; \overline{u})\|_{X_\alpha} \leq K\|u_0 - \overline{u}\|_{X_\alpha}, \quad 0 \leq t \leq \delta. \tag{7.1.25}$$

In case (ii), there are $r > 0$, $\delta > 0$, $K > 0$ such that if $\|u_0 - \overline{u}\| \leq r$, then $u(\cdot; u_0)$ is well defined in $[0, \delta]$, and

$$\|u(t; u_0) - u(t; \overline{u})\| + \|t^\alpha(u(t; u_0) - u(t; \overline{u}))\|_{X_\alpha} \leq K\|u_0 - \overline{u}\|, \quad 0 < t \leq \delta. \tag{7.1.26}$$

Similarly, in case (iii) there are $r > 0$, $\delta > 0$, $K > 0$ such that if $\|u_0 - \overline{u}\|_{D_A(\beta, \infty)} \leq r$, then $u(\cdot; u_0)$ is well defined in $[0, \delta]$, and

$$\|u(t; u_0) - u(t; \overline{u})\| + \|t^{\alpha-\beta}(u(t; u_0) - u(t; \overline{u}))\|_{X_\alpha} \leq K\|u_0 - \overline{u}\|_{D_A(\beta, \infty)}, \tag{7.1.27}$$

for $0 < t \leq \delta$. In all three cases, $u(\delta; \overline{u})$ and $u(\delta; u_0)$ belong to $\overline{D(A)}^\alpha$, and $u \in C([\delta, a]; X_\alpha)$. From now on, the statements follows from standard arguments, which we recall below.

Theorem 7.1.3(i) implies that for every $x_0 \in \overline{D(A)}^\alpha$ there are $\varepsilon = \varepsilon(x_0)$, $r = r(x_0)$, $K = K(x_0) > 0$ such that for every $a \in [0, T[$ and $x \in \overline{D(A)}^\alpha$ with $\|x - x_0\|_{X_\alpha} \leq r$, problem (7.0.5) has a unique mild solution $u(t; a, x)$ in $C([a, (a + \varepsilon) \wedge T]; X_\alpha)$, and moreover

$$\|u(t; a, x_0) - u(t; a, x)\|_{X_\alpha} \leq K\|x_0 - x\|_{X_\alpha}, \quad a \leq t \leq (a + \varepsilon) \wedge T.$$

Since the function $t \mapsto u(t; \overline{u})$ is continuous in $[\delta, b]$ with values in X_α, then the orbit

$$\{u(t; \overline{u}) : \delta \leq t \leq b\}$$

is compact, so that it may be covered by a finite number of open balls $B(\overline{u}_i, r(\overline{u}_i))$, with $\overline{u}_i = u(t_i; \overline{u})$, $i = 1, \ldots, n$. Set

$$\overline{\varepsilon} = \min\{\varepsilon(\overline{u}_i) : i = 1, \ldots, n\}, \quad \overline{K} = \max\{K(\overline{u}_i) : i = 1, \ldots, n\}.$$

There exists i_1 such that $u(\delta; \overline{u})$ belongs to $B(\overline{u}_{i_1}, r(\overline{u}_{i_1}))$. Due to estimates (7.1.25), (7.1.26), (7.1.27), if $\|u_0 - \overline{u}\|_{X_\alpha}$ (respectively, $\|u_0 - \overline{u}\|$, $\|u_0 - \overline{u}\|_{D_A(\beta,\infty)}$) is small enough, then also $u(\delta; u_0)$ belongs to the same ball $B(\overline{u}_{i_1}, r(\overline{u}_{i_1}))$. Therefore, $u(\cdot; u_0)$ is well defined in $[\delta, (\delta + \overline{\varepsilon}) \wedge T]$, and

$$\|u(t; u_0) - u(t; \overline{u})\|_{X_\alpha} \leq \overline{K} \|u(\delta; u_0) - u(\delta; \overline{u})\|_{X_\alpha}, \quad \delta \leq t \leq (\delta + \overline{\varepsilon}) \wedge T. \quad (7.1.28)$$

If $\delta + \overline{\varepsilon} \geq b$, the statements follow patching toghether estimates (7.1.28) and (7.1.25) (respectively, (7.1.26), (7.1.27)). If $\delta + \overline{\varepsilon} < b$, we repeat the above procedure, finding that if u_0 is sufficiently close to $\overline{u}$, then $u(\cdot; u_0)$ is well defined in $[\delta + \overline{\varepsilon}, (\delta + 2\overline{\varepsilon}) \wedge T]$, and

$$\|u(t; u_0) - u(t; \overline{u})\|_{X_\alpha} \leq \overline{K} \|u(\delta + \overline{\varepsilon}; u_0) - u(\delta + \overline{\varepsilon}; \overline{u})\|_{X_\alpha}$$
$$\leq \overline{K}^2 \|u(\delta; u_0) - u(\delta; \overline{u})\|_{X_\alpha}, \quad \delta + \overline{\varepsilon} \leq t \leq (\delta + 2\overline{\varepsilon}) \wedge T.$$

If $\delta + 2\overline{\varepsilon} \geq b$, the statements follow. Otherwise, we repeat the same procedure up to cover the whole interval $[\delta, b]$ by a finite numbers of intervals of length less or equal to $\overline{\varepsilon}$. ∎

7.1.3 Further regularity, classical and strict solutions

Now we give some further regularity results. In particular, we give sufficient conditions in order that the mild solution of (7.0.1) be classical or strict.

Proposition 7.1.10 *Let the hypotheses of any of the theorems in Subsection 7.1.1 hold. Assume in addition that there exists $\theta \in \,]0,1[$ such that for each $u_0 \in \mathcal{O}$[2] there are r, K such that*

$$\|f(t,x) - f(s,x)\| \leq K(t-s)^\theta, \quad 0 \leq s < t \leq T, \ \|x - u_0\|_{X_\alpha} \leq r. \quad (7.1.29)$$

Let $u = u(\cdot; u_0) : I(u_0) \mapsto X_\alpha$ be the maximally defined mild solution of problem (7.0.1), and fix any compact interval $[0, b] \subset I(u_0)$. Then u belongs to $C^\theta([\varepsilon, b]; D(A)) \cap C^{1+\theta}([\varepsilon, b]; X)$ for every $\varepsilon \in \,]0, b[$. Moreover,

(i) If $\alpha > 0$, or if $\alpha = 0$ and $u_0 \in \overline{D(A)}$, u is a classical solution of (7.0.1) in $I(u_0)$;

(ii) if $u_0 \in D_A(\sigma, \infty)$ for some $\sigma \in \,]\alpha, 1[$, then u belongs to $C^\sigma([0, b]; X) \cap B([0, b]; D_A(\sigma, \infty))$;

---
[2] If the assumptions of either Theorem 7.1.2 or Theorem 7.1.6 hold, we mean $\mathcal{O} = X_\alpha$.

1. LOCAL EXISTENCE AND REGULARITY

(iii) *if $u_0 \in D(A)$, then $u \in B([0,b]; D(A)) \cap \text{Lip}([0,b]; X)$; if in addition $Au_0 + f(0, u_0) \in \overline{D(A)}$, then $u \in C([0,b]; D(A)) \cap C^1([0,b]; X)$ and it is a strict solution of problem (7.0.1);*

(iv) *if $u_0 \in D(A)$ and $Au_0 + f(0, u_0) \in D_A(\sigma, \infty)$ for some $\sigma \in \,]0,1[$, then $u \in C^{\theta \wedge \sigma}([0,b]; D(A)) \cap C^{1+\theta \wedge \sigma}([0,b]; X)$.*

Proof — Let $0 < \varepsilon < b$. Since $u \in C([\varepsilon, b]; X_\alpha)$, then $t \mapsto f(t, u(t))$ belongs to $C([\varepsilon, b]; X)$. By Lemma 7.1.1, the function $t \mapsto \varphi(t) = u(t) - e^{tA}u_0$ belongs to $C^{1-\alpha}([\varepsilon, b]; X_\alpha)$ if $\alpha > 0$, and to $C^\gamma([\varepsilon, b]; X)$ for every $\gamma \in \,]0,1[$. Since $t \mapsto e^{tA}u_0$ belongs to $C^\infty(]0,b]; D(A))$, summing up we find that u belongs to $C^\gamma([\varepsilon, b]; X_\alpha)$, with $\gamma = 1 - \alpha$ if $\alpha > 0$, with any $\gamma \in \,]0,1[$ if $\alpha = 0$. Consequently, thanks to assumption (7.1.29), the function $t \mapsto f(t, u(t))$ belongs to $C^\gamma([\varepsilon, b]; X)$ with $\gamma = \min\{1 - \alpha, \theta\}$. Recalling that u satisfies

$$u(t) = e^{(t-\varepsilon)A}u(\varepsilon) + \int_\varepsilon^t e^{(t-s)A} f(s, u(s))ds, \quad \varepsilon \le t \le b, \tag{7.1.30}$$

we apply Theorem 4.3.1, which implies that u belongs to $C^\gamma([2\varepsilon, b]; D(A)) \cap C^{1+\gamma}([2\varepsilon, b]; X)$ for every $\varepsilon \in \,]0, b/2[$, and

$$u'(t) = Au(t) + f(t, u(t)), \quad \varepsilon < t \le b.$$

If $\gamma = \theta$, the proof of the first statement is finished. If $\gamma = 1 - \alpha$, we use a bootstrap argument: since $u \in C^{1-\alpha}([\varepsilon, b]; D(A)) \cap C^{2-\alpha}([\varepsilon, b]; X)$ for every $\varepsilon \in \,]0, b[$, from Proposition 1.1.5 one gets $u \in C^{2(1-\alpha)}([\varepsilon, b]; X_\alpha)$ if $2(1-\alpha) \ne 1$, $u \in C^\beta([\varepsilon, b]; X_\alpha)$ for every $\beta \in \,]0,1[$ if $2(1-\alpha) = 1$, so that $t \mapsto f(t, u(t))$ belongs to $C^{\gamma_1}([\varepsilon, b]; X)$, with $\gamma_1 = \min\{\theta, 2(1-\alpha)\}$. Due again to Theorem 4.3.1, $u \in C^{\gamma_1}([2\varepsilon, b]; D(A)) \cap C^{1+\gamma_1}([2\varepsilon, b]; X)$ for every $\varepsilon \in \,]0, b/2[$. If $\gamma_1 = \theta$ we have finished, otherwise we repeat the same procedure an arbitrary number of times, finding that $u \in C^{\gamma_n}([2\varepsilon, b]; D(A)) \cap C^{1+\gamma_n}([2\varepsilon, b]; X)$, with $\gamma_n = \min\{\theta, n(1-\alpha)\}$. For n large enough, $\gamma_n = \theta$, and the first part of the statement is proved.

Statement (i) is an obvious consequence of the continuity up to $t = 0$ of the function $t \mapsto e^{tA}u_0$ and of the convolution $t \mapsto e^{tA} * f(\cdot, u(\cdot))$.

Let us prove (ii). If $u_0 \in D_A(\sigma, \infty)$ with $\sigma > \alpha$, then Theorem 7.1.3(i) may be applied. It follows that $t \mapsto f(t, u(t)) \in L^\infty(0, b; X)$, and then, by Proposition 4.2.1, the function $t \mapsto \varphi(t) = u(t) - e^{tA}u_0$ belongs to $C^{1-\gamma}([0,b]; D_A(\gamma, 1))$, and hence to $C^{1-\sigma}([0,b]; D_A(\sigma, \infty)) \cap C^\sigma([0,b]; X)$, for every $\sigma \in \,]0,1[$. Moreover, $t \mapsto e^{tA}u_0$ belongs to $C^\sigma([0,b]; X) \cap L^\infty([0,b]; D_A(\sigma, \infty))$. Statement (ii) follows.

Let us prove (iii). We know from the proof of statement (ii) that the function $t \mapsto u(t) - e^{tA}u_0$ is Hölder continuous up to $t = 0$ with values in X_α (with exponent $1 - \alpha$ if $\alpha > 0$, with any exponent $\gamma \in \,]0,1[$ if $\alpha = 0$). Since $u_0 \in D(A)$, the same property is true for $t \mapsto e^{tA}u_0$. Therefore, u is Hölder continuous up to $t = 0$ with values in X_α, so that $t \mapsto f(t, u(t))$ is Hölder continuous in $[0, b]$ with values in X. Statement (iii) follows now from Theorem 4.3.1.

To prove statement (iv), the starting point is again the fact that $t \mapsto f(t, u(t))$ $\in C^\gamma([0, b]; X)$, with $\gamma = \min\{\theta, 1 - \alpha\}$ (see the proof of statement (iii)). Since $u_0 \in D(A)$ and $Au_0 + f(0, u_0) \in D_A(\sigma, \infty)$, by Theorem 4.3.1(iii) we get $u \in C^{\gamma \wedge \sigma}([0, b]; D(A)) \cap C^{1+\gamma \wedge \sigma}([0, b]; X)$. Statement (iv) follows now by the same bootstrap argument used in the proof of statement (i). ∎

Proposition 7.1.11 *Let the hypotheses of any of the theorems of Subsection 7.1.1 be satisfied, and assume in addition that f is continuous with values in $D_A(\theta, \infty)$, for some $\theta \in \;]0, 1[$. Let $u : I(u_0) \mapsto X_\alpha$ be the maximally defined solution of problem (7.0.3), and fix any compact interval $[0, b] \subset I(u_0)$. Then u belongs to $C^1(]0, b]; D_A(\theta, \infty)) \cap C(]0, b]; D_A(\theta + 1, \infty))$. Moreover,*

(i) *if $\alpha > 0$, or if $\alpha = 0$ and $u_0 \in \overline{D(A)}$, u is a classical solution of problem (7.0.3);*

(ii) *if $u_0 \in D_A(\sigma, \infty)$ for some $\sigma \in \;]\alpha, 1[$, then u belongs to $C^\sigma([0, b]; X) \cap B([0, b]; D_A(\sigma, \infty))$;*

(iii) *if $u_0 \in D(A)$ and $Au_0 \in \overline{D(A)}$, then $u \in C([0, b]; D(A)) \cap C^1([0, b]; X)$, and it is a strict solution of problem (7.0.3);*

(iv) *if $u_0 \in D(A)$ and $Au_0 \in D_A(\theta, \infty)$, then u' and $Au \in B([0, b]; D_A(\theta, \infty))$.*

Proof — We know that $u(\varepsilon) \in \overline{D(A)}$, and $u \in C([\varepsilon, b]; X_\alpha)$ for every $\varepsilon \in \;]0, b[$. Then $f(\cdot, u(\cdot)) \in C([\varepsilon, b]; D_A(\theta, \infty))$, and, since u satisfies (7.1.30), by Theorem 4.3.8(ii) u belongs to $C^1(]\varepsilon, b]; D_A(\theta, \infty)) \cap C(]\varepsilon, b]; D_A(\theta + 1, \infty))$, and $u'(t) = Au(t) + f(t, u(t))$ for $\varepsilon < t \leq b$. If in addition $u_0 \in \overline{D(A)}$, then $u \in C([0, b]; X)$ and it is a classical solution. So, statement (i) is proved.

The proof of statement (ii) is identical to the proof of statement (ii) of Proposition 7.1.8.

Let us prove that (iii) and (iv) hold. Since $u_0 \in D(A)$ and $Au_0 \in \overline{D(A)}$, then $t \mapsto e^{tA}u_0 \in C([0, b]; D(A)) \subset C([0, b]; X_\alpha)$. In addition, we know that $t \mapsto u(t) - e^{tA}u_0 \in C([0, b]; X_\alpha)$. Therefore u is continuous up to $t = 0$ with values in X_α, so that $f(\cdot, u(\cdot))$ is continuous in $[0, b]$ with values in $D_A(\theta, \infty)$. Statements (ii) and (iii) follow now from Corollary 4.3.9(ii)(iv). ∎

Remark 7.1.12 The last two propositions give sufficient criteria for the mild solution be classical or strict. Now we can discuss uniqueness of the classical and of the strict solution. It follows from Proposition 4.1.2 that any classical solution u of (7.0.1) in an interval $[0, \delta]$ such that $f(\cdot, u(\cdot)) \in L^1(0, \delta; X)$ is a mild solution. In particular, if f satisfies assumption (7.1.12), the classical solution is unique in $C([0, \delta]; X_\alpha)$; if f satisfies assumption (7.1.13) with $\alpha \gamma < 1$, the classical solution is unique in $C_\alpha(]0, \delta]; X_\alpha) \cap C([0, \delta]; X)$; if f satisfies assumption (7.1.13) with $\gamma(\alpha - \beta) < 1 - \beta$, the classical solution is unique in $C_{\alpha - \beta}(]0, \delta]; X_\alpha) \cap C([0, \delta]; X)$. In any case, the strict solution of (7.0.1) is unique.

7.2 A priori estimates and existence in the large

Through the section, $f : [0,T] \times \mathcal{O} \mapsto X$ is a continuous function satisfying (7.1.12), and $u_0 \in \mathcal{O} \cap \overline{D(A)}^\alpha$. This is not restrictive, as far as *a priori* estimates and existence in the large are concerned, because if u is any mild solution of (7.0.1) in an interval $[0,b]$ then $u(\varepsilon) \in D_A(\theta, 1)$ for every $\varepsilon \in \,]0,b]$ and $\theta \in \,]0,1[$; so that $u(\varepsilon) \in \overline{D(A)}^\alpha$ may be taken as initial value for problem (7.0.5), with $a = \varepsilon$.

As in the previous section, we denote by $I(u_0)$ or simply by I the maximal interval of existence of the solution. We have seen in subsection 7.1.2 that if $f(\cdot, u(\cdot))$ is bounded with values in X near $\tau = \sup I(u_0)$, and u remains far from the boundary of $\mathcal{O}$, then $\tau = T$ and $I(u_0) = [0,T]$. In most applications, if u remains far from the boundary of $\mathcal{O}$ and its X_α-norm is bounded, then $f(\cdot, u(\cdot))$ is bounded with values in X. Therefore, the estimates on the X_α-norm of u are of crucial importance to prove global existence. We shall give some sufficient conditions which let one find *a priori* estimates for $\|u(t)\|_{X_\alpha}$.

The simplest case is when f grows not more than linearly with respect to u.

Proposition 7.2.1 *Assume that there exists $C > 0$ such that*

$$\|f(t,x)\| \le C(1 + \|x\|_{X_\alpha}) \quad \forall x \in \mathcal{O}, \; t \in [0,T]. \tag{7.2.1}$$

Let $u : I \mapsto X_\alpha$ be a mild solution of (7.0.1). Then u is bounded in I with values in X_α.

Proof — For every $t \in I$

$$\begin{aligned}
\|u(t)\|_{X_\alpha} &\le M_\alpha \|u_0\|_{X_\alpha} + C_\alpha C \int_0^t (t-s)^{-\alpha}(1 + \|u(s)\|_{X_\alpha})ds \\
&= M_\alpha \|u_0\|_{X_\alpha} + C_\alpha C \left(\frac{T^{1-\alpha}}{1-\alpha} + \int_0^t \frac{\|u(s)\|_{X_\alpha}}{(t-s)^\alpha} ds \right).
\end{aligned}$$

Using Lemma 7.0.3 we get

$$\|u(t)\|_{X_\alpha} \le C_1 \left(M_\alpha \|u_0\|_{X_\alpha} + \frac{C_\alpha C T^{1-\alpha}}{1-\alpha} \right), \quad t \in I,$$

and the statement follows. ∎

Note that condition (7.2.1) is satisfied when f is globally Lipschitz continuous.

In the applications to parabolic PDE's it often happens to find *a priori* estimates in the norm of X (especially when X is endowed with an L^p norm, $1 \le p \le \infty$). Then, if $\alpha = 0$, $\mathcal{O} = X_\alpha = X$, Propositions 7.1.8 and 7.2.1 yield existence in the large of the solution. If $\alpha > 0$, one can find an *a priori* estimate in the norm of X_α if the solution is bounded in X and f satisfies suitable growth conditions, in such a way that one can argue as in the case of the linear growth: see the proof below.

Proposition 7.2.2 *Assume that there exists an increasing function $\mu : [0, +\infty[\mapsto [0, +\infty[$ such that*

$$\|f(t,x)\| \leq \mu(\|x\|)(1 + \|x\|_{X_\alpha}^\gamma), \quad 0 \leq t \leq T, \ x \in \mathcal{O}, \tag{7.2.2}$$

with $1 < \gamma < 1/\alpha$. Let $u : I \mapsto X_\alpha$ be a mild solution of (7.0.1). If u is bounded in I with values in X, then it is bounded in I with values in X_α and in $D_A(\theta, \infty)$, for every $\theta \in \,]0,1[$.

More precisely, for every $\theta \in \,]0,1[$ and for $0 < a < b$, $b \in I$, we have

$$\|u(t)\|_{D_A(\theta,\infty)} \leq C(1 + \|u(a)\|_{D_A(\theta,\infty)}), \quad a \leq t \leq b, \tag{7.2.3}$$

where C depends on θ, $\sup_{t \in I} \|u(t)\|$, and it does not depend on a, b.

Proof — It is sufficient to prove that (7.2.3) holds for some $\theta > \alpha$. Indeed, in this case u is bounded in $[a,b]$ with values in X_α, so that $f(\cdot, u(\cdot))$ is bounded in $[a,b]$ with values in X. Since u satisfies the variation of constants formula (7.0.6) in the interval $[a,b]$, with $x = u(a)$, then (7.2.3) holds for every $\theta \in \,]0,1[$, thanks to Proposition 4.2.1 and to estimates (2.2.20). Taking again $\theta > \alpha$, it holds $D_A(\theta, \infty) \subset X_\alpha$, so that u is bounded in $[a,b]$ with values in X_α. Moreover, since $u \in C([0,a]; X_\alpha)$, then u is bounded in $[0,a]$ with values in X_α. Therefore, u is bounded in $[0,b]$ with values in X_α by a constant independent of b. Since b is arbitrary, u is bounded in I with values in X_α.

We shall show that (7.2.3) holds for $\theta = \alpha\gamma$.

By the Reiteration Theorem, X_α belongs to the class $J_{\alpha/\theta}$ between X and $D_A(\theta, \infty)$. Therefore, there is a constant c such that

$$\|u(s)\|_{X_\alpha}^\gamma \leq c \|u(s)\|^{\gamma(1-\alpha/\theta)} \|u(s)\|_{D_A(\theta,\infty)}^{\alpha\gamma/\theta} \leq cK^{\gamma(1-\alpha/\theta)} \|u(s)\|_{D_A(\theta,\infty)},$$

so that

$$\|f(s, u(s))\| \leq \mu(K)(1 + cK^{\gamma(1-\alpha/\theta)} \|u(s)\|_{D_A(\theta,\infty)}).$$

Let M_θ be a constant such that $\|t^\theta e^{tA} x\|_{D_A(\theta,\infty)} \leq M_\theta \|x\|$ for $x \in X$, and $\|e^{tA} x\|_{D_A(\theta,\infty)} \leq M_\theta \|x\|_{D_A(\theta,\infty)}$ for $x \in D_A(\theta, \infty)$, $0 < t \leq T$. Then for $a \leq t \leq b$ we have

$$\|u(t)\|_{D_A(\theta,\infty)} \leq M_\theta \|u(a)\|_{D_A(\theta,\infty)}$$
$$+ M_\theta \mu(K) \int_\varepsilon^t (t-s)^{-\theta} (1 + cK^{\gamma(1-\alpha/\theta)} \|u(s)\|_{D_A(\theta,\infty)}) ds, \tag{7.2.4}$$

and (7.2.3) follows from Lemma 7.0.3. ∎

Estimate (7.2.3) holds also for certain values of γ greater than $1/\alpha$, provided that the solution is bounded with values in $D_A(\beta, \infty)$ for some $\beta \in \,]0, \alpha[$. In fact, following the proof of proposition 7.2.2 (with $\alpha\gamma$ replaced by $\alpha\gamma + \beta(1-\gamma)$) and recalling that X_α belongs to the class $J_{(\alpha-\beta)/(\theta-\beta)}$ between $D_A(\beta, \infty)$ and $D_A(\theta, \infty)$ for $\beta < \alpha < \theta$, one can prove the next proposition.

2. A priori ESTIMATES AND EXISTENCE IN THE LARGE

Proposition 7.2.3 *Assume that there exists an increasing function* $\mu : [0, +\infty[\mapsto [0, +\infty[$ *such that*

$$\|f(t,x)\| \leq \mu(\|x\|_{D_A(\beta,\infty)})(1 + \|x\|_{X_\alpha}^\gamma), \quad 0 \leq t \leq T, \ x \in \mathcal{O}, \tag{7.2.5}$$

with $\beta < \alpha$, $\gamma < (1-\beta)/(\alpha-\beta)$. *Let* $u : I \mapsto X_\alpha$ *be a mild solution of (7.0.1). If u is bounded in I with values in $D_A(\beta, \infty)$, then it is bounded in I with values in X_α and with values in $D_A(\theta, \infty)$, for every $\theta \in \,]0,1[$. More precisely, for every $\theta \in \,]0,1[$ and for every $a, b \in I$, $0 < a < b$, we have*

$$\|u(t)\|_{D_A(\theta,\infty)} \leq C(1 + \|u(a)\|_{D_A(\theta,\infty)}), \quad a \leq t \leq b, \tag{7.2.6}$$

where C does not depend on a and b.

However, in the applications finding *a priori* estimates in $D_A(\beta, \infty)$ is much harder than finding *a priori* estimates in X. Therefore, we come back to the case where the solution is bounded with values in X.

Note that the procedure of Proposition 7.2.2 fails in the case $\gamma = 1/\alpha$: one should replace $D_A(\alpha\gamma, \infty)$ by $D(A)$, and the integral in (7.2.4) would not make sense. What is behind this difficulty is the fact that one cannot expect to estimate the $D(A)$-norm of the convolution $(e^{tA} * \varphi)(t)$ in terms of sup $\|\varphi(t)\|$ (see counterexample 4.1.7). But we know from Theorem 4.3.1 that it is possible to estimate the norm $\|(e^{tA} * \varphi)\|_{L^\infty(D(A))}$ in terms of the Hölder norm $\|\varphi\|_{C^\sigma(X)}$, for any $\sigma \in \,]0,1[$. So, in the critical case $\gamma = 1/\alpha$ we use the results of Theorem 4.3.1(iii). To do this, boundedness of u with values in X is not enough: we need that u is Hölder continuous with values in X in its maximal interval of existence.

Theorem 7.2.4 *Assume that there are $\theta \in \,]0,1[$ and two increasing functions μ_1, $\mu_2 : [0, +\infty[\mapsto [0, +\infty[$ such that for $0 \leq t \leq T$, $x \in \mathcal{O}$*

$$\begin{cases} (i) & \|f(t,x) - f(s,x)\| \leq \mu_1(\|x\|)(t-s)^\theta (1 + \|x\|_{X_\alpha}^{1/\alpha}); \\ (ii) & \|f(t,x) - f(t,y)\| \leq \mu_2(\max\{\|x\|, \|y\|\})\big[(1 + \|x\|_{X_\alpha}^{1/\alpha-1} \\ & + \|y\|_{X_\alpha}^{1/\alpha-1})\|x-y\|_{X_\alpha} + (\|x\|_{X_\alpha}^{1/\alpha} + \|y\|_{X_\alpha}^{1/\alpha})\|x-y\|\big]. \end{cases} \tag{7.2.7}$$

Let $u : I \mapsto X_\alpha$ be a mild solution of (7.0.1) such that

$$\|u\|_{C^\theta(I;X)} = K < \infty.$$

Then u is bounded in I with values in X_α. More precisely, u is a classical solution of (7.0.1), and there is C such that for every $a, b \in I$, $0 < a < b$

$$\|u(t)\|_{D(A)} \leq C(1 + \|u(a)\|_{D(A)}), \quad a \leq t \leq b. \tag{7.2.8}$$

Proof — By Proposition 7.1.10, u belongs to $C^\theta([a,b]; D(A)) \cap C^{1+\theta}([a,b]; X)$. Moreover, for every $\delta \in \,]0, b-a]$ and for every $\sigma \in \,]0, \theta]$ we have, by estimate (4.3.2) of Theorem 4.3.1(ii),

$$\|u\|_{C([a,\delta];D(A))} \leq C(\|u(a)\|_{D(A)} + \|f(\cdot, u(\cdot))\|_{C^\sigma([a,\delta];X)}), \tag{7.2.9}$$

with $C = C(\sigma, T)$. For reasons which will be clear soon, we fix once and for all $\sigma \in\,]0, \theta(1-\alpha)[$.

Let us estimate $\|f(\cdot, u(\cdot))\|_{C^\sigma([a,\delta];X)}$. Due to assumption 7.2.7(ii), there exists an increasing function $\mu_3 : [0, +\infty[\mapsto [0, +\infty[$ such that

$$\|f(t,x)\| \leq \mu_3(\|x\|)(1 + \|x\|_{X_\alpha}^{1/\alpha}), \ 0 \leq t \leq T, \ x \in \mathcal{O}. \tag{7.2.10}$$

Let c_α be such that $\|x\|_{X_\alpha} \leq c_\alpha \|x\|_{D(A)}^\alpha \|x\|^{1-\alpha}$ for every $x \in D(A)$. Then, for $a \leq t \leq b$

$$\|f(a, u(a))\| \leq \mu_3(K)(1 + \|u(t)\|_{X_\alpha}^{1/\alpha}) \leq \mu_3(K)(1 + c_\alpha^{1/\alpha} K^{(1-\alpha)/\alpha} \|u(a)\|_{D(A)}), \tag{7.2.11}$$

and for $a \leq s \leq t \leq \delta$ we have $\|f(t, u(t)) - f(s, u(s))\| \leq \|f(t, u(t)) - f(s, u(t))\| + \|f(s, u(t)) - f(s, u(s))\|$, where

$$\|f(t, u(t)) - f(s, u(t))\| \leq \mu_1(K)(1 + \|u(t)\|_{X_\alpha}^{1/\alpha})(t-s)^\theta$$
$$\leq \mu_1(K)(1 + c^{1/\alpha} K^{(1-\alpha)/\alpha} \|u(t)\|_{D(A)})(t-s)^\theta,$$

and

$$\|f(s, u(t)) - f(s, u(s))\|$$
$$\leq \mu_2(K)\big[(1 + \|u(t)\|_{X_\alpha}^{1/\alpha - 1} + \|u(s)\|_{X_\alpha}^{1/\alpha - 1})\|u(t) - u(s)\|_{X_\alpha}$$
$$+ (\|u(t)\|_{X_\alpha}^{1/\alpha} + \|u(s)\|_{X_\alpha}^{1/\alpha})\|u(t) - u(s)\|\big]$$
$$\leq \mu_2(K)\big[(1 + 2c_\alpha^{1/\alpha - 1}\|u\|_{L^\infty(D(A))}^{1-\alpha} K^{(1-\alpha)^2/\alpha})c_\alpha K^{1-\alpha}(2\|u\|_{L^\infty(D(A))})^\alpha \cdot$$
$$\cdot (t-s)^{\theta(1-\alpha)} + 2c_\alpha^{1/\alpha} K^{(1-\alpha)/\alpha} \|u\|_{L^\infty(D(A))} K(t-s)^\theta\big]$$

so that

$$[f(\cdot, u(\cdot))]_{C^\sigma(X)} \leq M_1 \left[(1 + \|u\|_{L^\infty(D(A))})\delta^{\theta - \sigma} + (1 + \|u\|_{L^\infty(D(A))})\delta^{\theta(1-\alpha) - \sigma}\right], \tag{7.2.12}$$

with $M_1 = M_1(K, \alpha)$. Estimates (7.2.11) and (7.2.12) give

$$\|f(\cdot, u(\cdot))\|_{C^\sigma(X)} \leq \|f(a, u(a))\| + (T^\sigma + 1)[f(\cdot, u(\cdot))]_{C^\sigma(X)}$$
$$\leq M_2(1 + \|u(a)\|_{D(A)} + \|u\|_{L^\infty(D(A))}\delta^{\theta(1-\alpha) - \sigma}),$$

with $M_2 = M_2(K, T, \sigma, \alpha, \theta)$. By inserting this estimate in (7.2.9) we get

$$\|u\|_{C([a,\delta];D(A))} \leq C\left[(1 + M_2)\|u(a)\|_{D(A)} + M_2(1 + \delta^{\theta(1-\alpha) - \sigma}\|u\|_{C^\sigma([a,\delta];D(A))})\right].$$

If δ is sufficiently small, in such a way that

$$CM_2 \delta^{\theta(1-\alpha) - \sigma} \leq 1/2,$$

we obtain

$$\|u\|_{C([a,\delta];D(A))} \leq 2C[(1 + M_2)\|u(a)\|_{D(A)} + M_2],$$

3. SOME EXAMPLES

so that
$$\|u\|_{C([a,\delta];D(A))} \leq M_3(1+\|u(a)\|_{D(A)}),$$
with $M_3 = M_3(K,T,\sigma,\alpha,\theta)$. Arguing by recurrence, for every integer n we find
$$\|u\|_{C([a+n\delta,(a+(n+1)\delta)\wedge b];D(A))} \leq \sum_{j=1}^{n} M_3^j + M_3^{n+1}(1+\|u(a)\|_{D(A)}),$$
from which we obtain easily (7.2.8).

In particular, $\|u(t)\|_{D(A)}$ is bounded in $[a,b]$ by a constant independent of b. Since u is bounded in $[0,a]$ with values in X_α, then u is bounded in $[0,b]$ with values in X_α, by a constant independent of b. The statement follows. ∎

Theorem 7.2.4 and Proposition 7.1.9 give an existence in the large result.

Corollary 7.2.5 *Let $f : X_\alpha \mapsto X$ satisfy the assumptions of Theorem 7.2.4. Let $u : I(x_0) \mapsto X_\alpha$ be the maximally defined solution of problem (7.0.1). If there are $K > 0$ and $\theta \in \,]0,1[$ such that*
$$\|u(t) - u(s)\| \leq K|t-s|^\theta, \quad t,s \in I(x_0),$$
then $I(x_0) = [0,T]$.

7.3 Some examples

7.3.1 Reaction-diffusion systems

First we consider a reaction-diffusion system in $[0,T] \times \mathbb{R}^n$.

Let $D = diag(d_1,\ldots,d_m)$ be a diagonal $m \times m$ matrix, with positive, uniformly continuous and bounded entries $d_i : \mathbb{R}^n \mapsto \mathbb{R}$, such that
$$\inf_{x \in \mathbb{R}^n} d_i(x) > 0, \quad i=1,\ldots,n.$$

Consider the problem
$$u_t = D\Delta u + \varphi(t,x,u), \ t > 0, \ x \in \mathbb{R}^n; \quad u(0,x) = u_0(x), \ x \in \mathbb{R}^n, \quad (7.3.1)$$
with unknown $u = (u_1,\ldots,u_m)$.

Local existence and regularity for problem (7.3.1) may be studied using the results of Subsections 7.1.1 and 7.1.2. There are several choices of the space X, according to the regularity of the data D, φ and u_0. Here we choose
$$X = L^\infty(\mathbb{R}^n;\mathbb{R}^m).$$

It follows from Theorem 3.1.7 that the operator A defined by
$$\begin{cases} D(A) = \{u \in W^{2,p}_{loc}(\mathbb{R}^n;\mathbb{R}^m) \ \forall p \geq 1: \ u, \ \Delta u \in X\}, \\ A : D(A) \mapsto X, \ Au = D\Delta u, \end{cases}$$
is sectorial in X, and $\overline{D(A)} = UC(\mathbb{R}^n)$.

Proposition 7.3.1 *Let $\varphi : [0,T] \times \mathbb{R}^n \times \mathbb{R}^m \mapsto \mathbb{R}^m$ be a continuous function, and assume that there exists $\alpha \in \,]0,1[$ such that for every $r > 0$*

$$|\varphi(t,x,u) - \varphi(s,x,v)| \leq K((t-s)^\alpha + |u-v|), \qquad (7.3.2)$$

for $0 \leq s < t \leq T$, $x \in \mathbb{R}^n$, $u, v \in \mathbb{R}^m$, $|v| + |u| \leq r$, with $K = K(r)$. Let moreover $u_0 \in L^\infty(\mathbb{R}^n)$.

Then the following statements hold.

(i) *There are $\delta > 0$ and a solution $u \in L^\infty(\,]0, \delta[\times \mathbb{R}^n; \mathbb{R}^m)$ of (7.1.20) such that u, u_t, $D_i u$, Δu are continuous in $\,]0, \delta] \times \mathbb{R}^n$.*

(ii) *If $u_0 \in C(\mathbb{R}^n; \mathbb{R}^m)$, then $u \in C([0,\delta] \times \mathbb{R}^n; \mathbb{R}^m)$. If $u_0 \in UC(\mathbb{R}^n; \mathbb{R}^m)$, then $u(t,x) \to u_0(x)$ as $t \to 0$, uniformly for $x \in \mathbb{R}^n$. Moreover, u is the unique solution of (7.1.20) in the class of the functions v such that $t \mapsto v(t,\cdot)$ belongs to $C^1(\,]0, \delta]; L^\infty(\mathbb{R}^n; \mathbb{R}^m)) \cap C([0, \delta]; L^\infty(\mathbb{R}^n; \mathbb{R}^m))$.*

(iii) *If $u_0 \in C^\beta(\mathbb{R}^n; \mathbb{R}^m)$, with $0 < \beta < 2$, $\beta \neq 1$, then u belongs to $C^{\beta/2, \beta}([0, \delta] \times \mathbb{R}^n; \mathbb{R}^m)$.*

(iv) *If in addition the diffusion coefficients d_i and $\varphi(t, \cdot, u)$ belong to $C^{2\alpha}(\mathbb{R}^n)$ ($0 < \alpha < 1$), with $C^{2\alpha}$ norm locally independent of (t, u), then the derivatives $D_{ij} u$ are continuous in $\,]0, \delta] \times \mathbb{R}^n$; specifically, if $\alpha \neq 1/2$, $u \in C^{1+\alpha, 2+2\alpha}([\varepsilon, \delta] \times \mathbb{R}^n; \mathbb{R}^m)$ for every $\varepsilon < \delta$.*

(v) *u can be extended to a maximal time interval $I(u_0)$, which is relatively open in $[0,T]$. The mapping $(t, u_0) \mapsto u(t; u_0)$ is continuous from*

$$\{(t, u_0) : u_0 \in L^\infty(\mathbb{R}^n; \mathbb{R}^m), t \in I(u_0) \setminus \{0\}\}$$

to $L^\infty(\mathbb{R}^n; \mathbb{R}^m)$, and from

$$\{(t, u_0) : u_0 \in UC(\mathbb{R}^n; \mathbb{R}^m), t \in I(u_0)\}$$

to $UC(\mathbb{R}^n; \mathbb{R}^m)$. If u is bounded in $I(u_0) \times \mathbb{R}^n$, then $I(u_0) = [0, T]$.

Proof — By setting

$$f(t,u)(x) = \varphi(t, x, u(x)), \quad 0 \leq t \leq T, \ x \in \mathbb{R}^n, u \in X,$$

the function $f : [0,T] \times X \mapsto X$ is continuous and satisfies assumption (7.1.5), with $\alpha = 0$. Theorem 7.1.2 guarantees the existence of a unique mild solution $t \mapsto U(t) \in C(\,]0, \delta]; X) \cap L^\infty(0, \delta; X)$ of (7.0.1). Moreover, for every $a \in \,]0, \delta]$, $U(a)$ belongs to $\overline{D(A)}$, so that it is continuous and bounded. Take now $Y = C(\mathbb{R}^n; \mathbb{R}^m)$, the space of all continuous and bounded functions from $\mathbb{R}^n$ to $\mathbb{R}^m$. Corollary 3.1.9 implies that the part of A in Y is sectorial in Y. Moreover, the function f defined above satisfies assumption (7.1.5) with X replaced by Y and $\alpha = 0$. From Theorem 7.1.2 and Proposition 7.1.10, applied with initial time $a > 0$, we get that U, U', AU are continuous in $\,]a, \delta]$ with values in Y. Therefore, the function $(t, x) \mapsto u(t, x) = U(t)(x)$ is continuous and bounded in $]0, \delta] \times \mathbb{R}^n$, it is continuously differentiable in $]0, \delta] \times \mathbb{R}^n$, Δu is continuous in $]0, \delta] \times \mathbb{R}^n$, and u is a solution of (7.3.1). Statement (i) follows.

Let us prove statement (ii). The function $t \mapsto U(t) - e^{tA}u_0 = e^{tA} * f(\cdot, U(\cdot))$ belongs to $C([0, \delta]; Y)$. We have shown in the proof of Theorem 5.1.2 that if u_0 is continuous and bounded, then $(t, x) \mapsto (e^{tA}u_0)(x)$ is continuous and bounded in $[0, \delta] \times \mathbb{R}^n$. Summing up, we get that u is continuous and bounded in $[0, \delta] \times \mathbb{R}^n$.

If u_0 belongs to $UC(\mathbb{R}^n; \mathbb{R}^m) = \overline{D(A)}$, then U belongs to $C([0, \delta]; X)$ and it is a classical solution of (7.0.1). Moreover, $\lim_{t \to 0} u(t, \cdot) = u_0$ in $L^\infty(\mathbb{R}^n, \mathbb{R}^m)$. If v is a solution of (7.1.20) such that $t \mapsto V(t) = v(t, \cdot)$ belongs to $C^1(]0, \delta]; L^\infty(\mathbb{R}^n; \mathbb{R}^m))$ $\cap C([0, \delta]; L^\infty(\mathbb{R}^n; \mathbb{R}^m))$, then V, $f(\cdot, V(\cdot))$ belong to $C([0, \delta]; X)$, V' belongs to $C(]0, \delta]; X)$, so that V is a classical solution of (7.0.1). Since the classical solution is unique in $C([0, \delta]; X)$, it coincides with U. Statement (ii) follows.

From Theorem 3.1.12 we get

$$D_A(\alpha, \infty) = C^{2\alpha}(\mathbb{R}^n; \mathbb{R}^m), \quad \alpha \neq 1/2.$$

Therefore, if u_0 is β-Hölder continuous with $0 < \beta < 2$, $\beta \neq 1$, then by Proposition 7.1.10(ii) $U \in C^{\beta/2}([0, \delta]; X) \cap B([0, \delta]; D_A(\beta, \infty))$. This implies that $u \in C^{\beta/2, \beta}([0, \delta] \times \mathbb{R}^n; \mathbb{R}^m)$, and statement (iii) follows.

Statement (iv) is a consequence of Theorem 5.1.8. Indeed, since $t \mapsto f(t, U(t))$ belongs to $C^\alpha([\varepsilon, \delta]; X) \cap B([\varepsilon, \delta]; C^{2\alpha}(\overline{\Omega}))$, then $(t, x) \mapsto \varphi(t, x, u(t, x))$ belongs to $C^{\alpha, 2\alpha}([\varepsilon, \delta] \times \mathbb{R}^n; \mathbb{R}^m)$. Theorem 5.1.8 implies that $u \in C^{1+\alpha, 2+2\alpha}([\varepsilon, \delta] \times \mathbb{R}^n; \mathbb{R}^m)$.

Statement (v) follows from Proposition 7.1.9(i), recalling that if $u_0 \in \overline{D(A)}$ then $t \mapsto U(t; u_0)$ is continuous up to $t = 0$ with values in $\overline{D(A)}$. ∎

Similar results hold for reaction-diffusion systems in $[0, T] \times \overline{\Omega}$, where Ω is a bounded open set in $\mathbb{R}^n$ with C^2 boundary $\partial \Omega$. Precisely, let $\varphi : [0, T] \times \overline{\Omega} \times \mathbb{R}^m \mapsto \mathbb{R}^m$ be a regular function, and let $D = \text{diag}(d_1, \ldots, d_m)$ be a diagonal $m \times m$ matrix, with positive entries $d_i \in C(\overline{\Omega})$. Consider the problem

$$\begin{cases} u_t = D\Delta u + \varphi(t, x, u), & t > 0, \; x \in \overline{\Omega}, \\ u(0, x) = u_0(x) = (u_{01}(x), \ldots, u_{0n}(x)), & x \in \overline{\Omega}, \\ u_i(t, x) = 0, \; i \in I_1, \; \partial u_i/\partial \nu(t, x) = 0, & i \in I_2, \; x \in \partial \Omega, \end{cases} \quad (7.3.3)$$

with unknown $u = (u_1, \ldots, u_m)$. Here $I_1 \cup I_2 = \{1, \ldots, m\}$, $I_1 \cap I_2 = \emptyset$.

Arguing as in the proof of Proposition 7.3.1, the following results may be shown.

Proposition 7.3.2 *Assume that $\varphi : [0, T] \times \overline{\Omega} \times \mathbb{R}^m \mapsto \mathbb{R}^m$ is a continuous function satisfying (7.3.2) for $0 \leq s < t \leq T$, $x \in \overline{\Omega}$, $u, v \in \mathbb{R}^m$, $|u| + |v| \leq r$, with $K = K(r)$. Let moreover $u_0 \in C(\overline{\Omega}; \mathbb{R}^m)$ be such that $u_{0i} = 0$ on $\partial \Omega$ for $i \in I_1$. Then the following statements hold.*

(i) *There are $\delta > 0$ and a unique solution u of (7.3.3) such that $u \in C([0, \delta] \times \overline{\Omega}; \mathbb{R}^m)$, and u_t, $D_i u$, Δu are continuous in $]0, \delta] \times \overline{\Omega}$.*

(ii) If $u_0 \in C^\beta(\overline{\Omega}; \mathbb{R}^m)$, with $0 < \beta < 2$, $\beta \neq 1$, (and $\partial u_{0i}/\partial \nu = 0$ for $i \in I_2$, if $\beta > 1$), then u belongs to $C^{\beta/2,\beta}([0,\delta] \times \overline{\Omega}; \mathbb{R}^m)$.

(iii) If in addition the diffusion coefficients d_i and $\varphi(t, \cdot, u)$ belong to $C^{2\alpha}(\overline{\Omega})$, with $C^{2\alpha}$ norm locally independent of (t, u), and $\partial\Omega$ is uniformly $C^{2+2\alpha}$, then the derivatives $D_{ij}u$ are continuous in $]0,\delta] \times \overline{\Omega}$. More precisely, if $\alpha \neq 1/2$, u belongs to $C^{1+\alpha,2+2\alpha}([\varepsilon,\delta] \times \overline{\Omega}; \mathbb{R}^m)$ for every $\varepsilon \in]0,\delta[$.

(iv) u can be extended to a maximal time interval $I(u_0)$, which is relatively open in $[0,T]$. The mapping

$$\begin{cases} \{(t, u_0) : u_0 \in C(\overline{\Omega}), u_{0i|\partial\Omega} = 0, i \in I_1, t \in I(u_0)\} \mapsto C(\overline{\Omega}), \\ (t, u_0) \mapsto u(t; u_0) \end{cases}$$

is continuous. If u is bounded in $I(u_0) \times \overline{\Omega}$, then $I(u_0) = [0,T]$.

7.3.2 A general semilinear equation

Let Ω be a bounded open set in $\mathbb{R}^n$ with C^{2m} boundary $\partial\Omega$, and let $\mathcal{A}$, $\mathcal{B}_j$ ($j = 1, \ldots, m$) be differential operators of order $2m$, m_j respectively, satisfying (3.2.2), (3.2.3), (3.2.5), (3.2.6), (3.2.7). Let k be any integer in $[0, 2m-1]$, and consider the problem

$$\begin{cases} u_t = \mathcal{A}u + \varphi(t, x, u, Du, \ldots, D^k u), & t > 0, \ x \in \overline{\Omega}, \\ u(0, x) = u_0(x), & x \in \overline{\Omega}, \\ \mathcal{B}_j u = 0, & t > 0, \ x \in \partial\Omega, \ j = 1, \ldots, m. \end{cases} \qquad (7.3.4)$$

Here $D^h u$ denotes the (ordered) set of all space derivatives of u of order h. The function

$$(t, x, u, q) \mapsto \varphi(t, x, u, q)$$

is defined in $[0,T] \times \overline{\Omega} \times \Lambda$, where Λ is an open set in $\mathbb{R}^{1+n+\cdots+n^k}$, and has values in $\mathbb{R}$. We assume, as usual, that φ is continuous and satisfies a Hölder condition with respect to t, a Lipschitz condition with respect to (u, q). Precisely, we assume that there exists $\theta \in]0,1[$ such that for every $(\overline{u}, \overline{q}) \in \Lambda$ there are $r > 0$, $K > 0$ satisfying

$$|\varphi(t, x, u, p) - \varphi(s, x, v, q)| \leq K((t-s)^\theta + |u - v| + |p - q|), \qquad (7.3.5)$$

for $0 \leq s < t \leq T$, $(u, p), (v, q) \in B((\overline{u}, \overline{q}), r) \subset \mathbb{R}^{1+n+\cdots+n^k}$. The initial datum u_0 belongs to $C^k(\overline{\Omega})$, and it satisfies the boundary conditions

$$\mathcal{B}_j u_0 = 0, \ x \in \partial\Omega, \ m_j \leq k.$$

According to the notation of Section 3.2, this means that $u_0 \in C^k_\mathcal{B}(\overline{\Omega})$. Moreover, we assume that the range of $(u_0, Du_0, \ldots, D^k u_0)$ is contained in Λ.

3. SOME EXAMPLES

Proposition 7.3.3 *Let the above assumptions hold. Then:*

(i) *There is $\delta > 0$ such that problem (7.3.4) has a unique solution $u : [0, \delta] \times \overline{\Omega} \mapsto \mathbb{R}$, such that u and all the space derivatives $D^\beta u$, with $|\beta| \leq k$, are continuous in $[0, \delta] \times \overline{\Omega}$, and u_t, $\mathcal{A}u$, $D^\beta u$ with $k < |\beta| \leq 2m - 1$ are continuous in $]0, \delta] \times \overline{\Omega}$.*

(ii) *u can be extended to a maximally defined solution $u(t, x; u_0) : I(u_0) \times \overline{\Omega} \mapsto \mathbb{R}$, $I(u_0)$ being relatively open in $[0, T]$. The mapping*

$$\{(t, u_0) : u_0 \in C_{\mathcal{B}}^k(\overline{\Omega}), \ t \in I(u_0)\} \mapsto C_{\mathcal{B}}^k(\overline{\Omega}), \quad (t, u_0) \mapsto u(t, \cdot; u_0)$$

is continuous.

(iii) *If in addition the boundary $\partial\Omega$ is uniformly $C^{2m+2m\theta}$, the coefficients a_β belong to $C^{2m\theta}(\overline{\Omega})$, the coefficients $b_{j\beta}$ belong to $C^{2m-m_j+2m\theta}(\overline{\Omega})$, and the function $\varphi(t, \cdot, u, q)$ belongs to $C^{2m\theta}(\overline{\Omega})$, with Hölder norm locally independent of t, u, and q, then the derivatives $D^\beta u$, with $|\beta| = 2m$, exist continuous for $t > 0$; more precisely, if $2m\theta$ is not integer, $u \in C^{1+\theta, 2m+2m\theta}([a, b] \times \overline{\Omega})$ for $0 < a < b < \sup I(u_0)$.*

Proof — We choose $X = C(\overline{\Omega})$. The realization $A : D(A) \mapsto X$ of the operator $\mathcal{A}$ in X, with domain

$$D(A) = \{u \in W^{2m,p}(\Omega) \ \forall p \geq 1 : \mathcal{A}u \in C(\overline{\Omega}), \ \mathcal{B}_j u|_{\partial\Omega} = 0, \ j = 1, \ldots, m\}$$

is sectorial in X thanks to Theorem 3.2.5 (for the second order case, see Corollaries 3.1.21(ii), 3.1.24(ii)).

If $k = 0$, we choose $\alpha = 0$, $X_0 = X$. If $k > 0$, we choose $\alpha = k/2m$, and $X_\alpha = C_{\mathcal{B}}^k(\overline{\Omega})$. In both cases, $\mathcal{O}$ is the open set in X_α consisting of all the functions such that the range of $(u, \ldots, D^k u)$ is contained in Λ. The function

$$f : [0, T] \times \mathcal{O} \mapsto X, \quad f(t, u)(x) = \varphi(t, x, u(x), \ldots, D^k u(x))$$

is continuous, and satisfies assumption (7.1.12).

Let us check that, for $\alpha > 0$, the space $X_\alpha = C_{\mathcal{B}}^k(\overline{\Omega})$ satisfies (7.0.2). The embedding $D_A(\alpha/2m, 1) \subset C_{\mathcal{B}}^k(\overline{\Omega})$ holds thanks to Proposition 3.2.7, and the embedding $C_{\mathcal{B}}^k(\overline{\Omega}) \subset D_A(\alpha/2m, \infty)$ is true thanks to Theorem 2.4 of [1]. So, (7.0.2)(i) holds. Moreover, part of A in $C_{\mathcal{B}}^k(\overline{\Omega})$ is sectorial, thanks to [55], so that (7.0.2)(ii) holds. Therefore, (7.0.2) is satisfied.

In the case of second order equations, one does not need the results of Section 3.2, but it is sufficient to use Proposition 3.1.28, which states that $C_{\mathcal{B}}^1(\overline{\Omega})$ belongs to $J_{1/2}(X, D(A))$, Theorems 3.1.29, 3.1.30, which imply that $C_{\mathcal{B}}^1(\overline{\Omega})$ belongs to $K_{1/2}(X, D(A))$, and Theorems 3.1.25, 3.1.26, which imply that the part of A in $C_{\mathcal{B}}^k(\overline{\Omega})$ is sectorial in $C_{\mathcal{B}}^k(\overline{\Omega})$.

If either $k > 0$, or $k = 0$ and all the boundary differential operators $\mathcal{B}_j$ have order > 0, then $D(A)$ is dense in X_α, and hence $u_0 \in \overline{D(A)}^\alpha$. If one of the operators $\mathcal{B}_j$ has order 0 (that is, the solution must vanish on the boundary), then

$$\overline{D(A)} = \{u \in C(\overline{\Omega}) : u_{|\partial\Omega} = 0\},$$

so that, even if $k = 0$, $u_0 \in \overline{D(A)}$. In any case, the assumptions of Theorem 7.1.3(i) are satisfied, so that problem (7.0.1) has a unique local solution $u \in C([0,\delta]; C_\mathcal{B}^k(\overline{\Omega}))$, which belongs to $C^\theta([\varepsilon,\delta]; D(A)) \cap C^{\theta+1}([\varepsilon,\delta]; X)$ for every $\varepsilon \in \,]0,\delta[$, thanks to Proposition 7.1.10. It follows that the function

$$u(t, x; u_0) = u(t; u_0)(x), \quad 0 \le t \le \delta, \ x \in \overline{\Omega},$$

is a solution of (7.3.4), and it enjoys the regularity properties stated in (i).

Statement (ii) follows from Proposition 7.1.9(i). Since we have considered initial data belonging to $\overline{D(A)}^\alpha$, for every u_0 the function $u(\cdot; u_0)$ is continuous up to $t = 0$ with values in X_α.

Let us prove (iii). Theorem 3.2.6 states that if $0 < \theta < 1$ and $2m\theta$ is not integer then

$$D_A(\theta, \infty) = C_\mathcal{B}^{2m\theta}(\overline{\Omega}).$$

Since f satisfies the assumptions of Proposition 7.1.10, then for $0 < a < b < \tau(u_0)$, $t \mapsto u_t(t, \cdot)$ is bounded in $[a, b]$ with values in $D_A(\theta, \infty)$. This implies that $\sup_{a \le t \le b} \|u_t(t, \cdot)\|_{C^{2m\theta}(\overline{\Omega})} < \infty$. Since $u(t, \cdot)$ is bounded in $[a, b]$ with values in $D(A) \subset C^\beta(\overline{\Omega})$ for every $\beta < 2m$, then $t \mapsto \varphi(t, \cdot, u(t, \cdot), \ldots, D^k u(t, \cdot))$ is bounded with values in $C^\beta(\overline{\Omega})$, for every $\beta < 2m - k$, $\beta \le 2m\theta$. From the differential equation in (7.3.4) we get $\mathcal{A}u(t, \cdot) \in C^\beta(\overline{\Omega})$ for every $t > 0$, and $\sup_{a \le t \le b} \|\mathcal{A}u(t, \cdot)\|_{C^\beta(\overline{\Omega})} < \infty$. From the Schauder's estimates for higher order elliptic equations ([13]) it follows that $u(t, \cdot) \in C^{2m+\beta}(\overline{\Omega})$ for every $t > 0$, and $\sup_{a \le t \le b} \|u(t, \cdot)\|_{C^{2m+\beta}(\overline{\Omega})} < \infty$. From a bootstrap argument similar to the one of Proposition 7.1.10 we get then $u(t, \cdot) \in C^{2m+2m\theta}(\overline{\Omega})$ for every $t > 0$, and $\sup_{a \le t \le b} \|u(t, \cdot)\|_{C^{2m+2m\theta}(\overline{\Omega})} < \infty$. Moreover, thanks to Proposition 7.1.10, $u_t(\cdot, x)$ is θ-Hölder continuous in $[a, b]$, uniformly with respect to x. From an interpolation result proved in [156, Th. 2.2] it follows that $u \in C^{1+\theta, 2m+2m\theta}([a,b] \times (\overline{\Omega}))$. ∎

We consider now different growth and qualitative assumptions on φ, in order to study existence in the large of the solution. We write $q \in \mathbb{R}^{n + \ldots + n^k}$ as $q = (q_1, q_2, \ldots, q_k)$, with $q_j \in \mathbb{R}^{n^j}$, $j = 1, \ldots, k$. Consider the assumptions

$$|\varphi(t, x, u, q)| \le C(1 + |u| + |q|) \tag{7.3.6}$$

$$\varphi(t, x, u, q) \le \mu(|u|)\left(1 + \sum_{j=1}^k |q_j|^{(2m-\varepsilon)/j}\right) \tag{7.3.7}$$

3. SOME EXAMPLES

$$\begin{cases} (i) & |\varphi(t,x,u,q) - \varphi(t,x,v,q)| \\ & \leq \mu_1(|u|+|v|)\left(1 + \sum_{j=1}^k |q_j|^{2m/j}\right)|u-v|, \\ (ii) & |\varphi(t,x,u,q_1,\ldots,q_j,\ldots q_k) - \varphi(t,x,u,q_1,\ldots,\widetilde{q}_j,\ldots q_k)| \\ & \leq \mu_2(|u|)\left(1 + |q_j|^{2m/j-1} + |\widetilde{q}_j|^{2m/j-1}\right)|q_j - \widetilde{q}_j|, \quad j=1,\ldots,k, \\ (iii) & |\varphi(t,x,u,q) - \varphi(s,x,u,q)| \leq \\ & \leq \mu_3(|u|)\left(1 + \sum_{j=1}^k |q_j|^{2m/j}\right)(t-s)^\theta, \end{cases} \quad (7.3.8)$$

where the functions μ, μ_1, μ_2, μ_3 are increasing.

Let us make some comments on assumptions (7.3.6), (7.3.7), (7.3.8). (7.3.6) states that φ grows not more than linearly with respect to (u,q). (7.3.7) means that φ may grow arbitrarily fast with respect to u (which is immaterial once one knows that u is bounded), and subcritically with respect to q. (7.3.8) are standard critical growth conditions.

Proposition 7.3.4 *Let Ω, $\mathcal{A}$, $\mathcal{B}_j$ satisfy the assumptions stated at the beginning of the subsection, and let $\varphi : [0,T] \times \mathbb{R}^{1+n+\ldots n^k} \mapsto \mathbb{R}$ satisfy (7.3.5). For every $u_0 \in C_\mathcal{B}^k(\overline{\Omega})$ let $u : I(u_0) \times \overline{\Omega} \mapsto \mathbb{R}$ be the maximally defined solution of (7.3.4). Assume that one of the following conditions holds.*

(i) φ satisfies (7.3.6);

(ii) φ satisfies (7.3.7) and u is bounded in $I(x_0) \times \overline{\Omega}$;

(iii) φ satisfies (7.3.8) and u belongs to $C^{\sigma,0}(I(x_0) \times \overline{\Omega})$, i.e.

$$\sup_{t \in I(u_0),\, x \in \overline{\Omega}} |u(t,x)| + \sup_{s,t \in I(u_0),\, x \in \overline{\Omega}} \frac{|u(t,x) - u(s,x)|}{|t-s|^\sigma} < +\infty,$$

for some $\sigma > 0$.

Then $I(u_0) = [0,T]$.

Proof — Thanks to Proposition 7.1.8, u exists in the large provided

$$\|u(t,\cdot)\|_{C^k(\overline{\Omega})} \leq C, \quad (7.3.9)$$

with C independent of $t \in I(u_0)$.

(i) If φ satisfies (7.3.6), then f satisfies (7.2.1). Estimate (7.3.9) follows from Proposition 7.2.1.

(ii) If φ satisfies (7.3.7), then f satisfies (7.2.2), with $\gamma = (2m - \varepsilon)/k$, thanks to the interpolatory estimates

$$\|u\|_{C^j(\overline{\Omega})} \leq c_j \|u\|_\infty^{(1-j/k)} \|u\|_{C^k(\overline{\Omega})}^{j/k}, \quad (7.3.10)$$

which follow from Proposition 1.1.2(iii). If u is bounded, estimate (7.3.9) follows from Proposition 7.2.2.

(iii) If φ satisfies (7.3.8), then f satisfies (7.2.7), thanks again to (7.3.10). Estimate (7.3.9) follows from Theorem 7.2.4. ∎

A priori estimates for the sup norm or the time Hölder seminorm of u are available if problem (7.1.21) has a special structure. An example of a fourth order equation for which it is possible to bound the sup norm of the solution is the Cahn-Hilliard equation, see Subsection 7.3.4. The maximum principle is of great help in the second order case.

Proposition 7.3.5 *Let $m = 1$, and let the boundary operator $\mathcal{B}_1 = \mathcal{B}$ be either the trace operator,*
$$\mathcal{B}u = u,$$
or an oblique boundary differential operator,
$$\mathcal{B}u = \beta_0(x)u + \sum_{j=1}^{n} \beta_j(x) D_i u, \qquad (7.3.11)$$

with $\beta_j \in C^1(\overline{\Omega})$ for $j = 0, \ldots, n$, and either $\sum_{j=1}^{n} \beta_j(x)\nu_i(x) > 0$, $\beta_0(x) \geq 0$ for every $x \in \partial\Omega$, or $\sum_{j=1}^{n} \beta_j(x)\nu_i(x) < 0$, $\beta_0(x) \leq 0$ for every $x \in \partial\Omega$.
Let φ satisfy (7.3.5) with $k = 1$, and
$$\varphi(t, x, u, 0) u \leq C(1 + u^2), \quad 0 \leq t \leq T,\ x \in \overline{\Omega},\ u \in \mathbb{R}. \qquad (7.3.12)$$

For $u_0 \in C^1_\mathcal{B}(\overline{\Omega})$, let u be the maximally defined solution of (7.1.21). Then
$$\sup_{t \in I(u_0),\, x \in \overline{\Omega}} |u(t,x)| < +\infty.$$

If in addition φ is differentiable with respect to u, and
$$\begin{cases} (i) & \varphi_u(t, x, u, p) \leq 0, \\ (ii) & |\varphi(t, x, u, p) - \varphi(s, x, u, p)| \leq \mu_4(|u|)(t-s)^\theta, \end{cases} \qquad (7.3.13)$$

then for $\sigma < 1/2$, $\sigma \leq \theta$ it holds
$$\|u\|_{C^{\sigma,0}(I(u_0) \times \overline{\Omega})} < +\infty.$$

Sketch of the proof — The classical maximum principle (see e.g. [124, p. 23]) may be adapted to our case thanks to Proposition 3.1.10, and imply that the first statement holds. Concerning the second statement, (7.3.13)(i) yields (7.3.12), so that u is bounded. For $\sigma < 1/2$, $\sigma \leq \theta$, the estimate on the σ-Hölder seminorm of u with respect to time seminorm follows by using the maximum principle in the equation satisfied by $v(t,x) = u(t+h,x) - u(t,x)$. We get $\|v\|_\infty \leq \|v(0,\cdot)\|_\infty = \|u(h,\cdot) - u_0\|_\infty$. On the other hand, $\|u(h,\cdot) - u_0\|_\infty$ const. h^σ for h small, since $t \mapsto u(t,\cdot) \in C^{1/2}([0,\delta]; C(\overline{\Omega}))$ for δ small thanks to Theorem 7.1.3(i). ∎

We have considered so far problems with nonlinearities φ depending on the space derivatives of the solution up to the order k, and initial datum u_0 in $C^k(\overline{\Omega})$. In the case where φ satisfies certain growth or qualitative conditions, it is possible to consider less regular initial data. However, we do not develop here the most general theory. See next subsections for two examples.

7.3.3 Second order equations with nonlinearities in divergence form

Consider the initial value problem

$$\begin{cases} u_t = \mathcal{A}u + \sum_{i=1}^n D_i(\varphi_i(u)), & t > 0, \ x \in \overline{\Omega}, \\ u(0,x) = u_0(x), & x \in \overline{\Omega}, \\ \mathcal{B}u(t,x) = 0, & t > 0, \ x \in \partial\Omega, \end{cases} \quad (7.3.14)$$

where Ω is a bounded open set in $\mathbb{R}^n$ with C^2 boundary, $\mathcal{A}$ is an elliptic second order differential operator with continuous coefficients, and the functions $\varphi_i : \mathbb{R} \mapsto \mathbb{R}$, $i = 1, \ldots, n$, are twice continuously differentiable. $\mathcal{B}$ is either the trace operator, or the oblique boundary differential operator defined in (7.3.11), with coefficients satisfying the assumptions of Proposition 7.3.5. By Proposition 7.3.3, for every $u_0 \in C^1_{\mathcal{B}}(\overline{\Omega})$ problem (7.3.14) has a unique local regular solution. However, it is possible to get a global solution even if u_0 is merely continuous.

Proposition 7.3.6 *Let the above assumptions hold. Then for every $u_0 \in C(\overline{\Omega})$ (satisfying the compatibility condition $u_{0|\partial\Omega} = 0$ if $\mathcal{B}$ is the trace operator) problem (7.3.14) has a unique solution $u = u(t,x;u_0) \in C([0,+\infty[\times\overline{\Omega})$, such that $D_i u$, u_t, $\mathcal{A}u$ are continuous in $]0,+\infty[\times\overline{\Omega}$, and $t^{1/2} D_i u$ is bounded near $t = 0$ for every $i = 1, \ldots, n$. The mapping $[0,+\infty[\times C(\overline{\Omega}) \mapsto C(\overline{\Omega})$ (respectively, $[0,+\infty[\times C_0(\overline{\Omega}) \mapsto C_0(\overline{\Omega})$, if $\mathcal{B}$ is the trace operator), $(t, u_0) \mapsto u(t,\cdot;u_0)$, is continuous.*

Proof — We set as before $X = C(\overline{\Omega})$, $\alpha = 1/2$, $D(A) = \{u \in W^{2,p}(\Omega) \ \forall p \geq 1 : \mathcal{A}u \in C(\overline{\Omega}), \mathcal{B}u_{|\partial\Omega} = 0\}$, $X_{1/2} = C^1_{\mathcal{B}}(\overline{\Omega})$, and

$$f : X_{1/2} \mapsto X, \quad f(u) = \sum_{i=1}^n D_i(\varphi_i(u)) = \sum_{i=1}^n \varphi'_i(u) D_i u.$$

For every $u, v \in X_{1/2}$ such that $\|u\|_\infty \leq R$, $\|v\|_\infty \leq R$, it holds

$$\|f(u) - f(v)\|_\infty \leq \sum_{i=1}^n (\|\varphi'_i(u) - \varphi'_i(v)\|_\infty \|D_i u\|_\infty + \|\varphi'_i(v)\|_\infty \|D_i u - D_i v\|_\infty)$$

$$\leq \sum_{i=1}^n \sup_{|\xi|\leq R} |\varphi''_i(\xi)| \|u - v\|_\infty \|u\|_{C^1} + \sum_{i=1}^n \sup_{|\xi|\leq R} |\varphi'_i(\xi)| \|u - v\|_{C^1},$$

so that f satisfies assumption (7.1.15), with $\gamma = 1$. By Theorem 7.1.5(i) and Proposition 7.1.10 it follows that for every $u_0 \in \overline{D(A)}$ there exists $\delta > 0$ such that problem (7.0.1) has a unique local solution $u \in C([0,\delta];X) \cap C^1(]0,\delta];X) \cap C(]0,\delta];D(A))$, such that $t \mapsto t^{1/2}u(t)$ is bounded with values in $X_{1/2}$. Moreover, $\overline{D(A)} = \{u \in C(\overline{\Omega}) : u_{|\partial\Omega} = 0\}$ in the case of the Dirichlet boundary condition, and $\overline{D(A)} = C(\overline{\Omega})$ in the case of the oblique boundary condition. It follows that the function $u(t,x) = u(t)(x)$ is the unique solution of (7.3.14) in $[0,\delta] \times \overline{\Omega}$ with the regularity properties specified above.

Let us prove global existence. By applying the maximum principle, which is possible thanks to Proposition 3.1.10, we get $|u(t,x)| \leq \|u_0\|_\infty$ for every t in the maximal interval of existence $I(u_0)$ and for every $x \in \overline{\Omega}$. So, the mapping $t \mapsto u(t,\cdot)$ is bounded with values in X in $I(u_0)$. Moreover, since f satisfies (7.1.15), then it satisfies (7.2.2), with $\gamma = 1$. From Proposition 7.2.2 it follows that $\|u(t,\cdot)\|_{X_{1/2}}$ remains bounded in every bounded interval contained in $I(u_0)$, and consequently $I(u_0) = [0,+\infty[$.

The last statement follows from Theorem 7.1.5. ∎

We remark that if the functions φ_i are defined in a (possibly unbounded) interval $]a,b[$, it is still possible to find a local solution, which enjoys the regularity properties stated above, and leaving unchanged the rest of the proof. Of course, one needs that the initial datum belongs to $\mathcal{U} = \{u \in C(\overline{\Omega}) : u(x) \in]a,b[\ \forall x \in \overline{\Omega}\}$. Concerning existence in the large, the above arguments still hold, provided the solution takes values far away from a and b.

7.3.4 The Cahn-Hilliard equation

$$\begin{cases} u_t = \Delta(-\Delta u + \varphi(u)), & t > 0,\ x \in \overline{\Omega}, \\ \dfrac{\partial u}{\partial n} = \dfrac{\partial \Delta u}{\partial n} = 0, & t > 0,\ x \in \partial\Omega, \\ u(0,x) = u_0(x), & x \in \overline{\Omega}. \end{cases} \quad (7.3.15)$$

Here Ω is a bounded open set in $\mathbb{R}^n$ with C^4 boundary $\partial\Omega$, $n = 1,2$, or 3, $\varphi :]a,b[\mapsto \mathbb{R}$ is a smooth function, and $u_0 : \overline{\Omega} \mapsto \mathbb{R}$ is a continuous function. The physically relevant cases are $\varphi(u) = u^3 - u$, with $]a,b[= \mathbb{R}$, and $\varphi(u) = \log[(1+u)/(1-u)]$, with $]a,b[=]-1,1[$. See J.W. CAHN – J. E. HILLIARD [43, 44], C.M. ELLIOT – S. LUCKHAUS [80], A. DEBUSSCHE – L. DETTORI [70].

The nonlinearity in the right hand side of (7.3.5) may be rewritten as $\varphi'(u)\Delta u + \varphi''(u)|Du|^2$, and it is well defined if u is C^2 with respect to the space variables. Following the general procedure of Subsection 7.3.2, we set

$$X = C(\overline{\Omega}),\ X_{1/2} = C^2_{\partial/\partial\nu}(\overline{\Omega}) = \{u \in C^2(\overline{\Omega}) : \partial u/\partial \nu = 0\},$$
$$D(A) = \{u \in \bigcap_{p \geq 1} W^{4,p}(\Omega) : \Delta^2 u \in C(\overline{\Omega}),\ \partial u/\partial\nu = \partial \Delta u/\partial\nu = 0\},$$
$$Au = -\Delta^2 u,$$

3. Some examples

and
$$\mathcal{U} = \{u \in X : u(\xi) \in \,]a,b[\ \forall \xi \in \overline{\Omega}\}.$$

Note that $D(A)$ is dense in X, and that the operator A has a very special structure. Indeed, $A = -\Delta^2$, where $D(\Delta) = \{u \in W^{2,p}(\Omega) \ \forall p \geq 1 : \Delta u \in C(\overline{\Omega}), \ \partial u/\partial n = 0\}$. A is sectorial in X thanks to Proposition 2.4.4 and Corollary 3.1.24(ii).

Proposition 7.3.3 implies that for every $u_0 \in C^2_{\partial/\partial\nu}(\overline{\Omega}) \cap \mathcal{U}$ problem (7.3.15) has a unique maximally defined solution $u : [0, \tau[\times\overline{\Omega} \mapsto \mathbb{R}$, such that u and all the spatial derivatives $D^\beta u$, with $|\beta| \leq 2$, are continuous in $[0, \tau[\times\overline{\Omega}$, and u_t, $D^\beta u$ with $|\beta| = 3$, $\Delta^2 u$ are continuous in $]0, \tau[\times\overline{\Omega}$. However, using the results of Theorem 7.1.6, it is possible to get a solution of (7.3.15) even if u_0 is merely continuous.

Proposition 7.3.7 *Under the above assumptions, for every $u_0 \in C(\overline{\Omega})$ such that $u_0(\xi) \in \,]a,b[$ for every $\xi \in \overline{\Omega}$, problem (7.3.15) has a unique maximally defined solution $u : [0, \tau[\times\overline{\Omega} \mapsto \mathbb{R}$, such that u is continuous in $[0, \tau[\times\overline{\Omega}$, u_t, $D^\beta u$ with $1 \leq |\beta| \leq 3$, $\Delta^2 u$ are continuous in $]0, \tau[\times\overline{\Omega}$, and $t^{1/2}D^\beta u$ is bounded near $t = 0$ for $|\beta| = 2$. Consequently, $t^{1/4}D^\beta u$ is bounded near $t = 0$ for $|\beta| = 1$.*

Proof — The statements are easy consequences of Theorem 7.1.6(i) and Proposition 7.1.10, provided we show that the function
$$f : C^2_{\partial/\partial\nu}(\overline{\Omega}) \cap \mathcal{U} \mapsto C(\overline{\Omega}), \quad f(u) = \Delta\varphi(u),$$

satisfies assumption (7.1.15), with $\alpha = 1/2$, $X_\alpha = C^2_{\partial/\partial\nu}(\overline{\Omega})$, $\gamma = 1$.

It is clear that the first addendum $u \mapsto \varphi'(u)\Delta u$ satisfies (7.1.15) with $\gamma = 1$. Concerning the second addendum, for $u, v \in C^2(\overline{\Omega}) \cap \mathcal{U}$ it holds

$$\|\varphi''(u)|Du|^2 - \varphi''(v)|Dv|^2\|_\infty \leq \|\varphi''(u) - \varphi''(v)\|_\infty \|Du\|_\infty^2$$
$$+\|\varphi''(v)\|_\infty \|\langle D(u-v), D(u+v)\rangle\|_\infty$$
$$\leq \sup_{\xi \in K} |\varphi'''(\xi)| \|u-v\|_\infty \|Du\|_\infty^2$$
$$+ \sup_{\xi \in K} |\varphi''(\xi)| \|D(u-v)\|_\infty (\|Du\|_\infty + \|Dv\|_\infty),$$

where $K = \{\xi \in \,]a,b[: \xi = \theta u(x) + (1-\theta)v(x), \ 0 \leq \theta \leq 1, \ x \in \overline{\Omega}\}$. Taking into account the interpolatory estimate

$$\|Dz\|_\infty \leq c\|z\|_\infty^{1/2}\|D^2z\|_\infty^{1/2}, \quad \forall z \in C^2(\overline{\Omega})$$

(see Proposition 1.1.2(iii)), we get

$$\|D(u-v)\|_\infty(\|Du\|_\infty + \|Dv\|_\infty) \leq c^2\|u-v\|_{C^2}^{1/2}\|u-v\|_\infty^{1/2} \cdot$$
$$\cdot(\|u\|_\infty^{1/2}\|u\|_{C^2}^{1/2} + \|v\|_\infty^{1/2}\|v\|_{C^2}^{1/2})$$
$$\leq c^2/2(\|u-v\|_{C^2} + 2\|u-v\|_\infty(\|u\|_\infty\|u\|_{C^2} + \|v\|_\infty\|v\|_{C^2})).$$

Replacing in the above estimate, (7.1.15) follows. ∎

Concerning existence in the large, since f satisfies (7.2.2) with $\gamma = 1$, then Propositions 7.2.2 and 7.1.8 imply that u exists globally if $\|u\|_\infty$ is bounded and u takes values far away from a and b.

A priori estimates for the sup norm of u are available via Sobolev embedding and $W^{k,p}$ estimates when φ is a polynomial of odd degree,

$$\varphi(\xi) = \sum_{k=0}^{2q-1} a_k \xi^k,$$

with $a_{2q-1} > 0$, or when φ has a nonnegative primitive function. This is rather simple if $n = 1, 2$. If $n = 3$, a priori estimates guaranteeing existence in the large follow from nontrivial interpolation estimates. See [196, p. 154-158] for the case where φ is a polynomial of degree 3, and [205] for the case where φ has a nonnegative primitive and $|\varphi(x)| \leq c(1 + |x|)^{5-\varepsilon}$, $\varepsilon > 0$.

7.4 Bibliographical remarks for Chapter 7

Abstract semilinear parabolic problems have been studied since many years. See e.g. H. FUJITA – T. KATO [86], A. PAZY [166, Sect. 6.3], H. KIELHÖFER [109], W. VON WAHL [206], E. SINESTRARI, P. VERNOLE [178], H. AMANN [17, 18]. We have followed here the approach of A. LUNARDI [151] which unifies the previous ones.

The case of not regular initial data has been considered also by W. VON WAHL [206], H. HOSHINO, Y. YAMADA [101], S. RANKIN [171]. The present results are due to A LUNARDI [151]. They are refinements of the ones of H. HOSHINO - Y. YAMADA [101], who proved a local existence theorem similar to 7.1.5 taking $X_\alpha = D(-A)^\alpha$ and initial data in $D(-A)^\beta$ with $0 \leq \beta < \alpha$.

Comprehensive treatments of reaction-diffusion systems, from different points of view, may be found in E. ROTHE [172], C. COSNER – J. HERNANDEZ – E. MITIDIERI [56].

The results of Proposition 7.3.3 are similar to those obtained by X. MORA in [160], who considered also certain systems. The regularity assumptions on the coefficients in [160] are stronger than the present ones.

The results of Subsection 7.3.3 are comparable to those of S. RANKIN [171], who used different abstract techniques and worked in L^p spaces. Precisely, he considered the case where $\varphi_i(\xi) = \xi|\xi|^{\gamma_i - 1}$, and found a global solution for every $u_0 \in L^{p\gamma}(\Omega)$, where $n < p < +\infty$ and $\gamma = \max\{\gamma_i : i = 1, \ldots, n\}$.

The geometric theory of semilinear parabolic equations has been mostly developed by D. HENRY [99]. See also R. TEMAM [196], J. SMOLLER [180], D. DANERS – P. KOCH MEDINA [61].

Chapter 8

Fully nonlinear equations

Let D be a Banach space endowed with the norm $\|\cdot\|_D$, continuosly embedded in X, and let $\mathcal{O}$ be an open set in D. We consider the initial value problem

$$u'(t) = F(t, u(t)), \quad t > 0; \quad u(0) = u_0, \qquad (8.0.1)$$

where $F : [0, T] \times \mathcal{O} \mapsto X$ is a sufficiently smooth function, $T \in \,]0, +\infty[$, and $x_0 \in \mathcal{O}$. The simplest example is a fully nonlinear one dimensional problem,

$$\begin{cases} u_t = f(u, u_\xi, u_{\xi\xi}), & t \geq 0, \ 0 \leq \xi \leq 1, \\ u(t, 0) = u(t, 1) = 0, & t \geq 0, \\ u(0, \xi) = u_0(\xi), & 0 \leq \xi \leq 1, \end{cases} \qquad (8.0.2)$$

with regular data f, u_0, in which case it is in general convenient to choose $X = C([0,1])$, $D = \{v \in C^2([0,1]) : v(0) = v(1) = 0\}$, and obviously $F(t,v)(\xi) = f(v(\xi), v'(\xi), v''(\xi))$. The key assumption on problem (8.0.1) is

$$\begin{cases} \text{for every } t \in [0, T] \text{ and } v \in \mathcal{O}, \text{ the Fréchet derivative } F_v(t, v) \text{ is} \\ \text{sectorial in } X, \text{ and its graph norm is equivalent to the norm of } D. \end{cases} \qquad (8.0.3)$$

In the case of example (8.0.2), the Fréchet derivative $F_v(t, v)$ is the linear operator

$$\hat{v} \mapsto f_u \hat{v} + f_p \hat{v}' + f_q \hat{v}'',$$

where the derivatives of f are evaluated at $(v(\xi), v'(\xi), v''(\xi))$. Thanks to the generation theorems of Chapter 3, assumption (8.0.3) is satisfied if the parabolicity condition $f_q(u, p, q) > 0$ holds for every (u, p, q).

To solve (at least locally) problem (8.0.1) we linearize it near u_0, setting

$$A = F_x(0, u_0); \quad G(t, u) = F(t, u) - Au, \ t \in [0, T], u \in \mathcal{O} \qquad (8.0.4)$$

and writing it as

$$u'(t) = Au(t) + G(t, u(t)), \quad t > 0; \quad u(0) = u_0. \tag{8.0.5}$$

Equation (8.0.5) looks like (7.0.1), and the notions of strict, classical, and mild solution can be given as in Definition 7.0.1. In particular, a strict solution in an interval $[0, a]$ is a function $u \in C([0, a]; D) \cap C^1([0, a]; X)$ satisfying (8.0.1) pointwise. In the case of example (8.0.2), with the above choice of X and D, a strict solution is a function u such that u, u_x, u_{xx}, u_t are continuous in $[0, a] \times [0, 1]$, which means that $u \in C^{1,2}([0, a] \times [0, 1])$.

The main difference between fully nonlinear and semilinear equations is the fact that the nonlinear perturbation G is "of the same order" of the linear part A, in the sense that $G(t, x)$ is defined only for x in $\mathcal{O} \subset D$.

While showing local existence and uniqueness in a semilinear equation is relatively easy, the same is not trivial at all in the fully nonlinear case. Assume, for instance, that one looks for a strict solution of (8.0.1). As easily seen, a necessary condition is

$$u_0 \in \mathcal{O}, \quad F(0, u_0) \in \overline{D}. \tag{8.0.6}$$

In the applications (8.0.6) is usually a compatibility condition at the boundary for the initial datum. In example (8.0.2), where $\overline{D} = C_0([0, 1]) = \{u \in C([0, 1]) : u(0) = u(1) = 0\}$, (8.0.6) means that

$$u_0 \in C^2([0, 1]), \quad u_0(i) = 0, \quad f(u_0(i), u_0'(i), u_0''(i)) = 0, \quad i = 0, 1, \tag{8.0.7}$$

which is a necessary condition in order that u_t be continuous up to $t = 0$.

Fixed a small $a > 0$, it is natural to look for a solution in the interval $[0, a]$ as a fixed point of the operator Γ, defined in a small ball $B(u_0, r) \subset C([0, a]; D)$, where u_0 is the constant function $u_0(t) \equiv u_0$, as

$$\Gamma v(t) = e^{tA} u_0 + \int_0^t e^{(t-s)A} G(s, v(s)) ds, \quad 0 \le t \le a. \tag{8.0.8}$$

Now, if $v \in C([0, a]; D)$, then $G(\cdot, v(\cdot)) \in C([0, a]; X)$, and there is no hope, in general, that $\Gamma v \in C([0, a]; D)$ (see counterexample 4.2.1). Therefore we have to replace $C([0, a]; D)$ by a subspace $H([0, a]; D)$ of $C([0, a]; D)$ such that, if $v \in H([0, a]; D)$, then $\Gamma v \in H([0, a]; D)$ (provided, possibly, that some compatibility condition on u_0 holds). Since the D-norm is equivalent to the graph norm of A, this means that we need to find a subspace $H([0, a]; X)$ of $C([0, a]; X)$ such that if both v and Av belong to $H([0, a]; X)$, then $\Gamma v(t) \in D$ for every t, and both Γv and $A\Gamma v$ belong to $H([0, a]; X)$. In other words we need that the space $H([0, a]; X)$ enjoys the optimal regularity property. We can choose among the optimal regularity results of Section 4.3, and for any choice we get different regularity properties of the solution up to $t = 0$ and different compatibility conditions on the initial datum. For instance, if we choose $H([0, a]; X) = C^\alpha([0, a]; X)$, with $0 < \alpha < 1$, the compatibility condition is $F(0, u_0) \in D_A(\alpha, \infty)$ (see Theorem 4.3.1(iii)); if

INTRODUCTION

we choose $H([0,a]; X) = C([0,a]; D_A(\alpha))$, the compatibility condition is $u_0 \in D_A(\alpha)$ (see Corollary 4.3.10), and we have to assume in addition that F is smooth enough from $[0,T] \times D_A(\alpha+1)$ to $D_A(\alpha)$. In the case of example (8.0.2),, for $\alpha \neq 1/2$ the condition $F(0, u_0) \in D_A(\alpha, \infty)$ means that $f(u_0, u'_0, u''_0) \in C_0^{2\alpha}([0,1])$ and $u_0 \in D_A(\alpha+1)$ means that $u_0 \in h_0^{2+2\alpha}([0,1])$, $f_q(u_0(i), u'_0(i), u''_0(i))u''_0(i) + f_p(u_0(i), u'_0(i), u''_0(i))u'_0(i) = 0$ for $i = 0, 1$.

If we do not want to assume that some additional compatibility condition on u_0 holds except (8.0.6), we can choose $H([0,a]; X) = C([0,a]; X) \cap C_\alpha^\alpha(]0,a]; X)$, for some $\alpha \in \,]0,1[$ (see Corollary 4.2.6). We shall see in Sections 8.1 and 8.4 that many choices of the space $H([0,a]; X)$ enjoying the optimal regularity property lead to different results of local existence, uniqueness, and regularity of the solution to (8.0.1).

As in the case of semilinear equations, it is possible to define a unique noncontinuable strict solution, defined on a maximal time interval. However, in the present case boundedness of the solution with values in D is not enough to guarantee existence in the large. We are able to show that the solution exists in the large only if it is uniformly continuous with values in D. In the applications, finding a bound on the modulus of continuity with values in D is often very difficult. The solution may be shown to be uniformly continuous with values in D if its range is contained in a relatively compact subset of D. The properties of the noncontinuable solution of (8.0.1) are discussed in Section 8.2.

In Section 8.3 we treat further regularity properties of the solution, such as C^k and C^∞ regularity, and analyticity. More precisely, we consider a family of equations depending on a parameter,

$$u'(t) = F(t, u(t), \lambda), \quad t \geq 0; \quad u(0) = u_0,$$

and we study the dependence of u on t, u_0, λ.

In Section 8.4 we consider the case where X is an interpolation space. Then many proofs may be significantly simplified.

In Section 8.5 we give a number of applications of the abstract theory and of its methods to fully nonlinear parabolic problems arising in several fields, such as Detonation Theory, Stochastic Control, Differential Geometry, Free Boundary Problems. We also consider general fully nonlinear second order parabolic equations with nonlinear boundary condition.

8.1 Local existence, uniqueness and regularity

We make the following regularity assumptions on the nonlinear function $F : [0,T] \times \mathcal{O} \mapsto X$, $\mathcal{O}$ being an open set in D.

$$\begin{cases} (t,u) \mapsto F(t,u) \text{ is continuous with respect to } (t,u), \text{ and it is Fréchet} \\ \text{differentiable with respect to } u. \text{ There exists } \alpha \in \,]0,1[\text{ such that for all} \\ \overline{u} \in \mathcal{O} \text{ there are } R = R(\overline{u}), L = L(\overline{u}), K = K(\overline{u}) > 0 \text{ verifying} \\ \qquad \|F_u(t,v) - F_u(t,w)\|_{L(D,X)} \leq L\|v - w\|_D, \\ \qquad \|F(t,u) - F(s,u)\| + \|F_u(t,u) - F_u(s,u)\|_{L(D,X)} \leq K|t-s|^{\alpha}, \\ \text{for all } t, \ s \in [0,T], \ u, \ v, \ w \in B(\overline{u}, R) \subset D. \end{cases}$$

(8.1.1)

We now state the main local existence theorem. It is convenient to choose an arbitrary initial time $t_0 \in [0,T[$, and to consider the initial value problem

$$u'(t) = F(t,u(t)), \ t_0 \leq t \leq t_0 + \delta, \quad u(t_0) = x_0, \qquad (8.1.2)$$

with $\delta \in \,]0, T-t_0]$ and $x_0 \in \mathcal{O}$. We shall find a local strict solution, that is, a solution $u \in C([t_0, t_0+\delta]; D) \cap C^1([t_0, t_0+\delta]; X)$, with δ small, which in addition belongs to the weighted Hölder space $C_\alpha^\alpha(]t_0, t_0+\delta]; D)$. See the definition in Section 4.3.

Theorem 8.1.1 *Let $\mathcal{O} \subset D$ be an open set. Let $F : [0,T] \times \mathcal{O} \mapsto X$ satisfy assumptions (8.0.3) and (8.1.1). Fix $\bar{t} \in [0,T]$ and $\overline{u} \in \mathcal{O}$ such that $F(\bar{t}, \overline{u}) \in \overline{D}$. Then there are $\delta = \delta(\bar{t}, \overline{u}) > 0$, $r = r(\bar{t}, \overline{u}) > 0$ such that*

(i) For every $t_0 \in [\bar{t} - r, \bar{t} + r] \cap [0,T]$, and $x_0 \in \mathcal{O}$ such that $F(t_0, x_0) \in \overline{D}$ and $\|x_0 - \overline{u}\| \leq r$, there is a strict solution $u \in C([t_0, t_0+\delta]; D) \cap C^1([t_0, t_0+\delta]; D)$ of (8.1.2) in $[t_0, t_0+\delta]$.

(ii) u belongs to $C_\alpha^\alpha(]t_0, t_0+\delta]; D)$, u' belongs to $B_\alpha(]t_0, t_0+\delta]; (X,D)_{\alpha,\infty}$, and in addition[1]

$$\lim_{\varepsilon \to 0} \varepsilon^\alpha [u]_{C^\alpha([t_0+\varepsilon, t_0+2\varepsilon]; D)} = 0. \qquad (8.1.3)$$

Moreover, u is the unique solution of (8.1.2) belonging to

$$\bigcup_{0 < \beta < 1} C_\beta^\beta(]t_0, t_0+\delta]; D) \cap C([t_0, t_0+\delta]; D).$$

Proof — We look for a solution of (8.1.2) belonging to the metric space

$$\begin{aligned} Y \ = \ & \{u \in C_\alpha^\alpha(]t_0, t_0+\delta]; D) \cap C([t_0, t_0+\delta]; D) : \\ & u(t_0) = x_0, \ \|u(\cdot) - \overline{u}\|_{C_\alpha^\alpha(]t_0, t_0+\delta]; D)} \leq \rho \}, \end{aligned} \qquad (8.1.4)$$

---
[1]This fact will be used later, and precisely in Theorem 8.3.4.

1. LOCAL EXISTENCE, UNIQUENESS AND REGULARITY

endowed with the distance of $C_\alpha^\alpha(]t_0, t_0+\delta]; D)$, where $\delta < T-t_0$ and $\rho \leq R(\overline{u})$ are positive numbers to be chosen later. Clearly, Y is a closed set in $C_\alpha^\alpha(]t_0, t_0+\delta]; D)$. Set moreover

$$A = F_u(\overline{t}, \overline{u}),$$

It is easy to see that, due to (8.1.1), for every $u \in Y$ the function $t \mapsto F(t, u(t)) - Au(t)$ belongs to $C_\alpha^\alpha(]t_0, t_0 + \delta]; X) \cap C([t_0, t_0 + \delta]; X)$.

Define a nonlinear operator Γ on Y, by $\Gamma(u) = v$, where v is the solution of

$$v'(t) = Av(t) + [F(t, u(t)) - Au(t)], \quad t_0 \leq t \leq t_0 + \delta; \quad v(t_0) = x_0, \quad (8.1.5)$$

By Theorem 4.3.5(iii), for every $u \in Y$, $\Gamma(u)$ belongs to $C_\alpha^\alpha(]t_0, t_0 + \delta]; D) \cap C([t_0, t_0+\delta]; D)$. The compatibility condition $Ax_0 + [F(t_0, u(t_0)) - Au(t_0)] \in \overline{D}$ is satisfied because $u(t_0) = x_0$, and $F(t_0, x_0) \in \overline{D}$. It is clear that a function $u \in Y$ is a solution of (8.1.2) if and only if it is a fixed point of Γ. We shall show that Γ is a contraction and maps Y into itself, provided δ, ρ, $|t_0 - \overline{t}|$, and $\|x_0 - \overline{u}\|_D$ are suitably small. Let C be the constant given by Corollary 4.3.6(ii), and let γ, $\gamma_\alpha \geq 1$ be such that

$$\begin{cases} \gamma^{-1}\|y\|_D \leq \|y\| + \|Ay\| \leq \gamma\|y\|_D, \quad \forall y \in D, \\ \gamma_\alpha^{-1}\|y\|_{(X,D)_{\alpha,\infty}} \leq \|y\|_{D_A(\alpha,\infty)} \leq \gamma_\alpha \|y\|_{(X,D)_{\alpha,\infty}}, \quad \forall y \in (X,D)_{\alpha,\infty}. \end{cases} \quad (8.1.6)$$

Obviously, C, γ, γ_α depend only on $\overline{t}$ and $\overline{x}$. For $v_1, v_2 \in Y$ we have[2]

$$\begin{aligned} \|\Gamma(v_1) - \Gamma(v_2)\|_{C_\alpha^\alpha(D)} &\leq \gamma \|\Gamma(v_1) - \Gamma(v_2)\|_{C_\alpha^\alpha(D(A))} \\ &\leq \gamma C \|F(\cdot, v_1(\cdot)) - F(\cdot, v_2(\cdot)) - A(v_1(\cdot) - v_2(\cdot))\|_{C_\alpha^\alpha(X)}. \end{aligned} \quad (8.1.7)$$

For each $t \in [t_0, t_0 + \delta]$ we have

$$F(t, v_1(t)) - F(t, v_2(t)) - A(v_1(t) - v_2(t))$$
$$= \int_0^1 [F_x(t, \sigma v_1(t) + (1-\sigma)v_2(t)) - A]d\sigma(v_1(t) - v_2(t)),$$

so that

$$\begin{aligned} \|F(t, v_1(t)) - F(t, v_2(t)) &- A(v_1(t) - v_2(t))\| \\ &\leq (K(|t_0 - \overline{t}| + \delta)^\alpha + L\rho)\|v_1(t) - v_2(t)\|_D, \end{aligned} \quad (8.1.8)$$

---

[2] If B is any Banach space, we write as usual $C(B)$, $C_\alpha^\alpha(B)$ for $C([0,\delta]; B)$, $C_\alpha^\alpha(]t_0, t_0+\delta]; B)$, respectively.

and for $t_0 + \varepsilon \leq s \leq t \leq t_0 + \delta$

$$\|F(t, v_1(t)) - F(t, v_2(t)) - F(s, v_1(s)) + F(s, v_2(s)) \\ - A(v_1(t) - v_2(t) - v_1(s) + v_2(s))\|$$

$$\leq \left\| \int_0^1 [F_x(t, \sigma v_1(t) + (1-\sigma)v_2(t)) \right.$$

$$\left. - F_x(s, \sigma v_1(s) + (1-\sigma)v_2(s))]d\sigma(v_1(t) - v_2(t)) \right\|,$$

$$+ \left\| \int_0^1 (F_x(t, \sigma v_1(t) + (1-\sigma)v_2(t)) - A)d\sigma(v_1(t) - v_2(t) - v_1(s) + v_2(s)) \right\|$$

$$\leq (K + L\varepsilon^{-\alpha}\rho)(t-s)^\alpha \|v_1 - v_2\|_{C(D)}$$

$$+ (K(t-\bar{t})^\alpha + L\rho)\varepsilon^{-\alpha}[v_1 - v_2]_{C_\alpha^\alpha(D)}(t-s)^\alpha, \tag{8.1.9}$$

so that

$$[F(\cdot, v_1(\cdot)) - F(\cdot, v_2(\cdot)) - A(v_1(\cdot) - v_2(\cdot))]_{C_\alpha^\alpha(D)} \\ \leq (K(|t_0 - \bar{t}| + \delta)^\alpha + L\rho)\|v_1 - v_2\|_{C_\alpha^\alpha(D)}. \tag{8.1.10}$$

Using now (8.1.7), we find

$$\|\Gamma(v_1) - \Gamma(v_2)\|_{C_\alpha^\alpha(D)} \leq 2\gamma C(K(|t_0 - \bar{t}| + \delta)^\alpha + L\rho)\|v_1 - v_2\|_{C_\alpha^\alpha(D)}.$$

Therefore, Γ is a $1/2$-contraction provided

$$\begin{cases} (a) & |t_0 - \bar{t}| + \delta \leq \delta_0 = (8\gamma CK)^{-\frac{1}{\alpha}} \\ (b) & \rho \leq \min\{R(\bar{u}), (8\gamma CL)^{-1}\}. \end{cases} \tag{8.1.11}$$

Let t_0, δ, ρ satisfy (8.1.11). We now show that Γ maps Y into itself if δ, ρ are sufficiently small, t_0 is sufficiently close to $\bar{t}$, and x_0 is sufficiently close to $\bar{u}$. If $v \in Y$, we have, denoting by u_0 the constant function $u_0(t) = x_0$, $\forall t \in [t_0, t_0 + \delta]$,

$$\|\Gamma(v) - \bar{u}\|_{C_\alpha^\alpha(D)} \leq \|\Gamma(v) - \Gamma(u_0)\|_{C_\alpha^\alpha(D)} + \|\Gamma(u_0) - \bar{u}\|_{C_\alpha^\alpha(D)} \\ \leq \tfrac{1}{2}\|v - u_0\|_{C_\alpha^\alpha(D)} + \|\Gamma(u_0) - u_0\|_{C_\alpha^\alpha(D)} + \|x_0 - \bar{u}\|_D. \tag{8.1.12}$$

Since $\Gamma(u_0) - u_0$ is the solution w of

$$w'(t) = Aw(t) + F(t, x_0), \quad t_0 \leq t \leq t_0 + \delta; \quad w(t_0) = 0, \tag{8.1.13}$$

then

$$\Gamma(u_0)(t) - u_0 = \int_{t_0}^t e^{(t-s)A}[F(s, x_0) - F(\bar{t}, \bar{u})]ds \\ + \int_{t_0}^t e^{(t-s)A} F(\bar{t}, \bar{u})ds = I_1 + I_2, \quad t_0 \leq t \leq t_0 + \delta. \tag{8.1.14}$$

Using Theorem 3.2.5 and (8.1.6) we get

$$\|I_1\|_{C_\alpha^\alpha(D)} \leq \|F(\cdot, \bar{u}) - F(\bar{t}, \bar{u})\|_{C^\alpha(X)} \leq \gamma C[2K(|t_0 - \bar{t}| + \delta)^\alpha + L\|x_0 - \bar{u}\|_D],$$

1. LOCAL EXISTENCE, UNIQUENESS AND REGULARITY 293

so that if
$$|t_0 - \bar{t}| + \delta \leq \delta_1 = (12\gamma KC)^{-1/\alpha}\rho^{1/\alpha},$$

then
$$\|I_1\|_{C_\alpha^\alpha(]t_0,t_0+\delta];D)} \leq \rho/6.$$

Moreover, since $F(\bar{t}, \bar{u}) \in \overline{D}$, then $\lim_{\delta \to 0} \|I_2\|_{C_\alpha^\alpha(]t_0,t_0+\delta];D)} = 0$. Indeed, one checks easily that for every $y \in X$ it holds $\|\int_{t_0}^t e^{(t-s)A} y\|_{C_\alpha^\alpha(]t_0,t_0+\delta];D)} \leq \text{const.}\|y\|$, with constant independent of t_0, δ, y, and that for every $y \in D$

$$\lim_{\delta \to 0} \left\| \int_{t_0}^t e^{(t-s)A} y \, ds \right\|_{C_\alpha^\alpha(]t_0,t_0+\delta];D)} = 0. \tag{8.1.15}$$

Therefore (8.1.15) holds for every $y \in \overline{D}$. It yields $\lim_{\delta \to 0} \|I_2\|_{C_\alpha^\alpha(]t_0,t_0+\delta];D)} = 0$, so that there is $\delta_2 > 0$ (depending on $\bar{t}$ and $\bar{u}$) such that, if $\delta \leq \delta_2$, then

$$\|I_2\|_{C_\alpha^\alpha(]t_0,t_0+\delta];D)} \leq \rho/6.$$

Summing up we find that if

$$|t_0 - \bar{t}| + \delta \leq \min\{\delta_0, \delta_1\}, \ \delta \leq \delta_2, \ \|x_0 - \bar{u}\|_D \leq \rho(6\gamma CL + 1)^{-1},$$

then
$$\|\Gamma(u_0) - \bar{u}\|_{C_\alpha^\alpha(]t_0,t_0+\delta];D)} \leq \rho/2,$$

and using also (8.1.12) we see that Γ maps Y into itself. Therefore there is a unique fixed point u of Γ in Y, which is a solution of (8.1.2). Statement (i) follows.

Let us prove that statement (ii) holds. The derivatives of the functions in the range of Γ belong to $B_\alpha(]t_0, t_0 + \varepsilon]; (X, D)_{\alpha,\infty})$ thanks to Corollary 4.3.6(ii). Since u is a fixed point of Γ, then u' is in $B_\alpha(]t_0, t_0 + \varepsilon]; (X, D)_{\alpha,\infty})$.

Concerning (8.1.3), set

$$Y' = \{v \in Y : \lim_{\varepsilon \to 0} [v]_{C_\alpha^\alpha(]t_0,t_0+\varepsilon];D)} = 0\}.$$

Y' is closed in $C_\alpha^\alpha(]t_0, t_0 + \delta]; D)$, and $\Gamma(Y') \subset Y'$ (the proof is the same used above to show that $\Gamma(Y) \subset Y$). The fixed point u of Γ belongs to Y', which means that u satisfies (8.1.3).

Let us prove uniqueness. If u, v are solutions of (8.1.2) belonging to $C([t_0, t_0 + \delta]; D) \cap C_\beta^\beta(]t_0, t_0 + \delta]; D)$ for some $\beta \in \,]0, 1[$, set

$$t_1 = \sup\{t \in [t_0, t_0 + \delta] : u_{|[t_0,t]} = v_{|[t_0,t]}\}. \tag{8.1.16}$$

Assume by contradiction that $t_1 < t_0 + \delta$. Then $u(t_1) = v(t_1)$, because both u and v are continuous. Setting $x_1 = u(t_1) = v(t_1)$, both u and v are solutions of

$$u'(t) = F(t, u(t)), \ t_1 \leq t \leq t_0 + \delta; \ u(t_1) = x_1, \tag{8.1.17}$$

where $x_1 \in \mathcal{O}$ and $F(t_1, x_1) = u'(t_1) \in \overline{D}$. Taking $\bar{t} = t_1$, $\bar{u} = x_1$, and replacing α by any $\gamma < \theta = \min\{\alpha, \beta\}$ in point (i), we obtain the existence of $\delta_1 > 0$ such that problem (8.1.17) has a unique strict solution in the set

$$\begin{aligned} Y_1 &= \{u \in C_\gamma^\gamma(]t_1, t_1 + \delta_1]; D) \cap C([t_1, t_1 + \delta_1]; D) : \\ &\quad u(t_1) = x_1, \; \|u(\cdot) - x_1\|_{C_\gamma^\gamma(]t_1, t_1+\delta_1]; D} \leq \rho_1 \}, \end{aligned}$$

provided ρ_1 and δ_1 are sufficiently small. Now, both $u_{|[t_1, t_1+\delta_1]}$ and $v_{|[t_1, t_1+\delta_1]}$ belong to Y_1 if δ_1 is small: in particular, $\|u(t) - x_1\|_D \leq \rho_1$, $\|v(t) - x_1\|_D \leq \rho_1$ for δ_1 small, because u and v are continuous and they assume the value x_1 at $t = t_1$; moreover for $t_0 < t_0 + \varepsilon \leq s \leq t \leq t_0 + \delta_1$ and for every $a \in]0, 1[$ it holds

$$\|u(t) - u(s)\|_D = \|u(t) - u(s)\|_D^a \|u(t) - u(s)\|_D^{1-a}$$
$$\leq (\sup_{t_0 \leq s \leq t \leq t_0 + \delta_1} \|u(t) - u(s)\|_D)^a ([u]_{C_\theta^\theta(]t_0, t_0+\delta]; D)})(t-s)^\theta \varepsilon^{-\theta})^{1-a}.$$

Choosing $a = 1 - \gamma/\theta$ we get

$$[u]_{C_\gamma^\gamma(]t_0, t_0+\delta_1]; D)} \leq (\sup_{t_0 \leq s \leq t \leq t_0 + \delta_1} \|u(t) - u(s)\|_D)^a ([u]_{C_\theta^\theta(]t_0, t_0+\delta]; D)})^{1-a},$$

and, similarly,

$$[v]_{C_\gamma^\gamma(]t_0, t_0+\delta_1]; D)} \leq (\sup_{t_0 \leq s \leq t \leq t_0 + \delta_1} \|v(t) - v(s)\|_D)^a ([v]_{C_\theta^\theta(]t_0, t_0+\delta]; D)})^{1-a}.$$

Since both u and v are continuous with values in D, we get

$$[u]_{C_\gamma^\gamma(]t_0, t_0+\delta_1]; D)} \leq \rho_1, \; [v]_{C_\gamma^\gamma(]t_0, t_0+\delta_1]; D)} \leq \rho_1,$$

provided δ_1 is small enough.

Therefore, $u_{|[t_1, t_1+\delta_1]} = v_{|[t_1, t_1+\delta_1]}$, but this contradicts the definition of t_1. Hence $u = v$, and statement (ii) follows. ∎

Corollary 8.1.2 *Under the assumptions of Theorem 8.1.1, for every $\bar{t} \in [0, T]$ and $\bar{u} \in \mathcal{O}$ there is $k = k(\bar{t}, \bar{u}) > 0$ such that for every $t_0 \in [\bar{t} - r, \bar{t} + r]$, and $x_0, x_1 \in \mathcal{O}$ with $\|x_i - \bar{u}\| \leq r$, $F(t_0, x_i) \in \overline{D}$, we have, denoting by u_i the solution of (8.1.2) with initial value x_i, $i = 0, 1$,*

$$\|u_0 - u_1\|_{C_\alpha^\alpha(]t_0, t_0+\delta]; D)} + \|u_0' - u_1'\|_{C_\alpha^\alpha(]t_0, t_0+\delta]; X)}$$
$$+ \sup\{(t-t_0)^\alpha \|u_0'(t) - u_1'(t)\|_{(X, D)_{\alpha, \infty}} : t_0 \leq t \leq t_0 + \delta\} \leq k\|x_1 - u_0\|_D.$$

Proof — We cannot use directly the Principle of Contractions depending on a parameter, because the metric space Y introduced in the proof of Theorem 8.1.1 depends on x_0. However, the argument here is similar. Using notation from the

1. LOCAL EXISTENCE, UNIQUENESS AND REGULARITY

proof of Theorem 8.1.1 and estimate (4.3.12), for $\|x_i - \overline{u}\|_D \leq \rho/6\gamma CL$, $F(t_0, x_i) \in \overline{D}$, $i = 0, 1$, we get

$$\|u_0 - u_1\|_{C_\alpha^\alpha(]t_0,t_0+\delta];D)} + \|u_0' - u_1'\|_{C_\alpha^\alpha(]t_0,t_0+\delta];X)} + \|u_0' - u_1'\|_{B_\alpha(]t_0,t_0+\delta];(X,D)_{\alpha,\infty})}$$
$$\leq \|e^{tA}(x_0 - x_1)\|_{C_\alpha^\alpha(]t_0,t_0+\delta];D)} + \tfrac{1}{2}\|u_0 - u_1\|_{C_\alpha^\alpha(]t_0,t_0+\delta];D)},$$

so that

$$\|u_0 - u_1\|_{C_\alpha^\alpha(]t_0,t_0+\delta];D)} + 2\|u_0' - u_1'\|_{C_\alpha^\alpha(]t_0,t_0+\delta];X)}$$
$$+2\gamma_\alpha \|u_0' - u_1'\|_{B_\alpha(]t_0,t_0+\delta];(X,D)_{\alpha,\infty})}$$
$$\leq 2\|e^{tA}(x_0 - x_1)\|_{C_\alpha^\alpha(]t_0,t_0+\delta];D)} \leq 2\gamma^2 C \|x_0 - x_1\|_D.$$

∎

It is possible to prove local existence of the solution even if the compatibility condition $F(t_0, x_0) \in \overline{D}$ does not hold, but in this case we need x_0 to be close enough to some $\overline{u}$ such that $F(t_0, \overline{u}) \in \overline{D}$. It is sufficient to repeat the proof of Theorem 8.1.1, replacing Y by the set

$$\{u \in C_\alpha^\alpha(]t_0, t_0 + \delta]; D) : \|u(\cdot) - \overline{u}\|_{C_\alpha^\alpha(]t_0,t_0+\delta];D)} \leq \rho\}.$$

This fact will be discussed again in Section 8.3.

If the initial datum x_0 is "more regular", in the sense that

$$F(t_0, x_0) \in (X, D)_{\alpha, \infty} \tag{8.1.18}$$

then the local solution of (8.1.1) is α-Hölder continuous with values in D up to $t = 0$, and the corresponding continuous dependence result holds, as the following theorem shows.

Theorem 8.1.3 *Let the assumptions of Theorem 8.1.1 be satisfied, and let in addition (8.1.18) hold. If $u : [t_0, t_0 + \delta] \mapsto D$ is the solution of (8.1.2) given by Theorem 8.1.1, then*

$$u \in C^\alpha([t_0, t_0 + \delta]; D) \cap C^{1+\alpha}([t_0, t_0 + \delta]; X), u' \in B([t_0, t_0 + \delta]; (X, D)_{\alpha, \infty}).$$

Moreover, for every $\overline{t} \in [0, T]$, $\overline{u} \in \mathcal{O}$ such that $F(\overline{t}, \overline{u}) \in (X, D)_{\alpha, \infty}$, there are $r_0 = r_0(\overline{t}, \overline{u}) > 0$, $k_0 = k_0(\overline{t}, \overline{u}) > 0$ with the following property: for every $t_0 \in [\overline{t} - r_0, \overline{t} + r_0]$, and $x_0, x_1 \in \mathcal{O}$ such that

$$\|x_i - \overline{u}\| \leq r_0, \ \|F(t_0, x_i) - F(\overline{t}, \overline{u})\|_{(X,D)_{\alpha,\infty}} \leq r_0, \ i = 0, 1,$$

we have, denoting by u_i, $i = 0, 1$, the solution of (8.1.2) with initial value x_i,

$$\|u_0 - u_1\|_{C^\alpha([t_0,t_0+\delta];D)} + \|u_0' - u_1'\|_{C^\alpha([t_0,t_0+\delta];X)} + \|u_0' - u_1'\|_{B([t_0,t_0+\delta];(X,D)_{\alpha,\infty})}$$
$$\leq k_0(\|x_1 - x_0\|_D + \|F(t_0, u_0) - F(t_0, x_1)\|_{(X,D)_{\alpha,\infty}}).$$
$$\tag{8.1.19}$$

Proof — It suffices to prove that the statement holds in an interval $[t_0, t_0 + \delta_0]$, with $\delta_0 > 0$ possibly smaller than δ: the final statement will follow from Theorem 8.1.1. Fix $\bar{t} \in [0, T]$, $\bar{u} \in \mathcal{O}$ such that $F(\bar{t}, \bar{u}) \in D_A(\alpha, \infty)$, and set

$$Y_0 = \{u \in C^\alpha([t_0, t_0 + \delta_0]; D) : u(t_0) = x_0,\ \|u(\cdot) - x_0\|_{C^\alpha([t_0,t_0+\delta_0];D)} \leq \rho_0\},$$

where δ_0 will be chosen later, and

$$\rho_0 = 2\gamma C(\|F(\bar{t}, \bar{u})\|_{D_A(\alpha,\infty)} + K(1 + T^\alpha) + \|F(\bar{t}, \bar{u})\|),$$

γ, C, K being the constants in (8.1.6), (4.3.3), (8.1.1), respectively. Note that, if

$$\rho_0 \delta_0^\alpha \leq R(\bar{u})/2,\quad \|u_0 - \bar{u}\| \leq R(\bar{u})/2,$$

then for every $v \in Y_0$

$$\|v(t) - \bar{u}\|_D \leq R(\bar{u}),$$

so that $v(t) \in \mathcal{O}$.

Define the operator Γ in Y_0, by $\Gamma(u) = v$, where v is the solution of (8.1.5) in the interval $[t_0, t_0 + \delta_0]$. As in the proof of Theorem 8.1.1, we show that Γ is a contraction on Y_0 and $\Gamma(Y_0) \subset Y_0$, provided δ_0, $|t_0 - \bar{t}|$, $\|u_0 - \bar{u}\|_D$, and $\|F(t_0, x_0) - F(\bar{t}, \bar{u})\|_{D_A(\alpha,\infty)}$ are sufficiently small. First of all, for every $u \in Y_0$, the function $t \mapsto F(t, u(t)) - Au(t)$ belongs to $C^\alpha([t_0, t_0 + \delta_0]; D)$, so that Γ maps Y_0 into $C^\alpha([t_0, t_0 + \delta_0]; D) \cap C^{1+\alpha}([t_0, t_0 + \delta_0]; X)$, thanks to Theorem 4.3.1(iii) and to the compatibility condition (8.1.18). For $v_1, v_2 \in Y_0$, estimate (8.1.7) holds with C_α^α replaced by C^α, where now the constant C is given by Theorem 4.3.1(iii). So we can argue as in the proof of estimates (8.1.8) and (8.1.9), which have to be replaced by

$$\|F(t, v_1(t)) - F(t, v_2(t)) - A(v_1(t) - v_2(t))\|$$
$$\leq [K(t - \bar{t})^\alpha + \tfrac{1}{2}L(\|v_1(t) - \bar{u}\|_D + \|v_2(t) - \bar{u}\|)]\|v_1(t) - v_2(t)\|_D \quad (8.1.20)$$
$$\leq [K(|t_0 - \bar{t}| + \delta_0)^\alpha + L(\rho_0 \delta_0^\alpha + \|u_0 - \bar{u}\|_D)]\|v_1 - v_2\|_{C(D)}$$

and

$$\|F(t, v_1(t)) - F(t, v_2(t)) - F(s, v_1(s)) + F(s, v_2(s))$$
$$- A(v_1(t) - v_2(t) - v_1(s) + v_2(s))\|_D \leq (K + L\rho_0)(t - s)^\alpha \|v_1(t) - v_2(t)\|_D$$
$$+ [(K(s - \bar{t})^\alpha + \tfrac{1}{2}L(\|v_1(t) - \bar{u}\|_D + \|v_2(t) - \bar{u}\|_D)](t - s)^\alpha [v_1 - v_2]_{C^\alpha(D)}$$

$$\leq (K + L\rho_0)(t - s)^\alpha [v_1 - v_2]_{C^\alpha(D)} \delta_0^\alpha$$
$$+ [K(|t_0 - \bar{t}|) + \delta_0)^\alpha + L(\delta_0^\alpha \rho_0 + \|x_0 - \bar{u}\|_D)](t - s)^\alpha [v_1 - v_2]_{C^\alpha(D)}, \quad (8.1.21)$$

where we have used the fact that $v_1(t_0) = v_2(t_0) = x_0$. Summing up we find

$$\|F(\cdot, v_1(\cdot)) - F(\cdot, v_2(\cdot)) - A(v_1(\cdot) - v_2(\cdot))\|_{C^\alpha(X)}$$
$$\leq [2(K(|t_0 - \bar{t}| + \delta_0)^\alpha + 2L\delta_0^\alpha \rho_0 + L\|x_0 - \bar{u}\|_D)]\|v_1 - v_2\|_{C^\alpha(D)}.$$

1. Local existence, uniqueness and regularity

Then

$$\|\Gamma v_1 - \Gamma v_2\|_{C^\alpha(D)}$$
$$\leq \gamma C(2K(|t_0 - \bar{t}| + \delta_0)^\alpha + 2L\delta_0^\alpha \rho_0 + L\|x_0 - \bar{u}\|_D)\|v_1 - v_2\|_{C^\alpha(D)},$$

so that Γ is a $1/2$-contraction provided δ_0, $|t_0 - \bar{t}|$, and $\|u_0 - \bar{u}\|_D$ are so small that

$$\begin{cases} (a) & (|t_0 - \bar{t}| + \delta_0)^\alpha \leq (12\gamma CK)^{-1} \\ (b) & \delta_0^\alpha \leq (12CL\rho_0)^{-1} \\ (c) & \|x_0 - \bar{u}\|_D \leq r_1 = \min\{(6\gamma CL)^{-1}, R(\bar{u})/2\} \end{cases} \quad (8.1.22)$$

If (8.1.22) holds, for every $v \in Y_0$ we have

$$\|\Gamma v(\cdot) - x_0\|_{C^\alpha(D)} \leq \frac{1}{2}\|v(\cdot) - u_0\|_{C^\alpha(D)} + \|\Gamma u_0 - u_0\|_{C^\alpha(D)}, \quad (8.1.23)$$

where u_0 is the constant function $u_0(t) \equiv x_0$. Since $\Gamma u_0 - u_0$ is the solution w of

$$w'(t) = Aw(t) + F(t, x_0), \quad t_0 \leq t \leq t_0 + \delta_0; \quad w(t_0) = 0,$$

then, again by Theorem 3.2.3,

$$\|\Gamma u_0 - u_0\|_{C^\alpha(D)} \leq \gamma C(\|F(t_0, x_0)\|_{D_A(\alpha,\infty)} + \|F(\cdot, x_0)\|_{C^\alpha(X)})$$
$$\leq \gamma C(\|F(\bar{t}, \bar{u})\|_{D_A(\alpha,\infty)} + \|F(t_0, x_0) - F(\bar{t}, \bar{u})\|_{D_A(\alpha,\infty)}$$
$$+ K(1 + T^\alpha) + L\|x_0 - \bar{u}\|_D + \|F(\bar{t}, \bar{u})\|)$$
$$\leq \tfrac{1}{2}\rho_0 + \gamma C(\gamma_\alpha \|F(t_0, x_0) - F(\bar{t}, \bar{u})\|_{(X,D)_{\alpha,\infty}} + L\|x_0 - \bar{u}\|_D)$$

Therefore, if

$$\|x_0 - \bar{u}\|_D \leq r_0, \quad \|F(t_0, x_0) - F(\bar{t}, \bar{u})\|_{(X,D)_{\alpha,\infty}} \leq r_0,$$

where

$$r_0 = \min\{r_1, [2\rho_0 \gamma C(\gamma_\alpha + L)]^{-1}\}, \quad (8.1.24)$$

then Γ maps Y_0 into itself. Since Γ is a $1/2$-contraction, then it has a unique fixed point u in Y_0, which, due to the uniqueness part of Theorem 8.1.1, coincides with $u(\cdot, t_0, x_0)_{|[t_0, t_0+\delta_0]}$. Consequently, $u(\cdot, t_0, x_0)$ is α-Hölder continuous up to $t = t_0$, and, choosing $\bar{u} = x_0$, the first part of the statement follows.

Concerning the dependence on the initial value, let $x_0, x_1 \in \mathcal{O}$ be such that

$$\|x_i - \bar{u}\| \leq r_0, \quad \|F(t_0, x_i) - F(\bar{t}, \bar{u})\|_{(X,D)_{\alpha,\infty}} \leq r_0, \quad i = 0, 1,$$

where r_0 is defined by (8.1.24). Then the difference $z(t) = u(\cdot, t_0, x_0) - u(\cdot, t_0, x_1)$ satisfies

$$z'(t) = Az(t) + g(t), \quad t_0 \leq t \leq t_0 + \delta_0, \quad z(t_0) = x_0 - x_1,$$

with $g(t) = F(t, u(t, t_0, x_0)) - F(t, u(t, t_0, x_1)) - A(u(t, t_0, x_0) - u(t, t_0, x_1))$. By Theorem 4.3.1 we have

$$\|z\|_{C^\alpha(D)} + \|z'\|_{C^\alpha(X)} + \|z'\|_{B(D_A(\alpha,\infty))}$$
$$\leq \gamma C(\|x_0 - x_1\|_D + \|F(t_0, x_0) - F(t_0, x_1)\|_{D_A(\alpha,\infty)} + \|g\|_{C^\alpha(X)}).$$

Arguing as in estimates (8.1.20) and (8.1.21), and using (8.1.8), we find

$$\|g\|_{C^\alpha(X)} \leq \frac{1}{2\gamma C}\|z\|_{C^\alpha(D)} + (K + L\rho_0)\|x_0 - x_1\|_D,$$

and (8.1.19) follows. ∎

8.2 The maximally defined solution

Throughout the section, $F : [0, T] \times \mathcal{O} \mapsto X$ is a nonlinear function satisfying assumptions (8.1.1) and (8.0.3). For each $u_0 \in \mathcal{O}$ such that $F(0, u_0) \in \overline{D}$, Theorem 8.1.1 yields existence and uniqueness of a local solution u of (8.0.1) in the space $C^\alpha_\alpha(]0, \delta]; D)$, enjoying property (8.1.3). Since $u \in C^\alpha([\delta/2, \delta]; D) \cap C^{1+\alpha}([\delta/2, \delta]; X)$, then $u'(\delta) = F(\delta, u(\delta))$ belongs to $(X, D)_{\alpha,\infty}$ (see Proposition 2.2.10), so that, thanks to Theorem 8.1.3, u can be continued to some interval $[\delta, \delta_1]$ with $\delta_1 > \delta$, in such a way that the extension belongs to $C^\alpha_\alpha(]0, \delta_1]; D)$. So we define $u = u(\cdot; u_0)$ by

$$\begin{cases} I = I(u_0) = \bigcup\{[0, \delta] : \text{problem (8.1.1) has a solution } u_\delta \in \\ \qquad C([0, \delta]; D) \cap C^\beta_\beta(]0, \delta]; D) \text{ for some } \beta \in]0, 1[\}, \\ u : I \mapsto D, \ u(t) = u_\delta(t) \ \text{ for } t \in [0, \delta] \subset I. \end{cases} \quad (8.2.1)$$

u is well defined thanks to the uniqueness part of Theorem 8.1.1. We shall see in the next proposition that u has no continuous extension with values in D. So, it is the maximally defined solution of problem (8.0.1). We set

$$\tau = \tau(u_0) = \sup I(u_0).$$

Proposition 8.2.1 *If $u(\cdot; u_0)$ is uniformly continuous with values in D, then either*

$$\lim_{t \to \tau(u_0)} u(t; u_0) \in \partial\mathcal{O}, \quad (8.2.2)$$

or

$$I(u_0) = [0, T]. \quad (8.2.3)$$

Proof — Assume that (8.2.2) does not hold, so that $\lim_{t \to \tau} u(t) = u(\tau) \in \mathcal{O}$. The continuation of u belongs to $C([0, \tau]; D) \cap C^1([0, \tau]; X)$, so that $u'(\tau) = F(\tau, u(\tau)) \in \overline{D}$.

2. THE MAXIMALLY DEFINED SOLUTION

Fix $\bar{t} = \tau$ and $\bar{u} = u(\tau)$. By Theorem 8.1.1, there are positive numbers r, δ such that if
$$|t_0 - \tau| \leq r, \quad \|x_0 - \bar{u}\|_D \leq r, \quad F(t_0, x_0) \in \overline{D},$$
then the problem
$$v'(t) = F(t, v(t)), \quad t \geq t_0; \quad v(t_0) = x_0, \qquad (8.2.4)$$
has a unique solution $v \in C_\alpha^\alpha(]t_0, t_0 + \delta]; D) \cap C([t_0, t_0 + \delta]; D)$. Taking $t_0 = \tau - \varepsilon$, $x_0 = u(\tau - \varepsilon)$, where $\varepsilon \in]0, \min(\delta/2, r)[$ is so small that $\|u(t_0) - u(\tau)\|_D \leq r$, we get a solution v of (8.2.4) in the interval $[t_0, t_0 + \delta]$, which contains τ in its interior. Moreover, v coincides with u in the interval $[t_0, \tau]$. Let $\beta \in]0, \alpha]$ be such that $u \in C_\beta^\beta(]0, t_0]; D)$. Then u has an extension belonging to $C_\beta^\beta(]0, \tau_1]; D)$, with $\tau_1 > \tau$, a contradiction. Therefore, $\tau = T$. ∎

A sufficient condition for u be uniformly continuous in I is given by next lemma.

Lemma 8.2.2 *Let the assumptions of Theorem 8.1.1 hold, and let $u : I \mapsto D$ be the maximally defined solution of problem (8.1.1). If the orbit*
$$\{u(t) : 0 \leq t < \tau\}$$
is relatively compact in D, then $u : [0, \tau[\mapsto D$ is uniformly continuous.

Proof — The ω-limit set
$$\{y \in D : \exists \{t_n\} \to \tau \text{ such that } \|u(t_n) - y\|_D \to 0\}$$
is closed and not empty, since it is the intersection of the closed, nonempty sets $\overline{\{u(t) : s \leq t < \tau\}}$ with $0 \leq s < \tau$. Moreover, since the orbit is relatively compact, the set $\{F(t, u(t)) : 0 \leq t < \tau\}$ is bounded in X. Since $u'(t) = F(t, u(t))$, u is Lipschitz continuous with values in X and there exists $\lim_{t \to \tau} u(t)$ in X. This implies that the ω-limit set consists of a unique element y. By compactness, $\|u(t) - y\|_D \to 0$ as $t \to \tau$. Hence, u is uniformly continuous in $[0, \tau[$. ∎

Assume that $\mathcal{O} = D$. By Proposition 8.2.1, if u is uniformly continuous with values in D then $I = [0, T]$. Here is an important difference between semilinear (or quasilinear) and fully nonlinear equations. In semilinear equations the nonlinearity is defined in an intermediate space X_α between X and D, and an *a priori* estimate of the type $\|u(t)\|_{X_\alpha} \leq K \; \forall t \in I$ is enough to guarantee that $I = [0, T]$ (see Chapter 7). In fully nonlinear equations, where the nonlinearity is defined on D, the *a priori* estimate $\|u(t)\|_D \leq K \; \forall t \in I$ is not sufficient for existence in the large, but we need an estimate on the modulus of continuity of u. This fact produces additional difficulties when treating concrete examples. See Section 8.5, and in particular example 8.5.2.

The maximally defined solution depends continuously on the initial value, as the following proposition shows.

Proposition 8.2.3 Let $\bar{u} \in \mathcal{O}$ be such that $F(0, \bar{u}) \in \overline{D}$, and fix $\bar{\tau} \in \,]0, \tau(\bar{u})[$. Then there are $\epsilon = \epsilon(\bar{u}, \bar{\tau}) > 0$, $H = H(\bar{u}, \bar{\tau}) > 0$ such that if

$$u_0 \in \mathcal{O}, \ F(0, u_0) \in \overline{D}, \ \|u_0 - \bar{u}\|_D \leq \epsilon,$$

then

$$\tau(u_0) \geq \bar{\tau}$$

and

$$\|u(\cdot; u_0) - u(\cdot; \bar{u})\|_{C_\alpha^\alpha(]0,\bar{\tau}];D)} + \|u_t(\cdot; u_0) - u_t(\cdot; \bar{u})\|_{C_\alpha^\alpha(]0,\bar{\tau}];X)}$$
$$+ \sup\{t^\alpha \|u_t(t, u_0) - u_t(t; \bar{u})\|_{(X,D)_{\alpha,\infty}} : 0 < t \leq \bar{\tau}\} \leq H \|u_0 - \bar{u}\|_D.$$

Proof — The statement is a consequence of Theorem 8.1.1, via a suitable covering argument, similar to the one used in the proof of Proposition 7.1.9.

Since $u(\cdot; \bar{u})$ has values in D, then $u_t(s, \bar{u}) = F(s, u(s, \bar{u}))$ belongs to $\overline{D}$ for every $s \in [0, \bar{\tau}]$. By Theorem 8.1.1, there are r_s, δ_s, $k_s > 0$ such that if $|t_0 - s| \leq r_s$, $y \in D$ and $\|y - u(s, \bar{u})\|_D \leq r_s$, $F(s, y) \in \overline{D}$, then the problem

$$v'(t) = F(t, v(t)), \ t \geq s; \ v(s) = y, \tag{8.2.5}$$

has a unique solution $v = v(s; y) \in C_\alpha^\alpha(]s, s + \delta_s]; D) \cap C([s, s + \delta_s]; D)$, and

$$\|v(\cdot; y_1) - v(\cdot; y_2)\|_{C_\alpha^\alpha(]s,s+\delta_s];D)} + \|v'(\cdot; y_1) - v'(\cdot; y_2)\|_{C_\alpha^\alpha(]s,s+\delta_s];X)}$$
$$+ \sup_{s<t\leq s+\delta_s}(t - s)^\alpha \|v'(t; y_1) - v'(t; y_2)\|_{(X,D)_{\alpha,\infty}} \leq k_s \|y_1 - y_2\|_D.$$

Since the set

$$[0, \bar{\tau}] \times \{u(s; \bar{u}) : s \in [0, \bar{\tau}]\}$$

is compact in $\mathbb{R} \times D$, it can be covered by a finite number of open sets

$$O_i = \,]s_i - r_i, s_i + r_{s_i}[\times \overset{\circ}{B}(u(s_i; \bar{u}), r_i), \ i = 1, .., n,$$

where $\overset{\circ}{B}(y, r)$ denotes the open ball in D centered at y with radius r, and $r_i = r(s_i, u(s_i, \bar{u}))$. Set

$$\delta = \min\{\delta_{s_i} : i = 1, .., n\}, \ k = \max\{k_{s_i} : i = 1, .., n\}, \ r = \min\{r_{s_i} : i = 1, .., n\}.$$

There is $r_0 > 0$ such that

$$\|u_0 - \bar{u}\|_D \leq r_0 \Rightarrow (0, u_0) \in O_{i_0} \text{ for some } i_0 \in \{1, ..., n\}$$

Then $u(\cdot; u_0)$ belongs to $C_\alpha^\alpha(]0, \delta]; D)$, and

$$\|u(\cdot; u_0) - u(\cdot; \bar{u})\|_{C_\alpha^\alpha(]0,\delta];D)} + \|u_t(\cdot; u_0) - u_t(\cdot; \bar{u})\|_{C_\alpha^\alpha(]0,\delta];X)}$$
$$+ \sup\{t^\alpha \|u_t(t; u_0) - u_t(t; \bar{u})\|_{(X,D)_{\alpha,\infty}} : 0 < t \leq \delta\} \leq k \|u_0 - \bar{u}\|_D. \tag{8.2.6}$$

3. FURTHER REGULARITY PROPERTIES AND DEPENDENCE ON THE DATA

If $\delta \geq \bar{\tau}$ the proof is finished. If $\delta < \bar{\tau}$, there exists i_1 such that $(\delta/2, u(\delta/2, \bar{u})) \in O_{i_1}$. By (8.2.6) there is $r_1 \leq r_0$ such that

$$\|u_0 - \bar{u}\|_D \leq r_1 \Rightarrow (\delta/2, u(\delta/2, u_0)) \in O_{i_1}.$$

If $\|u_0 - \bar{u}\|_D \leq r_1$, then problem (8.2.5), with $y = u(\delta/2; u_0)$, is solvable in $[\delta/2, 3\delta/2]$. It follows that $\tau(u_0) \geq 3\delta/2$, and

$$\|u(\cdot; u_0) - u(\cdot; \bar{u})\|_{C^\alpha_\alpha(]\delta/2, 3\delta/2]; D)} + \|u_t(\cdot; u_0) - u_t(\cdot; \bar{u})\|_{C^\alpha_\alpha(]\delta/2, 3\delta/2]; X)}$$
$$+ \sup\{(t - \delta/2)^\alpha \|u_t(t; u_0) - u_t(t; \bar{u})\|_{(X,D)_{\alpha,\infty}} : \delta/2 \leq t \leq 3\delta/2\} \quad (8.2.7)$$
$$\leq k \|u(\delta/2; u_0) - u(\delta/2; \bar{u})\|_D.$$

By (8.2.6) and (8.2.7) there is $k_1 > 0$ such that

$$\|u(\cdot; u_0) - u(\cdot; \bar{u})\|_{C^\alpha_\alpha(]0, 3\delta/2]; D)} + \|u_t(\cdot; u_0) - u_t(\cdot; \bar{u})\|_{C^\alpha_\alpha(]0, 3\delta/2]; X)}$$
$$+ \sup\{t^\alpha \|u_t(t; u_0) - u_t(t; \bar{u})\|_{(X,D)_{\alpha,\infty}} : 0 < t \leq 3\delta/2\} \quad (8.2.8)$$
$$\leq k_1 \|u_0 - \bar{u}\|_D.$$

If $3\delta/2 \geq \bar{\tau}$, the proof is finished. Otherwise, we repeat the procedure a finite number of times, up to cover the whole interval $[0, \bar{\tau}]$. ∎

We remark that in the autonomous case $F = F(x)$, Proposition 8.2.3 implies that the solution u defines a local dynamical system on the set of the admissible initial values

$$C = \{x \in \mathcal{O} : F(x) \in \overline{D}\},$$

which is closed in D. We recall that a *local dynamical system* or *local semiflow* on a closed set C in a complete metric space is a map $T : \mathcal{A} = \{(t, x) : x \in C, t \in [0, \tau(x)[\} \mapsto C$, $(t, x) \mapsto T(t, x)$, such that $\mathcal{A}$ is relatively open in $[0, +\infty[\times C$, T is continuous, $T(0, x) = x$ for every $x \in C$, and if $s \in [0, \tau(x)[$, $t \in [0, \tau(T(s, x))[$ then $t + s < \tau(x)$ and $T(t, T(s, x)) = T(t + s, x)$.

8.3 Further regularity properties and dependence on the data

In applied mathematics one often encounters PDE's depending on one or several parameters, having specific physical meaning. For this reason, in this section we consider a family of equations depending on a parameter $\lambda \in \Lambda$, where Λ is any Banach space,

$$u'(t) = F(t, u(t), \lambda), \quad t \geq 0; \quad u(0) = x, \quad (8.3.1)$$

and for every λ the function $(t, x) \mapsto F(t, x, \lambda)$ satisfies the assumptions of Theorem 8.1.1. We give results of dependence of the maximal solution on λ, t, x.

8.3.1 C^k regularity with respect to (x, λ)

In view of assumption (8.1.1) it is useful to introduce the following notation.

Definition 8.3.1 *Let $\mathcal{O}_D \subset D$ and $\mathcal{O}_\Lambda \subset \Lambda$ be open sets, let Y be any Banach space, and let*
$$G : [0, T] \times \mathcal{O}_D \times \mathcal{O}_\Lambda \mapsto Y, \quad 0 < \alpha < 1.$$
G is said to belong to the class (C^α, Lip) if G is continuous, and for every $\bar{u} \in \mathcal{O}_D$, $\bar{\lambda} \in \mathcal{O}_\Lambda$, there exist $R = R(\bar{u}, \bar{\lambda})$, $L = L(\bar{u}, \bar{\lambda})$, $K = K(\bar{u}, \bar{\lambda}) > 0$ such that for $t, s \in [0, T]$, $x, y \in B(\bar{u}, R) \subset D$, $\lambda \in B(\bar{\lambda}, R) \subset \Lambda$ it holds
$$\|G(t, x, \lambda) - G(s, y, \lambda)\|_Y \le K|t - s|^\alpha + L\|x - y\|_D.$$

The basic assumptions on F are the following.
$$F : [0, T] \times \mathcal{O}_D \times \mathcal{O}_\Lambda \mapsto X$$
is a function such that

$$\begin{cases} (t, x, \lambda) \mapsto F(t, x, \lambda) \text{ is continuous with respect to } (t, x, \lambda), \\ \text{and it is Fréchet differentiable with respect to } (x, \lambda). \\ \exists \, \alpha \in \,]0, 1[\text{ such that } F, F_x, F_\lambda \text{ belong to the class } (C^\alpha, Lip) \\ \text{with values in } X, L(D, X), L(\Lambda, X), \text{ respectively,} \end{cases} \quad (8.3.2)$$

and

$$\begin{cases} \forall (t, x, \lambda) \in [0, T] \times \mathcal{O}_D \times \mathcal{O}_\Lambda, \text{ the operator } F_x(t, x, \lambda) : D \mapsto X \text{ is} \\ \text{sectorial in } X, \text{ and its graph norm is equivalent to the norm of } D. \end{cases} \quad (8.3.3)$$

From (8.3.2), (8.3.3) it follows that for every $\lambda \in \mathcal{O}_\Lambda$ the function $(t, x) \mapsto F(t, x, \lambda)$ satisfies the assumptions of Theorem 8.1.1. Therefore, for every $x \in \mathcal{O}_D$ satisfying the compatibility condition
$$F(0, x, \lambda) \in \overline{D} \quad (8.3.4)$$
problem (8.3.1) has a noncontinuable solution $u = u(t; x, \lambda)$ defined in a maximal interval $I(x, \lambda)$. Let $\mathcal{M}$ be the set of the initial data satisfying (8.3.4),
$$\mathcal{M} = \{(x, \lambda) \in \mathcal{O}_D \times \mathcal{O}_\Lambda : F(0, x, \lambda) \in \overline{D}\}. \quad (8.3.5)$$

Theorem 8.3.2 *Let $F : [0, T] \times \mathcal{O}_D \times \mathcal{O}_\Lambda \mapsto X$ satisfy assumptions (8.3.2), (8.3.3). Let $(\bar{u}, \bar{\lambda})$ in $\mathcal{M}$, and fix $\bar{\tau} \in I(\bar{u}, \bar{\lambda})$, $\bar{\tau} > 0$. Then there are $r = r(\bar{u}, \bar{\lambda}, \bar{\tau}) > 0$, $H = H(\bar{u}, \bar{\lambda}, \bar{\tau}) > 0$ such that if*
$$(u_0, \lambda) \in \mathcal{M}, \quad \|\lambda - \bar{\lambda}\|_\Lambda \le r, \quad \|u_0 - \bar{u}\|_D \le r,$$
then
$$\tau(\lambda, x) \ge \bar{\tau}$$

3. FURTHER REGULARITY PROPERTIES AND DEPENDENCE ON THE DATA

and

$$\|u(\cdot;u_0,\lambda)-u(\cdot;\bar{u},\bar{\lambda})\|_{C^\alpha_\alpha(]0,\bar{\tau}];D)} + \|u_t(\cdot;u_0,\lambda)-u_t(\cdot;\bar{u},\bar{\lambda})\|_{C^\alpha_\alpha(]0,\bar{\tau}];X)}$$
$$+\sup\{t^\alpha\|u_t(t;u_0,\lambda)-u_t(t;\bar{u},\bar{\lambda})\|_{(X,D)_{\alpha,\infty}} : 0\leq t\leq\bar{\tau}\} \quad (8.3.6)$$
$$\leq H(\|\lambda-\bar{\lambda}\|_\Lambda + \|u_0-\bar{u}\|_D).$$

Proof — The proof consists of two steps: first one shows a result of dependence on the data for the problem

$$v'(t) = F(t,v(t),\lambda), \quad t_0\leq t\leq t_0+\delta; \quad v(t_0)=x_0, \quad (8.3.7)$$

with $t_0\geq 0$ and small $\delta>0$. Then the final statement is proved by a covering argument as in Proposition 8.2.3.

Step 1) Following the proof of Theorem 8.1.1, one sees that for every $\bar{t}\geq 0$, $\bar{u}\in\mathcal{O}_D$, and $\bar{\lambda}$ in $\mathcal{O}_\Lambda$ such that $F(\bar{t},\bar{u},\bar{\lambda})\in\bar{D}$ there are $\delta=\delta(\bar{t},\bar{u},\bar{\lambda})>0$, $r=r(\bar{t},\bar{u},\bar{\lambda})>0, k=k(\bar{t},\bar{u},\bar{\lambda})>0$ with the property that for every $t_0\in[\bar{t}-r,\bar{t}+r]\cap[0,+\infty[$, $x_0\in\mathcal{O}_D$, and $\lambda\in\mathcal{O}_\Lambda$ such that $F(t_0,x_0,\lambda)\in\bar{D}$ and $\|\lambda-\bar{\lambda}\|_\Lambda\leq r$, $\|x_0-\bar{u}\|_D\leq r$, problem (8.3.7) has a unique solution u belonging to $C^\alpha_\alpha(]t_0,t_0+\delta];D)\cap C([t_0,t_0+\delta];D)$. u is the fixed point of the 1/2-contraction $\Gamma:Y\mapsto Y$ defined by $\Gamma(u)=v$, v being the solution of (8.1.5) with $F(t,u(t))$ replaced by $F(t,u(t),\lambda_0)$. The set Y has been defined in (8.3.5).

Let now $x_0,x_1\in\mathcal{O}_D$, $\lambda_0,\lambda_1\in\mathcal{O}_\Lambda$ be such that $F(t_0,x_0,\lambda_0)\in\bar{D}$, $F(t_0,x_1,\lambda_1)\in\bar{D}$. Using assumption (8.3.2), it is not hard to check that

$$\|F(\cdot,u(\cdot;x_0,\lambda_0),\lambda_0)-F(\cdot,u(\cdot;x_0,\lambda_0),\lambda_1)\|_{C^\alpha_\alpha(]t_0,t_0+\delta];D)}\leq K\|\lambda_0-\lambda_1\|_\Lambda.$$

The function $w=u(\cdot;x_0,\lambda_0)-u(\cdot;x_1,\lambda_1)$ satisfies

$$w'(t) = Aw(t) + \varphi(t), \quad t_0\leq t\leq t_0+\delta, \quad w(t_0)=x_0-x_1,$$

where $\varphi(t)=F(\cdot,u(\cdot;x_0,\lambda_0),\lambda_0)-F(\cdot,u(\cdot;x_1,\lambda_1),\lambda_1)-A(u(\cdot;x_0,\lambda_0)-u(\cdot;x_1,\lambda_1))$. It follows that

$$\|u(\cdot;x_0,\lambda_0)-u(\cdot;x_1,\lambda_1)\|_{C^\alpha_\alpha(]t_0,t_0+\delta];D)}\leq C\gamma(\|x_0-x_1\|_{D(A)}+\|\varphi\|_{C^\alpha_\alpha(X)})$$
$$\leq C\gamma(\gamma\|x_0-x_1\|_D+K\|\lambda_0-\lambda_1\|_\Lambda)+\tfrac{1}{2}\|u(\cdot;x_0,\lambda_0)-u(\cdot;x_1,\lambda_1)\|_{C^\alpha_\alpha(]t_0,t_0+\delta];D)},$$

which in its turn implies that

$$\|u_0-u_1\|_{C^\alpha_\alpha(]t_0,t_0+\delta];D)} + \|u'_0-u'_1\|_{C^\alpha_\alpha(]t_0,t_0+\delta];D)}$$
$$+ \sup\{(t-t_0)^\alpha\|u'_0(t)-u'_1(t)\|_{(X,D)_{\alpha,\infty}} : t_0\leq t<t_0+\delta\}$$
$$\leq \text{const.}(\|x_1-x_0\|_D + \|\lambda_0-\lambda_1\|_\Lambda).$$

Step 2) It is sufficient to follow closely the proof of Proposition 8.2.3, without important modifications. We omit the details. ∎

The following corollary is an obvious consequence of Theorem 8.3.2.

Corollary 8.3.3 *Let* $F : [0,T] \times \mathcal{O}_D \times \mathcal{O}_\Lambda \mapsto X$ *satisfy assumptions (8.3.2) and (8.3.3). Then the function*

$$\{(t,x,\lambda) : (x,\lambda) \in \mathcal{M}, \ t \in I(x,\lambda)\} \mapsto D, \quad (t,x,\lambda) \mapsto u(t,x,\lambda),$$

is continuous.

If F is more regular, one can show further regularity results with respect to (λ, x). We assume that

$$\begin{cases} \exists k \in \mathbb{N} \text{ such that } F \text{ is } k+1 \text{ times continuously differentiable} \\ \text{with respect to } (x,\lambda), \text{ and every derivative of order } \leq (k+1) \\ \text{with respect to } (x,\lambda) \text{ belongs to the class } (C^\alpha, Lip). \end{cases} \quad (8.3.8)$$

We shall see that under assumption 8.3.8 the function $(u,\lambda) \mapsto F(\cdot, u(\cdot), \lambda)$ is differentiable from $\{(u,\lambda) \in C_\alpha^\alpha(]0,a];D) \times \Lambda \ : \ u(t) \in \mathcal{O}_D, \ \lambda \in \mathcal{O}_\Lambda\}$ to $C_\alpha^\alpha(]0,a];X)$, for every $a > 0$. This will allow us to use the Implicit Function Theorem and to prove local existence of a solution even for initial data which do not satisfy the compatibility condition (8.3.4), provided they are close to initial data which satisfy (8.3.4). Of course, such a solution is not continuous at $t = 0$ with values in D, but it is continuous with values in X and in every interpolation space $(X,D)_{\theta,p}$, with $0 < \theta < 1$, $1 \leq p \leq \infty$.

Theorem 8.3.4 *Let* $F : [0,T] \times \mathcal{O}_D \times \mathcal{O}_\Lambda \mapsto X$ *satisfy assumptions (8.3.2), (8.3.3), and (8.3.8). Fix* $(\overline{u}, \overline{\lambda}) \in \mathcal{M}$, *and let* $u(\cdot; \overline{u}, \overline{\lambda}) : I(\overline{u}, \overline{\lambda}) \mapsto D$ *be the maximally defined solution of (8.3.1) with* $\lambda = \overline{\lambda}$, $x_0 = \overline{u}$. *For every* $a \in I(\overline{u}, \overline{\lambda})$, $a > 0$, *there are positive numbers* $r_0 = r_0(\overline{u}, \overline{\lambda}, a)$, $r_1 = r_1(\overline{u}, \overline{\lambda}, a)$, *such that for every* $x \in D$ *and* $\lambda \in \Lambda$ *with*

$$\|x - \overline{u}\|_D \leq r_0, \quad \|\lambda - \overline{\lambda}\|_\Lambda \leq r_0,$$

the problem

$$u'(t) = F(t, u(t), \lambda), \ 0 < t \leq a; \quad u(0) = x, \quad (8.3.9)$$

has a unique solution $u = u(\cdot; x, \lambda) \in C_\alpha^\alpha(]0,a];D) \cap C([0,a];X)$, *such that*

$$\|u - u(\cdot; \overline{u}, \overline{\lambda})\|_{C_\alpha^\alpha(]0,a];D)} \leq r_1.$$

The mapping $(x, \lambda) \mapsto u(\cdot; x, \lambda)$ *is* k *times continuously differentiable in the interior of* $B(\overline{u}, r_0) \times B(\overline{\lambda}, r_0) \subset D \times \Lambda$ *with values in* $C_\alpha^\alpha(]0,a];D)$.

Proof — Setting $u(t) = v(t) - x$, the initial value problem (8.3.9) is equivalent to

$$v'(t) = F(t, v(t) + x, \lambda), \ 0 < t \leq a; \quad v(0) = 0. \quad (8.3.10)$$

Problem (8.3.10) will be solved by the Implicit Function Theorem. Define

$$\begin{aligned} Z_\alpha([0,a]) &= \{v \in C_\alpha^\alpha(]0,a];D) \cap C([0,a];X), \ v(0) = 0, \ \exists v'(t) \\ & \quad \text{for } 0 < t \leq a, \ v' \in C_\alpha^\alpha(]0,a];X)\} \end{aligned}$$

3. FURTHER REGULARITY PROPERTIES AND DEPENDENCE ON THE DATA

and
$$\mathcal{A} = \{(v, x, \lambda) \in Z_\alpha([0, a]) \times \mathcal{O}_D \times \mathcal{O}_\Lambda : v(t) + x \in \mathcal{O}_D\},$$
$$\Phi : \mathcal{A} \mapsto C_\alpha^\alpha(]0, a]; X), \quad \Phi(v, x, \lambda)(t) = v'(t) - F(t, v(t) + x, \lambda).$$

To apply the Implicit Function Theorem depending on a parameter, we need to show that

(i) Φ is k times continuously differentiable in $\mathcal{A}$,

(ii) setting $\overline{v}(t) = u(t; \overline{u}, \overline{\lambda}) - \overline{u}$, the linear operator $\Phi_v(\overline{v}, \overline{u}, \overline{\lambda})$ is an isomorphism between $Z_\alpha([0, a])$ and $C_\alpha^\alpha(]0, a]; X)$.

(i) First we show that Φ is continuously Fréchet differentiable at every point $(v_0, x_0, \lambda_0) \in \mathcal{A}$. It is sufficient to show that $(v, x, \lambda) \mapsto F(\cdot, v(\cdot + x), \lambda)$ is differentiable at (v_0, x_0, λ_0), and that the derivative is the linear operator

$$(\hat{v}, \hat{x}, \hat{\lambda}) \mapsto F_x(\cdot, v_0(\cdot) + x_0, \lambda_0)(\hat{v}(\cdot) + \hat{x}) + F_\lambda(\cdot, v_0(\cdot) + x_0, \lambda_0)\hat{\lambda}. \quad (8.3.11)$$

For $(v, x, \lambda) \in \mathcal{A}$ set

$$\varphi(t) = F(t, v(t) + x, \lambda) - F(t, v_0(t), \lambda_0)$$
$$- F_x(t, v_0(t) + x_0, \lambda_0)(v(t) - v_0(t) + x - x_0) - F_\lambda(t, v_0(t) + x_0, \lambda_0)(\lambda - \lambda_0).$$

If (v, x, λ) is close to (v_0, x_0, λ_0), then for $0 < t \leq a$

$$\varphi(t) = \int_0^1 d\theta \int_0^1 \sigma d\sigma F_{uu}(t, y(t), \mu)(v(t) - v_0(t) + x - x_0)^2$$
$$+ \int_0^1 d\theta \int_0^1 \sigma d\sigma F_{u\lambda}(t, y(t), \mu)(v(t) - v_0(t) + x - x_0, \lambda - \lambda_0)$$
$$+ \int_0^1 d\theta \int_0^1 \sigma d\sigma F_{u\lambda}(t, y(t), \mu)(\lambda - \lambda_0)^2,$$

where

$$y(t) = \theta(\sigma(v(t) + x) + (1 - \sigma)(v_0(t) + x_0)) + (1 - \theta)(v_0(t) + x_0),$$
$$\mu = \theta(\sigma\lambda + (1 - \sigma)\lambda_0) + (1 - \theta)\lambda_0.$$

It is not difficult to check that

$$[\varphi]_{C_\alpha^\alpha(]0,a];X)} \leq \text{const.}(\|v - v_0\|_{C_\alpha^\alpha(]0,a];D)} + \|x - x_0\|_D + \|\lambda - \lambda_0\|_\Lambda)^2,$$

provided (v, x, λ) is sufficiently close to (v_0, x_0, λ_0). This implies that $(v, x, \lambda) \mapsto F(\cdot, v(\cdot) + x, \lambda)$ is Fréchet differentiable at (v_0, x_0, λ_0), and that the derivative is given by formula (8.3.11). Arguing as in the proof of Theorem 8.3.2, one sees that the derivative is locally Lipschitz continuous. It follows that Φ is continuously differentiable. Arguing by recurrence, we get that Φ is k times continuously differentiable.

(ii) Set $\overline{v}(t) = \overline{u}(t) - \overline{u}$. From step (i) we know that

$$[\Phi_v(\overline{v}, \overline{u}, \overline{\lambda})v](t) = v'(t) - F_u(t, \overline{v}(t) + \overline{u}, \overline{\lambda})v(t),$$

To show that $\Phi_v(\overline{v}, \overline{u}, \overline{\lambda}) : Z_\alpha([0, a]) \mapsto C_\alpha^\alpha(]0, a]; X)$ is an isomorphism, we have to prove that for every $f \in C_\alpha^\alpha(]0, a]; X)$ the problem

$$v'(t) = A(t)v(t) + f(t), \quad 0 < t \leq a; \quad v(0) = 0, \tag{8.3.12}$$

with $A(t) = F_u(t, \overline{v}(t) + \overline{u}, \overline{\lambda})$, has a unique solution $v \in Z_\alpha([0, a])$. The procedure is similar to the proofs of Proposition 6.1.3 and Theorem 6.1.4. First we consider problem (8.3.12) in a small time interval $[0, \delta]$, rewriting it as

$$v'(t) = A(0)v(t) + [A(t) - A(0)]v(t) + f(t), \quad 0 < t \leq \delta; \quad v(0) = 0,$$

and the solution is sought as a fixed point of the operator Γ, which maps any $v \in C_\alpha^\alpha(]0, \delta]; D)$ into the unique solution $u \in Z_\alpha([0, \delta])$ of

$$u'(t) = A(0)u(t) + [A(t) - A(0)]v(t) + f(t), \quad 0 < t \leq a; \quad u(0) = 0.$$

Since $\overline{v}$ is continuous up to $t = 0$ with values in D, and $\lim_{\varepsilon \to 0} [\overline{v}]_{C_\alpha^\alpha(]0,\varepsilon];D)} = 0$, then

$$\lim_{t \to 0} \|A(t) - A(0)\|_{L(D,X)} = 0, \quad \lim_{\varepsilon \to 0} [A(\cdot)]_{C_\alpha^\alpha(]0,\varepsilon];L(D,X))} = 0.$$

Using estimate (4.3.11) it is not difficult to see that Γ is a $1/2$-contraction, provided δ is small enough. So, it has a unique fixed point $u_1 \in Z_\alpha([0, \delta])$. Taking $u_1(\delta)$ as initial value, the problem

$$v'(t) = A(t)v(t) + f(t), \quad \delta \leq t \leq a; \quad v(\delta) = u_1(\delta),$$

has a unique solution $u_2 \in C^\alpha([\delta, a]; D) \cap C^{1+\alpha}([\delta, a]; X)$ thanks to Proposition 6.1.3, since $u_1(\delta) \in D$ and $A(\delta)u_1(\delta) + f(\delta) = u_1'(\delta) \in (X, D)_{\alpha, \infty}$. The function u defined by $u(t) = u_1(t)$ for $0 \leq t \leq \delta$, $u(t) = u_2(t)$ for $\delta < t \leq a$, is the unique solution of (8.3.12) belonging to $Z_\alpha([0, a])$. Therefore, $\Phi_v(\overline{v}, \overline{u}, \overline{\lambda})$ is an isomorphism, and the statement follows. ∎

Corollary 8.3.5 *Let F satisfy the assumptions of Theorem 8.3.4 for every $k \in \mathbb{N}$. Then for every $(\overline{u}, \overline{\lambda}) \in \mathcal{M}$ there is $r_0 > 0$ such that the mapping $(x, \lambda) \mapsto u(\cdot; x, \lambda)$ is C^∞ in the interior of $B(\overline{u}, r_0) \times B(\overline{\lambda}, r_0) \subset D \times \Lambda$ with values in $C_\alpha^\alpha(]0, a]; D)$.*

8.3.2 C^k regularity with respect to time

Now we consider further regularity with respect to time. In order to treat C^k regularity, we make the following assumption.

$$\begin{cases} \exists k \in \mathbb{N} \cup \{0\} \text{ such that } F \text{ is } k+1 \text{ times differentiable with respect to } (t, x), \\ \text{and each derivative of order } \leq k+1 \text{ belongs to the class } (C^\alpha, \text{Lip}). \end{cases} \tag{8.3.13}$$

Proposition 8.3.6 Let F satisfy (8.3.2), (8.3.3), and (8.3.13). For every $(\overline{u}, \overline{\lambda}) \in \mathcal{M}$, the maximally defined solution $u : I(\overline{u}, \overline{\lambda}) \mapsto D$ of problem (8.3.1) belongs to $C^{k+\alpha}([\varepsilon, a]; D) \cap C^{k+1+\alpha}([\varepsilon, a]; X)$ for every $\varepsilon \in]0, a[$.

If in addition F is $k+1$ times differentiable with respect to (t, x, λ), with derivatives up to the order $k+1$ in the class (C^α, Lip), then for every $(\overline{u}, \overline{\lambda}) \in \mathcal{M}$ and for $0 < \varepsilon < a < \tau(\overline{u}, \overline{\lambda})$, the mapping

$$(x, \lambda) \mapsto \frac{d^k}{dt^k} u(\cdot; x, \lambda)$$

is well defined and continuous in a neighborhood of $(\overline{u}, \overline{\lambda})$ with values in $C^\alpha([\varepsilon, a]; D)$.

Proof — Let $0 < a \in I(\overline{u}, \overline{\lambda})$. For $\mu > 0$ define $v(t) = u(\mu t; \overline{u}, \overline{\lambda})$. If μ is sufficiently small (precisely, $\mu < \tau(\overline{u}, \overline{\lambda})/a$), then v is well defined in $[0, a]$, and it satisfies

$$v'(t) = \mu F(\mu t, v(t), \overline{\lambda}), \quad 0 \leq t \leq a; \quad v(0) = \overline{u}.$$

The function

$$G(t, y, \mu) = \mu F(\mu t, y, \overline{\lambda})$$

is $k+1$ times continuously differentiable with respect to (y, μ) in $[0, T/a] \times \mathcal{O}_D$, and its derivatives up to the order $k+1$ belong to (C^α, Lip). Theorem 8.3.4 implies that the mapping $]0, \tau(\overline{u}, \overline{\lambda})/a[\mapsto C^\alpha_\alpha(]0, a]; D)$, $\mu \mapsto v$, is k times continuously differentiable. Recalling that $u(t, \overline{u}, \overline{\lambda}) = v(t/\mu)$, it follows that u is k times continuously differentiable with values in D in the interval $]0, a]$, and $u^{(k)} \in C^\alpha([\varepsilon, a]; D)$ for $0 < \varepsilon < a$.

If in addition F is $k+1$ times differentiable with respect to (t, x, λ), with derivatives up to the order $k+1$ in the class (C^α, Lip), then G is $k+1$ times differentiable with respect to (t, x, λ, μ), with derivatives up to the order $k+1$ in the class (C^α, Lip), so that, using again Theorem 8.3.4, we get that $d^k/d\mu^k v(\cdot; x, \lambda, \mu)$ is continuous with respect to (x, λ, μ) in a neighborhood of $(\overline{u}, \overline{\lambda}, 1)$, with values in $C^\alpha_\alpha(]0, a]; D)$. The last statement follows. ∎

Corollary 8.3.7 Let F satisfy assumptions (8.3.2), (8.3.3) and be of class C^∞ with respect to (t, x). If $u : [0, \tau[\mapsto D$ is any solution of (8.3.1) belonging to $C^\alpha([\varepsilon, \tau - \varepsilon]; D)$ for every $\varepsilon \in]0, \tau[$, then $u \in C^\infty(]0, \tau[; D)$.

Corollary 8.3.8 Let $F \in C^\infty([0, T] \times \mathcal{O}_D \times \mathcal{O}_\Lambda) \mapsto X$ satisfy assumptions (8.3.2), (8.3.3). Then the function

$$\mathcal{O} = \{(t, x, \lambda) : (x, \lambda) \in \mathcal{M}, \, 0 < t < \tau(x, \lambda)\} \mapsto D, \quad (t, x, \lambda) \mapsto u(t, x, \lambda),$$

has a C^∞ extension in an open neighborhood of $\mathcal{O}$. In particular, if $\mathcal{M}$ is a C^∞ Banach manifold, then $(t, x, \lambda) \mapsto u(t, x, \lambda)$ is C^∞ in $\mathcal{M}$.

We recall that a subset $\mathcal{M}$ of a Banach space Y is said to be a *Banach manifold* (or simply a *manifold*) of class C^k, with $k \in \mathbb{N} \cup \{\infty\}$, if there exist a Banach space Z, an open covering $\mathcal{U}_\alpha$ of $\mathcal{M}$, and a family of mappings $F_\alpha : \mathcal{U}_\alpha \mapsto Z$ such that $F_\alpha : \mathcal{U}_\alpha \mapsto F_\alpha(\mathcal{U}_\alpha)$ is a homeomorphism and $F_\alpha F_\beta^{-1} : F_\beta(\mathcal{U}_\alpha \cap \mathcal{U}_\beta) \mapsto F_\alpha(\mathcal{U}_\alpha \cap \mathcal{U}_\beta)$ is of class C^k.

Note that the set $\mathcal{M}$ defined in (8.3.5) is not necessarily a Banach manifold in $\mathcal{O}_D \times \mathcal{O}_\Lambda$ in the general case. However, in many applications it is indeed a manifold: for instance, it coincides with $\mathcal{O}_D \times \mathcal{O}_\Lambda$ when D is dense in X; it is the intersection of $\mathcal{O}_D \times \mathcal{O}_\Lambda$ with a subspace of $D \times \Lambda$ in some of the applications of Section 8.5.

8.3.3 Analyticity

We end this section with an analyticity result. If X and D are real Banach spaces, we denote by $\widetilde{X}$, $\widetilde{D}$ their complexifications.

Theorem 8.3.9 *Let $F : [0, +\infty[\times \mathcal{O}_D \times \mathcal{O}_\Lambda \mapsto X$ be analytic. Fix $(\overline{u}, \overline{\lambda}) \in \mathcal{M}$, and let $\overline{u} : [0, \tau(\overline{u}, \overline{\lambda})[\mapsto D$ be the maximal solution of (8.3.1), for $\lambda = \overline{\lambda}$, $x = \overline{u}$. Fix moreover $a \in \,]0, \tau(\overline{u}, \overline{\lambda})[$, and let $r_0(\overline{u}, \overline{\lambda}, a)$, be given by Theorem 8.3.4. Then there is a positive $r_0' \leq r_0(\overline{u}, \overline{\lambda}, a)$ such that for every $x \in D$ and $\lambda \in \Lambda$ with*

$$\|x - \overline{u}\|_D \leq r_0', \quad \|\lambda - \overline{\lambda}\|_\Lambda,$$

the solution $u(\cdot; x, \lambda)$ of problem (8.3.9) given by Theorem 8.3.4 is analytic in $]0, a[$ with values in D. Moreover:

(i) *The mapping $(x, \lambda) \mapsto u(\cdot; x, \lambda)$ is analytic in the interior of $B(\overline{u}, r_0') \times B(\overline{\lambda}, r_0') \subset D \times \Lambda$ with values in $C_\alpha^\alpha(]0, a]; D)$;*

(ii) *The function*

$$\mathcal{O} = \{(t, x, \lambda) : (x, \lambda) \in \mathcal{M}, \, 0 < t < \tau(x, \lambda)\} \mapsto D, \quad (t, x, \lambda) \mapsto u(t; x, \lambda),$$

has an analytic extension in a neighborhood of $\mathcal{O}$.

Proof — We follow the proof of Theorem 8.3.4. The analytic dependence on (x, λ) follows also in this case from the Implicit Function Theorem depending on a parameter, provided we show that the function

$$(v, x, \lambda) \mapsto \Psi(v, x, \lambda) = F(\cdot, v(\cdot) + x, \lambda)$$

has an analytic extension in a neighborhood of $(0, \overline{u}, \overline{\lambda}) \in C_\alpha^\alpha(]0, a]; \widetilde{D}) \times \widetilde{D} \times \widetilde{\Lambda}$ with values in $C_\alpha^\alpha(]0, a]; \widetilde{X})$. In fact, since F is analytic, it has a holomorphic extension $\widetilde{F}$ in a neighborhood of $[0, +\infty[\times \{\overline{u}\} \times \{\overline{\lambda}\}$ in $\mathbb{C} \times \widetilde{D} \times \widetilde{\Lambda}$. We may assume without loss of generality that all the derivatives of $\widetilde{F}$ up to the third order are bounded

4. THE CASE WHERE X IS AN INTERPOLATION SPACE

in such a neighborhood. Then, arguing as in the proof of the differentiability of Φ in Theorem 8.3.4, we see that the function

$$\widetilde{\Psi}(v, x, \lambda) = \widetilde{F}(\cdot, v(\cdot) + x, \lambda)$$

is a holomorphic extension of Ψ in a neighborhood of $(0, \overline{u}, \overline{\lambda}) \in C_\alpha^\alpha(]0, a]; \widetilde{D}) \times \widetilde{D} \times \widetilde{\Lambda}$, with values in $C_\alpha^\alpha(]0, a]; \widetilde{X})$. This implies that Ψ is analytic, and statement (i) follows.

To prove time analyticity we use the method of Proposition 8.3.6. We introduce a new real parameter ξ near 1, and we consider the problem satisfied by $w(t) = v(\xi t)$,

$$w'(t) = \xi F(\xi t, v(t) + x, \lambda), \quad 0 < t \leq a; \quad w(0) = 0.$$

Since F is analytic, then the function

$$G(t, v, x, \lambda, \xi) = \xi F(\xi t, v + x, \lambda)$$

is analytic with respect to (v, x, λ, ξ) in a neighborhood of $(0, \overline{u}, \overline{\lambda}, 1)$. Statement (i) yields that $(x, \lambda, \xi) \mapsto w$ is analytic in a neighborhood of $(\overline{u}, \overline{\lambda}, 1)$, with values in $C_\alpha^\alpha(]0, a]; D)$. Therefore, v is analytic with respect to time in $]0, a[$ with values in D, and statement (ii) follows. ∎

8.4 The case where X is an interpolation space

We have considered so far general Banach spaces X and D. So we have been led to look for a solution of (8.0.1) in $C^\alpha([0, \delta]; D)$ or in the weighted space $C_\alpha^\alpha(]0, \delta]; D)$, which created some technical difficulties in the proofs. Things can be considerably simplified if X and D are replaced by the continuous interpolation spaces $D_A(\theta)$, $D_A(\theta + 1)$, because then one can work in $C([0, \delta]; D_A(\theta + 1))$, using the optimal regularity property of $C([0, \delta]; D_A(\theta))$.

Let $E_1 \subset E_0 \subset E$ be Banach spaces, let $0 < \theta < 1$, and let Λ be another Banach space. We are given a number $T > 0$, two open sets $\mathcal{O}_1 \subset E_1$, $\mathcal{O}_\Lambda \subset \Lambda$, and a function

$$F : [0, T] \times \mathcal{O}_1 \times \mathcal{O}_\Lambda \mapsto E_0, \quad (t, x, \lambda) \mapsto F(t, x, \lambda)$$

such that

$$\begin{cases} F \text{ and } F_x \text{ are continuous in } [0, T] \times \mathcal{O}_1 \times \mathcal{O}_\Lambda, \\ \text{for every } (\overline{t}, \overline{u}, \overline{\lambda}) \in [0, T] \times \mathcal{O}_1 \times \mathcal{O}_\Lambda \text{ the operator} \\ F_x(\overline{t}, \overline{u}, \overline{\lambda}) : E_1 \mapsto E_0 \\ \text{is the part in } E_0 \text{ of a sectorial operator } A : D(A) \subset E \mapsto E, \text{ such that} \\ D_A(\theta) \simeq E_0, \quad D_A(\theta + 1) \simeq E_1. \end{cases}$$

(8.4.1)

Theorem 8.4.1 *Let F satisfy assumption (8.4.1). Then for every $\bar{t} \in [0,T]$, $(\bar{u}, \bar{\lambda}) \in \mathcal{O}_1 \times \mathcal{O}_\Lambda$ there are $\delta > 0$, $r > 0$, such that if $t_0 \in [0, T[$, $|t_0 - \bar{t}| \leq \delta$, $\|\lambda - \bar{\lambda}\|_\Lambda \leq r$, $\|x_0 - \bar{u}\|_{E_1} \leq r$, then problem (8.3.7) has a unique solution $v \in C([t_0, t_0 + \delta]; E_1) \cap C^1([t_0, t_0 + \delta]; E_0)$.*

Sketch of the proof — The proof is quite similar to the proofs of Theorems 8.1.1 and 8.1.3. Fix $(\bar{t}, \bar{u}, \bar{\lambda}) \in [0, T] \times \mathcal{O}_1 \times \mathcal{O}_\Lambda$, and consider problem (8.3.7) for t_0 close to $\bar{t}$, x_0 close to $\bar{u}$, λ close to $\bar{\lambda}$, say

$$\|x_0 - \bar{u}\|_{E_1} \leq r_0, \ \|\lambda - \bar{\lambda}\|_\Lambda \leq r_0.$$

Set, as usual,

$$A = F_u(\bar{t}, \bar{u}, \bar{\lambda})$$

and

$$Y = B(\bar{u}, \rho) \subset C([t_0, t_0 + \delta]; E_1),$$

with $\delta \in \,]0, 1]$ and $\rho > 0$ to be chosen suitably. Define the nonlinear operator Γ on Y as in (8.1.5). Using Corollary 4.3.10 and estimate (4.3.20) one sees that Γ is a $1/2$-contraction in Y and maps Y into itself, provided r_0, ρ, δ are sufficiently small. The fixed point of Γ is a solution of (8.3.7).

Uniqueness of the solution in $C([t_0, t_0+\delta]; E_1)$ may be shown in the customary way, see e.g. the proof of uniqueness in Theorem 7.1.3(i). ■

It is possible to extend the result of Theorem 8.4.1 to the case of initial data belonging to $D_A(\theta + 1, \infty)$ instead of $D_A(\theta + 1)$ (provided, however, they are sufficiently close to elements of $D_A(\theta + 1)$). Indeed, if F is continuously differentiable with respect to u for $u \in D_A(\theta + 1, \infty)$, one could work in $C(\,]t_0, t_0 + \delta], D_A(\theta + 1, \infty)) \cap L^\infty(t_0, t_0 + \delta, D_A(\theta + 1))$, thanks to the optimal regularity result of Corollary 4.3.9(iv). The ball $Y = B(\bar{u}, \rho) \subset C([t_0, t_0 + \delta], D_A(\theta + 1))$ should be replaced of course by $Y' = B(\bar{u}, \rho) \subset C(\,]t_0, t_0 + \delta], D_A(\theta + 1, \infty)) \cap L^\infty(t_0, t_0 + \delta, D_A(\theta + 1, \infty))$. The operator Γ is still a $1/2$-contraction, since $C([t_0, t_0+\delta], D_A(\theta+1))$ and $C(\,]t_0, t_0+\delta], D_A(\theta+1, \infty)) \cap L^\infty(t_0, t_0+\delta, D_A(\theta+1))$ have the same norm. Moreover, Γ maps Y' into itself if u_0 is sufficiently close to $\bar{u} \in D_A(\theta + 1)$, and δ is sufficiently small. So, the following proposition holds.

Proposition 8.4.2 *Let assumption (8.4.1) be satisfied, with $D_A(\theta)$, $D_A(\theta + 1)$ replaced respectively by $D_A(\theta, \infty)$, $D_A(\theta + 1, \infty)$. Let $\bar{u} \in E_1$, $0 \leq t_0 < T$, $\lambda_0 \in \mathcal{O}_\Lambda$ be such that setting $A = F_x(t_0, \bar{u}, \lambda_0)$ then $\bar{u} \in D_A(\theta+1)$. Then there exist positive numbers r, $R > 0$ such that for every $x_0 \in E_1 = D_A(\theta + 1, \infty)$, satisfying*

$$\|x_0 - \bar{u}\|_{E_1} \leq r,$$

problem (8.3.7) has a unique local solution $u \in C^1(\,]t_0, t_0+\delta], E_0) \cap C([t_0, t_0+\delta], E) \cap C(\,]t_0, t_0 + \delta], E_1) \cap L^\infty(t_0, t_0 + \delta, E_1)$, such that

$$\|u(t) - \bar{u}\|_{E_1} \leq R, \ \forall t \in [t_0, t_0 + \delta].$$

4. The case where X is an interpolation space

Note that, since the solution given by Proposition 8.4.2 belongs to $C([t_0, t_0 + \delta], E) \cap L^\infty(t_0, t_0+\delta, D_A(\theta+1))$, then it belongs also to $C([t_0, t_0+\delta], D_A(\theta+1-\varepsilon))$ for every $\varepsilon \in]0, \theta+1[$, thanks to Proposition 1.1.4(iii).

Proposition 8.4.2 is particularly meaningful in the study of the behavior near stationary solutions or periodic solutions. In the applications to fully nonlinear parabolic problems, it will let us work in spaces of Hölder continuous functions instead of spaces of little-Hölder continuous functions. See Chapter 9.

Take $t_0 = 0$. Then the local solution $u : [0, \delta] \mapsto E_1$ may be continued to a larger time interval, taking δ as initial time and $u(\delta)$ as initial value. The construction of the maximal interval of existence of the solution is similar to that of Section 8.2. Consider problem (8.3.1), with initial time 0. Under assumption (8.4.1), for every $x_0 \in \mathcal{O}_1$ and $\lambda \in \mathcal{O}_\Lambda$ define

$$\begin{cases} I = I(x, \lambda) = \bigcup \{[0, \delta] : \text{problem (8.3.1) has a solution} \\ \qquad u_\delta \in C([0, \delta]; E_1) \cap C^1([0, \delta]; E_0) \}, \\ u : I \mapsto E_1, \ u(t) = u_\delta(t) \text{ for } t \in [0, \delta] \subset I. \end{cases} \tag{8.4.2}$$

The result of Proposition 8.2.1 still holds: if u is uniformly continuous in $[0, \tau[$ with values in E_1, then either

$$\lim_{t \to \tau} u(t) \in \partial \mathcal{O}_1,$$

or

$$\tau = +\infty.$$

Also the result of Lemma 8.2.2 still holds: if the orbit $\{u(t) : 0 \le t < \tau\}$ is relatively compact in E_1, then u is uniformly continuous. The proofs of these statements are similar to the proofs of Proposition 8.2.1 and of Lemma 8.2.2, and we omit them.

By assumption, the space E_1 coincides with a certain continuous interpolation space $D_A(\theta+1)$. In many applications, if x belongs to $D_A(\theta+\varepsilon+1)$ and $\varepsilon > 0$ is small enough, then $F(t, x, \lambda)$ belongs to $D_A(\theta+\varepsilon)$. In these cases, it is possible to give another sufficient condition for u be uniformly continuous with values in D. To be precise, fix $(\overline{u}, \overline{\lambda}) \in \mathcal{O}_1 \times \mathcal{O}_\Lambda$. By assumption, $F_u(0, \overline{u}, \overline{\lambda})$ is the part in E_0 of the generator of an analytic semigroup $\overline{A} : D(\overline{A}) \mapsto E$. Fix $\varepsilon > 0$ and set $\mathcal{U} = \mathcal{O}_1 \cap D_{\overline{A}}(\theta+\varepsilon+1)$. Then $\mathcal{U}$ is an open set in $D_{\overline{A}}(\theta+\varepsilon+1)$.

Proposition 8.4.3 *Let F satisfy hypothesis (8.4.1), and assume in addition that for some $\lambda \in \mathcal{O}_\Lambda$ the restriction of $F(\cdot, \cdot, \lambda)$ to $[0, T] \times \mathcal{U}$ maps bounded subsets of $[0, T] \times \mathcal{U}$ into bounded subsets of $D_{\overline{A}}(\theta+\varepsilon)$. If there is $K > 0$ such that*

$$\|u(t; x, \lambda)\|_{D_{\overline{A}}(\theta+\varepsilon)} \le K, \ \forall t \in I,$$

then u is uniformly continuous in I with values in $E_1 = D_{\overline{A}}(\theta+1)$.

Proof — Since u is bounded with values in $D_{\overline{A}}(\theta+\varepsilon+1)$, then $\overline{A}u$ and $F(\cdot, u(\cdot), \lambda)$ are bounded with values in $D_{\overline{A}}(\theta+\varepsilon)$. From the equality

$$u'(t) = \overline{A}u(t) + [F(t, u(t), \lambda) - \overline{A}u(t)], \quad 0 \leq t < \tau,$$

we get

$$u(t) = e^{t\overline{A}}x + \int_0^t e^{(t-s)\overline{A}}[F(s, u(s), \lambda) - \overline{A}u(s)]ds, \quad 0 \leq t < \tau.$$

Since the function $s \mapsto F(s, u(s), \lambda) - \overline{A}u(s)$ is bounded with values in $D_{\overline{A}}(\theta+\varepsilon, \infty)$, Theorem 4.3.8 implies that the integral $\int_0^t e^{(t-s)\overline{A}}[F(s, u(s), \lambda) - \overline{A}u(s)]ds$ is bounded with values in $D_{\overline{A}}(\theta+\varepsilon+1, \infty)$ and uniformly continuous (in fact, Hölder continuous) with values in $D(\overline{A})$. By Proposition 1.1.4(iii), u is uniformly continuous with values in $D_{\overline{A}}(\theta+1, \infty)$. ∎

The dependence of the maximal solution on the data may be studied as in Section 8.3, with the aid of the Implicit Function Theorem depending on a parameter.

Theorem 8.4.4 *Let $F : [0, T] \times \mathcal{O}_1 \times \mathcal{O}_\Lambda$ satisfy (8.4.1).*

For every $(\overline{u}, \overline{\lambda}) \in \mathcal{O}_1 \times \mathcal{O}_\Lambda$ and for every $\overline{\tau} \in]0, \tau(\overline{u}, \overline{\lambda})[$ there is $r > 0$ such that if $\|x - \overline{u}\|_D \leq r$, $\|\lambda - \overline{\lambda}\|_\Lambda \leq r$, then

$$\tau(x, \lambda) \geq \overline{\tau},$$

and the mapping

$$\begin{cases} \Phi : B(\overline{u}, r) \times B(\overline{\lambda}, r) \mapsto C([0, \overline{\tau}]; E_1) \cap C^1([0, \overline{\tau}]; E_0), \\ \Phi(x, \lambda) = u(\cdot, x, \lambda), \end{cases} \quad (8.4.3)$$

is continuous with respect to (x, λ), continuously differentiable with respect to x.

If in addition F is k times continuously differentiable[3] *with respect to (x, λ), then Φ is k times continuously differentiable. If F is analytic, then Φ is analytic.*

The proof is omitted, since it is similar to the proofs of Theorems 8.3.4 and 8.3.9. Of course, in this case the set $Z_\alpha([0, a])$ has to be replaced by $\{v \in C([0, a]; E_1) \cap C^1([0, a]; E_0) : v(0) = 0\}$.

Let us consider further regularity with respect to time.

Proposition 8.4.5 *Let F satisfy (8.4.1), and in addition assume that for some $\lambda \in \mathcal{O}_\Lambda$, $F(\cdot, \cdot, \lambda)$ is k times continuously differentiable. Then for every $x \in \mathcal{O}_1$, $u(\cdot, x, \lambda)$ is k times continuously differentiable with values in E_1 and $k+1$ times continuously differentiable with values in E_0, in the open interval $]0, \tau(x, \lambda)[$.*

If F is analytic, then u is analytic in $]0, \tau(x, \lambda)[$ with values in D.

---
[3]We mean that all the derivatives of F with respect to (x, λ) up to the order k are continuous with respect to (t, x, λ).

5. EXAMPLES AND APPLICATIONS

Proof — Fixed any $a \in]0, \tau(x, \lambda)[$, for $\xi > 0$ define $v(t) = u(\xi t, x, \lambda)$. If ξ is sufficiently close to 1, v is defined in $[0, a]$, and

$$v'(t) = \xi F(\xi t, v(t), \lambda), \quad 0 \leq t \leq a; \quad v(0) = x.$$

If F is k times continuously differentiable, then the function

$$G(t, v, \xi) = \xi F(\xi t, v, \lambda)$$

is k times continuously differentiable with respect to (ξ, v), and its derivatives are continuous with respect to (t, v, ξ). By Theorem 8.4.4, for ε small enough the mapping $]1 - \varepsilon, 1 + \varepsilon[\mapsto C([0, a]; E_1) \cap C^1([0, a]; E_0)$, $\xi \mapsto v$, is k times continuously differentiable. Recalling that $u(t) = v(t/\xi)$, it follows that u is k times continuously differentiable with values in D and $k + 1$ times continuously differentiable with values in X in the interval $]0, a]$, with

$$\frac{\partial^h v}{\partial t^h}(t, x, \lambda) = \frac{1}{t^h} \frac{\partial^h v}{\partial \xi^h}(t, x, \xi), \quad h = 1, \ldots, k.$$

If $F(\cdot, \cdot, \lambda)$ is analytic, then by Theorem 8.4.4 v is analytic with respect to ξ. It follows that there are $K, M > 0$ such that

$$\sup_{0 < t \leq a} t^h \left\| \frac{\partial^h v}{\partial t^h}(t, x, \lambda)) \right\|_D = \sup_{0 < t \leq a} \left\| \frac{\partial^h v}{\partial \xi^h}(t, x, \xi) \right\|_1 \leq \frac{K h!}{M^h}, \quad h \in \mathbb{N},$$

which implies that u is analytic in $]0, a[$ with values in E_1. Since a is arbitrary, then u is analytic in $]0, \tau(x, \lambda)[$ with values in E_1. ∎

Corollary 8.4.6 *If F is k times continuously differentiable with respect to all its arguments, then the function*

$$\{(t, x, \lambda) : (x, \lambda) \in \mathcal{O}_1 \times \mathcal{O}_\Lambda, \ 0 < t < \tau(x, \lambda)\} \mapsto E_1, \quad (t, x, \lambda) \mapsto u(t; x, \lambda)$$

is k times continuously differentiable. If F is C^∞, then u is C^∞. If F is analytic, then u is analytic.

8.5 Examples and applications

8.5.1 An equation from detonation theory

In recent papers (see e.g. C.-M. BRAUNER, J. BUCKMASTER, J.W. DOLD, C. SCHMIDT-LAINÉ [34]) a fully nonlinear one dimensional equation has been introduced in the study of the shock displacement in a detonation phenomenon:

$$\begin{cases} g_t(t, x) = \log\left(\dfrac{\exp(cgg_{xx}) - 1}{cg_{xx}}\right) - \tfrac{1}{2}(g_x)^2, & t \geq 0, \ 0 \leq x \leq l, \\ g_x(t, 0) = g_x(t, l) = 0, & t \geq 0, \end{cases} \quad (8.5.1)$$

where c is a positive constant. Only positive solutions have physical meaning. In particular, the constant solution $g \equiv 1$ corresponds to the detonation wave.

The real function

$$f(u,p,q) = \begin{cases} \log\left(\dfrac{e^{cuq}-1}{cq}\right) - \tfrac{1}{2}p^2, & q \neq 0, \\ \log u - \tfrac{1}{2}p^2, & q = 0, \end{cases}$$

is well defined and analytic in the set $\{(u,p,q) \in \mathbb{R}^3 : u > 0\}$. Moreover, for every $u > 0$ and p, $q \in \mathbb{R}$, $f_q(u,p,q) > 0$. So, the differential equation in (8.5.1) is parabolic.

Problem (8.5.1) is supported with the initial condition

$$g(0,x) = g_0(x), \quad 0 \leq x \leq l, \tag{8.5.2}$$

where g_0 is a positive regular (at least C^2) initial datum, satisfying the compatibility condition

$$g_0'(0) = g_0'(l) = 0. \tag{8.5.3}$$

The results of Sections 8.1 and 8.3 yield existence and uniqueness of a local regular solution of (8.5.1)-(8.5.2). Set

$$X = C([0,l]), \quad D = \{g \in C^2([0,l]) : g'(0) = g'(1) = 0\},$$

and

$$\mathcal{O} = \{g \in D : g(x) > 0 \; \forall x \in [0,l]\}.$$

As usual X is endowed with the sup norm, and D is endowed with the C^2 norm. Then $\mathcal{O}$ is an open set in D, and the function

$$F : \mathcal{O} \mapsto X, \quad F(g)(x) = f(g(x), g'(x), g''(x)),$$

is well defined and analytic in $\mathcal{O}$. Moreover, for every $g_0 \in \mathcal{O}$ it holds

$$F'(g_0)g = f_u g(x) + f_p g'(x) + f_q g''(x), \quad g \in D, \; 0 \leq x \leq l,$$

where the derivatives f_u, f_p, f_q are evaluated at $(g_0(x), g_0'(x), g_0''(x))$. Since g_0 is twice continuously differentiable, then there is $\varepsilon > 0$ such that $f_q \geq \varepsilon$ in $[0,l]$. Therefore, $F'(g_0)$ is the realization in X of an elliptic operator with continuous coefficients. By Corollary 3.1.24, $F'(g_0)$ generates an analytic semigroup in X. Its graph norm is clearly equivalent to the norm of D. Moreover, D is dense in X.

By applying Theorem 8.1.1 and Theorem 8.3.9, we find a local existence, uniqueness, and regularity result. To be precise, Theorem 8.1.1 gives uniqueness of the solution in a class of weighted Hölder spaces, but uniqueness of the solution in $C^{1,2}([0,a] \times [0,l])$ follows from an obvious application of the Maximum Principle.

5. EXAMPLES AND APPLICATIONS

Proposition 8.5.1 *For every $g_0 \in \mathcal{O}$ there exist a time interval $[0, \tau[$ and a unique solution g of (8.5.1)-(8.5.2), which belongs to $C^{1,2}([0, a] \times [0, l])$ for every $a \in\,]0, \tau[$. The functions g, g_x, and g_{xx} are analytic with respect to t in the open interval $]0, \tau[$.*

There is no hope, in general, to find existence in the large. A counterexample is constructed by choosing a constant initial datum $g_0 \equiv \gamma$. Then the solution is independent of x, and it may be explicitly computed by integrating the ordinary differential equation

$$g'(t) = \log g(t), \quad t \geq 0; \quad g(0) = \gamma.$$

If $0 < \gamma < 1$, the solution is $g(t) = \varphi^{-1}(t)$, where $\varphi(s) = \int_\gamma^s \frac{d\sigma}{\log \sigma}$. It is defined in the interval $[0, \varphi(0)[$, and it goes to 0 as $t \to \varphi(0)$. Using comparison arguments, it is possible in fact to see that for every initial datum g_0 such that $0 < g_0(x) < 1$, the solution cannot exist in the large, since it goes to 0 in finite time.

A thorough study of problem (8.5.1)-(8.5.2), with detailed description of the dynamics near the stationary solution $g(t, x) \equiv 1$, may be found in C.-M. BRAUNER – C. SCHMIDT-LAINÈ – S. GERBI [36].

Similar arguments may be used to study the general one dimensional problem

$$\begin{cases} u_t(t, x) = f(t, x, u(t, x), u_x(t, x), u_{xx}(t, x)), & t \geq 0, \ 0 \leq x \leq 1, \\ \alpha_0 u_x(t, 0) + \beta_0 u(t, 0) = \alpha_1 u_x(t, 1) + \beta_1 u(t, 1) = 0, & t \geq 0, \\ u(0, x) = u_0(x), \ 0 \leq x \leq 1, \end{cases} \quad (8.5.4)$$

with $\alpha_i, \beta_i \in \mathbb{R}$, $\alpha_i^2 + \beta_i^2 > 0$, $i = 1, 2$. We need that the problem is parabolic, i.e.

$$f_q(t, x, u, p, q) > 0 \quad \forall (t, x, u, p, q),$$

and that $u_0 \in C^2([0, 1])$ satisfies the compatibility condition

$$\alpha_i u_0'(i) + \beta_i u_0(i) = 0, \quad i = 0, 1.$$

If α_0 or α_1 vanish, the domain $D = \{g \in C^2([0, 1]) : \alpha_i g'(i) + \beta_i g(i) = 0, \ i = 0, 1\}$ is not dense in $X = C([0, 1])$, and we need the further compatibility condition $f(0, 0, u_0(0), u_0'(0), u_0''(0)) = 0$ (respectively, $f(0, 1, u_0(1), u_0'(1), u_0''(1)) = 0$).

8.5.2 An example of existence in the large

Consider the initial boundary value problem

$$\begin{cases} u_t = f(\Delta u), & t \geq 0, \ x \in \overline{\Omega}, \\ u(t, x) = 0, & t \geq 0, \ x \in \partial \Omega, \\ u(0, x) = u_0(x), & x \in \overline{\Omega}. \end{cases} \quad (8.5.5)$$

where Ω is a bounded open set in $\mathbb{R}^n$ with regular boundary $\partial\Omega$, $f : \mathbb{R} \mapsto \mathbb{R}$ is a regular increasing function such that $f(0) = 0$, $u_0 : \overline{\Omega} \mapsto \mathbb{R}$ is a regular function satisfying the compatibility conditions

$$u_{0|\partial\Omega} = \Delta u_{0|\partial\Omega} = 0. \tag{8.5.6}$$

Local existence and uniqueness results follow from the results of Sections 8.1, 8.4, or else from the extension of next Theorem 8.5.4 to the case of the Dirichlet boundary condition. Due to the special character of the differential equation in (8.5.5), the application of the abstract results of Section 8.1 is particularly simple. Indeed, it is natural to see problem (8.5.5) as an initial value problem in the space $X = C(\overline{\Omega})$, with domain

$$D = D(\Delta) = \{\varphi \in W^{2,p}(\Omega) \, \forall p \geq 1 : \, \Delta\varphi \in C(\overline{\Omega}), \, \varphi_{|\partial\Omega} = 0\}.$$

As usual, X is endowed with the sup norm, and D is endowed with the graph norm of the Laplace operator. Setting

$$F : D \mapsto X, \quad F(\varphi) = f(\Delta\varphi),$$

then F (as a function from D to X) enjoys the same regularity properties of f (as a function from $\mathbb{R}$ to $\mathbb{R}$). In particular, if f is differentiable, then F is differentiable, and for every $u_0 \in D$

$$F'(u_0)v(x) = f'(\Delta u_0(x))\Delta v(x),$$

so that $F'(u_0)$ is an elliptic operator with continuous coefficients, which is sectorial in X thanks to Corollary 3.1.21(ii). Moreover, its graph norm is equivalent to the norm of D. So, according to the regularity of u_0 and f, the existence and regularity results of Sections 8.1 and 8.3 may be applied. We state in the next proposition only some properties of the solution which will be useful to prove existence in the large, leaving to the reader the application of the other results.

Proposition 8.5.2 *Let $0 < \alpha < 1$. Assume that $\partial\Omega$ is uniformly $C^{2+\alpha}$, that $u_0 \in D$ satisfies (8.5.6), and that $f : \mathbb{R} \mapsto \mathbb{R}$ is an increasing C^3 function, with $f(0) = 0$. Then there are a maximal $\tau > 0$ and a unique $u \in C([0, \tau[\times\overline{\Omega})$, with u_t and Δu in $C([0, \tau[\times\overline{\Omega})$, which satisfies (8.5.6) in $[0, \tau[\times\overline{\Omega}$. Moreover, $u \in C^{1+\alpha/2, 2+\alpha}([\varepsilon, a] \times \overline{\Omega})$ for $0 < \varepsilon < a < \tau$, $u_t(t, \cdot)$ belongs to D for $0 < t < \tau$, and u_{tt}, Δu_t exist and belong to $C^{\alpha/2,\alpha}([\varepsilon, a] \times \overline{\Omega})$ for $0 < \varepsilon < a < \tau$.*

Proof — Uniqueness of the local solution $u \in C([0, \tau[\times\overline{\Omega})$, with u_t and Δu in $C([0, \tau[\times\overline{\Omega})$ is a simple consequence of the Maximum Principle, trough Proposition 3.1.10. Existence of such a solution follows immediately from Theorem 8.1.1 and from the arguments at the beginning of Section 8.2: indeed, since f is twice continuously differentiable, then $F : D \mapsto X$ is twice continuously Fréchet differentiable. Moreover, the assumption $\Delta u_{0|\partial\Omega} = 0$ implies that $F(u_0) \in \overline{D} = \{\varphi \in C(\overline{\Omega}) : \varphi_{|\partial\Omega} = 0\}$. So, the problem

$$U'(t) = F(U(t)), \quad t \geq 0; \quad U(0) = u_0,$$

5. EXAMPLES AND APPLICATIONS   317

has a unique local solution $U \in C([0, \tau[; D) \cap C^1([0, \tau[; X)$, such that $u \in C_\beta^\beta(]0, a]$; D), for every $a \in]0, \tau[$ and $\beta \in]0, 1[$. Setting

$$u(t, x) = U(t)(x), \quad 0 \le t < \tau, \; x \in \overline{\Omega},$$

then u is a solution of (8.5.5), with u, u_t, and Δu in $C([0, \tau[\times\overline{\Omega})$. Fix $a < \tau$ and take $\beta = \alpha/2$. The fact that $U \in C_{\alpha/2}^{\alpha/2}(]0, a]; D)$ implies that $U \in C^{\alpha/2}([\varepsilon, a]; D) \cap C^{1+\alpha/2}([\varepsilon, a]; X)$ for every $\varepsilon \in]0, a[$, so that $U' \in B([\varepsilon, a]; (X, D)_{\alpha/2,\infty})$. Thanks to the characterization of Theorem 3.1.29,

$$(X, D)_{\alpha/2,\infty} = C^\alpha(\overline{\Omega}).$$

Therefore, $\Delta u(t, \cdot) = f^{-1}(u_t(t, \cdot)) \in C^\alpha(\overline{\Omega})$ for $0 < t \le T$. Since $\partial \Omega$ is uniformly $C^{2+\alpha}$, from the Schauder Theorem 3.1.24(i) it follows that $u(t, \cdot) \in C^{2+\alpha}(\overline{\Omega})$ for $t > 0$, and in addition $\|u(t, \cdot)\|_{C^{2+\alpha}(\overline{\Omega})}$ is bounded in $[\varepsilon, a]$ for every $\varepsilon \in]0, a[$. By Lemma 5.1.1, the second order space derivatives $D_{ij}u$ belong to $C^{(1+\alpha)/2,0}([\varepsilon, a] \times \overline{\Omega})$, and hence to $C^{\alpha/2,\alpha}([\varepsilon, a] \times \overline{\Omega})$. Therefore, $u \in C^{1+\alpha/2,2+\alpha}([\varepsilon, a] \times \overline{\Omega})$, for every $\varepsilon \in]0, a[$.

From Proposition 8.3.6, with $k=1$, it follows that U belongs to $C^{1+\alpha/2}([\varepsilon, a]; D) \cap C^{2+\alpha/2}([\varepsilon, a]; X)$, and hence $U'' \in B([\varepsilon, a]; (X, D)_{\alpha,\infty}) \subset B([\varepsilon, a]; C^\alpha(\overline{\Omega}))$, for every $\varepsilon \in]0, a[$. This implies that u_{tt} exists and belongs to $C^{\alpha/2,\alpha}([\varepsilon, a] \times \overline{\Omega})$, that $u_t(t, \cdot) \in W^{2,p}(\Omega)$ for every p for $t > 0$, and $\Delta u_t \in C^{\alpha/2,0}([\varepsilon, a] \times \overline{\Omega})$. Moreover, since $u_{tt} = f'(\Delta u)\Delta u_t$, then $\Delta u_t = u_{tt}/f'(\Delta u)$ belongs to $B([\varepsilon, a]; C^\alpha(\overline{\Omega}))$. Therefore, $\Delta u_t \in C^{\alpha/2,\alpha}([\varepsilon, a] \times \overline{\Omega})$, and the statement follows. ∎

Proposition 8.5.3 *Let the hypotheses of Proposition 8.5.2 hold, and assume in addition that*

$$f'(\xi) \ge \nu > 0 \text{ for every } \xi \in \mathbb{R}.$$

Then $\tau = +\infty$, i.e. the solution u of (8.5.5) exists in the large.

Proof — We shall show that the function $U(t) = u(t, \cdot)$ is uniformly continuous in $[0, \tau[$ with values in D. Then the statement will follow from Proposition 8.2.1.

Fix ε, $a \in]0, \tau[$, $\varepsilon < a$. From Proposition 8.5.2 we know that u_t and Δu are continuously differentiable with respect to time in $[\varepsilon, a] \times \overline{\Omega}$. Setting $u_t = v$, it holds

$$\begin{cases} v_t = f'(\Delta u)\Delta v, & \varepsilon \le t \le a, \; x \in \overline{\Omega}, \\ v(t, x) = 0, & \varepsilon \le t \le a, \; x \in \partial\Omega, \\ v(\varepsilon, x) = f(\Delta u(\varepsilon, x)), & x \in \overline{\Omega}. \end{cases} \quad (8.5.7)$$

Using the Maximum Principle we see that v is bounded, and $\|v\|_\infty \le \|f(\Delta u(\varepsilon, \cdot))\|_\infty$. Letting $\varepsilon \to 0$ we get $|u_t(t, x)| \le \|f(\Delta u_0)\|_\infty$. Therefore,

$$\|U(t) - U(s)\|_\infty \le \|f(\Delta u_0)\|_\infty (t - s), \quad 0 \le s < t < \tau,$$

so that U is uniformly continuous with values in X. Let us prove that ΔU is uniformly continuous with values in X. From the equality $\Delta u = f^{-1}(u_t)$ it follows that Δu is bounded, and precisely

$$|\Delta u(t,x)| \leq \frac{1}{\nu}\|f(\Delta u_0)\|_\infty.$$

From the regularity part of Proposition 8.5.2 we know that $u_t(t,\cdot) = f(\Delta u(t,\cdot))$ belongs to D for $t > 0$. In particular, $f(\Delta u(t,\cdot)) \in W^{2,p}(\Omega)$ for every $p \in [1,\infty[$, so that $\Delta u \in W^{2,p}(\Omega)$ for every $p \in [1,\infty[$. The function $z = \Delta u$ satisfies

$$\begin{cases} z_t = \Delta f(\Delta u) = \sum_{i=1}^n D_i(f'(\Delta u)D_i z), & \varepsilon \leq t < \tau,\ x \in \overline{\Omega}, \\ z(t,x) = 0, & \varepsilon \leq t < \tau,\ x \in \partial\Omega, \\ z(\varepsilon,x) = \Delta u(\varepsilon,x), & x \in \overline{\Omega}, \end{cases}$$

where the coefficient $f'(\Delta u)$ belongs to $L^\infty(]0,\tau[\times\Omega)$, and the initial datum $\Delta u(\varepsilon,\cdot)$ is continuously differentiable in $\overline{\Omega}$. By the Nash-Moser Theorem it follows that there are $K > 0$, $\beta \in]0,1[$ such that

$$\|\Delta u\|_{C^{\beta/2,\beta}([\varepsilon,a]\times\overline{\Omega})} \leq K, \quad \forall a \in]\varepsilon,\tau[. \tag{8.5.8}$$

Since Δu is continuous in $[0,\varepsilon] \times \overline{\Omega}$, then Δu is uniformly continuous in $[0,\tau[\times\overline{\Omega}$, and the statement follows. ∎

It is clear from the proof that the existence in the large result of Proposition 8.5.3 may be extended to more general equations. The argument works as well if the Laplace operator Δ is replaced by any elliptic operator in divergence form $\mathcal{A} = \sum_{i=1}^n D_i(a_{ij}(x)D_j)$ with Hölder continuous coefficients.

In the proof of Proposition 8.5.3, instead of the Nash-Moser Theorem (for a proof, see O.A. LADYZHENSKAJA – V.A. SOLONNIKOV – N.N. URAL'CEVA [124, Thm. 10.1] or the original papers by J. NASH [164] and J. MOSER [162]) it is possible to use the Krylov-Safonov Theorem, and more precisely the Krylov-Safonov estimates up to the boundary, which may be found for instance in the paper by M. GRUBER [96] (the original paper by N.V. KRYLOV – M.V. SAFONOV [123] gives only interior estimates). Such a theorem has to be applied to the equation satisfied by $v = u_t$, and it gives

$$\|u_t\|_{C^{\beta'/2,\beta'}([\varepsilon,a]\times\overline{\Omega})} \leq K', \quad \forall a \in]\varepsilon,\tau[,$$

for some $K' > 0$, $\beta' \in]0,1[$. Recalling that $\Delta u = f^{-1}(u_t)$, one gets (8.5.8).

If Δ is replaced by an elliptic operator in nondivergence form $\mathcal{A} = \sum_{i,j=1}^n a_{ij}(x) D_{ij}$, with Hölder continuous coefficients, then the result of Proposition 8.5.3 still holds, but in the proof one has to use the Krylov-Safonov Theorem instead of the Nash-Moser Theorem.

A remarkable feature of these existence in the large results is the fact that no growth condition is imposed on f. Such results have been pointed out by W. VON

5. EXAMPLES AND APPLICATIONS

WAHL [207] and N. KIKUCHI [110]. They have been generalized to equations of the type $u_t = f(t, x, u, Du, \mathcal{A}u)$ by A. LUNARDI [146].

Existence in the large for classical solutions of second order fully nonlinear problems of general type with Dirichlet boundary condition has been considered by N. KRYLOV in the papers [119], [120], [121], and in the book [122]. He gives sufficient growth and qualitative conditions on f in order to find *a priori* estimates on the solution in the $C^{1+\alpha/2, 2+\alpha}$ norm, for some $\alpha > 0$. Once *a priori* estimates in the $C^{1+\alpha/2, 2+\alpha}$ norm are established, it is not hard to prove existence in the large, using for instance the continuity method such as in [122, Thm. 6.2.5].

Global weak solutions of certain second order equations may be found by the nonlinear semigroup approach (see e.g. P. BENILAN – K.S. HA [31], Y. KONISHI [113], and C.-Y. LIN [126] for the case where the space dimension is 1), and by the viscosity solution approach (see H. ISHII – P.L. LIONS [104] and the survey paper M.G. CRANDALL – H. ISHII – P.-L. LIONS [57]). Some regularity results for the solutions provided by the above methods may be found respectively in M.I. HAZAN [98] and in L. WANG [208].

8.5.3 A general second order problem

Let Ω be a bounded open set in $\mathbb{R}^n$ with regular boundary $\partial\Omega$. Consider a second order fully nonlinear equation in $[0, T] \times \overline{\Omega}$, supported with first order nonlinear condition at the boundary and initial value,

$$\begin{cases} u_t(t, x) = f(t, x, u(t, x), Du(t, x), D^2 u(t, x)), & t \geq 0, \ x \in \overline{\Omega}, \\ u(0, x) = u_0(x), & x \in \overline{\Omega}, \\ g(t, x, u(t, x), Du(t, x)) = 0, & t \geq 0, \ x \in \partial\Omega, \end{cases} \quad (8.5.9)$$

where f, u_0, g are given regular functions, Du denotes the gradient, and $D^2 u$ is the ordered set of all second order spatial derivatives of u. The real function $(t, x, u, p, q) \mapsto f(t, x, u, p, q)$ is defined in $Q = [0, T] \times \overline{\Omega} \times B((\overline{u}, \overline{p}, \overline{q}), R_0)$, where[4] $(\overline{u}, \overline{p}, \overline{q}) \in \mathbb{R} \times \mathbb{R}^n \times \mathbb{R}_S^{n^2}$ satisfies the ellipticity condition

$$\sum_{i,j=1}^n \frac{\partial f}{\partial q_{ij}}(t, x, u, p, q)\xi_i\xi_j > 0, \quad \forall (t, x, u, p, q) \in Q, \ \xi = (\xi_1, \ldots, \xi_n) \in \mathbb{R}^n \setminus \{0\}, \tag{8.5.10}$$

and $g : S = [0, T] \times \overline{\Omega} \times B((\overline{u}, \overline{p}), R_0)$, $(t, x, u, p) \mapsto g(t, x, u, p)$ satisfies the non-tangentiality condition

$$\sum_{i=1}^n \frac{\partial g}{\partial p_i}(t, x, u, p)\nu_i(x) \neq 0, \quad \forall (t, x, u, p, q) \in S \text{ such that } x \in \partial\Omega. \tag{8.5.11}$$

The abstract theory developed in the previous sections cannot be applied directly to problem (8.5.9), because of the nonlinear boundary condition. Instead

---

[4] $\mathbb{R}_S^{n^2}$ is the subset of $\mathbb{R}^{n^2}$ consisting of all symmetric $n \times n$ matrices.

of developing an abstract theory for problems with nonlinear boundary condition, it seems simpler to work directly on the problem (8.5.9), using the linearization procedure of Sections 8.1 and 8.4, and some of the optimal regularity results of Subsection 5.1.2.

The most familiar optimal regularity result for linear problems with oblique derivative boundary condition is certainly Corollary 5.1.22, dealing with $C^{1+\alpha/2, 2+\alpha}$ regularity. To use such a result we need certain regularity assumptions on $\partial\Omega$, f and g, listed below.

(i) There exists $\alpha \in\ ]0,1[$ such that $\partial\Omega$ is uniformly $C^{2+\alpha}$;

(ii) $f : Q \mapsto \mathbb{R}$ is differentiable with respect to $z = (u, p, q)$, moreover f, f_{p_i}, $f_{q_{ij}}$ are locally Lipschitz continuous with respect to z and locally $C^{\alpha/2, \alpha}$ with respect to (t, x), uniformly with respect to the other variables: i.e., for every $\bar{t} \geq 0$ we have

$$\sup\{\|D_z^\beta f(\cdot, \cdot, z)\|_{C^{\alpha/2,\alpha}([0,\bar{t}]\times\overline{\Omega})} : z \in B((\bar{u},\bar{p},\bar{q}), R_0),\ |\beta| = 0, 1\} = K < \infty, \tag{8.5.12}$$

and there exists $L > 0$ such that

$$\begin{aligned}&|D_z^\beta f(t, x, z_1) - D_z^\beta f(t, x, z_2)| \leq L|z_1 - z_2|,\\ &\forall (t, x) \in [0, \bar{t}] \times \overline{\Omega},\ z_1,\ z_2 \in B((\bar{u},\bar{p},\bar{q}), R_0),\ |\beta| = 0, 1.\end{aligned} \tag{8.5.13}$$

Similarly,

(iii) $g : S \mapsto \mathbb{R}$ is twice differentiable with respect to $w = (u, p)$, each derivative up to the second order is locally Lipschitz continuous with respect to w and locally $C^{(1+\alpha)/2, 1+\alpha}$ with respect to (t, x), uniformly with respect to the other variables: i.e., for every $\bar{t} \geq 0$ we have

$$\sup\{\|D_w^\beta g(\cdot, \cdot, w)\|_{C^{(1+\alpha)/2, 1+\alpha}([0,\bar{t}]\times\overline{\Omega})} : w \in B((\bar{u},\bar{p}), R_0),\ |\beta| = 0, 1, 2\} = H < \infty \tag{8.5.14}$$

and there exists $M > 0$ such that

$$\begin{aligned}&|D_w^\beta g(t, x, w_1) - D_w^\beta g(t, x, w_2)| \leq M|w_1 - w_2|,\\ &\forall (t, x) \in [0, \bar{t}] \times \overline{\Omega},\ w_1,\ w_2 \in B((\bar{u},\bar{p}), R_0),\ |\beta| = 0, 1, 2.\end{aligned} \tag{8.5.15}$$

Assumptions (8.5.12) and (8.5.13) are satisfied if f is twice continuously differentiable with respect to all its arguments, and assumptions (8.5.14), (8.5.15) are satisfied if g is thrice continuously differentiable with respect to all its arguments.

We are now in position of stating a local existence and uniqueness result.

Theorem 8.5.4 *Let (8.5.10),…,(8.5.15) hold. Assume that $u_0 \in C^{2+\alpha}(\overline{\Omega})$ verifies the compatibility condition*

$$g(0, x, u_0(x), Du_0(x)) = 0,\ x \in \partial\Omega, \tag{8.5.16}$$

5. EXAMPLES AND APPLICATIONS

and that the range of (u_0, Du_0, D^2u_0) is contained in the ball centered at $(\bar{u}, \bar{p}, \bar{q})$ with radius $R_0/2$. Then there are $\delta > 0$ and a unique $u \in C^{1+\alpha/2, 2+\alpha}([0, \delta] \times \overline{\Omega})$ satisfying (8.5.9) in $[0, \delta] \times \overline{\Omega}$.

Proof — We use a method similar to the one employed in the proof of Theorem 5.1.21. In fact, the proof below reduces to the proof of Theorem 5.1.21 and Corollary 5.1.22 in the case where f and g are linear with respect to (u, p, q).

Set

$$\mathcal{A}v = \sum_{i,j=1}^{n} f_{q_{ij}} D_{ij}v + \sum_{i=1}^{n} f_{p_i} D_i v + f_u v, \quad \mathcal{B}v = \sum_{i=1}^{n} g_{p_i} D_i v + g_u v,$$

where the derivatives of f are evaluated at the point $(0, x, u_0(x), Du_0(x), D^2u_0(x))$, and the derivatives of g are evaluated at the point $(0, x, u_0(x), Du_0(x))$. Let $\delta \leq 1$, R be positive numbers to be precised later, satisfying

$$(\delta + C'_\alpha \delta^{(1+\alpha)/2} + \delta^{\alpha/2})R \leq R_0/2, \quad (8.5.17)$$

where C'_α is the constant given by Lemma 5.1.1. We seek the solution of (8.5.9) as a fixed point of the operator Γ defined in the set

$$Y = \{u \in C^{1+\alpha/2, 2+\alpha}([0, \delta] \times \overline{\Omega}) : u(0, \cdot) = u_0, \ \|u - u_0\|_{C^{1+\alpha/2, 2+\alpha}} \leq R\}$$

by $\Gamma(u) = w$, where w is the solution of

$$\begin{cases} w_t(t, x) = \mathcal{A}v(t, x) + f(t, x, u(t, x), Du(t, x), D^2u(t, x)) - \mathcal{A}u(t, x), \\ \qquad\qquad 0 \leq t \leq \delta, \ x \in \overline{\Omega}, \\ \mathcal{B}w(t, x) = -g(t, x, u(t, x), Du(t, x)) + \mathcal{B}u(t, x), \ 0 \leq t \leq \delta, \ x \in \partial\Omega, \\ w(0, x) = u_0(x), \ x \in \overline{\Omega}. \end{cases}$$

Note that for every $u \in Y$

$$\sum_{i,j=1}^{n} \|D_{ij}u - D_{ij}u_0\|_\infty + \sum_{i=1}^{n} \|D_i u - D_i u_0\|_\infty + \|u - u_0\|_\infty$$
$$\leq (\delta^{\alpha/2} + C'_\alpha \delta^{(1+\alpha)/2} + \delta)R,$$

so that, due to (8.5.17), the range of (u, Du, D^2u) is contained in the ball $B((\bar{u}, \bar{p}, \bar{q}), R_0)$.

We shall show that for every $u, v \in Y$

$$\|\Gamma(u) - \Gamma(v)\|_{C^{1+\alpha/2, 2+\alpha}} \leq C(R)\delta^{\alpha/2} \|u - v\|_{C^{1+\alpha/2, 2+\alpha}}. \quad (8.5.18)$$

So, if δ and R satisfy

$$C(R)\delta^{\alpha/2} \leq 1/2 \quad (8.5.19)$$

then Γ is a 1/2-contraction, and we get also, for every $u \in Y$,

$$\|\Gamma(u) - u_0\|_{C^{1+\alpha/2, 2+\alpha}} \leq R/2 + \|\Gamma(u_0) - u_0\|_{C^{1+\alpha/2, 2+\alpha}}. \tag{8.5.20}$$

The function $v = \Gamma(u_0) - u_0$ is the solution of

$$\begin{cases} v_t(t,x) = \mathcal{A}v(t,x) + f(t,x,u_0(x),Du_0(x),D^2u_0(x)), & 0 \leq t \leq \delta, \ x \in \overline{\Omega}, \\ \mathcal{B}v(t,x) = -g(t,x,u_0(x),Du_0(x)), & 0 \leq t \leq \delta, \ x \in \partial\Omega, \\ v(0,x) = 0, & x \in \overline{\Omega}. \end{cases}$$

By Corollary 5.1.22 there is $C > 0$, independent of δ, such that

$$\|v\|_{C^{1+\alpha/2, 2+\alpha}} \leq C(\|f(\cdot,\cdot,u_0,Du_0,D^2u_0)\|_{C^{\alpha/2,\alpha}} + \|g(\cdot,\cdot,u_0,Du_0)\|_{C^{(1+\alpha)/2, 1+\alpha}})$$
$$= C'.$$

Summing up, we find

$$\|\Gamma(u) - u_0\|_{C^{1+\alpha/2, 2+\alpha}} \leq R/2 + C'.$$

Therefore for R suitably large Γ is a contraction mapping Y into itself, and it has a unique fixed point in Y. Uniqueness of the solution in $C^{1+\alpha/2, 2+\alpha}([t_0, t_0+\delta] \times \overline{\Omega})$ follows easily, arguing as in the proof of Theorem 7.1.2, or else applying the Maximum Principle to the problem satisfied by the difference of two solutions.

The verification of (8.5.18) is rather lenghty, although staightforward.

Proof of (8.5.18).

Let $u, v \in Y$. Then $w = \Gamma(u) - \Gamma(v)$ satisfies

$$\begin{cases} w_t = \mathcal{A}w + f(t,x,u,Du,D^2u) - f(t,x,v,Dv,D^2v) - \mathcal{A}(u-v), \\ \qquad 0 \leq t \leq \delta, \ x \in \overline{\Omega}, \\ \mathcal{B}w = \mathcal{B}(u-v) - g(t,x,u,Du) + g(t,x,v,Dv), \\ \qquad 0 \leq t \leq \delta, \ x \in \overline{\Omega}, \\ w(0,x) = 0, \ x \in \overline{\Omega}. \end{cases}$$

By Corollary 5.1.22, there is $C > 0$, independent of δ, such that

$$\|w\|_{C^{1+\alpha/2, 2+\alpha}} \leq C(\|\varphi\|_{C^{\alpha/2,\alpha}} + \|\psi\|_{C^{(1+\alpha)/2, 1+\alpha}})$$

where

$$\varphi(t,x) = f(t,x,u(t,x),Du(t,x),D^2u(t,x)) \\ - f(t,x,v(t,x),Dv(t,x),D^2v(t,x)) - (\mathcal{A}u(t,x) - \mathcal{A}v(t,x)),$$

$$\psi(t,x) = \mathcal{B}u(t,x) - \mathcal{B}v(t,x) - g(t,x,u(t,x),Du(t,x)) + g(t,x,v(t,x),Dv(t,x)).$$

5. Examples and applications

To estimate $\|\varphi\|_{C^{\alpha/2,\alpha}}$ it is convenient to write $\varphi(t,x)$ as

$$\varphi(t,x) = \int_0^1 (f_u(t,x,\xi_\sigma(t,x)) - f_u(0,x,\xi_0(x)))\,(u(t,x) - v(t,x))d\sigma$$

$$+ \int_0^1 \sum_{i=1}^n (f_{p_i}(t,x,\xi_\sigma(t,x)) - f_{p_i}(0,x,\xi_0(x)))\,(D_i u(t,x) - D_i v(t,x))d\sigma$$

$$+ \int_0^1 \sum_{i,j=1}^n (f_{q_{ij}}(t,x,\xi_\sigma(t,x)) - f_{q_{ij}}(0,x,\xi_0(x)))\,(D_{ij} u(t,x) - D_{ij} v(t,x))d\sigma,$$

with

$$\xi_\sigma(t,x) = \sigma(u(t,x), Du(t,x), D^2 u(t,x)) + (1-\sigma)(v(t,x), Dv(t,x), D^2 v(t,x)),$$
$$\xi_0(x) = (u_0(x), Du_0(x), D^2 u_0(x)).$$

There is $C > 0$ such that

$$|\xi_\sigma(t,x) - \xi_\sigma(t,y)| \leq C(\|u_0\|_{C^{2+\alpha}(\overline{\Omega})} + R)|x-y|^\alpha,$$

$$|\xi_0(x) - \xi_0(y)| \leq C\|u_0\|_{C^{2+\alpha}(\overline{\Omega})}|x-y|^\alpha$$

To estimate $|\varphi(t,x) - \varphi(s,x)|$ for $0 \leq s \leq t \leq \delta$, and $x \in \overline{\Omega}$, it is convenient to add and subtract

$$\int_0^1 f_u(s,x,\xi_\sigma(s,x))(u(t,x) - v(t,x))d\sigma$$

$$+ \int_0^1 \sum_{i=1}^n f_{p_i}(s,x,\xi_\sigma(s,x))(D_i u(t,x) - D_i v(t,x))d\sigma$$

$$+ \int_0^1 \sum_{i,j=1}^n f_{q_{ij}}(s,x,\xi_\sigma(s,x))(D_{ij} u(t,x) - D_{ij} v(t,x))d\sigma.$$

So, we have to estimate

$$|D^\beta f(t,x,\xi_\sigma(t,x)) - D^\beta f(s,x,\xi_\sigma(s,x))|$$

and

$$|D^\beta f(s,x,\xi_\sigma(s,x)) - D^\beta f(0,x,\xi_0(x))|,$$

where $D^\beta f$ is any first order derivative of f with respect to (u,p,q). Thanks to (8.5.11) and (8.5.12) it holds

$$|D^\beta f(t,x,\xi_\sigma(t,x)) - D^\beta f(s,x,\xi_\sigma(s,x))| \leq \left(K + \frac{L}{2}([u]_{C^{\alpha/2,0}} + [v]_{C^{\alpha/2,0}}\right.$$

$$\left. +[Du]_{C^{\alpha/2,0}} + [Dv]_{C^{\alpha/2,0}} + [D^2 u]_{C^{\alpha/2,0}} + [D^2 v]_{C^{\alpha/2,0}}\right)(t-s)^{\alpha/2}$$

$$\leq C_1(R)(t-s)^{\alpha/2},$$

and

$$|D^\beta f(s,x,\xi_\sigma(s,x)) - D^\beta f(0,x,\xi_0(x))|$$
$$\leq K|s-t_0|^{\alpha/2} + \frac{L}{2}\big([u-u_0]_{C^{\alpha/2,0}} + [v-u_0]_{C^{\alpha/2,0}} + [D(u-u_0)]_{C^{\alpha/2,0}}$$
$$+[D(v-u_0)]_{C^{\alpha/2,0}} + [D^2(u-u_0)]_{C^{\alpha/2,0}} + [D^2(v-u_0)]_{C^{\alpha/2,0}}\big)\delta^{\alpha/2}$$
$$\leq C_2(R)\delta^{\alpha/2}.$$

So, we get

$$|\varphi(t,x) - \varphi(s,x)| \leq C_1(R)(t-s)^{\alpha/2}\Big(\delta\|u_t - w_t\|_\infty$$
$$+\delta^{(1+\alpha)/2}\sum_{i=1}^n [D_i(u-w)]_{C^{(1+\alpha)/2,0}} + \delta^{\alpha/2}\sum_{i,j=1}^n [D_{ij}(u-w)]_{C^{\alpha/2,0}}\Big)$$
$$+C_2(R)\delta^{\alpha/2}\Big((t-s)\|u_t - v_t\|_\infty + (t-s)^{(1+\alpha)/2}\sum_{i=1}^n [D_i(u-v)]_{C^{(1+\alpha)/2,0}}$$
$$+(t-s)^{\alpha/2}\sum_{i,j=1}^n [D_{ij}(u-v)]_{C^{\alpha/2,0}}\Big)$$
$$\leq C_3(R)\delta^{\alpha/2}(t-s)^{\alpha/2}\|u-v\|_{C^{1+\alpha/2,2+\alpha}}, \tag{8.5.21}$$

which implies also (since $\varphi(0,\cdot) \equiv 0$)

$$\|\varphi\|_\infty \leq C_3(R)\delta^\alpha \|u-v\|_{C^{1+\alpha/2,2+\alpha}}. \tag{8.5.22}$$

To estimate $|\varphi(t,x) - \varphi(t,y)|$ it is convenient to add and subtract

$$\int_0^1 \big(f_u(t,y,\xi_\sigma(t,y)) - f_u(0,y,\xi_0(y))\big)(u(t,x) - v(t,x))d\sigma$$

$$+\int_0^1 \sum_{i=1}^n \big(f_{p_i}(t,y,\xi_\sigma(t,y)) - f_{p_i}(0,y,\xi_0(y))\big)(D_i u(t,x) - D_i v(t,x))d\sigma$$

$$+\int_0^1 \sum_{i=1}^n \big(f_{q_{ij}}(t,y,\xi_\sigma(t,y)) - f_{q_{ij}}(0,y,\xi_0(y))\big)(D_{ij}u(t,x) - D_{ij}v(t,x))d\sigma.$$

Thanks to (5.1.11) and (8.5.13), for any first order derivative $D^\beta f$ with respect to (u,p,q) it holds

$$|D^\beta f(t,x,\xi_\sigma(t,x)) - D^\beta f(t,y,\xi_\sigma(t,y))| + |D^\beta f(0,x,\xi_0(x)) - D^\beta f(0,y,\xi_0(y))|$$
$$\leq \Big(K + \frac{L}{2}([u]_{C^{0,\alpha}} + [v]_{C^{0,\alpha}} + [Du]_{C^{0,\alpha}} + [Dv]_{C^{0,\alpha}} + [D^2 u]_{C^{0,\alpha}} + [D^2 v]_{C^{0,\alpha}})$$
$$+K + L([u_0]_{C^{0,\alpha}} + [Du_0]_{C^{0,\alpha}} + [D^2 u_0]_{C^{0,\alpha}})\Big)|x-y|^\alpha \leq C_4(R)|x-y|^\alpha,$$

5. EXAMPLES AND APPLICATIONS

and for every $y \in \overline{\Omega}$,

$$|D^\beta f(t, y, \xi_\sigma(t,y)) - D^\beta f(0, y, \xi_0(y))|$$
$$\leq K\delta^{\alpha/2} + \frac{L}{2}\Big([u-u_0]_{C^{\alpha/2,0}} + [v-u_0]_{C^{\alpha/2,0}} + [D(u-u_0)]_{C^{\alpha/2,0}}$$
$$+ [D(v-u_0)]_{C^{\alpha/2,0}} + [D^2(u-u_0)]_{C^{\alpha/2,0}} + [D^2(v-u_0)]_{C^{\alpha/2,0}}\Big)\delta^{\alpha/2}$$
$$\leq C_5(R)\delta^{\alpha/2}.$$

So, we get

$$|\varphi(t,x) - \varphi(t,y)| \leq C_4(R)(\delta\|u_t - v_t\|_\infty + \delta^{(1+\alpha)/2}\sum_{i=1}^n [D_i(u-v)]_{C^{(1+\alpha)/2,0}}$$
$$+ \delta^{\alpha/2}\sum_{i,j=1}^n [D_{ij}(u-v)]_{C^{\alpha/2,0}})|x-y|^\alpha$$
$$+ C_5(R)\delta^{\alpha/2}([u-v]_{C^{0,\alpha}} + \sum_{i=1}^n [D_i(u-v)]_{C^{0,\alpha}}$$
$$+ \sum_{i,j=1}^n [D_{ij}(u-v)]_{C^{0,\alpha}})|x-y|^\alpha$$
$$\leq C_6(R)\delta^{\alpha/2}\|u-v\|_{C^{1+\alpha/2,2+\alpha}}|x-y|^\alpha.$$
(8.5.23)

From (8.5.21), (8.5.22) and (8.5.23) we get

$$\|\varphi\|_{C^{\alpha/2,\alpha}} \leq C_7(R)\delta^{\alpha/2}\|u-v\|_{C^{1+\alpha/2,2+\alpha}}.$$

Arguing similarly, we find that there exists $C_8(R)$ such that

$$\|\psi\|_{C^{(1+\alpha)/2,1+\alpha}} \leq C_8(R)\delta^{\alpha/2}\|u-v\|_{C^{1+\alpha/2,2+\alpha}}.$$

Summing up, (8.5.18) follows. ∎

Theorem 8.5.4 gives local existence of a solution $u \in C^{1+\alpha/2,2+\alpha}([0,\delta] \times \overline{\Omega})$ of (8.5.9). Taking then δ as initial time and $u(\delta,\cdot)$ as initial datum, one can continue the solution to a larger time interval. The procedure may be repeated indefinitely, up to construct a maximally defined solution $u : [0,\tau[\times\overline{\Omega} \mapsto \mathbb{R}$, belonging to $C^{1+\alpha/2,2+\alpha}([0,T] \times \overline{\Omega})$ for every $T < \tau$. The interval $[0,\tau[$ is maximal in the sense that if $\tau < \infty$ then there does not exist any solution of (8.5.9) belonging to $C^{1+\alpha/2,2+\alpha}([0,\tau] \times \overline{\Omega})$.

Further regularity results will be used later. To prove them, we shall use a lemma about Hölder continuous functions, whose proof is left to the reader.

Lemma 8.5.5 *Let Ω be an open set in $\mathbb{R}^n$ with uniformly $C^{2+\alpha}$ boundary, and let $a < b$. Let $\{u_n\}_{n\in\mathbb{N}}$ be a sequence of $C^{1+\alpha/2,2+\alpha}$ functions defined in $[a,b] \times \overline{\Omega}$. Assume that*

$$\|u_n\|_{C^{1+\alpha/2,2+\alpha}([a,b]\times\overline{\Omega})} \leq M,$$

with M independent on n, and that u_n converges to a function u in $L^\infty([a,b] \times \overline{\Omega})$. Then u belongs to $C^{1+\alpha/2,2+\alpha}([a,b] \times \overline{\Omega})$, and $\|u\|_{C^{1+\alpha/2,2+\alpha}([a,b]\times\overline{\Omega})} \leq M$.

Proposition 8.5.6 Let the assumptions of Theorem 8.5.4 hold, and let $u : [0, T] \times \overline{\Omega}$ be any solution of (8.5.9) belonging to $u \in C^{1+\alpha/2, 2+\alpha}([0, T] \times \overline{\Omega})$. Then the first order space derivatives $D_k u$, $k = 1, \ldots, n$, belong to $C^{1+\alpha/2, 2+\alpha}([\varepsilon, T] \times \overline{\Omega}')$, for $0 < \varepsilon < T$ and for every open set Ω' such that $\overline{\Omega}' \subset \Omega$. If in addition $u_0 \in C^{3+\alpha}(\overline{\Omega})$, then $D_k u \in C^{1+\alpha/2, 2+\alpha}([0, T] \times \overline{\Omega}')$, for $k = 1, \ldots, n$.

Proof — Let $B(x_0, R) \subset \Omega$. Fix any integer $k = 1, \ldots, n$, and for small $h \in \mathbb{R}$ set

$$u_h(t, x) = \frac{u(t, x + he_k) - u(t, x)}{h}, \quad 0 \leq t \leq T, \ x \in B(x_0, R),$$

where e_k is the vector in $\mathbb{R}^n$ whose k-th component is 1 and the others are 0. Then u_h satisfies

$$\begin{cases} D_t u_h = \sum_{i,j=1}^n a_{ij} D_{ij} u_h + \sum_{i=1}^n b_i u_h + c u_h + f_h, \\ \qquad\qquad 0 \leq t \leq T, \ x \in B(x_0, R), \\ u_h(0, x) = \dfrac{u_0(x + he_k) - u_0(x)}{h}, \quad x \in B(x_0, R), \end{cases}$$

where

$$a_{ij}(t, x) = \int_0^1 f_{q_{ij}}(t, \xi_\sigma(t, x)) d\sigma, \quad b_i(t, x) = \int_0^1 f_{p_i}(t, \xi_\sigma(t, x)) d\sigma,$$

$$c(t, x) = \int_0^1 f_u(t, \xi_\sigma(t, x)) d\sigma, \quad f_h(t, x) = \int_0^1 f_{x_k}(t, \xi_\sigma(t, x)) d\sigma,$$

and

$$\xi_\sigma(t, x) = (x + \sigma h e_k, \sigma(u, Du, D^2 u)(t, x + h e_k) + (1 - \sigma)(u, Du, D^2 u)(t, x)).$$

If the initial datum u_0 belongs to $C^{3+\alpha}(\overline{\Omega})$, choose $\theta \in C^\infty(\mathbb{R}^n)$ such that

$$\theta \equiv 1 \text{ in } B(x_0, R/2), \quad \theta \equiv 0 \text{ outside } B(x_0, R).$$

If u_0 belongs only to $C^{2+\alpha}(\overline{\Omega})$, choose $\theta \in C^\infty([0, T] \times \mathbb{R}^n)$ such that

$$\theta \equiv 1 \text{ in } [\varepsilon, T] \times B(x_0, R/2), \quad \theta \equiv 0 \text{ outside } [\varepsilon/2, T] \times B(x_0, R),$$

with $0 < \varepsilon < T$.

In both cases, the function v defined by

$$v(t, x) \begin{cases} = \theta u_h, & \text{in } [0, T] \times B(x_0, R), \\ = 0, & \text{outside } [0, T] \times B(x_0, R), \end{cases}$$

satisfies

$$\begin{cases} v_t = \sum_{i,j=1}^n a_{ij} D_{ij} v + \sum_{i=1}^n b_i v + cv + \widetilde{f}, \ 0 \leq t \leq T, \ x \in \overline{\Omega}, \\ v(0, x) = v_0(x), \ x \in \overline{\Omega}, \\ v(t, x) = 0, \ 0 \leq t \leq T, \ x \in \partial\Omega, \end{cases}$$

5. EXAMPLES AND APPLICATIONS

where

$$\widetilde{f} = \theta f_h - u_h \theta_t - u_h \sum_{i=1}^n b_i D_i \theta - \sum_{i,j=1}^n a_{ij}(D_j u_h D_i \theta + D_i u_h D_j \theta + u_h D_{ij}\theta),$$

$$v_0(x) \begin{cases} = \theta(0,x)\dfrac{u_0(x+he_k) - u_0(x)}{h}, & \text{in } [0,T] \times B(x_0,R), \\ = 0, & \text{outside } [0,T] \times B(x_0,R). \end{cases}$$

The coefficients a_{ij}, b_i, c, and the function θf_h belong to $C^{\alpha/2,\alpha}([0,T] \times \overline{\Omega})$, with norm independent of h, and thanks to (8.5.10) the highest order coefficients a_{ij} satisfy the ellipticity condition

$$\sum_{i,j=1}^n a_{ij}\xi_i\xi_j \geq \nu |\xi|^2, \quad \forall \xi \in \mathbb{R}^n,$$

with ν independent of h. Moreover, since $u \in C^{1+\alpha/2,2+\alpha}([0,T] \times \overline{\Omega})$, then u_h and its derivatives $D_i u_h$ belong to $C^{\alpha/2,\alpha}([0,T] \times \overline{\Omega})$, with norm independent of h. It follows that $\widetilde{f}$ belongs to $C^{\alpha/2,\alpha}([0,T] \times \overline{\Omega})$, with norm independent of h.

Due to the choice of θ, if the initial datum u_0 belongs to $C^{3+\alpha}(\overline{\Omega})$, then v_0 belongs to $C^{2+\alpha}(\overline{\Omega})$ with norm independent of h; if u_0 belongs to $C^{2+\alpha}(\overline{\Omega})$, then $v_0 \equiv 0$. In both cases v_0 satisfies any compatibility conditions at $\partial\Omega$, because everything vanishes in a neighborhood of $\partial\Omega$. So, Theorem 5.1.16 implies that $v \in C^{1+\alpha/2,2+\alpha}([0,T] \times \overline{\Omega})$, with norm independent of h. Therefore, the incremental ratio u_h belongs to $C^{1+\alpha/2,2+\alpha}([0,T] \times B(u_0,R/2))$ if $u_0 \in C^{3+\alpha}(\overline{\Omega})$, to $C^{1+\alpha/2,2+\alpha}([\varepsilon,T] \times B(u_0,R/2))$ if $u_0 \in C^{2+\alpha}(\overline{\Omega})$. Of course $u_h \to D_k u$ in $C([0,T] \times B(u_0,R))$ as $h \to 0$. From Lemma 8.5.5 it follows that $D_k u$ belongs to $C^{1+\alpha/2,2+\alpha}([0,T] \times B(u_0,R/2))$ if $u_0 \in C^{3+\alpha}(\overline{\Omega})$, to $C^{1+\alpha/2,2+\alpha}([\varepsilon,T] \times B(u_0,R/2))$ if $u_0 \in C^{2+\alpha}(\overline{\Omega})$. ∎

An unpleasant aspect of Hölder spaces is the fact that, in general, the mapping $[0,\tau[\mapsto C^{2+\alpha}(\overline{\Omega})$, $t \mapsto u(t,\cdot)$ is not continuous, but only bounded, at $t = 0$. In the linear case with smooth coefficients and boundary, for instance, we know that $u(t,\cdot) \to u_0$ in the $C^{2+\alpha}$ norm as $t \to 0$ if and only if $u_0 \in h^{2+\alpha}(\overline{\Omega})$.

This problem may be overcome working in little Hölder spaces. Of course, the Hölder continuity assumptions on f, g have to be replaced by little-Hölder continuity assumptions. However, in the sequel we will not need that $t \mapsto u(t,\cdot)$ is continuous with values in $C^{2+\alpha}(\overline{\Omega})$. So, we do not go into details.

By using the techniques of Theorem 8.5.4, it is possible to treat fully nonlinear problems with nonlinear condition at the boundary also working in the space $C^{1,2+\alpha}([0,T] \times \overline{\Omega})$ instead of $C^{1+\alpha/2,2+\alpha}([0,T] \times \overline{\Omega})$, and in the space of the functions u such that $t \mapsto u(t,\cdot)$ belongs to $C([0,\delta];h^{2+\alpha}(\overline{\Omega})) \cap C^1([0,\delta];h^\alpha(\overline{\Omega}))$. See A. LUNARDI [144].

Results similar to the ones of Theorem 8.5.4 hold for fully nonlinear equations supported with Dirichlet boundary condition, such as

$$\begin{cases} u_t(t,x) = f(t,x,u(t,x),Du(t,x),D^2u(t,x)), & t \geq 0, \ x \in \overline{\Omega}, \\ u(0,x) = u_0(x), & x \in \overline{\Omega}, \\ u(t,x) = g(t,x), & t \geq 0, \ x \in \partial\Omega, \end{cases} \quad (8.5.24)$$

and for fully nonlinear equations in the whole space $\mathbb{R}^n$,

$$\begin{cases} u_t(t,x) = f(t,x,u(t,x),Du(t,x),D^2u(t,x)), & t \geq 0, \ x \in \mathbb{R}^n, \\ u(0,x) = u_0(x), & x \in \mathbb{R}^n. \end{cases} \quad (8.5.25)$$

The proofs are similar to the ones above and even simpler, thanks to the lack of the nonlinear boundary condition. The result of Theorem 8.5.4 holds also for certain systems. See B. TERRENI [9]. It may be extended to higher order equations, thanks to the optimal regularity results in Hölder spaces for higher order linear equations.

Hystorically, the so called "quasilinearization method" was the first method used to study local existence for fully nonlinear parabolic problems. It goes back to S.D. EIDEL'MAN [79], who considered problem (8.5.25). The method consists in differentiating problem (8.5.25) with respect to x_i, $i = 1, \ldots, n$, to get a quasilinear parabolic system for (u, Du): then such a system may be solved by local existence theorems for quasilinear systems. This method and a variant of it was used by S.I. HUDJAEV [102] and N.N. SOPOLOV [185], also for higher order equations: they differentiate (8.5.3) with respect to time, in order to get a linear problem (with coefficients depending on u) for $v = u_t$:

$$\begin{cases} v_t = f_t + f_u v + \sum_{i=1}^n f_{p_i} D_i v + \sum_{i,j=1}^n f_{q_{ij}} D_{ij} v, & t \geq 0, \ x \in \overline{\Omega}, \\ g_t + g_u v + \sum_{i=1}^n g_{p_i} D_i v = 0, & t \geq 0, \ x \in \partial\Omega, \\ v(0,x) = f(0,x,u_0(x),Du_0(x),D^2u_0(x)), & x \in \overline{\Omega}, \end{cases}$$

where the derivatives of f are evaluated at $(t,x,u(t,x),Du(t,x),D^2u(t,x))$, and the derivatives of g are evaluated at $(t,x,u(t,x),Du(t,x))$. Fixed any u in a closed subset C of a suitable Banach space of functions, this problem has a unique solution v, and one looks at a fixed point in C of the nonlinear operator Γ defined by

$$(\Gamma u)(t,x) = \int_0^t v(s,x)ds, \ x \in \overline{\Omega}.$$

The set C considered in [102] is a ball in $\{u \in C^{1+\alpha,2+\alpha}([0,T] \times \overline{\Omega}) : \exists u_t \in C^{1+\alpha,2,2+\alpha}([0,T] \times \overline{\Omega})\}$ with $0 < \alpha < 1$ and small $T > 0$. The quasilinearization method requires strong regularity properties for f and g; it requires also unnecessary compatibility conditions between u_0 and the boundary data.

The linearization method, which we have used here, seems to have been introduced by S.N. KRUZHKOV – A. CASTRO – M. LOPES [116, 117].

8.5.4 Motion of hypersurfaces by mean curvature

At initial time $t = 0$, a smooth connected hypersurface Γ_0 is the boundary of an open bounded set $U \subset \mathbb{R}^n$. The surface evolves in time via mean curvature. This means that there exists a family of hypersurfaces $\{\Gamma_t\}_{0 \leq t \leq T}$, such that for $0 \leq t_0 \leq t \leq T$ we have $\Gamma_t = \{x(t, x_0) : x_0 \in \Gamma_{t_0}\}$, and $x(\cdot, x_0)$ satisfies the ODE

$$\dot{x}(t) = -[div(\nu(t, x(t))] \nu(t, x(t)), \quad t_0 \leq t \leq T; \quad x(t_0) = x_0. \tag{8.5.26}$$

Here $\nu(t, x)$ is the exterior normal vector to Γ_t at the point x. The name "motion by mean curvature" comes from the identity

$$div(\nu) = -(\kappa_1 + \ldots + \kappa_{n-1}),$$

where $\kappa_1, \ldots, \kappa_{n-1}$ are the principal curvatures of Γ_t computed with respect to ν, and

$$H = \frac{1}{n-1}(\kappa_1 + \ldots + \kappa_{n-1}) \tag{8.5.27}$$

is the mean curvature.

The problem of determining whether a given smooth initial hypersurface evolves (at least for short time) by mean curvature into a family of smooth hypersurfaces, has been studied by several points of view: see M. GAGE – R.S. HAMILTON [88], Y-G. CHEN – Y. GIGA – S. GOTO [53], L.C. EVANS – J. SPRUCK [81], [82]. We follow here the approach of L.C. EVANS – J. SPRUCK [82], who reduced such a problem to the study of a fully nonlinear parabolic problem with nonlinear boundary condition of the type considered in Subsection 8.5.3.

Assume that Γ_0 evolves into the family $\{\Gamma_t\}_{0 \leq t \leq T}$, Γ_t being the smooth boundary of an open bounded set U_t. It is not difficult to deduce an equation for the signed distance function

$$d(t, x) = \begin{cases} dist\,(x, \Gamma_t), & x \in \mathbb{R}^n \setminus \overline{U}_t, \\ -dist\,(x, \Gamma_t), & x \in U_t. \end{cases}$$

Indeed, since Γ_t is assumed to be smooth, then d is smooth in the set

$$Q^+ = \{(t, x) : 0 \leq t \leq T, \ 0 \leq d(t, x) < \delta_0\},$$

and in the set

$$Q^- = \{(t, x) : 0 \leq t \leq T, \ -\delta_0 < d(t, x) \leq 0\},$$

provided $\delta_0 > 0$ is small enough. Moreover, if δ_0 is sufficiently small, for every $(t, x) \in Q^+$ there exists a unique $y \in \Gamma_t$ such that $d(t, x) = |y - x|$. Due to (8.5.26), it holds

$$d_t(t, x) = div(\nu(t, y)).$$

On the other hand, the eigenvalues of $D^2 d(t,x)$ are given by

$$\lambda_i = -\frac{\kappa_i(t,y)}{1-\kappa_i(t,y)d}, \quad i=1,\ldots,n-1, \quad \lambda_n = 0,$$

so that

$$\kappa_i = \frac{\lambda_i}{\lambda_i d - 1}, \quad i=1,\ldots,n-1.$$

From (8.5.27) we get

$$d_t = f(d, D^2 d),$$

where

$$f(u,q) = \sum_{i=1}^n \frac{\lambda_i}{1-\lambda_i u}, \quad u \in \mathbb{R}, \ q \in \mathbb{R}_S^{n^2}, \ \lambda_i u \neq 1 \ \forall i=1,\ldots,n,$$

λ_i being the eigenvalues of q. The same equation may be deduced if $(t,x) \in Q^-$.

Since $|d|$ is a distance, then the spatial gradient Dd should have modulus 1 at any point. This provides a nonlinear first order boundary condition for d. So, we are led to study the evolution equation

$$\begin{cases} v_t = f(v, D^2 v), \ t \geq 0, \ x \in \overline{\Omega}, \\ |Dv|^2 = 1, \ t \geq 0, \ x \in \partial\Omega, \\ v(0,x) = d_0(x), \ x \in \overline{\Omega}, \end{cases} \quad (8.5.28)$$

where $\Omega = Q^+ \cup Q^- = \{x \in \mathbb{R}^n : -\delta_0 < d_0(x) < \delta_0\}$, d_0 is the signed distance function from Γ_0, and f is the function defined above. The number δ_0 has been chosen so small that $\lambda_i(D^2 d_0)\delta_0 \neq 1$ for every i, so that f is well defined in a neighborhood of the range of $(d_0, D^2 d_0)$.

Note that, although the eigenvalues $\lambda_i(q)$ do not depend smoothly on q near the region where two or more of them coincide, the function f is analytic in its domain. Indeed, it is easy to see that

$$f(u,q) = Tr(q(I-uq)^{-1}),$$

(where $Tr\, q$ denotes the trace of the matrix q), so that f is analytic, and

$$\frac{\partial f}{\partial q}(u,q)s = Tr((I-uq)^{-2} s). \quad (8.5.29)$$

Therefore, for every $\xi = (\xi_1, \ldots, \xi_n) \in \mathbb{R}^n$ we have

$$\sum_{i,j=1}^n f_{q_{ij}}(u,q)\xi_i \xi_j = Tr\left(\frac{\partial f}{\partial q}(u,q)\xi \otimes \xi\right) = \sum_{i=1}^n \frac{1}{(1-\lambda_i u)^2}\langle \xi, \overline{x}_i\rangle^2,$$

5. EXAMPLES AND APPLICATIONS

where $\{\bar{x}_1, \ldots, \bar{x}_n\}$ is an orthonormal basis in $\mathbb{R}^n$ such that each $\bar{x}_i$ is an eigenvector of q with eigenvalue λ_i, and the symbol $\eta \otimes \xi$ denotes as usual the matrix $[\eta_i \xi_j]_{i,j=1,\ldots,n}$. It follows that

$$\sum_{i,j=1}^{n} f_{q_{ij}}(u,q) \xi_i \xi_j \geq \nu |\xi|^2,$$

with $\nu = \min_{i=1,\ldots,n}(1 - \lambda_i u)^{-2}$, and the ellipticity condition (8.5.10) is satisfied in a neighborhood of the range of $(d_0, D^2 d_0)$. Then the results of the previous subsection may be applied to problem (8.5.28).

Proposition 8.5.7 *Let ∂U_0 be uniformly $C^{2+\alpha}$, with $0 < \alpha < 1$, and let d_0 be the signed distance from $\Gamma_0 = \partial U_0$. Then there are $T > 0$ and a unique $v \in C^{1+\alpha/2, 2+\alpha}([0,T] \times \overline{\Omega})$, solution of (8.5.28). If in addition ∂U_0 is uniformly $C^{3+\alpha}$, setting*

$$\Gamma_t = \{x \in \overline{\Omega} : v(t,x) = 0\}, \tag{8.5.30}$$

the family $\{\Gamma_t\}_{0 \leq t \leq T}$ is a family of $C^{3+\alpha}$ hypersurfaces evolving by mean curvature from Γ_0.

Proof — Existence and uniqueness of a local $C^{1+\alpha/2, 2+\alpha}$ solution v of (8.5.28) follows from Theorem 8.5.4. To prove the second part of the statement, we show preliminarly that

$$|Dv| \equiv 1. \tag{8.5.31}$$

Thanks to the regularity results of Proposition 8.5.4, the function $w = |Dv|^2 - 1$ belongs to $C^{1+\alpha/2, 2+\alpha}([0,T] \times \overline{\Omega'})$, for every open set Ω' such that $\overline{\Omega'} \subset \Omega$. It satisfies

$$\begin{aligned} w_t &= 2\sum_{k=1}^{n} v_{x_i} v_{t x_i} \\ &= 2\sum_{i,j,k=1}^{n} f_{q_{ij}}(v, D^2 v) v_{x_i} v_{x_i x_j x_k} + 2 f_u(v, D^2 v)|Dv|^2 \\ &= \sum_{i,j=1}^{n} f_{q_{ij}}(v, D^2 v) w_{x_i x_j} - 2\sum_{i,j,k=1}^{n} f_{q_{ij}}(v, D^2 v) v_{x_i x_k} v_{x_k x_j} \\ &\quad + 2 f_u(v, D^2 v)|Dv|^2 \end{aligned}$$

in $[0,T] \times \Omega$. On the other hand, by (8.5.29),

$$\sum_{i,j,k=1}^{n} f_{q_{ij}}(v, D^2 v) v_{x_i x_k} v_{x_k x_j} = \sum_{i=1}^{n} \frac{(\lambda_i(D^2 v))^2}{(1 - \lambda_i(D^2 v)v)^2} = f_u(v, D^2 v).$$

By replacing in (8.5.31), we get

$$D_t w = \sum_{i,j=1}^{n} f_{q_{ij}}(v, D^2 v) D_{ij} w + 2 f_u(v, D^2 v) w, \quad 0 \leq t \leq T, \; x \in \overline{\Omega}.$$

Since w vanishes on the parabolic boundary of $[0,T] \times \overline{\Omega}$, then $w \equiv 0$, and (8.5.31) holds.

Since $|Dv| = 1$, the sets Γ_t defined by (8.5.30) are hypersurfaces of class $C^{2+\alpha}$, and $\nu(t, \cdot) = Dv(t, \cdot)$ is a unit normal vector field to Γ_t.

Consider now the problem

$$\begin{cases} \dot{x}(t) = -f(v(t, x(t)), D^2(v(t, x(t))), \\ x(t_0) = x_0, \end{cases} \quad (8.5.32)$$

where $0 \le t_0 \le T$, and $x_0 \in \Gamma_{t_0}$. Since every first order space derivative of v belongs to $C^{1+\alpha/2, 2+\alpha}([0,T] \times \overline{\Omega'})$, then the second order space derivatives are $(\alpha+1)/2$-Hölder continuous with respect to t, $C^{\alpha+1}$ with respect to x, hence Lipschitz continuous. It follows that problem (8.5.32) is uniquely solvable, and precisely there is $\delta > 0$ such that for every $t_0 \in [0,T]$, $x_0 \in \Gamma_{t_0}$, (8.5.32) has a unique solution $x \in C^{(3+\alpha)/2}([t_0 - \delta, t_0 + \delta] \cap [0,T])$. Moreover, $v(t, x(t)) \equiv 0$, since, by construction, $d/dt\, v(t, x(t)) = 0$, and $v(t_0, x_0) = 0$. Then $x(t) \in \Gamma_t$, and $t \mapsto x(t)$ may be extended to the whole $[0,T]$. The statement follows. ∎

8.5.5 Bellman equations

In Stochastic Control Theory one encounters problems of the type

$$\begin{cases} v_t(t,x) + \inf_{\alpha \in B}(L^\alpha v(t, \cdot)(x) + f(x, \alpha)) = 0, \quad t \le T,\ x \in \mathbb{R}^n, \\ v(T, x) = v_0(x),\ x \in \mathbb{R}^n. \end{cases} \quad (8.5.33)$$

where the parameter α belongs to a Banach space B (or to a closed set in a Banach space B), and L^α is a second order differential operator,

$$L^\alpha u(x) = \sum_{i,j=1}^n a_{ij}(x, \alpha) \frac{\partial^2 u}{\partial x_i \partial x_j} + \sum_{i=1}^n b_i(x, \alpha) \frac{\partial u}{\partial x_i}$$

We refer to N.V. KRYLOV [118] for the derivation of (8.5.33). In certain cases, problem (8.5.33) is a fully nonlinear parabolic problem. Consider, for instance, $B = \mathbb{R}^n$ and

$$a_{ij}(x, \alpha) = \frac{1}{2} a_{ij}^0(x) + \frac{1}{2} \alpha_i \alpha_j, \quad b(x, \alpha) = b^0(x) + \alpha, \quad f(x, \alpha) = \frac{1}{2}|\alpha|^2,$$

where the matrix $[a_{ij}^0]_{i,j=1,\ldots,n}$ satisfies the ellipticity condition

$$\sum_{i,j=1}^n a_{ij}^0(x) \xi_i \xi_j \ge \nu |\xi|^2, \quad \forall x,\ \xi \in \mathbb{R}^n. \quad (8.5.34)$$

5. EXAMPLES AND APPLICATIONS

Then
$$L^\alpha u(x) = \frac{1}{2}\sum_{i,j=1}^n (a_{ij}^0(x) + \alpha_i\alpha_j)D_{ij}u(x) + \sum_{i=1}^n (b_i^0(x) + \alpha_i)D_i u(x)$$
$$= \frac{1}{2}\,Tr\,[(A^0(x) + (\alpha \otimes \alpha))D^2 u(x)] + \langle b^0(x) + \alpha, Du(x)\rangle$$

The infimum of the function
$$F(\alpha) = \frac{1}{2}\,Tr\,[(\alpha \otimes \alpha)D^2 u] + \langle \alpha, Du\rangle + \frac{1}{2}|\alpha|^2, \quad \alpha \in \mathbb{R}^n,$$
is attained at the point $\alpha \in \mathbb{R}^n$ such that
$$\sum_{j=1}^n \alpha_j D_{ij}u + D_i u + \alpha_i = 0, \quad i = 1, ..., n,$$
which means that
$$\alpha = -(I + D^2 u)^{-1}(Du).$$

On the other hand,
$$F(-(I + D^2 u)^{-1}Du) = \frac{1}{2}\langle D^2 u(I + D^2 u)^{-1}Du, (I + D^2 u)^{-1}Du\rangle$$
$$-\langle (I + D^2 u)^{-1}Du, Du\rangle + \frac{1}{2}|(I + D^2 u)^{-1}Du| = -\frac{1}{2}(I + D^2 u)^{-1}Du.$$

Replacing in (8.5.33) we find
$$\begin{cases} v_t = \frac{1}{2}\langle (I + D^2 v)^{-1}Dv, Dv\rangle - \frac{1}{2}Tr[A^0 D^2 v] - \langle b^0, Dv\rangle, \quad t \geq 0, \; x \in \mathbb{R}^n, \\ v(T, x) = v_0(x), \quad x \in \mathbb{R}^n. \end{cases}$$

Setting $u(t, x) = v(T - t, x)$, we get a forward problem for u,
$$\begin{cases} u_t = \frac{1}{2}Tr(A_0 D^2 u) - \frac{1}{2}\langle (I + D^2 u)^{-1}Du, Du\rangle + \langle b^0, Du\rangle, \; 0 \leq t \leq T, \; x \in \mathbb{R}^n, \\ u(0, x) = v_0(x), \quad x \in \mathbb{R}^n. \end{cases}$$
(8.5.35)

Since the coefficients a_{ij} satisfy the ellipticity condition (8.5.34), then problem (8.5.35) is parabolic. Indeed, the function
$$f(p, q) = \sum_{i,j=1}^n a_{ij}^0(x)q_{ij} - \langle (I + q)^{-1}p, p\rangle$$
satisfies
$$\frac{\partial f}{\partial q_{ij}}\xi_i \xi_j \geq \nu|\xi|^2 + \langle (I + q)^{-1}\xi, p\rangle^2$$
for every $(u, p) \in \mathbb{R}^{n+1}$ and $q \in \mathbb{R}^{n^2}_S$ such that $I + q$ is invertible.

Arguing as in the proof of Theorem 8.5.4, one gets

Proposition 8.5.8 *Let (8.5.34) hold, and let the coefficients a_{ij}^0, b_i^0 belong to $C^\theta(\mathbb{R}^n)$, for some $\theta \in {]0,1[}$. Then for every $v_0 \in C^{2+\theta}(\mathbb{R}^n)$ such that*

$$\det(I + D^2 v_0(x)) \geq \varepsilon > 0, \quad x \in \mathbb{R}^n,$$

there exist $\delta > 0$ and a unique local solution $v \in C^{1+\theta/2, 2+\theta}([T-\delta, T] \times \mathbb{R}^n)$ of (8.5.31).

8.6 Bibliographical remarks

In Sections 8.1, 8.2, 8.3 we have gathered and developed previous results spread in several papers by A. LUNARDI [139, 141, 142, 148, 149].

The results of Section 8.4, dealing with equations in continuous interpolation spaces, are older. The local existence and uniqueness theorem 8.4.1, as well as Proposition 8.4.3 and Theorem 8.4.4, are essentially due to G. DA PRATO – P. GRISVARD [63]. A proof of Proposition 8.4.2 may be found in A. LUNARDI [152]. The analiticity result of Proposition 8.4.5 has been proved in A. LUNARDI [132].

An exposition of the Da Prato — Grisvard theory may be found in S. ANGENENT [26], who simplified and clarified the original proofs, especially as far as further regularity and dependence on the data are concerned. We have followed in fact the method of [26] in the proof of Propositions 8.4.5 and 8.3.6. The analyticity result has been applied to prove analyticity of the free boundary in certain free boundary problems. See [25], [35], [15].

Some of the results of this chapter have been extended to fully nonlinear integrodifferential problems. See E. SINESTRARI [175], A. LUNARDI – E. SINESTRARI [155].

The bibliographical references for the examples of Section 8.5 are contained in Section 8.5 itself.

An important class of nonlinear problems are the quasilinear ones,

$$u' = A(t,u)u + f(t,u),$$

where the nonlinear functions A and f are defined for $t \in [0,T]$ and x in an open set $\mathcal{O}$ of an intermediate space X_θ in the class $J_\theta \cap K_\theta$ between X and D, $0 < \theta < 1$, and have values respectively in $L(D,X)$ and in X. Moreover, for every $t \in [0,T]$ and $x \in \mathcal{O}$ the operator $A(t,x): D \mapsto X$ is sectorial, and its graph norm is equivalent to the norm of D.

Under suitable regularity assumptions on A and f, the function $F(t,x) = A(t,x)x + f(t,x)$ satisfies (8.1.1) and (8.1.3), so that for every $u_0 \in D$ such that $F(0, u_0) = A(0, u_0)u_0 + f(0, u_0) \in \overline{D}$ the initial value problem (8.0.1) has a maximally defined solution $u \in C(I(u_0); D)$ (see Sections 8.1, 8.2).

Local existence and regularity for quasilinear problems can be studied without using optimal regularity techniques. Indeed, a local solution may be sought as a

6. BIBLIOGRAPHICAL REMARKS

solution in $C^\alpha([0,\delta]; X_\theta)$ of the integral equation

$$u(t) = G_u(t,0)u_0 + \int_0^t G_u(t,s)f(s,u(s))ds, \quad 0 \leq t \leq \delta,$$

where $G_u(t,s)$ is the evolution operator associated to the family $\{A(t,u(t)) : 0 \leq t \leq \delta\}$. See P.E. SOBOLEVSKIĬ [181], A. FRIEDMAN [84], A. LUNARDI [133], H. AMANN [22]. The advantage of this approach is threefold: first, one can allow initial data not belonging to D, but to an intermediate space between X_θ and D (however, not to X_θ); secondly, if a solution u is Hölder continuous with values X_θ in its maximal interval of existence $[0, \tau(u_0)[$ and its orbit is far from the boundary of $\mathcal{O}$, then it exists in the large; third, one can consider also certain cases in which the domains $D(A(t,x))$ are not constant. See H. AMANN [21, 24], A. YAGI [212]. However, the proofs are rather long, technical, and complicated. We have not room here to develop such approach.

Chapter 9

Asymptotic behavior in fully nonlinear equations

In the previous chapter we have studied local solvability and regularity in fully nonlinear equations. Now we proceed further, studying asymptotic behavior and in particular stability.

First of all, we consider an autonomous problem

$$u'(t) = F(u(t)), \quad t \geq 0, \tag{9.0.1}$$

where $F : D \mapsto X$ satisfies the assumptions of one of the local existence theorems 8.1.1, 8.4.1, 8.4.2, and we study the stability of the stationary solutions. A stationary solution $\bar{u}$ of (9.0.1) is a solution independent of time. It is said to be stable if for each $\varepsilon > 0$ there is $\delta > 0$ such that for $\|u_0 - \bar{u}\|_D \leq \delta$ then $\tau(u_0) = +\infty$ and $\|u(t; u_0) - \bar{u}\|_D \leq \varepsilon$ for every $t > 0$[1]. It is said to be asymptotically stable if it is stable and in addition $\lim_{t \to \infty} \|u(t; u_0) - \bar{u}\|_D = 0$, uniformly for u_0 in a neighborhood of $\bar{u}$. It is said to be unstable if it is not stable.

Without any loss of generality we may assume that $F(0) = 0$, and that F is defined in a neighborhood of 0 in D. We show that the Principle of Linearized Stability holds, that is, with the exception of the critical case of stability discussed below, the null solution of (9.0.1) enjoys the same stability properties of the null solution of the linearized equation

$$v'(t) = F'(0)v(t), \quad t \geq 0. \tag{9.0.2}$$

Specifically, if

$$\sup\{\operatorname{Re}\lambda : \lambda \in \sigma(F'(0))\} = -\omega_0 < 0,$$

---
[1] We recall that $u(\cdot; u_0) : [0, \tau(u_0)[\mapsto D$ is the maximally defined solution of (9.0.1) such that $u(0) = u_0$.

then the null solution of (9.0.1) is exponentially asymptotically stable in D: for every $\omega \in [0, \omega_0[$ there are $r, M > 0$ such that if $\|u_0\|_D \leq r$, then $\tau(u_0) = +\infty$, and
$$\|u(t; u_0)\|_D \leq M e^{-\omega t} \|u_0\|_D, \quad t \geq 0.$$
Moreover, if the intersection σ_+ between the spectrum of $F'(0)$ and the half plane $\{\operatorname{Re} \lambda > 0\}$ is not empty, and
$$\inf\{\operatorname{Re} \lambda : \lambda \in \sigma_+\} > 0,$$
then the null solution of (9.0.1) is unstable and it is possible to find nontrivial backward solutions going to 0 as $t \to -\infty$. The local unstable manifold, (consisting, roughly, of the orbits of all the small backward solutions going to 0 as t goes to $-\infty$) is readily constructed. In the saddle point case
$$\sigma_+ \neq \emptyset, \quad \sigma(F'(0)) \cap i\mathbb{R} = \emptyset,$$
also the local stable manifold (consisting, roughly, of the orbits of all the small solutions going to 0 as t goes to $+\infty$) is constructed. The Principle of Linearized Stability, the stable and unstable manifolds and their regularity are treated in Section 9.1.

The critical case of stability
$$\sup\{\operatorname{Re} \lambda : \lambda \in \sigma(F'(0))\} = 0$$
is much more difficult and challenging from the mathematical point of view. There are two kind of assumptions under which we are able to study the stability of the null solution. The first one is when the elements of the spectrum of A with null real part are a finite number of isolated eigenvalues, with finite algebraic multiplicity. The second one is when decay estimates such as
$$\|e^{tF'(0)}\|_{L(X)} \leq M_0, \quad \|tF'(0)e^{tF'(0)}\|_{L(X)} \leq M_1, \quad \forall t > 0$$
hold. Such kinds of assumptions are independent: it is easy to construct examples in which only one of them, or both of them, or none of them, are satisfied.

The first case is discussed in Subsections 9.2.1 and 9.2.2, which deal with center and center-unstable manifolds. We assume, more generally, that the elements of the spectrum of A with nonnegative real part are a finite number of isolated eigenvalues with finite algebraic multiplicity, and we construct a local regular invariant manifold[2] $\mathcal{M}$, the so called center-unstable manifold, which attracts all the orbits corresponding to small initial values. Then we show that $\mathcal{M}$ is asymptotically stable, which lets one show that the null solution of (9.0.1) is stable (respectively, asymptotically stable, unstable) if and only if the null solution is

---

[2] A set M is said to be invariant if for every $u_0 \in \mathcal{M}$ we have $u(t; u_0) \in \mathcal{M}$ for every $t \in [0, \tau(u_0)[$.

INTRODUCTION

stable (respectively, asymptotically stable, unstable) with respect to the flow on $\mathcal{M}$.

For the construction of $\mathcal{M}$ we use the Lyapunov-Perron method as in ordinary differential equations. Due to the fully nonlinear character of (9.0.1) such a construction is rather complicated, and we work only in the context of $D_A(\theta, \infty) - D_A(\theta + 1, \infty)$, or $D_A(\theta) - D_A(\theta + 1)$ spaces.

Concerning the second critical case of stability, treated in Subsection 9.2.3, we take advantage of the polynomial decay of the derivatives of $e^{tF'(0)}$ as $t \to +\infty$ to show that the null solution of (9.0.1) is stable provided F satisfies certain growth conditions near $u = 0$.

Other stability problems may be reduced to the problem of stability of stationary solutions. For instance, in Subsection 9.1.6 the study of the orbital stability of the travelling waves in a free boundary parabolic problem is reduced to the study of the stability of the null solution of an abstract equation.

To prove existence of stationary solutions, one has to solve an elliptic fully nonlinear equation. Here we consider only the case of equations depending on a parameter, where, once a stationary solution is known to exist, other stationary solutions near the estabilished one may be found by bifurcation arguments.

Then we consider the problem of existence and stability of periodic solutions, both for nonautonomous time periodic equations and for autonomous equations, for which a Hopf bifurcation theorem is proved. Concerning the stability of periodic solutions, the Linearized Stability Principle holds. Let $F(t,u)$ be either T-periodic with respect to time, or independent of time, and let $\bar{u}$ be a T-periodic solution of

$$u'(t) = F(t, u(t)), \quad t \in \mathbb{R}. \tag{9.0.3}$$

Under suitable assumptions it turns out that $\bar{u}$ has the same stability properties of the null solution of the linearized problem

$$v'(t) = F_u(t, \bar{u}(t))v(t), \quad t \geq 0. \tag{9.0.4}$$

To be more precise, let $G(t,s)$ be the evolution operator associated to the family $\{F_u(t, \bar{u}(t)) : t \in \mathbb{R}\}$, and define

$$V(s) = G(s+T, s), \quad s \in \mathbb{R}.$$

According to the results of Section 6.3, if

$$\sup\{|\lambda| : \lambda \in \sigma(V(0))\} < 1,$$

then the null solution of (9.0.4) is exponentially asymptotically stable, and so is the periodic solution $\bar{u}$ of (9.0.3). If

$$\sigma_+ = \sigma(V(0)) \cap \{|\lambda| > 1\} \neq \emptyset, \quad \inf\{|\lambda| : \lambda \in \sigma_+\} > 1,$$

then the null solution of (9.0.4) is unstable, and so is $\bar{u}$.

If F is independent of time, then 1 is an eigenvalue of $V(s)$, for every s. So, the Principle of Linearized Stability may work only in the unstable case. However, it is possible to discuss the stability of the periodic orbits obtained by Hopf bifurcation. Note that if $\bar{u}$ is a solution of any autonomous problem, then also $\bar{u}(\cdot + t_0)$ is a solution, so that it is natural to deal with orbital stability rather than stability. Let $\gamma = \{\bar{u}(t) : t \in \mathbb{R}\}$ be the orbit of $\bar{u}$. $\bar{u}$ is said to be *orbitally stable* if for every $\varepsilon > 0$ there is $\delta > 0$ such that

$$dist(u_0, \gamma) \leq \delta \Rightarrow \tau(u_0) = +\infty, \ dist(u(t; u_0), \gamma) \leq \varepsilon \ \forall t \geq 0.$$

$\bar{u}$ is said to be *orbitally unstable* if it is not orbitally stable. It is said to be *asymptotically orbitally stable* if it is orbitally stable, and moreover

$$\lim_{t \to +\infty} dist(u(t; u_0), \gamma) = 0,$$

uniformly for u_0 in a neighborhood of γ.

9.1 Behavior near stationary solutions

It is convenient to rewrite (9.0.1) in the form

$$u'(t) = Au(t) + G(u(t)), \ t \geq 0. \tag{9.1.1}$$

Here $A : D(A) \subset X \mapsto X$ is a linear operator such that

$$\begin{cases} A : D(A) \mapsto X \text{ is sectorial, and the graph} \\ \text{norm of } A \text{ is equivalent to the norm of } D. \end{cases} \tag{9.1.2}$$

Moreover, $\mathcal{O}$ is a neighborhood of 0 in D, and $G : \mathcal{O} \mapsto X$ is a C^1 function with locally Lipschitz continuous derivative, such that

$$G(0) = 0, \ G'(0) = 0, \tag{9.1.3}$$

so that equation (9.1.1) admits the stationary solution $u \equiv 0$. The linearization of (9.1.1) near $\bar{u} = 0$ is

$$v'(t) = Av(t), \ t \geq 0. \tag{9.1.4}$$

Problem (9.1.1) is supported with the initial condition

$$u(0) = u_0, \tag{9.1.5}$$

with $u_0 \in \mathcal{O}$.

Since G' is continuous, by the perturbation result of Proposition 2.4.2 there exists a neighborhood $\mathcal{O}'$ of 0 in D such that for every $u_0 \in \mathcal{O}'$ the operator $A + G'(u_0) : D \mapsto X$ is sectorial, and due to the continuity of G' its graph norm

1. BEHAVIOR NEAR STATIONARY SOLUTIONS

is equivalent to the norm of D. As far as stability is concerned we may assume without loss of generality that $\mathcal{O}' = \mathcal{O}$.

In the previous chapter we have shown that for every $u_0 \in \mathcal{O}$ satisfying the compatibility condition

$$Au_0 + G(u_0) \in \overline{D}, \qquad (9.1.6)$$

there is $\tau = \tau(u_0)$ such that problem (9.1.1)-(9.1.5) has a unique noncontinuable solution $u \in C([0, \tau[; D) \cap C^1([0, \tau[; X) \cap C_\beta^\beta(]0, \tau - \varepsilon]; D)$ for every $\beta \in]0, 1[$ and $\varepsilon \in]0, \tau[$. A local solution exists also when the compatibility condition (9.1.6) is not satisfied, provided u_0 is sufficiently close to 0. In this case, however, the solution is not continuous but only bounded near $t = 0$ with values in D, and it is not C^1 but only Lipschitz continuous near $t = 0$ with values in X. See Theorems 8.1.1, 8.3.4, Proposition 8.4.2.

Concerning existence in the large for small initial data, the result of Proposition 8.2.1 may be improved. Indeed, it is not necessary to bound the modulus of continuity of the solution, but an *a priori* estimate in the D-norm is sufficient to guarantee existence in the large.

Proposition 9.1.1 *Let (9.1.2), (9.1.3) hold, and let $u(\cdot; u_0) : [0, \tau(u_0)[\mapsto D$ be the maximally defined solution of the initial value problem (9.1.1)-(9.1.5). There exists $M > 0$, independent of u_0, such that if*

$$\|u(t)\|_D \leq M, \ \ 0 \leq t < \tau, \qquad (9.1.7)$$

then $\tau = +\infty$.

Proof — The statement is a consequence of the continuous dependence on the initial datum proved in Proposition 8.2.3. Indeed, $\tau(0) = +\infty$, so that, by Proposition 8.2.3, for every $T > 0$ there is $\varepsilon = \varepsilon(T) > 0$ such that if $\|u_0\|_D \leq \varepsilon$ then $\tau(u_0) \geq T$. Set

$$M = \varepsilon(1).$$

If (9.1.7) holds, then $\|u_0\|_D \leq M$, so that $u(\cdot, u_0)$ is defined at least in $[0, 1]$. Since $\|u(1)\|_D \leq M$, the solution of the problem with initial value $u(1)$ is defined at least in $[0, 1]$, so that, by uniqueness, $u(\cdot, u_0)$ is defined at least in $[0, 2]$. Proceeding in this way, one shows by recurrence that $u(\cdot, u_0)$ is defined in $[0, +\infty[$. ∎

9.1.1 Stability and instability by linearization

The stability assumption is

$$\sup\{\operatorname{Re}\lambda : \lambda \in \sigma(A)\} = -\omega_0 < 0. \qquad (9.1.8)$$

Theorem 9.1.2 *Let $A : D(A) \mapsto X$ be a linear operator satisfying (9.1.2) and (9.1.8). Let $\mathcal{O}$ be a neighborhood of the origin in D, and let $G : \mathcal{O} \mapsto X$ be a*

C^1 function with locally Lipschitz continuous derivative, satisfying (9.1.3). Fix $\omega \in [0, \omega_0[$. Then there exist $r > 0$, $M > 0$ such that for each $u_0 \in B(0,r) \subset D$ we have $\tau(u_0) = +\infty$, and

$$\|u(t)\|_D + \|u'(t)\|_X \leq Me^{-\omega t}\|u_0\|_D, \quad \forall t \geq 0. \tag{9.1.9}$$

Proof — Fix $\alpha \in]0,1[$. We shall find a solution in the weighted space[3]

$$Y = C([0,\infty[;X) \cap C_\alpha^\alpha(]0,1];D) \cap C^\alpha([1,+\infty[;D,-\omega).$$

Y is endowed with the norm

$$\|u\|_Y = \|u\|_{C_\alpha^\alpha(]0,1];D)} + \|u\|_{C^\alpha([1,+\infty[;D,\omega)}.$$

We look for a fixed point of the operator Γ defined in a ball $B(0,R) \subset Y$ by $\Gamma(u) = w$, where w is the solution of

$$w'(t) = Aw(t) + G(u(t)), \quad t > 0; \quad w(0) = u_0.$$

The number R is so small that $\mathcal{O}$ contains the ball centered at 0 with radius R, and $R \leq R_0$, where

$$[G']_{Lip} = \sup\left\{\frac{\|G'(x) - G'(y)\|_{L(D,X)}}{\|x - y\|_D} : \|x\|_D, \|y\|_D \leq R_0\right\} < \infty. \tag{9.1.10}$$

This choice of R implies that G' is bounded in the ball $B(0,R) \subset D$, with

$$\|G'(x)\|_{L(D,X)} \leq [G']_{Lip}R, \quad \forall x \in B(0,R) \subset D.$$

If $u \in B(0,R) \subset Y$, then $t \mapsto G(u(t))$ belongs to $C_\alpha^\alpha(]0,1];X) \cap C^\alpha([1,+\infty[;X,-\omega)$. By Corollary 4.3.6(ii) and Proposition 4.4.10(i), $\Gamma(u)$ belongs to Y. We prove now that Γ is a contraction, provided R is small enough. Estimates (4.3.11) and (4.4.34) imply

$$\|\Gamma(u) - \Gamma(v)\|_{C_\alpha^\alpha(]0,1];D)} + \|\Gamma(u) - \Gamma(v)\|_{C^\alpha([1,+\infty[;D,-\omega)}$$
$$\leq C(\|G(u) - G(v)\|_{C_\alpha^\alpha(]0,1];X)} + \|G(u) - G(v)\|_{C^\alpha([1,+\infty[;X,-\omega)}).$$

For $t > 0$ and for every $u, v \in B(0,R) \subset Y$ we have

$$\varphi(t) = G(u(t)) - G(v(t)) = \int_0^1 G'(\sigma u(t) + (1-\sigma)v(t))d\sigma \, (u(t) - v(t)).$$

Arguing as in the proof of Theorem 8.1.1 one gets

$$\|\varphi(t)\| \leq [G']_{Lip}R\|u(t) - v(t)\|_D, \quad t \geq 0,$$

---
[3]The definitions of the weighted spaces $C_\alpha^\alpha(]0,1];D)$ and $C^\alpha([1,+\infty[;D,-\omega)$ may be found in the introduction of Chapter 4 and in Section 4.4, respectively.

1. Behavior near stationary solutions

and
$$[\varphi]_{C^\alpha_\alpha(]0,1];X)} \leq [G']_{Lip} R \|u - v\|_{C^\alpha_\alpha(]0,1];D)},$$
whereas for $t \geq s \geq 1$
$$\|e^{\omega t}\varphi(t) - e^{\omega s}\varphi(s)\|$$
$$\leq [G']_{Lip} \tfrac{1}{2} (\|u(t) - u(s)\|_D + \|v(t) - v(s)\|_D) \|u - v\|_{C([1,+\infty[;D,-\omega)}$$
$$+ [G']_{Lip} \tfrac{1}{2} (\|u(s)\|_D + \|v(s)\|_D) [u - v]_{C^\alpha([1,+\infty[;D,-\omega)}$$
$$\leq [G']_{Lip} \Big[(c(\omega,\alpha)R + R) \|u - v\|_{C([1,+\infty[;D,-\omega)}$$
$$+ R[u - v]_{C^\alpha([1,+\infty[;D,-\omega)} \Big] (t - s)^\alpha,$$
with $c(\omega,\alpha) = [e^{-\omega\cdot}]_{C^\alpha([1,+\infty[)} \leq \omega^\alpha (1 - \alpha)^{1-\alpha}$. Therefore,
$$[\varphi]_{C^\alpha([1,+\infty[;D,-\omega)} \leq [G']_{Lip} (c(\omega,\alpha) + 1) R \|u - v\|_{C^\alpha([1,+\infty[;D,-\omega)},$$
so that
$$\|\Gamma(u) - \Gamma(v)\|_Y \leq C[G']_{Lip} (c(\omega,\alpha) + 2) R \|u - v\|_Y.$$
Then, Γ is a $1/2$-contraction provided
$$R \leq (2C(c(\omega,\alpha) + 2)[G']_{Lip})^{-1}. \tag{9.1.11}$$
If (9.1.11) holds, for every $u \in B(0,R) \subset Y$ we have
$$\|\Gamma(u)\|_Y \leq \|\Gamma(u) - \Gamma(0)\|_Y + \|\Gamma(0)\|_Y \leq \tfrac{1}{2}\|u\|_Y + \|e^{\cdot A} u_0\|_Y \leq \frac{R}{2} + M_0 \|u_0\|_D, \tag{9.1.12}$$
with M_0 independent of u_0. Therefore, Γ maps $B(0,R) \subset Y$ into itself provided (9.1.11) holds, and
$$\|u_0\|_D \leq r = \frac{R}{2M_0},$$
in which case there exists a unique fixed point u of Γ in Y. Thanks to (9.1.12), u satisfies $\|u\|_Y \leq \tfrac{1}{2}\|u\|_Y + M_0\|u_0\|_D$, so that $\|u\|_Y \leq 2M_0\|u_0\|_D$, and (9.1.9) holds with $M = 2M_0$. ∎

We shall prove an instability result under the assumption
$$\begin{cases} \sigma_+(A) = \sigma(A) \cap \{\lambda \in \mathbb{C} : \operatorname{Re}\lambda > 0\} \neq \emptyset, \\ \inf\{\operatorname{Re}\lambda : \lambda \in \sigma_+(A)\} = \omega_+ > 0. \end{cases} \tag{9.1.13}$$
Then the linearized problem (9.1.4) has nontrivial backward solutions, which decay exponentially as $t \to -\infty$. We refer to notation and results of Subsection 4.3.4. (9.1.13) implies that for every $\omega \in]0, \omega_+[$
$$\sigma(A) \cap \{\lambda \in \mathbb{C} : \operatorname{Re}\lambda = \omega\} = \emptyset.$$

Theorem 9.1.3 *Let the assumptions of Theorem 9.1.2 be satisfied, with (9.1.8) replaced by (9.1.13). Then problem (9.1.1) has a nontrivial backward solution $v \in C^\alpha(]-\infty,0]; D, \omega)$, with $v' \in C^\alpha(]-\infty,0]; X, \omega)$, for every $\alpha \in \,]0,1[$ and $\omega \in \,]0, \omega_+[$. It follows that the null solution of (9.1.1) is unstable.*

Proof — Let P be the projection associated to the spectral set $\sigma_+(A)$. Fix $\alpha \in \,]0,1[$, $\omega \in \,]0, \omega_+[$, and a small $x \in P(X), x \neq 0$. We recall that there is $M(\alpha, \omega) > 0$, independent of x, such that $t \mapsto e^{tA}x \in C^\alpha(]-\infty,0]; D, \omega)$, and

$$\|e^{\cdot A}x\|_{C^\alpha(]-\infty,0];D,\omega)} \leq M(\alpha, \omega)\|x\|.$$

Choose $R > 0$ as in the proof of Theorem 9.1.2, that is such that $\mathcal{O}$ contains the ball centered at 0 with radius R, and $R \leq R_0$, where R_0 is such that (9.1.10) holds.

Define an operator Φ in the ball $B = B(0,R) \subset C^\alpha(]-\infty,0]; D, \omega)$ by $\Phi u = v$, where v is the unique solution in $C^\alpha(]-\infty,0]; D, \omega)$ of

$$v'(t) = Av(t) + G(u(t)), \quad t \leq 0; \quad Pv(0) = x.$$

B is endowed with the metrics generated by the norm of $C^\alpha(]-\infty,0]; D, \omega)$. By using estimate (4.4.37), one shows easily that if R is sufficiently small, say $R \leq R(\alpha,\omega)$, then Φ is a 1/2-contraction. Therefore, for every $u \in B(0,R)$

$$\|\Phi(u)\|_B \leq \|\Phi(u) - \Phi(0)\|_B + \|\Phi(0)\|_B$$
$$\leq \tfrac{1}{2}\|u\|_B + \|e^{\cdot A}x\|_B \leq R/2 + M(\alpha,\omega)\|x\|,$$

so that Γ maps $B(0,R) \subset Y$ into itself provided

$$\|x\| \leq \frac{R}{2M(\alpha,\omega)},$$

in which case it has a unique fixed point v in $B(0,R)$. Moreover, $Pv(0) = P(\Phi v)(0) = x \neq 0$, so that $v \not\equiv 0$.

Let us prove further regularity. Fix any $\beta \in [\alpha, 1[$, $\omega' \in \,]\omega, \omega_+[$, and let $R(\beta, \omega')$ be such that Φ is a 1/2-contraction in $B(0,R) \subset C^\beta(]-\infty,0]; D, \omega')$ if $R \leq R(\beta, \omega')$ and $\|x\| \leq r = R/2M(\beta, \omega')$. Let $R' \leq R(\beta, \omega')$ be such that $\|v\|_{C^\beta(]-\infty,0];D,\omega')} \leq R'$ implies $\|v\|_{C^\alpha(]-\infty,0];D,\omega)} \leq R(\alpha, \omega)$.

Since $\lim_{t \to -\infty} v(t) = 0$, then there exists $t_0 \leq 0$ such that $\|v(t_0)\| \leq \frac{R'}{2M(\beta,\omega')}$. The backward problem

$$z' = Az + G(z), \quad t \leq t_0; \quad Pz(t_0) = Pv(t_0),$$

has a unique solution $z \in B(0, R') \subset C^\beta(]-\infty, t_0]; D, \omega')$. Due to the choice of R', v belongs also to $B(0,R) \subset C^\alpha(]-\infty, t_0]; D, \omega)$, so that it coincides with $v_{|]-\infty,t_0]}$, by the uniqueness of the solution in $B(0,R) \subset C^\alpha(]-\infty,t_0]; D, \omega)$. Therefore, the restriction $v_{|]-\infty,t_0]}$ belongs to $C^\beta(]-\infty, t_0]; D, \omega')$. By the further regularity result of Proposition 8.3.6, for every $\bar{t} < 0$ the restriction $v_{|[\bar{t},0]}$ belongs to $C^\beta([\bar{t}, 0]; D)$. The statement follows. ∎

1. Behavior near stationary solutions

9.1.2 The saddle point property

We now assume that
$$\sigma(A) \cap i\mathbb{R} = \emptyset. \tag{9.1.14}$$

Then the sets
$$\sigma_-(A) = \sigma(A) \cap \{\operatorname{Re}\lambda < 0\}, \quad \sigma_+(A) = \sigma(A) \cap \{\operatorname{Re}\lambda > 0\}$$

are spectral sets, and
$$\sup\{\operatorname{Re}\lambda : \lambda \in \sigma_-(A)\} = -\omega_- < 0,$$
$$\inf\{\operatorname{Re}\lambda : \lambda \in \sigma_+(A)\} = \omega_+ > 0.$$

Let again P be the projection associated with the spectral set $\sigma_+(A)$.

Theorem 9.1.4 *Let $A : D(A) \mapsto X$ be a linear operator satisfying (9.1.2), (9.1.13), and let $G : \mathcal{O} \mapsto X$ be a C^1 function with locally Lipschitz continuous derivative, satisfying assumption (9.1.3). Then for every $\alpha \in\,]0,1[$ there are positive numbers r_0, r_1, such that*

(i) There exist $R_0 > 0$ and a Lipschitz continuous function
$$\varphi : B(0, r_0) \subset P(X) \mapsto (I - P)(D),$$

differentiable at 0 with $\varphi'(0) = 0$, such that for every u_0 belonging to the graph of φ problem (9.1.1) has a unique backward solution v in $C^\alpha(]-\infty, 0]; D)$, satisfying
$$\|v\|_{C^\alpha(]-\infty,0];D)} \le R_0. \tag{9.1.15}$$

Moreover, $v \in C^\beta(]-\infty, 0]; D, \omega)$ for every $\beta \in\,]0,1[$ and $\omega \in\,]0, \omega_+[$. In particular, $\lim_{t\to-\infty} v(t) = 0$.

Conversely, if problem (9.1.1) has a backward solution v which satisfies (9.1.15) and $\|Pv(0)\| \le r_0$, then $v(0) \in \operatorname{graph}\varphi$.

(ii) There exist $R_1 > 0$ and a Lipschitz continuous function
$$\psi : B(0, r_1) \subset (I - P)(D) \mapsto P(X),$$

differentiable at 0 with $\psi'(0) = 0$, such that for every u_0 belonging to the graph of ψ problem (9.1.1) has a unique solution u in $C_\alpha^\alpha(]0,1]; D) \cap C([0,1]; X) \cap C^\alpha([1,+\infty[; D)$, such that
$$\|u\|_{C_\alpha^\alpha(]0,1];D)} + \|u\|_{C^\alpha([1,+\infty[;D)} \le R_1. \tag{9.1.16}$$

Moreover, $u \in C^\beta([1,+\infty[; D, -\omega)$ for every $\beta \in\,]0,1[$ and $\omega \in\,]0, \omega_-[$.

Conversely, if problem (9.1.1) has a solution u which satisfies (9.1.16), and $\|(I-P)u(0)\|_D \le r_1$, then $u(0) \in \operatorname{graph}\psi$.

(iii) If in addition $G \in C^k(\mathcal{O}; X)$ is such that $G^{(k)}$ is locally Lipschitz continuous, with $k \in \mathbb{N}$, $k \ge 2$, then ψ and φ are $k-1$ times differentiable, with Lipschitz continuous derivatives.

Proof — (i) Due to assumption (9.1.14), the statement of Theorem 9.1.3 holds also with $\omega = 0$. Let $R = R(\alpha, 0)$, $M = M(\alpha, 0)$, $r_0 = R/2M$ be the constants given by Theorem 9.1.3 for $\omega = 0$. Let $x \in P(X)$ be such that $\|x\| \leq r_0$, and let $\Phi = \Phi_x$ be the operator defined in the proof of Theorem 9.1.3. By Proposition 4.4.12, for every $v \in B(0, R) \subset C^\alpha(]-\infty, 0]; D)$ and $t \leq 0$ it holds

$$(\Phi_x(v))(t) = e^{tA}x + \int_0^t e^{(t-s)A} PG(v(s)) ds + \int_{-\infty}^t e^{(t-s)A}(I-P)G(v(s)) ds.$$

Let $v(\cdot; x)$ be the unique fixed point of Φ_x in $B(0, R)$, and set

$$\varphi(x) = (I - P)v(0; x) = \int_{-\infty}^0 e^{-sA}(I-P)G(v(s; x)) ds.$$

Since the mapping $(x, v) \mapsto \Phi_x(v)$ is Lipschitz continuous in $B(0, r_0) \times B(0, R) \subset P(X) \times C^\alpha(]-\infty, 0]; D)$, then the mapping $x \mapsto v(\cdot; x)$ is Lipschitz continuous, and consequently φ is Lipschitz continuous.

Let us prove that φ is differentiable at 0 with null derivative. Choose a small $r \leq R(\alpha, 0)$, let $x \in P(X)$ be so small that Φ_x has a unique fixed point in $B(0, r) \subset C^\alpha(]-\infty, 0]; D)$. From the proof of Theorem 9.1.3 we know that

$$\|u(\cdot; x)\|_{C^\alpha(]-\infty,0];D)} \leq 2M\|x\|.$$

Using estimate (4.4.37) we get easily

$$\|(I-P)u(\cdot; x)\|_{C^\alpha(]-\infty,0];D)} \leq C\|G(u(\cdot; x))\|_{C^\alpha(]-\infty,0];X)}$$
$$\leq C[G']_{Lip}\|u(\cdot; x)\|_{C^\alpha(]-\infty,0];D)}r \leq 2MC[G']_{Lip}\|x\|r.$$

Therefore,

$$\|\varphi(x)\|_D \leq 2MC[G']_{Lip}\|x\|r.$$

Letting $r \to 0$ we see that φ is Fréchet differentiable at $x = 0$, and $\varphi'(0) = 0$.

Let us prove the uniqueness statement. If z is a backward solution of (9.1.1) which satisfies (9.1.15), and such that $\|Pz(0)\| \leq r_0$, then z is a fixed point of the operator Φ_x belonging to the ball $B(0, R) \subset C^\alpha(]-\infty, 0]; D)$. By uniqueness of the fixed point, $(I - P)z(0) = \varphi(z(0))$.

Let us prove that $\lim_{t \to -\infty} v(t) = 0$, provided $\|x\|$ is small enough. Fix any $\omega' \in]0, \omega_+[$, and let $R' \leq R(\alpha, \omega')$ be such that for every $u \in C^\alpha(]-\infty, 0]; D, \omega')$ with $\|u\|_{C^\alpha(]-\infty,0];D,\omega')} \leq R'$ it holds $\|u\|_{C^\alpha(]-\infty,0];D)} \leq R$. In the proof of Theorem 9.1.3 we have shown that if $x \in P(X)$ is such that $\|x\| \leq r' = R'/2M(\alpha, \omega')$, then the operator Φ_x has a unique fixed point in $B(0, R') \subset C^\alpha(]-\infty, 0]; D, \omega')$, which belongs to $B(0, R) \subset C^\alpha(]-\infty, 0]; D)$ and hence coincides with v. In particular, $\lim_{t \to -\infty} v(t) = 0$. Arguing now as in the proof of Theorem 9.1.3, one sees that v

1. BEHAVIOR NEAR STATIONARY SOLUTIONS

belongs to $C^\beta(]-\infty, 0]; D, \omega)$ for every $\beta \in \,]0,1[$ and $\omega \in \,]0, \omega_+[$. So, statement (i) follows, with $R_0 = \min\{R, R'\}$.

(ii) The proof is similar to the one of statement (i). So we sketch it, leaving the details to the reader.

Fix any $\omega \in [0, \omega_-[$ and a small $x \in (I-P)(D)$. In view of Proposition 4.4.10, we seek a globally defined solution of (9.1.1) as a fixed point of the operator Ψ_x defined by

$$(\Psi_x(u))(t) = e^{tA}x + \int_0^t e^{(t-s)A}(I-P)G(u(s))ds - \int_t^{+\infty} e^{(t-s)A}PG(u(s))ds,$$

for every $u \in B(0,R) \subset C_\alpha^\alpha(]0,1];D) \cap C([0,1];X) \cap C^\alpha([1,+\infty[;D,-\omega)$, and $t \geq 0$. Using Proposition 4.4.10(i) and arguing as in the proof of Theorem 9.1.3, one finds that Ψ_x is a 1/2-contraction and maps $B(0,R)$ into itself provided $R \leq R(\alpha, \omega)$ and $\|x\|_D \leq R/2M(\alpha, \omega)$, where $M(\alpha, \omega)$ is any number such that

$$\|e^{\cdot A}y\|_{C_\alpha^\alpha(]0,1];D)} + \|e^{\cdot A}y\|_{C^\alpha([1,+\infty[;D,-\omega)} \leq M(\alpha,\omega)\|y\|_D, \quad \forall y \in (I-P)(D).$$

Take $\omega = 0$ and define $\psi(x) = Pu(0;x)$, where $u(\cdot; x)$ is the fixed point of Ψ_x. Arguing again as in the proof of Theorem 9.1.3, one shows that u belongs in fact to $C_\beta^\beta(]0,1];D) \cap C^\beta([1,+\infty[;D,-\omega')$ for every $\beta \in \,]0,1[$ and $\omega' \in [0,\omega_-[$, provided $\|x\|_D$ is small enough. The rest of the proof follows as the proof of statement (i), with obvious modifications.

(iii) If G is k times differentiable, and $G^{(k)}$ is locally Lipschitz continuous, then the mappings $(x,u) \mapsto \Phi_x(u)$ and $(x,u) \mapsto \Psi_x(u)$, introduced in the proofs of statements (i) and (ii), are $k-1$ times differentiable, with $(k-1)$-th Lipschitz continuous derivative, so that their fixed points are $k-1$ times differentiable with respect to x, with $(k-1)$-th Lipschitz continuous derivative. The statement follows. ∎

The graphs of φ and ψ are called *local unstable manifold* and *local stable manifold*, respectively.

9.1.3 The case where X is an interpolation space

In the case where X is an interpolation space $D_A(\theta, \infty)$ or $D_A(\theta)$ the results of Subsection 9.1.2 may be improved as far as the regularity of G and the uniqueness of the solution are concerned, since the results of Section 9.4 allow to work in spaces of continuous functions, requiring less regularity on G than the spaces of weighted Hölder continuous functions which we have used up to now.

Let $A: D(A) \subset E \mapsto E$ be a sectorial operator. Let $0 < \theta < 1$.

We consider two sets of assumptions.

$$\begin{cases} \mathcal{O} \text{ is a neighborhood of } 0 \text{ in } D_A(\theta+1,\infty),\ G \in C^1(\mathcal{O}; D_A(\theta+1,\infty)) \\ \text{is a nonlinear function such that } G(0) = 0,\ G'(0) = 0; \end{cases} \quad (9.1.17)$$

$$\begin{cases} \mathcal{O} \text{ is a neighborhood of } 0 \text{ in } D_A(\theta+1),\ G \in C^1(\mathcal{O}; D_A(\theta+1)) \\ \text{is a nonlinear function such that } G(0) = 0,\ G'(0) = 0. \end{cases} \quad (9.1.18)$$

From Theorem 8.4.1 and Proposition 8.4.2 local existence results follow.

Theorem 9.1.5 *Let assumption (9.1.17) (respectively, (9.1.18)) hold. Then for every $T > 0$ there are $R = R(T)$, $R_0 = R_0(T)$, $M = M(T) > 0$ such that if $u_0 \in D_A(\theta+1,\infty)$ (respectively, $u_0 \in D_A(\theta+1)$) is such that $\|u_0\|_{D_A(\theta+1,\infty)} \leq R$, then the initial value problem $u(0) = u_0$ for equation (9.1.1) has a solution $u \in C(]0,T]; D_A(\theta+1,\infty)) \cap C^\theta([0,T]; D(A)) \cap C^1(]0,T]; D_A(\theta,\infty))$, (respectively, $u \in C([0,T]; D_A(\theta+1)) \cap h^\theta([0,T]; D(A)) \cap C^1([0,T]; D_A(\theta)))$, such that*

$$\sup_{0 \leq t \leq T} \|u(t)\|_{D_A(\theta+1,\infty)} + \|u\|_{C^\theta([0,T]; D(A))} \leq M \|u_0\|_{D_A(\theta+1,\infty)}. \quad (9.1.19)$$

Moreover, u is the unique solution of (9.1.1) such that $u(0) = u_0$ and $\sup_{0 \leq t \leq T} \|u(t)\|_{D_A(\theta+1,\infty)} \leq R_0$.

Note that the result of Theorem 9.1.5 is stronger than a usual local existence result, because the interval of existence $[0,T]$ is arbitrarily large. It can be interpreted as a theorem about continuous dependence of the solution on the initial datum, at $u_0 = 0$. As a corollary, a continuation result follows.

Corollary 9.1.6 *Let either (9.1.17) or (9.1.18) hold. Set*

$$R = \sup_{T>0} R(T),$$

where $R(T)$ is given by Theorem 9.1.5. Let u be a solution of (9.1.1) and set

$$\tau = \sup\{t > 0 : u \in C(]0,t]; D_A(\theta+1,\infty)) \text{ solves (9.1.1) in }]0,t]\ \},$$

if (9.1.17) holds; set

$$\tau = \sup\{t > 0 : u \in C([0,t]; D_A(\theta+1)) \text{ solves (9.1.1) in } [0,t]\ \},$$

if (9.1.18) holds. If there exists $\varepsilon > 0$ such that $\|u(t)\|_{D_A(\theta+1,\infty)} < R - \varepsilon$ for every $t < \tau$, then $\tau = +\infty$.

The principle of linearized stability reads as follows.

1. BEHAVIOR NEAR STATIONARY SOLUTIONS

Theorem 9.1.7 *Let either (9.1.17) or (9.1.18) hold. Then*

(i) If A satisfies (9.1.8), then for every $\omega \in {]}0, \omega_0{[}$ there are r, M such that if $\|u_0\|_{D_A(\theta+1,\infty)} \leq r$ then the solution u of (9.1.1) with initial value u_0, whose existence is stated in Theorem 9.1.5, is defined in $[0, +\infty[$, and

$$\|u(t)\|_{D_A(\theta+1,\infty)} + \|u'(t)\|_{D_A(\theta,\infty)} \leq Me^{-\omega t}\|u_0\|_{D_A(\theta+1,\infty)}, \quad t \geq 0.$$

(ii) If A satisfies (9.1.13), then the null solution of (9.1.1) is unstable in $D_A(\theta+1, \infty)$ (respectively, in $D_A(\theta+1)$). Specifically, there exist nontrivial backward solutions of (9.1.1) going to 0 as t goes to $-\infty$.

In the saddle point case a result similar to Theorem 9.1.3 holds.

Theorem 9.1.8 *Let A satisfy (9.1.14), and let $G : \mathcal{O} \mapsto X$ satisfy assumption (9.1.17) (respectively, (9.1.18)). Then for every $\alpha \in {]}0, 1{[}$ there are positive numbers r_0, r_1, such that*
(i) There exists $R_0 > 0$ and a Lipschitz continuous function

$$\varphi : B(0, r_0) \subset P(D_A(\theta, \infty)) \mapsto (I - P)(D_A(\theta + 1, \infty)),$$

(respectively, $\varphi : B(0, r_0) \subset P(D_A(\theta)) \mapsto (I - P)(D_A(\theta + 1)))$ differentiable at 0 with $\varphi'(0) = 0$, such that for every u_0 belonging to the graph of φ problem (9.1.1) has a unique backward solution v in $C({]}-\infty, 0]; D_A(\theta + 1, \infty))$ (respectively, in $C({]}-\infty, 0]; D_A(\theta + 1))$), such that

$$\|v\|_{C({]}-\infty,0];D_A(\theta+1,\infty))} \leq R_0. \tag{9.1.20}$$

Moreover $v \in C({]}-\infty, 0]; D_A(\theta + 1, \infty), \omega)$ for every $\omega \in {]}0, \omega_+{[}$.

Conversely, if problem (9.1.1) has a backward solution v which satisfies (9.1.19) and $\|Pv(0)\|_{D_A(\theta,\infty)} \leq r_0$, then $v(0) \in \text{graph } \varphi$.

(ii) There exist R_1, $r_1 > 0$ and a Lipschitz continuous function

$$\psi : B(0, r_1) \subset (I - P)(D_A(\theta + 1, \infty)) \mapsto P(D_A(\theta, \infty)),$$

(respectively, $\psi : B(0, r_1) \subset (I - P)(D_A(\theta + 1)) \mapsto P(D_A(\theta, \infty)))$, differentiable at 0 with $\psi'(0) = 0$, such that for every u_0 belonging to the graph of ψ problem (9.1.1) has a unique solution u in $L^\infty(0, +\infty; D_A(\theta + 1, \infty)) \cap C({]}0, +\infty[; D_A(\theta + 1, \infty))$ (respectively, in $C([0, +\infty[; D_A(\theta + 1)))$, such that

$$\|u\|_{L^\infty(0,+\infty;D_A(\theta+1,\infty))} \leq R_1. \tag{9.1.21}$$

Moreover, $u \in C([1, +\infty[; D_A(\theta + 1, \infty), -\omega)$ for every $\omega \in {]}0, \omega_-{[}$.

Conversely, if problem (9.1.1) has a solution u which satisfies (9.1.21), and $\|(I - P)u(0)\|_{D_A(\theta+1,\infty)} \leq r_1$, then $u(0) \in \text{graph } \psi$.

(iii) If in addition $G \in C^k(\mathcal{O}; D_A(\theta, \infty))$ (respectively, $G \in C^k(\mathcal{O}; D_A(\theta)))$ and $G^{(k)}$ is Lipschitz continuous for some $k \in \mathbb{N}$, then ψ and φ are k times differentiable, with Lipschitz continuous k-th order derivatives.

9.1.4 Bifurcation of stationary solutions

In this subsection we discuss the problem of existence of stationary solutions to fully nonlinear equations depending on a parameter λ,

$$F(x, \lambda) = 0, \qquad (9.1.22)$$

where $F : \mathcal{O}_D \times \mathcal{O}_\Lambda \mapsto X$ is a regular function, satisfying the assumptions of the local existence Theorem 8.3.2. We assume that for $\lambda = \lambda_0$ equation (9.1.22) has a stationary solution x_0, and we look for stationary solutions of (9.1.22) near x_0, for λ near λ_0. Without loss of generality we may assume that $\lambda_0 = 0 \in \mathcal{O}_\Lambda$, $x_0 = 0 \in \mathcal{O}_D$. We set as usual

$$A = F_x(0, 0).$$

If $0 \in \rho(A)$, the Local Inversion Theorem implies that for every λ sufficiently close to 0 there is a unique small stationary solution $x(\lambda)$ of (9.1.22), depending continuously on λ. If in addition A satisfies the stability condition (9.1.8) (i.e., if $\sup\{\operatorname{Re} z : z \in \sigma(A)\} < 0$) then by the perturbation lemma A.3.1 also the operator $F_x(0, \lambda)$ satisfies (9.1.8) for λ close to 0, so that the stationary solution $x(\lambda)$ is exponentially asymptotically stable. Similarly, if A satisfies the instability condition (9.1.13) then $F_x(0, \lambda)$ satisfies (9.1.13) for λ close to 0, so that the stationary solution $x(\lambda)$ is unstable. So, the case where A is nonsingular is rather trivial.

From now on we consider the case where 0 is in the spectrum of A. Precisely, we assume that X and D are real Banach spaces, $\Lambda = \mathbb{R}$, $\mathcal{O}_\Lambda = \,]-1, 1[$, and

$$0 \text{ is an algebraically simple isolated eigenvalue of } A. \qquad (9.1.23)$$

Let P be the projection associated with the spectral set $\{0\}$, and let x_0 be a generator of the kernel of A. Then there is x_0^* in the dual space X' such that $\langle x_0, x_0^* \rangle = 1$, $\langle y, x_0^* \rangle = 0$ for every $y \in (I - P)(X)$, and the projection P may be expressed as

$$Px = \langle x, x_0^* \rangle x_0, \quad x \in X.$$

The operators

$$A(\lambda) = F_x(0, \lambda)$$

have real simple eigenvalues for λ close to 0, as the next lemma shows.

Lemma 9.1.9 *Let $F : \mathcal{O}_D \times \,]-1, 1[\,\mapsto X$ be a C^k function, with $k \geq 2$. Then there are $\lambda_0 \in \,]0, 1[$ and C^{k-1} functions $\gamma : \,]-\lambda_0, \lambda_0[\, \mapsto \mathbb{R}$, $x : \,]-\lambda_0, \lambda_0[\, \mapsto D$ such that $\gamma(0) = 0$, $x(0) = x_0$, and*

$$A(\lambda)x(\lambda) = \gamma(\lambda)x(\lambda).$$

Moreover, $\gamma(\lambda)$ is an algebraically simple eigenvalue of $A(\lambda)$, and it is the unique element of the spectrum of $A(\lambda)$ close to 0; $x(\lambda)$ is the unique element x in the kernel of $\gamma(\lambda)I - A$ such that $x - x_0 \in (I - P)(X)$.

1. Behavior near stationary solutions

Proof — Since the function $\lambda \mapsto A(\lambda)$ is $k-1$ times continuously differentiable with values in $L(D, X)$, most of the statements follow from Proposition A.3.2 concerning perturbation of simple eigenvalues. Since it is clear that $x(\lambda)$ may be chosen uniquely in such a way that $x(\lambda) - x_0 \in (I - P)(X)$, we have only to prove the regularity statements, and the fact that $\gamma(\lambda)$ is real.

To see that the functions $\lambda \mapsto \gamma(\lambda)$, $\lambda \mapsto x(\lambda)$ are C^{k-1} one can follow the proof of Proposition A.3.2 and check that at any step regularity is preserved. Otherwise, one can follow the procedure below, which shows also that $\gamma(\lambda)$ is real.

Define the nonlinear function

$$\mathcal{F} : (I - P)(D) \times\,]-1,1[\, \times \mathbb{R} \mapsto X, \quad \mathcal{F}(z, \lambda, r) = A(\lambda)(x_0 + z) - r(x_0 + z),$$

$\mathcal{F}$ is $k-1$ times continuously differentiable, $F(0,0,0) = 0$, and the derivative of $\mathcal{F}$ with respect to (z, r) at $(0,0,0)$ is the linear operator $(\hat{z}, \hat{r}) \mapsto A\hat{z} - \hat{r}x_0$, which is an isomorphism from $(I - P)(D) \times \mathbb{R}$ onto X. By the Implicit Function Theorem, there are C^{k-1} functions $\lambda \mapsto z(\lambda)$, $\lambda \mapsto r(\lambda)$ defined for λ near 0, such that $r(0) = 0$, $z(0) = 0$, and $\mathcal{F}(z(\lambda), \lambda, r(\lambda)) = 0$. By uniqueness, $r(\lambda) = \gamma(\lambda)$, and $x_0 + z(\lambda) = x(\lambda)$. ∎

Now we are able to state the existence theorem.

Theorem 9.1.10 *Let $k \geq 2$ and let $F : \mathcal{O}_D \times\,]-1,1[\, \mapsto X$ be a k times continuously differentiable function satisfying $F(0, \lambda) \equiv 0$. Assume that $A = F_x(0,0) : D(A) = D \mapsto X$ is a sectorial operator satisfying (9.1.23). Assume moreover that*

$$\gamma'(0) = \langle F_{x\lambda}(0,0)x_0, x_0^* \rangle \neq 0. \tag{9.1.24}$$

Then there are $c_0 > 0$ and C^{k-1} functions $\lambda :\,]-c_0, c_0[\, \mapsto \mathbb{R}$, $u :\,]-c_0, c_0[\, \mapsto D$, such that

$$\lambda(0) = 0, \quad u(0) = 0, \quad u(\lambda) \neq 0 \text{ for } \lambda \neq 0,$$

and

$$F(u(c), \lambda(c)) = 0.$$

Moreover, there is $\varepsilon_0 > 0$ such that if $|\lambda| \leq \varepsilon_0$, $\|u\|_D \leq \varepsilon_0$ and $F(u, \lambda) = 0$, then there exists $c \in\,]-c_0, c_0[$ such that $u = u(c)$, $\lambda = \lambda(c)$.

Proof — It is convenient to rewrite (9.1.22) in the form

$$Au + G(u, \lambda) = 0, \tag{9.1.25}$$

where $G(u, \lambda) = F(u, \lambda) - Au$ is such that $G_u(0,0) = 0$. We look for a solution u,

$$u = c(x_0 + v),$$

with $c \in \mathbb{R}$ and $v \in (I - P)(D)$. Replacing in (9.1.25) we get

$$Av + \mathcal{G}(v, \lambda, c) = 0, \tag{9.1.26}$$

where

$$\mathcal{G}(v,\lambda,c) = \begin{cases} \dfrac{1}{c} G(c(x_0+v),\lambda), & \text{if } c \neq 0, \\ G_x(0,\lambda)(x_0+v), & \text{if } c = 0. \end{cases}$$

Applying P and $(I-P)$, (9.1.26) is equivalent to the system

$$\begin{cases} (i) & Av + (I-P)\mathcal{G}(v,\lambda,c) = 0, \\ (ii) & P\mathcal{G}(v,\lambda,c) = 0. \end{cases} \quad (9.1.27)$$

Since $\mathcal{G}_v(0,0,0) = 0$, from the Implicit Function Theorem it follows that if c, λ are sufficiently small there is a unique small solution $v = v(\lambda,c)$ of (9.1.27)(i) in $(I-P)(D)$. Replacing in (9.1.27)(ii) we get the *bifurcation equation*

$$\eta(\lambda,c) = \langle \mathcal{G}(v(\lambda,c),\lambda,c), x_0^* \rangle = 0. \quad (9.1.28)$$

The function η is $k-1$ times continuously differentiable; moreover $\eta(0,0)=0$, and

$$\eta_\lambda(0,0) = \langle \mathcal{G}_{x\lambda}(0,0)x_0, x_0^* \rangle = \langle F_{x\lambda}(0,0)x_0, x_0^* \rangle,$$

$$\eta_c(0,0) = \frac{1}{2}\langle \mathcal{G}_{xx}(0,0)(x_0,x_0), x_0^* \rangle = \frac{1}{2}\langle F_{xx}(0,0)(x_0,x_0), x_0^* \rangle.$$

Since $\langle F_{x\lambda}(0,0)x_0, x_0^* \rangle \neq 0$ by assumption, by the Implicit Function Theorem there is a C^{k-1} function $c \mapsto \lambda(c)$, defined for c small, such that

$$\lambda(0)=0, \quad \lambda'(0) = -\frac{\langle F_{xx}(0,0)(x_0,x_0), x_0^* \rangle}{2\langle F_{x\lambda}(0,0)x_0, x_0^* \rangle} = -\frac{\langle F_{xx}(0,0)(x_0,x_0), x_0^* \rangle}{2\gamma'(0)} \quad (9.1.29)$$

and $\eta(\lambda(c),c) = 0$. Then

$$u(c) = c(x_0 + v(\lambda(c),c)) \quad (9.1.30)$$

is a solution of (9.1.22) with $\lambda = \lambda(c)$. Note that for $c \neq 0$ u does not vanish, because $x_0 \in P(D)$ and $v(\lambda(c),c) \in (I-P)(D)$.

We prove now uniqueness. By the Implicit Function Theorem there is a neighborhood of $(0,0)$ in $(I-P)(D) \times \mathbb{R}$, say $B(0,\delta_0) \times]-\delta_0,\delta_0[$, such that for every small $c \in \mathbb{R}$, say $|c| < \delta_1$, the couple $(v(\lambda(c),c), \lambda(c))$ is the unique solution of

$$F(c(x_0+v), \lambda) = 0$$

in $B(0,\delta_0) \times]-\delta_0,\delta_0[$. Let $(\bar{u},\bar{\lambda})$ be a solution of $F(\bar{u},\bar{\lambda})=0$ satisfying

$$\|u\|_D < \varepsilon, \quad |\lambda| < \varepsilon,$$

with $\varepsilon \leq \delta_0$. $\bar{u}$ may be written uniquely as

$$\bar{u} = cx_0 + \bar{v} = c(x_0 + \bar{v}/c),$$

1. BEHAVIOR NEAR STATIONARY SOLUTIONS   353

with $c \in \mathbb{R}$ and $\overline{v} \in (I - P)(D)$. Since $c = \langle \overline{u}, x_0^* \rangle$, if $\varepsilon < \delta_1/\|x_0^*\|_{X'}$ then $|c| < \delta_1$. Since $\overline{v} = (I - P)\overline{u}$, if ε is small then $\|v\|_D$ is small. So, to prove uniqueness it is sufficient to show that $\|\overline{v}\|_D/c < \delta_0$ if ε is small enough. This will be a consequence of the assumption $F(0, \lambda) \equiv 0$. Indeed,

$$\begin{aligned} 0 &= F(cx_0 + \overline{v}, \overline{\lambda}) = F(cx_0 + \overline{v}, \overline{\lambda}) - F(0, \overline{\lambda}) - F(0, \overline{\lambda}) - F_x(0, \overline{\lambda})(cx_0 + \overline{v}) \\ &+ [F_x(0, \overline{\lambda}) - F_x(0, 0) - \overline{\lambda} F_{x\lambda}(0, 0)](cx_0 + \overline{v}) \\ &+ \overline{\lambda} F_{x\lambda}(0, 0)\overline{v} + F_x(0, 0)\overline{v} + c\overline{\lambda} F_{x\lambda}(0, 0)x_0. \end{aligned}$$

Since the operator $(\hat{v}, \hat{r}) \mapsto F_x(0, 0)\hat{v} + \hat{r} F_{x\lambda}(0, 0)x_0$ is an isomorphism from $(I - P)(D) \times \mathbb{R}$ onto X, it follows that

$$\|\overline{v}\|_D + |c\overline{\lambda}| \leq const.(\|\overline{v}\|^2 + c^2 + \overline{\lambda}^2(|c| + \|\overline{v}\|_D) + \overline{\lambda}\|\overline{v}\|_D\|c\overline{v}\|_D + |c|^2 + |c|o(|\overline{\lambda}|))$$

if ε is small enough. Therefore, for ε small,

$$\|\overline{v}\|_D + |c\overline{\lambda}| \leq const.|c|(\|\overline{v}\|_D + |c|),$$

and the statement follows. ∎

From (9.1.29) one deduces that if $\langle F_{xx}(0,0)(x_0, x_0), x_0^* \rangle \neq 0$, then the range of the function $\lambda(c)$ contains a neighborhood of 0, so that problem (9.1.22) admits stationary solutions both for $\lambda < 0$ and for $\lambda > 0$. If $\langle F_{xx}(0,0)(x_0, x_0), x_0^* \rangle = 0$, in order to know the sign of $\lambda(c)$ one may compute the derivatives of the function $\lambda(c)$ at $c = 0$ differentiating the identity $\eta(\lambda(c), c) \equiv 0$.

To study the stability properties of the bifurcating stationary solutions of (9.1.22) we have to know the position of the spectrum of $F_x(u(c), \lambda(c))$ with respect to the imaginary axis. If A satisfies the instability condition (9.1.13), then for c small the operator $F_x(u(c), \lambda(c))$ too satisfies (9.1.13), thanks to the perturbation lemma A.3.1. So, by Theorem 9.1.3, $u(c)$ is unstable. In the critical case of stability

$$\sup\{\operatorname{Re} z : z \in \sigma(A)\} = 0,$$

the operator $F_x(u(c), \lambda(c))$ has an isolated eigenvalue $\mu(c)$ near 0 thanks to Proposition A.3.2, and the rest of the spectrum lies in the left complex half plane. If $\mu(c)$ is not purely imaginary, the stability of $u(c)$ is decided by the sign of the real part of $\mu(c)$. In the next Theorem we show that $\mu(c)$ is real for c small, and we study the behavior of $\mu(c)$ for c near 0.

Theorem 9.1.11 *Let the assumptions of Theorem 9.1.10 hold. Then there are $c_1 > 0$ and a C^{k-1} function $\mu :]-c_1, c_1[\mapsto \mathbb{R}$ such that $\mu(0) = 0$ and $\mu(c)$ is an eigenvalue of $F_x(u(c), \lambda(c))$ for every $c \in]-c_1, c_1[$. Moreover, $\mu(c)$ is the unique element of the spectrum of $F_x(u(c), \lambda(c))$ near 0, and*

$$|\mu(c) + \gamma'(0)c\lambda'(c)| = \epsilon(c)|c\lambda'(c)|, \qquad (9.1.31)$$

where $\lim_{c\to 0} \epsilon(c) = 0$. It follows that there is a neighborhood of 0 in which $\mu(c)$ and $-\gamma'(0)c\lambda'(c)$ have the same zeroes, and if they do not vanish, they have the same sign.

Proof — The proof of the existence of $\mu(c)$ is identical to the proof of Lemma 9.1.9, and it is omitted. Precisely, one shows that there is a C^{k-1} function $c \mapsto w(c)$, defined in a neighborhood of 0 and with values in D, such that

$$F_x(u(c), \lambda(c))w(c) = \mu(c)w(c),$$

and $w(c)$ is the unique eigenfunction w of $F_x(u(c), \lambda(c))$ with eigenvalue $\mu(c)$ such that $w - x_0 \in (I - P)(X)$.

As a second step, we show that

$$\|u'(c) - w(c)\|_D \leq \text{const.} \, (|c\lambda'(c)| + |\mu(c)|), \qquad (9.1.32)$$

for c small. Differentiating the identity $F(u(c), \lambda(c)) \equiv 0$, we get

$$F_x(u(c), \lambda(c))u'(c) + F_\lambda(u(c), \lambda(c))\lambda'(c) = 0.$$

Subtracting the equality $F_x(u(c), \lambda(c))w(c) - \mu(c)w(c) = 0$ we obtain

$$F_x(u(c), \lambda(c))(u'(c) - w(c)) + F_\lambda(u(c), \lambda(c))\lambda'(c) + \mu(c)w(c) = 0. \qquad (9.1.33)$$

On the other hand,

$$F_x(u(c), \lambda(c)) = F_x(0,0) + c(F_{xx}(0,0)u'(0) + F_{x\lambda}(0,0)\lambda'(0)) + o(c) = A + o(1),$$

$$F_\lambda(u(c), \lambda(c)) = F_\lambda(0,0) + c(F_{x\lambda}(0,0)u'(0) + F_{\lambda\lambda}(0,0)\lambda'(0)) + o(c)$$
$$= cF_{x\lambda}(0,0)cx_0 + o(c).$$

Since $w(0) = x_0$, then $w(c) = x_0 + o(1)$. By replacing in (9.1.33) we get

$$A(u'(c) - w(c)) = (-c + o(c))F_{x\lambda}(0,0)x_0\lambda'(c) - \mu(c)x_0 \\ + o(1)(u'(c) - w(c)) + o(1)\mu(c). \qquad (9.1.34)$$

We claim that $u'(c) - w(c) \in (I - P)(D)$. Indeed, $u'(c) - x_0 = v(c) + cv(c) \in (I - P)(D)$ (see the proof of Theorem 9.1.10), and $w(c) - x_0 \in (I - P)(D)$ by construction; therefore $u'(c) - w(c) \in (I-P)(D)$. Since $A_{|(I-P)(D)} : (I-P)(D) \mapsto (I-P)(X)$ is invertible, (9.1.34) yields (9.1.32).

Once (9.1.32) is proved, by applying P to both sides of (9.1.34) one gets

$$0 = (c + o(c))\langle F_{x\lambda}(0,0)x_0, x_0^*\rangle\lambda'(c) + (1 + o(1))\mu(c) + \langle o(1)(u'(c) - w(c)), x_0^*\rangle.$$

Recalling that $\langle F_{x\lambda}(0,0)x_0, x_0^*\rangle = \gamma'(0)$ and using (9.1.32) we prove the last statement. Note that the "$o(1)$" before $u'(c) - w(c)$ is not a scalar but it is an operator,

1. Behavior near stationary solutions

so that $\langle o(1)(u'(c) - w(c)), x_0^* \rangle$ does not necessarily vanish, and we really need estimate (9.1.32). ∎

Theorem 9.1.11 implies that the sign of $\mu(c)$ is opposit to the sign of $c\lambda'(c)\gamma'(0)$ for c small. Formula (9.1.29) gives $\lambda'(0)$. If $\lambda'(0) \neq 0$, the sign of $\mu(c)$ is equal to the sign of

$$-c\lambda'(0)\gamma'(0) = c\langle F_{xx}(0,0)(x_0, x_0), x_0^* \rangle,$$

which coincides with the sign of $\pm\lambda\langle F_{xx}(0,0)(x_0, x_0), x_0^* \rangle$ if $\lambda'(0) \gtreqless 0$.

9.1.5 Applications to nonlinear parabolic problems, I

We consider a nonlinear evolution problem in $[0, +\infty[\times\overline{\Omega}$, Ω being a bounded open set in $\mathbb{R}^n$ with uniformly $C^{2+\theta}$ boundary $\partial\Omega$ $(0 < \theta < 1)$,

$$\begin{cases} u_t = \Delta u + cu + g(x, u, Du, D^2u), & t \geq 0, \; x \in \overline{\Omega}, \\ \partial u(t,x)/\partial\nu = 0, & t \geq 0, \; x \in \partial\Omega, \end{cases} \quad (9.1.35)$$

$$u(0, x) = u_0(x), \quad x \in \overline{\Omega}. \quad (9.1.36)$$

Here c is a real number, $g = g(x, u, p, q)$ is a sufficiently regular function defined for $x \in \overline{\Omega}$ and (u, p, q) in a neighborhood of 0 in $\mathbb{R} \times \mathbb{R}^n \times \mathbb{R}_S^{n^2}$, satisfying[4]

$$g(x, 0) = g_u(x, 0) = g_{p_i}(x, 0) = g_{q_{ij}}(x, 0) = 0, \quad i, j = 1, \ldots, n.$$

This implies that problem (9.1.35) is parabolic near $u \equiv 0$.

Local existence and regularity for problem (9.1.35)-(9.1.36) with initial datum $u_0 \in C^{2+\theta}(\overline{\Omega})$ have been discussed in Subsection 8.5.3. As in Subsection 8.5.3 we assume here that g is twice continuously differentiable with respect to (u, p, q), with derivatives up to the second order θ-Hölder continuous with respect to x and Lipschitz continuous with respect to (u, p, q).

We discuss now the behavior of the solution for small initial data, applying the results of this Section.

By Corollary 3.1.24 the realization A of $\Delta + cI$ with homogeneous Neumann boundary condition in $E = C(\overline{\Omega})$ is sectorial, and by Theorem 3.1.29 and Corollary 3.1.35(ii)

$$D_A(\theta/2, \infty) = C^\theta(\overline{\Omega}), \quad D_A(\theta/2 + 1, \infty) = C^{2+\theta}_{\partial/\partial\nu}(\overline{\Omega}),$$

with equivalence of the respective norms. We recall that $C^{2+\theta}_{\partial/\partial n}(\overline{\Omega})$ denotes the subspace of $C^{2+\theta}(\overline{\Omega})$ consisting of the functions with vanishing normal derivative at the boundary.

---

[4]We recall that $\mathbb{R}_S^{n^2}$ is the set of the symmetric matrices $[q_{ij}]_{i,j=1,\ldots,n}$ endowed with the topology of $\mathbb{R}^{n^2}$.

The mapping $\psi \mapsto G(\psi)$ defined by
$$(G(\varphi))(x) = g(x, \varphi(x), D\varphi(x), D^2\varphi(x)), \quad x \in \overline{\Omega} \tag{9.1.37}$$
is continuously differentiable in a neighborhood of 0 in $D_A(\theta/2+1, \infty)$, with values in $D_A(\theta/2, \infty)$. Therefore, setting $u(t) = u(t, \cdot)$, problem (9.1.35) may be seen as an evolution equation of the type (9.1.1), such that assumptions (9.1.2), (9.1.3), (9.1.17) are satisfied.

Applying Theorem 9.1.5 with θ replaced by $\theta/2$, one proves local existence of a classical solution u of (9.1.35) for small initial data. More precisely, for every $T > 0$ there exists $r_0 > 0$ such that if $u_0 \in C^{2+\theta}_{\partial/\partial\nu}(\overline{\Omega})$ and $\|u_0\|_{C^{2+\theta}(\overline{\Omega})} \leq r_0$ then problem (9.1.35)-(9.1.36) has a solution $u : [0, T] \times \overline{\Omega} \mapsto \mathbb{R}$. The regularity properties of the solution are the following: u_t and $D_{ij}u$ are continuous in $[0, T] \times \overline{\Omega}$, and they are θ-Hölder continuous with respect to the space variables, with Hölder constant independent of t; Δu is θ-Hölder continuous with respect to time, with Hölder constant independent of the space variables. Moreover, $t \mapsto u(t, \cdot)$ is continuous in $]0, T]$ with values in $C^{\theta+2}(\overline{\Omega})$.

If we denote by $\{-\lambda_n\}_{n \in \mathbb{N}}$ the ordered sequence of the eigenvalues of Δ with homogeneous Neumann boundary condition, then $\sigma(A) = \{-\lambda_n + c\}_{n \in \mathbb{N}}$. In the case where $c < 0$, the spectrum of A is contained in the left complex halfplane. Then the principle of linearized stability (Theorem 9.1.7(i)) may be applied, and it gives existence in the large and exponential decay of the solution for small initial data. Precisely, since the first eigenvalue of A is c, then for every $\omega \in]0, c[$ there are $M(\omega)$, $R(\omega)$ such that if $\|u_0\|_{C^{\theta+2}(\overline{\Omega})} \leq R(\omega)$ then
$$\|u(t, \cdot)\|_{C^{\theta+2}(\overline{\Omega})} \leq M(\omega)e^{-\omega t}\|u_0\|_{C^{\theta+2}(\overline{\Omega})}, \quad t \geq 0.$$

In the case where $c > 0$ the assumptions of Theorem 9.1.7(ii) are satisfied, so that the null solution is unstable. If $-\lambda_n + c \neq 0$ for every $n \in \mathbb{N}$, then 0 is a saddle point: there exist an infinite dimensional Lipschitz continuous stable manifold and a finite dimensional Lipschitz continuous unstable manifold.

The critical case of stability $c = 0$ will be discussed in the next Section.

Similar considerations hold for the Dirichlet problem
$$\begin{cases} u_t = \Delta u + cu + g(x, u, Du, D^2u), & t \geq 0, \; x \in \overline{\Omega}, \\ u(t, x) = 0, & t \geq 0, \; x \in \partial\Omega, \end{cases} \tag{9.1.38}$$
where the critical value of the parameter c is $c = \lambda_1$, $-\lambda_1$ being the first eigenvalue of the Laplace operator with Dirichlet boundary condition. Moreover we have to assume that $g(x, 0, p, q) = 0$ for $x \in \partial\Omega$ to work in the space $D_A(\theta/2, \infty) = C_0^{2\theta}(\overline{\Omega})$.

In the case where the nonlinear function g in (9.1.35) depends on a real parameter λ,
$$g = g(\lambda, x, u, Du, D^2u), \tag{9.1.39}$$
existence of stationary solutions of problem (9.1.35) may be proved by applying the results of Subsection 9.1.4.

1. Behavior near stationary solutions

We assume that the function g is defined for $\lambda \in \,]-1,1[$, $x \in \overline{\Omega}$, and (u,p,q) in a neighborhood of 0 in $\mathbb{R} \times \mathbb{R}^n \times \mathbb{R}_S^{n^2}$, and that g is thrice continuously differentiable with respect to (λ, u, p, q), with derivatives up to the third order θ-Hölder continuous with respect to x and Lipschitz continuous with respect to (u,p,q). Moreover, we assume that

$$g(\lambda, x, 0) = g_u(0, x, 0) = g_{p_i}(0, x, 0) = g_{q_{ij}}(0, x, 0) = 0, \quad i,j = 1, \ldots, n.$$

The problem

$$\Delta u(x) + cu(x) + g(\lambda, x, u(x), Du(x), D^2 u(x)) = 0, \quad x \in \overline{\Omega}, \quad \frac{\partial u}{\partial n} = 0, \quad x \in \partial\Omega \tag{9.1.40}$$

is seen as an equation of the type (9.1.22) in the space $D = C_{\partial/\partial\nu}^{2+\theta}(\overline{\Omega})$. The function

$$F(u, \lambda) = \Delta u + cu + g(\lambda, \cdot, u(\cdot), Du(\cdot), D^2 u(\cdot))$$

is well defined and twice continuously differentiable in a neighborhood of 0 in $\mathbb{R} \times C_{\partial/\partial\nu}^{2+\theta}(\overline{\Omega})$ with values in $C^\theta(\overline{\Omega})$, and

$$F_u(0, \lambda) = \Delta + cI.$$

If $c \neq \lambda_n$ for every n, the operator $A = \Delta + cI : C_{\partial/\partial n}^{2+\theta}(\overline{\Omega}) \mapsto C^\theta(\overline{\Omega})$ is invertible, and existence of a unique small solution of (9.1.40) for every small λ follows easily from the considerations made at the beginning of Subsection 9.1.4.

In the case $c = 0$ the result of Theorem 9.1.10 may be applied. Indeed, $A = \Delta$ admits the simple eigenvalue 0, with eigenspace consisting of all the constant functions. We denote by ξ_0 the constant function equal to $1/(\operatorname{meas}\Omega)^{1/2}$ over $\overline{\Omega}$. Thanks to Lemma A.2.8, the associated projection P is given by

$$P\varphi(x) = \frac{1}{\operatorname{meas}\Omega}\int_\Omega \varphi(y)\,dy, \quad \varphi \in C^\theta(\overline{\Omega}), \ x \in \overline{\Omega}, \tag{9.1.41}$$

So, the element ξ_0^* of the dual space $(C^\theta(\overline{\Omega}))'$ such that $P\varphi = \langle\varphi, \xi_0^*\rangle \xi_0$ is the functional $\varphi \mapsto (\operatorname{meas}\Omega)^{-1/2} \int_\Omega \varphi(y)\,dy$. Therefore,

$$\langle F_{u\lambda}(0,0)\xi_0, \xi_0^*\rangle = \frac{1}{(\operatorname{meas}\Omega)^{1/2}}\int_\Omega g_{u\lambda}(0, y, 0, 0, 0)\,dy.$$

Setting $\alpha = \langle F_{u\lambda}(0,0)\xi_0, \xi_0^*\rangle$, if $\alpha \neq 0$ Theorem 9.1.10 guarantees existence of small solutions of $F(u, \lambda) = 0$. More precisely, if also

$$\beta = \langle F_{uu}(0,0)(\xi_0, \xi_0), \xi_0^*\rangle = \frac{1}{(\operatorname{meas}\Omega)^{1/2}}\int_\Omega g_{uu}(0, y, 0, 0, 0)\,dy \neq 0, \tag{9.1.42}$$

then for every λ close to 0 the problem (9.1.35) (with $c = 0$) has a unique small stationary solution $u(\lambda) \in C_{\partial/\partial\nu}^{2+\theta}(\overline{\Omega})$.

With the aid of Theorem 9.1.11 it is possible to study the stability of the bifurcating stationary solutions $u(\lambda)$ for λ small. Indeed, the linearized operator $F_u(u(\lambda), \lambda)$ is close to the Laplace operator Δ in $L(C^{2+\theta}_{\partial/\partial\nu}(\overline{\Omega}), C^\theta(\overline{\Omega}))$, so that its spectrum consists of a sequence of negative eigenvalues, plus an isolated real eigenvalue μ, the sign of which may be found using Theorem 9.1.11. The remarks after Theorem 9.1.11 imply that the sign of μ is equal to the sign of $\lambda\beta$, if α and β have opposite sign, and it is equal to the sign of $-\lambda\beta$, if α and β have the same sign.

If μ is negative, then the Principle of Linearized Stability (Theorem 9.1.7(i)) may be applied, finding that the stationary solution $u(\lambda)$ is exponentially asymptotically stable in the space $C^{2+\theta}_{\partial/\partial\nu}(\overline{\Omega})$. If μ is positive, part (ii) of Theorem 9.1.7 implies that $u(\lambda)$ is unstable in $C^{2+\theta}_{\partial/\partial\nu}(\overline{\Omega})$.

9.1.6 Stability of travelling waves in two-phase free boundary problems

Consider a free boundary problem, with unknown $\xi : [0, +\infty[\mapsto \mathbb{R}, \; u : \mathbb{R} \mapsto \mathbb{R}$,

$$\begin{cases} u_t(t,x) = u_{xx}(t,x) + u(t,x)u_x(t,x), \; t \geq 0, \; x \in \mathbb{R}, x \neq \xi(t), \\ u(t,\xi(t)) = u_*, \; [u_x(t,\xi(t))] = u_x(t,\xi(t)^+) - u_x(t,\xi(t)^-) = -1, \; t \geq 0, \\ u(t,-\infty) = 0, \; u(t,+\infty) = u_\infty, \; t \geq 0. \end{cases}$$
(9.1.43)

Problem (9.1.43) is the simplest example of a class of free boundary problems with jump condition on the gradient at the free boundary arising in Combustion Theory. See G.S.S. LUDFORD and D.S. STEWART [130, 131, 186], B.J. MATKOWSKY – A. VAN HARTEN [159].

$u(t,x)$ represents a normalized temperature, and it satisfies the Burgers equation on both sides of the free boundary, where the combustion reaction takes place. u_∞ is the normalized temperature of the burnt mixture, and u_* is the normalized combustion temperature.

Of physical relevance are travelling wave solutions and their perturbations. We recall that a travelling wave solution of a one dimensional problem is a solution of the type

$$u(t,x) = U(x + ct), \quad \xi(t) = -ct + \xi_0,$$

where $c, \xi_0 \in \mathbb{R}$ and $U : \mathbb{R} \mapsto \mathbb{R}$ is a given function.

We present here the approach of C.-M. BRAUNER–A. LUNARDI–C. SCHMIDT-LAINÉ [35] to the stability of the travelling wave solutions.

It is not difficult to see that a (unique up to translations) travelling wave solution exists if and only if

$$u_\infty > \sqrt{2}, \; 2/u_\infty < u_* < u_\infty + 2/u_\infty,$$

1. Behavior near stationary solutions

in which case we have

$$c = \frac{1}{u_\infty} + \frac{u_\infty}{2} > \sqrt{2},$$

$$\begin{cases} U(y) = \dfrac{2cu_* e^{cy}}{u_* e^{cy} + 2c - u_*}, & y < 0, \\ U(y) = \dfrac{u_\infty(u_* + u_\infty - 2c) + \frac{2}{u_\infty}(u_\infty - u_*)e^{-(u_\infty/2 - 1/u_\infty)y}}{u_* + u_\infty - 2c + (u_\infty - u_*)e^{-(u_\infty/2 - 1/u_\infty)y}}, & y \geq 0. \end{cases}$$

The wave decays exponentially, as $y \to \pm\infty$, to its limits 0 and u_∞. Precisely,

$$\begin{cases} U(y) \sim \dfrac{2cu_*}{2c - u_*} e^{cy} = k_- e^{2\alpha - y} \text{ as } y \to -\infty, \\ U(y) - u_\infty \sim \dfrac{2(u_\infty - u_*)}{u_* + u_\infty - 2c} e^{-(u_\infty/2 - 1/u_\infty)y} = k_+ e^{-2\alpha + y}, \text{ as } y \to +\infty. \end{cases}$$

It is convenient to rewrite problem (9.1.43) in a frame attached to the front, setting $y = x - \xi(t)$ and getting

$$\begin{cases} u_t = \dot\xi(t) u_y + u_{yy} + (\tfrac{1}{2} u^2)_y, & t \geq 0, \ y \neq 0, \\ u(0,t) = u_*, \ [u_y(t)(0)] = u_y(t, 0^+) - u_y(t, 0^-) = -1, & t \geq 0, \\ u(t, -\infty) = 0, \ u(t, +\infty) = u_\infty, & t \geq 0. \end{cases} \quad (9.1.44)$$

Here and in the following we use the subscript t or the superdot to denote differentiation with respect to time, and the subscript y or the prime to denote differentiation with respect to the space variable y.

We introduce the perturbations

$$v(x, t) = u(x, t) - U(x), \quad s(t) = \xi(t) + ct.$$

Problem (9.1.44) is equivalent to

$$\begin{cases} v_t = v_{yy} - cv_y - \dot s(t) U'(y) + \dot s(t) v_y + (Uv)_y + vv_y, & t \geq 0, \ y \neq 0, \\ v(t, 0) = [v_y(t, 0)] = 0, \ v(t, -\infty) = v(t, +\infty) = 0, & t \geq 0, \\ v(0, y) = v_0(y) = u_0(y) - U(y), \ y \in \mathbb{R}, \ s(0) = 0. \end{cases} \quad (9.1.45)$$

A suggestion on the way to decouple (9.1.45) is given by the linearized problem,

$$\begin{cases} \hat v_t = \hat v_{yy} - c\hat v_y + (U\hat v)_y + \dot s(t) U' = \mathcal{A} \hat v + \dot s(t) U', & t \geq 0, \ y \neq 0, \\ \hat v(t, 0) = [\hat v_y(t, 0)] = 0, \ \hat v(t, -\infty) = \hat v(t, +\infty) = 0, \\ \hat v(0, y) = \hat v_0(y) = u_0(y) - U(y), \ y \in \mathbb{R}, \ s(0) = 0, \end{cases}$$

which is easily decoupled by setting $v_1(t, y) = \hat v(t, y) - s(t) U'(y)$. Indeed, since $\mathcal{A} U' = 0$ by construction, we get

$$\partial v_1/\partial t = \hat v_t - \dot s(t) U' = \mathcal{A} \hat v = \mathcal{A} v_1, \quad t \geq 0, \ y \neq 0,$$

and

(i) $v_1(t, 0^+) = -s(t)U'(0^+)$, $v_1(t, 0^-) = -s(t)U'(0^-)$, $t \geq 0$,

(ii) $\dfrac{\partial v_1}{\partial y}(t, 0^+) = -s(t)U''(0^+)$, $\dfrac{\partial v_1}{\partial y}(t, 0^-) = -s(t)U''(0^-)$, $t \geq 0$.

The boundary conditions (i) are equivalent to

$$[v_1(t, 0)] = s(t), \quad v_1(t, 0^+)U'(0^-) = v_1(t, 0^-)U'(0^+),$$

which decouple completely the system. Then the boundary conditions (ii) may be written as

$$[\partial v_1/\partial y(t, 0)] = [v_1(t, 0)](c - u_*).$$

So, we have associated "natural" boundary conditions to the operator $\mathcal{A}$, which were not immediately deducible from the homogeneous problem (9.1.45).

We now read the nonlinear problem (9.1.45) as an evolution problem in the weigted space

$$Y = \{v : \mathbb{R} \mapsto \mathbb{R} : y \mapsto e^{-\alpha - y}v(y) \in UC(\mathbb{R}_-), \ y \mapsto e^{\alpha + y}v(y) \in UC(\mathbb{R}_+)\}$$

The realization of the operator $\mathcal{A}$ with the above boundary conditions in Y is defined by

$$\begin{cases} D(A) = \{v \in Y : v', v'' \in Y, [v'(0)] = (c - u_*)[v(0)], \\ \qquad v(0^-)U'(0^+) = v(0^+)U'(0^-)\}, \\ A : D(A) \mapsto Y, \ Av = \mathcal{A}v = (v' - cv + Uv)'. \end{cases}$$

The reason for choosing a weighted space instead of a nonweighted one will be clear soon. Note that the derivative of the travelling wave solution U' is in $D(A)$.

The properties of the operator A are summarized in the next Proposition.

Proposition 9.1.12 *A is a sectorial operator in Y. Its spectrum has three components:*

(i) the half-line $]-\infty, -c^2/4 + 1/2]$, with $-c^2/4 + 1/2 < 0$;

(ii) the simple eigenvalue 0, the kernel of A being spanned by U';

(iii) possibly an additional eigenvalue λ given by the dispersion relation

$$p_1(\lambda) - u_*/2 - u_*(c - u_*/2)(p_1(\lambda) + p_2(\lambda) - c) = 0,$$

where $p_1(\lambda) = (c + \sqrt{c^2 + 4\lambda})/2$, $p_2(\lambda) = (c + \sqrt{c^2 + 4\lambda - 2})/2$.

Proposition 9.1.12 has been proved by elementary methods in [35]. The general theorems of Chapter 3 about generation of analytic semigroups by elliptic operators may be avoided, since the problem is one dimensional and the resolvent operator can be explicitly computed.

1. Behavior near stationary solutions

We may explain now the reason of the choice of the space Y. Since $c > \sqrt{2}$, the continuous spectrum $]-\infty, -c^2/4 + 1/2]$ of A is far from the imaginary axis. If Y is replaced by the nonweighted space $\tilde{Y} = \{v : \mathbb{R} \mapsto \mathbb{R} : y_{|\mathbb{R}_-} \in UC(\mathbb{R}_-), y_{|\mathbb{R}_+} \in UC(\mathbb{R}_+)\}$, then the continuous spectrum of the realization of $\mathcal{A}$ in $\tilde{Y}$ reaches 0, and this fact prevents us from using the Principle of Linearized Stability as stated in Subsection 9.1.1. Some results about stability when 0 is not an isolated element of the spectum may be found in Subsection 9.2.3, but they are not applicable to the present example.

The dispersion relation may be completely solved (see e.g. [165]) as far as the existence and the sign of the solutions are concerned. Indeed, setting

$$u_0 = c - \sqrt{c^2 - 1 - \sqrt{2c^2 - 3}}, \quad u_c = c + \sqrt{c^2 - 1}$$

it is possible to see that

- If $2/u_\infty < u_* \leq u_0$ the dispersion relation has a unique root λ_D such that $(2-c^2)/2 \leq \lambda_D < 0$;

- if $u_0 < u_* \leq u_c$ it has no solution;

- if $u_c < u_* < u_\infty + 2/u_\infty = 2c$ it admits a unique solution $\lambda_D > 0$.

Problem (9.1.45) is written in abstract form as an evolution system in Y,

$$\dot{v} = Av + \dot{s}U' + \dot{s}v' + vv', \quad t \geq 0; \quad v(0) = v_0, \quad s(0) = 0. \tag{9.1.46}$$

To decouple it, in view of Proposition 9.1.12 we apply the projection P associated to the spectral set $\{0\}$. Using Lemma A.2.8 one sees that P is given by

$$Pv(y) = \frac{1}{u_\infty} \int_{-\infty}^{+\infty} v(x) dx \, U'(y), \quad v \in Y.$$

Splitting v as $v(t) = Pv(t) + (I - P)v(t) = p(t)U' + w(t)$ we get

$$p(t) = [w(t)(0)], \quad t \geq 0,$$

and then applying P and $I - P$ to (9.1.46) we obtain

$$\dot{p}(t) = \dot{s}(t), \quad t \geq 0,$$

$$\dot{w}(t) = (I - P)Aw(t) + \dot{s}(t)v'(t) + v(t)v'(t), \quad t \geq 0; \quad w(0) = w_0 = (I - P)v_0.$$

The latter equation is used to evaluate the jump of $\dot{w}(t)$ at $y = 0$:

$$\begin{aligned}\dot{s}(t) &= [\dot{w}(t)(0)] = [(I - P)Aw(t)(0)] + [(\dot{s}(t) + v(t))v_y(t)(0)] \\ &= [(I - P)Aw(t)(0)].\end{aligned}$$

By replacing $\dot{s}$ in the equation for w we obtain

$$\begin{cases} \dot{w} = (I-P)Aw + [(I-P)Aw]([w]U' + w') + \tfrac{1}{2}\tfrac{d}{dy}([w]U' + w')^2, \\ w(0) = w_0, \end{cases} \quad (9.1.47)$$

which is an initial value problem in the space $(I-P)(Y)$, with linear operator $(I-P)A$ and nonlinear function

$$G(w) = [(I-P)Aw]([w]U' + w') + \frac{1}{2}\frac{d}{dy}([w]U' + w')^2.$$

$G : (I-P)(D(A)) \mapsto (I-P)(Y)$ is analytic, so that the local existence theorem 8.1.1 and the regularity theorems of Section 8.3 may be applied. One finds that for every $T > 0$ there is $\delta_1 > 0$ such that for each $w_0 \in (I-P)D(A)$ with $\|w_0\|_{D(A)} \leq \delta_1$, problem (9.1.47) has a unique small solution $w \in C_\alpha^\alpha(]0,T]; (I-P)(D(A))) \cap \overline{C([0,T];Y)}$, which is analytic in $]0,T[$ with values in $D(A)$. If in addition $Aw_0 + G(w_0) \in \overline{D(A)}$, then w is continuous up to $t=0$ with values in $D(A)$, and $\dot{w}$ is continuous up to $t=0$ with values in Y.

The spectrum of the linear operator $(I-P)A : D(A) \cap (I-P)(Y) \mapsto (I-P)(Y)$ consists of the half line $]-\infty, -c^2/4 + 1/2]$ plus the eigenvalue λ_D given by the dispersion relation. Therefore it does not intersect the imaginary axis, and the Principle of Linearized Stability may be applied. In the case where

$$2/u_\infty < u_* \leq u_c, \quad (9.1.48)$$

the spectrum is contained in the left complex halfplane, and Theorem 9.1.2 implies that the null solution of (9.1.47) is exponentially asymptotically stable. If

$$u_c < u_* < u_\infty + 2/u_\infty, \quad (9.1.49)$$

the spectrum has a unique element with positive real part, and Theorem 9.1.3 implies that the null solution of (9.1.47) is unstable.

Coming back to problem (9.1.43), one finds

$$\begin{cases} u(t,\cdot) = [w(t)]U' + w(t) + U, \\ s(t) = \xi(t) - ct = \int_0^t \dot{p}(\sigma)d\sigma = p(t) - p(0) = [w(t)(0)] - \dfrac{1}{u_\infty}\int_{-\infty}^{+\infty} v_0(x)dx. \end{cases}$$

If (9.1.48) holds, the travelling wave U is orbitally stable with respect to perturbations belonging to the weighted space $D(A)$, and the perturbation of the front $s(t)$ converges exponentially to the limiting shift

$$s_\infty = -\frac{1}{u_\infty}\int_{-\infty}^{+\infty} v_0(x)dx.$$

If (9.1.49) holds, the travelling wave U is unstable.

The method described above is effective in a larger class of singular free boundary problems, see the paper [35]. It may be extended to systems with nonlinear conditions at the free boundary, such as the one considered in [15].

9.2 Critical cases of stability

9.2.1 The center-unstable manifold

While the construction of the stable and unstable manifolds in the saddle point case is relatively easy, the construction of finite dimensional invariant manifolds near stationary solutions in the critical case of stability $\sup\{\operatorname{Re}\lambda : \lambda \in \sigma(A)\} = 0$, or, more generally, when $\sigma(A) \cap i\mathbb{R} \neq \emptyset$, is rather complicated. We are able to treat only the case where the underlying Banach space is an interpolation space $D_A(\theta, \infty)$ or $D_A(\theta)$, and, correspondingly, the domain of A is $D_A(\theta + 1, \infty)$ or $D_A(\theta+1)$. To be definite, we assume that (9.1.17) holds. There are not substantial differences if assumption (9.1.17) is replaced by (9.1.18). To simplify notation, we set here

$$E_i = D_A(\theta + i, \infty), \quad \|\cdot\|_i = \|\cdot\|_{D_A(\theta+i,\infty)}, \quad i = 0, 1.$$

Moreover, we assume that

$$\begin{cases} \text{the set } \sigma_+(A) = \{\lambda \in \sigma(A) : \operatorname{Re}\lambda \geq 0\} \text{ consists of a finite} \\ \text{number of isolated eigenvalues with finite algebraic multiplicity.} \end{cases} \quad (9.2.1)$$

We define as before $\sigma_-(A) = \{\lambda \in \sigma(A) : \operatorname{Re}\lambda < 0\}$, and $\omega_- = -\sup\{\operatorname{Re}\lambda : \lambda \in \sigma_-(A)\}$. P is again the projection associated with the spectral set $\sigma_+(A)$. Thanks to assumption (9.2.1), $P(E_0) \subset E_1$ is finite dimensional. This fact is of crucial importance in what follows.

Problem (9.1.1) is equivalent to the system

$$\begin{cases} x'(t) = A_+ x(t) + PG(x(t) + y(t)), & t \geq 0, \\ y'(t) = A_- y(t) + (I - P)G(x(t) + y(t)), & t \geq 0, \end{cases} \quad (9.2.2)$$

with $x(t) = Pu(t)$, $y(t) = (I-P)u(t)$, $A_+ = A_{|P(E_0)} : P(E_0) \mapsto P(E_0)$, $A_- = A_{|(I-P)(E_1)} : (I-P)(E_1) \mapsto (I-P)(E_0)$.

We modify G by introducing a smooth cutoff function $\rho : P(E_0) \mapsto \mathbb{R}$ such that

$$0 \leq \rho(x) \leq 1, \quad \rho(x) = 1 \text{ if } \|x\|_0 \leq 1/2, \quad \rho(x) = 0 \text{ if } \|x\|_0 \geq 1.$$

Since $P(E_0)$ is finite dimensional, the existence of such a function is obvious. For small $r > 0$ we consider the system

$$x'(t) = A_+ x(t) + f(x(t), y(t)), \quad y'(t) = A_- y(t) + g(x(t), y(t)), \quad t \geq 0, \quad (9.2.3)$$

$$x(0) = x_0 \in P(E_0), \quad y(0) = y_0 \in (I-P)(E_0), \quad (9.2.4)$$

where

$$f(x, y) = PG(\rho(x/r)x + y), \quad g(x, y) = (I-P)G(\rho(x/r)x + y).$$

System (9.2.3) coincides with (9.2.2) if $\|x(t)\|_0 \leq r/2$. In particular, they are equivalent as far as stability of the null solution is concerned.

Throughout the subsection, for every $r > 0$ we denote by $L(r)$ the maximum between the Lipschitz constants of f and g over $P(E_0) \times B(0,r) \subset (I-P)(E_1)$ (of course, $P(E_0)$ is endowed with the norm of E_0 and $(I-P)(E_1)$ is endowed with the norm of E_1). Then

$$\lim_{r \to 0} L(r) = 0.$$

Moreover, setting

$$M(r) = \sup_{x \in P(E_0), \|y\|_1 \leq r} \|f(x,y)\|_0 + \|g(x,y)\|_0,$$

it holds

$$\lim_{r \to 0} M(r)/r = 0.$$

Theorem 9.1.5 and Corollary 9.1.6 may be applied to problem (9.2.3)-(9.2.4), getting local existence for small initial data and existence in the large provided the solution remains small enough. In fact, while it is easy to find an *a priori* estimate on $\|y(t)\|_{D_A(\theta+1,\infty)}$ for r small (see Proposition 9.2.1 below), in general $x(t)$ is not necessarily bounded. For instance, in the case where $f \equiv 0$ and $\|e^{tA}P\|_{L(X)}$ is not bounded there are arbitrarily small x_0 such that $x(t)$ is not bounded. Therefore, in the general case Corollary 9.1.6 is not of help. However, due to the truncation in f and g, we can show that if r and the initial data are small enough, then the solution of (9.2.3)-(9.2.4) exists in the large.

Proposition 9.2.1 *There exists $r_0 > 0$ and a function $C_0 :]0, r_0] \mapsto]0, \infty[$ such that if $0 < r \leq r_0$ and $\|x_0\|_0 + \|y_0\|_1 \leq C_0$ then problem (9.2.3)-(9.2.4) has a solution (x,y) such that x belongs to $C^1([0,+\infty[; E_0)$ and y belongs to $C(]0,+\infty[; E_1) \cap B([0,+\infty[; E_1) \cap C([0,+\infty[; E_0)$. The solution is unique in the class of the functions (x,y) enjoying the above properties of regularity and such that $\|y(t)\|_1 \leq r$.*

Proof — The proof is in two steps. As a first step we show that for any fixed $T > 0$, system (9.2.3)-(9.2.4) has a solution in $[0,T]$ provided r is small and $\|y_0\|_1 \leq r/2M$, where

$$M = \sup_{t>0} \|e^{tA_-}\|_{L((I-P)(E_1))},$$

no matter how $\|x_0\|_0$ is large. As a second step, we prove that $\|y(t)\|_1 \leq r/2M$ for every t in the maximal interval of existence, provided $\|x_0\|_0$ and $\|y_0\|_1$ are small enough.

(i) As a consequence of Proposition 4.4.10, for every $\varphi \in L^\infty(0,T; P(E_0))$ we have

$$\sup_{0 \leq t \leq T} \left\| \int_0^t e^{(t-s)A_-} \psi(s) ds \right\|_1 \leq C_1 \sup_{0 \leq t \leq T} \|\psi(t)\|_0,$$

2. CRITICAL CASES OF STABILITY

with C_1 independent of T and ψ. Let moreover $C_2 = C_2(T)$ be such that

$$\sup_{0\leq t\leq T}\left\|\int_0^t e^{(t-s)A_+}\varphi(s)ds\right\|_1 \leq C_2 \sup_{0\leq t\leq T}\|\varphi(t)\|_0,$$

for every $\varphi \in L^\infty(0, T; P(E_0))$.

Any solution of (9.2.3)-(9.2.4) in the interval $[0, T]$ is a fixed point of the operator Γ defined by

$$\Gamma(x,y)(t) = \left(e^{tA_+}x_0 + \int_0^t e^{(t-s)A_+}f(x(s),y(s))ds ,\right.$$

$$\left. e^{tA_-}y_0 + \int_0^t e^{(t-s)A_-}g(x(s),y(s))ds\right).$$

If r is small enough Γ is well defined on the set $Y = B(e^{tA_+}x_0, r) \times B(0, r) \subset (C([0,T]; P(E_0)) \times C(]0,T]; (I-P)(E_1)) \cap B([0,T]; E_1))$. Y is endowed with the product norm $\|(x,y)\|_Y = \|x\|_{B([0,T];E_0)} + \|y\|_{B([0,T];E_1)}$.

For (x, y), $(\overline{x}, \overline{y})$ in Y we get easily

$$\|f(x(s),y(s)) - f(\overline{x}(s),\overline{y}(s))\|_0 \leq L(r)(\|x(s) - \overline{x}(s)\|_0 + \|y(s) - \overline{y}(s)\|_1),$$

$$\|g(x(s),y(s)) - g(\overline{x}(s),\overline{y}(s))\|_0 \leq L(r)(\|x(s) - \overline{x}(s)\|_0 + \|y(s) - \overline{y}(s)\|_1).$$

Therefore, Γ is a $1/2$-contraction provided

$$(C_1 + C_2)L(r) \leq 1/2.$$

For $(x, y) \in Y$ we get

$$\|\Gamma(x,y) - (e^{tA_+}x_0, 0)\|_Y \leq (C_1 + C_2)M(r) + M\|y_0\|_1,$$

so that Γ maps Y into itself if r is so small that $(C_1 + C_2)M(r)/r \leq 1/2$ and $\|y_0\|_1 \leq r/2M$. Therefore, if r is small enough and $\|y_0\|_1 \leq r/2M$, problem (9.2.3)-(9.2.4) has a unique solution in Y. Indeed, it is the unique solution of (9.2.3)-(9.2.4) such that $\|y(t)\|_1 \leq r$.

(ii) Let us prove now that $\|y(t)\|_1 \leq r/2M$ for every t in the interval of existence if the initial data are small enough. By Theorem 9.1.5, there is $C(r)$ such that if $\|x_0\|_0 + \|y_0\|_1 \leq C(r)$, then the solution of (9.2.3)-(9.2.4) is defined at least in $[0,1]$, and $\|y(t)\|_D \leq r/2M$ for $0 \leq t \leq 1$. Take

$$\|x_0\|_0 + \|y_0\|_1 \leq C(r),$$

and set

$$\tau = \sup\{t > 0 : y(s) \text{ exists and } \|y(s)\|_1 \leq r/2M \text{ for } 0 \leq s \leq t\}.$$

Then $\tau \geq 1$. If $\tau < \infty$, y is well defined and continuous at $t = \tau$ with values in E_1, so that

$$r/2M = \|y(\tau)\|_1 \leq M\|y_0\|_1 + C_1 \sup_{s \leq \tau} \|g(x(s), y(s))\|_0 \leq M\|y_0\|_1 + C_1 M(r),$$

which is impossible if r is so small that $C_1 M(r) < r/4M$ and

$$\|y_0\|_1 \leq r/4M.$$

For such values of r and of the norms of the initial data, $\|y(t)\|_1$ remains bounded by $r/2M$, as far as it exists. By Step 1, the solution exists in the large. ■

We shall prove that there exists a finite dimensional invariant manifold $\mathcal{M}$ for system (9.2.3)-(9.2.4), provided r is sufficiently small. Then we shall see that such a manifold attracts exponentially all the orbits starting from an initial datum sufficiently close to the manifold itself. As a consequence, we shall see that the null solution of (9.1.1) is stable, asymptotically stable, or unstable, if and only if it is stable, asymptotically stable, or unstable, with respect to the restriction of the flow to $\mathcal{M}$.

The manifold $\mathcal{M}$ is sought as the graph of a bounded, Lipschitz continuous function $\gamma : P(E_0) \mapsto (I - P)(E_1)$. Let us derive heuristically an equation which should be satisfied by γ. Invariance of the graph of γ means that if $y_0 = \gamma(x_0)$, then

$$x'(t) = A_+ x(t) + f(x(t), \gamma(x(t))), \quad y'(t) = A_- y(t) + g(x(t), \gamma(x(t))),$$

for every t such that $(x(t), y(t))$ exists. For every $x_0 \in P(E_0)$, let $z = z(s; x_0, \gamma)$ be the solution of the finite dimensional system

$$z' = A_+ z + f(z + \gamma(z)); \quad z(0) = x_0. \tag{9.2.5}$$

Since γ is Lipschitz continuous, then z exists in the large. If $\sup_{x \in P(E_0)} \|\gamma(x)\|$ is small enough, then $g(z(\cdot), \gamma(z(\cdot)))$ is bounded, and $y = \gamma(z(\cdot))$ is also bounded. Thanks to Proposition 4.4.12(ii), with A replaced by A_-

$$y(t) = \int_{-\infty}^{t} e^{(t-s)A_-} g(z(s), \gamma(z(s))) ds, \quad t \leq 0,$$

and in particular, for $t = 0$,

$$y(0) = \gamma(x_0) = \int_{-\infty}^{0} e^{-sA_-} g(z(s; x_0, \gamma), \gamma(z(s; x_0, \gamma))) ds.$$

Theorem 9.2.2 *Let A satisfy (9.2.1) and let (9.1.17) hold. Then there exists $r_1 > 0$ such that for $r \leq r_1$ there is a Lipschitz continuous function $\gamma : P(E_0) \mapsto (I - P)(E_1)$ such that the graph of γ is invariant for system (9.2.3)-(9.2.4). If in*

2. Critical cases of stability

addition G is k times continuously differentiable, with $k \geq 2$, then there exists $r_k > 0$ such that if $r \leq r_k$ then $\gamma \in C^{k-1}$, $\gamma^{(k-1)}$ is Lipschitz continuous, and

$$\gamma'(x)(A_- x + f(x, \gamma(x))) = A_+ \gamma(x) + g(x, \gamma(x)), \quad x \in P(X). \tag{9.2.6}$$

Proof — Set

$$\begin{aligned} Y = \{ \gamma : P(E_0) \mapsto (I-P)(E_1) : \gamma(0) = 0, \, \|\gamma(x)\|_1 \leq r, \\ \|\gamma(x) - \gamma(\overline{x})\|_1 \leq a\|x - \overline{x}\|_0 \; \forall x \in P(E_0) \}. \end{aligned}$$

Y is closed in the space of the bounded functions from $P(E_0)$ to $(I-P)(E_1)$, endowed with the sup norm. Define a nonlinear operator Γ on Y by

$$(\Gamma \gamma)(x) = \int_{-\infty}^{0} e^{-sA_-} g(z(s), \gamma(z(s))) ds, \quad x \in P(E_0), \tag{9.2.7}$$

where $z = z(s; x, \gamma)$ is the solution of (9.2.5). We are going to prove that Γ is well defined in Y and it is a strict contraction mapping Y into itself, provided r and a are suitably chosen.

Since $z \mapsto f(z + \gamma(z))$ is globally Lipschitz continuous for every $\gamma \in Y$, then the solution z of (9.2.5) exists in the large. Let us give precise estimates on z. Fix once and for all a number $\varepsilon_0 \in \,]0, \omega_-[$, and set

$$M_0 = \sup_{t<0} e^{\varepsilon_0 t} \|e^{tA}\|_{L(P(E_0))}. \tag{9.2.8}$$

Moreover, for $\mu \in [0, \omega_-[$ let $C(\mu)$ be such that

$$\left\| \int_{-\infty}^{0} e^{-sA_-} \varphi(s) ds \right\|_1 \leq C(\mu) \sup_{s \leq 0} e^{\mu s} \|\varphi(s)\|_0$$

for each $\varphi \in C(]-\infty, 0]; E_0, -\mu)$.

By the Gronwall Lemma we get easily

$$\|z(s; x, \gamma) - z(s, \overline{x}, \overline{\gamma})\|_0 \leq M_0 e^{-\mu s} \|x - \overline{x}\|_0 + \frac{1 - e^{-\mu s}}{1 + a} \|\gamma - \overline{\gamma}\|_\infty, \tag{9.2.9}$$

where

$$\mu = \mu(r) = M_0 L(r)(1 + a) + \varepsilon_0. \tag{9.2.10}$$

Let $\gamma \in Y$ and $x \in P(E_0)$. Then

$$\|g(z(s; \gamma, x), \gamma(z(s; \gamma, x)))\|_0 \leq M(r),$$

so that $g(z(\cdot; \gamma, x), \gamma(z(\cdot; \gamma, x)))$ belongs to $C(]-\infty, 0]; E_0)$. Therefore, Γ is well defined in Y, and since $\lim_{r \to 0} M(r)/r = 0$, then $\|\Gamma(\gamma)(x)\|_1 \leq r$ for every x, if r is small. Fix now $\gamma \in Y$ and $x, \overline{x} \in P(E_0)$. Then

$$\|g(z(s; \gamma, x), \gamma(z(s; \gamma, x))) - g(z(s; \gamma, \overline{x}), \gamma(z(s; \gamma, \overline{x})))\|_0$$
$$\leq M_0 L(r)(1+a) e^{-\mu s} \|x - \overline{x}\|_0, \; s \leq 0.$$

Choosing r so small that $\mu = M_0 L(r)(1+a) + \varepsilon_0 < \omega_-$, we get

$$\|\Gamma\gamma(x) - \Gamma\gamma(\overline{x})\|_1 \le C(\mu) M_0 L(r)(1+a)\|x - \overline{x}\|_0,$$

so that $\|\Gamma\gamma(x) - \Gamma\gamma(\overline{x})\|_1 \le a\|x - \overline{x}\|_0$, provided r is small enough. So, Γ maps Y into itself. Let us prove that it is a contraction: for $\gamma, \overline{\gamma} \in Y$ and $x \in P(E_0)$ we have, for every $s \le 0$,

$$\|g(z(s;\gamma,x),\gamma(z(s;\gamma,x))) - g(z(s;\overline{\gamma},x),\gamma(z(s;\overline{\gamma},x)))\|_0 \le L(r)e^{-\mu s}\|\gamma - \overline{\gamma}\|_\infty.$$

Therefore,

$$\|\Gamma\gamma(x) - \Gamma\overline{\gamma}(x)\|_\infty \le C(\mu) L(r) \|\gamma - \overline{\gamma}\|_\infty, \quad x \in P(E_0),$$

so that Γ is a contraction on Y if r is small. In that case there exists a unique fixed point of Γ in Y.

Let us prove now that the graph of Γ is invariant. Let $y_0 = \gamma(x_0)$, and set

$$x(t) = z(t; x_0, \gamma), \quad y(t) = \gamma(x(t)), \quad t \ge 0.$$

Then

$$\begin{aligned}
\gamma(x(t)) &= \int_{-\infty}^{0} e^{-sA_-} g(z(s; x(t), \gamma), \gamma(z(s; x(t), \gamma))) ds \\
&= \int_{-\infty}^{0} e^{-sA_-} g(z(s+t; x_0, \gamma), \gamma(z(s+t; x_0, \gamma))) ds \\
&= \int_{-\infty}^{t} e^{(t-s)A_-} g(z(s; x_0, \gamma), \gamma(z(s; x_0, \gamma))) ds \\
&= e^{tA_-} y_0 + \int_{0}^{t} e^{(t-s)A_-} g(x(s), y(s)) ds,
\end{aligned}$$

so that (x, y) is the solution of (9.2.3)-(9.2.4). The first part of the statement follows.

If in addition G is k times continuously differentiable, one looks for a fixed point of Γ in the set

$$\begin{aligned}
Y_k = \{&\gamma \in C^{k-1}(P(E_0), (I-P)(E_1)) : \gamma(0) = 0, \|\gamma(x)\|_1 \le r, \\
&\|\gamma^{(h)}(x) - \gamma^{(h)}(\overline{x})\|_1 \le a_h \|x - \overline{x}\|_0, \ h = 1, \ldots, k-1\}.
\end{aligned}$$

Using the same technique as above, one shows that Γ maps Y_k into itself, if r and the constants a_h are suitably chosen. Since Y_k is closed in the space of the bounded functions from $P(E_0)$ to $(I-P)(E_1)$, the fixed point of Γ in Y belongs in fact to Y_k.

In the case where the fixed point of Γ is continuously differentiable, equality (9.2.6) follows by replacing $y = \gamma(x)$ in (9.2.3)-(9.2.4). ∎

2. CRITICAL CASES OF STABILITY

$\mathcal{M}$ is called *center-unstable manifold*. In the case where γ is differentiable, its derivatives may be computed recursively by differentiating (9.2.7). One gets for instance

$$\gamma'(0) = 0, \quad \gamma''(0) = \int_{-\infty}^{0} e^{-sA_-} g''(0) ds = -A_-^{-1}(I-P)G''(0). \quad (9.2.11)$$

Let us prove a property of attractivity of $\mathcal{M}$.

Proposition 9.2.3 *Let G be twice continuously differentiable. For every $\omega \in]0, \omega_-[$ there are $r(\omega)$, $M(\omega)$ such that if $\|x_0\|_0$ and $\|y_0\|_1$ are sufficiently small, then the solution of (9.2.3)-(9.2.4) exists in the large and satisfies*

$$\|y(t) - \gamma(x(t))\|_1 \le M(\omega) e^{-\omega t} \|y_0 - \gamma(x_0)\|_1, \quad t \ge 0. \quad (9.2.12)$$

Proof — By Proposition 9.2.1 for every sufficiently small r there is $C_0(r)$ such that if $\|x_0\|_0 + \|y_0\|_1 \le C_0(r)$ then the solution of (9.2.3)-(9.2.4) exists in the large, and $\|y(t)\|_1 \le r$. Let $v(t) = y(t) - \gamma(x(t))$. Since γ satisfies (9.2.6), then

$$\begin{aligned} v'(t) &= A_- v(t) + g(x(t), y(t)) - g(x(t), \gamma(x(t))) \\ &\quad - \gamma'(x(t))(f(x(t), y(t)) - f(x(t), \gamma(x(t)))) \\ &= A_- v(t) + H(x(t), y(t)), \quad t \ge 0. \end{aligned}$$

Recalling that $\|H(x(s), y(s))\| \le L(r)(1+a)\|v(s)\|_1$, from Proposition 4.4.10(ii) with A replaced by A_- we get

$$\sup_{t \ge 0} e^{\omega t} \|v(t)\|_1 \le C(\|v(0)\|_1 + L(r)(1+a) \sup_{t \ge 0} e^{\omega t} \|v(t)\|_1), \quad (9.2.13)$$

where $C = C(\omega)$ is independent of T and v. Taking r so small that $CL(r)(1+a) \le 1/2$ the statement follows, with $M(\omega) = 2C$. ∎

Once exponential attractivity of $\mathcal{M}$ is established, one can prove that it is asymptotically stable with asymptotic phase, in the sense specified by next proposition.

Proposition 9.2.4 *For every $\omega \in]0, \omega_-[$ there is $C(\omega) > 0$ such that if $\|x_0\|_0$ and $\|y_0\|_1$ are small enough there exists $\overline{x} \in P(E_0)$ such that*

$$\|x(t) - \overline{z}(t)\|_0 + \|y(t) - \gamma(\overline{z}(t))\|_1 \le C(\omega) e^{-\omega t} \|y_0 - \gamma(x_0)\|_0, \quad t \ge 0, \quad (9.2.14)$$

where $\overline{z}(t) = z(t; \gamma, \overline{x})$ is the solution of (9.2.5) with $x_0 = \overline{x}$.

Proof — It is convenient to introduce the function $w(s, t) = z(s-t; x(t), \gamma)$, which satisfies

$$w_s(s, t) = A_+ w(s, t) + f(w(s, t), \gamma(w(s, t))), \quad s \in \mathbb{R}; \quad w(t, t) = x(t).$$

Using (9.2.8) we get for $0 \leq s \leq t$

$$\|x(s) - w(s,t)\|_0 \leq \int_s^t M_0 e^{\varepsilon_0(\sigma-s)} \|f(x(\sigma), y(\sigma)) - f(w(\sigma,t), \gamma(w(\sigma,t)))\|_0 d\sigma$$

$$\leq M_0 \int_s^t e^{\varepsilon_0(\sigma-s)} L(r) \big(\|x(\sigma) - w(\sigma,t)\|_0$$
$$+ \|\gamma(x(\sigma)) - \gamma(w(\sigma,t))\|_1 + \|y(\sigma) - \gamma(x(\sigma))\|_1\big) d\sigma$$

$$\leq M_0 L(r) \int_s^t e^{\varepsilon_0(\sigma-s)} (\|v(\sigma)\|_1 + (1+a)\|x(\sigma) - w(\sigma,t)\|_0) d\sigma,$$

where $v(t) = y(t) - \gamma(x(t))$ is the function considered in Proposition 9.2.3. Using the Gronwall Lemma we get

$$\|x(s) - w(s,t)\|_0 \leq M_0 L(r) \int_s^t e^{\mu(\sigma-s)} \|v(\sigma)\|_1 d\sigma, \quad 0 \leq s \leq t,$$

where μ is given by (9.2.10). For every $\omega \in]\mu, \omega_-[$ estimate (9.2.12) yields

$$\sup_{s < \sigma < t} e^{\omega(\sigma-s)} \|v(\sigma)\|_1 \leq M(\omega)\|v(s)\|_1,$$

so that

$$\|x(s)-w(s,t)\|_0 \leq M(\omega)M_0 L(r) \int_s^t e^{(\mu-\omega)(\sigma-s)} d\sigma \|v(s)\|_1 \leq \frac{M(\omega)M_0 L(r)}{\omega - \mu} \|v(s)\|_1. \tag{9.2.15}$$

Choosing $s = 0$ in (9.2.15) we get

$$\|x_0 - w(0,t)\|_0 \leq \frac{M(\omega)M_0 L(r)}{\omega - \mu} \|y_0 - \gamma(x_0)\|_1, \quad t \geq 0,$$

so that the set $\{w(0,t) : t \geq 0\}$ is bounded in $P(E_0)$. Since $P(E_0)$ is finite dimensional, there is a sequence $t_n \to +\infty$ such that $w(0, t_n)$ goes to $\bar{x} \in P(E_0)$ as $n \to \infty$. Let $\bar{z}$ be the solution of (9.2.5) when x_0 is replaced by $\bar{x}$. Thanks to the continuous dependence theorem for ODE's, for every $s \in \mathbb{R}$ we have

$$\bar{z}(s) = \lim_{n \to \infty} z(s; w(0,t_n), \gamma) = \lim_{n \to \infty} z(s; z(-t_n; x(t_n), \gamma), \gamma)$$
$$= \lim_{n \to \infty} z(s - t_n; x(t_n), \gamma) = \lim_{n \to \infty} w(s, t_n).$$

It follows that

$$\|x(s) - \bar{z}(s)\|_0 = \lim_{n \to \infty} \|x(s) - w(s, t_n)\|_0.$$

Using (9.2.15) and then (9.2.12), we get, for $s \geq 0$,

$$\|x(s) - \bar{z}(s)\|_0 \leq \frac{M(\omega)M_0 L(r)}{\omega - \mu} \|v(s)\|_1 \leq \frac{M(\omega)^2 M_0^2 L(r)}{\omega - \mu} e^{-\omega s} \|y_0 - \gamma(x_0)\|_1. \tag{9.2.16}$$

2. CRITICAL CASES OF STABILITY

Moreover

$$\|y(s) - \gamma(\bar{z}(s))\|_1 \leq \|y(s) - \gamma(x(s))\|_1 + a\|x(s) - \bar{z}(s)\|_0, \quad s \geq 0, \qquad (9.2.17)$$

and the statement follows from (9.2.16), (9.2.17), (9.2.12). ∎

From Proposition 9.2.4 we get easily the following corollary.

Corollary 9.2.5 *Let A be a sectorial operator satisfying (9.2.1) and let (9.1.17) hold. Assume in addition that $\sigma_+(A) \subset i\mathbb{R}$. Then the null solution of (9.1.1) is stable (respectively, asymptotically stable, unstable) in $D_A(\theta + 1, \infty)$ if and only if the null solution of the finite dimensional system (9.2.5) is stable (respectively, asymptotically stable, unstable).*

9.2.2 Applications to nonlinear parabolic problems, II

Let us consider again problem (9.1.35), in the critical case of stability $c = 0$. We refer to Subsection 9.1.5 for notation and assumptions. The nonlinear function g is assumed here to be C^∞, since we will need to compute many derivatives of g at $(x, 0, 0, 0)$ to discuss the stability of the null solution. The part of the spectrum of A in the imaginary axis consists only of the simple eigenvalue 0, and the corresponding eigenspace is the set of the constant functions. The operator A_+ is the null operator, and the projection P is given by

$$P\psi(x) = \frac{1}{\text{meas } \Omega} \int_\Omega \psi(y) dy = \langle \psi, \xi_0 \rangle \xi_0(x), \quad x \in \overline{\Omega},$$

where $\langle \cdot, \cdot \rangle$ denotes the scalar product in $L^2(\Omega)$, and ξ_0 is the constant function equal to $(\text{meas } \Omega)^{-1/2}$ over $\overline{\Omega}$.

By Corollary 9.2.5 the null solution of (9.1.35) is stable (respectively, asymptotically stable, unstable) in $C^{2+\theta}(\overline{\Omega})$ if and only if the null solution of the one dimensional equation

$$z'(t) = PG(z(t) + \gamma(z(t))) = \Psi(z(t)), \quad t \geq 0, \qquad (9.2.18)$$

is stable (respectively, asymptotically stable, unstable). Here G is defined by (9.1.37), and γ is the function given by Theorem 9.2.2. Setting $z(t) = \zeta(t)\xi_0$, problem (9.2.18) is equivalent to the scalar equation

$$\zeta'(t) = \psi(\zeta(t)), \quad t \geq 0, \qquad (9.2.19)$$

where, for $\zeta \in \mathbb{R}$,

$$\psi(\zeta) = \frac{1}{\text{meas } \Omega} \int_\Omega g(x, \zeta + \gamma(\zeta\xi_0)(x), D\gamma(\zeta\xi_0)(x), D^2\gamma(\zeta\xi_0)(x)) dx.$$

Sufficient conditions for the stability of the null solution of (9.1.38) may be given by computing the derivatives of ψ at 0. Using (9.2.11) we get

$$\psi'(0) = 0, \quad \psi''(0) = \langle G''(0)(\xi_0, \xi_0), \xi_0^* \rangle,$$

$$\begin{aligned}\psi'''(0) &= \langle G'''(0)(\xi_0, \xi_0, \xi_0^*) + 3\langle G''(0)(\xi_0, \gamma''(0)(\xi_0, \xi_0)), \xi_0^* \rangle \\ &= \langle g'''(0)(\xi_0, \xi_0, \xi_0), \xi_0^* \rangle - 3\langle G''(0)(\xi_0, A_-^{-1}(I - P)G''(0)(\xi_0, \xi_0)), \xi_0^* \rangle.\end{aligned}$$

In particular,

$$\psi''(0) = \frac{1}{(\operatorname{meas} \Omega)^{3/2}} \int_\Omega g_{uu}(x, 0) dx.$$

So, if

$$\int_\Omega g_{uu}(x, 0) dx \neq 0,$$

then the null solution of (9.2.19) is unstable, and hence the null solution of (9.1.35) is unstable. If $\int_\Omega g_{uu}(x, 0) dx = 0$, it holds

$$\psi'''(0)(\operatorname{meas} \Omega)^2 = \int_\Omega g_{uuu}(x, 0) dx - 3 \int_\Omega g_{uu}(x, 0)(A_-^{-1} g_{uu}(\cdot, 0))(x) dx.$$

If $\psi'''(0) < 0$, the null solution of (9.2.19) is asymptotically stable, and hence the null solution of (9.1.35) is asymptotically stable; if $\psi'''(0) > 0$, the null solution of (9.2.19) is unstable, and hence the null solution of (9.1.35) is unstable.

In the one dimensional case it is possible to perform the computation of $\psi'''(0)$, because of the explicit knowledge of A_-^{-1}. Take for instance $\Omega =]0, 1[$. Then for every $\varphi \in (I - P)(D_A(\theta/2, \infty))$, that is, for every $\varphi \in C^\theta([0, 1])$ with zero mean value,

$$A_-^{-1}\varphi(x) = \int_0^x (x - \sigma)\varphi(\sigma) d\sigma - \int_0^1 \frac{(1-\sigma)^2}{2}\varphi(\sigma) d\sigma, \quad 0 \leq x \leq 1.$$

In the case where $g = g(u, p, q)$ does not depend explicitly on the space variables x, then $\gamma \equiv 0$. Indeed, G maps $P(X)$ into itself, so that for any r the unique small solution of (9.2.7) is the null function. It follows that

$$\psi^{(n)}(0) = \frac{\partial^n g}{\partial u^n}(0), \quad n \in \mathbb{N}.$$

Assume that there exists $n \in \mathbb{N}$, $n \geq 2$ such that

$$\frac{\partial^k g}{\partial u^k}(0) = 0, \quad k = 1, \ldots, n-1, \quad \frac{\partial^n g}{\partial u^n}(0) \neq 0.$$

If either n is even, or n is odd and $\partial^n g/\partial u^n(0) > 0$, then the null solution of (9.2.19) is unstable, and hence the null solution of (9.1.35) is unstable; if n is odd and $\partial^n g/\partial u^n(0) < 0$, then the null solution of (9.2.19) is asymptotically stable, and hence the null solution of (9.1.35) is asymptotically stable.

2. CRITICAL CASES OF STABILITY

The case where $g = g(u,p,q)$ depends explicitly on u, but all the derivatives $\partial^n g/\partial u^n(0)$ vanish, will be discussed in Subsection 9.2.4.

If $g = g(p,q)$ does not depend explicitly on (x,u), then G vanishes on $P(X)$, and this fact lets one decouple system (9.2.3), which becomes

$$x'(t) = PG(y(t)), \quad y'(t) = A_- y(t) + (I - P)G(y(t)), \quad t \geq 0.$$

The second equation satisfies the assumptions of the principle of linearized stability, with X replaced of course by $(I-P)(X)$, because the spectrum of A_- consists of the negative eigenvalues $-\lambda_n$, $n \geq 1$. By applying the Linearized Stability Theorem 9.1.7(i) we find that for every $\omega \in \,]0, \lambda_1[$ there are $R(\omega)$, $M(\omega)$ such that if $\|(I-P)u_0\|_{C^{2+\theta}(\overline{\Omega})} \leq R(\omega)$ then the solution of the y-equation with initial value $y(0) = (I-P)u_0$ exists in the large and satisfies

$$\|y(t)\|_{C^{2+\theta}(\overline{\Omega})} \leq M(\omega)e^{-\omega t}\|(I-P)u_0\|_{C^{2+\theta}(\overline{\Omega})}, \quad t \geq 0.$$

Replacing in the x-equation, we find that $x(t)$ remains bounded, and

$$\|x(t)\|_{C^{2+\theta}(\overline{\Omega})} = \|x(t)\|_{C(\overline{\Omega})} \leq const.\|u_0\|_{C^{2+\theta}(\overline{\Omega})}, \quad t \geq 0.$$

Therefore, the null solution is stable in $C^{2+\theta}(\overline{\Omega})$.

The same procedure may be performed in the case of the Dirichlet boundary condition, or in the case of periodic conditions in $\mathbb{R}^n$. We leave the details to the interested reader.

9.2.3 The case where the linear part generates a bounded semigroup

Now we consider a critical case of stability for equation (9.1.1) which cannot be treated by the center manifold theory developed in the previous subsection because the intersection of the spectrum of A with the imaginary axis is not necessarily isolated in $\sigma(A)$. This happens, for instance, when A is the realization of the Laplace operator in spaces of functions defined in $\mathbb{R}^n$ or, more generally, in unbounded domains.

Specifically, we consider the case where the solution of the linearized problem enjoys some decay properties, such as

$$\|e^{tA}\|_{L(X)} \leq M_0, \quad \|Ae^{tA}\|_{L(X)} \leq \frac{M_1}{t}, \quad t > 0. \tag{9.2.20}$$

The estimates (9.2.20) hold if and only if A satisfies assumption (2.0.1) with $\omega = 0$. In particular, they are satisfied in the noncritical case $\sup\{\operatorname{Re}\lambda : \lambda \in \sigma(A)\} < 0$. Then the Principle of Linearized Stability holds for the nonlinear equation (9.1.1), and the result of next Theorem 9.2.21 is weaker than the result of Theorem 9.1.7(i): see Subsection 9.1.3. Theorem 9.2.21 is meaningful in the critical case $\omega_0 = 0$.

374   CHAPTER 9. ASYMPTOTIC BEHAVIOR IN FULLY NONLINEAR EQUATIONS

(9.2.20) implies that the null solution of the linear problem $v'(t) = Av(t)$ is stable, but it is not necessarily asymptotically stable.

Also in this critical case of stability we work in the context of the spaces $D_A(\theta, \infty)$, $D_A(\theta+1, \infty)$.

We recall (see Subsection 4.4.1) that the norms $x \mapsto \|x\|_{D_A(\theta,\infty)}$ and

$$x \mapsto \|x\| + |x|_{D_A(\theta,\infty)} = \|x\| + \sup_{\xi>0} \|\xi^{1-\theta} A e^{\xi A} x\|$$

are equivalent. In what follows, we shall use the seminorm $|\cdot|_{D_A(\theta,\infty)}$. For every $T > 0$ and $u \in L^\infty(0, T; D_A(\theta+1, \infty))$ we set

$$\|u\|_T = \sup_{0 \le t \le T} \left(\|u(t)\|_X + \max(1, t) \|Au(t)\| + \max(1, t^{1+\theta}) |Au(t)|_{D_A(\theta,\infty)} \right)$$

Note that if $T \le 1$ the $\|\cdot\|_T$-norm coincides with the $L^\infty(0, T; D_A(\theta+1, \infty))$ norm. For every $T > 1$ the $\|\cdot\|_T$-norm is equivalent to the norm of $L^\infty(0, T; D_A(\theta+1, \infty))$, but one of the equivalence constants blows up as $T \to +\infty$.

Theorem 9.2.6 *Let $A : D(A) \subset X \mapsto X$ be a sectorial operator satisfying (9.2.20). Let $\mathcal{O}$ be a neighborhood of 0 in $D_A(\theta+1, \infty)$, $0 < \theta < 1$, and let $G : \mathcal{O} \mapsto D_A(\theta, \infty)$ be a C^1 function such that*

$$\|G(u)\|_{D_A(\theta,\infty)} \le K(\|Au\|_{D_A(\theta,\infty)})^p, \qquad (9.2.21)$$

with $K > 0$ and $p \ge \theta+1$. Then there are R_1, $M_1 > 0$ such that if $\|u_0\|_{D_A(\theta+1,\infty)} \le R_1$, the solution of problem (9.1.1)-(9.1.5) exists in the large, and for every $t \ge 0$

$$\|u(t)\| + \max(1, t) \|Au(t)\| + \max(1, t^{1+\theta}) |Au(t)|_{D_A(\theta,\infty)} \le M_1 \|u_0\|_{D_A(\theta+1,\infty)}. \qquad (9.2.22)$$

Proof — Let us fix some positive T, say $T = 1$. Theorem 9.1.5 states that there are R, $M > 0$ such that if $\|u_0\|_{D_A(\theta+1,\infty)} \le R$ then problem (9.1.1)-(9.1.5) has a unique small solution u, defined at least in $[0, 1]$, such that

$$\sup_{0 \le t \le 1} \|u(t)\|_{D_A(\theta+1,\infty)} = \|u\|_1 \le M \|u_0\|_{D_A(\theta+1,\infty)}.$$

We are going to prove an *a priori* estimate on u (provided the initial datum is sufficiently small) yielding existence in the large and estimate (9.2.22). Let C be the constant of estimate (4.4.17). Fix any positive number $\overline{R} \le R$ such that

$$KC(2 + 1/\theta)\overline{R}^p < \frac{\overline{R}}{2}, \qquad (9.2.23)$$

and set

$$\overline{T} = \sup\{T > 0 : u \text{ exists in } [0, T] \text{ and } \|u\|_T \le \overline{R}\}.$$

2. CRITICAL CASES OF STABILITY

Then $\overline{T} \geq 1$. If $\overline{T} < \infty$, u is continuous at $t = \overline{T}$ and $\|u\|_{\overline{T}} = \overline{R}$. From the equality

$$u(t) = e^{tA}u_0 + \int_0^t e^{(t-s)A}G(u(s))ds, \quad 0 \leq t \leq \overline{T},$$

using estimates (4.4.15) and (4.4.17) we get

$$\overline{R} = \|u\|_{\overline{T}}$$
$$\leq C(\|u_0\|_{D_A(\theta+1,\infty)} + \|G(u)\|_{L^1(0,\overline{T};X)} + \sup_{0<t\leq\overline{T}} \max(1, t^{\theta+1})|G(u)|_{D_A(\theta,\infty)}).$$
(9.2.24)

Let us estimate $\|G(u)\|_{L^1(0,\overline{T};X)}$ and $|G(u(t))|_{D_A(\theta,\infty)}$. Using (9.2.21) we get

$$\|G(u(t))\|_{D_A(\theta,\infty)} \leq K(\|Au(t)\| + |Au(t)|_{D_A(\theta,\infty)})^p \leq \frac{K}{\max(t^{\theta+1}, 1)} \overline{R}^p,$$

so that

$$\|G(u)\|_{L^1(0,\overline{T};X)} \leq K(1 + 1/\theta)\overline{R}^p, \quad \sup_{0<t\leq\overline{T}} \max(1, t^{\theta+1})|G(u)|_{D_A(\theta,\infty)} \leq K\overline{R}^p.$$

Replacing in (9.2.24) and using (9.2.23) we get

$$\overline{R} = C\left(\|u_0\|_{D_A(\theta+1,\infty)} + K(2 + 1/\theta)\overline{R}^p\right) < C\|u_0\|_{D_A(\theta+1,\infty)} + \overline{R}/2,$$

which is impossible if $\|u_0\|_{D_A(\theta+1,\infty)} \leq \overline{R}/2C$. For such small initial data, $\|u\|_T \leq \overline{R}$ for every interval $[0, T]$ contained in the interval of existence of u. Thanks to Corollary 9.1.6, this implies that u exists in the large and it satisfies estimate (9.2.22). ∎

Theorem 9.2.21 implies that the null solution of (9.1.1) is stable, but not necessarily asymptotically stable.

9.2.4 Applications to nonlinear parabolic problems, III

Let us consider an initial value problem in $[0, +\infty[\times \mathbb{R}^n$,

$$u_t = \Delta u + g(x, u(t,x), Du(t,x), D^2u(t,x), u(t,x_0), Du(t,x_1), D^2u(t,x_2)),$$
(9.2.25)
$$u(0, x) = u_0(x), \quad x \in \mathbb{R}^n,$$
(9.2.26)

where $x_0, x_1, x_2 \in \mathbb{R}^n$, the function $(x, u, p, q, u_0, p_0, q_0) \mapsto g(x, u, p, q, u_0, p_0, q_0)$ is defined in $\mathbb{R}^n \times \Omega$, Ω being a neighborhood of 0 in $(\mathbb{R} \times \mathbb{R}^n \times \mathbb{R}^{n^2}_S)^2$, and it is twice differentiable with respect to (u, p, q, u_0, p_0, q_0) with Lipschitz continuous derivatives. We assume moreover that

$$|g(x, u, p, q, u_0, p_0, q_0)| \leq K(|p|^{2+\varepsilon} + |p_0|^{2+\varepsilon} + |q|^2 + |q_0|^2),$$
(9.2.27)

with $\varepsilon > 0$, for all $x \in \mathbb{R}^n$, $(u, p, q, u_0, p_0, q_0) \in \Omega$.

Let A be the realization of the Laplace operator Δ in the space $X = C(\mathbb{R}^n)$. It is easy to see that e^{tA} satisfies the bounds (9.2.20): it is sufficient to represent $e^{tA}\varphi$ by a convolution integral with the Gauss-Weierstrass kernel, for every $\varphi \in X$. See formula (3.1.40). However, the eigenvalue 0 is not isolated in the spectrum of A, so that the results of Subsection 9.2.1 cannot be applied to problem (9.2.25)-(9.2.26).

Proposition 9.2.7 *Let $g : \mathbb{R}^n \times \Omega \mapsto \mathbb{R}$ satisfy the above assumptions. For every $\alpha \in \,]0, 2\varepsilon/(4+\varepsilon)]$ there are R_1, $M_1 > 0$ such that if $\|u_0\|_{C^{\alpha+2}(\mathbb{R}^n)} \leq R_1$, the solution u of problem (9.2.25)-(9.2.26) exists in the large, and*

$$\|u(t,\cdot)\|_{C^\alpha(\mathbb{R}^n)} + \max(1,t)\|D^2 u(t,\cdot)\|_{C^\alpha(\mathbb{R}^n)} \leq M_1 \|u_0\|_{C^{\alpha+2}(\mathbb{R}^n)}, \quad t \geq 0.$$

Proof — By Theorem 3.1.12 and Theorem 3.1.15,

$$C^\alpha(\mathbb{R}^n) = D_A(\alpha/2, \infty), \quad C^{\alpha+2}(\mathbb{R}^n) = D_A(\alpha/2 + 1, \infty). \tag{9.2.28}$$

We have to check that the function $G : C^{\alpha+2}(\mathbb{R}^n) \mapsto C^\alpha(\mathbb{R}^n)$ defined by

$$G(u)(x) = g(x, u, Du(x), D^2 u(x), u(x_0), Du(x_1), D^2 u(x_2)), \quad x \in \mathbb{R}^n,$$

satisfies the assumptions of Theorem 9.2.21, with $\theta = \alpha/2$. We know already that G is continuously differentiable. Moreover, due to Proposition 1.1.3(ii), for every $\varphi \in C^{\alpha+2}(\mathbb{R}^n)$ it holds

$$\|D\varphi\|_{C^\alpha} \leq C \|\varphi\|_{C^\alpha}^{1/2} \|D^2\varphi\|_{C^\alpha}^{1/2}.$$

So, we get

$$\|G(\varphi)\|_{C^\alpha(\mathbb{R}^n)} \leq C\|D^2\varphi\|_{C^\alpha(\mathbb{R}^n)}^{(2+\varepsilon)/2} \tag{9.2.29}$$

for $\|\varphi\|_{C^{2+\alpha}(\mathbb{R}^n)}$ small. Estimate (3.1.41) yields

$$\|D^2\varphi\|_\infty \leq C(\lambda^{1-\alpha/2}\|\varphi\|_{C^\alpha(\mathbb{R}^n)} + \lambda^{\alpha/2}\|\Delta\varphi\|_{C^\alpha(\mathbb{R}^n)}), \quad \forall \lambda > 0.$$

Taking the minimum for $\lambda \in \,]0, +\infty[$ we get

$$\|D^2\varphi\|_\infty \leq C' \|\varphi\|_{C^\alpha(\mathbb{R}^n)}^{\alpha/2} \|\Delta\varphi\|_{C^\alpha(\mathbb{R}^n)}^{1-\alpha/2}. \tag{9.2.30}$$

Estimate (3.1.42) gives

$$[D^2\varphi]_{D_A(\alpha/2,\infty)} \leq C''(\lambda\|\varphi\|_{C^\alpha(\mathbb{R}^n)} + \|\Delta\varphi\|_{C^\alpha(\mathbb{R}^n)}), \quad \forall \lambda > 0.$$

Letting $\lambda \to 0$, adding to (9.2.30) and taking into account (9.2.28), we get

$$\|D^2\varphi\|_{D_A(\alpha/2,\infty)} \leq C(\|\varphi\|_{D_A(\alpha/2,\infty)}^{\alpha/2} \|\Delta\varphi\|_{D_A(\alpha/2,\infty)}^{1-\alpha/2} + \|\Delta\varphi\|_{D_A(\alpha/2,\infty)}).$$

Replacing in (9.2.29) we see that (9.2.21) is satisfied, provided $(1-\alpha/2)(1+\varepsilon/2) \geq 1 + \alpha/2$, i.e. $\alpha \leq 2\varepsilon/(4+\varepsilon)$. Theorem 9.2.21 may be applied, and the statement follows. ∎

2. CRITICAL CASES OF STABILITY

Proposition 9.2.7 implies that the null solution of (9.2.25) is stable in $C^{\alpha+2}(\mathbb{R}^n)$. It is not asymptotically stable, since the solution is constant when the initial datum u_0 is constant.

Let us make some comments on the growth condition (9.2.27). If the right hand side of (9.2.27) is replaced by $K(|u|^m+|u_0|^m)$, with $m \geq 2$ arbitrarily large, the null solution is not necessarily stable: indeed, the solutions of the initial value problem for $u_t = u_{xx} + u^m$ and $u_t = u_{xx} + u(t,x_0)^m$ are independent of the space variable x if the initial data are constant, and they blow up in finite time if such constants are positive. These counterexamples rely on the fact that 0 is an eigenvalue of the Laplace operator in $C^{\theta+2}(\mathbb{R}^n)$ and the kernel consists of the constant functions. Examples of blowing up in finite time for $u_t = \Delta u + u^m$, with compactly supported initial data, can be found in [85, 209] in the case $n(m-1)/2 \leq 1$.

Let us show by a counterexample that the exponent $2+\varepsilon$ in (9.2.27) cannot be replaced by 2: consider problem

$$u_t(t,x) = u_{xx}(t,x) + (u_x(t,0))^2, \quad t \geq 0, \ x \in \mathbb{R}; \quad u(0,x) = u_0(x), \ x \in \mathbb{R},$$

where $u_0' \geq 0$ and u_0 is not constant. Then

$$u(t,x) = \frac{1}{(4\pi t)^{1/2}} \int_{-\infty}^{+\infty} \exp(-y^2/4t) u_0(x-y) dy + \int_0^t u_x(s,0)^2 ds, \quad (9.2.31)$$

so that

$$u_x(t,0) = \frac{1}{(4\pi t)^{1/2}} \int_{-\infty}^{+\infty} \exp(-y^2/4t) u_0'(-y) dy, \quad t > 0,$$

and for $t \geq 1$

$$u_x(t,0) \geq \frac{1}{(4\pi t)^{1/2}} \int_{-\infty}^{+\infty} \exp(-y^2) u_0'(-y) dy = \frac{C}{t^{1/2}},$$

with $C > 0$. Replacing in (9.2.31), we find

$$u(t,x) \geq (e^{tA} u_0)(x) + C^2 \int_1^t \frac{1}{s} ds, \quad t \geq 1.$$

Since $\|e^{tA} u_0\|_\infty \leq \|u_0\|_\infty$, it follows that $\|u(t,\cdot)\|_\infty \geq -\|u_0\|_\infty + C^2 \log t$ for $t \geq 1$, so that the null solution is unstable.

If the nonlinearity $g = g(u,p,q)$ does not depend on nonlocal terms,

$$\begin{cases} u_t(t,x) = \Delta u(t,x) + g(u(t,x), Du(t,x), D^2 u(t,x)), & t \geq 0, \ x \in \mathbb{R}^n, \\ u(0,x) = u_0(x), & x \in \mathbb{R}^n. \end{cases}$$

the growth condition (9.2.27) may be weakened. In the case where $g = g(Du, D^2 u)$, S. ZHENG [214] chose $H^s(\mathbb{R}^n)$ as a phase space, s integer $> n/2+3$. He proved a local existence result for $\|u_0\|_{H^s}$ small, and he obtained energy estimates, guaranteeing existence in the large and decay rates of the type $\|D^j u(t,\cdot)\|_{L^2} \leq C(1+t)^{-j/2}$, for $j = 1, ..., s$.

In the case where g depends explicitly on u, and $g(u,p,q) = O(|u|^\gamma + |p|^\gamma + |q|^\gamma)$, the Hörmander-Nash-Moser iteration scheme, together with energy estimates in higher norms, has been used in S. KLAINERMAN [111], S. ZHENG – Y. CHEN [215]. Some restriction on (γ, n) are assumed: if $\gamma = 2$, one requires that $n \geq 3$, if $\gamma = 3$ one requires that $n \geq 2$, if $\gamma = 4$ any n is allowed. Then one shows that there exists a large integer s_0 such that if $\|u_0\|_{H^{s_0}}$ is small enough, the solution exists in the large. A. LUNARDI [152] and Z. CHEN [51, 52] consider the case $|g(u, p, q)| \leq K(|u|^2 + |p|^2 + |q|^2)$. The first author considers initial data in $C^{\theta+2}(\mathbb{R}^n)$, and assumes some conditions on g in order that the maximum principle holds for a suitable combination of the solution and its derivatives. The second author takes initial data u_0 in the intersection of $C^{\theta+2}(\mathbb{R}^n)$ with a suitable Besov space (an interpolation space between $W^{1,1}(\mathbb{R}^n)$ and $W^{3,1}(\mathbb{R}^n)$), and such that $\lim_{|x|\to+\infty} u_0(x) = \lim_{|x|\to+\infty} \Delta u_0 = 0$. Then he uses the hypercontractivity of the semigroup $e^{t\Delta}$ an a procedure similar to the one of Theorem 9.3.7. Both of them prove existence in the large and polynomial decay of the derivatives of the solution.

9.3 Periodic solutions

Existence of periodic solutions to fully nonlinear equations will be stated for equations depending on a parameter λ. The procedure is similar to the one employed in Subsection 9.1.4 to find stationary solutions, but there are additional technical difficulties. So, we consider a fully nonlinear equation,

$$u'(t) = F(t, u(t), \lambda), \quad t \in \mathbb{R}, \tag{9.3.1}$$

where $F : \mathbb{R} \times \mathcal{O}_D \times \mathcal{O}_\Lambda \mapsto X$ satisfies assumptions (8.3.2), (8.3.3), guaranteeing existence and uniqueness of a local solution to the initial value problem for equation (9.3.1), see Section 8.3. Here $\mathcal{O}_D$ is an open neighborhood of 0 in D, and $\mathcal{O}_\Lambda$ is a neighborhood of 0 in the parameter space Λ. Moreover, F is T-periodic with respect to time,

$$F(t, x, \lambda) = F(t+T, x, \lambda), \quad t \in \mathbb{R}, x \in \mathcal{O}_D, \lambda \in \mathcal{O}_\Lambda, \tag{9.3.2}$$

with minimum period $T > 0$, and

$$F(t, 0, 0) = 0, \quad t \in \mathbb{R}. \tag{9.3.3}$$

Due to the regularity assumptions on F, the family

$$\{B(t) = F_x(t, 0, 0) : t \in \mathbb{R}\}$$

generates an evolution operator $G(t, s)$ in X. See Chapter 6.

The simplest case is when a nonresonance condition holds. Then existence of periodic solutions is a straightforward consequence of the Local Inversion Theorem.

3. Periodic solutions

Proposition 9.3.1 *Let $F : \mathbb{R} \times \mathcal{O}_D \times \mathcal{O}_\Lambda \mapsto X$ satisfy (9.3.2), (9.3.3), (9.3.2), (9.3.3). Assume moreover that*

$$1 \in \rho(G(T,0)).$$

Then there are λ_0, $r_0 > 0$ such that for $\|\lambda\|_\Lambda \leq \lambda_0$ problem (9.3.1) has a unique T-periodic solution u such that

$$\|u\|_{C^\alpha(\mathbb{R};D)} + \|u\|_{C^{1+\alpha}(\mathbb{R};X)} \leq r_0.$$

Proof — Consider the Banach spaces

$$Y = \{u \in C^\alpha(\mathbb{R}; D) \cap C^{1+\alpha}(\mathbb{R}; X) : u(t) = u(t+T), \ \forall t \in \mathbb{R}\},$$

$$Z = \{u \in C^\alpha(\mathbb{R}; X) : u(t) = u(t+T), \ \forall t \in \mathbb{R}\},$$

and define a mapping $\Phi : Y \mapsto Z$ by

$$\Phi(u, \lambda) = u' - F(\cdot, u(\cdot), \lambda).$$

A function $u \in Y$ is a T-periodic solution of (9.3.1) if and only if $\Phi(u, \lambda) = 0$.
In the proof of Theorem 8.3.4 we have shown that Φ is differentiable, and

$$\Phi_u(0,0)v = v' - F_x(\cdot, 0, 0)v = v' - B(\cdot)v.$$

Thanks to Proposition 6.3.12, $\Phi_u(0,0)$ is an isomorphism. Since $\Phi(0,0) = 0$, the statement follows from the Local Inversion Theorem. ∎

In the autonomous case $F = F(x, \lambda)$, denoting by $A = F_x(0,0)$ we have $G(t, s) = e^{(t-s)A}$, and hence $\rho(G(T,0)) = \rho(e^{TA})$. Sufficient conditions for $1 \in \rho(e^{TA})$ may be found in Proposition 2.3.6.

9.3.1 Hopf bifurcation

Now we consider a resonance condition in an autonomous problem,

$$u'(t) = F(u(t), \lambda), \ t \in \mathbb{R}, \tag{9.3.4}$$

where $F : \mathcal{O}_D \times [-1, 1] \mapsto X$ is a regular function such that $A = F_x(0,0) : D \mapsto X$ is sectorial. We assume that X and D are real Banach spaces, we denote their complexifications by $\widetilde{X}$, $\widetilde{D}$ respectively, and we denote by $\widetilde{A}$ the complexification of the operator A. The resonance condition is the following: there exists $\omega \in \mathbb{R}$, $\omega \neq 0$, such that

$$\begin{cases} \pm\omega i \text{ are simple isolated eigenvalues of } \widetilde{A}, \\ k\omega i \in \rho(\widetilde{A}), \ k \in \mathbb{Z}, \ k \neq \pm 1. \end{cases} \tag{9.3.5}$$

We know from Proposition 2.3.9 that for every $t \in \mathbb{R}$, $e^{\pm i\omega t}$ are semisimple isolated eigenvalues of $e^{t\widetilde{A}}$, which implies that the linear problem $v' = Av$ has nontrivial $2\pi/\omega$-periodic solutions. There are x_0, $y_0 \in D$ such that

$$\widetilde{A}(x_0 \pm iy_0) = \pm i\omega(x_0 \pm iy_0),$$

which means

$$Ax_0 = -\omega y_0, \quad Ay_0 = \omega x_0. \tag{9.3.6}$$

Moreover, $e^{t\widetilde{A}}(x_0 \pm iy_0) = e^{\pm i\omega t}(x_0 \pm iy_0)$, i.e.

$$e^{tA}x_0 = x_0 \cos(\omega t) - y_0 \sin(\omega t), \quad e^{tA}y_0 = x_0 \sin(\omega t) + y_0 \cos(\omega t), \quad t \in \mathbb{R}. \tag{9.3.7}$$

Let P be the projection associated to the spectral set $\{\omega i, -\omega i\}$,

$$P = \frac{1}{2\pi i}\int_{C(\omega i,\varepsilon)} (zI - \widetilde{A})^{-1}dz + \frac{1}{2\pi i}\int_{C(-\omega i,\varepsilon)} (zI - \widetilde{A})^{-1}dz,$$

where ε is so small that the set $\{z \in \mathbb{C} : 0 < |z \pm \omega i| \leq \varepsilon\}$ is contained in the resolvent set of $\widetilde{A}$. Although P is defined through complex integrals, it holds $P(X) \subset X$: indeed, after some computation one gets

$$\frac{2\pi}{\varepsilon}P = \int_0^{2\pi}(\varepsilon + \omega\sin\alpha - 2\cos\alpha A)(\omega i + \varepsilon e^{i\alpha} - \widetilde{A})^{-1}(-\omega i + \varepsilon e^{-i\alpha} - \widetilde{A})^{-1}d\alpha,$$

and the operator $(\omega i + \varepsilon e^{i\alpha} - \widetilde{A})^{-1}(-\omega i + \varepsilon e^{-i\alpha} - \widetilde{A})^{-1}$ maps X into X, since $\overline{-\omega i + \varepsilon e^{-i\alpha}} = \overline{\omega i + \varepsilon e^{i\alpha}}$.

Since $P(X)$ is the subspace of X spanned by x_0 and y_0, there are x_0^*, y_0^* in the dual space X' such that

$$Px = \langle x, x_0^*\rangle x_0 + \langle x, y_0^*\rangle y_0, \quad x \in X,$$

and

$$\langle x_0, x_0^*\rangle = \langle y_0, y_0^*\rangle = 1, \quad \langle x_0, y_0^*\rangle = \langle y_0, x_0^*\rangle = 0.$$

It follows that

$$\begin{cases} A^*x_0^* = \omega y_0^*, \quad A^*y_0^* = -\omega x_0^*, \\ (e^{tA})^*x_0^* = x_0^*\cos(\omega t) + y_0^*\sin(\omega t), \quad t \in \mathbb{R}, \\ (e^{tA})^*y_0^* = -x_0^*\sin(\omega t) + y_0^*\cos(\omega t), \quad t \in \mathbb{R}. \end{cases} \tag{9.3.8}$$

In the following, we shall need a transversality assumption on the eigenvalues of the operator

$$A(\lambda) = F_x(0,\lambda).$$

The next lemma deals with such eigenvalues.

3. PERIODIC SOLUTIONS

Lemma 9.3.2 *Let $F : \mathcal{O}_D \times]-1,1[\mapsto X$ be k times continuously differentiable ($k \geq 2$), and such that (9.3.5) holds. Then there are $\lambda_0 \in]0,1[$ and C^{k-1} functions $a, b :]-\lambda_0, \lambda_0[\mapsto \mathbb{R}$, $x, y :]-\lambda_0, \lambda_0[\mapsto D$ such that*

$$A(\lambda)x(\lambda) = a(\lambda)x(\lambda) - b(\lambda)y(\lambda), \quad A(\lambda)y(\lambda) = b(\lambda)x(\lambda) + a(\lambda)y(\lambda), \quad (9.3.9)$$

and $a(0) = 0$, $b(0) = \omega$.

Proof — Consider the simple eigenvalue ωi of $\widetilde{A}(0)$. By Proposition A.3.2, there are $\delta > 0$, $r > 0$ such that for λ so small that $\|\widetilde{A}(\lambda) - \widetilde{A}(0)\|_{L(D,X)} \leq \delta$, the operator $\widetilde{A}(\lambda)$ has a simple eigenvalue $z(\lambda) = a(\lambda) + ib(\lambda)$, with eigenfunction $w(\lambda) = x(\lambda) + iy(\lambda)$ ($a(\lambda), b(\lambda) \in \mathbb{R}$, $x(\lambda), y(\lambda) \in D$), and $z(\lambda)$ is the unique element in the spectrum of $\widetilde{A}(\lambda)$ such that $|z - \omega i| \leq r$. Following step by step the procedure of Proposition A.3.2, one sees that the functions $\lambda \mapsto z(\lambda)$, $\lambda \mapsto w(\lambda)$, $\lambda \mapsto \widetilde{A}(\lambda)$ are $k-1$ times continuously differentiable. Equalities (9.3.9) follow from $\widetilde{A}(\lambda)\omega(\lambda) = z(\lambda)w(\lambda)$, separating real and imaginary parts. ■

Theorem 9.3.3 *Let $F : \mathcal{O}_D \times]-1,1[\mapsto X$ be k times continuously differentiable, with $\partial^k F/\partial x^k$ Lipschitz continuous with respect to x, $k \geq 2$. Assume that $A = F_x(0,0)$ is a sectorial operator satisfying (9.3.5). Assume moreover that the nontrasversality condition*

$$a'(0) \neq 0 \quad (9.3.10)$$

holds. Then for every $\alpha \in]0,1[$ there are $c_0 > 0$ and C^{k-1} functions $\lambda :]-c_0, c_0[\mapsto \mathbb{R}$, $\rho :]-c_0, c_0[\mapsto \mathbb{R}$, $u :]-c_0, c_0[\mapsto C^\alpha(\mathbb{R}; X) \cap C^{\alpha+1}(\mathbb{R}; D)$, such that

$$\lambda(0) = 0, \quad \rho(0) = 1, \quad u(0) \equiv 0, \quad u(c) \text{ is nonconstant for } c \neq 0,$$

and $u(c)$ is a $2\pi\rho(c)/\omega$-periodic solution of

$$u' = F(u, \lambda(c)), \quad t \in \mathbb{R}.$$

Moreover there exists ε_0 such that if $\overline{\lambda} \in]-1,1[$, $\overline{\rho} \in \mathbb{R}$, and $\overline{u} \in C^\alpha(\mathbb{R}; D) \cap C^{\alpha+1}(\mathbb{R}; X)$ is a $2\pi\overline{\rho}/\omega$-periodic solution of

$$\overline{u}'(t) = F(\overline{u}(t), \overline{\lambda})$$

such that

$$\|u\|_{C^\alpha(\mathbb{R};D)} + \|u\|_{C^{\alpha+1}(\mathbb{R};X)} + |\overline{\lambda}| + |1 - \overline{\rho}| \leq \varepsilon_0,$$

then there are $c \in]-c_0, c_0[$, $t_0 \in \mathbb{R}$, such that

$$\overline{\lambda} = \lambda(c), \quad \overline{\rho} = \rho(c), \quad \overline{u}(t) = u(c)(t + t_0).$$

Proof — The proof is similar to the one of Theorem 9.1.10. It is divided in three steps. First we show that a branch of periodic solutions depending on a real parameter c exists, provided the determinant D of a certain 2×2 matrix is nonzero. Then we verify that $a'(0) \neq 0$ implies that $D \neq 0$, and finally we prove uniqueness.

(i) Since we look for periodic solutions of (9.3.4), with period close to $2\pi/\omega$, it is convenient to rescale time changing t with t/ρ, ρ close to 1, and look for $2\pi/\omega$-periodic solutions of

$$u'(t) = \rho F(u(t), \lambda), \quad t \in \mathbb{R}. \tag{9.3.11}$$

So, we define the nonlinear function

$$\Phi : (C^\alpha_\sharp(\mathbb{R}; D) \cap C^{\alpha+1}_\sharp(\mathbb{R}; X)) \times \,]-1,1[\times \mathbb{R} \mapsto C^\alpha_\sharp(\mathbb{R}; X),$$

$$\Phi(u, \lambda, \rho) = u' - \rho F(u(\cdot), \lambda),$$

where the subscript $\sharp$ stands for $2\pi/\omega$-periodic, and we look for a solution (u, λ, ρ) of

$$\Phi(u, \lambda, \rho) = 0. \tag{9.3.12}$$

Φ is $k-1$ times continuously differentiable, and

$$(\Phi_u(u, \lambda, \rho)v)(t) = v'(t) - \rho F_x(u(t), \lambda)v(t),$$
$$\Phi_\lambda(u, \lambda, \rho)(t) = -\rho F_\lambda(u(t), \lambda), \quad \Phi_\rho(u, \lambda, \rho)(t) = -F(u(t), \lambda).$$

In particular, $\Phi_u(0, 0, 1)v = v' - Av$ is not an isomorphism. By Propositions 4.4.8(ii) and 4.4.9(ii)

$$\mathcal{N} = \operatorname{Ker} \Phi_u(0, 0, 1) = \{e^{tA}x : x \in P(X)\},$$

$$\mathcal{R} = \operatorname{Range} \Phi_u(0, 0, 1) = \left\{ h \in C^\alpha_\sharp(\mathbb{R}; X) : \int_0^{2\pi/\omega} e^{(2\pi/\omega - s)A} Ph(s)\,ds = 0 \right\}.$$

Projections on $\mathcal{N}$ and on $\mathcal{R}$ are given respectively by

$$(P_\mathcal{N} u)(t) = \frac{\omega}{2\pi} \int_0^{2\pi/\omega} \langle e^{(2\pi/\omega - s)A} u(s), x_0^* \rangle ds\, e^{tA} x_0$$
$$+ \frac{\omega}{2\pi} \int_0^{2\pi/\omega} \langle e^{(2\pi/\omega - s)A} u(s), y_0^* \rangle ds\, e^{tA} y_0,$$

for $u \in C^\alpha_\sharp(\mathbb{R}; D) \cap C^{\alpha+1}_\sharp(\mathbb{R}; X)$, and

$$(P_\mathcal{R} h)(t) = h(t) - \frac{\omega}{2\pi} \int_0^{2\pi/\omega} \langle e^{(2\pi/\omega - s)A} h(s), x_0^* \rangle ds\, e^{tA} x_0$$
$$- \frac{\omega}{2\pi} \int_0^{2\pi/\omega} \langle e^{(2\pi/\omega - s)A} h(s), y_0^* \rangle ds\, e^{tA} y_0,$$

for $h \in C^\alpha_\sharp(\mathbb{R}; X)$.

3. Periodic solutions

Every element $u \in C^\alpha_\sharp(\mathbb{R}; D) \cap C^{\alpha+1}_\sharp(\mathbb{R}; X)$ may be written uniquely as

$$u(t) = c_1 e^{tA} x_0 + c_2 e^{tA} y_0 + v(t), \quad t \in \mathbb{R},$$

with

$$c_1, c_2 \in \mathbb{R}, \quad v \in \mathcal{V} = (I - P_\mathcal{N})(C^\alpha_\sharp(\mathbb{R}; D) \cap C^{\alpha+1}_\sharp(\mathbb{R}; X)).$$

Thanks to (9.3.7), there are $c \in \mathbb{R}$, $t_0 \in [0, 2\pi/\omega[$ such that

$$u(t) = c e^{tA} x_0 + v(t - t_0), \quad t \in \mathbb{R},$$

where $v(\cdot - t_0)$ still belongs to $\mathcal{V}$. In other words, neglecting translations in the time variable, every element of $C^\alpha_\sharp(\mathbb{R}; D) \cap C^{\alpha+1}_\sharp(\mathbb{R}; X)$ may be written as

$$u(t) = c e^{tA} x_0 + v(t), \quad t \in \mathbb{R},$$

with $c \in \mathbb{R}$ and $v \in \mathcal{V}$.

It is convenient to look for the solution of (9.3.11) in the form

$$u(t) = c(e^{tA} x_0 + v(t)), \quad t \in \mathbb{R}.$$

Replacing in (9.3.12), we are led to solve

$$v'(t) - Av(t) = \mathcal{G}(v(t), \lambda, \rho, c)(t), \quad t \in \mathbb{R}, \tag{9.3.13}$$

where

$$\mathcal{G}(x, \lambda, \rho, c)(t) \begin{cases} = \dfrac{\rho}{c} F(c(e^{tA} x_0 + x), \lambda) - A(e^{tA} x_0 + x), & \text{if } c \neq 0, \\ = \rho F_x(0, \lambda)(e^{tA} x_0 + x) - A(e^{tA} x_0 + x), & \text{if } c = 0. \end{cases}$$

Applying $P_\mathcal{R}$ and $I - P_\mathcal{R}$ we find that (9.3.13) is equivalent to the system

$$\begin{cases} (i) & v' - Av = P_\mathcal{R} \mathcal{G}(v(\cdot), \lambda, \rho, c), \\ (ii) & 0 = (I - P_\mathcal{R}) \mathcal{G}(v(\cdot), \lambda, \rho, c) \end{cases} \tag{9.3.14}$$

Since $\mathcal{G}_v(0, 0, 1, 0) = 0$ and $v \mapsto v' - Av$ is an isomorphism from $\mathcal{V}$ to $P_\mathcal{R}(C^\alpha_\sharp(\mathbb{R}; X))$, by the Implicit Function Theorem if λ, $\rho - 1$, c are small enough there is a unique small solution $v = v(\lambda, \rho, c) \in \mathcal{V}$ of (9.3.14)(i), say

$$\|v\|_{C^\alpha(\mathbb{R};D)} + \|v\|_{C^{\alpha+1}(\mathbb{R};X)} \le \delta_0. \tag{9.3.15}$$

Replacing in (9.3.14)(ii) we get the bifurcation equation

$$\eta(\lambda, \rho, c) = (I - P_\mathcal{R}) \mathcal{G}(v(\lambda, \rho, c)(\cdot), \lambda, \rho, c) = 0. \tag{9.3.16}$$

For every small c we shall find $\lambda = \lambda(c)$, $\rho = \rho(c)$ such that $\eta(\lambda(c), \rho(c), c) = 0$. Setting then

$$u(c)(t) = c(e^{tA} x_0 + v(\lambda(c), \rho(c), c)(t)), \quad t \in \mathbb{R}, \tag{9.3.17}$$

u will be a solution of (9.3.11), with $\lambda = \lambda(c)$, $\rho = \rho(c)$. Note that if $c \neq 0$ then u is not constant.

Problem (9.3.16) is equivalent to the two-dimensional system in the unknown (λ, ρ),
$$\eta_1(\lambda, \rho, c) = 0, \quad \eta_2(\lambda, \rho, c) = 0,$$
where
$$\eta_1(\lambda, \rho, c) = \int_0^{2\pi/\omega} \langle e^{(2\pi/\omega - s)A} \mathcal{G}(v(\lambda, \rho, c)(s), \lambda, \rho, c), x_0^* \rangle ds,$$
$$\eta_2(\lambda, \rho, c) = \int_0^{2\pi/\omega} \langle e^{(2\pi/\omega - s)A} \mathcal{G}(v(\lambda, \rho, c)(s), \lambda, \rho, c), y_0^* \rangle ds.$$

Since $\eta_1(0,0,0) = \eta_2(0,0,0) = 0$, to solve (9.3.16) it is sufficient to check that
$$D = \det \begin{pmatrix} \eta_{1\lambda}(0,1,0) & \eta_{1\rho}(0,1,0) \\ \eta_{2\lambda}(0,1,0) & \eta_{2\rho}(0,1,0) \end{pmatrix} \neq 0,$$

and then to apply the Implicit Function Theorem. If $D \neq 0$ there is $r_0 > 0$ such that for every sufficiently small $c > 0$, say $|c| \leq c_0$, the problem
$$Ae^{tA}x_0 + v'(t) = \rho F(c(e^{tA}u_0 + v), \lambda)$$
has a unique solution $(\lambda, \rho, v) \in]-1, 1[\times \mathbb{R} \times \mathcal{V}$ such that
$$\|v\|_{C^\alpha(\mathbb{R}, D)} + \|v\|_{C^{\alpha+1}(\mathbb{R}, X)} + |\lambda| + |1 - \rho| \leq r_0.$$

(ii) We show now that if $a'(0) \neq 0$, then $D \neq 0$. After some computation one gets
$$\eta_{1\lambda}(0,1,0) = \int_0^{2\pi/\omega} \langle e^{(2\pi/\omega - s)A} F_{x\lambda}(0,0) e^{sA} x_0, x_0^* \rangle ds,$$
$$\eta_{1\rho}(0,1,0) = \int_0^{2\pi/\omega} \langle e^{(2\pi/\omega - s)A} A e^{sA} x_0, x_0^* \rangle ds = 0,$$
$$\eta_{2\lambda}(0,1,0) = \int_0^{2\pi/\omega} \langle e^{(2\pi/\omega - s)A} F_{x\lambda}(0,0) e^{sA} x_0, y_0^* \rangle ds,$$
$$\eta_{2\rho}(0,1,0) = \int_0^{2\pi/\omega} \langle e^{(2\pi/\omega - s)A} A e^{sA} x_0, y_0^* \rangle ds = -2\pi.$$

Therefore
$$D = -2\pi \int_0^{2\pi/\omega} \langle e^{(2\pi/\omega - s)A} F_{x\lambda}(0,0) e^{sA} x_0, x_0^* \rangle ds. \tag{9.3.18}$$

With the notation of Lemma 9.3.2, we have $F_{x\lambda}(0,0) = A'(0)$. Differentiating with respect to λ the identity (9.3.9) at $\lambda = 0$, we get
$$A'(0)x_0 = -Ax'(0) + a'(0)x_0 - b'(0)y_0 - \omega y'(0),$$
$$A'(0)y_0 = -Ay'(0) + b'(0)x_0 + a'(0)y_0 + \omega x'(0).$$

3. Periodic solutions

Replacing in (9.3.18) and using (9.3.7), (9.3.8), we get

$$D = -2\pi \int_0^{2\pi/\omega} \langle A'(0)(x_0\cos\omega s - y_0\sin\omega s), e^{(2\pi/\omega - s)A^*} x_0^* \rangle ds = -\frac{4\pi}{\omega} a'(0).$$

By assumption, $a'(0) \neq 0$. Therefore the bifurcation equation is solvable, and the function u given by (9.3.17) is a nonconstant $2\pi/\omega$-periodic solution of (9.3.11).

(iii) Let us prove the uniqueness statement. Let $\overline{u} \in C^\alpha(\mathbb{R}; D) \cap C^{\alpha+1}(\mathbb{R}; X)$ be a small $2\pi/\omega$-periodic solution of

$$\overline{u}' = \overline{\rho} F(\overline{u}, \overline{\lambda}),$$

satisfying

$$\|\overline{u}\|_{C^\alpha(\mathbb{R}; D)} + \|\overline{u}\|_{C^{\alpha+1}(\mathbb{R}; X)} + |\overline{\lambda}| + |1 - \overline{\rho}| \leq \varepsilon_0.$$

Split $\overline{u}$ into the sum

$$\overline{u}(t) = c_1 e^{tA} x_0 + c_2 e^{tA} y_0 + \overline{v}(t),$$

where $\overline{v} = (I - P_\mathcal{R})\overline{u}$, and let $t_0 \in [0, 2\pi/\omega[$, $c \in \mathbb{R}$ be such that

$$\overline{u}(t - t_0) = c e^{tA} x_0 + \overline{v}(t - t_0), \quad t \in \mathbb{R}.$$

Set $\hat{v}(t) = \overline{v}(t - t_0)$, $\hat{u}(t) = c(e^{tA} x_0 + \hat{v}(t)/c)$. Out aim is to show that if ε_0 is small enough then

$$|c| \leq c_0, \quad |\overline{\lambda}| + |1 - \overline{\rho}| + \frac{1}{c}\|\hat{v}\|_{C^\alpha(\mathbb{R}; D) \cap C^{\alpha+1}(\mathbb{R}; X)} \leq r_0, \tag{9.3.19}$$

(see the last part of Step 1). To prove that (9.3.19) holds, we show preliminarily that the linear operator

$$\Psi : \mathbb{R}^2 \times \mathcal{V} \mapsto C_\sharp^\alpha(\mathbb{R}; X),$$

$$\Psi(\lambda, \rho, v)(t) = -\lambda A'(0) e^{tA} x_0 + \rho e^{tA} y_0 + v'(t) - Av(t)$$

is an isomorphism. We claim that it is one to one: otherwise, there would exist $(\lambda, \rho, v) \neq (0, 0, 0)$ such that $\Psi(\lambda, \rho, v) = 0$, so that

$$\langle e^{(2\pi/\omega - s)A} \Psi(\lambda, \rho, v)(s), x_0^* \rangle = 0, \quad 0 \leq s \leq 2\pi/\omega,$$

and integrating over $[0, 2\pi/\omega]$ one would get $a'(0) = 0$, contradicting the transversality assumption (9.3.10). Let us prove that it is onto: the range of Ψ is spanned by $\mathcal{R}$, which has codimension 2, and by the functions $t \mapsto A'(0)e^{tA}x_0$, $t \mapsto e^{tA}y_0$, which are linearly independent due to the above argument. So, Ψ is an isomorphism, and there is $K > 0$ such that

$$|c\overline{\lambda}| + |c(1-\overline{\rho})| + \|\hat{v}\|_{C^\alpha(\mathbb{R};D) \cap C^{\alpha+1}(\mathbb{R};X)} \leq K \|\Psi(c\overline{\lambda}, c(1-\overline{\rho}), \hat{v})\|_{C^\alpha(\mathbb{R};X)}$$
$$\leq K \big(\|A'(0)e^{tA}x_0\|_{C^\alpha(\mathbb{R};X)} |c||\overline{\lambda}| + \|e^{tA}y_0\|_{C^\alpha(\mathbb{R};X)} |c||1-\overline{\rho}| + \|\hat{v}' - A\hat{v}\|_{C^\alpha(\mathbb{R};X)}\big).$$

Since
$$\hat{v}' - A\hat{v} = \overline{\rho}(F(\hat{v} + ce^{tA}x_0, \overline{\lambda}) - A(ce^{tA}x_0 + \hat{v})) + c(\overline{\rho} - 1)Ae^{tA}x_0,$$

and $Ae^{tA}x_0 = -e^{tA}y_0$, then, after some computation one finds

$$\|\hat{v}' - A\hat{v}\|_{C^\alpha(\mathbb{R};X)} \leq K_1\left(\|\hat{v}\|^2_{C^\alpha(\mathbb{R};D)} + \|\overline{\lambda}\hat{v}\|^2_{C^\alpha(\mathbb{R};D)} + |c|^2 + |c\overline{\lambda}| + |c(1-\overline{\rho})|\right),$$

provided $\|\hat{v}\|_{C^\alpha(\mathbb{R};D)}$, $|\overline{\lambda}|$, $|c|$ are small enough. It follows that

$$\|\hat{v}\|_{C^\alpha(\mathbb{R};D) \cap C^{\alpha+1}(\mathbb{R};X)} \leq K_2\big(|c(1-\overline{\rho})| + |c\overline{\lambda}| + c^2$$
$$+ \|(1-\overline{\rho})\hat{v}\|^2_{C^\alpha(\mathbb{R};D)} + \|\overline{\lambda}\hat{v}\|_{C^\alpha(\mathbb{R};D)} + \|\hat{v}\|^2_{C^\alpha(\mathbb{R};D)}\big),$$

which concludes the proof. ∎

To know whether the periodic orbits given by the previous theorem occur for $\lambda < 0$ or for $\lambda > 0$, one has to study the function $c \mapsto \lambda(c)$. The next lemma is of help.

Lemma 9.3.4 *Under the assumptions of Theorem 9.3.3, it holds*

$$\lambda'(0) = 0, \quad v'(0) = 0.$$

If F is four times continuously differentiable, then $c \mapsto \lambda(c)$ is twice continuously differentiable, and

$$\lambda''(0) = -\frac{1}{8a'(0)}\Big(\langle F_{xxx}(0,0)((x_0,x_0,x_0) + (x_0,y_0,y_0)), x_0^*\rangle$$
$$- \langle F_{xxx}(0,0)((x_0,x_0,y_0) + (y_0,y_0,y_0)), y_0^*\rangle\Big).$$

Proof — Let $c \mapsto v(c)$ be the function considered in the proof of Theorem 9.3.3. We know already that $v'(0) = 0$ and that

$$\left(\frac{d}{dt} - A\right)v(c) = \mathcal{G}(v(c)(\cdot), \lambda(c), \rho(c), c)(\cdot), \tag{9.3.20}$$

where the function in right hand side $(x,c) \mapsto \mathcal{G}(x, \lambda(c), \rho(c), c)$ is twice continuously differentiable with respect to x, and continuously differentiable with respect to c, with values in $C^\alpha_\sharp(\mathbb{R}, X)$. By Theorem (9.3.3), $c \mapsto v(c)$ is continuously differentiable with values in $C^\alpha_\sharp(\mathbb{R}, D) \cap C^{1+\alpha}_\sharp(\mathbb{R}, X)$, and differentiating (9.3.20) with respect to c we get

$$\left(\frac{d}{dt} - A\right)v'(c) = \mathcal{G}_x v'(c) + \mathcal{G}_\lambda \lambda'(c) + \mathcal{G}_\rho \rho'(c) + \mathcal{G}_c,$$

3. Periodic solutions

where the derivatives of g are evaluated at $(v(c)(\cdot), \lambda(c), \rho(c), c)$. Recall that

$$\mathcal{G}_x(0,0,1,0) = 0, \quad \mathcal{G}_\lambda(0,0,1,0)(t) = F_{x\lambda}(0,0)e^{tA}x_0,$$
$$\mathcal{G}_\rho(0,0,1,0)(t) = Ae^{tA}x_0, \quad \mathcal{G}_c(0,0,1,0)(t) = F_{xx}(0,0)(e^{tA}x_0)^2/2.$$

Applying $I - P_\mathcal{R}$ and taking $c = 0$, one gets

$$0 = (I - P_\mathcal{R})\left(\lambda'(0)F_{x\lambda}(0,0)e^{tA}x_0 + \rho'(0)Ae^{tA}x_0 + \frac{1}{2}F_{xx}(0,0)(e^{tA}x_0, e^{tA}x_0)\right),$$

so that

$$0 = \frac{\omega}{2\pi}\int_0^{2\pi/\omega} \langle e^{(2\pi/\omega - s)A}F_{x\lambda}(0,0)e^{sA}x_0, x_0^*\rangle ds\, \lambda'(0)$$
$$+ \frac{\omega}{2\pi}\int_0^{2\pi/\omega} \langle e^{(2\pi/\omega - s)A}(Ae^{sA}x_0\rho'(0) + \frac{1}{2}F_{xx}(0,0)(e^{tA}x_0, e^{tA}x_0)), x_0^*\rangle ds$$
$$= a'(0)\lambda'(0).$$

Since $a'(0) \neq 0$, then $\lambda'(0) = 0$.

By Theorem 9.3.5, if F is four times continuously differentiable then $c \mapsto v(c)$ is twice continuously differentiable with values in $C^\alpha_\#(\mathbb{R}, D) \cap C^{1+\alpha}_\#(\mathbb{R}, X)$. Differentiating twice (9.3.20) with respect to c, taking $c = 0$ and using the equalities

$$g_{\rho\rho}(0,0,1,0) = 0, \quad g_{\rho c}(0,0,1,0) = F_{xx}(0,0)(e^{tA}x_0)^2/2,$$
$$g_{cc}(0,0,1,0) = F_{xxx}(0,0)(e^{tA}x_0)^3/3$$

one gets

$$(d/dt - A)v''(0) = F_{x\lambda}(0,0)e^{tA}x_0\lambda''(0) + F_{xx}(0,0)(e^{tA}x_0)^2$$
$$+ Ae^{tA}x_0\rho''(0) + \tfrac{1}{3}F_{xxx}(0,0)(e^{tA}x_0)^3.$$

Applying $I - P_\mathcal{R}$, one obtains

$$0 = a'(0)\lambda''(0) + \frac{\omega}{6\pi}\int_0^{2\pi/\omega} \langle F_{xxx}(0,0)(e^{sA}x_0)^3, (e^{(2\pi/\omega - s)A})^* x_0^*\rangle ds,$$

and the last statement follows, recalling formulas (9.3.7), (9.3.8). ∎

9.3.2 Stability of periodic solutions

Consider again the problem

$$u'(t) = F(t, u(t)), \tag{9.3.21}$$

where $F : \mathbb{R} \times \mathcal{O}_D \mapsto X$ satisfies assumptions (8.0.3), (8.1.1) and it is periodic with respect to time, with minimum period $T > 0$. We assume that (9.3.21) has a T-periodic solution $\bar{u} \in C^\alpha(\mathbb{R}, D) \cap C^{\alpha+1}(\mathbb{R}; X)$ and we study its stability properties. To begin with, we state the Principle of Linearized Stability.

Let $t_0 \in \mathbb{R}$, $u_0 \in \mathcal{O}$, and consider the initial value problem

$$u(t_0) = u_0 \qquad (9.3.22)$$

for equation (9.3.21). If either $F(t_0, u_0) \in \overline{D}$ or u_0 is sufficiently close to $\bar{u}(t_0)$, Theorems 8.1.1 and 8.3.4 imply existence and uniqueness of a maximally defined solution $u = u(t; t_0, u_0)$ belonging to $C([t_0, t_0 + \tau[; X) \cap C^1(]t_0, t_0 + \tau[; X) \cap C_\alpha^\alpha(]t_0, t_0 + \tau - \varepsilon]; D)$ for every $\varepsilon \in]0, \tau[$ (if $F(t_0, u_0) \in \overline{D}$ then u belongs also to $C([t_0, t_0 + \tau[; D))$. Here $\tau = \tau(t_0, x_0) > 0$. We set

$$A(t) = F_x(t, \bar{u}(t)), \quad t \in \mathbb{R}.$$

Since $\bar{u} \in C^\alpha(\mathbb{R}, D)$ and F_x is locally α-Hölder continuous with respect to t, then $A(\cdot)$ belongs to $C^\alpha(\mathbb{R}, L(D, X))$, so that it generates a parabolic evolution operator $G(t, s)$, see Chapter 6. Set

$$V(s) = G(s + T, s), \quad s \in \mathbb{R},$$

and recall that $\sigma(V(s)) \setminus \{0\}$ is independent of s.

Theorem 9.3.5 *Let $F : \mathbb{R} \times \mathcal{O}_D \mapsto X$ satisfy assumptions (8.0.3), (8.1.1) and be T-periodic with respect to time. Assume that*

$$\rho_0 = \sup\{|\lambda| : \lambda \in \sigma(V(0))\} < 1. \qquad (9.3.23)$$

Then $\bar{u}$ is exponentially asymptotically stable. Specifically, for every ω in the interval $[0, -T^{-1} \log \rho_0[$ there are $r > 0$, $M > 0$ such that for each $t_0 \in \mathbb{R}$ and $u_0 \in B(\bar{u}(t_0), r) \subset D$ we have $\tau(t_0, u_0) = +\infty$, and

$$\|u(t; t_0, u_0) - \bar{u}(t)\|_D \leq M e^{-\omega(t-t_0)} \|u_0 - \bar{u}(t_0)\|_D, \quad \forall t \geq t_0. \qquad (9.3.24)$$

Sketch of the proof — From Theorem 8.3.4 we know that for every $a > 0$, if u_0 is sufficiently close to $\bar{u}(t_0)$ then problem (9.3.21)–(9.3.22) has a unique local solution $u(\cdot, t_0, u_0) \in C([t_0, t_0 + a]; X) \cap C^1(]t_0, t_0 + a]; X) \cap C_\alpha^\alpha(]t_0, t_0 + a]; D)$. Let z be the difference

$$z(t) = u(t; t_0, u_0) - \bar{u}(t), \quad t \geq t_0.$$

Then z satisfies

$$z'(t) = A(t)z(t) + G(t, z(t)), \quad t_0 < t \leq a; \quad z(t_0) = u_0 - \bar{u}(t_0),$$

where

$$G(t, x) = F(t, \bar{u}(t) + x) - F(t, \bar{u}(t)) - F_x(t, \bar{u}(t))x.$$

To prove that u is defined in $[t_0, +\infty[$, one has to find a fixed point of the operator Γ defined by $\Gamma(u) = z$, where z is the solution of

$$z'(t) = A(t)z(t) + G(t, u(t)), \quad t > t_0; \quad z(t_0) = u_0 - \bar{u}(t_0).$$

3. Periodic solutions

Γ is well defined in the ball $B(0,r) \subset Y$, where

$$Y = C([t_0, \infty[; X) \cap C_\theta^\theta(]t_0, t_0 + 1]; D) \cap C^\theta([t_0+1, +\infty[; D, -\omega),$$

provided $0 < \theta < \alpha$, and r is small enough. Following the procedure of Theorem 9.1.2, using Proposition 6.3.10 instead of Proposition 4.4.10, one shows that Γ maps $B(0,r) \subset Y$ into itself and it is a $1/2$-contraction if r and $\|u_0 - \overline{u}(t_0)\|_D$ are sufficiently small. The proof is close to the one of Theorem 9.1.2, and we leave it to the reader. ∎

We now give an instability result.

Theorem 9.3.6 *Let the assumptions of Theorem 9.3.5 be satisfied, with (9.3.23) replaced by*

$$\sigma_+ = \{\lambda \in \sigma(V(0)) : |\lambda| > 1\} \neq \emptyset, \ \inf\{|\lambda| : \lambda \in \sigma_+\} = \rho_+ > 1. \qquad (9.3.25)$$

Then $\overline{u}$ is unstable. Specifically, for every $t_0 \in \mathbb{R}$ there exists an initial datum $u_0 \neq \overline{u}(t_0)$ such that problem (9.3.21) has a backward solution u with $u(t_0) = u_0$ and

$$\|u(t) - \overline{u}(t)\|_D \leq M e^{\omega t}, \ t \leq 0.$$

We omit the proof, which is similar to the proof of Theorem 9.1.3. Of course, one has to use the results of Proposition 6.3.11 instead of Proposition 4.4.12.

In the autonomous case with smooth nonlinearity

$$u'(t) = F(u(t)), \qquad (9.3.26)$$

assumption (9.3.23) is never satisfied. Indeed, from Proposition 8.3.6 it follows that any periodic solution $\overline{u}$ belongs to $C^{2+\theta}(\mathbb{R}; X) \cap C^{1+\theta}(\mathbb{R}; D)$ for every $\theta \in]0,1[$, and the derivative $v(t) = \overline{u}'$ satisfies

$$v'(t) = F'(\overline{u}(t))v(t), \ t \in \mathbb{R},$$

so that 1 is an eigenvalue of $V(s)$ for every s.

Theorem 9.3.7 *Let $F : \mathcal{O}_D \mapsto X$ be a thrice continuously differentiable function. Let $\overline{u}$ be a nonconstant T-periodic solution of (9.3.26) being Hölder continuous with values in D. Assume that*

$$\begin{cases} (i) & 1 \text{ is a simple eigenvalue of } V(0), \\ (ii) & \sup\{|\lambda| : \lambda \in \sigma(V(0)) \setminus \{1\}\} = \rho_- < 1. \end{cases} \qquad (9.3.27)$$

Then $\overline{u}$ is orbitally asymptotically stable with asymptotic phase. Precisely, for every $\omega \in [0, -T^{-1} \log \rho_-[$ there are $r_0 > 0$, $M > 0$ such that if $\text{dist}(u_0, \gamma) = r \leq r_0$ then $u(\cdot; u_0)$ exists in the large and there is $\theta = \theta(u_0)$ such that

$$\|u(t, u_0) - \overline{u}(t + \theta)\|_D \leq M e^{-\omega t} r, \ t \geq 0.$$

Proof — Fix $\rho \in]\rho_-, 1[$, and let $P_-(s)$, $P_+(s)$, $s \in \mathbb{R}$, be the projections defined in (6.3.8). Replacing possibly $\overline{u}(t)$ by $\overline{u}(t + t_0)$, with $0 < t_0 < T$, we may assume that u_0 is close to $\overline{u}(0)$.

For every $\theta \in \mathbb{R}$ the difference
$$z(t, \theta) = u(t; u_0) - \overline{u}(t + \theta)$$
satisfies
$$z_t(t, \theta) = A(t)z(t, \theta) + g(t, z(t, \theta), \theta), \quad 0 < t < \tau, \tag{9.3.28}$$
where
$$A(t) = F'(\overline{u}(t)), \quad g(t, x, \theta) = F(\overline{u}(t + \theta) + x) - F(\overline{u}(t + \theta)) - F'(\overline{u}(t))x.$$

We seek for a globally defined, exponentially decaying solution of (9.3.28). In view of Proposition 6.3.10, fixed any small $y \in P_-(0)(D)$, we look for a solution z of

$$\begin{aligned}
z(t) &= G(t, 0)y + \int_0^t G(t, s)P_-(s)g(s, z(s), \theta)ds \\
&\quad - \int_t^{+\infty} G(t, s)P_+(s)g(s, z(s), \theta)ds = (\Lambda_\theta)z(t)
\end{aligned} \tag{9.3.29}$$

in a ball $B(0, r) \subset Y$, where
$$Y = C_\alpha^\alpha(]0, 1]; D) \cap C^\alpha([1, +\infty[; D, -\omega),$$

and $\alpha \in]0, 1[$. We recall that, thanks to Proposition 6.1.5(ii), Corollary 6.1.6(ii) and Proposition 6.3.10, for every $\alpha \in]0, 1[$ there exists a constant $C > 0$ such that for every $y \in P_-(0)(D)$ and $f \in C_\alpha^\alpha(]0, 1]; X) \cap C^\alpha([1, +\infty[; X, -\omega)$, then the function
$$z(t) = G(t, 0)y + \int_0^t G(t, s)P_-(s)f(s)ds - \int_t^{+\infty} G(t, s)P_+(s)f(s)ds$$
belongs to $C_\alpha^\alpha(]0, 1]; D) \cap C^\alpha([1, +\infty[; D, -\omega)$, and
$$\|z\|_{C_\alpha^\alpha(]0,1];D)} + \|z\|_{C^\alpha([1,+\infty[;D,-\omega)} \leq C(\|y\|_D + \|f\|_{C_\alpha^\alpha(]0,1];X)} + \|f\|_{C^\alpha([1,+\infty[;X,-\omega)})$$

Using the assumption $F \in C^3(\mathcal{O}_D, X)$ and recalling that $\overline{u} \in C^1(\mathbb{R}; D)$, one checks that $g : \mathbb{R} \times \mathcal{O}_D \times \mathbb{R} \mapsto X$ satisfies

(i) $g(t + T, x, \theta) = g(t, x, \theta)$, $g(t, 0, 0) \equiv 0$,

(ii) $\sup_{t \in \mathbb{R}, \|x\|_D \leq r} \|g_x(t, x, \theta)\|_{L(D, X)} = K_1(r, \theta) \to 0$ as $(r, \theta) \to 0$,

(iii) $\sup_{\|x\|_D \leq r} [g_x(\cdot, x, \theta)]_{C^\alpha(\mathbb{R}; L(D, X))} = K_2(r, \theta) \to 0$ as $(r, \theta) \to 0$,

(iv) $\sup_{t, \theta \in \mathbb{R}, \|x\|_D \leq r} \|g_\theta(t, x, \theta)\| = K_3(r) \to 0$ as $r \to 0$,

(v) $\sup_{\theta \in \mathbb{R}} [g_\theta(\cdot, x, \theta)]_{C^\alpha(\mathbb{R}; X)} \leq K_4(r)\|x\|_D$ for all $x \in B(0, r) \subset D$.

3. Periodic solutions

For $r > 0$ define

$$\begin{cases} K_5(r) = \sup_{t, \theta \in \mathbb{R}, \|x\|_D \leq r} \|g_{xx}(t, x, \theta)\|_{L(D, L(D, X))} \\ K_6(r) = \sup_{t, \theta \in \mathbb{R}, \|x\|_D \leq r} \|g_{x\theta}(t, x, \theta)\|_{L(D, X)}. \end{cases}$$

Arguing as in the proof of Theorem (9.1.2) one sees that Λ_θ is Lipschitz continuous if r is small enough, with Lipschitz constant less or equal to

$$\varphi(r, \theta) = C[K_1(r, \theta) + K_2(r, \theta) + (\omega^\alpha (1-\alpha)^{1-\alpha} + 1) r K_5(r)]. \tag{9.3.30}$$

If $r \leq r_0$, $|\theta| \leq \theta_0$, where r_0, θ_0 are so small that $\varphi(r_0, \theta_0) \leq 1/2$, then Λ_θ is a contraction with constant $1/2$ in $B(0, r) \subset Y$. Moreover, Λ_θ maps $B(0, r)$ into itself provided $\|y\|_D \leq r/2C$. So, Λ_θ has a unique fixed point $z = z(\cdot, y, \theta)$ in $B(0, r)$, and it satisfies the inequality

$$\|z(\cdot, y, \theta)\|_{C^\alpha_\alpha(]0,1];D)} + \|z(\cdot, y, \theta)\|_{C^\alpha([1,+\infty[;D,-\omega)} \leq 2C \|y\|_D.$$

We shall show that if u_0 is sufficiently close to $\bar{u}(0)$ there are $y \in B(0, r_0/2C) \subset D$, $\theta \in \mathbb{R}$ near 0 and $\bar{t} > 0$ such that

$$u(t; u_0) = \bar{u}(t + \theta) + z(t, y, \theta), \quad 0 \leq t \leq \bar{t}. \tag{9.3.31}$$

This will prove the statement: indeed, if $u(t, u_0)$ coincides with $\bar{u}(t + \theta) + z(t, y, \theta)$ in an interval, then it exists in the large and coincides with $\bar{u}(t + \theta) + z(t, y, \theta)$ in $[0, +\infty[$, thanks to the uniqueness part of Theorem 8.1.1. In its turn, (9.3.31) is equivalent to

$$u_0 = \bar{u}(\theta) + z(0, y, \theta). \tag{9.3.32}$$

This is because the initial value problem for equation (9.3.26) has a unique solution in $B(\bar{u}(0), \rho_0) \subset C^\alpha_\alpha(]0, \bar{t}]; D)$ for ρ_0 small (see Theorem 8.3.4) provided u_0 is sufficiently close to $\bar{u}(0)$, and $\bar{u}(\theta + \cdot) + z(\cdot, y, \theta)$ belongs to $B(\bar{u}(0), \rho_0)$ if $\bar{t}$ and $\|y\|_D$ are small enough.

Let $x^* \in X'$ be such that

$$P_+(0)x = \langle x, x^* \rangle \bar{u}'(0), \quad \forall x \in X.$$

Then equation (9.3.32) is equivalent to

$$H(y, \theta) = (y, \theta),$$

where

$$H : B(0, \varepsilon) \subset P_-(0)(D) \times [-\delta, \delta] \mapsto P_-(0)(D) \times \mathbb{R},$$

$$H(y, \theta) = (P_-(0)(u_0 - \bar{u}(\theta)), \langle (u_0 - \bar{u}(\theta)) - z(0, y, \theta) + \theta \bar{u}'(0), x^* \rangle),$$

and $\varepsilon \in\,]0, r_0/2C]$, $\delta \in\,]0, \theta_0]$. $B(0, \varepsilon) \times [-\delta, \delta]$ is endowed with the usual product norm. We shall show that H is a contraction and maps $B(0, \varepsilon) \times [-\delta, \delta]$ into itself, provided ε, δ, and $\|u_0 - \overline{u}(0)\|_D$ are sufficiently small. It holds

$$H(y_1, \theta_1) - H(y_2, \theta_2) =$$
$$= \big(-P_-(0)(\overline{u}(\theta_1) - \overline{u}(\theta_2)), \langle -\overline{u}(\theta_1) + \overline{u}(\theta_2) + (\theta_1 - \theta_2)\overline{u}'(0)$$
$$-z(0, y_1, \theta_1) + z(0, y_2, \theta_2), x^* \rangle \big)$$

Since $P_-(0)\overline{u}'(0) = 0$, then

$$P_-(0)(\overline{u}(\theta_1) - \overline{u}(\theta_2)) = P_-(0) \int_0^1 [\overline{u}'(\sigma\theta_1 + (1-\sigma)\theta_2) - \overline{u}'(0)]d\sigma(\theta_1 - \theta_2),$$

so that

$$\|P_-(0)(\overline{u}(\theta_1) - \overline{u}(\theta_2))\|_D \leq c\delta^\alpha |\theta_1 - \theta_2|,$$

and similarly

$$\langle -\overline{u}(\theta_1) + \overline{u}(\theta_2) + (\theta_1 - \theta_2)\overline{u}'(0), x^* \rangle \leq c\delta^\alpha |\theta_1 - \theta_2|.$$

Moreover, since $z_i(\cdot, y_i, \theta_i)$, $i = 1, 2$, belong to $B(0, r) \subset Y$ and $r = 2C\varepsilon \leq r_0$, $|\theta_i| \leq \theta_0$,

$$\|z(\cdot, y_2, \theta_2) - z(\cdot, y_1, \theta_1)\|_Y$$
$$\leq C(\|y_2 - y_1\|_D + \|g(z(\cdot, y_2, \theta_2), \theta_2) - g(z(\cdot, y_1, \theta_1), \theta_1)\|_Y) \quad (9.3.33)$$
$$\leq C\|y_2 - y_1\|_D + \tfrac{1}{2}\|z(\cdot, y_2, \theta_2) - z(\cdot, y_1, \theta_1)\|_Y$$
$$+ C\|g(z(\cdot, y_2, \theta_2), \theta_2) - g(z(\cdot, y_2, \theta_2), \theta_1)\|_Y,$$

and

$$\|g(z(\cdot, y_2, \theta_2), \theta_2) - g(z(\cdot, y_2, \theta_2), \theta_1)\|_Y$$
$$\leq K_3(r)|\theta_1 - \theta_2| + [g(z(\cdot, y_2, \theta_2), \theta_2) - g(z(\cdot, y_2, \theta_2), \theta_1)]_{C^\alpha(]0,1];X)}$$
$$+ [g(z(\cdot, y_2, \theta_2), \theta_2) - g(z(\cdot, y_2, \theta_2), \theta_1)]_{C^\alpha([1,+\infty[;X,-\omega)}$$
$$\leq K_3(r)|\theta_1 - \theta_2| + r(K_4(r) + K_6(r))|\theta_1 - \theta_2|$$
$$+ r(\omega K_3(r) + K_4(r) + K_6(r)(\omega^\alpha(1-\alpha)^{1-\alpha} + 1))|\theta_1 - \theta_2|$$
$$= K_7(r)|\theta_1 - \theta_2|,$$

where

$$\lim_{r \to 0} K_7(r) = 0.$$

Replacing in (9.3.33) we get

$$\|z(\cdot, y_2, \theta_2) - z(\cdot, y_1, \theta_1)\|_Y \leq 2C(\|y_2 - y_1\|_D + K_7(r)|\theta_1 - \theta_2|),$$

so that

$$\|P_+(0)(z(0, y_2, \theta_2) - z(0, y, \theta_1))\|$$
$$\leq C\|g(z(\cdot, y_2, \theta_2), \theta_2) - g(z(\cdot, y_2, \theta_2), \theta_1)\|_Y$$
$$\leq C(\varphi(2C\varepsilon, \delta)(2C(\|y_2 - y_1\|_D + K_7(r)|\theta_1 - \theta_2|))$$

3. Periodic solutions

where φ is defined in (9.3.30). Recalling that $\lim_{(r,\delta)\to 0} \varphi(r,\delta) = 0$, we see that H is a 1/2-contraction provided ε and δ are sufficiently small. Fix such ε and δ. Since
$$g(0,0) = (P_-(0)(u_0 - \overline{u}(0)), \langle u_0 - \overline{u}(0), x^* \rangle),$$
then H maps $B(0,\varepsilon) \subset P_-(0)(D) \times [-\delta, \delta]$ into itself, provided
$$\|u_0 - \overline{u}(0)\|_D \leq \max\left(\frac{\varepsilon}{2\|P_-(0)\|_{L(D)}}, \frac{\delta}{2\|x^*\|_{X'}}\right).$$
Then, H has a unique fixed point in $B(0,\varepsilon) \times [-\delta, \delta]$, and the proof is complete. ∎

Since orbital stability is weaker than stability, then the next result about orbital instability is stronger than the result of Theorem 9.3.6.

Theorem 9.3.8 *Let the assumptions of Theorem 9.3.7 hold, with (9.3.27) replaced by*
$$\begin{cases} (i) & \sigma_+ = \{\lambda \in \sigma : |\lambda| > 1\} \neq \emptyset, \\ (ii) & \inf\{|\lambda| : \lambda \in \sigma_+\} = \rho_+ > 1. \end{cases} \quad (9.3.34)$$
Then $\overline{u}$ is orbitally unstable. Specifically, there is an initial value u_0 not belonging to the orbit γ such that problem (9.3.26) has a backward solution u with $u(0) = u_0$ and $\|u(t) - \overline{u}(t)\|_D$ goes to 0 as $t \to -\infty$.

Proof — Fix $\rho \in]1, \rho_+[$ and define $P_-(s)$, $P_+(s)$ by (6.3.8). If u is any backward solution of (9.3.26), the difference $w(t) = u(t) - \overline{u}(t)$ satisfies
$$w'(t) = A(t)w(t) + g(t, w(t), 0), \quad t \leq 0,$$
with G defined in (9.3.29). In view of Proposition 6.3.11, we look for a solution of
$$\begin{aligned} w(t) = & \; G(t,0)x_+ + \int_{-\infty}^t G(t,s)P_-(s)g(s,w(s),0)ds \\ & + \int_0^t G(t,s)P_+(s)g(s,w(s),0)ds, \quad t \leq 0, \end{aligned} \quad (9.3.35)$$
with $x_+ \in P_+(s)(X)$ fixed. w is sought in the ball $B(0,r) \subset C^\alpha(]-\infty, 0]; D, \omega)$, with fixed $\alpha \in]0,1[$ and small $r > 0$.

By Proposition 6.3.11, there is $C > 0$ such that for every $x \in P_+(0)(X) = P_+(0)(D)$ and $g \in C^\alpha(]-\infty, 0]; X, \omega)$, the function
$$w(t) = G(t,0)x + \int_{-\infty}^t G(t,s)P_-(s)g(s)ds + \int_0^t G(t,s)P_+(s)g(s)ds, \quad t \leq 0,$$
belongs to $C^\alpha(]-\infty, 0]; D, \omega)$ and satisfies
$$\|w\|_{C^\alpha(]-\infty,0]; D,\omega)} \leq C(\|x\| + \|g\|_{C^\alpha(]-\infty,0]; X,\omega)}).$$

Arguing as in the proof of the stability theorem, one finds that if r is sufficiently small and $0 < \|x\| \leq r/2C$, then equation (9.3.35) has a unique solution $w \in B(0, r)$, and moreover

$$\|w(0) - x\|_D \leq K(r)\|x\|, \tag{9.3.36}$$

with

$$\lim_{r \to 0} K(r) = 0. \tag{9.3.37}$$

Set now

$$u(t) = \overline{u}(t) + w(t), \quad t \leq 0.$$

Then u is a backward solution of (9.3.26) such that $\|u(t) - \overline{u}(t)\|_D$ converges exponentially to 0 as $t \to -\infty$.

It remains to show that $u(0)$ does not belong to the orbit γ. For every $t \in [0, T]$ it holds

$$\|u(0) - \overline{u}(t)\|_D = \|\overline{u}(0) + w(0) - \overline{u}(t)\|_D$$
$$\geq \|x - \overline{u}'(0)t\|_D - \|w(0) - x\|_D - \|\overline{u}(t) - \overline{u}(0) - \overline{u}'(0)t\|_D.$$

Since $x \in P_+(0)(D)$ and $\overline{u}'(0) \in P_-(0)(D)$, then

$$\inf_{0 \leq t \leq T} \|x - \overline{u}'(0)t\|_D > 0.$$

Recalling (9.3.36) and (9.3.37) we get that there are ε, r_1, δ_1 such that if $r \leq r_1$ then

$$\|u(0) - \overline{u}(t)\|_D \geq \delta_1, \quad t \in [0, \varepsilon] \cup [T - \varepsilon, T].$$

Moreover, since $\overline{u}$ is nonconstant, then

$$\inf_{\varepsilon \leq t \leq T-\varepsilon} \|\overline{u}(t) - \overline{u}(0)\|_D > 0.$$

Since

$$\|u(0) - \overline{u}(t)\|_D = \|\overline{u}(0) + w(0) - \overline{u}(t)\|_D$$
$$\geq \|\overline{u}(0) - \overline{u}(t)\|_D - \|w(0)\|_D \geq \|\overline{u}(0) - \overline{u}(t)\|_D - r,$$

then there are r_2, δ_2 such that if $r \leq r_2$ then

$$\|u(0) - \overline{u}(t)\|_D \geq \delta_2, \quad t \in [\varepsilon, T - \varepsilon].$$

Therefore, $\operatorname{dist}(u(0), \gamma) \geq \delta = \min\{\delta_1, \delta_2\}$ if r is small enough, and the statement follows. ∎

Theorems 9.3.7 and 9.3.8 are applicable provided the position of the spectrum of $V(0)$ with respect to the unit circle is known. In the case of the periodic orbits $\overline{u} = u(c)$ arising from Hopf bifurcation (see Theorem 9.3.3) it is in fact possible to study the relation between the spectrum of A and the spectrum of $V(0)$. In the applications, computing the spectrum of A is generally easier than computing the

3. Periodic solutions

spectrum of $V(0)$. Note that from the continuous dependence theorem 8.3.2, if c is close to 0 then $V(0)$ is close to $e^{2\pi A/\omega}$ in $L(X)$. By the perturbation lemma A.3.1, the spectrum of $V(0)$ is close to the spectrum of $e^{2\pi A/\omega}$. So, if the spectrum of A contains elements with positive real part, then by Corollary 2.3.7 the spectrum of $e^{2\pi A/\omega}$ contains elements with modulus greater than 1, and so does the spectrum of $V(0)$, if c is small enough.

The case where the spectrum of A does not have elements with positive real part is more difficult to handle. Indeed, since $\pm\omega i$ are eigenvalues of A, then 1 is a double eigenvalue of e^{tA}. We know already that 1 is an eigenvalue of $V(0)$. By the above perturbation argument, there is another eigenvalue of $V(0)$ close to 1, for c close to 0. In the next theorem we study the behavior of the second eigenvalue.

Theorem 9.3.9 *Let the assumptions of Theorem 9.3.3 hold, and in addition let $F \in C^4(\mathcal{O}_D \times] -1, 1[; X)$. Then there are $c_1 > 0$ and a continuous function $[-c_1, c_1] \mapsto \mathbb{R}$, $c \mapsto \kappa(c)$, such that $\kappa(0) = 0$, and for every $c \in [-c_1, c_1]$, the problem*

$$w' = \rho(c) F_x(u(c), \lambda(c)) w(c) - \kappa(c) w \qquad (9.3.38)$$

has a nontrivial 2π-periodic solution. Moreover,

$$|\kappa(c) + a'(0) c \lambda'(c)| = \epsilon(c) |c \lambda'(c)|, \qquad (9.3.39)$$

where $\lim_{c \to 0} \epsilon(c) = 0$.

Proof — The proof is similar to the proof of Theorem 9.1.11. We use the notation of Theorem 9.3.3. Recalling that the function $h(t) = u'(c)(t) = c\, d/dt(e^{tA} x_0 + v(c))$ satisfies

$$h' = \rho(c) F_x(u(c), \lambda(c)) h, \quad t \in \mathbb{R},$$

we look for a solution of (9.3.38) in the form

$$w(t) = e^{tA} x_0 + z(t) + \frac{\eta}{\kappa c} \frac{d}{dt} u(c)(t), \quad t \in \mathbb{R},$$

where $z \in \mathcal{V}$, and $\eta \in \mathbb{R}$. So, we define the nonlinear function

$$\mathcal{F} : [-c_0, c_0] \times \mathbb{R}^2 \times \mathcal{V} \mapsto C_\#^\alpha(\mathbb{R}; X),$$

$$\mathcal{F}(c, \kappa, \eta, z)(t) = d/dt\, (e^{tA} x_0 + z(t))$$
$$- \rho(c) F_x(u(c), \lambda(c))(e^{tA} x_0 + z(t)) + \kappa(e^{tA} x_0 + z(t)) + \frac{\eta}{c} d/dt\, u(c)(t).$$

Since F is four times continuously differentiable, by Theorem 9.3.3 the mapping $c \mapsto d/dt\, u(c)$ is continuously differentiable with values in $C_\#^\alpha(\mathbb{R}; D) \cap C_\#^{\alpha+1}(\mathbb{R}; X)$. It follows that $\mathcal{F}$ is continuously differentiable. The derivative of $\mathcal{F}$ with respect to (κ, η, z) at $(0, 0, 0, 0)$ is the linear operator

$$(\hat{\kappa}, \hat{\eta}, \hat{z}) \mapsto e^{tA} x_0 \hat{\kappa} + A e^{tA} x_0 \hat{\eta} + \hat{z}'(t) - A \hat{z}(t),$$

396   CHAPTER 9. ASYMPTOTIC BEHAVIOR IN FULLY NONLINEAR EQUATIONS

and it is an isomorphism, as we have shown in the proof of Theorem 9.3.3. So, there are continuously differentiable functions $\kappa(c)$, $\eta(c)$, $z(c)$, defined near $c = 0$, such that
$$\mathcal{F}(c, \kappa(c), \eta(c), z(c)) = 0.$$
If $\kappa(c) = 0$, then $t \mapsto e^{tA}x_0 + z(c)(t)$ and $d/dt\, u(c)$ are linearly independent solutions of (9.3.38), so that 1 is a double eigenvalue of $V(0)$. If $\kappa(c) \neq 0$, then the function $w(t) = e^{tA}x_0 + z(t) + \eta/(\kappa c)\, d/dt\, u(c)(t)$ is a nonzero solution of (9.3.38). The first part of the statement is so proved.

Let us prove that (9.3.39) holds. Differentiate with respect to c the identity
$$\frac{d}{dt}(c(e^{tA}x_0 + v(c))) - \rho(c)F(c(e^{tA}x_0 + v(c)), \lambda(c)) = 0, \quad t \in \mathbb{R},$$
and subtract $\mathcal{F}(c, \kappa(c), \eta(c), z(c)) = 0$ from the resulting equality, to get
$$\kappa(c)(e^{tA}x_0 + z(c)) + \left(\frac{c\rho'(c)}{\rho} + \eta(c)\right)\frac{d}{dt}(e^{tA}x_0 + v(c)) + \\ \lambda'(c)F_\lambda(c(e^{tA}x_0 + v(c)), \lambda(c)) + \Lambda(c)\,(d/dc(cv(c)) - z(c)) = 0, \quad (9.3.40)$$
where
$$\Lambda(c)y = \rho(c)F_x(u(c), \lambda(c))y - y'.$$
The linear mapping
$$(h, \xi, y) \mapsto h(e^{tA}x_0 + z(c)) + \xi\frac{d}{dt}(e^{tA}x_0 + v(c)) + \Lambda(c)y$$
depends continuously on c, and at $c = 0$ it is an isomorphism from $\mathbb{R}^2 \times \mathcal{V}$ onto $C^\alpha_\sharp(\mathbb{R}; X)$. Moreover,
$$\lim_{c \to 0} \frac{F_\lambda(c(e^{tA}x_0 + v(c)), \lambda(c))}{c} = F_{x\lambda}(0, 0)e^{tA}x_0,$$
in $C^\alpha_\sharp(\mathbb{R}; X)$. Therefore, there is a constant C such that for c small enough
$$|\kappa(c)| + \left|\frac{c\rho'(c)}{\rho} + \eta(c)\right| + \|d/dc(cv(c)) - z(c)\|_{C^\alpha_\sharp(\mathbb{R};D)} \leq C|c\lambda'(c)|. \quad (9.3.41)$$
Set now
$$\begin{aligned}g(c) &= \kappa(c)e^{tA}x_0 + \frac{c\rho'(c)}{\rho} + \eta(c)Ae^{tA}x_0 + \lambda'(c)F_\lambda(c(e^{tA}x_0 + v(c)), \lambda(c)) \\ &= -\Lambda(c)\,(d/dc(cv(c)) - z(c)) + \kappa(c)z(c) + c\rho'(c)/\rho + \eta(c)\tfrac{d}{dt}v(c).\end{aligned}$$
Estimate (9.3.41) and the equalities $v(0) = 0$, $z(0) = 0$ imply that
$$\|g(c) + \Lambda(c)(d/dc(cv(c)))\|_{C^\alpha_\sharp(\mathbb{R};X)} = \epsilon(c)|c\lambda'(c)|,$$

3. PERIODIC SOLUTIONS

with $\lim_{c \to 0} \epsilon(c) = 0$. Applying $I - P_{\mathcal{R}}$ we find

$$\left| \frac{\omega}{2\pi} \int_0^{2\pi/\omega} \langle e^{(2\pi/\omega - s)A} g(c)(s), x_0^* \rangle ds \right| \leq \epsilon(c) |c\lambda'(c)|,$$

which yields

$$|\kappa(c) + c\lambda'(c)a'(0)| = \epsilon(c)|c\lambda'(c)|,$$

and the statement follows. ∎

Estimate (9.3.39) implies that there is a neighborhood of $c = 0$ in which $\kappa(c)$ and $c\lambda'(c)$ have the same zeroes, and in which $\kappa(c)$ and $-a'(0)c\lambda'(c)$ have the same sign, if they do not vanish. Therefore, information on the sign of $\kappa(c)$ may be found by computing the derivatives $\lambda^{(k)}(0)$. See Lemma 9.3.4.

9.3.3 Applications to nonlinear parabolic problems, IV

Let us consider a one dimensional fully nonlinear parabolic equation,

$$u_t = f(u, u_x, u_{xx}, \lambda), \quad (t, x) \in \mathbb{R}^2, \tag{9.3.42}$$

where $f : \mathbb{R}^4 \mapsto \mathbb{R}$ is a smooth function such that

$$f(0, 0, 0, \lambda) = 0, \quad f_q(0, 0, 0, \lambda) > 0, \quad \forall \lambda \in \mathbb{R}. \tag{9.3.43}$$

We look for periodic solutions, both with respect to x and with respect to t. We fix the period with respect to x, say $T_x = 2\pi$, and we set

$$X = C_\sharp(\mathbb{R}) = \{\varphi \in C(\mathbb{R}) : \varphi(x) = \varphi(x + 2\pi)\},$$
$$D = C^2(\mathbb{R}) \cap X.$$

Setting as usual $u(t) = u(t, \cdot)$, problem (9.3.42) can be written in the abstract form (9.3.4), with $F(\varphi, \lambda)(x) = f(\varphi(x), \varphi'(x), \varphi''(x), \lambda)$. So, the function $F : D \mapsto X$ is smooth. The operator $A(\lambda) = F_u(0, \lambda)$ is given by

$$A(\lambda)\varphi(x) = f_u(0, 0, 0, \lambda)\varphi(x) + f_p(0, 0, 0, \lambda)\varphi'(x) + f_q(0, 0, 0, \lambda)\varphi''(x).$$

The spectrum of its complexification $\widetilde{A}(\lambda)$ consists of the simple eigenvalues

$$\lambda_k = f_u(0, 0, 0, \lambda) + ik f_p(0, 0, 0, \lambda) - k^2 f_q(0, 0, 0, \lambda), \quad k \in \mathbb{Z},$$

so that $\widetilde{A}(0)$ satisfies assumption (9.3.5) provided there is $h \in \mathbb{N}$ such that

$$f_u(0, 0, 0, 0) = h^2 f_q(0, 0, 0, 0). \tag{9.3.44}$$

If (9.3.44) holds the unique elements of the spectrum of $\widetilde{A}(0)$ on the imaginary axis are the simple eigenvalue $\pm hi$. The nontransversality condition (9.3.10) is satisfied if

$$f_{u\lambda}(0, 0, 0, 0) \neq h^2 f_{q\lambda}(0, 0, 0, 0). \tag{9.3.45}$$

If (9.3.43) and (9.3.45) hold, Theorem 9.3.3 is applicable. It implies that problem (9.3.42) has small periodic solutions for suitable values of λ near 0, with period 2π with respect to x and period near $2h\pi$ with respect to time. To know whether the periodic orbits occur for $\lambda > 0$ or for $\lambda < 0$, we have to compute $\lambda''(0)$, using Lemma 9.3.4, taking into account the fact that $x_0(x) = \cos hx/\sqrt{\pi}$, $y_0(x) = \sin hx/\sqrt{\pi}$.

However, the periodic solutions are orbitally unstable, because (9.3.43) and (9.3.44) imply that $f_u(0,0,0,0) > 0$, so that, taking $k = 0$, $\widetilde{A}(0)$ has at least the eigenvalue $f_u(0,0,0,0)$ with positive real part. Then the spectrum of $V(0)$ has at least an element with modulus greater than 1, see the discussion after Theorem 9.3.8. Since D is compactly embedded in X, then the part of the spectrum of $V(0)$ outside the unit circle consists of isolated eigenvalues. The instability theorem 9.3.8 is so applicable, and it implies that the periodic orbits obtained are orbitally unstable.

9.4 Bibliographical remarks

The geometric theory of this chapter is the natural extension to fully nonlinear problems of the results of D. HENRY [99] about semilinear problems.

The Principle of Linearized Stability for the stationary solutions of fully nonlinear problems has been proved in the present form by A. LUNARDI [149]. The case where X is an interpolation space $D_A(\theta)$ had been considered before by G. DA PRATO – A. LUNARDI [66]. Some stability results have been extended to fully nonlinear integrodifferential equations by A. LUNARDI [138], A. LUNARDI – E. SINESTRARI [154].

The bifurcation results of Subsection 9.2.4 are due to M.G. CRANDALL – P.H. RABINOWITZ [58, 59]. For the Implicit Function Theorem and the Local Inversion Theorem in Banach spaces we refer to J.T. SCHWARTZ [173].

Concerning the center-unstable manifold, the results of G. DA PRATO – A. LUNARDI [66], dealing with the case $E_0 = D_A(\theta)$, $E_1 = D_A(\theta + 1)$, were extended by A. LUNARDI [150] to the case $E_0 = D_A(\theta, \infty)$, $E_1 = D_A(\theta + 1, \infty)$.

Theorem 9.2.6 is due to A. LUNARDI [152].

In the theory of Hopf bifurcation we have followed G. DA PRATO – A. LUNARDI [65]. The study of the stability of the periodic solutions in fully nonlinear problems is due to A. LUNARDI [150].

The bibliographical references for the examples of Subsections 9.1.6, 9.2.4 are contained in Subsections 9.1.6, 9.2.4.

In the quasilinear case some geometric theory results have been proved without the use of optimal regularity. See M. POITIER-FERRY [168], A. LUNARDI [136], A. DRANGEID [77] for the Principle of Linearized Stability, H. AMANN [22] for Hopf bifurcation. G. SIMONETT [174] constructed center manifolds for quasilinear equations with nonconstant domains by means of extrapolation techniques and the optimal regularity results of [63], [26].

Appendix A
Spectrum and resolvent

Throughout this section $X \neq \{0\}$ is a real or complex Banach space. Even in the case where X is a real vector space, we need to deal with complex spectrum and resolvent: so we introduce the complexification of X, defined as

$$\tilde{X} = \{x + iy : x, y \in X\}; \quad \|x + iy\|_{\tilde{X}} = \sup_{0 \leq \theta \leq 2\pi} \|x\cos\theta + y\sin\theta\|.$$

If $A : D(A) \subset X \mapsto X$ is a linear operator, the complexification of A is defined by

$$D(\tilde{A}) = \{x + iy : x, y \in D(A)\}, \quad \tilde{A}(x + iy) = Ax + iAy.$$

In the sequel if no confusion will arise we shall drop out all the tildes, and by spectrum and resolvent of A we shall mean spectrum and resolvent of $\tilde{A}$.

Definition A.0.1 *Let $A : D(A) \subset X \mapsto X$ be a linear operator. The resolvent set $\rho(A)$ and the spectrum $\sigma(A)$ of A are defined by*

$$\rho(A) = \{\lambda \in \mathbb{C} : \exists\, (\lambda I - A)^{-1} \in L(X)\}, \quad \sigma(A) = \mathbb{C}\backslash\rho(A). \tag{A.0.1}$$

The complex numbers $\lambda \in \rho(A)$ such that $\lambda I - A$ is not one to one are called eigenvalues. *The set $\sigma_p(A)$ consisting of all eigenvalues of A is called* point spectrum.

If $\lambda \in \rho(A)$, we set

$$(\lambda I - A)^{-1} = R(\lambda, A). \tag{A.0.2}$$

$R(\lambda, A)$ is called *resolvent operator* or simply *resolvent*.

We state below some properties of the spectrum and the resolvent set.

First, it is clear that if $A : D(A) \subset X \mapsto X$ and $B : D(B) \subset X \mapsto X$ are linear operators such that $R(\lambda_0, A) = R(\lambda_0, B)$ for some $\lambda_0 \in \mathbb{C}$, then $D(A) = D(B)$ and $A = B$. Indeed, $D(A) = \text{Range } R(\lambda_0, A) = \text{Range } R(\lambda_0, B) = D(B)$, and for every $x \in D(A) = D(B)$ we have

$$R(\lambda_0, A)(\lambda_0 x - Ax) = R(\lambda_0, B)(\lambda_0 x - Ax) = R(\lambda_0, B)(\lambda_0 x - Bx)$$

so that $\lambda_0 x - Ax = \lambda_0 x - Bx$, which implies $Ax = Bx$.

Next formula is called *"resolvent identity"*, or else *"first resolvent identity"*, its verification is straightforward:

$$R(\lambda, A) - R(\mu, A) = (\mu - \lambda)R(\lambda, A)R(\mu, A), \quad \forall \, \lambda, \, \mu \in \rho(A). \tag{A.0.3}$$

The resolvent identity characterizes the resolvent operators, in the sense specified by the following proposition.

Proposition A.0.2 *Let $\Omega \subset \mathbb{C}$ be an open set, and let $\{F(\lambda) : \lambda \in \Omega\} \subset L(X)$ be a family of linear operators satisfying the resolvent identity*

$$F(\lambda) - F(\mu) = (\mu - \lambda)F(\lambda)F(\mu), \quad \forall \lambda, \, \mu \in \Omega.$$

Assume that for some $\lambda_0 \in \Omega$, the operator $F(\lambda_0)$ is invertible. Then there exists a linear operator $A : D(A) \subset X \mapsto X$ such that $\rho(A) \supset \Omega$, and $R(\lambda, A) = F(\lambda)$ for $\lambda \in \Omega$.

Proof — Fix any $\lambda_0 \in \Omega$, and set

$$D(A) = \text{Range } F(\lambda_0), \quad Ax = \lambda_0 x - F(\lambda_0)^{-1} x \quad \forall x \in D(A).$$

For $\lambda \in \Omega$ and $y \in X$ the resolvent equation $\lambda x - Ax = y$ is equivalent to $(\lambda - \lambda_0)x + F(\lambda_0)^{-1}x = y$. Applying $F(\lambda)$ we get $(\lambda - \lambda_0)F(\lambda)x + F(\lambda)F(\lambda_0)^{-1}x = F(\lambda)y$. From the resolvent identity we obtain easily $F(\lambda)F(\lambda_0)^{-1} = (\lambda_0 - \lambda)F(\lambda) + I$. Therefore the resolvent equation is uniquely solvable, with $x = F(\lambda)y$. Then $\lambda \in \rho(A)$, and $R(\lambda, A) = F(\lambda)$. ∎

Proposition A.0.3 *Let $\lambda_0 \in \rho(A)$. Then the ball*

$$\{\lambda \in \mathbb{C} : \, |\lambda - \lambda_0| < \|R(\lambda_0, A)\|_{L(X)}^{-1}\}$$

is contained in $\rho(A)$, and

$$\begin{aligned} R(\lambda, A) &= \sum_{n=0}^{\infty} (-1)^n (\lambda - \lambda_0)^n R^{n+1}(\lambda_0, A) \\ &= R(\lambda_0, A)[I + (\lambda - \lambda_0)R(\lambda_0, A)]^{-1}. \end{aligned} \tag{A.0.4}$$

Therefore, the resolvent set $\rho(A)$ is open in $\mathbb{C}$ and $\lambda \mapsto R(\lambda, A)$ is analytic in $\rho(A)$.

Proof — For every $y \in X$ the equation $\lambda x - Ax = y$ is equivalent to $(\lambda - \lambda_0)x + (\lambda_0 - A)x = y$, and setting $z = (\lambda_0 - A)x$, to $z + (\lambda - \lambda_0)R(\lambda_0, A)z = y$. If $\|(\lambda - \lambda_0)R(\lambda_0, A)\|_{L(X)} < 1$, then $I + (\lambda - \lambda_0)R(\lambda_0, A)$ is invertible with bounded inverse, so that

$$z = [I + (\lambda - \lambda_0)R(\lambda_0, A)]^{-1} = \sum_{n=0}^{\infty} (-1)^n (\lambda - \lambda_0)^n R^n(\lambda_0, A),$$

and the statement follows. ∎

Spectrum and resolvent

Corollary A.0.4 *The function $\lambda \mapsto R(\lambda, A)$ has not removable singularities.*

Proof — Assume by contradiction that $\lambda_0 \in \mathbb{C}$ is a removable singularity of $R(\cdot, A)$. Then there is a neighborhood Ω of λ_0 such that $\Omega \setminus \{\lambda_0\}$ is contained in $\rho(A)$, and $R(\cdot, A)$ has an analytic extension to the whole of Ω. In particular, $R(\cdot, A)$ is bounded near λ_0. From Proposition A.0.3 we get, for every $\lambda \in \rho(A)$,

$$\|R(\lambda, A)\|_{L(X)} \geq \frac{1}{\operatorname{dist}(\lambda, \sigma(A))},$$

which is impossible if λ_0 belongs to $\sigma(A)$. ∎

As a consequence of Propositions A.0.2, A.0.3, and Corollary A.0.4, one finds easily that the resolvent set is the biggest domain of analyticity of function $\lambda \mapsto R(\lambda, A)$.

Let us study now the spectral properties of the bounded operators.

Proposition A.0.5 *Let $A \in \mathcal{L}(X)$. The following statements hold.*

(i) $\sigma(A)$ is contained in the circle $C(0, r(A))$ centered at 0 with radius

$$r(A) = \limsup_{n \to \infty} \sqrt[n]{\|A^n\|_{L(X)}}.$$

If $|\lambda| > r(A)$ then $\lambda \in \rho(A)$, and

$$R(\lambda, A) = \sum_{k=0}^{\infty} A^k \lambda^{-k-1}. \tag{A.0.5}$$

If $|\lambda| > \|A\|_{L(X)}$ then

$$\|R(\lambda, A)\|_{L(X)} \leq \frac{1}{|\lambda| - \|A\|_{L(X)}}. \tag{A.0.6}$$

(ii) $\sigma(A)$ is not empty.

Proof — To prove statement (i), define $F(\lambda) = \sum_{k=0}^{\infty} A^k \lambda^{-k-1}$. Since the series $\sum_{k=1}^{\infty} \|A^k\| |\lambda|^{-k-1}$ converges for $|\lambda| > r(A)$, then $F(\lambda)$ does, and it is not difficult to see that $(\lambda I - A) F(\lambda) = F(\lambda)(\lambda I - A) = I$. Therefore, $\lambda \in \rho(A)$ and $F(\lambda) = R(\lambda, A)$. Estimate (A.0.6) follows from the inequality

$$\|F(\lambda)\| \leq \sum_{k=0}^{\infty} |\lambda|^{-k-1} \|A\|^k.$$

Let us prove (ii). Assume by contradiction $\sigma(A) = \emptyset$; then $R(\cdot, A)$ is analytic in $\mathbb{C}$. For every $x \in X$ and $x' \in X'$ the complex function $\langle R(\cdot, A)x, x' \rangle$ is analytic in $\mathbb{C}$ and bounded thanks to (A.0.6), so that it is constant. Hence there exists a linear operator $L \in \mathcal{L}(X)$ such that $R(\lambda, A) = L$ for each $\lambda \in \mathbb{C}$, a contradiction. ∎

$r(A)$ is called *spectral radius* of A.

A.1 Spectral sets and projections

Definition A.1.1 *A subset $\sigma_1 \subset \sigma(A)$ is said to be a* spectral set *if both σ_1 and $\sigma(A)\setminus\sigma_1$ are closed in $\mathbb{C}$.*

Let σ_1 be a bounded spectral set, and let $\sigma_2 = \sigma(A) \setminus \sigma_1$. The distance between σ_2 and σ_1 is positive, so that there exists a bounded open set Ω containing σ_1 and such that its closure is disjoint from σ_2. We may assume that the boundary γ of Ω consists of a finite number of rectifiable closed Jordan curves, oriented counterclockwise. We define a linear bounded operator P by

$$P = \frac{1}{2\pi i} \int_\gamma R(\xi, A) d\xi. \tag{A.1.1}$$

The following proposition holds.

Proposition A.1.2 *Let σ_1 be a bounded spectral set. Then the operator P defined by (A.1.1) is a projection, and $P(X)$ is contained in $D(A^n)$ for every $n \in \mathbb{N}$. Moreover if we set*

$$X_1 = P(X), \quad X_2 = (I-P)(X), \tag{A.1.2}$$

$$\begin{cases} A_1 : X_1 \mapsto X_1, \; A_1 x = Ax \;\; \forall x \in X_1, \\ A_2 : D(A_2) = D(A) \cap X_2 \mapsto X_2, \; A_2 x = Ax \;\; \forall x \in D(A_2), \end{cases} \tag{A.1.3}$$

then $A_1 \in L(X_1)$, and

$$\sigma(A_1) = \sigma_1, \quad \sigma(A_2) = \sigma_2, \tag{A.1.4}$$

$$R(\lambda, A_1) = R(\lambda, A)_{|X_1}, \quad R(\lambda, A_2) = R(\lambda, A)_{|X_2}, \;\; \forall \lambda \in \rho(A). \tag{A.1.5}$$

If $\lambda \in \mathbb{C} \setminus \sigma_1$, and $\Omega \supset \sigma_1$ is an open set with the above properties and such that $\lambda \notin \overline{\Omega}$, then

$$R(\lambda, A_1) = \frac{1}{2\pi i} \int_\gamma R(\xi, A)(\lambda - \xi)^{-1} d\xi. \tag{A.1.6}$$

If $\lambda \in \mathbb{C} \setminus \sigma_2$, and $\Omega \supset \sigma_1$ is an open set with the above properties and such that $\lambda \in \Omega$, then

$$R(\lambda, A_2) = -\frac{1}{2\pi i} \int_\gamma R(\xi, A)(\lambda - \xi)^{-1} d\xi. \tag{A.1.7}$$

Proof — First we show that P is a projection. There exists an open set Ω' whose boundary γ' consists of a finite number of closed rectifiable Jordan curves, such that $\overline{\Omega} \subset \Omega'$, and $\overline{\Omega'} \cap \sigma_2 = \emptyset$. Then we have

$$\int_{\gamma'} (\xi - \lambda)^{-1} d\xi = 2\pi i, \;\; \forall \lambda \in \gamma, \quad \int_\gamma (\xi - \lambda)^{-1} d\xi = 0, \;\; \forall \xi \in \gamma',$$

so that, by (A.0.3),

$$\begin{aligned}P^2 &= \left(\frac{1}{2\pi i}\right)^2 \int_{\gamma'} R(\xi, A)d\xi \int_{\gamma} R(\lambda, A)d\lambda \\ &= \left(\frac{1}{2\pi i}\right)^2 \int_{\gamma' \times \gamma} [R(\lambda, A) - R(\xi, A)](\xi - \lambda)^{-1}d\xi d\lambda \\ &= \left(\frac{1}{2\pi i}\right)^2 \int_{\gamma} R(\lambda, A)d\lambda \int_{\gamma'} (\xi - \lambda)^{-1}d\xi \\ &\quad - \left(\frac{1}{2\pi i}\right)^2 \int_{\gamma'} R(\xi, A)d\xi \int_{\gamma} (\xi - \lambda)^{-1}d\lambda \\ &= P.\end{aligned}$$

From (A.1.1) it follows that X_1 is contained in $D(A^n)$ for every $n \in \mathbb{N}$, and

$$A^n P = \frac{1}{2\pi i}\int_{\gamma} \xi^n R(\xi, A)d\xi. \qquad (A.1.8)$$

It is easy to see that $\rho(A) \subset \rho(A_1) \cap \rho(A_2)$ and that (A.1.5) holds.

For every $\lambda \in \mathbb{C} \setminus \sigma_1$, define $F(\lambda)$ to be the right hand side of (A.1.6). Then

$$\begin{aligned}(\lambda I - A)F(\lambda) &= \frac{1}{2\pi i}\int_{\gamma} [\lambda R(\xi, A) - \xi R(\xi, A) + I](\lambda - \xi)^{-1}d\xi \\ &= \frac{1}{2\pi i}\int_{\gamma} R(\xi, A)d\xi = P.\end{aligned}$$

Similarly,
$$F(\lambda)(\lambda I - A) = P \quad \text{on } D(A).$$

This implies that λ belongs to the resolvent set of A_1, and that $F(\lambda) = R(\lambda, A_1)$.

Let now $\lambda \in \mathbb{C} \setminus \sigma_2$, and define $G(\lambda)$ by the right hand side of (A.1.7). Then

$$\begin{aligned}(\lambda - A)G(\lambda) &= -\frac{1}{2\pi i}\int_{\gamma} [\lambda R(\xi, A) - \xi R(\xi, A) + I](\lambda - \xi)^{-1}d\xi \\ &= \left(\frac{1}{2\pi i}\int_{\gamma} R(\xi, A)d\xi + I\right) = I - P,\end{aligned}$$

and similarly,
$$G(\lambda)(\lambda - A) = I - P \quad \text{on } D(A).$$

Therefore λ belongs to $\rho(A_2)$, and $R(\lambda, A_2) = G(\lambda)$, i.e. (A.1.7) holds. Let now λ belong to $\rho(A_1) \cap \rho(A_2)$. Then it is easy to check that $\lambda \in \rho(A)$, with $R(\lambda, A) = R(\lambda, A)(I - P) + R(\lambda, A)P$. This implies that $\sigma(A_1) = \sigma_1$, $\sigma(A_2) = \sigma_2$, and (A.1.4) is proved. ∎

A.2 Isolated points of the spectrum

We are interested here in the behavior of $R(\lambda, A)$ near isolated points of the spectrum of A.

Proposition A.2.1 *Let $\lambda_0 \in \sigma(A)$ be isolated in $\sigma(A)$, and let $\varepsilon > 0$ be such that the open ball centered at λ_0 with radius ε is contained in $\rho(A)$. Then for $0 < \lambda - \lambda_0 < \varepsilon$*

$$R(\lambda, A) = \sum_{n=0}^{\infty} (-1)^n S^{n+1} (\lambda - \lambda_0)^n + P(\lambda - \lambda_0)^{-1} + \sum_{n=1}^{\infty} D^n (\lambda - \lambda_0)^{-n-1}, \quad (A.2.1)$$

where

$$\begin{cases} P = \dfrac{1}{2\pi i} \displaystyle\int_{\gamma} R(\xi, A) d\xi, \quad D = (A - \lambda_0 I) P, \\ S = \dfrac{1}{2\pi i} \displaystyle\int_{\gamma} R(\xi, A)(\xi - \lambda_0)^{-1} d\xi = \lim_{\lambda \to \lambda_0} (I - P) R(\lambda, A) \end{cases} \quad (A.2.2)$$

and $\gamma = C(\lambda_0, r)$, with any $r < \varepsilon$.

Proof — First we show that

$$\lim_{\lambda \to \lambda_0} (I - P) R(\lambda, A) = \frac{1}{2\pi i} \int_{\gamma} R(\xi, A)(\xi - \lambda_0)^{-1} d\xi.$$

For $0 < |\lambda - \lambda_0| < \varepsilon$, choosing $r \in \,]|\lambda - \lambda_0|, \varepsilon[$ we get

$$\begin{aligned} PR(\lambda, A) &= \frac{1}{2\pi i} \int_{\gamma} R(\xi, A) R(\lambda, A) d\xi \\ &= \frac{1}{2\pi i} \int_{\gamma} R(\lambda, A)(\xi - \lambda)^{-1} d\xi - \frac{1}{2\pi i} \int_{\gamma} R(\xi, A)(\xi - \lambda)^{-1} d\xi \\ &= R(\lambda, A) - \frac{1}{2\pi i} \int_{\gamma} R(\xi, A)(\xi - \lambda)^{-1} d\xi \end{aligned} \quad (A.2.3)$$

and the last integral goes to S as λ goes to λ_0. Moreover one can show by a recurrence argument that for every $n \in \mathbb{N} \cup \{0\}$

$$\frac{1}{2\pi i} \int_{\gamma} R(\xi, A)(\xi - \lambda_0)^{-n-1} d\xi = (-1)^n S^{n+1}, \quad (A.2.4)$$

and for every $n \in \mathbb{N}$

$$\frac{1}{2\pi i} \int_{\gamma} R(\xi, A)(\xi - \lambda_0)^n d\xi = [(A - \lambda_0) P]^n. \quad (A.2.5)$$

Recall now that $\lambda \mapsto R(\lambda, A)$ is analytic with values in $L(X)$ in the set $\{\lambda \in \mathbb{C} : 0 < |\lambda - \lambda_0| < \varepsilon\}$. The Laurent development of $R(\lambda, A)$ near $\lambda = \lambda_0$ is

$$R(\lambda, A) = \sum_{n=0}^{\infty} (\lambda - \lambda_0)^n \frac{1}{2\pi i} \int_\gamma R(\xi, A)(\xi - \lambda_0)^{-n-1} d\xi$$
$$+ (\lambda - \lambda_0)^{-1} \frac{1}{2\pi i} \int_\gamma R(\xi, A) d\xi + \sum_{n=0}^{\infty} (\lambda - \lambda_0)^{-n-1} \frac{1}{2\pi i} \int_\gamma R(\xi, A)(\xi - \lambda_0)^n d\xi,$$

and (A.2.1) follows from (A.2.4), (A.2.5). ∎

Formula (A.2.1) is therefore the Laurent development of $R(\cdot, A)$ near $\lambda = \lambda_0$. The projection P is the residue of $R(\lambda, A)$ at $\lambda = \lambda_0$. Let the subspaces X_1, X_2 be defined by (A.1.2), with $\sigma_1 = \{\lambda_0\}$, and let the operators A_1, A_2 be defined by (A.1.3). By Proposition A.1.2 we get

$$\sigma(A_1) = \{\lambda_0\}, \quad \sigma(A_2) = \sigma(A) \setminus \{\lambda_0\}; \tag{A.2.6}$$

$$R(\lambda_0, A_2) = S, \tag{A.2.7}$$

where S is defined in (A.2.2).

In general, the subspace X_1 does not coincide with the kernel of $\lambda_0 I - A$, but the following proposition holds.

Proposition A.2.2 *Let λ_0 be an isolated point of $\sigma(A)$. Then $X_1 \supset \mathrm{Ker}(\lambda_0 I - A)$, and $X_2 \subset \mathrm{Range}(\lambda_0 I - A)$. Moreover, the following conditions are equivalent.*

(i) $X_1 = \mathrm{Ker}\,(\lambda_0 I - A)$;

(ii) $X_2 = \mathrm{Range}\,(\lambda_0 I - A)$;

(iii) λ_0 is a simple pole of $\lambda \mapsto R(\lambda, A)$ (that is, $D = 0$);

(iv) $\mathrm{Range}\,(\lambda_0 I - A)$ is closed, and

$$X = \mathrm{Ker}(\lambda_0 I - A) \oplus \mathrm{Range}(\lambda_0 I - A).$$

Proof — Let $x \in \mathrm{Ker}\,(\lambda_0 I - A)$. Then $\lambda_0 x - Ax = 0$, so that $R(\xi, A)x = (\xi - \lambda_0)^{-1} x$ for $\xi \in \rho(A)$. It follows that

$$Px = \frac{1}{2\pi i} \int_\gamma (\xi - \lambda_0)^{-1} x \, d\xi = x.$$

Therefore, $\mathrm{Ker}\,(\lambda_0 I - A) \subset X_1$. Let now $x \in X_1$: then $x = Px$, so that $(\lambda_0 I - A)x = (\lambda_0 I - A)Px$, and x belongs to $\mathrm{Ker}\,(\lambda_0 I - A)$ if and only if $Dx = 0$. This implies that $X_1 = \mathrm{Ker}\,(\lambda_0 I - A) \Leftrightarrow D = 0$, that is $(i) \Leftrightarrow (ii)$. Since $\lambda_0 \in \rho(A_2)$ due to Proposition A.1.2, then $\mathrm{Range}\,(\lambda_0 I - A) \supset X_2$. Let now $x = \lambda_0 y - Ay$, with $y \in D(A)$. Then $Px = (\lambda_0 - A)Py = Dy$, so that x belongs to X_2 if and only if $Dy = 0$. This implies that $X_2 = \mathrm{Range}\,(\lambda_0 I - A) \Leftrightarrow D = 0$, that is $(ii) \Leftrightarrow (iii)$.

Let (iii) hold. Then Range $(\lambda_0 I - A) =$ Range $(I - P)$ is closed because $I - P$ is a projection. Moreover, due to (i) and (ii), we have $X = $ Ker $(\lambda_0 I - A) \oplus$ Range $(\lambda_0 I - A)$, so that (iv) holds.

Conversely, if (iv) holds, then Range $(\lambda_0 I - A)$, endowed with the norm of X, is a Banach space, and $D(A) \cap$ Range $(\lambda_0 I - A)$, endowed with the graph norm of A, is a Banach space. Moreover, the part of $\lambda_0 I - A$ in Range $(\lambda_0 I - A)$, defined by $A_1 : D(A) \cap$ Range $(\lambda_0 I - A) \mapsto$ Range $(\lambda_0 I - A)$, $A_1 x = \lambda_0 x - Ax$, is one to one, so that its inverse is bounded. So, $\lambda \mapsto R(\lambda, A)_{|\text{Range}(\lambda_0 I - A)}$ is holomorphic at $\lambda = \lambda_0$. The part of $\lambda_0 I - A$ in Ker $(\lambda_0 I - A)$ is the null operator, and for λ close to λ_0 we have $R(\lambda, A)_{|\text{Ker}(\lambda_0 I - A)} = (\lambda - \lambda_0)^{-1} I_{|\text{Ker}(\lambda_0 I - A)}$. Since $X = $ Ker $(\lambda_0 I - A) \oplus$ Range $(\lambda_0 I - A)$, then λ_0 is a simple pole of $R(\cdot, A)$, that is (iii) holds. ∎

Definition A.2.3 *An isolated eigenvalue $\lambda_0 \in \sigma(A)$ is said to be a* semisimple eigenvalue *if one of the equivalent conditions of Proposition A.2.2 is satisfied.*

Remark A.2.4 From Proposition A.2.2 it follows that if λ_0 is a semisimple eigenvalue, then Ker $(\lambda_0 I - A) \cap$ Range $(\lambda_0 I - A) = \{0\}$, so that

$$(v) \quad \text{Ker}\,(\lambda_0 I - A) = \text{Ker}\,(\lambda_0 I - A)^2.$$

Let λ_0 be an isolated eigenvalue. Then condition (v) is not equivalent to (i), (ii), (iii), (iv) of Proposition A.2.2. See next example. However, if λ_0 is a pole of $R(\cdot, A)$ and (v) holds, then it is a simple pole: actually, if λ_0 is a pole of $R(\cdot, A)$ of order n then $D^{n+1} \neq 0$ and $D^n = 0$, so that there is $x \in X$ such that $(A - \lambda_0)^{n-1} Px \neq 0$, $(A - \lambda_0)^n Px = 0$, and this contradicts (v) unless $n = 1$. Hence, if λ_0 is a pole of $R(\cdot, A)$ (which is true, for instance, if X is finite dimensional) then conditions (i) through (v) are equivalent.

Example A.2.5 Let $X = \{f \in C([0,1]; \mathbb{C}) : f(0) = 0\}$ be endowed with the sup norm, and let $A : X \mapsto X$ be defined by

$$Af(t) = \int_0^t f(s)ds, \ 0 \leq t \leq 1.$$

As easily seen, 0 is the unique element of the spectrum of A, and Ker $A = $ Ker $A^2 = \{0\}$, so that $0 \in \mathbb{C}$ is not an eigenvalue of A. If 0 would be a simple pole of $R(\cdot, A)$, then it would be an eigenvalue of A. Therefore, (iii) does not hold.

Set now $Y = \{u \in C([0,1]; \mathbb{C}) : u(0) = 0\} \times \mathbb{C}$ be endowed with the product norm, and let $B : Y \mapsto Y$ be defined as

$$B(f, z) = (Af, 0).$$

Again, $\sigma(B) = \{0\}$, and Ker $B = $ Ker $B^2 = \{0\} \times \mathbb{C}$, so that 0 is an isolated eigenvalue and condition (v) holds. If 0 would be a simple pole of $R(\cdot, B)$ then it would be a simple pole of $R(\cdot, A)$. So, condition (iii) is not satisfied, and 0 is not a semisimple eigenvalue of B. Moreover, Range $B = \{(g, 0) : g \in C^1([0,1];\mathbb{C}),\ g(0) = 0\}$ is not closed in Y, and its closure is the set $\{(g,0) : g \in C([0,1];\mathbb{C}),\ g(0) = 0\}$, so that $Y \neq \mathrm{Ker}\, B \oplus \overline{\mathrm{Range}\, B}$.

Definition A.2.6 *Let $\lambda_0 \in \sigma(A)$ be an eigenvalue. Then*

(i) *If x belongs to $D(A^k)$ for some $k \in \mathbb{N}$, and $(\lambda_0 I - A)^k x = 0$, then x is called a* generalized eigenvector *of A.*

(ii) *If there is $k \in \mathbb{N}$ such that $\mathrm{Ker}\,(\lambda_0 I - A)^k = \mathrm{Ker}\,(\lambda_0 I - A)^{k+1}$, the minimum of such k is called the* index *of λ_0.*

(iii) *The dimension of $\mathrm{Ker}\,(\lambda_0 I - A)$ is called the* geometric multiplicity *of λ_0. The dimension of the space of the generalized eigenvectors is called the* algebraic multiplicity *of λ_0.*

Therefore, the algebraic multiplicity of any eigenvalue is greater or equal to its geometric multiplicity. By Remark 0.3.10, the index of any semisimple eigenvalue is 1, and the geometric multiplicity of λ_0 is equal to its algebraic multiplicity. The name "algebraic multiplicity" comes from the fact that if $\dim X < \infty$, the dimension of the space of generalized eigenvectors is equal to the multiplicity of λ_0 as a root of the characteristic polynomial $\lambda \mapsto \det(A - \lambda I)$.

Definition A.2.7 *A semisimple eigenvalue λ_0 is said to be an (algebraically)* simple eigenvalue *if its algebraic multiplicity (and hence also its geometric multiplicity) is equal to 1.*

In the applications, the explicit computation of P is often not very easy. Next lemma may be of help.

Lemma A.2.8 *Let λ_0 be a simple eigenvalue of A. Then P is the unique projection onto $\mathrm{Ker}\,(\lambda_0 I - A)$ which commutes with A.*

Proof — Let Q be any projection onto $\mathrm{Ker}(\lambda_0 I - A)$ which commutes with A, that is
$$AQx = QAx \quad \forall x \in D(A).$$
Let $x_0 \neq 0 \in \mathrm{Ker}(\lambda_0 I - A)$, and for every $x \in X$ set
$$Px = \alpha(x)x_0, \quad Qx = \beta(x)x_0.$$
Since Q commutes with A, then for every $x \in D(A)$ it holds $\lambda_0 \beta(x) x_0 = \beta(Ax)x_0$, so that $\beta(\lambda_0 x - Ax) = 0$. Therefore Q vanishes on $\mathrm{Range}(\lambda_0 I - A)$. Moreover, since Q is a projection, then $\beta(x_0) = 1$.

Let $x \in X$. Then
$$Qx = QPx + Q(I-P)(X) = QPx = \beta(\alpha(x)x_0) = \alpha(x) = Px.$$

∎

A.3 Perturbation results

We shall see that some spectral properties depend continuously on the operator A. So, we fix once and for all a linear closed operator $A_0 : D \subset X \mapsto X$. Its domain D is endowed with the graph norm.

Lemma A.3.1 *Let $\lambda \in \rho(A_0)$. There is $\delta > 0$ such that $\lambda \in \rho(A)$ for every operator $A \in L(D, X)$ such that $\|A - A_0\|_{L(D,X)} \leq \delta$. Moreover, $R(\lambda, A) - R(\lambda, A_0) \to 0$ in $L(X)$ and in $L(X, D)$ as $A \to A_0$ in $L(D, X)$.*

Proof — For every $y \in X$ the equation

$$\lambda x - Ax = y \tag{A.3.1}$$

is equivalent to

$$\lambda x - A_0 x = (A - A_0)x + y,$$

and, setting $u = \lambda x - A_0 x$, to

$$u = (A - A_0)R(\lambda, A_0)u + y. \tag{A.3.2}$$

If A is so close to A_0 that $\|A - A_0\|_{L(D,X)} \leq 1/2 \|R(\lambda, A_0)\|_{L(X,D)}$, then equation (A.3.1) has a unique solution u, and $\|u\| \leq 2\|y\|$. Consequently, (A.3.1) has a unique solution $x = R(\lambda, A_0)u$, and $\|x\| \leq 2\|R(\lambda, A_0)\|_{L(X)}\|y\|$. Therefore λ belongs to $\rho(A)$.

The last assertion follows from the equality

$$R(\lambda, A) - R(\lambda, A_0) = R(\lambda, A)(A - A_0)R(\lambda, A_0).$$

∎

Also the simple isolated eigenvalues depend continuously on A, in the sense of next proposition.

Proposition A.3.2 *Let λ_0 be an algebraically simple isolated eigenvalue of A_0. There are r, $\delta > 0$ such that for every $A \in L(D, X)$ with $\|A - A_0\|_{L(D,X)} \leq \delta$, $\sigma(A)$ has a unique isolated element $\lambda = \lambda(A)$ in the circle $\{z \in \mathbb{C} : |z - \lambda_0| \leq r\}$. Moreover, λ is an algebraically simple eigenvalue of A, and*

$$\lim_{A \to A_0} \lambda(A) = \lambda_0.$$

Proof — Since λ_0 is isolated in $\sigma(A_0)$, then there is $r > 0$ such that the set $\{r/2 \leq |z - \lambda_0| \leq 2r\}$ is contained in $\rho(A_0)$. By Lemma A.3.1, if A is sufficiently close to A_0 in $L(D, X)$, say $\|A - A_0\|_{L(D,X)} \leq \delta_0$, then the circumference $C(\lambda_0, r)$ is contained in $\rho(A)$. So, the projection

$$P_A = \frac{1}{2\pi i} \int_{C(\lambda_0, r)} R(z, A)dz$$

is well defined, and by Lemma A.3.1 $P_A \to P_{A_0}$ in $L(X, D)$ as $A \to A_0$.

Set now
$$U_A = P_A P_{A_0} + (I - P_A)(I - P_{A_0}).$$
Since $\lim_{A \to A_0} U_A = I$, then U_A is invertible for A close to A_0, say for $\|A - A_0\|_{L(D,X)} \leq \delta_1$. From the equality
$$U_A P_{A_0} = P_A P_{A_0} = P_A U_A$$
one gets
$$P_A = U_A P_{A_0} (U_A)^{-1}$$
for $\|A - A_0\|_{L(D,X)} \leq \delta_2$. It follows that $P_A(X)$ is one dimensional, and if $\operatorname{Ker}(\lambda_0 I - A_0) = P_{A_0}(X)$ is spanned by x_0, then $P_A(X)$ is spanned by $x = U_A x_0 = P_A x_0$. Since $A(P_A(X))$ is contained in $P_A(X)$, then there is $\lambda \in \mathbb{C}$ such that
$$Ax = \lambda x,$$
so that x is an eigenvector of A with eigenvalue λ.

The complex number λ satisfies $|\lambda - \lambda_0| < r$. Indeed, from the equality $Ax = \lambda x$ it follows that $R(z, A)x = x/(z - \lambda)$ for $|z - \lambda_0| = r$, and hence $x = P_A x = \frac{1}{2\pi i} \int_{C(\lambda_0, r)} x/(z - \lambda) dz$ implies that $|\lambda - \lambda_0| < r$. Applying now Proposition A.1.2, we get that λ is the unique element of the spectrum of A in the circle $\{|z - \lambda_0| < r\}$.

The last assertion follows from Lemma A.3.1. ■

Bibliography

[1] P. ACQUISTAPACE: *Zygmund Classes with Boundary Conditions as Interpolation Spaces*, Proceedings of the conference "Trends in Semigroup Theory and Applications" (Trieste 1987), P. Clément, E. Mitidieri, S. Vrabie Eds. M. Dekker, New York (1989), 1–19.

[2] P. ACQUISTAPACE: *Evolution operators and strong solutions of abstract linear parabolic equations*, Diff. Int. Eqns. **1** (1988), 433–457.

[3] P. ACQUISTAPACE: *Abstract Linear Nonautonomous Parabolic Equations: A Survey*, in: "Differential Equations in Banach Spaces", Proceedings, Bologna 1991, G. Dore, A. Favini, E. Obrecht, A. Venni Eds. Lect. Notes in Pure and Applied Math. **148**, M. Dekker, New York (1993), 1–19.

[4] P. ACQUISTAPACE, B. TERRENI: *Some existence and regularity results for abstract non-autonomous parabolic equations*, J. Math. Anal. Appl. **99** (1984), 9–64.

[5] P. ACQUISTAPACE, B. TERRENI: *Maximal Space Regularity for Abstract Linear Non-autonomous Parabolic Equations*, J. Funct. Anal. **60** (1985), 168–210.

[6] P. ACQUISTAPACE, B. TERRENI: *Linear Parabolic Equations in Banach Spaces with Variable Domains but Constant Interpolation Spaces*, Ann. Sc. Norm. Sup. Pisa, Serie IV, **13** (1986), 75–107.

[7] P. ACQUISTAPACE, B. TERRENI: *Hölder classes with boundary conditions as interpolation spaces*, Math. Z. **195** (1987), 451–471.

[8] P. ACQUISTAPACE, B. TERRENI: *A unified approach to abstract linear non-autonomous parabolic equations*, Rend. Sem. Mat. Univ. Padova **78** (1987), 47–107.

[9] P. ACQUISTAPACE, B. TERRENI: *Fully nonlinear parabolic systems*, in "Recent Advances in Nonlinear Elliptic and Parabolic Problems", Nancy 1988, Ph. Bénilan, M. Chipot, L. Evans, M. Pierre Eds. Pitman Res. Notes in Math. Series **208**, 97–111, Longman, Harlow (1989).

[10] R.A. ADAMS: *Sobolev spaces*, Academic Press, New York (1985).

[11] J.O.-O. ADEYEYE: *Characterization of real interpolation spaces between the domain of the Laplace operator and $L_p(\Omega)$; Ω polygonal and applications*, J. Maths. Pures Appl. **67** (1988), 263–290.

[12] S. AGMON: *On the eigenfunctions and the eigenvalues of general elliptic boundary value problems*, Comm. Pure Appl. Math. **15** (1962), 119–147.

[13] S. AGMON, A. DOUGLIS, L. NIRENBERG: *Estimates near the boundary for solutions of elliptic partial differential equations satisfying general boundary conditions*, Comm. Pure Appl. Math. **12** (1959), 623–727.

[14] M.S. AGRANOVICH, M.I. VISHIK: *Elliptic problems with a parameter and parabolic problems of general type*, Uspehi Mat. Nauk **19** (1964), 53–161 (Russian); English translation: Russian Math. Surveys **19** (1964), 53–157.

[15] F. ALABAU, A. LUNARDI: *Behavior near the travelling wave solution of a free boundary system in Combustion Theory*, Dyn. Syst. Appl. **1** (1992), 391–417.

[16] H. AMANN: *Dual semigroups and second order linear elliptic value problems*, Israel J. Math. **45** (1983), 25–54.

[17] H. AMANN: *Existence and regularity for semilinear parabolic evolution equations*, Ann. Sc. Norm. Sup. Pisa, Serie IV, **XI** (1984), 593–676.

[18] H. AMANN: *Global existence for semilinear parabolic systems*, J. Reine Ang. Math. **360** (1985), 47–83.

[19] H. AMANN: *Quasilinear evolution equations and parabolic systems*, Trans. Amer. Math. Soc. **293** (1986), 191–227.

[20] H. AMANN: *On abstract parabolic fundamental solutions*, J. Math. Soc. Japan **39** (1987), 93–116.

[21] H. AMANN: *Parabolic Evolution Equations in Interpolation and Extrapolation Spaces*, J. Funct. Anal. **78** (1988), 233–270.

[22] H. AMANN: *Hopf bifurcation in quasilinear reaction-diffusion systems*, in "Delay Differential Equations and Dynamical Systems", Proceedings. Lect. Notes in Math. **1455**, Springer Verlag, Berlin (1991).

[23] H. AMANN: *Highly degenerate quasilinear parabolic systems*, Ann. Sc. Norm. Sup. Pisa, Serie IV **18** (1991), 135–166.

[24] H. AMANN: *Linear and Quasilinear Parabolic Problems*, Birkhäuser Verlag, to appear.

[25] S. ANGENENT: *Analyticity of the interface of the porous media equation after the waiting time*, Proc. Amer. Math. Soc. **102** (1988), 329–336.

[26] S. ANGENENT: *Nonlinear Analytic Semiflows*, Proc. Royal Soc. Edinb. **115A** (1990), 91–107.

[27] A. ARDITO, P. RICCIARDI: *Existence and regularity for linear delay partial differential equations*, Nonlinear Analysis T.M.A. **4** (1980), 411–414.

[28] J.B. BAILLON: *Caractère borné de certains générateurs de semigroupes linéaires dans les espaces de Banach*, C. R. Acad. Sci. Paris **290** (1980), 757–760.

[29] A.V. BALAKRISHNAN: *Applied functional analysis*, Springer Verlag, New York (1976).

[30] B. BEAUZAMY: *Espaces d'interpolation réels: topologie et géometrie*, Springer Verlag, Berlin (1978).

[31] P. BENILAN, K.S. HA: *Equations d'évolution du type $du/dt + \beta \partial \varphi(u) \ni 0$ dans $L^\infty(\Omega)$*, C.R. Acad. Sci. Paris **281** (1975), 947–950.

[32] J. BERGH, J. LÖFSTRÖM: *Interpolation Spaces. An Introduction*, Springer Verlag, Berlin (1976).

[33] H. BERENS, P.L. BUTZER: *Approximation theorems for semigroup operators in intermediate spaces*, Bull. Amer. Math. Soc. **70** (1964), 689–692.

[34] C.-M. BRAUNER, J. BUCKMASTER, J.W. DOLD, C. SCHMIDT-LAINÉ: *On an evolution equation arising in Detonation Theory*, in: "Fluid Dynamical Aspects of Combustion Theory", M. Onofri & A. Tesei Eds. Pitman Res. Notes Math. Series **223** (1991), 196–210.

[35] C.-M. BRAUNER, A. LUNARDI, C. SCHMIDT-LAINÉ: *Stability of travelling waves with interface conditions*, Nonlinear Analysis T.M.A. **19** (1992), 455–474.

[36] C.-M. BRAUNER, S. GERBI, C. SCHMIDT-LAINÉ: *A stability analysis for a fully nonlinear parabolic problem in detonation theory*, Applic. Anal. (to appear).

[37] H. BREZIS: *Opérateurs maximaux monotones et semi-groupes de contractions dans les espaces de Hilbert*, North-Holland, Amsterdam (1973).

[38] F.E. BROWDER: *On the spectral theory of elliptic differential operators. I*, Math. Annalen **142** (1961), 22–130.

[39] YU. BRUDNYI, N. KRUGLJAK: *Interpolation functors and interpolation spaces*, North-Holland, Amsterdam (1991).

[40] A. BUTTU: *On the evolution operator for a class of parabolic nonautonomous equations*, J. Math. Anal. Appl. **170** (1992), 115–137.

[41] A. BUTTU: *A construction of the evolution operator for a class of abstract parabolic equations*, Dyn. Syst. Appl. **3** (1994), 221–234.

[42] P.L. BUTZER, H. BERENS: *Semigroups of Operators and Approximation*, Springer Verlag, Berlin (1967).

[43] J.W. CAHN: *On spinodal decomposition*, Acta Metall. **9** (1961), 795–801.

[44] J.W. CAHN, J. E. HILLIARD: *Free energy of a non-uniform system, I. Interfacial free energy*, J. Chem. Phys. **2** (1958), 258–267.

[45] S. CAMPANATO: *Generation of analytic semigroups, in the Hölder topology, by elliptic operators of second order with Neumann boundary condition*, Le Matematiche **35** (1980), 61–72.

[46] S. CAMPANATO: *Generation of analytic semigroups by elliptic operators of second order in Hölder spaces*, Ann. Sc. Norm. Sup. Pisa **8** (1981), 495–512.

[47] P. CANNARSA, V. VESPRI: *Analytic semigroups generated on Hölder spaces by second order elliptic systems under Dirichlet boundary conditions*, Ann. Mat. Pura Appl. (IV) **CXL** (1985), 393–415.

[48] P. CANNARSA, V. VESPRI: *Generation of analytic semigroups by elliptic operators with unbounded coefficients*, SIAM J. Math. Anal. **18** (1987), 857–872.

[49] P. CANNARSA, V. VESPRI: *Generation of analytic semigroups in the L^p topology by elliptic operators in $\mathbb{R}^n$*, Israel J. Math. **61** (1988), 235–255.

[50] P. Cannarsa, B. Terreni, V. Vespri, *Analitic semigroups generated by non-variational elliptic systems of second order under Dirichlet boundary conditions*, J. Math. Anal. Appl. **112** (1985), 56–103.

[51] Z. Chen: *Long time small solutions to nonlinear parabolic equations*, Arkiv för Matematik **28** (1990), 371–381.

[52] Z. Chen: *A specifically fully nonlinear perturbation of the heat equations in exterior domains*, Nonlinear Analysis T.M.A. **19** (1992), 475–485.

[53] Y-G. Chen, Y. Giga, S. Goto: *Uniqueness and existence of viscosity solutions of generalized mean curvature flow equations*, J. Diff. Geom. **33** (1991), 749–786.

[54] Ph. Clément, H.J.A.M. Heijmans, S. Angenent, C.J. van Duijn, B. de Pagter: *One-Parameter Semigroups*, North-Holland, Amsterdam (1987).

[55] F. Colombo, V. Vespri: *Generation of analytic semigroups in C^k and $W^{k,p}$ spaces*, Diff. Int. Eqns. (to appear).

[56] C. Cosner, J. Hernandez, E. Mitidieri: *Reaction diffusion systems and applications*, Birkhauser Verlag, to appear.

[57] M.G. Crandall, H. Ishii, P.-L. Lions: *A user's guide to viscosity solutions of nonlinear PDE's*, Bull. Amer. Math. Soc. **27** (1992), 1–67.

[58] M.G. Crandall, P.H. Rabinowitz: *Bifurcation from simple eigenvalues*, J. Funct. Anal. **8** (1971), 321–340.

[59] M.G. Crandall, P.H. Rabinowitz: *Bifurcation, Perturbation of Simple Eigenvalues, and Linearized Stability*, Arch. Rat. Mech. Anal. **52** (1973), 161–180.

[60] M.G. Crandall, P.H. Rabinowitz: *The Hopf Bifurcation Theorem in Infinite Dimensions*, Arch. Rat. Mech. Anal. **68** (1978), 53–72.

[61] D. Daners, P. Koch Medina: *Abstract evolution equations, periodic problems and applications*, Pitman Res. Notes in Math. Ser. **275**, Longman, Harlow (1992).

[62] G. Da Prato, P. Grisvard: *Sommes d'opérateurs linéaires et équations différentielles opérationelles*, J. Maths. Pures Appliquées **54** (1975), 305–387.

[63] G. Da Prato, P. Grisvard: *Equations d'évolution abstraites non linéaires de type parabolique*, Ann. Mat. Pura Appl. (IV) **120** (1979), 329–396.

[64] G. Da Prato, P. Grisvard: *Maximal regularity for evolution equations by interpolation and extrapolation*, J. Funct. Anal. **58** (1984), 107–124.

[65] G. Da Prato, A. Lunardi: *Hopf bifurcation for fully nonlinear equations in Banach space*, Ann. Inst. Henri Poincaré **3** (1986), 315–329.

[66] G. Da Prato, A. Lunardi: *Stability, Instability, and Center Manifold Theorem for Fully Nonlinear Autonomous Parabolic Equations in Banach Space*, Arch. Rat. Mech. Anal. **101** (1988), 115–141.

[67] G. Da Prato, E. Sinestrari: *Hölder regularity for nonautonomous abstract parabolic equations*, Israel J. Math. **42** (1982), 1–19.

[68] G. Da Prato, E. Sinestrari: *Differential Operators with Nondense Domains*, Ann. Sc. Norm. Sup. Pisa **14** (1987), 285–344.

BIBLIOGRAPHY

[69] E.B. DAVIES: *Heat kernels and spectral theory*, Cambridge Univ. Press, Cambridge (1989).

[70] A. DEBUSSCHE, L. DETTORI: *On the Cahn-Hilliard equation with a logarithmic free energy*, Nonlinear Analysis T.M.A. (to appear).

[71] W. DESCH, W. SCHAPPACHER: *Some Perturbation Results for Analytic Semigroups*, Math. Ann. **281** (1988), 157–162.

[72] G. DI BLASIO: *Linear Parabolic Evolution Equations in L^p-Spaces*, Ann. Mat. Pura Appl. **138** (1984), 55–104.

[73] G. DI BLASIO: *Non autonomous functional differential equations in Hilbert spaces*, Nonlinear Analysis T.M.A. **12** (1985), 1367–1380.

[74] G. DI BLASIO, K. KUNISCH, E. SINESTRARI: L^2-*regularity for parabolic partial integrodifferential equations with delay in the highest order derivatives*, J. Math. Anal. Appl. **102** (1984), 38–57.

[75] G. DI BLASIO, K. KUNISCH, E. SINESTRARI: *Stability for abstract linear functional differential equations*, Israel J. Math. **50** (1985), 231–263.

[76] G. DORE, A. VENNI: *On the Closedness of the Sum of two closed Operators*, Math. Z. **196** (1987), 189–201.

[77] A.K. DRANGEID: *The principle of linearized stability for quasilinear parabolic evolution equations*, Nonlinear Analysis T.M.A. **13** (1989), 1091–1113.

[78] B. EBERHARDT, G. GREINER: *Baillon's Theorem on Maximal Regularity*, Acta Appl. Math. **27** (1992), 47–54.

[79] S. D. EIDEL'MAN: *Parabolic systems*, Nauka, Moskow (1964) (Russian); English translation: North-Holland, Amsterdam (1969).

[80] C.M. ELLIOT, S. LUCKHAUS: *A generalized equation for phase separation of a multicomponent mixture with interfacial free energy*, preprint.

[81] L.C. EVANS, J. SPRUCK: *Motion of level sets by mean curvature. I*, J. Diff. Geom. **33** (1991), 635–681.

[82] L.C. EVANS, J. SPRUCK: *Motion of level sets by mean curvature. II*, Trans. Amer. Math. Soc. **330** (1992), 321–332.

[83] A. FRIEDMAN: *Partial Differential Equations of Parabolic Type*, Prentice-Hall, Englewood Cliffs, NJ (1964).

[84] A. FRIEDMAN: *Partial Differential Equations*, Holt, Rinehart & Winston, New York (1969).

[85] H. FUJITA: *On the blowing up of solutions of the Cauchy problem for $u_t = \Delta u + u^{1+\alpha}$*, J. Fac. Sci. Univ. Tokyo, Sect. 1, **13** (1966), 109–124.

[86] H. FUJITA, T. KATO: *On the Navier-Stokes initial value problem I*, Arch. Rat. Mech. Anal. **16** (1964), 269–315.

[87] M. FUHRMAN: *Bounded solutions for abstract time-periodic parabolic equations with non-constant domains*, Diff. Int. Eqns. **4** (1991), 493–518.

[88] M. GAGE, R.S. HAMILTON: *The heat equation shrinking convex plane curves*, J. Diff. Geom. **23** (1986), 69–96.

[89] G. GEYMONAT, P. GRISVARD: *Alcuni risultati di teoria spettrale per i problemi ai limiti lineari ellittici*, Rend. Sem. Mat. Univ. Padova **38** (1967), 121–173.

[90] Y. GIGA, H. SOHR: *Abstract L^p estimates for the Cauchy problem with applications to the Navier-Stokes equations in exterior domains*, J. Funct. Anal. **102** (1991), 72–94.

[91] D. GILBARG, N.S. TRUDINGER, *Elliptic partial differential equations of second order*, 2nd Edition, Springer-Verlag, Berlin (1983).

[92] J. GOLDSTEIN: *Semigroups of linear operators and applications*, Oxford Mathematical Monographs, Oxford (1985).

[93] P. GRISVARD: *Commutativité de deux functeurs d'interpolation et applications*, J. Maths. Pures Appl. **45** (1966), 143–290.

[94] P. GRISVARD: *Équations differentielles abstraites*, Ann. Scient. Éc. Norm. Sup., 4e série **2** (1969), 311–395.

[95] P. GRISVARD: *Caractérisation de quelques espaces d'interpolation*, Arch. Rat. Mech. Anal. **25** (1967), 40–63.

[96] M. GRUBER: *Harnack Inequalities for Solutions of General Second Order Parabolic Equations and Estimates of their Hölder Constants*, Math. Z. **185** (1984), 23–43.

[97] G.H. HARDY, J.E. LITTLEWOOD, G. PÒLYA: *Inequalities*, Cambridge Univ. Press, Cambridge (1934).

[98] M.I. HAZAN: *Differentiability of nonlinear semigroups and the classical solvability of nonlinear boundary value problems for the equation $f(u_t) = u_{xx}$*, Dokl. Akad. Nauk SSSR **228** (1976) (Russian). English transl.: Soviet Math. Dokl. **17** (1976), 839–843.

[99] D. HENRY: *Geometric theory of semilinear parabolic equations*, Lect. Notes in Math. **840**, Springer-Verlag, New York (1981).

[100] E. HILLE, R.S. PHILLIPS: *Functional Analysis and Semigroups*, Amer. Math. Soc., Providence (1957).

[101] H. HOSHINO, Y. YAMADA: *Solvability and smoothing effect for Semilinear Parabolic Equations*, Funkt. Ekv. **34** (1991), 475–494.

[102] S.I. HUDJAEV: *The First Boundary Value Problem for Non-Linear Parabolic Equations*, Dokl. Akad. Nauk SSSR **149** (1963), 535–538 (Russian). English transl.: Soviet Math. Dokl. **4** (1963), 441–445.

[103] A.M. ILIN, A.S. KALASHNIKOV, O.A. OLEINIK: *Linear equations of the second order of parabolic type*, Russian Math. Surveys **17** (1962), 1–143.

[104] H. ISHII, P.L. LIONS: *Viscosity Solutions of Fully Nonlinear Second-Order Elliptic Partial Differential Equations*, J. Diff. Eqns. **83** (1990), 26–78.

[105] F. KAPPEL, W. SCHAPPACHER: *Strongly continuous semigroups – an introduction*, preprint.

[106] T. KATO: *Abstract evolution equations of parabolic type in Banach and Hilbert spaces*, Nagoya Math. J. **19** (1961), 93–125.

[107] T. KATO: *Perturbation Theory for Linear Operators*, Springer Verlag, New York (1966).

Bibliography

[108] T. Kato, H. Tanabe: *On the abstract evolution equations*, Osaka Math. J. **14** (1962), 107–133.

[109] H. Kielhöfer: *Global solutions of semilinear evolution equations satisfying an energy inequality*, J. Diff. Eqns. **36** (1980), 188–222.

[110] N. Kikuchi: *A construction of a solution for $\partial u/\partial t = a(\Delta u)$ ($a \in C^1(-\infty, \infty)$, $a(0) = 0$, $a' > 0$) with initial and boundary condition of class $C^{3+\alpha}$* ($0 < \alpha < 1$), preprint.

[111] S. Klainerman: *Long-time behavior of solutions to nonlinear evolution equations*, Arch. Rat. Mech. Anal. **78** (1982), 73–98.

[112] H. Komatsu: *Fractional powers of operators*, Pacific J. Math. **19** (1966), 285–346; *Fractional powers of operators. II*, Pacific J. Math. **21** (1967), 89–111; *Fractional powers of operators. III. Negative powers*, J. Math. Soc. Japan **21** (1969), 205–220; *Fractional powers of operators. IV. Potential operators*, J. Math. Soc. Japan **21** (1969), 221–228; *Fractional powers of operators. V. Dual operators*, J. Fac. Sci. Univ. Tokyo, Sec. I A, Math. **17** (1970), 373–396; *Fractional powers of operators. VI. Interpolation of non-negative operators and imbedding theorems*, J. Fac. Sci. Univ. Tokyo, Sec. I A, Math. **19** (1972), 1–63.

[113] Y. Konishi: *On the nonlinear semi-groups associated with $u_t = \Delta b(u)$ and $f(u_t) = \Delta u$*, J. Math. Soc. Japan **25** (1973), 622–628.

[114] S.G. Krein: *Linear Differential Equations in Banach Space*, Nauka, Moskow (1963) (Russian); English translation: Transl. Amer. Math. Soc. **29**, Providence (1971).

[115] S.G. Krein, Yu. Petunin, E.M. Semenov: *Interpolation of linear operators*, Amer. Math. Soc., Providence (1982).

[116] S.N. Kruzhkov, A. Castro, M. Lopes: *Schauder type estimates and theorems on the existence of the solution of fundamental problem for linear and nonlinear parabolic equations*, Dokl. Akad. Nauk SSSR **20** (1975), 277–280 (Russian). English transl.: Soviet Math. Dokl. **16** (1975), 60–64.

[117] S.N. Kruzhkov, A. Castro, M. Lopes: *Mayoraciones de Schauder y teorema de existencia de las soluciones del problema de Cauchy para ecuaciones parabolicas lineales y no lineales*, (I) Ciencias Matemáticas **1** (1980), 55–76; (II) Ciencias Matemáticas **3** (1982), 37–56.

[118] N.V. Krylov: *Controlled Diffusion Processes*, Nauka, Moskow (1977) (Russian). English transl.: Springer-Verlag, New York (1980).

[119] N.V. Krylov: *Boundedly Nonhomogeneous Elliptic and Parabolic Equations*, Izv. Akad. Nauk SSSR, Ser. Mat. **46** (1982), 487–523 (Russian). English transl.: Math. USSR, Izv. **20** (1983), 459–492.

[120] N.V. Krylov: *Boundedly Nonhomogeneous Elliptic and Parabolic Equations in a Domain*, Izv. Akad. Nauk SSSR, Ser. Mat. **47** (1983), 75–108 (Russian). English transl.: Math. USSR, Izv. **21** (1984), 67–98.

[121] N.V. Krylov: *Estimates for the Derivatives of Solutions of Nonlinear Parabolic Equations*, Dokl. Akad. Nauk SSSR **274** (1984), 23–26 (Russian). English transl.: Soviet Math. Dokl. **29** (1984), 14–17.

[122] N.V. Krylov: *Nonlinear Elliptic and Parabolic Equations of the Second Order*, Nauka, Moscow (1985). English transl.: D. Reidel Publishing Co., "Mathematics and Its Applications", Dordrecht (1987).

[123] N.V. KRYLOV, M.V. SAFONOV: *On a certain property of solutions of parabolic equations with measurable coefficients*, Izv. Akad. Nauk SSSR, Ser. Mat. **44** (1980), 161–175 (Russian). English transl.: Math. USSR Izv. **16** (1981), 151–164.

[124] O.A. LADYZHENSKAJA, V.A. SOLONNIKOV, N.N. URAL'CEVA: *Linear and quasilinear equations of parabolic type*, Nauka, Moskow 1967 (Russian). English transl.: Transl. Math. Monographs, AMS, Providence (1968).

[125] I. LASIECKA: *A unified theory for abstract parabolic boundary problems: a semigroup approach*, Appl. Math. Optim. **6** (1980), 287–333.

[126] C.-Y. LIN: *Fully nonlinear parabolic boundary value problem in one space dimension*, J. Diff. Eqns. **87** (1990), 62–69.

[127] J.L. LIONS: *Théoremes de traces et d'interpolation*, (I),...,(V); (I), (II) Ann. Sc. Norm. Sup. Pisa **13** (1959), 389–403; **14** (1960), 317–331; (III) J. Maths. Pures Appl. **42** (1963), 196–203; (IV) Math. Annalen **151** (1963), 42–56; (V) Anais de Acad. Brasileira de Ciencias **35** (1963), 1–110.

[128] J.L. LIONS, E. MAGENES: *Problèmes aux limites non homogènes et applications*, Dunod, Paris, vol. 1, 2 (1968), vol. 3 (1970).

[129] J.L. LIONS, J. PEETRE: *Sur une classe d'espaces d'interpolation*, Publ. I.H.E.S. **19** (1964), 5–68.

[130] G.S.S. LUDFORD, D.S. STEWART: *Fast deflagration waves*, J. Méc. Th. Appl., **2** (1983), 463–487.

[131] G.S.S. LUDFORD, D.S. STEWART: *The acceleration of fast deflagration waves*, Z.A.M.M., **63** (1983), 291–302.

[132] A. LUNARDI: *Analyticity of the maximal solution of an abstract nonlinear parabolic equation*, Nonlinear Analysis T.M.A. **6** (1982), 503–521.

[133] A. LUNARDI: *Abstract Quasilinear Parabolic Equations*, Math. Annalen **267** (1984), 405–425.

[134] A. LUNARDI: *An extension of Schauder's Theorem to little-Hölder continuous functions*, Boll. U.M.I., Sez. Anal. Funz. Appl., Serie VI, **3-C** (1984), 25–35.

[135] A. LUNARDI: *Global solutions of abstract quasilinear parabolic equations*, J. Diff. Eqns. **58** (1985), 228–243.

[136] A. LUNARDI: *Asymptotic exponential stability in quasilinear parabolic equations*, Nonlinear Analysis T.M.A. **9** (1985), 563–586.

[137] A. LUNARDI: *Characterization of Interpolation Spaces between Domains of Elliptic Operators and Spaces of Continuous Functions with Applications to Fully Nonlinear Parabolic Equations*, Math. Nachr. **121** (1985), 295–318.

[138] A. LUNARDI: *Laplace transform methods in integrodifferential equations*, in "Integrodifferential evolution equations and applications", G. Da Prato, M. Iannelli Eds., J. Int. Eqns. **10** (1985), 185–211.

[139] A. LUNARDI: C^∞ *regularity for fully nonlinear abstract evolution equations*, in : "Differential Equations in Banach Spaces", Proceedings, A. Favini, E. Obrecht Eds., Lect. Notes in Math. **1223**, Springer Verlag, New York (1986), 176–185.

[140] A. LUNARDI: *On the evolution operator for abstract parabolic equations*, Israel J. Math. **60** (1987), 281–314.

[141] A. LUNARDI: *On the local dynamical system associated to a fully nonlinear parabolic equation*, in : "Nonlinear Analysis and Applications", V. Lakshmikantham Ed., Marcel Dekker Publ., 1987, 319–326.

[142] A. LUNARDI: *Time analyticity for solutions of nonlinear abstract parabolic evolution equations*, J. Funct. Anal. **71** (1987), 294–308.

[143] A. LUNARDI: *Bounded solutions of periodic abstract parabolic equations*, Proc. Royal Soc. Edinb. **110 A** (1988), 135–159.

[144] A. LUNARDI: *Regular solutions for time dependent abstract integrodifferential equations with singular kernel*, J. Math. Anal. Appl. **130** (1988), 1–21.

[145] A. LUNARDI: *Stability of the periodic solutions to fully nonlinear parabolic equations in Banach spaces*, Diff. Int. Eqns. **1** (1988), 253–279.

[146] A. LUNARDI: *Maximal space regularity in nonhomogeneous initial boundary value parabolic problems*, Numer. Funct. Anal. Optimiz. **10** (1989), 323–439.

[147] A. LUNARDI: *Differentiability with respect to (t,s) of the parabolic evolution operator*, Israel J. Math. **68** (1989), 161–184.

[148] A. LUNARDI: *On a class of fully nonlinear parabolic equations*, Comm. P.D.E.'s **16** (1991), 145–172.

[149] A. LUNARDI: *An introduction to geometric theory of fully nonlinear parabolic equations*, in "Qualitative aspects and applications of nonlinear evolution equations", T.T. Li, P. de Mottoni Eds., World Scientific, Singapore (1991),107–131.

[150] A. LUNARDI: *Stability and local invariant manifolds in fully nonlinear parabolic equations*, in "Semigroups of Linear and Nonlinear Operations and Applications", Curaçao (August 10–14, 1992), Proceedings, G.& J. Goldstein Editors, Kluwer Academic Publishers, Dordrecht (1993), 185–203.

[151] A. LUNARDI: *A unified approach to semilinear parabolic problems*, Dyn. Syst. Appl. (to appear).

[152] A. LUNARDI: *Stability in fully nonlinear parabolic equations*, Arch. Rat. Mech. Anal. (to appear).

[153] A. LUNARDI, E. SINESTRARI: *Fully nonlinear integrodifferential equations in general Banach space*, Math. Z. **190** (1985), 225–248.

[154] A. LUNARDI, E. SINESTRARI: *Existence in the large and stability for nonlinear Volterra equations*, in "Integrodifferential evolution equations and applications", G. Da Prato, M. Iannelli Eds., J. Int. Eqns. **10** (1985), 213–239.

[155] A. LUNARDI, E. SINESTRARI: C^α-*regularity for non autonomous linear integrodifferential equations of parabolic type*, J. Diff. Eqns. **63** (1986), 88–116.

[156] A. LUNARDI, E. SINESTRARI, W. VON WAHL: *A semigroup approach to the time-dependent parabolic initial boundary value problem*, Diff. Int. Eqns. **5** (1992), 1275–1306.

[157] A. LUNARDI, V. VESPRI: *Hölder Regularity in Variational Parabolic Non-homogeneous Equations*, J. Diff. Eqns. **94** (1991), 1–40.

[158] R.H. MARTIN: *Nonlinear Operators and Differential Equations in Banach Spaces*, John Wiley & Sons, New York (1976).

[159] B.J. MATKOWSKI, A. VAN HARTEN: *A new model in flame theory*, SIAM J. Appl. Math. **42** (1982), 850–867.

[160] X. MORA: *Semilinear parabolic problems define semiflows on C^k spaces*, Trans. Amer. Math. Soc. **278** (1983), 21–55.

[161] B.S. MITJAGIN, E.M. SEMENOV: C^k *is not an interpolation space between C and C^n*, $0 < k < n$, Dokl. Akad. Nauk SSSR **228** (1976) (Russian). English transl.: Soviet Math. Dokl. **17** (1976), 778–782.

[162] J. MOSER: *A Harnack inequality for parabolic differential equations*, Comm. Pure Appl. Math. **17** (1964), 101–134.

[163] R. NAGEL, ED.: *One-parameter Semigroups of Positive Operators*, Lect. Notes in Math. **1184**, Springer-Verlag, Berlin (1986).

[164] J. NASH: *Continuity of solutions of parabolic and elliptic equations*, Amer. J. Math. **80** (1958), 931–954.

[165] S. NOOR EBAD, C. SCHMIDT-LAINÉ, *Further numerical results on a deflagration to detonation transition model*, preprint.

[166] A. PAZY: *Semigroups of linear operators and applications to partial differential equations*, 2nd Edition, Springer-Verlag, New York (1992).

[167] J. PEETRE: *New thoughts on Besov spaces*, Duke University (1976).

[168] M. POITIER-FERRY: *The linearization principle for the stability of solutions of quasilinear parabolic equations, I*, Arch. Rat. Mech. Anal. **77** (1981), 301–320.

[169] J. PRÜSS, H. SOHR: *On Operators with Bounded Imaginary Powers in Banach Spaces*, Math. Z. **203** (1990), 429–452.

[170] J. PRÜSS, H. SOHR: *Imaginary powers of elliptic second order differential operators in L^p-spaces*, Hiroshima Math. J. **23**, 161–192.

[171] S.M. RANKIN III: *Semilinear Evolution Equations in Banach Spaces with Applications to Parabolic Partial Differential Equations*, Trans. Amer. Math. Soc. **336** (1993), 523–535.

[172] F. ROTHE: *Global Solutions of Reaction-Diffusion Systems*, Lect. Notes in Math. **1072**, Springer Verlag, Berlin (1984).

[173] J.T. SCHWARTZ: *Nonlinear Functional Analysis*, Gordon & Breach, New York (1953).

[174] G. SIMONETT: *Invariant Manifolds and bifurcation for Quasilinear Reaction-Diffusion Systems*, Nonlinear Analysis T.M.A. **23** (1994), 515–544.

[175] E. SINESTRARI: *Continuous interpolation spaces and spatial regularity in non linear Volterra integrodifferential equations*, J. Int. Eqns. **5** (1983), 287–308.

[176] E. SINESTRARI: *On a Class of Retarded Partial Differential Equations*, Math. Z. **186** (1984), 223–246.

[177] E. SINESTRARI: *On the abstract Cauchy problem in spaces of continuous functions*, J. Math. Anal. Appl. **107** (1985), 16–66.

Bibliography

[178] E. SINESTRARI, P. VERNOLE: *Semilinear evolution equations in interpolation spaces*, Nonlinear Analysis T.M.A. **1** (1977), 249–261.

[179] E. SINESTRARI, W. VON WAHL: *On the solutions of the first boundary value problem for the linear parabolic equations*, Proc. Royal Soc. Edinburgh **108A** (1988), 339–355.

[180] J. SMOLLER: *Shock Waves and Reaction-Diffusion Equations*, Springer Verlag, Berlin (1983).

[181] P.E. SOBOLEVSKIĬ: *Equations of parabolic type in a Banach space*, Trudy Moskow Math. Obsc. **10** (1961), 297–350 (Russian). English transl.: Amer. Math. Soc. Transl. **49** (1964), 1–62.

[182] P.E. SOBOLEVSKIĬ: *Coerciveness inequalities for abstract parabolic equations*, Soviet Math. **5** (1964), 894–897.

[183] P.E. SOBOLEVSKIĬ: *Parabolic equations in a Banach space with an unbounded variable operator, a fractional power of which has a constant domain of definition*, Dokl. Akad. Nauk **138** (1961), 59–62 (Russian); English transl.: Soviet Math. Dokl. **2** (1961), 545–548.

[184] V.A. SOLONNIKOV: *On the boundary value problems for linear parabolic systems of differential equations of general form*, Proc. Steklov Inst. Math. **83** (1965), O. A. Ladyzenskaja Ed., Amer. Math. Soc. (1967).

[185] N.N. SOPOLOV: *The first boundary value problem for nonlinear parabolic equations of arbitrary order*, C. R. Acad. Bulgare Sci. **23** (1970), 899–902 (Russian).

[186] D.S. STEWART: *Transition to detonation on a model problem*, J. Méc. Th. Appl., **4** (1985), 103–137.

[187] H.B. STEWART: *Generation of analytic semigroups by strongly elliptic operators*, Trans. Amer. Math. Soc. **199** (1974), 141–162.

[188] H.B. STEWART: *Generation of analytic semigroups by strongly elliptic operators under general boundary conditions*, Trans. Amer. Math. Soc. **259** (1980), 299–310.

[189] M.H. TAIBLESON: *On the theory of Lipschitz spaces of distributions on Euclidean n-space, I*, J. Math. Mech. **13** (1964), 407–479.

[190] H. TANABE: *On the equations of evolution in a Banach space*, Osaka Math. J. **12** (1960), 363–376.

[191] H. TANABE: *Note on singular perturbation for abstract differential equations*, Osaka J. Math. **1** (1964), 239–252.

[192] H. TANABE: *On semilinear equations of elliptic and parabolic type*, in: Functional Analysis and Numerical Analysis. Japan-French Seminar, Tokyo and Kyoto, 1976. H. Fujita Editor. Japan Society for the Promotion of Science (1978), 455–473.

[193] H. TANABE: *Equations of evolution*, Pitman, London (1979).

[194] H. TANABE: *Functional Analysis, II*, Jikkyo Shuppan Publ. Comp., Tokyo (1981) (Japanese).

[195] H. TANABE: *Note on Volterra integrodifferential equations of parabolic type*, Proc. Japan Acad. Sci. (to appear).

[196] R. TEMAM: *Infinite Dimensional Dynamical Systems in Mechanics and Physics*, Springer-Verlag, New York (1988).

[197] B. TERRENI: *Nonhomogeneous initial-boundary value problems for linear parabolic systems*, Studia Math. **92** (1989), 141–175.

[198] C.C. TRAVIS: *Differentiability of weak solutions to an abstract inhomogeneous differential equation*, Proc. Amer. Math. Soc. **82** (1981), 425–430.

[199] R. TRIGGIANI: *On the stabilizability problem in Banach space*, J. Math. Anal. Appl. **52** (1975), 384–403.

[200] H. TRIEBEL: *Interpolation Theory, Function Spaces, Differential Operators*, North-Holland, Amsterdam (1978).

[201] V. VESPRI: *The functional space $C^{-1,\alpha}$ and analytic semigroups*, Diff. Int. Eqns. **1** (1988), 473–493.

[202] V. VESPRI: *Analytic semigroups generated by ultraweak operators*, Proc. Royal Soc. Edinburgh **119A** (1991), 87–105.

[203] W. VON WAHL: *Gebrochene Potenzen eines elliptischen Operators und parabolische Differentialgleichungen in Räumen Hölderstetiger Funktionen*, Nachr. Akad. Wiss. Göttingen, II. Mathem. Physik. Klasse (1972), 231–258.

[204] W. VON WAHL: *Lineare und semilineare parabolische Differentialgleichungen in Räumen Hölderstetiger Funktionen*, Abh. Math. Sem. Univ. Hamburg **43** (1975), 234–262.

[205] W. VON WAHL: *On the Cahn-Hilliard Equation $u' + \Delta^2 u - \Delta f(u) = 0$*, Delft Progr. Rep. **10** (1985), 291–310.

[206] W. VON WAHL: *The Equations of Navier-Stokes and Abstract Parabolic Equations*, Vieweg, Braunschweig/Wiesbaden (1985).

[207] W. VON WAHL: *On the Equation $u_t = f(\Delta u)$*, Boll. U.M.I. 7, 1-A (1987), 437–441.

[208] L. WANG: *On the regularity theory of fully nonlinear parabolic equations*, Bull. Amer. Math. Soc. **22** (1990), 107–114.

[209] F.B. WEISSLER: *Existence and nonexistence of global solutions for a semilinear heat equation*, Israel J. Math. **38** (1981), 29–40.

[210] A. YAGI: *On the abstract linear evolution equations in Banach spaces*, J. Math. Soc. Japan **28** (1976), 290–303.

[211] A. YAGI: *On the abstract evolution equation of parabolic type*, Osaka J. Math. **14** (1977), 557–568.

[212] A. YAGI: *Abstract Quasilinear Evolution Equations of Parabolic type in Banach Spaces*, Boll. U.M.I. **5-B** (1991), 341–368.

[213] K. YOSIDA: *Functional Analysis*, Springer Verlag, Berlin (1965).

[214] S. ZHENG: *Remarks on global existence for nonlinear parabolic equations*, Nonlinear Analysis T.M.A. **10** (1986), 107–114.

[215] S. ZHENG, Y. CHEN: *Global existence for nonlinear equations*, Chinese Ann. Math. **6B** (1986), 57–73.

[216] A. ZYGMUND: *Trigonometric Series*, Cambridge Univ. Press., 2nd Edition Reprinted (1968).

Index

$B_\mu(]a,b];X)$, 122
$(C_\sharp^\alpha(\mathbb{R};D)$, 382
$(X,Y)_\theta$, 15
$(X,Y)_{\theta,p}$, 15
$B(I;X)$, 1
$B(I;Y,\omega)$, 153
$C(I;X)$, 1
$C(I;Y,\omega)$, 153
$C([a,b]\times\overline{\Omega})$, 175
$C^\infty(I;X)$, 1
$C^{0,\alpha}([a,b]\times\overline{\Omega})$, 175
$C_0^1(\overline{\Omega})$, 101
$C_\mathcal{B}^{1+\alpha}(\overline{\Omega})$, 107
$C^{1,2+\alpha}([a,b]\times\overline{\Omega})$, 176
$C^{1,2}([a,b]\times\overline{\Omega})$, 176
$C_\sharp^{\alpha+1}(\mathbb{R};X)$, 382
$C^{\alpha/2,\alpha}([a,b]\times\overline{\Omega})$, 177
$C^\alpha(I;X)$, 3
$C^\alpha(I;Y,\omega)$, 153
$C_0^\alpha(\overline{\Omega})$, 107
$C_\beta^\alpha(]a,b];X)$, 123
$C_\sharp^\alpha(\mathbb{R};X)$, 382
$C^{\alpha,0}([a,b]\times\overline{\Omega})$, 175
$C^\theta(\overline{\Omega})$, 7
$C^{k+\alpha}(I;X)$, 3
$C^{k+\alpha}(I;Y,\omega)$, 153
$C^m(I;X)$, 1
$C_*(\overline{\Omega})$, 97
$C_0(\mathbb{R}^n)$, 81
$C_b(I;X)$, 1
$C_\mu(]a,b];X)$, 122
$C_\theta(]0,\delta];Y)$, 262
$C_b^m(I;X)$, 1
$D_A(\alpha)$, 46

$D_A(\alpha,p)$, 46
$K(t,x,X,Y)$, 15
$S_{\theta,\omega}$, 33
$UC([a,b]\times\overline{\Omega})$, 175
$V(p,\theta,Y,X)$, 20
$W^{\alpha,p}(\Omega)$, 115
$W_{loc}^{k,p}(\mathbb{R}^n)$, 76
$W_\mathcal{B}^{s,p}(\Omega)$, 113
$\Omega_{x_0,r}$, 94
$\eta\otimes\xi$, 331
ω-limit set, 299
$h_\mathcal{B}^{1+\alpha}(\overline{\Omega})$, 107
$h^\alpha(I;X)$, 4
$h_0^\alpha(\overline{\Omega})$, 107
$h^\theta(\overline{\Omega})$, 7
$\mathcal{C}_0^1(\overline{\Omega})$, 107
$\mathcal{C}_\mathcal{B}^1(\overline{\Omega})$, 109
$\mathcal{C}^\alpha(I;X)$, 5

Agmon-Douglis-Nirenberg estimates, 70, 71
algebraic multiplicity, 407
algebraically simple eigenvalue, 407
analytic semigroup, 34, 40
asymptotic orbital stability, 340
asymptotic phase, 369, 389
asymptotic stability, 337

Banach manifold, 308
Bernstein polynomials, 2
bifurcation equation, 352

Caccioppoli type inequality, 76
center-unstable manifold, 338, 369
characteristic multiplier, 236
characteristic polynomial, 407

class J_α, 12
classical solution, 124, 255
complementing condition, 113
complexification, 38, 399
continuous interpolation spaces, 32

eigenvalue, 399
Euler Γ function, 54
evolution operator, 211, 219
exponential asymptotic stability, 338

Floquet exponent, 236
Floquet multiplier, 236

generalized eigenvector, 407
geometric multiplicity, 407
graph norm, 33

Hölder continuous functions, 3
Hardy-Young inequalities, 20

index, 407
integral solution, 125
intermediate space, 11
interpolation space, 11
invariant manifold, 338

Laplace operator, 173
Laurent development of the resolvent, 405
$Lip(I;X)$, 3
Lipschitz continuous functions, 3
little- Hölder continuous functions, 4
local dynamical system, 301
local semiflow, 301

method of continuity, 88, 118
mild solution, 125, 212, 255

nontangentiality condition, 71
normality condition, 112

orbital instability, 340
orbital stability, 340
orbitally asymptotically stable, 389

parabolic Hölder space, 174, 177
part of A in X_0, 40
period map, 235

periodic solutions, 165
Poincaré map, 235
point spectrum, 62, 399
Principle of Linearized Stability, 337
Principle of Linearized Stability , 339

quenching, 256

real interpolation space, 15
resolvent identity, 400
resolvent operator, 399
resolvent set, 399

saddle point, 338
sectorial operator, 33
semigroup, 34
semisimple eigenvalue, 406
simple eigenvalue, 407
Sobolev embedding, 76
spectral determining condition, 56
Spectral Mapping Theorem, 62
spectral radius, 401
spectral set, 58, 60, 402
spectrum, 399
stable manifold, 338, 347
stable periodic solution, 388
stable stationary solution, 337
stationary solution, 337
strict solution, 123, 255, 290
strong ellipticity, 72
strong solution, 123
strongly continuous, 34

travelling waves, 358
type of A, 169
type of e^{tA}, 58

uniform ellipticity, 71
uniformly $C^{m+\alpha}$ boundary, 8
uniformly C^m boundary, 2
uniformly $h^{m+\alpha}$ boundary, 8
unstable manifold, 338, 347
unstable periodic solution, 389
unstable stationary solution, 337

Yosida approximation, 38

Progress in Nonlinear Differential Equations and Their Applications

Editor

Haim Brezis
Département de Mathématiques, Université P. et M. Curie, 4, Place Jussieu, 75252 Paris Cedex 05, France, and
Department of Mathematics, Rutgers University, New Brunswick, NJ 08903, U.S.A.

Progress in Nonlinear and Differential Equations and Their Applications is a book series that lies at the interface of pure and applied mathematics. Many differential equations are motivated by problems arising in such diversified fields as Mechanics, Physics, Differential Geometry, Engineering, Control Theory, Biology, and Economics. This series is open to both the theoretical and applied aspects, hopefully stimulating a fruitful interaction between the two sides. It will publish monographs, polished notes arising from lectures and seminars, graduate level texts, and proceedings of focused and refereed conferences.

PNLDE 1   Partial Differential Equations and the Calculus of Variations, Volume I
           Essays in Honor of Ennio De Giorgi
           F. Colombini, A. Marino, L. Modica, and S. Spagnolo, editors
PNLDE 2   Partial Differential Equations and the Calculus of Variations, Volume II
           Essays in Honor of Ennio De Giorgi
           F. Colombini, A. Marino, L. Modica, and S. Spagnolo, editors
PNLDE 3   Propagation and Interaction of Singularities in Nonlinear Hyperbolic Problems
           Michael Beals
PNLDE 4   Variational Methods
           Henri Berestycki, Jean-Michel Coron, and Ivar Ekeland, editors
PNLDE 5   Composite Media and Homogenization Theory
           Gianni Dal Maso and Gian Fausto Dell'Antonio, editors
PNLDE 6   Infinite Dimensional Morse Theory and Multiple Solution Problems
           Kung-ching Chang
PNLDE 7   Nonlinear Differential Equations and their Equilibrium States, 3
           N.G. Lloyd, W.M. Ni, L.A. Peletier, J. Serrin, editors
PNLDE 8   An Introduction to Γ–Convergence
           Gianni Dal Maso
PNLDE 9   Differential Inclusions in Nonsmooth Mechanical Problems. Shocks and Dry Friction
           Manuel D.P. Monteiro Marques
PNLDE 10   Periodic Solutions of Singular Lagrangian Systems
           Antonio Ambrosetti and Vittorio Coti Zelati
PNLDE 11   Nonlinear Waves and Weak Turbulence with Applications in Oceanography and
           Condensed Matter Physics
           N. Fitzmaurice, D. Gurarie, F. McCaughan and W.A. Woyczynski, editors
PNLDE 12   Semester on Dynamical Systems
           S. Kuksin, V. Lazutkin and J. Pöschel, editors
PNLDE 13   Ginzberg-Landau Vortices
           F. Bethuel, H. Brezis and F. Hélein
PNLDE 14   Variational Methods in Image Segmentation
           J.-M. Morel, S. Solimini
PNLDE 15   Topological Nonlinear Analysis: Degree, Singularity, and Variations
           M. Matzeu, A. Vignoli, editors
PNLDE 16   Analytic Semigroups and Optimal Regularity in Parabolic Problems
           A. Lunardi

MMA 84

H. Triebel, Friedrich-Schiller-Universität Jena, Germany

Theory of Function Spaces II

1992. 380 pages. Hardcover
ISBN 3-7643-2639-5

Theory of Function Spaces II deals with the theory of function spaces of type B^s_{pq} and F^s_{pq} as it stands at present. These two scales of spaces cover many well-known function spaces such as Hölder-Zygmund spaces, (fractional) Sobolev spaces, Besov spaces, inhomogeneous Hardy spaces, spaces of BMO-type and local approximation spaces which are closely connected with Morrey-Campanato spaces.
Theory of Function Spaces II monograph is self-contained, although it may be considered an update of the author's earlier book of the same title. The book's 7 chapters start with a historical survey of the subject, and then analyze the theory of function spaces in R^n and in domains, applications to (exotic) pseudodifferential operators, and function spaces on Riemannian manifolds.

MMA 87

J. Prüss, Universität-GH Paderborn, FB 17 Mathematik-Informatik, Germany

Evolutionary Integral Equations and Applications

1993. 294 pages. Softcover
ISBN 3-7643-5071-7

This book deals with evolutionary systems whose equation of state can be formulated as a linear Volterra equation in a Banach space. The main feature of the kernels involved is that they consist of unbounded linear operators. The aim is a coherent presentation of the state of art of the theory including detailed proofs and its applications to problems from mathematical physics, such as viscoelasticity, heat conduction, and electrodynamics with memory. The importance of evolutionary integral equations – which form a larger class than do evolution equations– stems from such applications and therefore special emphasis is placed on these. A number of models are derived and, by means of the developed theory, discussed thoroughly. An annotated bibliography containing 450 entries increases the book's value as an incisive reference text.

Please order through your bookseller or write to:
Birkhäuser Verlag AG
P.O. Box 133
CH-4010 Basel / Switzerland
FAX: ++41 / 61 / 271 76 66
e-mail:
100010.2310@compuserve.com

For orders originating in the USA or Canada:
Birkhäuser
333 Meadowlands Parkway
Secaucus, NJ 07094-2491
USA

Birkhäuser
Birkhäuser Verlag AG
Basel · Boston · Berlin